Zika Virus Impact, Diagnosis, Control, and Models

The Neuroscience of Zika Virus

Zika Virus Impact, Diagnosis, Control, and Models

The Neuroscience of Zika Virus

Volume 2

Edited by

Colin R. Martin

Institute for Clinical and Applied Health Research (ICAHR),
University of Hull
Hull, United Kingdom

Caroline J. Hollins Martin

Edinburgh Napier University
Edinburgh, Scotland, United Kingdom

Victor R. Preedy

King's College London,
London, United Kingdom

Rajkumar Rajendram

Department of Medicine, King Abdulaziz Medical City,
Ministry of National Guard Health Affairs,
Riyadh, Saudi Arabia

ELSEVIER

ACADEMIC PRESS
An imprint of Elsevier

Academic Press is an imprint of Elsevier
125 London Wall, London EC2Y 5AS, United Kingdom
525 B Street, Suite 1650, San Diego, CA 92101, United States
50 Hampshire Street, 5th Floor, Cambridge, MA 02139, United States
The Boulevard, Langford Lane, Kidlington, Oxford OX5 1GB, United Kingdom

Library of Congress Cataloging-in-Publication Data
A catalog record for this book is available from the Library of Congress

British Library Cataloguing-in-Publication Data
A catalogue record for this book is available from the British Library

ISBN: 978-0-12-820267-8

SET ISBN: 978-0-323-85864-9

For information on all Academic Press publications
visit our website at https://www.elsevier.com/books-and-journals

Publisher: Nikki Levy
Acquisitions Editor: Natalie Farra
Editorial Project Manager: Timothy Bennett
Production Project Manager: Paul Prasad Chandramohan
Cover Designer: Mark Rogers

Typeset by SPi Global, India

Dedication

This book is dedicated to my daughter Dr. Caragh Brien of whom I am so proud and also to my wonderful late father.

Colin R. Martin

Contents

Section F
Control, vaccines, and treatments

Contributors

Numbers in parentheses indicate the pages on which the authors' contributions begin.

Tawfik Aboellail (429), Department of Microbiology, Immunology and Pathology, Colorado State University, Fort Collins, CO, United States

Jônatas Santos Abrahão (417), Department of Microbiology, Federal University of Minas Gerais, Belo Horizonte, Minas Gerais, Brazil

Talita Adelino (29), Department of Genetics, Ecology and Evolution, Federal University of Minas Gerais; Central Laboratory of Public Health of Minas Gerais, Ezequiel Dias Foundation, Belo Horizonte, MG, Brazil

Ramesh Akkina (429), Department of Microbiology, Immunology and Pathology, Colorado State University, Fort Collins, CO, United States

Ayman Alboudi (201), Department of Neurology, Hauenstein Neuroscience Center, Mercy Health Saint Mary's Hospital, Grand Rapids, MI, United States

Luiz Carlos Junior Alcantara (29), Department of Genetics, Ecology and Evolution, Federal University of Minas Gerais, Belo Horizonte, MG; Oswaldo Cruz Institute (IOC), Oswaldo Cruz Foundation (Fiocruz), Rio de Janeiro, RJ, Brazil

Henning Andersen (143), Department of Neurology, Aarhus University Hospital, Aarhus N, Denmark

Teresinha De Jesus Aguiar Dos Santos Andrade (327), Nucleus of Applied Research to Sciences, Federal Institute of Maranhão, Timon, MA, Brazil

Masashi Arakawa (49), Department of Biochemistry and Molecular Biology, Faculty of Agriculture and Life Science, Hirosaki University, Hirosaki-shi, Aomori, Japan

Josélio Maria Galvão de Araújo (3), Department of Microbiology and Parasitology, Post-Graduate Program in Parasite Biology, Biosciences Center, Federal University of Rio Grande do Norte, Natal, RN, Brazil

Pamella Nunes Azevedo (85), Brain Institute of Rio Grande do Sul (BraIns), Medical School, Pontifical Catholic University of Rio Grande do Sul, Rio Grande do Sul, Brazil

Omar Bagasra (97,377), Department of Biology, South Carolina Center for Biotechnology, School of Natural Sciences and Mathematics, Claflin University, Orangeburg, SC, United States

Mark J. Bailey (367), Department of Microbiology, Graduate School of Biomedical Sciences, Icahn School of Medicine at Mount Sinai, New York, NY, United States

Serdar Baraklı (135), Department of Neurology, Faculty of Medicine, Ankara Yildirim Beyazit University, Ankara, Turkey

Alison Jane Basile (273), Diagnostic and Reference Laboratory, Arboviral Diseases Branch, Division of Vector-Borne Diseases, Centers for Disease Control and Prevention, Fort Collins, CO, United States

Luis Federico Bátiz (111), School of Medicine, Facultad de Medicina; Neuroscience Program, Center for Biomedical Research and Innovation (Centro de Investigación e Innovación Biomédica [CiiB]), Universidad de los Andes, Santiago, Chile

Benan Bayrakci (125), Department of Pediatrics, Division of Pediatric Critical Care Medicine, Hacettepe University, İhsan Doğramacı Children's Hospital, Altındağ/Ankara, Turkey

Aline Almeida Bentes (75), Departamento de Pediatria, Faculdade de Medicina, Universidade Federal de Minas Gerais, Belo Horizonte, MG, Brazil

Jean A. Bernatchez (385), Centre for Discovery and Innovation in Parasitic Diseases, Skaggs School of Pharmacy and Pharmaceutical Sciences, University of California, San Diego, La Jolla, CA, United States

A.B. Blázquez (409), Department of Biotechnology, National Agricultural and Food Research and Technology Institute (INIA), Madrid, Spain

Viola Borchardt-Lohölter (293), Institute for Experimental Immunology, EUROIMMUN Medizinische Labordiagnostika AG, Lübeck, Germany

Ana Luiza Vilela Borges (39), Department of Public Health Nursing, School of Nursing, University of São Paulo, São Paulo, SP, Brazil

Maria Sole Burali (349), Department of Pharmaceutical Sciences, University of Perugia, Perugia, Italy

Felipe A. Bustamante-Barrientos (111), Cells for Cells; Immunology Program, Center for Biomedical Research and Innovation (Centro de Investigación e Innovación Biomédica [CiiB]), Universidad de los Andes, Santiago, Chile

Paulo E. Cabral Filho (283), Biophysics and Radiobiology Department, Federal University of Pernambuco (UFPE), Biomedical Nanotechnology Research Group (NanoBio), Recife, Pernambuco, Brazil

Steven Vargas Cañas (193), Nerve and Muscle Clinic, Department of Neurology, National Institute of Neurology and Neurosurgery Manuel Velasco Suárez, Tlalpan, Mexico City, Mexico

Talita Castro (255), School of Dentistry, University of Santo Amaro, Sao Paulo, SP, Brazil

Gwong-Jen J. Chang (303), Division of Vector-Borne Diseases, Centers for Disease Control and Prevention, US Department of Health and Human Services, Fort Collins, CO, United States

Day-Yu Chao (303), Graduate Institute of Microbiology and Public Health, College of Veterinary Medicine, National Chung Hsing University, Taichung, Taiwan

Ameya Chaudhari (263), Department of Biomedical Engineering, Duke University, Durham, NC, United States

Duverney Chaverra-Rodriguez (317), Department of Cell and Developmental Biology, University of California, San Diego, La Jolla, CA, United States

Raissa R. Christoff (451), Institute of Biomedical Sciences, Federal University of Rio de Janeiro, Rio de Janeiro, Brazil

Luiz Felipe Leomil Coelho (417), Vaccine Laboratory, Federal University of Alfenas, Alfenas, Minas Gerais, Brazil

Diogo Goulart Corrêa (179,185,215), Department of Radiology, Clinic of Diagnostic Imaging/DASA, Rio de Janeiro, Rio de Janeiro, Brazil

Michael Coste (385), Department of Chemistry and Biochemistry and the Viral Information Institute, San Diego State University, San Diego, CA, United States

Raquel Zanatta Coutinho (39), Department of Demography, Center for Development and Regional Planning, Federal University of Minas Gerais, Belo Horizonte, MG, Brazil

Joaquim Soares da Costa Júnior (327), Department of Chemistry, Federal Institute of Piauí, Teresina, PI, Brazil

Anna Carolina Toledo da Cunha Pereira (417), Federal University of the Delta of Parnaíba, Parnaíba, Piauí, Brazil

Paulo Marcos da Matta Guedes (3), Department of Microbiology and Parasitology, Post-Graduate Program in Parasite Biology, Biosciences Center, Federal University of Rio Grande do Norte, Natal, RN, Brazil

Guilherme Liberato da Silva (85), Medical Science Center, Universidade do Vale do Taquari—Univates, Lajeado, Rio Grande do Sul, Brazil

Jaderson Costa DaCosta (85), Brain Institute of Rio Grande do Sul (BraIns), Medical School, Pontifical Catholic University of Rio Grande do Sul, Rio Grande do Sul, Brazil

Prajakta Dandekar (263), Department of Pharmaceutical Sciences and Technology, Institute of Chemical Technology, Mumbai, India

Amos Danielli (231), Faculty of Engineering, The Institute of Nanotechnology and Advanced Materials, Bar-Ilan University, Ramat Gan, Israel

Maria das Dores Alves de Oliveira (327), Chemistry Department, Federal University of Piauí, Teresina, PI, Brazil

Orhan Deniz (135), Department of Neurology, Faculty of Medicine, Ankara Yildirim Beyazit University, Ankara, Turkey

Philippe Desprès (397), University of Reunion Island, UMR PIMIT, INSERM U1187, CNRS UMR 9192, IRD UMR 249, Technological platform CYROI, Sainte Clotilde, France

Betânia Paiva Drumond (417), Department of Microbiology, Federal University of Minas Gerais, Belo Horizonte, Minas Gerais, Brazil

Jonny Duque (317), Center for Research on Tropical Diseases (Centro de Investigaciones en Enfermedades Tropicales-CINTROP), Faculty of Health, School of Medicine, Department of Basic Sciences, Universidad Industrial de Santander, Bucaramanga, Colombia

Patrícia e Silva Alves (327), Department of Chemistry, Chemistry Graduate Program, State University of Piauí, Teresina, PI, Brazil

Chaker El Kalamouni (397), University of Reunion Island, UMR PIMIT, INSERM U1187, CNRS UMR 9192, IRD UMR 249, Technological platform CYROI, Sainte Clotilde, France

José Veríssimo Fernandes (3), Department of Microbiology and Parasitology, Post-Graduate Program in Parasite Biology, Biosciences Center, Federal University of Rio Grande do Norte, Natal, RN, Brazil

Gustavo Portela Ferreira (417), Federal University of the Delta of Parnaíba, Parnaíba, Piauí, Brazil

T. Foiadelli (151), Pediatric Clinic, Fondazione IRCCS Policlinico "San Matteo", University of Pavia, Pavia (PV), Italy

Vagner Fonseca (29), Department of Genetics, Ecology and Evolution, Federal University of Minas Gerais, Belo Horizonte, MG, Brazil; Kwazulu-Natal Research Innovation and Sequencing Platform (KRISP), Department of Virology, National Health Laboratory Service (NHLS), University of KwaZulu-Natal—Nelson R. Mandela School of Medicine, Berea, Durban, South Africa

Adriana Fontes (283), Biophysics and Radiobiology Department, Federal University of Pernambuco (UFPE), Biomedical Nanotechnology Research Group (NanoBio), Recife, Pernambuco, Brazil

Lawrence Frenkel (61), Departments of Pediatrics and Microbiology, University of Illinois College of Medicine, Rockford, IL; Department of Pediatrics, Division of Immunology, Allergy, and Infectious Disease, The Children's Hospital at Saint Peter's University Hospital, New Brunswick, NJ, United States

Gilles Gadea (397), University of Reunion Island, UMR PIMIT, INSERM U1187, CNRS UMR 9192, IRD UMR 249, Technological platform CYROI, Sainte Clotilde, France

Patricia P. Garcez (451), Institute of Biomedical Sciences, Federal University of Rio de Janeiro, Rio de Janeiro, Brazil

Marta Giovanetti (29), Department of Genetics, Ecology and Evolution, Federal University of Minas Gerais, Belo Horizonte, MG; Oswaldo Cruz Institute (IOC), Oswaldo Cruz Foundation (Fiocruz), Rio de Janeiro, RJ, Brazil

Fernando Gomez (61), Department of Specialty Medicine, Rocky Vista University, Parker, CO, United States

Bonnie E. Gulas-Wroblewski (243), Department of Pediatrics, Section of Tropical Medicine, Baylor College of Medicine and Texas Children's Hospital, Houston, TX, United States

Şadiye Gümüşyayla (135), Department of Neurology, Faculty of Medicine, Ankara Yildirim Beyazit University, Ankara, Turkey

Sunam Gurung (463), Dean Myers Lab, Department of Obstetrics and Gynecology, The University of Oklahoma Health Sciences Center, Oklahoma City, OK, United States

Juliano G. Haddad (397), University of Reunion Island, UMR PIMIT, INSERM U1187, CNRS UMR 9192, IRD UMR 249, Technological platform CYROI, Sainte Clotilde, France

Thomas Harbo (143), Department of Neurology, Aarhus University Hospital, Aarhus N, Denmark

Cecília Hedin-Pereira (207), Vice Presidency of Research and Biological Collections (VPPCB), Oswaldo Cruz Foundation (FIOCRUZ), Rio de Janeiro, RJ, Brazil

Roberto Henzi (111), Neuroscience Program, Center for Biomedical Research and Innovation (Centro de Investigación e Innovación Biomédica [CiiB]), Universidad de los Andes, Santiago, Chile

Holly R. Hughes (273), Diagnostic and Reference Laboratory, Arboviral Diseases Branch, Division of Vector-Borne Diseases, Centers for Disease Control and Prevention, Fort Collins, CO, United States

Luiz Celso Hygino da Cruz Júnior (179,185,215), Department of Radiology, Clinic of Diagnostic Imaging/DASA, Rio de Janeiro, Rio de Janeiro, Brazil

Andrew Jameson (201), Michigan State University—College of Human Medicine, Grand Rapids, MI, United States

Rachel Jordan (463), The University of Oklahoma Health Sciences Center, College of Medicine, Oklahoma City, OK, United States

Rebecca B. Kairis (243), Department of Pediatrics, Section of Tropical Medicine, Baylor College of Medicine and Texas Children's Hospital, Houston, TX, United States

Taruna Kaura (337), Department of Medical Parasitology, Postgraduate Institute of Medical Education and Research, Chandigarh, India

Selman Kesici (125), Department of Pediatrics, Division of Pediatric Critical Care Medicine, Hacettepe University, İhsan Doğramacı Children's Hospital, Altındağ/Ankara, Turkey

Ahmad Suhail Khazali (439), Faculty of Applied Sciences, MARA University of Technology (UiTM) Perlis Branch, Arau, Malaysia

Julia Maria Klemens (293), Institute for Experimental Immunology, EUROIMMUN Medizinische Labordiagnostika AG, Lübeck, Germany

Erna Geessien Kroon (75), Laboratório de Vírus, Departamento de Microbiologia, Instituto de Ciências Biológicas, Universidade Federal de Minas Gerais, Belo Horizonte, MG, Brazil

Walter Sze Tung Lam (15), Yong Loo Lin School of Medicine, National University of Singapore, Singapore

Erik Lattwein (293), Institute for Experimental Immunology, EUROIMMUN Medizinische Labordiagnostika AG, Lübeck, Germany

Vladimir V. Lazarev (207), Laboratory of Neurobiology and Clinical Neurophysiology, National Institute of Women, Children and Adolescents Health Fernandes Figueira (IFF), Oswaldo Cruz Foundation (FIOCRUZ), Rio de Janeiro, Brazil

Túlio César Rodrigues Leite (417), Biological Sciences Research Center, Federal University of Ouro Preto, Ouro Preto, Minas Gerais, Brazil

Cui Li (359), State Key Laboratory of Molecular Developmental Biology, Institute of Genetics and Developmental Biology, Chinese Academy of Sciences, Beijing, China

Nerilson Marques Lima (327), Department of Chemistry—ICE, Federal University of Juiz de Fora, Juiz de Fora, Brazil

Maria Elizabeth Lopes Moreira (207), Unit of Clinical Research, National Institute of Women, Children and Adolescents Health Fernandes Figueira (IFF), Oswaldo Cruz Foundation (FIOCRUZ), Rio de Janeiro, RJ, Brazil

Karina Carrillo Loza (193), Neurological Emergency Service, National Institute of Neurology and Neurosurgery Manuel Velasco Suarez, Tlalpan, México City, Mexico

Denise Cantarelli Machado (85), Brain Institute of Rio Grande do Sul (BraIns), Medical School, Pontifical Catholic University of Rio Grande do Sul, Rio Grande do Sul, Brazil

Fernanda Majolo (85), Brain Institute of Rio Grande do Sul (BraIns), Medical School, Pontifical Catholic University of Rio Grande do Sul; Post-Graduate Program in Biotechnology, Universidade do Vale do Taquari—Univates, Lajeado, Rio Grande do Sul, Brazil

Luiz Cosme Cotta Malaquias (417), Vaccine Laboratory, Federal University of Alfenas, Alfenas, Minas Gerais, Brazil

Giuseppe Manfroni (349), Department of Pharmaceutical Sciences, University of Perugia, Perugia, Italy

Daniel Rodrigo Marinowic (85), Brain Institute of Rio Grande do Sul (BraIns), Medical School, Pontifical Catholic University of Rio Grande do Sul, Rio Grande do Sul, Brazil

Dimitri Marques Abramov (207), Laboratory of Neurobiology and Clinical Neurophysiology, National Institute of Women, Children and Adolescents Health Fernandes Figueira (IFF), Oswaldo Cruz Foundation (FIOCRUZ), Rio de Janeiro, Brazil

G.L. Marseglia (151), Pediatric Clinic, Fondazione IRCCS Policlinico "San Matteo", University of Pavia, Pavia (PV), Italy

Ewen McLean (97,377), Aqua Cognoscenti, West Columbia, SC, United States

Breno de Mello Silva (417), Department of Biological Sciences, Federal University of Ouro Preto, Ouro Preto, Minas Gerais, Brazil

Stelia Mendez-Sanchez (317), Group for Research in Biochemistry and Microbiology (Grupo de Investigación en Bioquímica y Microbiología-GIBIM). School of Chemistry, Universidad Industrial de Santander, Bucaramanga, Colombia

Abhishek Mewara (337), Department of Medical Parasitology, Postgraduate Institute of Medical Education and Research, Chandigarh, India

Eiji Morita (49), Department of Biochemistry and Molecular Biology, Faculty of Agriculture and Life Science, Hirosaki University, Hirosaki-shi, Aomori, Japan

Kristy O. Murray (243), Department of Pediatrics, Section of Tropical Medicine, Baylor College of Medicine and Texas Children's Hospital, Houston, TX, United States

Dean Myers (463), Harold Hamm Diabetes Center, Department of Obstetrics and Gynecology, The University of Oklahoma Health Sciences Center, Oklahoma City, OK, United States

Manuela Sales Lima Nascimento (3), Department of Microbiology and Parasitology, Post-Graduate Program in Parasite Biology, Biosciences Center, Federal University of Rio Grande do Norte, Natal, RN, Brazil

Osvaldo J.M. Nascimento (215), Department of Neurology, Federal Fluminense University, Niterói, Rio de Janeiro, Brazil

Marciano Viana Paes (165), Laboratory of Interdisciplinary Medical Research, Oswaldo Cruz Institute, Rio de Janeiro, Brazil

James Papin (463), Division of Comparative Medicine, The University of Oklahoma Health Sciences Center, Oklahoma City, OK, United States

Giovannia A.L. Pereira (283), Department of Fundamental Chemistry, Federal University of Pernambuco (UFPE), Biomedical Nanotechnology Research Group (NanoBio), Recife, Pernambuco, Brazil

Goreti Pereira (283), Department of Fundamental Chemistry, Federal University of Pernambuco (UFPE), Biomedical Nanotechnology Research Group (NanoBio), Recife, Pernambuco, Brazil

Maria I.A. Pereira (283), Biophysics and Radiobiology Department, Federal University of Pernambuco (UFPE), Biomedical Nanotechnology Research Group (NanoBio), Recife, Pernambuco, Brazil

Naveed Pervaiz (337), Department of Zoology, Panjab University, Chandigarh, India

Byron W. Purse (385), Department of Chemistry and Biochemistry and the Viral Information Institute, San Diego State University, San Diego, CA, United States

Kíssila Rabelo (165), Laboratory of Ultrastructure and Tissue Biology, Rio de Janeiro State University, Rio de Janeiro, Brazil

Jéssika F.F. Ribeiro (283), Biophysics and Radiobiology Department, Federal University of Pernambuco (UFPE), Biomedical Nanotechnology Research Group (NanoBio), Recife, Pernambuco, Brazil

Shannon E. Ronca (243), Department of Pediatrics, Section of Tropical Medicine, Baylor College of Medicine and Texas Children's Hospital, Houston, TX, United States

Shira Roth (231), Faculty of Engineering, The Institute of Nanotechnology and Advanced Materials, Bar-Ilan University, Ramat Gan, Israel

Tania Regina Saad Salles (207), Laboratory of Neurobiology and Clinical Neurophysiology, National Institute of Women, Children and Adolescents Health Fernandes Figueira (IFF), Oswaldo Cruz Foundation (FIOCRUZ), Rio de Janeiro, Brazil

Lívia Sacchetto (417), Department of Microbiology, Federal University of Minas Gerais, Belo Horizonte, Minas Gerais, Brazil

J.C. Saiz (409), Department of Biotechnology, National Agricultural and Food Research and Technology Institute (INIA), Madrid, Spain

Natália Gedeão Salomão (165), Laboratory of Interdisciplinary Medical Research, Oswaldo Cruz Institute, Rio de Janeiro, Brazil

Beate S. Santos (283), Pharmaceutical Sciences Department, Federal University of Pernambuco (UFPE), Biomedical Nanotechnology Research Group (NanoBio), Recife, Pernambuco, Brazil

Sandra Saschenbrecker (293), Institute for Experimental Immunology, EUROIMMUN Medizinische Labordiagnostika AG, Lübeck, Germany

S. Savasta (151), Pediatric Clinic, Fondazione IRCCS Policlinico "San Matteo", University of Pavia, Pavia (PV), Italy

Wolfgang Schlumberger (293), Institute for Experimental Immunology, EUROIMMUN Medizinische Labordiagnostika AG, Lübeck, Germany

Kimberly Schmitt (429), Department of Microbiology, Immunology and Pathology, Colorado State University, Fort Collins, CO, United States

Jair L. Siqueira-Neto (385), Centre for Discovery and Innovation in Parasitic Diseases, Skaggs School of Pharmacy and Pharmaceutical Sciences, University of California, San Diego, La Jolla, CA, United States

Derek Tuck Loong Soon (15), Division of Neurology, University Medicine Cluster, National University Health Service, Singapore

José Luis Soto-Hernández (193), Department of Infectious Diseases, National Institute of Neurology and Neurosurgery Manuel Velasco Suárez, Tlalpan, México City, Mexico

Gabriel Augusto Pires de Souza (417), Department of Microbiology, Federal University of Minas Gerais, Belo Horizonte, Minas Gerais, Brazil

Katja Steinhagen (293), Institute for Experimental Immunology, EUROIMMUN Medizinische Labordiagnostika AG, Lübeck, Germany

Konstanze Stiba (293), Institute for Experimental Immunology, EUROIMMUN Medizinische Labordiagnostika AG, Lübeck, Germany

Paul Ananth Tambyah (15), Division of Infectious Diseases, University Medicine Cluster, National University Health Service, Singapore

Gene S. Tan (367), J. Craig Venter Institute; Division of Infectious Diseases, Department of Medicine, University of California San Diego, La Jolla, CA, United States

Stephane Tosta (29), Department of Genetics, Ecology and Evolution, Federal University of Minas Gerais, Belo Horizonte, MG, Brazil

C. Trabatti (151), Pediatric Clinic, Fondazione IRCCS Policlinico "San Matteo", University of Pavia, Pavia (PV), Italy

Gönül Vural (135), Department of Neurology, Faculty of Medicine, Ankara Yildirim Beyazit University, Ankara, Turkey

Heron Werner (185,215), Department of Radiology, Clinic of Diagnostic Imaging/DASA, Rio de Janeiro, Rio de Janeiro, Brazil

Joilson Xavier (29), Department of Genetics, Ecology and Evolution, Federal University of Minas Gerais, Belo Horizonte, MG, Brazil

Zhiheng Xu (359), State Key Laboratory of Molecular Developmental Biology, Institute of Genetics and Developmental Biology, Chinese Academy of Sciences, Beijing, China

Tay Wei Xuan (15), Yong Loo Lin School of Medicine, National University of Singapore, Singapore

Rohana Yusof (439), Department of Molecular Medicine, Faculty of Medicine, University of Malaya, Kuala Lumpur, Malaysia

Foreword

Zika virus is a disease of great concern for a great many reasons, although the fundamental issues relate to the current lack of medication or vaccination and the potentially devastating birth defects impacting babies following transmission of the virus during pregnancy. Even though spread mostly by the Aedes species mosquito, there are also other known transmission routes, for example, through unprotected sexual intercourse with an infected partner and potential transmission routes such as blood donation. Symptomatically, as with other viral infections, Zika may be accompanied by fever, headache, joint and muscular pain among other symptoms. However, again, similar to several other viral infections, there may be no symptoms whatsoever and thus the infected person would not be aware of his/her infection status. As a result of the recent Covid-19 pandemic international interest and awareness has been focussed on the devastating effects of viral infections at the population level, but interestingly, in the context of Zika, the impact of birth defects associated with this particular virus can be devastating and lifelong. Fortunately, Zika can be diagnosed by a blood or urine test and these offer some confidence in both a confirmatory diagnosis, the context for the development of interventions for those affected, and a fertile basis for proactive research enquiry and the nurturing of a robust evidence base.

There is still so much that we don't know about Zika and it is thus a great pleasure for me to write this foreword for this excellent work by Professors Martin, Hollins Martin, Preedy, and Dr. Rajendram in bringing together leading experts in the Zika disease field to provide an evidence-based and state-of-the-art account of the latest research insights into the diagnostic, control, and model aspects of the Zika virus. Understanding these factors is critical in the provision of evidence-informed clinical care as well as innovation in the treatment for those experiencing this potentially devastating infection. I thus commend these editors, and authors, in producing *Zika Virus Impact, Diagnosis, Control, and Models*, a comprehensive "one-stop shop" book, which is accessible to both the clinician and researcher alike and relevant to all those working in this area.

Finally, we should always borne in mind that as we innovate in treatment, particularly in the case of the more extreme consequences of Zika infection, there are persons and individuals impacted by the diagnosis and the potential consequential impact on their quality of life and life course development and aspirations, and it is hoped that this book will help in treatment at both the global and individual level.

Professor Alan St Clair Gibson MBChB PhD MD

Professor St Clair Gibson is an associate dean (research) at the University of Hull. He is a former head of the School of Medicine at the University of the Free State, South Africa, dean of the School of Health and Human Performance Sciences at the University of Waikato, New Zealand, and a research fellow at the National Institute of Neurological Disorders and Stroke, NIH, Washington DC, USA. He received MBChB, PhD, and MD degrees from the University of Cape Town, South Africa, and believes in the importance of all of teaching and training, clinical work, and research in best combatting diseases and pathology which afflict the world's population.

Preface

The Zika virus is a member of the virus family Flaviviridae and was first identified in the 1940s. In 2007 the first outbreak of Zika virus disease was reported; this was followed by substantial outbreaks of the viral infection in 2013. In 2015 a large outbreak was reported in Brazil. Concomitant associations with Guillain-Barré syndrome and microcephaly (currently termed as congenital ZIKV syndrome) were also reported. The gravity of the situation was recognized by the World Health Organization and declared the Zika virus a *Public Health Emergency of International Concern*. In 2018 the Zika virus was added to the World Health Organization's list of diseases which merits priority attention and research.

The Zika virus is transmitted by Aedes mosquitoes and clinical symptoms can be mild. However, in some adults there may be adverse consequences of Zika virus infection. Inflammatory conditions, nerve damage, and triggering of Guillain-Barré syndrome as a consequence of Zika virus infection can also arise. Zika virus infection during pregnancy can have an impact on the developing fetus resulting in offspring with microcephaly. These conditions are only one end of the spectrum of disorders and conditions associated with Zika virus. There is also the psychological impact on the mother carrying the child with microcephaly to term. The family unit and community at large are also adversely affected. From a public health point of view there needs to be strategies to prevent and treat Zika virus infection using emerging and established scientific knowledge and methods. For example, recently it has been shown that the Zika virus can be transmitted sexually and via breast milk. However, very little is known about the long-term consequences of Zika virus infection.

In order to provide a framework for the broader understanding of Zika virus it is necessary to gather all the relevant information into one single place. However, hitherto information on the Zika virus and associated conditions have been sporadic and within different scientific domains. Much of the material is targeted toward experts working within specific disciplines without regard for the novice or those newly embarking upon Zika virus-related research. To address these issues we have compiled two books on Zika virus:

Book 1: Zika Virus Biology, Transmission, and Pathways

Book 2: Zika Virus Impact, Diagnosis, Control, and Models

In the book **Zika Virus Impact, Diagnosis, Control, and Models** we have over 40 chapters divided into 7 sections.

Section A: Zika virus: Setting the scene
Section B: Microcephaly and congenital syndromes
Section C: Guillain-Barré syndrome
Section D: Case studies and short reports
Section E: Methods, biomarkers, and diagnosis
Section F: Control, vaccines, and treatments
Section G: Models and modeling

There are of course always difficulties in ascribing chapters to different sections and placing them in order. Some chapters are equally at home in more than one section. However, the excellent indexing system allows material to be rapidly located.

To aid understanding across the different disciplines we have introduced the following features within each chapter.

An Abstract: *published on line.*
A Mini-dictionary of terms: *to aid the novice and to bridge the transdisciplinary and transintellectual divides.*
Key Facts: *focused areas of interest.*
Policy and Procedure: *recommendations or suggestions for research, strategies, guidelines, prevention, and treatment.*
Summary Points: *encapsulating the whole chapter in a series of single sentences.*

Zika Virus Impact, Diagnosis, Control, and Models is designed for research and teaching purposes. It is suitable for neurologists, virologists, health scientists, public health workers, doctors, pharmacologists, and research scientists. It is valuable as a personal reference book and also for academic libraries that covers the domains of virology, neurology, and health sciences. Contributions are from leading national and international experts including those from world renowned institutions. It is suitable for undergraduates, postgraduates, lecturers, and academic professors.

The Editors

Section A

Zika virus: Setting the scene

Chapter 1

How Zika virus emerged and spread worldwide

Josélio Maria Galvão de Araújo, Manuela Sales Lima Nascimento, Paulo Marcos da Matta Guedes, and José Veríssimo Fernandes

Department of Microbiology and Parasitology, Post-Graduate Program in Parasite Biology, Biosciences Center, Federal University of Rio Grande do Norte, Natal, RN, Brazil

Abbreviations

CNS	central nervous system
GBS	Guillain-Barré syndrome
IgG	immunoglobulin G
PAHO	Pan American Health Organization
RNA	ribonucleic acid
RT-PCR	reverse transcription polymerase chain reaction
WHO	World Health Organization
ZIKV	Zika virus

African origin of Zika virus and its dispersion to Asia

The Zika virus (ZIKV) emerged in Uganda, Eastern Africa, around 1920, from where it spread westward in two different moments. First, the strain Zika MR766 Uganda 1947 migrated around 1935 from Uganda to the Central African Republic and then to Nigeria (MacNamara, 1954). The another introduction occurred when the virus migrated from Uganda to Ivory Coast, around 1940, and from there to Senegal (Cornet, Robin, Chateau, et al., 1979). This resulted in the strain Zika DakAr 41,524 Senegal 1984, which continued to circulate in the region and was detected from 1988 to 1993 (Monlun, Zeller, Le Guenno, et al., 1993).

ZIKV was first isolated in Uganda in April 1947 during a research supported by the Rockefeller Foundation to study the enzootic or wild cycle of the yellow fever virus and to identify new circulating arboviruses (Weaver, Costa, Garcia-Blanco, et al., 2016). During this research, one of the rhesus monkeys (*Macaca mulatta*) used as sentinels in the Zika Forest developed fever and viremia. A blood sample was collected from this animal during the fever spike, and the serum obtained was inoculated in newborn mice intracerebrally. These mice developed disease 10 days after inoculation of the unknown virus, which was then called Zika virus in reference to the forest of its first isolation.

One year later the same virus was isolated from *Aedes africanus* mosquitoes also collected in the Zika Forest (Dick, Kitchen, & Haddow, 1952. Again, mice were used to evaluate the pathogenesis of the disease caused by ZIKV. They found that animals of all ages tested were susceptible to the virus when it was inoculated intracerebrally. Mice less than 2 weeks old were highly susceptible to ZIKV intraperitoneal inoculation. In addition, the virus proved to be highly neurotropic, and any viral particle was recovered from other tissues. Guinea pigs and rabbits showed no clinical signs after ZIKV intracerebral inoculation, and the subcutaneous infection in monkeys resulted in asymptomatic infection, although all inoculated animals showed viremia during the first week and produced specific antibodies detected 14 days post infection. In the same study, from 99 human serum samples analyzed, six presented neutralizing antibodies against ZIKV, indicating the virus circulation in the local population (Dick, 1952).

Zika Virus Impact, Diagnosis, Control, and Models. https://doi.org/10.1016/B978-0-12-820267-8.00001-7

The first case reported and laboratory-confirmed ZIKV infection in humans was in Tanzania, Africa, in 1948, 1 year after the virus isolation from monkeys (Smithburn, 1952). In 1953 ZIKV was found in Nigeria, where three cases of the disease were diagnosed: one based on viral isolation and two based on the seroconversion. These cases were clinically characterized with a feverish condition, headache, and diffuse joint pain, and one of the patients had mild jaundice (MacNamara, 1954). In the same period, neutralizing antibodies were also detected in humans residing in Egypt, Northeast Africa (Smithburn, Taylor, Rizk, et al., 1954), and then in Mozambique in Southern Africa in 1957 (Kokernot, Smithburn, Gandara, et al., 1960).

As early as the 1950s, probably a descendant of the Zika MR766 Uganda 1947 lineage migrated from East Africa to the Asian continent in British Malaysia in 1954. The term British Malaysia is often used to refer to the Federated Malay States and Non-Federated, located on the Malay Peninsula and the island of Singapore that were colonized by the British from the 18th century to the second half of the 20th century. ZIKV circulation in this region was evidenced by the detection of neutralizing antibodies in the sera from long-term resident individuals in Malaysia and Borneo (Smithburn, 1954). In 1951 ZIKV neutralizing antibodies were also detected in humans in Indonesia (Olson, Ksiazek, Gubler, et al., 1983) and in Philippines in 1953 (Hammon, Schrack, & Sather, 1985), both located in Southeast Asia. In India, South Asia, it was detected in 1954 (Smithburn, Kerr, & Gatne, 1954). In the same period, neutralizing antibodies against the ZIKV were also found in humans living in Thailand and Vietnam in Southeast Asia (Pond, 1963; Smithburn, 1954).

The ability of ZIKV to infect humans was proven through an experimental infection in a human volunteer in 1956 (Bearcroft, 1956). According to Gudo, Falk, Ali, et al. (2016), although it has been omitted in several studies on the historical past of ZIKV, there is evidence of human cases in Mozambique since 1957, as shown by the presence of neutralizing antibodies in nine different localities of that country, with overall prevalence rates of 4% (Kokernot et al., 1960).

The first isolation of ZIKV from human serum only occurred in the early 1960s in Uganda (Simpson, 1964) and in 1966 ZIKV was isolated *A. aegypti* mosquito in Malaysia (Marchette, Garcia, & Rudnick, 1969). In the late 1960s Kenya demonstrated seropositivity to ZIKV at 52% of overall population, but a wide variation between areas was observed (Geser, Henderson, & Christensen, 1970).

Between 1964 and 1970, ZIKV was isolated from both humans and mosquitoes in Nigeria, characterizing the second introduction of the ZIKV into Nigerian territory (Moore, Causey, Carey, et al., 1975). The isolated strain during this period was Zika IbH 30,656 Nigeria 1968. From 1971 to 1975 the presence of neutralizing antibodies and the virus was also detected in human sera samples, showing that the virus continued circulating in the local population during this period (Fagbami, 1979).

From Nigeria the virus spread to other Gulf of Guinea countries, where neutralizing antibodies were detected in humans residing in several countries in the region, including Cameroon and Togo (1964), Liberia, and Benin (1967) and neighboring countries such as Central African Republic (1961) and Burkina Faso (1963), Mali (1964), and Niger (1965) (Brès, 1970). Neutralizing antibodies have also been found in humans residing in Atlantic coast countries such as Angola (1960) (Kokernot, Casaca, Weinbren, et al., 1965), Senegal (1961), Guinea-Bissau (1964) (Pinto, 1967), Morocco (1966), and Gabon (1967) (Brès, 1970). The same is true for the Ethiopia population, located in the Horn of Africa (1961) (Sérié, Casals, Panthier, et al., 1968) and neighboring countries such as Somalia, Kenya (1966) (Henderson, Metselaar, Cahill, et al., 1968), and Sierra Leone (1972) (Robin & Mouchet, 1975).

An epidemiological study carried out in Senegal from 1972 to 1975 found neutralizing antibodies against ZIKV in individuals residing in five different areas of the country, proving its circulation during this period in the local population (Renaudet, Jan, Ridet, et al., 1978). From 1969 to 1983, ZIKV circulated in Pakistan in South Asia, where it was isolated from mosquitoes and sporadic cases in humans were confirmed (Darwish, Hoogstraal, Roberts, et al., 1983).

Reports of human cases of ZIKV infection occurred in 1977 in Java, Indonesia (Olson, Ksiazek, & Triwibowo, 1981). In this period, detection of neutralizing antibodies in human and animal sera was reported in Indonesia (Olson et al., 1983). Subsequently, virus isolation was obtained from human serum in Senegal in 1990 (Monlun et al., 1993) and in Cameroon (Fokam et al., 2010). Then the virus reemerged in the Gulf of Guinea region reaching Côte d'Ivoire in 1999 (Akoua-Koffi, Diarrassouba, Bénié, et al., 2001).

In a retrospective seroprevalence study of anti-ZIKV IgG antibodies among Vietnam and Laos blood donors covering the period 2003–2004 and 2015, the presence of neutralizing antibodies was detected in the population of these two countries. The prevalence of these antibodies was low in the sera collected in 2003 and 2004, with a significant increase observed in 2015, with no difference between men and women with positive serology. This indicates that ZIKV has been circulating in these countries since the early 2000s as an intensification of the number of cases in 2015 (Pastorino, Sengvilaipaseuth, Chanthongthip, et al., 2019). Fig. 1 shows the year and countries where the ZIKV was reported.

FIG. 1 The history of Zika virus in Africa (A), Asia and Oceania (B), and Americas (C) (WHO, 2020).

Continued

Introduction of ZIKV in Oceania

For decades, ZIKV infection has been neglected because it is a benign and self-limiting disease with sporadic cases in Africa and Asia (Brogueira & Miranda, 2017). The introduction of the ZIKV in the Oceania resulted in the first outbreak documented in humans by this virus, reported in 2007 in a North Pacific island called Yap, which is part of the Federated States of Micronesia. The disease went with rash, fever, arthralgia, and conjunctivitis. Involvement ZIKV in the etiology was confirmed by detection of viral RNA or the presence of virus-specific antibodies. The epidemic lasted about 3 months

FIG. 1, CONT'D

and reached 73% of the local population, but no hospitalizations, bleeding manifestations, or deaths were reported. Phylogenetic and molecular analysis showed that the ZIKV strain isolated during the Yap outbreak was divergent of Uganda prototype strain and the Senegal strain that represent the East and West African strains, respectively (Lanciotti, Kosoy, Laven, et al., 2008).

Although ZIKV has been circulating in Southeast Asia since the early 1950s, as noted earlier, the first case of laboratory-confirmed disease on the Asian continent came only in 2010 in Cambodia. ZIKV RNA was detected by reverse transcription polymerase chain reaction (RT-PCR) in the serum of a patient with clinical symptoms compatible with the disease, and the virus was isolated and identified by sequencing (Heang, Yasuda, Sovann, et al., 2012). This isolate gave rise to the Zika FSS13025 Cambodia 2010 strain, considered the reference Asian strain of ZIKV. Given the geographical proximity, it is possible that the ZIKV migrated from Vietnam and Laos in the early 2000s to Cambodia.

In October 2013 ZIKV reemerged in Oceania causing an explosive outbreak in French Polynesia, reaching several islands in the region, with an estimated occurrence of 19,000 cases (Cao-Lormeau, Roche, Teissier, et al., 2014). During the year 2014 it spread rapidly to several South Pacific islands, with outbreaks reported in New Caledonia, Cook Islands, Solomon Islands, Vanuatu, Samoa, Fiji (Musso & Gubler, 2016), and Easter Island (Tognarelli, Ulloa, Villagra, et al., 2016). The strain isolated during this outbreak in the South Pacific was characterized as being from the Asian genotype and registered as Zika H/PF/2013. Phylogenetic analysis showed that this strain showed higher genetic identity with the Zika strain FSS13025 Cambodia 2010 (Cao-Lormeau et al., 2014; Heang et al., 2012). During this outbreak the association of ZIKV infection with neurological complications was first noticed, such as Guillain-Barré syndrome (GBS) (Oehler, Watrin, Leparc-Goffart, et al., 2014). Table 1 shows historical events associated with ZIKV.

Introduction of ZIKV in the Americas

A set of evidence suggests that the introduction of ZIKV into the American continent was through the northeast of Brazil, from May to November 2013, coming from French Polynesia. Several hypotheses have been proposed to explain the introduction of the ZIKV in Brazil, including the Football Confederations Cup in 2013, the World Cup FIFA in 2014, an international canoeing event in 2015, and even the possible successful deployment of reservoir hosts or infected mosquitoes carried by humans through vehicles or boats (Ai, Zhang, & Zhang, 2016). However, the most likely hypothesis is that the virus entered in Brazil during the Football Confederations Cup, a global sports event held in June 2013 that caused an increase in the flow of people from all over the world, including French Polynesia, where an outbreak of ZIKV was taking place (Musso, Nilles, & Cao-Lormeau, 2014).

A strong indication of this is the fact that the Tahiti football team, one of the countries affected by the ZIKV epidemic, played in Recife, Pernambuco, Northeastern Brazil, which means that not only the athletes and members of the technical

TABLE 1 Chronological list of events related to the distribution of Zika virus around the world.

Year	Country/place	Event	References
1947	Uganda	First isolation of the virus from rhesus monkeys	Dick et al. (1952)
1948	Uganda	Isolation of the virus from *Aedes africanus* mosquitoes	Dick (1952)
1948	Tanzania	First isolation of the virus from human sera	Smithburn (1952)
1953	Nigeria	Isolation of the virus from human sera	MacNamara (1954)
1954	Egypt	Detection of neutralizing antibodies in humans	Smithburn, Taylor, et al. (1954)
1954	India	Detection of neutralizing antibodies in humans	Smithburn, Kerr, et al. (1954)
1957	Mozambique	Detection of neutralizing antibodies in humans	Kokernot et al. (1960)
1964	Uganda	Isolation of the virus from humans	Simpson (1964)
1964	Malaysia	Isolation of the virus from mosquitoes	Marchette et al. (1969)
1970	Nigeria	Isolation of the virus from human sera	Monath, Wilson, and Casals (1973)
1977	Indonesia	First human case report outside Africa	Olson et al. (1981)
1978	Senegal	Detection of neutralizing antibodies in humans	Renaudet et al. (1978)
1975	Gabon	Detection of neutralizing antibodies in humans	Jan, Languillat, Renaudet, and Robin (1978)
1979	Central African Republic	Detection of neutralizing antibodies in humans	Saluzzo, Gonzalez, Hervé, and Georges (1981)
1983	Indonesia	Detection of neutralizing antibodies in humans	Olson et al. (1983)
1990	Senegal	Isolation of the virus from human sera	Monlun et al. (1993)
2007	Micronesia	First virus epidemic reported on Yap Island	Duffy et al. (2009)
2010	Cameroon	Isolation of the virus from human sera	Fokam et al. (2010)
2013	French Polynesia	Second report of epidemic and first reports of neurological complications	Hancock, Marfel, and Bel (2014)
2014	Chile	The virus reaches Easter Island in Polynesia and belongs to Chile	Tognarelli et al. (2016)
2015	Brazil	Beginning of the epidemic in Brazil with the first case reports in Rio Grande do Norte and Pernambuco	Zanluca, Melo, Mosimann, et al. (2015)
2015	Cabo Verde	Return of the virus to the African continent	Lourenço, Monteiro, Valdez, et al. (2018)
2016	Brazil	First case report of microcephaly associated with the Zika virus	Heukelbach, Alencar, Kelvin, et al. (2016)
2016	Americas	Broad spread of the virus throughout the Americas	Musso and Gubler (2016)

commission but also potentially infected supporters, coming from that country, circulated throughout the region. For all that, it's very likely that this was the way of introduction of the virus into Brazil (Albuquerque, de Souza, Araújo, et al., 2018). In addition, viral genome analyses with ecological and epidemiological data suggest that ZIKV was already circulating in northeastern Brazil in February 2014 (Faria, Quick, Claro, et al., 2017).

In 2014 the virus hit several other South Pacific islands including New Caledonia, Cook Islands, Solomon Islands, Vanuatu, Samoa and Fiji (Musso & Gubler, 2016), and Easter Island, which despite being in Polynesia belongs to Chile (Tognarelli et al., 2016).

In the second half of 2014, a disease characterized by low fever, rash, arthralgias, and conjunctivitis appeared simultaneously in Natal (capital of the state of Rio Grande do Norte) and Recife (capital of the state of Pernambuco), both in northeastern Brazil. Clinical manifestations did not fulfill the criteria for classical exanthematic diseases such as measles, rubella, scarlet fever, or sudden rash. In addition, although it is an endemic area for dengue, some serological evaluations were negative, and the patients did not present platelet reduction, very common in dengue. The possibility of infection with the chikungunya virus was also excluded by reverse transcription polymerase chain reaction (RT-PCR) (Luz, Glauco dos Santos, & Vieira, 2015), tests performed at the Infectious Diseases and Cancer Laboratory, Department of Microbiology and Parasitology, Federal University of Rio Grande do Norte.

The autochthonous transmission of ZIKV in the Americas was proven for the first time in March 2015 when the first cases of the disease were documented in the state of Rio Grande do Norte, Brazil (Zanluca et al., 2015), marking officially the arrival of ZIKV to the American continent, although since 2014 there were reports of acute exanthematic disease in the region (Heukelbach et al., 2016). The wide distribution of the mosquitoes *A. aegypti* and *A. albopictus* and the highly favorable climate conditions for the vector proliferation in the northeast region of Brazil contributed to the emergence of the ZIKV and its rapid spread to almost the entire country.

In March and April 2015 an outbreak of a disease with the same characteristics was reported in Salvador (Cardoso, Paploski, Kikuti, et al., 2015) and Camaçari (Campos, Bandeira, & Sardi, 2015), both in the state of Bahia, northeast of Brazil. The detection of ZIKV RNA by RT-PCR in June of the same year (Zanluca et al., 2015) confirmed that such outbreak was caused by the ZIKV. In December 2015 cases of autochthonous transmission of ZIKA were confirmed in all nine states of the northeastern region (Alagoas, Bahia, Ceará, Maranhão, Paraíba, Pernambuco, Piauí, Rio Grande do Norte, and Sergipe), in five states of the northern region (Amazonas, Pará, Rondônia, Roraima, and Tocantins), in three states of the southeast region (Espírito Santo, Rio de Janeiro, and São Paulo), in one state of southern region (Paraná), and in one state of the midwest region (Mato Grosso) (Musso & Gubler, 2016).

Brazil had the highest number of reported cases of ZIKV infection in the world, with an estimated 440,000–1,300,000 people infected by December 2016. During 2015 an increasing number of babies born with reduced head circumference, characterized as microcephaly, was observed in northeastern Brazil, with 29 cases recorded between August and October. In that year, there were 4783 suspected cases of microcephaly due to ZIKV congenital infection in the country, most of them in the northeast region, and 76 deaths of babies born from ZIKV-infected mothers during pregnancy were reported. ZIKV has been identified in blood and tissue samples from a newborn and of mothers who have given birth to babies with microcephaly and ophthalmic abnormalities (Heukelbach et al., 2016).

From Brazil the ZIKV spreads rapidly across the American continent. In October 2015 cases of the infection were confirmed in Colombia in the state of Bolivar from where it spread to other states. By the end of 2015, autochthonous transmission of the virus was already reported in 12 other countries and territories of the Americas and the Caribbean: Suriname, Venezuela, Guatemala, Honduras, Mexico, Paraguay, Panama, French Guiana, El Salvador, Haiti, Puerto Rico, and Martinique. In January 2016 ZIKV circulation was also proven in Bolivia, Nicaragua, Guyana, and Ecuador in South America and Barbados, Dominican Republic, Guadalupe, and Virgin Islands in the Caribbean (Musso & Gubler, 2016) (Fig. 1).

In northeastern Brazil, observational studies conducted in the states of Rio Grande do Norte, Pernambuco, and Bahia revealed increasing cases of GBS and that most patients had a clinical history of past symptoms compatible with fever of ZIKV (Araújo, Ferreira, & Nascimento, 2016). The first cases of neurological manifestations associated with ZIKV in Brazil were six patients seen at the Neurology Service of the *Hospital da Restauração* in Recife, Pernambuco, between December 2014 and June 2015, with four cases of GBS and one case of acute disseminated encephalomyelitis all in adults and one case of encephalitis in a 2-year-old child. ZIKV RNA was detected by RT-PCR in the serum of the five adult individuals and cerebrospinal fluid of the child with encephalitis. In addition, virus isolation was obtained from the serum of one of the GBS patients and the serum of the patient with acute disseminated encephalomyelitis (Brito Ferreira, Antunes de Brito, Moreira, et al., 2017).

In Brazil, only in 2015, there were 3530 cases of microcephaly, distributed by 20 states of the federation of the five regions of the country, mainly affecting the northeast, notably the states of Pernambuco, Rio Grande do Norte, Paraíba, and Bahia. This number of cases is extraordinarily high considering that the average annual incidence of microcephaly in the period from 2000 to 2014 was 157.3 cases per year. As this alarming growth in the number of cases of microcephaly and other congenital neuropathies coincided with the ZIKV epidemic in the region, the hypothesis of a possible association with maternal ZIKV infection during pregnancy and fetus microcephaly started to be considered (de Oliveira, Juan Cortez-Escalante, de Oliveira, et al., 2016).

A study conducted in Pernambuco revealed that only in the period from August to October 2015 were confirmed 40 cases of microcephaly distributed by eight municipalities of the metropolitan region of Recife, with the largest concentration of cases in the municipalities of Jaboatão dos Guararapes, Olinda, and Paulista. The children had a median head circumference of 29 cm, and 21 of 25 cases analyzed presented cerebral calcification, ventriculomegaly, or lissencephaly.

Twenty-seven of the 40 mothers reported rash with pruritus, headache, myalgia, and absence of fever during pregnancy, of which 20 are in the first trimester and seven in the second trimester (Vargas, Saad, Dimech, et al., 2016).

An observational study conducted in Brazil between September 2015 and May 2016 analyzed 345 pregnant women and found that among 117 live infants born to 116 ZIKV-positive women, 42% had congenital malformations, in most cases, of the central nervous system (CNS), evidenced clinically or by imaging exams in the first month of life, of which four had microcephaly. None of the newborns in the control group (children of mothers with no evidence of infection) had microcephaly (Brasil, Pereira, Moreira, et al., 2016).

Another study performed a retrospectively analysis of data from 1501 live births from November 2015 to February 2016 with suspected ZIKV-associated microcephaly, in which 602 were considered definitive or highly probable. It was found that most of the reported cases were newborns whose mothers had a clinical history of rash during pregnancy, with a directly proportional relationship between gestation time and head circumference size at birth. Neurological changes have also been observed in newborns with normal head circumference (França, Schuler-Faccini, Oliveira, et al., 2016).

The proof of vertical transmission of ZIKV and its direct relationship to congenital anomalies presented by newborns of mothers who had the infection during pregnancy, especially in first trimester, were well documented by detection of the virus in serum, cerebrospinal fluid, and fetal tissues. At least three important studies have contributed in this regard, all conducted with cases of microcephaly occurring during the ZIKV epidemic in Brazil (Araújo, Rodrigues, Alencar Ximenes, et al., 2016; Martines, Bhatnagar, Oliveira Ramos, et al., 2016; Mlakar, Korva, Tul, et al., 2016).

Given the magnitude of the epidemic and the risk of vertical transmission, with an alarming increase in the number of cases of microcephaly and other neurological manifestations resulting from congenital ZIKV infection, in addition to the association with GBS, the World Health Organization has declared a Public Health Emergency of Global Interest on September 1, 2016 (Chitti, Prasad, & Saxena, 2016).

Between 2015 and 2017, ZIKV spread rapidly throughout the populations of the Americas, reaching 51 countries and territories, especially in Latin America and the Caribbean, being reported over 800,000 cases of the disease (Pan American Health Organization/World Health Organization, 2018). The lack of prior exposure of the population to this virus resulted in the explosive increase in the number of cases of the disease in a short while (Harris, Caldwell, & Mordecai, 2019). Transmission of the virus was limited in 2015, but in 2016 and 2017 there were enough opportunities for widespread transmission of the virus in most cities, resulting in the exhaustion of susceptible individuals (O'Reilly, Lowe, Edmunds, et al., 2018).

Reintroduction of ZIKV in Africa

In addition to spreading widely across Asia and Americas, ZIKV returned back to its origin, the African continent, where it was responsible for the major outbreak of the disease in Cape Verde, where it reported the first cases of microcephaly associated with this virus in the African continent, with 18 cases diagnosed between July 2015 and May 2016 Lourenço et al. (2018) and then to Angola (Hill et al., 2019). Because the virus responsible for this outbreak is the same strain of Asian origin that caused the epidemic in the Americas, the most reasonable hypothesis is that the ZIKV was carried out from American countries, including Brazil, to the island by tourists. ZIKV remains endemic in other African countries near Cape Verde, such as Senegal. It is possible that the holding of the Olympic and Paralympic Games in Rio de Janeiro, Brazil, in August 2016, during which around 500,000 people from all over the world circulated throughout the city, contributed, at least in part, to dissemination of ZIKV around the world (Musso & Gubler, 2016).

During this period of dispersion, ZIKV underwent to processes of genetic diversification and molecular evolution, giving rise to genetically different strains (Shi, Zhang, Ling, et al., 2016). Genetic studies have revealed that since its discovery, ZIKV has evolved into three distinct genotypes: the East African genotype, which encompasses the descending cluster of the Zika MR766 Uganda 1947 prototype; the West African genotype, the descendant cluster of the Nigerian Zika IbH30656 Nigeria 1968 prototype; and the Asian genotype, which had as a common ancestor the East African genotype (Faye, Freire, Iamarino, et al., 2014). This Asian genotype has as a reference strain Zika FS13025 Cambodia 2010, isolated in Cambodia in Southeast Asia in 2010 (Haddow, Schuh, Yasuda, et al., 2012). These three genotypes share many common characteristics, because they all descend from a common ancestor, which is the prototype lineage isolated in Uganda 1947. Finally the history of the ZIKV teaches that surveillance of arboviruses in humans, nonhuman primates, and mosquitoes must persist for decision-making in public health, health education, and vector control, as vaccine against ZIKV is not yet available.

Policy and procedures

Not applied.

Mini-dictionary of terms

Arboviruses. An epidemiological name used to refer to viruses that are transmitted by arthropod vectors.

Genotypes. A term used regarding the genetic constitution of an organism. The ZIKV, for example, is a species of virus that currently has three known genotypes.

Guillain-Barré syndrome. A possible outcome of ZIKV infection, when the immune system attacks nerves causing neurological disorder.

Lissencephaly. An uncommon brain formation disorder associated with microcephaly and an absence of normal brain circumvolutions.

Microcephaly. When the brain is smaller than normal size for age, due to malformation. It may be the result of a number of factors, such as radiation, chemical substances, and congenital infections.

Neurotropic. What it has affinity with the nervous system.

Pathogenicity. The ability of a microorganism to cause disease.

Strain. A group of descendants with a common ancestor who share genetic and morphological similarities.

Ventriculomegaly. Dilation of the cerebral ventricles.

Vertical transmission. The transmission of an infection or disease from the mother to her fetus in the womb or newborn during childbirth. The main routes of infection are pregnancy, childbirth, and breastfeeding.

Viremia. The presence of viruses in the blood.

Key facts

Not applied.

Summary points

- This chapter describes the origin and spread of the ZIKV worldwide.
- ZIKV was first isolated in Uganda in April 1947.
- The first evidence of human exposure to ZIKV was obtained as early as 1948 in Tanzania.
- In Asia, ZIKV was first isolated in 1966 from the *Aedes aegypti* mosquito in Malaysia.
- The first reports of human cases of infection caused by this virus outside Africa only occurred in 1977 in Java, Indonesia.
- The introduction of the ZIKV on the Oceania resulted in the first outbreak documented in human caused, reported in 2007 at North Pacific island called Yap, which is part of the Federated States of Micronesia.
- ZIKV officially arrived in the American continent in March 2015 in the state of Rio Grande do Norte, Brazil. In this outbreak, an increased number of babies born with reduced head circumference were observed, called microcephaly.
- Given the magnitude of the epidemic and the risk of vertical transmission, with a large number of microcephaly cases and other neurological manifestations, the World Health Organization had declared a Public Health Emergency of Global Interest on September 1, 2016.
- In 2015, ZIKV went back to its origin continent, Africa.

References

Ai, J. W., Zhang, Y., & Zhang, W. (2016). Zika virus outbreak: 'A perfect storm'. *Emerging Microbes and Infections*, *5*, e21.

Akoua-Koffi, C., Diarrassouba, S., Bénié, V. B., et al. (2001). Investigation surrounding a fatal case of yellow fever in Côte d'Ivoire in 1999. *Bulletin de la Société de Pathologie Exotique*, *94*, 227–230.

Albuquerque, M. F. P. M., de Souza, W. V., Araújo, T. V. B., et al. (2018). The microcephaly epidemic and Zika virus: Building knowledge in epidemiology. *Cadernos de Saúde Pública*, *34*(10), e00069018.

Araújo, L. M., Ferreira, M. L. B., & Nascimento, O. J. M. (2016). Guillain-Barré syndrome associated with the Zika virus outbreak in Brazil. *Arquivos de Neuro-Psiquiatria*, *74*, 253–255.

Araújo, T. V., Rodrigues, L. C., Alencar Ximenes, R. A., et al. (2016). Association between Zika virus infection and microcephaly in Brazil, January to May, 2016: Preliminary report of a case-control study. *Lancet Infectious Diseases*, *16*, 1356–1363.

Bearcroft, W. G. (1956). Zika virus infection experimentally induced in a human volunteer. *Transactions of the Royal Society of Tropical Medicine and Hygiene*, *50*, 442–448.

Brasil, P., Pereira, J. P., Jr., Moreira, M. E., et al. (2016). Zika virus infection in pregnant women in Rio de Janeiro. *New England Journal of Medicine*, *375*, 2321–2334.

Brito Ferreira, M. L., Antunes de Brito, C. A., Moreira, Á. J. P., et al. (2017). Guillain-Barré syndrome, acute disseminated encephalomyelitis and encephalitis associated with Zika virus infection in Brazil: Detection of viral RNA and isolation of virus during late infection. *American Journal of Tropical Medicine and Hygiene*, *97*, 1405–1409.

Brogueira, P., & Miranda, A. C. (2017). Virus Zika: Emergência de um Velho Conhecido. *Medicină Internă*, *24*(2), 146–153.

Brès, P. (1970). Données récentes apportées par les enquêtes sérologiques sur la prévalence des arbovirus en Afrique, avec référence spéciale à la fièvre jaune. *Bulletin of the World Health Organization*, *43*, 223–267.

Campos, G. S., Bandeira, A. C., & Sardi, S. I. (2015). Zika virus outbreak, Bahia, Brazil. *Emerging Infectious Diseases*, *21*, 1885–1886.

Cao-Lormeau, V. M., Roche, C., Teissier, A., et al. (2014). Zika virus, French polynesia, South Pacific, 2013. *Emerging Infectious Diseases*, *20*, 1085–1086.

Cardoso, C. W., Paploski, I. A. D., Kikuti, M., et al. (2015). Outbreak of exanthematous illness associated with Zika, chikungunya, and dengue viruses, Salvador, Brazil. *Emerging Infectious Diseases*, *21*, 2274–2276.

Chitti, S. V., Prasad, A. K., & Saxena, S. K. (2016). Emerging Zika virus disease: A public health emergency of global concern. *Virus Disease*, *27*, 211–214.

Cornet, M., Robin, Y., Chateau, R., et al. (1979). Isolement d'arbovirus au Se ́ne ́gal oriental a ́partir de moustiques (1972–1977) et notes sur l'e ́pide ́ miologiedes virus par les Ae en particulier du virus amaril. *Cah OR-STOM, Ser Entomology Medical Parasitology*, *17*, 149–163.

Darwish, M. A., Hoogstraal, H., Roberts, T. J., et al. (1983). A sero-epidemiological survey for certain arboviruses (*Togaviridae*) in Pakistan. *Transactions of the Royal Society of Tropical Medicine and Hygiene*, *77*, 442–445.

de Oliveira, W. K., Juan Cortez-Escalante, J., de Oliveira, W. T. G. H., et al. (2016). Increase in reported prevalence of microcephaly in infants born to women living in areas with confirmed Zika virus transmission during the first trimester of pregnancy – Brazil, 2015. *CDC Weekly*, *65*, 242–247.

Dick, G. W. A. (1952). Zika virus (II) pathogenicity and physical properties. *Transactions of the Royal Society of Tropical Medicine and Hygiene*, *46*, 521–534.

Dick, G. W. A., Kitchen, S. F., & Haddow, A. J. (1952). Zika Virus (I). Isolations and serological specificity. *Transactions of the Royal Society of Tropical Medicine and Hygiene*, *46*, 509–520.

Duffy, M. R., Chen, T.-H., Hancock, W. T., Powers, A. M., Kool, J. L., Lanciotti, R. S., et al. (2009). Zika virus outbreak on Yap Island, Federated States of Micronesia. *The New England Journal of Medicine*, *360*, 2536–2543.

Fagbami, A. H. (1979). Zika virus infections in Nigeria: Virological and seroepidemiological investigations in Oyo State. *Journal of Hygiene (Lond)*, *83*, 213–219.

Faria, N. R., Quick, J., Claro, I. M., et al. (2017). Establishment and cryptic transmission of Zika virus in Brazil and the Americas. *Nature*, *546*, 406–410.

Faye, O., Freire, C. C., Iamarino, A., et al. (2014). Molecular evolution of Zika virus during its emergence in the 20th century. *PLoS Neglected Tropical Diseases*, *8*, e2636.

Fokam, E., Levai, L., Guzman, H., Amelia, P., Titanji, V., et al. (2010). Silent circulation of arboviruses in Cameroon. *East African Medical Journal*, *6*, 262–268.

França, G. V., Schuler-Faccini, L., Oliveira, W. K., et al. (2016). Congenital Zika virus syndrome in Brazil: A case series of the first 1501 livebirths with complete investigation. *Lancet*, *388*, 891–897.

Geser, A., Henderson, B. E., & Christensen, S. (1970). A multipurpose serological survey in Kenya. 2. Results of arbovirus serological tests. *Bulletin of the World Health Organization*, *43*, 539–552.

Gudo, E. S., Falk, K. I., Ali, S., et al. (2016). Historic report of Zika in Mozambique: Implications for assessing current risk. *PLoS Neglected Tropical Diseases*, *10*, e0005052.

Haddow, A. D., Schuh, A. J., Yasuda, C. Y., et al. (2012). Genetic characterization of Zika virus strains: Geographic expansion of the Asian lineage. *PLoS Neglected Tropical Diseases*, *6*, e1477.

Hammon, W. M., Schrack, W. D., Jr., & Sather, G. E. (1985). Serological survey for arthropodborne virus infections in the Philippines. *American Journal of Tropical Medicine and Hygiene*, *7*, 323–328.

Hancock, W. T., Marfel, M., & Bel, M. (2014). Zika virus, French Polynesia, South Pacific, 2013. *Emerging Infectious Diseases*, *20*, 1960.

Harris, M., Caldwell, J. M., & Mordecai, E. A. (2019). Climate drives spatial variation in Zika epidemics in Latin America. *Proceedings of the Royal Society*, *286*, 20191578.

Heang, V., Yasuda, C. Y., Sovann, L., et al. (2012). Zika virus infection, Cambodia, 2010. *Emerging Infectious Diseases*, *18*, 349–351.

Henderson, B. E., Metselaar, D., Cahill, K., et al. (1968). Yellow fever immunity surveys in northern Uganda and Kenya and eastern Somalia, 1966–67. *Bulletin of the World Health Organization*, *38*, 229–237.

Heukelbach, J., Alencar, C. H., Kelvin, A. A., et al. (2016). Zika virus outbreak in Brazil. *Journal of Infection in Developing Countries*, *10*(2), 116–120.

Hill, S. C., Vasconcelos, J., Neto, Z., Jandondo, D., Zé-Zé, L., Aguiar, R. S., et al. (2019). Emergence of the Asian lineage of Zika virus in Angola: an outbreak investigation. *Lancet Infectious Diseases*, *19*, 1138–1147.

Jan, C., Languillat, G., Renaudet, J., & Robin, Y. (1978). A serological survey of arboviruses in Gabon. *Bulletin de la Société de Pathologie Exotique*, *71*, 140–146.

Kokernot, R. H., Smithburn, K. C., Gandara, A. F., et al. (1960). Neutralization tests with sera from individuals residing in Mozambique against specific viruses isolated in Africa, transmitted by arthropods. *Anais do Instituto de Higiene e Medicina Tropical*, *17*, 201–230.

Kokernot, R. H., Casaca, V. M. R., Weinbren, M. P., et al. (1965). Survey for antibodies against arthropod-borne viruses in the sera of indigenous residents of Angola. *Transactions of the Royal Society of Tropical Medicine and Hygiene*, *59*, 563–570.

Lanciotti, R. S., Kosoy, O. L., Laven, J. J., et al. (2008). Genetic and serologic properties of Zika virus associated with an epidemic, Yap State, Micronesia, 2007. *Emerging Infectious Diseases*, *14*, 1232–1239.

Lourenço, J., Monteiro, M. L., Valdez, T., et al. (2018). Epidemiology of the Zika virus outbreak in the Cabo Verde Islands, West Africa. *PLOS Currents*, *10*. https://doi.org/10.1371/currents.outbreaks.19433b1e4d007451c691f138e1e67e8c.

Luz, K. G., Glauco dos Santos, I. V., & Vieira, R. M. (2015). Zika virus efver. *Revista Epidemiologia e Serviços de Saúde*, *24*, 4.

MacNamara, E. M. (1954). Zika virus: A report on three cases of human infection during an epidemic of jaundice in Nigeria. *Transactions of the Royal Society of Tropical Medicine and Hygiene*, *48*, 139–145.

Marchette, N. J., Garcia, R., & Rudnick, A. (1969). Isolation of Zika virus from *Aedes aegypti* mosquitoes in Malaysia. *American Journal of Tropical Medicine and Hygiene*, *18*, 411–415.

Martines, R. B., Bhatnagar, J., Oliveira Ramos, A. M., et al. (2016). Pathology of congenital Zika syndrome in Brazil: A case series. *Lancet*, *388*, 898–904.

Mlakar, J., Korva, M., Tul, N., et al. (2016). Zika virus associated with microcephaly. *New England Journal of Medicine*, *374*, 951–958.

Monath, T. P., Wilson, D. C., & Casals, J. (1973). The 1970 yellow fever epidemic in Okwoga District, Benue Plateau State, Nigeria. 3. Serological responses in persons with and without pre-existing heterologous group B immunity. *Bulletin World Health Organization*, *49*, 235–244.

Monlun, E., Zeller, H., Le Guenno, B., et al. (1993). Surveillance of the circulation of arbovirus of medical interest in the region of eastern Senegal. *Bulletin de la Société de Pathologie Exotique*, *86*, 21–28.

Moore, D. L., Causey, O. R., Carey, D. E., et al. (1975). Arthropod-borne viral infections of man in Nigeria, 1964–1970. *Annals of Tropical Medicine and Parasitology*, *69*, 49–64.

Musso, D., & Gubler, D. J. (2016). Zika virus. *Clinical Microbiology Reviews*, *29*, 487–524.

Musso, D., Nilles, E. J., & Cao-Lormeau, V. M. (2014). Rapid spread of emerging Zika virus in the Pacific area. *Clinical Microbiology and Infection*, *20*, O595–O596.

Oehler, E., Watrin, L., Leparc-Goffart, I., et al. (2014). Zika virus infection complicated by Guillain–Barré syndrome—Case report, French Polynesia, December 2013. *Euro Surveillance*, *19*, 20720.

Olson, J. G., Ksiazek, T. G., & Triwibowo, S. (1981). Zika virus, a cause of fever in Central Java, Indonesia. *Transactions of the Royal Society of Tropical Medicine and Hygiene*, *75*, 389–393.

Olson, J. G., Ksiazek, T. G., Gubler, D. J., et al. (1983). A survey for arboviral antibodies in sera of humans and animals in Lombok, Republic of Indonesia. *Annals of Tropical Medicine and Parasitology*, *77*, 131–137.

O'Reilly, K. M., Lowe, R., Edmunds, W. J., et al. (2018). Projecting the end of the Zika virus epidemic in Latin America: A modelling analysis. *BMC Medicine*, *16*, 180.

Pan American Health Organization/World Health Organization. (2018). *Zika suspected and confirmed cases reported by countries and territories in the Americas: Cumulative cases, 2015–2017*. http://ais.paho.org/phip/viz/ed_zika_cases.asp.

Pastorino, B., Sengvilaipaseuth, O., Chanthongthip, A., et al. (2019). Low Zika virus seroprevalence in Vientiane, Laos, 2003–2015. *American Journal of Tropical Medicine and Hygiene*, *100*, 639–642.

Pinto, M. R. (1967). Survey for antibodies to arboviruses in the sera of children in Portuguese Guinea. *Bulletin of the World Health Organization*, *37*, 101–108.

Pond, W. L. (1963). Arthropod-borne virus antibodies in sera from residents of South-East Asia. *Transactions of the Royal Society of Tropical Medicine and Hygiene*, *57*, 364–371.

Renaudet, J., Jan, C., Ridet, J., et al. (1978). A serological survey of arboviruses in the human population of Senegal. *Bulletin de la Societé de Pathologie Exotique et de ses Filiales*, *71*, 131–140.

Robin, Y., & Mouchet, J. (1975). Serological and entomological study on yellow fever in Sierra Leone. *Bulletin de la Societé de Pathologie Exotique et de ses Filiales*, *68*, 249–258.

Saluzzo, J. F., Gonzalez, J. P., Hervé, J. P., & Georges, A. J. (1981). Serological survey for the prevalence of certain arboviruses in the human population of the south-east area of Central African Republic. *Bulletin de la Société de Pathologie Exotique*, *74*, 490–499.

Shi, W., Zhang, Z., Ling, C., et al. (2016). Increasing genetic diversity of Zika virus in the Latin American outbreak. *Emerging Microbes and Infections*, *5*, e68.

Simpson, D. I. H. (1964). Zika virus infection in man. *Transactions of the Royal Society of Tropical Medicine and Hygiene*, *58*, 335–338.

Smithburn, K. C. (1952). Neutralizing antibodies against certain recently isolated viruses in the sera of human beings residing in East Africa. *Journal of Immunology*, *69*, 223–234.

Smithburn, K. C. (1954). Neutralizing antibodies against arthropod-borne viruses in the sera of long-time residents of Malaya and Borneo. *American Journal of Hygiene*, *59*, 157–163.

Smithburn, K. C., Taylor, R. M., Rizk, F., et al. (1954). Immunity to certain arthropod-borne viruses among indigenous residents of Egypt. *American Journal of Tropical Medicine and Hygiene*, *3*, 9–18.

Smithburn, K. C., Kerr, J. A., & Gatne, P. B. (1954). Neutralizing antibodies against certain viruses in the sera of residents of India. *Journal of Immunology*, *72*, 248–257.

Sérié, C., Casals, J., Panthier, R., et al. (1968). Etudes sur la fièvre jaune en Ethiopie. 2. Enquête sérologique sur la population humaine. *Bulletin of the World Health Organization*, *38*, 843–854.

Tognarelli, J., Ulloa, S., Villagra, E., et al. (2016). A report on the outbreak of Zika virus on Easter Island, South Pacific, 2014. *Archives of Virology*, *161*, 665–668.

Vargas, A., Saad, E., Dimech, G. S., et al. (2016). Characteristics of the first cases of microcephaly possibly related to Zika virus reported in the Metropolitan Region of Recife, Pernambuco State, Brazil. *Revista Epidemiologia e Serviços de Saúde*, *25*, 691–700.

Weaver, S. C., Costa, C., Garcia-Blanco, M. A., et al. (2016). Zika virus: History, emergence, biology, and prospects for control. *Antiviral Research*, *130*, 69–80.

World Health Organization (WHO). (2020). *The history of Zika virus*. Available from https://www.who.int/emergencies/zika-virus/timeline/en/.

Zanluca, C., Melo, V. C. A., Mosimann, A. L. P., et al. (2015). First report of autochthonous transmission of Zika virus in Brazil. *Memórias do Instituto Oswaldo Cruz*, *110*, 09.

Chapter 2

Clinical neurological spectrum of adult and congenital ZIKV infection: An overview of virology, pathogenesis, and management

Walter Sze Tung Lam[a], Tay Wei Xuan[a], Paul Ananth Tambyah[b], and Derek Tuck Loong Soon[c]

[a]*Yong Loo Lin School of Medicine, National University of Singapore, Singapore,* [b]*Division of Infectious Diseases, University Medicine Cluster, National University Health Service, Singapore,* [c]*Division of Neurology, University Medicine Cluster, National University Health Service, Singapore*

Abbreviations

AMSAN	acute motor sensory axonal neuropathy
CNS	central nervous system
CSF	cerebrospinal fluid
CT	computed tomography
DENV	dengue virus
ELISA	enzyme-linked immunosorbent assay
EUA	Emergency Use Authorization
GBS	Guillain-Barré syndrome
hNPC	human neural progenitor cells
IgM	immunoglobulin M
JEV	Japanese encephalitis virus
MRI	magnetic resonance imaging
NAAT	nucleic acid amplification tests
NSAID	nonsteroidal antiinflammatory drug
PCR	polymerase chain reaction
PHEIC	Public Health Emergency of International Concern
PRNT	plaque reduction neutralization test
RNA	ribonucleic acid
RT-PCR	reverse transcription polymerase chain reaction
TBEV	tick-borne encephalitis virus
YFV	yellow fever virus
ZIKV	Zika virus

Virology of Zika virus

Introduction to Zika virus

Zika virus (ZIKV) is an enveloped nonsegmented, single-stranded, positive-sense RNA mosquito-borne flavivirus, most notably associated with the *Aedes* mosquitoes. The *Flavivirus* genus includes many other significant human pathogens, such as dengue virus (DENV), yellow fever virus (YFV), and Japanese encephalitis virus (JEV).

ZIKV was first discovered in the Zika Forest of Uganda in 1947 from a febrile rhesus macaque; subsequent serological studies performed on human serum samples showed the presence of neutralizing antibodies against the virus, revealing ZIKV's pathogenicity in humans (Dick, Kitchen, & Haddow, 1952).

Zika Virus Impact, Diagnosis, Control, and Models. https://doi.org/10.1016/B978-0-12-820267-8.00002-9

Current genetic and phylogenetic analyses reveal two major lineages of ZIKV—African and Asian (Wang et al., 2016), with <5% amino acid variation among the different strains. Immune serum from ZIKV-infected humans and animals was shown to neutralize ZIKV strains and inhibit infections from both lineages equivalently, thereby demonstrating that ZIKV circulates as a single serotype (Dowd et al., 2016).

Epidemiology of Zika virus

Until 2007, ZIKV was believed to be an infection of mild clinical significance and limited geographical distribution among flaviviruses (Liu, Shi, & Qin, 2019). ZIKV clinically presents with mild symptoms including fever, headache, rash, conjunctivitis, arthralgia, and myalgia (Simpson, 1964). Furthermore, these symptoms only occurred in an estimated 18% of laboratory-confirmed infections (Duffy et al., 2009), with the majority being asymptomatic. The large outbreak on the island of Yap brought attention to this previously neglected tropical disease. This was further highlighted by the large outbreak in the Americas in 2015–16 (Fig. 1).

On February 1, 2016, the World Health Organization (WHO) declared ZIKV infection and its associated congenital malformations and neurological disorders as a Public Health Emergency of International Concern (PHEIC) (World Health Organisation, 2016).

Pathogenesis of Zika virus

Mechanisms of infection

ZIKV is primarily a mosquito-borne virus and is transmitted by the *Aedes* mosquito. Nonvector-borne transmission has also been recorded, with transmission through bodily fluids such as semen, amniotic fluid, and breast milk being clinically confirmed (Asif et al., 2017).

FIG. 1 Brief history of ZIKV infection.

Infection via mosquitoes

ZIKV is transmitted by mosquitoes of the genus *Aedes*, mainly of *Aedes aegypti* and *A. albopictus*. The virus has also been isolated from other genus such as *Anopheles*, *Culex*, and *Mansonia* spp. (Saiz et al., 2016).

ZIKV natural transmission cycle involves *Cercopithecus aethiops* and *Erythrocebus patas* monkeys in Africa (Faye et al., 2013) and semicaptive and wild orangutans in Asia (Wolfe et al., 2001), where these primates serve as reservoir hosts for ZIKV. There is no other animal other than humans and nonhuman primates known to be amplifying hosts for ZIKV.

Horizontal transmission within humans

A study on 72 pregnant women by Brasil et al. found that 26 of them tested positive for ZIKV RNA in blood samples (Brasil et al., 2016). ZIKV RNA and replicative virus have also been found in semen (Atkinson et al., 2016), saliva (Musso et al., 2015), and urine (Gourinat, O'Connor, Calvez, Goarant, & Dupont-Rouzeyrol, 2015). In February 2016 the CDC received 14 cases of suspected sexual transmission of ZIKV, of which all reported cases belonged to women whose only known risk factor was sexual intercourse with symptomatic partners who had recently returned from areas with ongoing ZIKV circulation (Hills et al., 2016).

While direct horizontal transmission remains unlikely to play a major role in ZIKV spread, these data have demonstrated that ZIKV spread can occur through blood transfusion and sexual transmission.

Vertical transmission and mechanisms of entry into the nervous system

ZIKV has a unique capacity to enter immunoprivileged sites such as the central nervous system and the placenta, due to its ability to cross the blood-brain barrier and placental barrier, respectively (Asif et al., 2017). Not only is understanding the mechanism through which it enters immunoprivileged sites important for understanding the pathogenesis of the congenital ZIKV syndrome, but its presence in these sites and shedding into bodily fluids have important implications for the diagnosis of ZIKV infection and subsequent treatment and prevention options.

Vertical transmission of ZIKV was first suspected when ZIKV was detected in the amniotic fluid of two pregnant women who had presented with clinical manifestations suspicious of ZIKV infection and whose fetuses were diagnosed with microcephaly (Calvet et al., 2016). RT-PCR confirmed ZIKV in sera from the mothers and fetuses. A subsequent autopsy performed on a fetus diagnosed with microcephaly revealed micrencephaly and dystrophic calcifications, and RT-PCR and electron microscopy confirmed ZIKV in the microcephalic tissue of the fetus (Štrafela et al., 2017). Subsequent studies in Brazil have reported the presence of ZIKV RNA in amniotic fluids and fetuses. These findings subsequently catalyzed further studies regarding the teratogenic effects of ZIKV in developing fetuses.

An in vitro model study investigated the tropism of ZIKV and showed that the virus readily infected human neural progenitor cells (hNPCs) derived from induced pluripotent stem cells. Apoptosis and cell-cycle dysregulation were induced in the infected hNPCs, which resulted in attenuated population growth, and infected cells further released infectious ZIKV particles (Tang et al., 2016). These findings provided the scientific basis for the epidemiological link between congenital ZIKV infection and fetal neurological abnormalities.

Cross-reactivity and cross-protection between ZIKV and DENV infections

In many ZIKV epidemic regions, there is an overlapping presence of DENV due to widespread and overlapping geographical distributions, coupled with an increase in vaccination coverage over the years. As such, there is an increased likelihood of exposure to multiple flaviviruses in these endemic regions.

Antibodies to flaviviruses can be protective and impart lifelong immunity to reinfection by homologous virus. However, the antigenic relationships between flaviviruses may complicate diagnostics. Cross-reactivity may also play a role in cross-protection or promote more severe pathologies, as demonstrated by antibody-dependent enhancement in subsequent DENV reinfection, thereby leading to severe disease—dengue hemorrhagic fever or dengue shock syndrome.

The endemic circulation of DENV and ZIKV in overlapping geographical regions has raised concerns that preexisting flavivirus immunity may modulate ZIKV disease and transmission potential.

A study done by Maurizio et al. evaluated the correlation between preexisting immunity in DENV experienced mothers and its role in immunological cross-protection for ZIKV with the use of novel assays that allowed distinction between the two flavivirus infections.

The study evaluated three cases of ZIKV infection in DENV-immune mothers. All three cases presented with the absence of ZIKV transmission to the fetus. Preexisting DENV infection was confirmed in these women, and ZIKV infection was diagnosed as well. None of the fetuses had developed congenital ZIKV syndrome and cerebral malformations. Two other cases of ZIKV infection in DENV-naïve mothers were studied. In both cases, ZIKV transmission to the fetuses was reported, and the presence of cerebral malformations such as intracranial calcifications was detected. In these two women, previous DENV infection was excluded serologically.

This supports the idea that prior DENV infections may protect against ZIKV transmission to the fetus. DENV-specific effector memory T cells characterized by rapid effector function in response to antigenic stimulation were demonstrated in ZIKV nontransmitter mothers.

A study by Swanstrom et al. demonstrated that sera from patients with secondary DENV infection can strongly neutralize ZIKV and yielded stronger cross-neutralizing activity than patients with primary DENV infection (Swanstrom et al., 2016). In vivo studies showed that individuals develop multitypic neutralizing antibodies, reactive against the exposed serotype, and also against serotypes to which they have not yet been exposed, after secondary DENV infection (Halstead, 2003). These antibodies could potentially play a role in DENV-ZIKV immunological cross-reactivity and cross-protection.

Clinical presentation of ZIKV infection

Prior to 2007, ZIKV infections typically clinically presented with a self-resolving illness, and symptoms only occurred in an estimated 18% of infected individuals (Duffy et al., 2009). Symptoms include low-grade fever, malaise, generalized erythematous maculopapular rashes, retroorbital headache, bilateral nonpurulent conjunctivitis, myalgia, and polyarticular arthralgia of the small joints in the hands and feet (Duffy et al., 2009; Musso et al., 2015). Following the global ZIKV epidemic, ZIKV infection has been found to be associated with more sinister, life-threatening disease, including multiorgan failure (Swaminathan, Schlaberg, Lewis, Hanson, & Couturier, 2016), neurological complications, and thrombocytopenic purpura (Karimi et al., 2016).

While the majority of ZIKV infection manifests with mild symptoms, there have been reports of deaths from severe ZIKV infection—a case of fatal ZIKV infection in a 15-year-old Colombian girl with sickle cell disease was reported (Arzuza-Ortega et al., 2016) and at least three other cases of fatal ZIKV infection in Brazil in individuals with comorbidities or weakened immune systems (Chan, Choi, Yip, Cheng, & Yuen, 2016). This highlights the importance of timely recognition of ZIKV infection and active intervention, especially in individuals with comorbidities.

There are no pathognomonic features—clinical, radiological, nor serological—that distinguish ZIKV infection from other viral infections (Chan et al., 2016) (Fig. 2). Therefore it must be noted that definitive diagnosis of ZIKV infection requires laboratory confirmation—with successful isolation of ZIKV in viral culture as the gold standard of laboratory diagnosis.

Neurological manifestations of ZIKV infection

Similar to other flaviviruses, there is an association between molecularly confirmed or serologically confirmed ZIKV infection and acute neuroinflammatory diseases, including Guillain-Barré syndrome, encephalitis, and transverse myelitis. Acute ZIKV infection is shown to be associated with an increased incidence of a spectrum of significant neurological manifestations (da Silva et al., 2017; Oehler et al., 2014).

ZIKV-associated Guillain-Barré syndrome

ZIKV has also been observed to induce autoimmunity in some patients (Monsalve et al., 2017). This was first observed in the ZIKV epidemic in French Polynesia of 2013 where a cluster of Guillain-Barré syndrome (GBS) cases were identified (Oehler et al., 2014; Cao-Lormeau et al., 2014). Subsequently a study by James et al. reported that the incidence rate of GBS cases during the ZIKV outbreak was approximately 20-fold higher than expected given the size of the FP population and the established incidence rates of GBS (1–2/100,000 population per year) (Sejvar, Baughman, Wise, & Morgan, 2011). Subsequently, ZIKV infection has been associated with GBS in outbreaks in Central and Latin American countries such as Brazil and the Dominican Republic (do Rosário et al., 2016; Dos Santos et al., 2016).

GBS is an autoimmune polyradiculoneuropathy triggered by antecedent bacterial or viral infections. It usually manifests as ascending weakness and numbness associated with diminution of deep tendon reflexes. The disease course is over 1–3 weeks. About two-thirds of ZIKV-associated GBS patients reported paresthesia, and approximately half of patients had cranial nerve palsies, most commonly facial palsy (Muñoz et al., 2017).

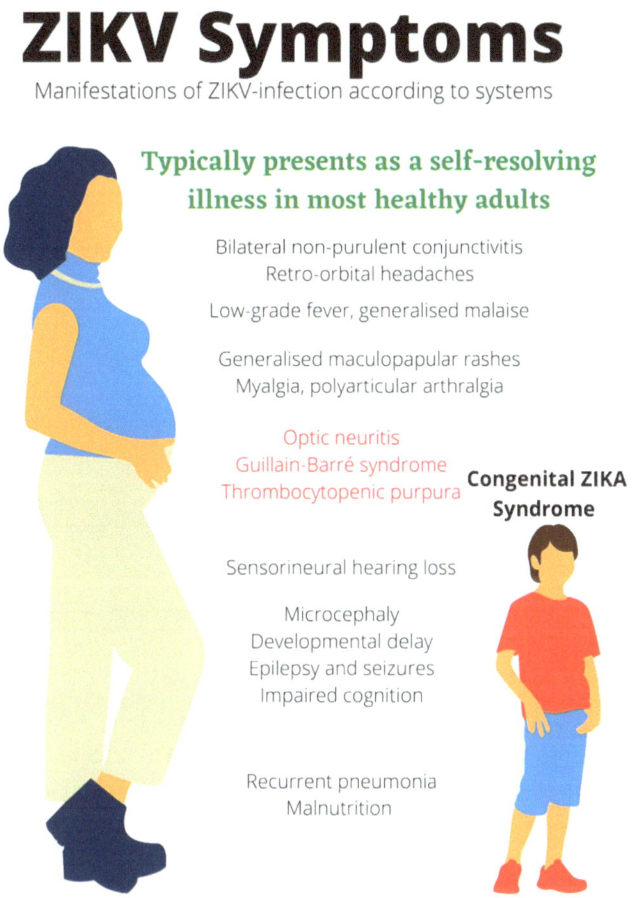

FIG. 2 Manifestations of ZIKV infection.

ZIKV infection is also sometimes associated with Miller-Fisher subtype of GBS, marked by a triad of ataxia, areflexia, and ophthalmoplegia. A study by Parra et al. reported four cases of Miller-Fisher syndrome (Parra et al., 2016). Other GBS variants reported with ZIKV include AMSAN (acute motor sensory axonal neuropathy), Bickerstaff's encephalitis, and pharyngeal-cervical-brachial variants (Anaya et al., 2017).

In patients with ZIKV-associated GBS, it was found that neurological symptoms followed quickly after a viral prodrome, with the median time between viral illness and GBS onset being 5–10 (Muñoz et al., 2017). This is in contrast to GBS not associated with ZIKV where neurological symptoms typically develop 2–4 weeks after infection (Willison, Jacobs, & van Doorn, 2016). This suggests a parainfectious temporal profile with regard to GBS associated with ZIKV, as opposed to the postinfectious profile reported in classical GBS cases (van den Berg et al., 2014).

Several case series have also showed that the prognosis of ZIKV-associated GBS was similar to GBS associated with other processes (Muñoz et al., 2017).

The incidence of ZIKV-associated GBS is estimated to be between 2 and 3 cases per 10,000 ZIKV infections, similar to that associated with *Campylobacter* infections (Halpin et al., 2018; Mier-Y-Teran-Romero, Delorey, Sejvar, & Johansson, 2018) (Fig. 3). Other autoimmune disorders such as thrombocytopenic purpura have also been associated with ZIKV infections (Karimi et al., 2016).

ZIKV-associated central nervous system involvement

Despite the recognized association with the Guillain-Barré syndrome, ZIKV rarely causes meningitis and encephalitis, as it is generally less neuroinvasive in adults (Carteaux et al., 2016), and the mature adult brain has shown to be less susceptible. However, ZIKV has a proclivity for infecting and impairing neural progenitor cells, which explains the pathogenesis of microcephaly and other neurodevelopmental disorders in congenital Zika syndrome.

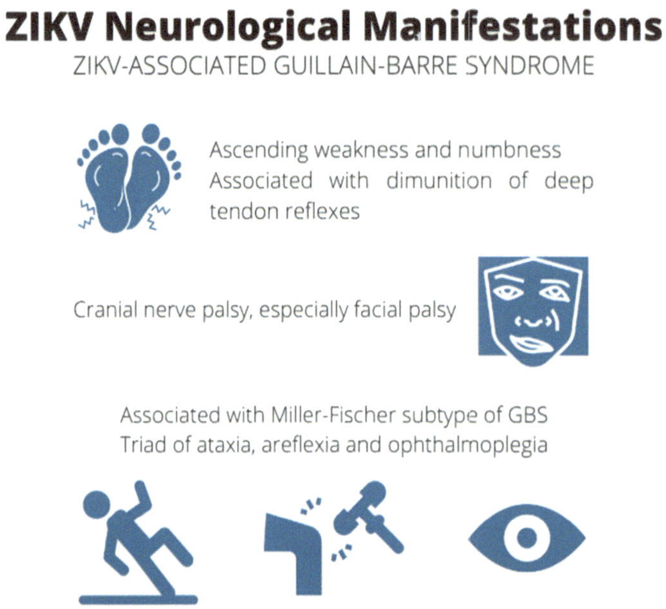

FIG. 3 ZIKV-associated Guillain-Barré syndrome.

CNS manifestations of ZIKV include encephalopathy, encephalitis, meningitis, myelitis, and seizures. However, such cases did not reveal consistent features on ancillary imaging and workup. An example would be how magnetic resonance imaging displayed variation in findings, where meningeal inflammation was reported in some cases, and in others, parenchymal lesions of restricted diffusion, which were suggestive of ischemia, were found (Carteaux et al., 2016; Sebastián et al., 2017).

Like ZIKV-associated GBS cases, there appears to be a predominance of neuroinflammation in ZIKV-associated CNS complications. This is further corroborated by the finding of pleocytosis and elevated protein in the CSF of ZIKV patients with CNS involvement (Mécharles et al., 2016; Rozé et al., 2016; Soares et al., 2016). Further research is however needed to elucidate the exact immunological mechanisms involved.

Congenital ZIKA syndrome

The link between maternal ZIKV infection and microcephaly was first suspected when there was a reported increase in the incidence of newborns with microcephaly in the state of Pernambuco, Brazil, from six cases of microcephaly between January and July 2015 to 39 in October 2015 alone. This alarming increase in cases of microcephaly caused the State Secretariat for Health of Pernambuco to raise an alert and notify the national surveillance system (Teixeira, Costa, de Oliveira, Nunes, & Rodrigues, 2016).

Subsequently a rapid assessment task force was set up to investigate and elucidate the epidemiology underlying the increased incidence of microcephaly. The mothers of 38 newborns with microcephaly were interviewed, among whom 24 reported experiencing an exanthematous rash during their pregnancy. A physician subsequently made the link that these women who delivered their babies in October would have been in the first trimester of their pregnancy from January to March 2015, during which there was an outbreak of an exanthematous disease, later identified to be ZIKV (which was previously not a reportable disease) (Teixeira et al., 2016). This raised suspicion that these cases of microcephaly were caused by maternal ZIKV infection.

A 2015 cohort study was conducted among pregnant women, regardless of week of gestation, who presented to the acute febrile illness clinic in Rio de Janeiro, with a rash that had developed in the preceding 5 days. A total of 345 women were enrolled and were tested for ZIKV by qualitative RT-PCR (reverse transcriptase polymerase chain reaction), with 182 of these women testing positive for ZIKV. Among the 117 live births in the cohort of ZIKV-positive women, 49 of the neonates were born with either grossly abnormal clinical or radiological findings, and 4 of the neonates were born with microcephaly. None of the infants in the cohort of ZIKV-negative women had microcephaly. These results suggested a substantial causal link between maternal ZIKV infection and neonatal malformations, especially regarding CNS and overall development (Brasil et al., 2016).

Another 2016 prospective case-control study was conducted among neonates born to women resident in the region Recife, state of Pernambuco, Brazil, and delivered in eight public hospitals in the same area. Thirty-two neonates with microcephaly were matched to 62 neonates without microcephaly by expected date of delivery and residential location. The cases and controls were analyzed for ZIKV infection, which was defined as a positive RT-PCR result (which detected ZIKV genomes) and/or detection of IgM antibodies against ZIKV. The analysis revealed a strong association (crude OR 55.5 [95% CI 8.6%–∞]) between microcephaly and laboratory confirmation of ZIKV by RT-PCR or by ZIKV-specific IgM antibodies in neonates, with laboratory-confirmed ZIKV infection present only in neonates with microcephaly (de Araújo et al., 2018). These two studies confirmed the hypothesis between in utero ZIKV infection and microcephaly.

ZIKV has a proclivity for infecting and impairing neural progenitor cells, thus explaining the pathogenesis of microcephaly and other neurodevelopmental disorders seen in neonates with congenital Zika syndrome as a result of ZIKV intrauterine infection.

Congenital ZIKV syndrome typically manifests as microcephaly, in which the neonate's brain is reduced in size with poorly differentiated cortex and few gyri (Klase et al., 2016; Moore et al., 2017). Congenital ZIKV infection halts cerebral growth, especially of the frontal, temporal, and parietal lobes. This leads to an abnormally collapsed cranial bone plate, with bilateral depression and occipital prominence (occipitoparietal crest) (Dain Gandelman Horovitz, da Silva Pone, Moura Pone, Dias Saad Salles, & Bastos Boechat, 2016) and abnormal skin folds (cutis gyrate) manifesting as a result of disproportionate scalp growth. Depending on the severity of microcephaly, the head circumference of infected neonates can range from normal size to severe microcephaly with head circumference of three standard deviation (SD) below the mean. In severe cases, infected neonates may present with significant craniofacial disproportion.

These CNS disruptions as a result of congenital ZIKV syndrome can cause multiple neurological manifestations of varying severity.

Infected neonates tend to be extremely irritable, partially due to postural disorganization. Neonates may present with developmental delay from the first days of life (Oliveira Melo et al., 2016; World Health Organisation, 2016). Hyperexcitability, increased muscle tonus, hyperreflexia, and clonus, which may be observed in patients, may stimulate epileptic seizures. Over time, these seizures may become increasingly evident, complex, and resistant to antiepileptic drugs. Some patients exhibit flexion spasms, typical of infantile spasms (West syndrome), with or without hypsarrhythmia. Others may develop Lennox-Gastaut syndrome. Around 4 months of age, as sucking and swallowing reflexes disappear and motor incoordination worsens, infants may present with dysphagia and microaspirations. As a result, these patients are more vulnerable to recurrent pneumonias and malnutrition.

Neonates with congenital ZIKV syndrome may have impaired cognitive functions, such as deficits in social interaction and communication skills. Impaired development of brainstem neurons can also lead to sensorineural hearing loss or hypoacusia in infected neonates.

In some cases, arthrogryposis and other osteoarticular malformations may be seen (van der Linden et al., 2016) (Fig. 4).

Ophthalmic manifestations of ZIKV

While the ophthalmic manifestations of congenital ZIKV are well described, in which ZIKV-infected neonates may present with ophthalmological abnormalities such as pale optic discs, chorioretinal atrophy, and colobomas (Zin et al., 2017), ophthalmic complications are relatively rare in ZIKV-infected adults. Reported manifestations include uveitis (Kodati et al., 2017), maculopathy (Parke et al., 2016), retinal vasculopathy (Singh et al., 2018), and optic neuritis (PACTRIMS 2017, 2018; Zaidi et al., 2018).

While the disease mechanism remains unclear, the finding of ZIKA virus in conjunctival fluid of infected individuals (Sun et al., 2016) and in the tears of mice (Miner et al., 2016) raises the possibility of damage arising from direct viral infection. However, a parainfectious or postinfectious inflammatory process, analogous to the processes involved in GBS, should not be discounted. Optic neuritis is a relatively common complication of other arboviruses, including DENV and West Nile virus (Zaidi et al., 2018). Prognosis in such infection-related optic neuritis tends to be good, typically with full recovery of vision.

Clinical management of ZIKA infections

Diagnosis of ZIKA infections

Until the recent epidemic, there was only a limited capacity for diagnostic testing of ZIKV infection due to the infrequency of ZIKV cases prior to 2016. Only certain public health and research laboratories that had their own reagents were equipped to do so (Landry & St George, 2017). After the surge in microcephaly cases and birth defects associated with congenital

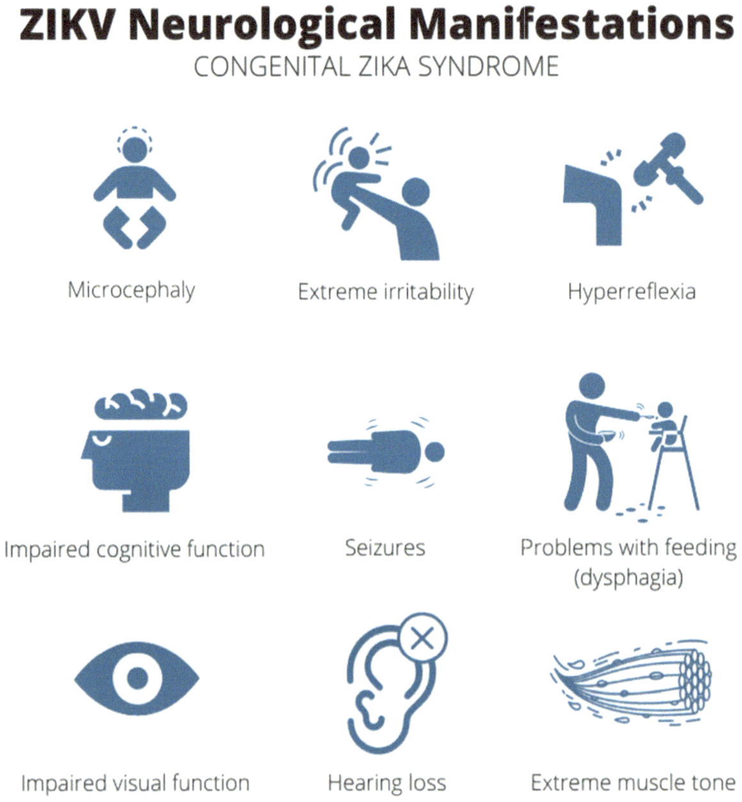

FIG. 4 Congenital ZIKA syndrome.

ZIKV infection was discovered in Brazil, there was an urgent need for increased accessibility and capabilities for ZIKV testing, especially among pregnant women.

Although ZIKV is no longer a Public Health Emergency of International Concern, there still remains a strong necessity for ZIKV screening during pregnancy in endemic regions, as it has the potential to infect the fetus and result in congenital infection (Duarte et al., 2020), with defects occurring more frequently among neonates whose mothers had been infected in the earlier stages of pregnancy (Hoen et al., 2018). Furthermore, diagnostic testing for ZIKV remains limited in many endemic regions despite the availability of commercial assays.

The clinical signs of Zika fever are not pathognomonic, and therefore confirmation of ZIKV infection requires isolation and identification of the virus. As mentioned earlier, successful isolation of ZIKV in viral culture is the gold standard for laboratory diagnosis of ZIKV infection. However, isolation remains a challenging and resource-intensive task, as it requires adequate biocontainment facilities to handle the samples. As such, diagnosis of ZIKV infections are made based on apparent pathological signs, results of serological and molecular assays, and epidemiological factors such as the presence of vectors in the geographical area (Singh et al., 2016).

Serological assays such as IgM and IgG enzyme-linked immunosorbent assay (ELISA) (Shan et al., 2017), as well as molecular assays like reverse transcription polymerase chain reaction (RT-PCR), which can be performed using urine, blood, and cerebrospinal fluid (CSF) from an early stage of infection, have been developed. There are currently 5 serological assays and 14 molecular assays for ZIKV with FDA Emergency Use Authorization (EUA) (Theel & Hata, 2018).

However, as ZIKV is genetically similar to other flaviviruses, there is the limitation of cross-reactivity in particular with dengue. As such, molecular diagnosis in particular by PCR from urine samples has been the most widely relied upon as in our case (Ho et al., 2017).

Follow-up in pregnant women

In pregnant women, molecular assays can be performed on amniotic fluid and placenta, in addition to urine, nasopharyngeal swabs, saliva, and CSF. Similarly, serological assays can be performed as well for diagnostic purposes. Differential diagnosis is essential in accurate and successful ZIKV detection.

To detect fetal microcephaly during pregnancy, prenatal ultrasound testing is used (Lazear & Diamond, 2016), in which the occipitofrontal circumference is measured.

Given the variety of neurological manifestations and the risk of severe developmental impairment, and also for diagnostic purposes, neuroimaging should be obtained in all patients. Postnatally, neuroimaging including computed tomography (CT) and magnetic resonance imaging (MRI) scans is used to detect congenital microcephaly (de Fatima Vasco Aragao et al., 2016). Infants with microcephaly will present with cortical and subcortical atrophy especially in the frontal, parietal, and occipital lobes and cerebellum, ventriculomegaly as a result of ventricular dilation, intracranial (juxtacortical and basal ganglia) calcifications, cortical development malformations such as polymicrogyria and pachygyria, and hypoplasia of the corpus callosum, cerebellum, and brainstem. ZIKV can also be identified by immunohistochemical staining of the antigen in the umbilical cord and placenta (Staples et al., 2016) in affected infants. Diagnosis of ZIKV-infected stillborn fetuses can be made by histopathological and immunohistochemical staining (Landry and St George, 2017). Cranial ultrasound, in case of an open anterior fontanel, can also be performed, but it has a more limited resolution.

Treatment of ZIKV infections

As evidenced in the clinical vignette, most patients with symptomatic ZIKV infection do not require active medical intervention. Most cases of symptomatic ZIKV infection resolve spontaneously. A modest proportion for patients may require symptomatic treatment after ruling out more serious conditions such as dengue, malaria, or bacterial infection.

The Pan-American Health Organization (WHO) recommends symptomatic treatment such as hydration, rest, antipyretics such as paracetamol, and antihistamines to control the itch associated with the rash. Use of aspirin and nonsteroidal antiinflammatory drugs (NSAIDs) is not advised, the former due to the risk of bleeding and Reye's syndrome in children under 12 and the latter due to contraindications in pathologies such as dengue and chikungunya, which are probable differential diagnoses for ZIKV infection and might precipitate hemorrhagic complications (Pan American Health Organization/World Health Organization (PAHO/WHO), 2015).

Possibility of a vaccine for ZIKA?

The severity of congenital Zika syndrome in infected pregnant women, coupled with the fact that ZIKV infection can be transmitted sexually to pregnant women, creates a strong need for the development of a vaccine against ZIKV infection. With the vaccine, both women of childbearing age and their male partners can be immunized to reduce the likelihood of vertical transmission, especially in individuals who live or travel frequently to ZIKV-endemic regions—a goal supported by the World Health Organization.

Licensed vaccines for four other flaviviruses—yellow fever virus (YFV), tick-borne encephalitis virus (TBEV), Japanese encephalitis virus (JEV), and dengue virus (DENV)—support the feasibility and development of a ZIKV vaccine (Shan, Xie, & Shi, 2018). Current ZIKV vaccine candidates belong to one of three categories, namely, inactivated vaccines, subunit vaccines, and live-attenuated vaccines, with more than 10 vaccine candidates being evaluated in phase 1 and 2 clinical trials (World Health Organisation, 2018). Unfortunately, there are no approved vaccines for ZIKV yet. The adverse consequences of the DENV vaccine have also raised the bar for the development of a safe and effective ZIKV vaccine.

For the ZIKV vaccine to be successful, not only should it have an acceptable safety profile across age ranges and pregnancy statuses, but it must also be economical and cost-effective to ensure high uptake in developing and underdeveloped countries that are highly susceptible to further ZIKV outbreaks (Garg, Mehmetoglu-Gurbuz, Ruddy, & Joshi, 2019). In addition, there should be no difference between flavivirus-naïve and flavivirus-experienced recipients, which may potentially enhance other flavivirus infections. This phenomenon, antibody-dependent enhancement of a clinical disease, presents a serious theoretical concern for ZIKV and other flavivirus vaccines (Barouch, Thomas, & Michael, 2017).

One of the first human clinical trials for the ZIKV vaccine were the VRC 319, which assessed the VRC5288 ZIKV vaccine candidate, and VRC 320, which assessed the VRC5283 ZIKV vaccine candidate (Gaudinski et al., 2018). Both the vaccine candidates were previously shown to be safe and immunogenic in nonhuman primate studies and induced strong neutralizing antibody responses after two doses (Dowd, DeMaso, et al., 2016; Dowd, Ko, et al., 2016), which prompted their assessment in the aforementioned phase I clinical trials. Both VRC5288 and VRC5283 were safe and well tolerated with mild adverse effects. However, the greatest effects in neutralizing antibody responses and CD4 and CD8 T-cell responses were seen with the VRC5283 vaccine candidate, which advanced it to an international placebo-controlled phase 2 clinical trial (Gaudinski et al., 2018).

Subsection policy and procedures

Screening for ZIKA virus among pregnant women

According to CDC's updated guidelines (https://www.cdc.gov/zika/hc-providers/testing-guidance.html), diagnostic testing should be performed as soon as possible within 12 weeks of symptom onset for symptomatic pregnant women who have had recent travel to areas with active dengue transmission and a risk of Zika. Diagnostic tests should include dengue and Zika virus NAAT (nucleic acid amplification tests) testing on a serum specimen, Zika virus NAAT on a urine specimen, and IgM testing for dengue only.

Zika virus IgM testing is not recommended for symptomatic pregnant women as Zika IgM antibodies can persist for months to years following infection, and its presence might not indicate a recent infection. Moreover the notable cross-reactivity between dengue IgM and Zika IgM antibodies in serologic tests could cause a false-positive Zika IgM result.

Diagnostic testing should also be done for pregnant women who live in or traveled to areas with endemic Zika infections during the period of her pregnancy and have a fetus with prenatal ultrasound findings consistent with congenital Zika virus infection.

If the patient tests negative for Zika virus NAATs but positive for ZIKV IgM, confirmatory plaque reduction neutralization tests (PRNTs) should be performed against Zika and dengue. Zika virus NAAT testing of amniocentesis specimens should also be performed should amniocentesis be performed as part of clinical care. Alternatively, testing of placental and fetal tissues could also be considered.

Diagnostic tests are currently not recommended for asymptomatic pregnant women.

Screening for congenital microcephaly

Congenital microcephaly is described as a structural defect of the cranium in which the fetal head circumference is smaller than expected, when compared with other fetuses or infants of the same gestational age, sex, and ethnic background.

Microcephaly can be diagnosed either postnatally or prenatally. A measuring tape is placed over the most prominent part on the posterior aspect of the head (occiput) and just above the eyebrows (supraorbital ridges) and above the ears. This allows for the measurement of the occipital-frontal circumference (head circumference).

Microcephaly is defined as an occipital-frontal circumference measurement that is more than 2 standard deviations (SDs) below the mean for age and sex or less than the third percentile for age and sex. Severe microcephaly is defined as head circumference more than 3 SDs below the mean for age and sex.

Mini-dictionary of terms

Antipyretics. Class of medications given to resolve fever (pyrexia).
Areflexia. Neuromuscular condition resulting in reduced responses of muscles to external stimuli.
Arthralgia. Inflammation of the joints.
Arthrogryposis. Congenital joint contracture in two or more areas of the body.
Ataxia. Neuromuscular condition resulting in reduced control of fine movements.
Cerebrospinal fluid. A bodily fluid that coats the brain and spinal cord.
Clonus. Involuntary and rhythmic muscle contractions caused by a permanent lesion in descending motor neurons.
Coloboma. Congenital malformation in which failure of fusion of intraocular fissure causes defects in the lens, iris, or retina.
Conjunctivitis. Inflammation of the conjunctiva (eyes).
Cross-reactivity. Reactivity of an observed agent that initiates reactions outside the main reaction expected.
Dysphagia. Difficulty in swallowing.
Encephalitis. Inflammation of the brain.
Horizontal transmission. Transmission of an infection from one individual to another, both of whom are in the same generation.
Hypsarrhythmia. Chaotic and disorganized brain electrical activity with no recognizable pattern.
Hypoacousia. Impaired sense of hearing.
Hypoplasia. Impaired development of a tissue or an organ.
Immune thrombocytopenic purpura. Bleeding disorder in which the immune system destroys platelets.
Induced pluripotent stem cells. Stem cells derived from skin or blood cells that have been reprogrammed back into an embryonic-like state in which they have two properties: self-renewal and potency.

Meningitis. Inflammation of the meninges (tissues covering the central nervous system).
Micrencephaly. Pathological condition of having an unusually small brain.
Microaspiration. Unintentional aspiration of small amounts of reflux material.
Microcephaly. Pathological condition of having an unusually small head.
Myalgia. Muscle aches and pains.
Nucleic acid amplification tests. Assay used for detection of a particular nucleic acid sequence to detect and identify a particular organism.
Ophthalmoplegia. Paralysis or weakness of the muscles controlling the eye.
Optic neuritis. Inflammation of the optic nerve.
Pachygyria. Congenital malformation of the cerebral hemisphere resulting in unusually thick convolutions of the cerebral cortex.
Plaque reduction neutralization test. Assay to quantify the titer of virus-neutralizing and protective antibodies. The serum sample is diluted and mixed with a viral suspension. The suspension is further incubated to allow the antibody to react with the virus.
Polymicrogyria. Cerebral cortical malformation characterized by excessive cortical folding and shallow sulci.
Polyneuropathy. A medical condition where multiple peripheral nerves in an individual are damaged.
Polyradiculoneuropathy. A medical condition comprising polyneuropathy and polyradiculopathy.
Polyradiculopathy. A medical condition where multiple nerve roots in an individual are damaged.
Serotype. Distinct variation within a species of bacteria or virus.
Teratogenic. Producing deleterious and harmful effects on the developing fetus.
Transverse myelitis. Neurological condition in which a section of the spinal cord is inflamed.
Vertical transmission. Transmission of an infection from an infected mother to her child.

Key facts of Zika virus neurological manifestations

1. There is an association between molecularly confirmed or serologically confirmed Zika virus infection and acute neuroinflammatory disease.
2. Zika virus can induce autoimmunity and is associated with Guillain-Barré syndrome outbreaks, Miller-Fisher subtype of Guillain-Barré syndrome, and other autoimmune disorders such as thrombocytopenic purpura.
3. It usually manifests as ascending weakness and numbness associated with diminution of deep tendon reflexes.
4. Zika virus is less neuroinvasive in adults.
5. Central nervous system manifestations of Zika virus include encephalopathy, encephalitis, meningitis, myelitis, and seizures.

Key facts of congenital ZIKV syndrome

1. The link between maternal Zika virus infection and microcephaly was first suspected when there was a reported increase in the incidence of newborns with microcephaly in the state of Pernambuco, Brazil.
2. Zika virus has a proclivity for infecting and impairing neural progenitor cells, thus leading to microcephaly and other neurodevelopmental disorders in neonates with congenital Zika syndrome.
3. Congenital Zika virus syndrome typically manifests as microcephaly.
4. Congenital Zika virus syndrome can manifest as multiple neurological manifestations of varying severity.
5. Infected neonates may present with extreme irritability, developmental delay, hyperexcitability, hyperreflexia, epileptic seizures, impaired cognitive functions, and sensorineural hearing loss.

Summary points

- This chapter focuses on the clinical neurological spectrum of adult and congenital Zika virus infection, in which Zika virus is a member of the virus family *Flaviviridae* and is transmitted primarily by *Aedes* mosquitoes.
- Zika virus was first discovered in the Zika Forest of Uganda in 1947 and was until recently believed to be an infection of mild clinical significance.

- Infection with Zika virus presents with mild clinical symptoms including fever, headache, rash, conjunctivitis, arthralgia, and myalgia—only a small proportion of laboratory-confirmed infections manifested clinically detectable features.
- In 2016, the World Health Organization declared Zika virus infection and its associated congenital malformations and neurological disorders as a Public Health Emergency of International Concern.
- Although Zika virus is no longer a Public Health Emergency of International Concern, Zika virus screening during pregnancy in endemic regions remains relevant and necessary, due to the severe long-term sequelae and cost to the healthcare systems involved.
- In summary, Zika virus is an important emerging infectious disease that can have a range of neurological complications in adults and in utero. Vector control remains the primary mode of control with good clinical management to reduce morbidity and even mortality.

References

Anaya, J. M., Rodríguez, Y., Monsalve, D. M., Vega, D., Ojeda, E., González-Bravo, D., … Acosta-Ampudia, Y. (2017). A comprehensive analysis and immunobiology of autoimmune neurological syndromes during the Zika virus outbreak in Cúcuta, Colombia. *Journal of Autoimmunity*, *77*, 123–138.

Arzuza-Ortega, L., Polo, A., Pérez-Tatis, G., López-García, H., Parra, E., Pardo-Herrera, L. C., … Rodríguez-Morales, A. J. (2016). Fatal sickle cell disease and Zika virus infection in girl from Colombia. *Emerging Infectious Diseases*, *22*(5), 925–927.

Asif, A., Manzoor, S., Tuz-Zahra, F., Saalim, M., Ashraf, M., Ishtiyaq, J., & Khalid, M. (2017). Zika virus: Immune evasion mechanisms, currently available therapeutic regimens, and vaccines. *Viral Immunology*, *30*(10), 682–690.

Atkinson, B., Hearn, P., Afrough, B., Lumley, S., Carter, D., Aarons, E. J., … Hewson, R. (2016). Detection of Zika virus in semen. *Emerging Infectious Diseases*, *22*(5), 940.

Barouch, D. H., Thomas, S. J., & Michael, N. L. (2017). Prospects for a Zika virus vaccine. *Immunity*, *46*(2), 176–182.

Brasil, P., Pereira, J. P., Jr., Moreira, M. E., Ribeiro Nogueira, R. M., Damasceno, L., Wakimoto, M., … Nielsen-Saines, K. (2016). Zika virus infection in pregnant women in Rio de Janeiro. *The New England Journal of Medicine*, *375*(24), 2321–2334.

Calvet, G., Aguiar, R. S., Melo, A., Sampaio, S. A., de Filippis, I., Fabri, A., … de Filippis, A. (2016). Detection and sequencing of Zika virus from amniotic fluid of fetuses with microcephaly in Brazil: A case study. *The Lancet Infectious Diseases*, *16*(6), 653–660.

Cao-Lormeau, V. M., Roche, C., Teissier, A., Robin, E., Berry, A. L., Mallet, H. P., … Musso, D. (2014). Zika virus, French polynesia, South pacific, 2013. *Emerging Infectious Diseases*, *20*(6), 1085–1086.

Carteaux, G., Maquart, M., Bedet, A., Contou, D., Brugières, P., Fourati, S., … Mekontso Dessap, A. (2016). Zika virus associated with meningoencephalitis. *The New England Journal of Medicine*, *374*(16), 1595–1596.

Chan, J. F., Choi, G. K., Yip, C. C., Cheng, V. C., & Yuen, K. Y. (2016). Zika fever and congenital Zika syndrome: An unexpected emerging arboviral disease. *The Journal of Infection*, *72*(5), 507–524.

da Silva, I., Frontera, J. A., Bispo de Filippis, A. M., Nascimento, O., & RIO-GBS-ZIKV Research Group. (2017). Neurologic complications associated with the Zika virus in Brazilian adults. *JAMA Neurology*, *74*(10), 1190–1198.

Dain Gandelman Horovitz, D., da Silva Pone, M. V., Moura Pone, S., Dias Saad Salles, T. R., & Bastos Boechat, M. C. (2016). Cranial bone collapse in microcephalic infants prenatally exposed to Zika virus infection. *Neurology*, *87*(1), 118–119.

de Araújo, T., Ximenes, R., Miranda-Filho, D. B., Souza, W. V., Montarroyos, U. R., de Melo, A., … State Health Department of Pernambuco. (2018). Association between microcephaly, Zika virus infection, and other risk factors in Brazil: Final report of a case-control study. *The Lancet Infectious Diseases*, *18*(3), 328–336.

de Fatima Vasco Aragao, M., van der Linden, V., Brainer-Lima, A. M., Coeli, R. R., Rocha, M. A., Sobral da Silva, P., … Valenca, M. M. (2016). Clinical features and neuroimaging (CT and MRI) findings in presumed Zika virus related congenital infection and microcephaly: Retrospective case series study. *BMJ (Clinical research ed.)*, *353*, i1901.

Dick, G. W., Kitchen, S. F., & Haddow, A. J. (1952). Zika virus. I. Isolations and serological specificity. *Transactions of the Royal Society of Tropical Medicine and Hygiene*, *46*(5), 509–520.

do Rosário, M. S., de Jesus, P. A., Vasilakis, N., Farias, D. S., Novaes, M. A., Rodrigues, S. G., … de Siqueira, I. C. (2016). Guillain-Barré syndrome after Zika virus infection in Brazil. *The American Journal of Tropical Medicine and Hygiene*, *95*(5), 1157–1160.

Dos Santos, T., Rodriguez, A., Almiron, M., Sanhueza, A., Ramon, P., de Oliveira, W. K., … Espinal, M. A. (2016). Zika virus and the Guillain-Barré syndrome – Case series from seven countries. *The New England Journal of Medicine*, *375*(16), 1598–1601.

Dowd, K. A., DeMaso, C. R., Pelc, R. S., Speer, S. D., Smith, A., Goo, L., … Pierson, T. C. (2016). Broadly neutralizing activity of Zika virus-immune sera identifies a single viral serotype. *Cell Reports*, *16*(6), 1485–1491.

Dowd, K. A., Ko, S. Y., Morabito, K. M., Yang, E. S., Pelc, R. S., DeMaso, C. R., … Graham, B. S. (2016). Rapid development of a DNA vaccine for Zika virus. *Science (New York, N.Y.)*, *354*(6309), 237–240.

Duarte, A. O., Oliveira, J. V., Carvalho, T., Pessoa, L. B., Filho, C. M., Lima, J., … de Siqueira, I.C. (2020). Maternal and congenital infections arising from Zika, dengue and chikungunya arboviruses in Salvador, Brazil. *Transactions of the Royal Society of Tropical Medicine and Hygiene*, *114*(3), 222–225.

Duffy, M. R., Chen, T. H., Hancock, W. T., Powers, A. M., Kool, J. L., Lanciotti, R. S., … Hayes, E. B. (2009). Zika virus outbreak on Yap Island, Federated States of Micronesia. *The New England Journal of Medicine*, *360*(24), 2536–2543.

Faye, O., Faye, O., Diallo, D., Diallo, M., Weidmann, M., & Sall, A. A. (2013). Quantitative real-time PCR detection of Zika virus and evaluation with field-caught mosquitoes. *Virology Journal*, *10*, 311.

Garg, H., Mehmetoglu-Gurbuz, T., Ruddy, G. M., & Joshi, A. (2019). Capsid containing virus like particle vaccine against Zika virus made from a stable cell line. *Vaccine*, *37*, 7123–7131.

Gaudinski, M. R., Houser, K. V., Morabito, K. M., Hu, Z., Yamshchikov, G., Rothwell, R. S., … VRC 320 study teams. (2018). Safety, tolerability, and immunogenicity of two Zika virus DNA vaccine candidates in healthy adults: randomised, open-label, phase 1 clinical trials. *Lancet (London, England)*, *391*(10120), 552–562.

Gourinat, A. C., O'Connor, O., Calvez, E., Goarant, C., & Dupont-Rouzeyrol, M. (2015). Detection of Zika virus in urine. *Emerging Infectious Diseases*, *21*(1), 84–86.

Halpin, A. L., Gu, W., Wise, M. E., Sejvar, J. J., Hoekstra, R. M., & Mahon, B. E. (2018). Post-campylobacter Guillain Barré syndrome in the USA: Secondary analysis of surveillance data collected during the 2009-2010 novel influenza A (H1N1) vaccination campaign. *Epidemiology and Infection*, *146*(13), 1740–1745.

Halstead, S. B. (2003). Neutralization and antibody-dependent enhancement of dengue viruses. *Advances in Virus Research*, *60*, 421–467.

Hills, S. L., Russell, K., Hennessey, M., Williams, C., Oster, A. M., Fischer, M., & Mead, P. (2016). Transmission of Zika virus through sexual contact with travelers to areas of ongoing transmission – Continental United States, 2016. *MMWR Morbidity and Mortality Weekly Report*, *65*(8), 215–216.

Ho, Z. J. M., Hapuarachchi, H. C., Barkham, T., Chow, A., Ng, L. C., Lee, J. M. V., … Sessions, O. M. (2017). Outbreak of Zika virus infection in Singapore: An epidemiological, entomological, virological, and clinical analysis. *The Lancet Infectious Diseases*, *17*(8), 813–821.

Hoen, B., Schaub, B., Funk, A. L., Ardillon, V., Boullard, M., Cabié, A., … Douine, M. (2018). Pregnancy outcomes after ZIKV infection in French territories in the Americas. *New England Journal of Medicine*, *378*(11), 985–994.

Karimi, O., Goorhuis, A., Schinkel, J., Codrington, J., Vreden, S., Vermaat, J. S., … Grobusch, M. P. (2016). Thrombocytopenia and subcutaneous bleedings in a patient with Zika virus infection. *Lancet (London, England)*, *387*(10022), 939–940.

Klase, Z. A., Khakhina, S., Schneider, A., Callahan, M. V., Glasspool-Malone, J., & Malone, R. (2016). Zika fetal Neuropathogenesis: Etiology of a viral syndrome. *PLoS Neglected Tropical Diseases*, *10*(8), e0004877.

Kodati, S., Palmore, T. N., Spellman, F. A., Cunningham, D., Weistrop, B., & Sen, H. N. (2017). Bilateral posterior uveitis associated with Zika virus infection. *Lancet (London, England)*, *389*(10064), 125–126.

Landry, M. L., & St George, K. (2017). Laboratory diagnosis of Zika virus infection. *Archives of Pathology & Laboratory Medicine*, *141*(1), 60–67.

Lazear, H. M., & Diamond, M. S. (2016). Zika virus: New clinical syndromes and its emergence in the Western Hemisphere. *Journal of Virology*, *90*, 4864–4875. https://doi.org/10.1128/JVI.00252-16.

Liu, Z. Y., Shi, W. F., & Qin, C. F. (2019). The evolution of Zika virus from Asia to the Americas. *Nature Reviews Microbiology*, *17*(3), 131–139.

Mécharles, S., Herrmann, C., Poullain, P., Tran, T. H., Deschamps, N., Mathon, G., … Lannuzel, A. (2016). Acute myelitis due to Zika virus infection. *Lancet (London, England)*, *387*(10026), 1481.

Mier-Y-Teran-Romero, L., Delorey, M. J., Sejvar, J. J., & Johansson, M. A. (2018). Guillain-Barré syndrome risk among individuals infected with Zika virus: A multi-country assessment. *BMC Medicine*, *16*(1), 67.

Miner, J. J., Sene, A., Richner, J. M., Smith, A. M., Santeford, A., Ban, N., … Apte, R. S. (2016). Zika virus infection in mice causes panuveitis with shedding of virus in tears. *Cell Reports*, *16*(12), 3208–3218.

Monsalve, D. M., Pacheco, Y., Acosta-Ampudia, Y., Rodríguez, Y., Ramírez-Santana, C., & Anaya, J.-M. (2017). Zika virus and autoimmunity. One-step forward. *Autoimmunity Reviews*, *16*(12), 1237–1245.

Moore, C. A., Staples, J. E., Dobyns, W. B., Pessoa, A., Ventura, C. V., Fonseca, E. B., … Rasmussen, S. A. (2017). Characterizing the pattern of anomalies in congenital Zika syndrome for pediatric clinicians. *JAMA Pediatrics*, *171*(3), 288–295.

Muñoz, L. S., Parra, B., Pardo, C. A., & Neuroviruses Emerging in the Americas Study. (2017). Neurological implications of Zika virus infection in adults. *The Journal of Infectious Diseases*, *216*(Suppl_10), S897–S905.

Musso, D., Roche, C., Nhan, T. X., Robin, E., Teissier, A., & Cao-Lormeau, V. M. (2015). Detection of Zika virus in saliva. *Journal of Clinical Virology: The Official Publication of the Pan American Society for Clinical Virology*, *68*, 53–55.

Oehler, E., Watrin, L., Larre, P., Leparc-Goffart, I., Lastere, S., Valour, F., … Ghawche, F. (2014). Zika virus infection complicated by Guillain-Barre syndrome–Case report, French Polynesia, December 2013. *Euro surveillance: bulletin Europeen sur les maladies transmissibles = European communicable disease bulletin*, *19*(9), 20720.

Oliveira Melo, A. S., Malinger, G., Ximenes, R., Szejnfeld, P. O., Alves Sampaio, S., & Bispo de Filippis, A. M. (2016). Zika virus intrauterine infection causes fetal brain abnormality and microcephaly: Tip of the iceberg? *Ultrasound in Obstetrics & Gynecology: The Official Journal of the International Society of Ultrasound in Obstetrics and Gynecology*, *47*(1), 6–7.

PACTRIMS 2017. (2018). *Multiple Sclerosis Journal*, *24*(3), 368–423.

Pan American Health Organization/World Health Organization (PAHO/WHO). (2015). *Epidemiological alert: Neurological syndrome, congenital malformations, and Zika virus infection*. Implications for Public Health in the Americas.

Parke, D. W., Almeida, D. R. P., Albini, T. A., Ventura, C. V., Berrocal, A. M., & Mittra, R. A. (2016). Serologically confirmed Zika-related unilateral acute maculopathy in an adult. *Ophthalmology*, *123*(11), 2432–2433.

Parra, B., Lizarazo, J., Jiménez-Arango, J. A., Zea-Vera, A. F., González-Manrique, G., Vargas, J., & Pardo, C. A. (2016). Guillain–Barré syndrome associated with Zika virus infection in Colombia. *New England Journal of Medicine*, *375*(16), 1513–1523.

Rozé, B., Najioullah, F., Signate, A., Apetse, K., Brouste, Y., Gourgoudou, S., … Neuro-Zika Working Group of Martinique. (2016). Zika virus detection in cerebrospinal fluid from two patients with encephalopathy, Martinique, February 2016. *Euro surveillance: bulletin Europeen sur les maladies transmissibles = European communicable disease bulletin*, *21*(16). https://doi.org/10.2807/1560-7917.ES.2016.21.16.30205.

Saiz, J. C., Vázquez-Calvo, Á., Blázquez, A. B., Merino-Ramos, T., Escribano-Romero, E., & Martín-Acebes, M. A. (2016). Zika virus: The latest newcomer. *Frontiers in Microbiology*, *7*, 496.

Sebastián, U. U., Ricardo, A., Alvarez, B. C., Cubides, A., Luna, A. F., Arroyo-Parejo, M., … LACCTIN Group. (2017). Zika virus-induced neurological critical illness in Latin America: Severe Guillain-Barre Syndrome and encephalitis. *Journal of Critical Care*, *42*, 275–281.

Sejvar, J. J., Baughman, A. L., Wise, M., & Morgan, O. W. (2011). Population incidence of Guillain-Barré syndrome: A systematic review and meta-analysis. *Neuroepidemiology*, *36*(2), 123–133.

Shan, C., Xie, X., Ren, P., Loeffelholz, M. J., Yang, Y., Furuya, A., … Shi, P. Y. (2017). A rapid Zika diagnostic assay to measure neutralizing antibodies in patients. *eBioMedicine*, *17*, 157–162.

Shan, C., Xie, X., & Shi, P. Y. (2018). Zika virus vaccine: Progress and challenges. *Cell Host & Microbe*, *24*(1), 12–17.

Simpson, D. I. (1964). Zika virus infection in man. *Transactions of the Royal Society of Tropical Medicine and Hygiene*, *58*, 335–338.

Singh, R. K., Dhama, K., Malik, Y. S., Ramakrishnan, M. A., Karthik, K., Tiwari, R., … Joshi, S. K. (2016). Zika virus – Emergence, evolution, pathology, diagnosis, and control: Current global scenario and future perspectives – A comprehensive review. *Veterinary Quarterly*, *36*, 150–175.

Singh, M. S., Marquezan, M. C., Omiadze, R., Reddy, A. K., Belfort, R., Jr., & May, W. N. (2018). Inner retinal vasculopathy in Zika virus disease. *American Journal of Ophthalmology Case Reports*, *10*, 6–7.

Soares, C. N., Brasil, P., Carrera, R. M., Sequeira, P., de Filippis, A. B., Borges, V. A., … Solomon, T. (2016). Fatal encephalitis associated with Zika virus infection in an adult. *Journal of Clinical Virology: The Official Publication of the Pan American Society for Clinical Virology*, *83*, 63–65.

Staples, J. E., Dziuban, E. J., Fischer, M., Cragan, J. D., Rasmussen, S. A., Cannon, M. J., … Moore, C. A. (2016). Interim guidelines for the evaluation and testing of infants with possible congenital Zika virus infection United States. *MMWR Morbidity and Mortality Weekly Report*, *65*, 63–67.

Štrafela, P., Vizjak, A., Mraz, J., Mlakar, J., Pižem, J., Tul, N., … Popović, M. (2017). Zika virus-associated micrencephaly: A thorough description of neuropathologic findings in the fetal central nervous system. *Archives of Pathology & Laboratory Medicine*, *141*(1), 73–81.

Sun, J., Wu, D., Zhong, H., Guan, D., Zhang, H., Tan, Q., & Ke, C. (2016). Presence of Zika virus in conjunctival fluid. *JAMA Ophthalmology*, *134*(11), 1330–1332.

Swaminathan, S., Schlaberg, R., Lewis, J., Hanson, K. E., & Couturier, M. R. (2016). Fatal Zika virus infection with secondary nonsexual transmission. *The New England Journal of Medicine*, *375*(19), 1907–1909.

Swanstrom, J. A., Plante, J. A., Plante, K. S., Young, E. F., McGowan, E., Gallichotte, E. N., … Baric, R. S. (2016). Dengue virus envelope dimer epitope monoclonal antibodies isolated from dengue patients are protective against Zika virus. *mBio*, *7*(4). e01123-16.

Tang, H., Hammack, C., Ogden, S. C., Wen, Z., Qian, X., Li, Y., … Ming, G. L. (2016). Zika virus infects human cortical neural progenitors and attenuates their growth. *Cell Stem Cell*, *18*(5), 587–590.

Teixeira, M. G., Costa, M., de Oliveira, W. K., Nunes, M. L., & Rodrigues, L. C. (2016). The epidemic of Zika virus-related microcephaly in Brazil: Detection, control, etiology, and future scenarios. *American Journal of Public Health*, *106*(4), 601–605.

Theel, E. S., & Hata, D. J. (2018). Diagnostic testing for Zika virus: A postoutbreak update. *Journal of Clinical Microbiology*, *56*(4). e01972-17.

van den Berg, B., Walgaard, C., Drenthen, J., Fokke, C., Jacobs, B. C., & van Doorn, P. A. (2014). Guillain-Barré syndrome: Pathogenesis, diagnosis, treatment and prognosis. *Nature Reviews Neurology*, *10*(8), 469–482.

van der Linden, V., Filho, E. L., Lins, O. G., van der Linden, A., Aragão, M., Brainer-Lima, A. M., … Ramos, R. C. (2016). Congenital Zika syndrome with arthrogryposis: Retrospective case series study. *BMJ (Clinical research ed.)*, *354*, i3899.

Wang, L., Valderramos, S. G., Wu, A., Ouyang, S., Li, C., Brasil, P., … Cheng, G. (2016). From mosquitos to humans: Genetic evolution of Zika virus. *Cell Host & Microbe*, *19*(5), 561–565.

Willison, H. J., Jacobs, B. C., & van Doorn, P. A. (2016). Guillain-Barré syndrome. *Lancet (London, England)*, *388*(10045), 717–727.

Wolfe, N. D., Kilbourn, A. M., Karesh, W. B., Rahman, H. A., Bosi, E. J., Cropp, B. C., … Gubler, D. J. (2001). Sylvatic transmission of arboviruses among Bornean orangutans. *The American Journal of Tropical Medicine and Hygiene*, *64*(5–6), 310–316.

World Health Organisation. (2016). *Zika situation report. Neurological syndrome and congenital abnormalities.*

World Health Organisation. (2018). *Immunization, vaccines and biologicals: WHO vaccine pipeline tracker.*

Zaidi, M. B., De Moraes, C. G., Petitto, M., Yepez, J. B., Sakuntabhai, A., Simon-Loriere, E., … Maestre, G. E. (2018). Non-congenital severe ocular complications of Zika virus infection. *JMM Case Reports*, *5*(6), e005152.

Zin, A. A., Tsui, I., Rossetto, J., Vasconcelos, Z., Adachi, K., Valderramos, S., … Moreira, M. (2017). Screening criteria for ophthalmic manifestations of congenital Zika virus infection. *JAMA Pediatrics*, *171*(9), 847–854.

Chapter 3

Classification of Zika virus sequences with respect to their species and subspecies

Joilson Xavier[a], Stephane Tosta[a], Talita Adelino[a,b], Vagner Fonseca[a,c], Marta Giovanetti[a,d], and Luiz Carlos Junior Alcantara[a,d]

[a]*Department of Genetics, Ecology and Evolution, Federal University of Minas Gerais, Belo Horizonte, MG, Brazil,* [b]*Central Laboratory of Public Health of Minas Gerais, Ezequiel Dias Foundation, Belo Horizonte, MG, Brazil,* [c]*Kwazulu-Natal Research Innovation and Sequencing Platform (KRISP), Department of Virology, National Health Laboratory Service (NHLS), University of KwaZulu-Natal—Nelson R. Mandela School of Medicine, Berea, Durban, South Africa,* [d]*Oswaldo Cruz Institute (IOC), Oswaldo Cruz Foundation (Fiocruz), Rio de Janeiro, RJ, Brazil*

Abbreviations

DNA	deoxyribonucleic acid
ICTV	International Committee on Taxonomy of Viruses
NS5	nonstructural protein 5
ORF	open reading frame
RNA	ribonucleic acid
ViPR	virus pathogen resource

Introduction

Zika virus (ZIKV) is an arthropod-borne virus belonging to the family *Flaviviridae*, genus *Flavivirus*. Its genome is a positive-sense, single-stranded, and nonsegmented RNA of approximately 11 kilobases (Kuno and Chang, 2007).

ZIKV was first identified in 1947 in a rhesus monkey from the Zika Forest in Uganda (Dick, Kitchen, & Haddow, 1952a). For many years the virus was responsible for causing sporadic infections in countries of Africa and Asia, such as Tanzania, Nigeria, and Malaysia (Marchette, Garcia, & Rudnick, 1969; Dick, Kitchen, & Haddow, 1952b; Macnamara, 1954).

Until now the largest epidemic caused by ZIKV was reported in 2015 in Brazil and, subsequently, in several other countries in South, North, and Central America (Pan American Health Organization/World Health Organization, 2016). Estimates indicate that more than 800,000 cases of ZIKV infection were reported in the Americas during that epidemic (Pan American Health Organization/World Health Organization, 2017). A considerable increase in the number of microcephaly cases was reported during the 2015 ZIKV epidemic in Brazil, suggesting for the first time a possible association between ZIKV infection and congenital abnormalities (Schuler-Faccini, 2016; Ministério Da Saúde, Brasil, 2015).

Transmission of ZIKV in urban settings occurs mainly via mosquitoes from the genus *Aedes* spp., also responsible for the transmission of other mosquito-borne viruses. Other nonvector modes of ZIKV transmission include congenital, perinatal, and sexual (Foy et al., 2011; Musso et al., 2015; Calvet et al., 2016; Hoen et al., 2018; Pomar et al., 2018). Possible transmission by blood transfusion has also been described (Musso et al., 2014; Barjas-Castro et al., 2016).

In humans the incubation period from a mosquito bite to symptom onset is ~3–12 days (Ioos et al., 2014). Infection is likely asymptomatic in ~80% of cases (Ioos et al., 2014). When symptoms occur, they are typically mild, self-limiting, and nonspecific; similarity to other arbovirus infections (e.g., dengue and chikungunya viruses) may confound the diagnosis. This superimposition of symptoms, combined with the possibility of cross-reactions with other arboviruses in serological tests, makes the ZIKV diagnosis a complicated task. Commonly reported symptoms include fever, headache, joint pain, muscle pain, conjunctivitis, and maculopapular eruption (Ioos et al., 2014; de Paula Freitas et al., 2016). Severe neurologic sequelae have also been described in adults, including meningitis, meningoencephalitis, and Guillain-Barré syndrome

Zika Virus Impact, Diagnosis, Control, and Models. https://doi.org/10.1016/B978-0-12-820267-8.00003-0

(WHO, 2016). A surge in Guillain-Barré syndrome cases has been observed in Brazil, Colombia, El Salvador, Suriname, Venezuela, and French Polynesia during outbreaks; however, ZIKV has been laboratory confirmed in only some of these cases (WHO, 2016).

Viruses are characterized in terms of sequence differentiation, and viral species are defined as a monophyletic group of viruses whose properties can be distinguished from those of other species by multiple criteria (Adams, Lefkowitz, King, & Carstens, 2013). According to the International Committee on Taxonomy of Viruses (ICTV), the nucleotide and deduced amino acid sequence relatedness is a relevant criterion for the demarcation of species. Other criteria such as antigenic and ecological characteristics, geographic, vector, host, and disease associations are also important to resolve the demarcation of genetically closely related viruses (Simmonds et al., 2017).

Zika virus sequences from African countries

ZIKV is currently taxonomically classified according to the ICTV Online (10th) Report (Simmonds et al., 2017), and the latest change into ZIKV species classification occurred in 2018, after the creation of the Riboviria realm, a taxon that includes RNA viruses (ICTV, 2019).

The first investigative studies on ZIKV epidemiology, mostly through serological surveys, found a wide distribution in African countries, including Gabon, Côte d'Ivoire, Central African Republic, Egypt, Kenya, Nigeria, Senegal, Sierra Leone, Somalia, and Tanzania (Jan, Languillat, Renaudet, & Robin, 1978; Akoua-Koffi et al., 2001; Saluzzo, Gonzalez, Herve, & Georges, 1981; Geser, Henderson, & Christensen, 1970; Fagbami, 1979; Robin & Mouchet, 1975).

The first report of ZIKV was published in 1952, when Dick, Kitchen, and Haddow reported two distinct isolation events in time and from different hosts at Zika Forest, Uganda, Africa (Dick et al., 1952a). Firstly, in April 1947, ZIKV was isolated from sentinel rhesus monkey placed in cages at the Zika Forest. The second isolation occurred 9 months later, in January 1948, from *Aedes africanus* mosquitoes at the same location (Dick et al., 1952a).

The ZIKV prototype (strain MR766 from Uganda) used for the NCBI Reference Sequence (NC_012532.1) was fully sequenced and genomically characterized by Kuno and Chang (2007). The sequence of this African strain comprises 10,794 nucleotides, and its ORF encodes 3419 amino acids. Later, Lanciotti et al. (2008) generated partial *NS5* sequences from three other African strains (strains 41,524, 41,525, and 41,662) isolated in 1984, in Senegal, from *Aedes* spp. mosquitoes.

Zika virus sequences from the Asian region

Following the first cases of ZIKV infection registered in the African continent, ZIKV transmission has been identified in other regions by virus isolation and/or serological studies. Pieces of evidence of ZIKV infection in humans in the Asian region have been registered since 1954 by serological studies performed in Malaysia, India, Philippines, Vietnam, Thailand, and Indonesia (Hammon et al., 1958; Olson et al., 1981; Pond, 1963; Smithburn, Kerr, & Gatne, 1954). However, a strain of the virus (P6-740) was isolated for the first time in Asia only in 1966 from infected mosquitoes *A. aegypti* collected in Malaysia, and the first human case from that region was confirmed in Central Java, Indonesia, in 1977 (Marchette et al., 1969; Olson et al., 1981).

Despite the existence of evidence of ZIKV circulation in Southeast Asia since 1954, an epidemic caused by this virus was reported for the first time in 2007 after an outbreak of a dengue-like illness in the Yap Island, Federated States of Micronesia (Duffy et al., 2009). Strains associated with this epidemic were sequenced, and phylogenetic analysis on the generated *NS5* gene fragment (strain ZIKV 2007 EC) showed that this strain is distantly related to the African sequences and reflected its geographical origin in Southeast Asia (Lanciotti et al., 2008). Moreover the authors of that study also pointed out to the tree topology displaying three subclades, where the first two subclades represent two lineages of African origin (referred to as the West and East African lineages, strains Senegal/1984 and MR766, respectively) and the third subclade represents the Asian lineage comprising the Yap strain (ZIKV 2007 EC) (Lanciotti et al., 2008).

Evolution of the Asian lineage

A different study from 2012, after performing phylogenetic analysis on sequences derived from the Yap Island and from a 2010 ZIKV case in Cambodia, corroborated Lanciotti et al.'s (2008) results by showing two geographically distinct lineages of ZIKV (Asian and African lineages), where the Asian lineage was represented by strains from the Yap epidemic and Cambodia, suggesting that the Asian strain was circulating in Southeast Asia until reaching and triggering the 2007 epidemic in Yap Island (Haddow et al., 2012). A 2014 study found similar results and also estimated the sequences of events that might have represented the emergence and international spread of the Asian lineage (Faye et al., 2014).

Faye et al. (2014) used ZIKV *NS5* and *envelope* genes generated from 43 strains to estimate the time of origin of ZIKV to early 20th century, suggesting the emergence of the virus in Uganda, Africa, around 1920, and from there the virus dispersed to Malaysia, probably around 1945, and then Micronesia (Faye et al., 2014). Another performed Bayesian analysis estimated an alternative earlier time of the most recent common ancestor of ZIKV to around 1837 (Gong et al., 2017). The time of emergence of the Asian lineage in Southeast Asia, in turn, was estimated to be around 1950, supporting the hypothesis of the African origin of ZIKV (Pettersson et al., 2018).

Spread of the Asian lineage to Pacific and Americas

Expansion of ZIKV to the South Pacific was determined after a ZIKV outbreak was reported in French Polynesia in 2013 (Cao-Lormeau et al., 2014). Phylogenetic analysis of partial ZIKV sequences from that outbreak showed that the French Polynesia sequences are closely related to Cambodia 2010 and Yap State 2007 strains (Cao-Lormeau et al., 2014). This expansion to South Pacific was also supported by phylogenetic analysis using the complete coding sequence of the ZIKV strain (H/PF/2013) from the outbreak in French Polynesia in 2013 (Baronti et al., 2014).

Following this spread to South Pacific, it has been argued that a strain from the ZIKV Asian lineage circulating in French Polynesia was likely introduced in the Americas around early 2014 (Faria et al., 2017; Metsky et al., 2017; Hu et al., 2019). Epidemiological and genomic data supported the hypothesis that this introduction might have occurred in the northeast region of Brazil, where the ZIKV caused a large outbreak in 2015 before spreading to other countries in the Americas (Faria et al., 2017; Metsky et al., 2017; Pettersson et al., 2018). The first ZIKV sequences from the 2015 outbreak in Brazil were partial *envelope* sequences generated from patients presenting dengue-like symptoms, and these sequences were closely related to the French Polynesia strain (H/PF/2013) (Zanluca et al., 2015; Campos et al., 2015).

A complete ZIKV genome sequence was obtained from fetal brain tissue of an associated microcephaly case originated in northeast Brazil, and this strain (Natal RGN) has been used as a provisional NCBI Reference Sequence (NC_035889.1) along with the MR 766 African strain (see Fig. 3) (Mlakar et al., 2016). In the following years, more ZIKV complete genome sequences were generated from the epidemic in the American continent (see Fig. 1 and Table 1), and two simultaneous studies from 2017 provided together 164 ZIKV genomes from Brazil and other countries from the Americas to understand the genomic epidemiology of ZIKV (Faria et al., 2017; Metsky et al., 2017). The results from these two studies

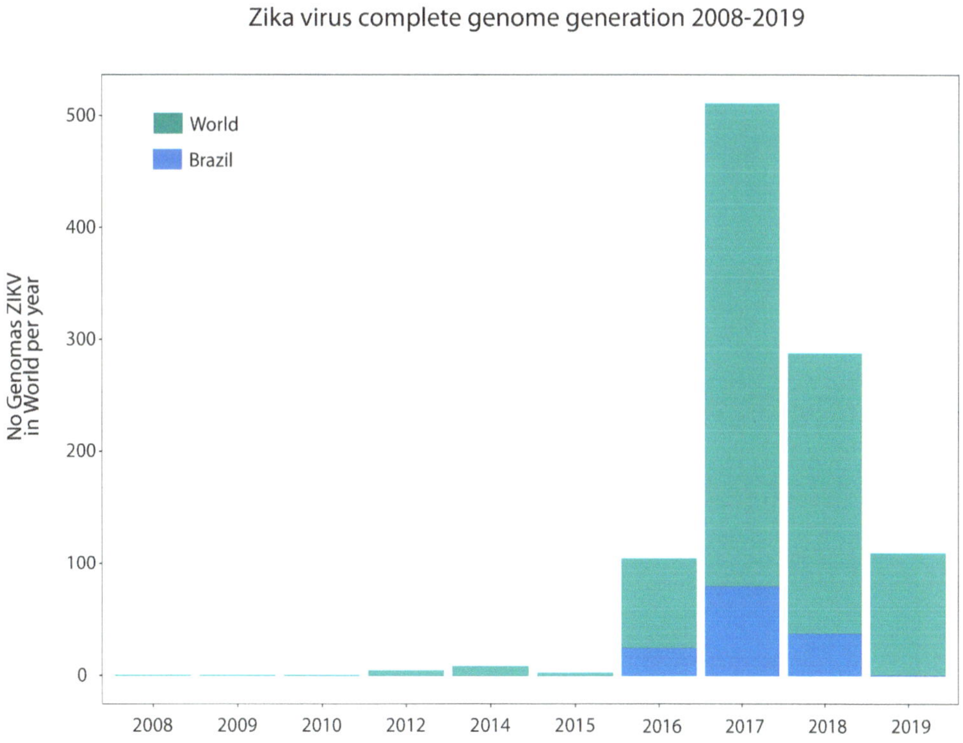

FIG. 1 Zika virus complete genome generation, 2008–19. A bar chart showing the number of Zika virus complete genomic sequences generated in each year. *Blue bars* represent the sequences from Brazilian isolates. The numbers were retrieved from the NCBI public database (www.ncbi.nlm.nih.gov).

TABLE 1 American ZIKV isolates sequenced between 2015 and 2017.

Accession number	Isolate	Lineage/ genotype	Country	Collection date	Host	Sample type	Sequence length (bp)
MH513599	BR/Sinop/H366/2015	Asian	Brazil	2015	Human	Plasma	10,658
KY785450	Zika virus/H. sapiens-wt/BRA/2016/ FC-6863-SER	Asian	Brazil	2016	Human	Serum	10,455
KU527068	Natal RGN	Asian	Brazil	2015	Human	Fetus' brain autopsy	10,808
KY631493	MEX_ENCB165	Asian	Mexico	2015	Human	Serum	10,683
MH063262	ZIKV/Homo_sapiens/Cuba/2017/ FL051Se	Asian	Cuba	2017	Human	Serum	10,619
MF783072	Zika virus/mosquito/Haiti/1855/2016	Asian	Haiti	2016	Mosquito	Mosquito	10,808
KX827268	ZIKV/Homo sapiens/USA/UT-1/2016	Asian	United States	2016	Human	Serum	10,787
KY785451	Zika virus/H. sapiens-wt/MTQ/2016/ FL-001-SAL	Asian	Martinique	2016	Human	Saliva	10,345
KY325467	Zika virus/H. sapiens-wt/USA/2016/ FLUR006	Asian	United States	2016	Human	Urine	10,742
MK028858	Zika virus/H. sapiens-tc/Panama/2015/ 259249	Asian	Panama	2015	Human	Serum	10,772
KY348640	SL1602	Asian	Suriname	2016	Human	Plasma	10,807
KX879604	ZIKV/EC/Esmeraldas/089/2016	Asian	Ecuador	2016	Human	Serum	10,810
KY348640	SL1602	Asian	Suriname	2016	Human	Plasma	10,807
KX893855	Zika virus/Homo sapiens/VEN/UF-2/ 2016	Asian	Venezuela	2016	Human	Serum	10,808
KU758877	17,271	Asian	French Guiana	2015	Human	Urine	10,364
MK049247	Aedes_aegypti/COL/FH04/2016	Asian	Colombia	2016	Mosquito	Mosquito	10,385
KY989971	FLA	Asian	Colombia	2015	Human	Serum	10,807
KY328289	HN16	Asian	Honduras	2016	Human	Serum	10,807
MH675628	CTS-223-16p	Asian	Puerto Rico	2016	Human	Blood	10,807
MK028858	Zika virus/H. sapiens-tc/Panama/2015/ 259249	Asian	Panama	2015	Human	Serum	10,772

A list of some ZIKV isolates from the Americas sequenced between the years 2015 and 2017. Sequence information were retrieved from the GenBank and Virus Pathogen Resource (ViPR) website.

are similar, and molecular clock analyses suggested that the virus emerged in the Americas, specifically in northeast Brazil, around early 2014, and from there the Asian strain disseminated internationally to other regions of the Americas, where thevirus might have circulated undetected for months before the first cases were locally confirmed (Faria et al., 2017; Metsky et al., 2017).

The proposition of three lineages

Since phylogenetic analyses have shown that the ZIKV sequences from the American epidemic clustered to form a single clade, these sequences have been classified as belonging to an American subclade derived from the Asian lineage circulating in Southeast Asia and the Pacific (see Fig. 2) (Faria et al., 2017; Gubler et al., 2017). The ZIKV phylogeny reconstructed by Gubler et al. (2017) showed a single clade comprising 93 ZIKV American strains from different countries from the Americas such as Brazil, Dominican Republic, Colombia, Ecuador, Guatemala, Haiti, Honduras, Mexico, Panama, Puerto Rico, Suriname, the United States, and Venezuela.

Besides the proposition of two major ZIKV lineages (African and Asian lineages), it has been argued that there might be a second African lineage sister to the previously defined African and Asian lineages (Gong et al., 2016; Shen et al., 2016). Using ZIKV *envelope* and *NS5* gene sequences available in GenBank, Shen et al. (2016) reconstructed the ZIKV phylogeny, whose tree topology revealed three ZIKV lineages (African lineage 1, African lineage 2, and Asian/American lineage) designated according to the geographical origin of the clustered strains. The African lineage 2 comprises strains from Senegal and Cote d'Ivoire, and it seems to have separated early from the other two lineages. Since there are no complete genomic sequences of the African lineage 2 and these three lineages appeared only in the *envelope* and *NS5* trees, it has been argued that sampling bias might explain why the African lineage 2 is absent in ZIKV phylogenies based on complete genomic sequences.

Regarding differences between ZIKV lineages, Smith et al. (2018) argued that mutations might influence phenotypic differences observed between ZIKV isolates from the Asian and African lineages, since this study showed approximately 75–100 amino acid residue variations between Asian and African isolates, while approximately 10–30 amino acid variations were observed between American and Asian isolates. A different study from 2018, by analyzing codon usage and nucleotide substitution rates in ZIKV genomes, identified that strains from the African lineage were more diverse than those from the Asian lineage and differential usage of codons was identified in genes from the Asian lineage. Moreover, that study also identified 131 nonsynonymous substitutions in their ZIKV dataset, and a higher selection pressure on the Asian lineage was also identified (Lin et al., 2018) (Figs. 3 and 4).

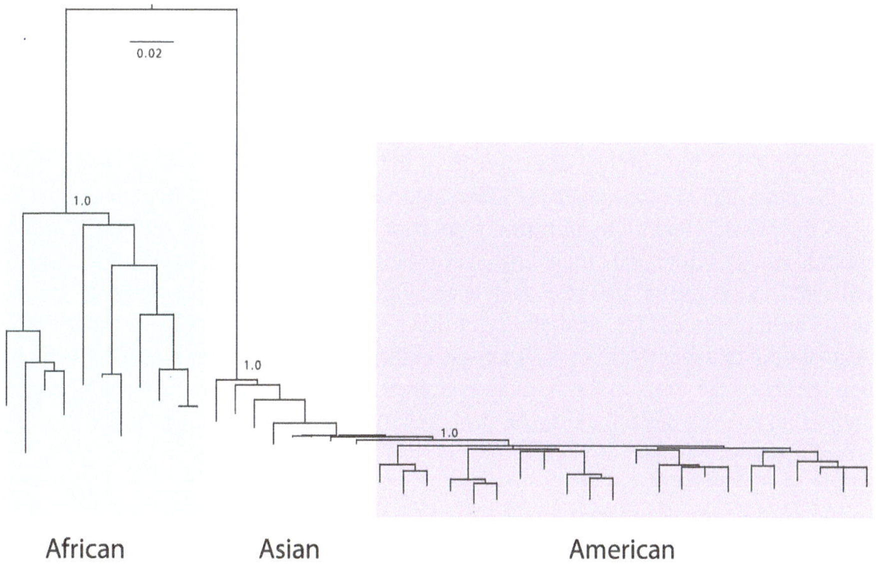

FIG. 2 Phylogenetic tree of Zika virus strains. A maximum likelihood tree reconstructed using complete genomic sequences of Zika virus strains from different countries. Closely related strains clustered to form three distinct groups (clades) that represent the proposed Zika virus lineages discussed in the main text (African, Asian, and American lineages). Sequences were retrieved from the GenBank public database (www.ncbi.nlm.nih.gov/genbank/). Phylogenetic analysis was performed using the IQ-TREE software.

FIG. 3 Alignment of two Zika virus sequences. Alignment of the NS1 gene of the Zika virus reference sequence (accession number: NC_012532) from NCBI database to another Zika virus sequence from Brazil set as provisional reference sequence from NCBI (accession number: NC_035889). This alignment shows a partial region of the NS1 gene, and each nucleotide is represented by a different color. Alignment was performed using MAFFT free alignment software and visualized using AliView.

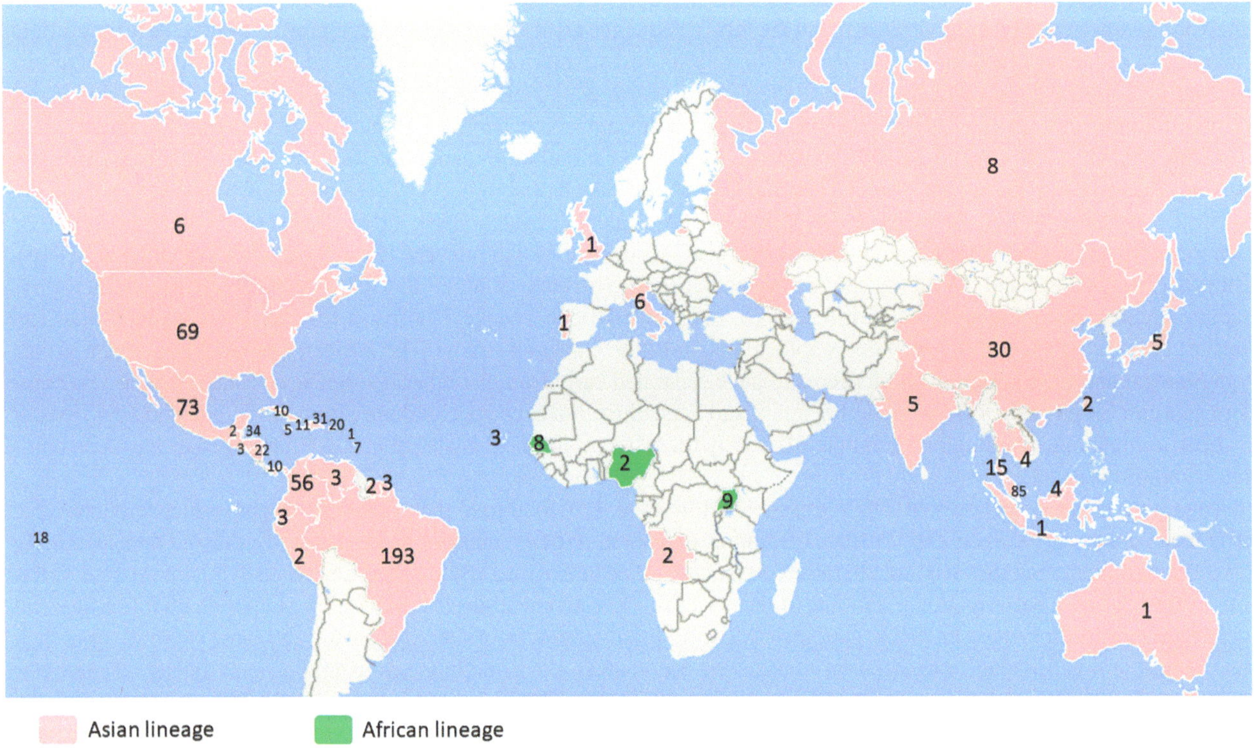

FIG. 4 Number of complete Zika virus sequences according to lineage by country. Geochart showing the number of Zika virus complete genomic sequences generated in each country (here, sequences >40% size of reference genome). Areas in *pink color* represent the sequences classified as Asian lineage, and the *green color* represents sequences classified as African lineage. The numbers were retrieved from the Virus Pathogen Resource (ViPR) public database (https://www.viprbrc.org/).

Conclusion

Recent epidemics caused by ZIKV have drawn attention from the international public due to its rapid spread, the high number of cases, and the neurological consequences associated with the infection. Such threat to public health posed by this virus influenced the World Health Organization's decision to include Zika in the list of priority diseases of the 2018 agenda for research and development, including surveillance and diagnostics. After the ZIKV epidemic in the Americas, an expressive increase in the number of complete and partial ZIKV genomic sequences generated from genomic epidemiology studies has been observed. The availability of this genomic information has allowed the characterization of the evolution and spread of viral lineages. Despite this increase in the amount of genomic data, little information is available on the genomic epidemiology of the virus in the African continent. Thus more genomic surveillance efforts in currently affected areas are required for a better understanding of viral evolution and dispersion dynamics that might help actions of prediction and control of future outbreaks.

Policy and procedures

Sequencing

To obtain genomic sequences from ZIKV, sequencing experiments are performed on clinical samples used as the source of ZIKV genetic material. Viral RNA can be extracted and purified from biological samples, such as saliva, serum, plasma, and semen, using commercial RNA extraction kits. For the sequencing experiments, it has been commonly used protocols from DNA sequencer machines from Illumina or Oxford Nanopore. A detailed protocol for ZIKV sequencing was published in Quick et al. (2017)

Classification of sequences

Genomic sequences can be classified according to its evolutionary relationship with other isolates or specimens by using Bioinformatics software that calculate similarity among nucleotide sequences from a given dataset and perform serial groupings of those more closely related sequences.

Genetic sequence database

ZIKV sequences can be stored in public databases such as GenBank and Virus Pathogen Resource (ViPR). The GenBank is an example of an online public database that stores DNA sequences and related information. Researchers can store their genetic sequences in GenBank and get an accession number (sequence ID) unique for each sequence. This number allows a user to access information about sample collection and download a file containing the nucleotide sequence of interest.

Mini-dictionary of terms

Bayesian analysis. Here, it refers to a statistical method that uses probability to calculate the chance of a hypothesis.
Guillain-Barré syndrome. A disease in which the immune system attacks the nerves.
Maculopapular eruption. Skin redness with small, high, and soft lumps in the skin.
Meningoencephalitis. Inflammation of the brain and surrounding tissues, usually caused by infection.
Microcephaly. It is a medical condition in which there is a malformation of the brain of a newborn.
Molecular clock analyses. It is a method for estimating the evolutionary rate between species.
Monophyletic. A group that includes the common ancestor and all their descendants.
Phylogenetic analysis. A tool for estimating evolutionary relationships between genes, genomes, and species.
Phylogeny. Representation of the evolutionary relationship between groups of related species.
Subclades. A subgroup formed by evolutionarily related strains.

Key facts of Zika virus lineage

- The Zika virus reference strain MR766 was isolated from mosquitoes in Uganda in 1948, and its complete genomic sequence was published in 2007.
- A Zika virus strain from the Asian lineage was isolated for the first time in Asia in 1966.
- The first epidemic in Southeast Asia caused by an Asian lineage strain was reported in the Yap Island, Federated States of Micronesia, in 2007.
- Expansion of the Asian lineage to the South Pacific was identified when an epidemic was reported in French Polynesia in 2013.
- The Asian lineage reached the Americas and caused a large epidemic in 2015, and 164 complete genomic sequences were generated from that epidemic.
- Zika virus sequences from the epidemic in the Americas form a single distinct clade, named the American subclade.

Summary points

- Zika virus is an arthropod-borne virus first identified in Uganda, Africa, that has caused large epidemics in the western hemisphere.
- The use of nucleotide sequence relatedness is a relevant criterion for the demarcation of viral species.
- Reports of Zika virus infection have been observed throughout many African countries.
- There is evidence of the Zika virus circulation in the Asian region since 1954 although the first epidemic in that region was reported only in 2007.
- It is argued that the Asian lineage diverged from the African lineage around 1950 and then spread to South Asia, the Pacific, and the Americas.
- A 2016 study suggested that Zika virus sequences can be classified into three lineages.

References

Adams, M. J., Lefkowitz, E. J., King, A. M. Q., & Carstens, E. B. (2013). Recently agreed changes to the International Code of Virus Classification and Nomenclature. *Archives of Virology*, *158*(12), 2633–2639.

Akoua-Koffi, C., Diarrassouba, S., Benie, V. B., Ngbichi, J. M., Bozoua, T., Bosson, A., ... Ehouman, A. (2001). Investigation surrounding a fatal case of yellow fever in Cote d'Ivoire in 1999. *Bulletin de la Societe de pathologie exotique (1990)*, *94*(3), 227–230.

Barjas-Castro, M. L., Angerami, R. N., Cunha, M. S., Suzuki, A., Nogueira, J. S., Rocco, I. M., ... Stucchi, R. S. (2016). Probable transfusion-transmitted Zika virus in Brazil. *Transfusion*, *56*(7), 1684–1688.

Baronti, C., Piorkowski, G., Charrel, R. N., Boubis, L., Leparc-Goffart, I., & de Lamballerie, X. (2014). Complete coding sequence of Zika virus from a French polynesia outbreak in 2013. *Genome Announcements*, *2*(3). e00500-14.

Calvet, G., Aguiar, R. S., Melo, A. S., Sampaio, S. A., De Filippis, I., Fabri, A., ... Tschoeke, D. A. (2016). Detection and sequencing of Zika virus from amniotic fluid of fetuses with microcephaly in Brazil: A case study. *The Lancet Infectious Diseases*, *16*(6), 653–660.

Campos, G. S., Bandeira, A. C., & Sardi, S. I. (2015). Zika virus outbreak, Bahia, Brazil. *Emerging Infectious Diseases*, *21*(10), 1885.

Cao-Lormeau, V. M., Roche, C., Teissier, A., Robin, E., Berry, A. L., Mallet, H. P., ... Musso, D. (2014). Zika virus, French polynesia, South Pacific, 2013. *Emerging Infectious Diseases*, *20*(6), 1085.

de Paula Freitas, B., de Oliveira Dias, J. R., Prazeres, J., Sacramento, G. A., Ko, A. I., Maia, M., & Belfort, R. (2016). Ocular findings in infants with microcephaly associated with presumed Zika virus congenital infection in Salvador, Brazil. *JAMA Ophthalmology*, *134*(5), 529–535.

Dick, G. W. A., Kitchen, S. F., & Haddow, A. J. (1952a). Zika virus (I). Isolations and serological specificity. *Transactions of the Royal Society of Tropical Medicine and Hygiene*, *46*(5), 509–520.

Dick, G. W. A., Kitchen, S. F., & Haddow, A. J. (1952b). Zika virus (II). Pathogenicity and physical properties. *Transactions of the Royal Society of Tropical Medicine and Hygiene*, *46*(5), 521–534.

Duffy, M. R., Chen, T. H., Hancock, W. T., Powers, A. M., Kool, J. L., Lanciotti, R. S., ... Guillaumot, L. (2009). Zika virus outbreak on Yap Island, Federated States of Micronesia. *New England Journal of Medicine*, *360*(24), 2536–2543.

Fagbami, A. H. (1979). Zika virus infections in Nigeria: Virological and seroepidemiological investigations in Oyo State. *Epidemiology and Infection*, *83*(2), 213–219.

Faria, N. R., Quick, J., Claro, I. M., Theze, J., de Jesus, J. G., Giovanetti, M., ... Franco, L. C. (2017). Establishment and cryptic transmission of Zika virus in Brazil and the Americas. *Nature*, *546*(7658), 406.

Faye, O., Freire, C. C., Iamarino, A., Faye, O., de Oliveira, J. V. C., Diallo, M., & Zanotto, P. M. (2014). Molecular evolution of Zika virus during its emergence in the 20th century. *PLoS Neglected Tropical Diseases*, *8*(1), e2636.

Foy, B. D., Kobylinski, K. C., Foy, J. L. C., Blitvich, B. J., da Rosa, A. T., Haddow, A. D., ... Tesh, R. B. (2011). Probable non-vector-borne transmission of Zika virus, Colorado, USA. *Emerging Infectious Diseases*, *17*(5), 880.

Geser, A., Henderson, B. E., & Christensen, S. (1970). A multipurpose serological survey in Kenya: 2. Results of arbovirus serological tests. *Bulletin of the World Health Organization*, *43*(4), 539.

Gong, Z., Gao, Y., & Han, G. Z. (2016). Zika virus: Two or three lineages? *Trends in Microbiology*, *24*(7), 521–522.

Gong, Z., Xu, X., & Han, G. Z. (2017). The diversification of Zika virus: Are there two distinct lineages? *Genome Biology and Evolution*, *9*(11), 2940–2945.

Gubler, D. J., Vasilakis, N., & Musso, D. (2017). History and emergence of Zika virus. *Journal of Infectious Diseases*, *216*(Suppl_10), S860–S867.

Haddow, A. D., Schuh, A. J., Yasuda, C. Y., Kasper, M. R., Heang, V., Huy, R., ... Weaver, S. C. (2012). Genetic characterization of Zika virus strains: Geographic expansion of the Asian lineage. *PLoS Neglected Tropical Diseases*, *6*(2), e1477.

Hammon, W. M., Schrack, W. D., Jr., & Sather, G. E. (1958). Serological survey for arthropod-borne virus infections in the Philippines. *American Journal of Tropical Medicine and Hygiene*, *7*(3), 323–328.

Hoen, B., Schaub, B., Funk, A. L., Ardillon, V., Boullard, M., Cabié, A., ... Douine, M. (2018). Pregnancy outcomes after ZIKV infection in French territories in the Americas. *New England Journal of Medicine*, *378*(11), 985–994.

Hu, T., Li, J., Carr, M. J., Duchêne, S., & Shi, W. (2019). The Asian lineage of Zika virus: Transmission and evolution in Asia and the Americas. *Virologica Sinica*, *34*(1), 1–8.

ICTV. (2019). *ICTV taxonomy history: Zika virus*. Available from https://talk.ictvonline.org/taxonomy/p/taxonomy-history?taxnode_id=201853123. (Accessed 13 December 2019).

Ioos, S., Mallet, H. P., Goffart, I. L., Gauthier, V., Cardoso, T., & Herida, M. (2014). Current Zika virus epidemiology and recent epidemics. *Medecine et maladies infectieuses*, *44*(7), 302–307.

Jan, C., Languillat, G., Renaudet, J., & Robin, Y. (1978). A serological survey of arboviruses in Gabon. *Bulletin de la Societe de pathologie exotique et de ses filiales*, *71*(2), 140–146.

Kuno, G., & Chang, G. J. (2007). Full-length sequencing and genomic characterization of Bagaza, Kedougou, and Zika viruses. *Archives of Virology*, *152* (4), 687–696.

Lanciotti, R. S., Kosoy, O. L., Laven, J. J., Velez, J. O., Lambert, A. J., Johnson, A. J., ... Duffy, M. R. (2008). Genetic and serologic properties of Zika virus associated with an epidemic, Yap State, Micronesia, 2007. *Emerging Infectious Diseases*, *14*(8), 1232.

Lin, D., Li, L., Xie, T., Yin, Q., Saksena, N., Wu, R., ... Chen, X. (2018). Codon usage variation of Zika virus: The potential roles of NS2B and NS4A in its global pandemic. *Virus Research*, *247*, 71–83.

Macnamara, F. N. (1954). Zika virus: A report on three cases of human infection during an epidemic of jaundice in Nigeria. *Transactions of the Royal Society of Tropical Medicine and Hygiene*, *48*(2), 139–145.

Marchette, N. J., Garcia, R., & Rudnick, A. (1969). Isolation of Zika virus from Aedes aegypti mosquitoes in Malaysia. *American Journal of Tropical Medicine and Hygiene*, *18*(3), 411–415.

Metsky, H. C., Matranga, C. B., Wohl, S., Schaffner, S. F., Freije, C. A., Winnicki, S. M., … Lin, A. E. (2017). Zika virus evolution and spread in the Americas. *Nature*, *546*(7658), 411.

Ministério Da Saúde, Brasil. (2015). *Procedimentos preliminares a serem adotados para a vigilância dos casos de microcefalia no Brasil.* Nota Informativa n^0 01/2015. Available from https://portalarquivos2.saude.gov.br/images/pdf/2015/novembro/18/microcefalia-nota-informativa-17nov2015-c.pdf. (Accessed 31 October 2019).

Mlakar, J., Korva, M., Tul, N., Popović, M., Poljšak-Prijatelj, M., Mraz, J., … Vizjak, A. (2016). Zika virus associated with microcephaly. *New England Journal of Medicine*, *374*(10), 951–958.

Musso, D., Nhan, T., Robin, E., Roche, C., Bierlaire, D., Zisou, K., … Broult, J. (2014). Potential for Zika virus transmission through blood transfusion demonstrated during an outbreak in French Polynesia, November 2013 to February 2014. *Eurosurveillance*, *19*(14), 20761.

Musso, D., Roche, C., Robin, E., Nhan, T., Teissier, A., & Cao-Lormeau, V. M. (2015). Potential sexual transmission of Zika virus. *Emerging Infectious Diseases*, *21*(2), 359.

Pan American Health Organization/World Health Organization. (2016). *Timeline of the emergence of Zika virus in the Americas*. Washington, DC: PAHO/WHO. Available from https://www.paho.org/hq/index.php?option=com_content&view=article&id=11959:timeline-of-emergence-of-zika-virus-in-the-americas&Itemid=41711&lang=en. (Accessed 13 December 2019).

Pan American Health Organization/World Health Organization. (2017). *Zika suspected and confirmed cases reported by countries and territories in the Americas cumulative cases, 2015–2017.* Available from https://www.paho.org/hq/dmdocuments/2017/2017-ago-10-phe-ZIKV-casess.pdf. (Accessed 28 October 2019).

Pettersson, J. H. O., Bohlin, J., Dupont-Rouzeyrol, M., Brynildsrud, O. B., Alfsnes, K., Cao-Lormeau, V. M., … Musso, D. (2018). Re-visiting the evolution, dispersal and epidemiology of Zika virus in Asia. *Emerging Microbes and Infections*, *7*(1), 1–8.

Pomar, L., Vouga, M., Lambert, V., Pomar, C., Hcini, N., Jolivet, A., … Panchaud, A. (2018). Maternal-fetal transmission and adverse perinatal outcomes in pregnant women infected with Zika virus: Prospective cohort study in French Guiana. *BMJ*, *363*, k4431.

Pond, W. L. (1963). Arthropod-borne virus antibodies in sera from residents of South-East Asia. *Transactions of the Royal Society of Tropical Medicine and Hygiene*, *57*(5), 364–371.

Robin, Y., & Mouchet, J. (1975). Serological and entomological study on yellow fever in Sierra Leone. *Bulletin de la Societe de Pathologie Exotique et de ses Filiales*, *68*(3), 249–258.

Saluzzo, J. F., Gonzalez, J. P., Herve, J. P., & Georges, A. J. (1981). Serological survey for the prevalence of certain arboviruses in the human population of the south-east area of Central African Republic (author's transl). *Bulletin de la Societe de pathologie exotique et de ses filiales*, *74*(5), 490–499.

Schuler-Faccini, L. (2016). Possible association between Zika virus infection and microcephaly—Brazil, 2015. *MMWR Morbidity and Mortality Weekly Report*, *65*, 59–62.

Shen, S., Shi, J., Wang, J., Tang, S., Wang, H., Hu, Z., & Deng, F. (2016). Phylogenetic analysis revealed the central roles of two African countries in the evolution and worldwide spread of Zika virus. *Virologica Sinica*, *31*(2), 118–130.

Simmonds, P., Becher, P., Bukh, J., Gould, E. A., Meyers, G., Monath, T., … Stapleton, J. T. (2017). ICTV virus taxonomy profile: Flaviviridae. *The Journal of General Virology*, *98*(1), 2.

Smith, D. R., Sprague, T. R., Hollidge, B. S., Valdez, S. M., Padilla, S. L., Bellanca, S. A., … Haddow, A. D. (2018). African and Asian Zika virus isolates display phenotypic differences both in vitro and in vivo. *American Journal of Tropical Medicine and Hygiene*, *98*(2), 432–444.

World Health Organization (WHO). (2016). *Zika situation report: Neurological syndrome and congenital anomalies.* Available from http://www.who.int/emergencies/zika-virus/situation-report/5-february-2016/en/. (Accessed 13 December 2019).

Zanluca, C., Melo, V. C. A. D., Mosimann, A. L. P., Santos, G. I. V. D., Santos, C. N. D. D., & Luz, K. (2015). First report of autochthonous transmission of Zika virus in Brazil. *Memórias do Instituto Oswaldo Cruz*, *110*(4), 569–572.

Olson, J. G., et al. (1981). Zika virus, a cause of fever in Central Java, Indonesia. *Transactions of The Royal Society of Tropical Medicine and Hygiene*. https://doi.org/10.1016/0035-9203(81)90100-0.

Smithburn, K. C., Kerr, J. A., & Gatne, P. B. (1954). Neutralizing antibodies against certain viruses in the sera of residents of India. *The Journal of Immunology*, *72*(4), 248–257.

Quick, J., et al. (2017). Multiplex PCR method for MinION and Illumina sequencing of Zika and other virus genomes directly from clinical samples. *Nature Protocols*. https://doi.org/10.1038/nprot.2017.066.

Chapter 4

Health knowledge about Zika virus: Brazil aspects

Ana Luiza Vilela Borges[a] and Raquel Zanatta Coutinho[b]

[a]*Department of Public Health Nursing, School of Nursing, University of São Paulo, São Paulo, SP, Brazil,* [b]*Department of Demography, Center for Development and Regional Planning, Federal University of Minas Gerais, Belo Horizonte, MG, Brazil*

Abbreviations

CDC	Centers for Disease Control
FWAT	Free Words Association Technique
HDI	Human Development Index
LARC	long-acting reversible contraceptive
LIRA	Larval Index Rapid Assay
PAHO	Pan-American Health Organization
RESP	Registro de Eventos em Saúde Pública (Public Health Events Registration)
SES	socioeconomic status
SINASC	Sistema de Informação sobre Nascidos Vivos (Brazilian Information System on Live Births)
WHO	World Health Organization

Introduction

At the end of 2014, bedside physicians in the northeastern region of Brazil, Latin America (Fig. 1), noticed an increase in reports of skin rash, a condition that came together with low fever and conjunctivitis. Attentive medical doctors noticed that the symptoms differed from those caused by dengue and chikungunya viruses, both transmitted by the mosquito *Aedes aegypti*.

Few months later, in April 2015, the presence of a foreign virus, the Zika, was confirmed in Brazil, followed by the reports of autochthonous infections and transmission (see timeline of Zika virus epidemics in Brazil, Fig. 2).

In September 2015 a sudden increase in cases of microcephaly was reported in the same region. Researching a peak of occurrence among infants born in late November 2015, researchers found that the incidence of microcephaly, which had been around 15 cases per month, had increased many times (França et al., 2016; Oliveira et al., 2017). The association between Zika virus and the increase in microcephaly, first suspected by Brazilian doctors, had no precedent in the medical literature investigating past upsurges of Zika virus elsewhere.

A rush of research including much international attention trying to link both conditions and prove the causality is what followed. Donation of biospecimen allowed the identification of the congenital Zika virus infection, proving the link. Calvet et al. (2016) were the first to isolate the genome of the Zika virus from amniotic fluid of a pregnant woman before delivery. Mlakar et al. (2016) were the first to find Zika virus in the brain tissue of a severely microcephalic baby.

In the same month, November 2015, the Brazilian Ministry of Health recognized the possible association of the virus and the increase in microcephaly cases, so a surveillance system for microcephaly (RESP[a]) was set up by the Ministry of Health to register cases of microcephaly. But it was not until April 2016 that the World Health Organization (WHO) finally documented the associations and declared Zika virus a Public Health Emergency of International Concern (França et al., 2016). The reports of transmission were appearing throughout other countries in the Americas, Africa, and many others of the world. Nevertheless, Brazil was considered the epicenter of the outbreak, more intensively the northeast region. Since the beginning of the notifications until the end of 2019, 3474 cases of congenital Zika virus syndrome were confirmed in the

a. Brasil, Ministério da Saúde, Registro de Eventos em Saúde Pública (RESP, Microcefalia) 2016. http://resp.saude.gov.br.

Zika Virus Impact, Diagnosis, Control, and Models. https://doi.org/10.1016/B978-0-12-820267-8.00004-2

FIG. 1 Brazil and regions. *(Source: Authors own elaboration.)*

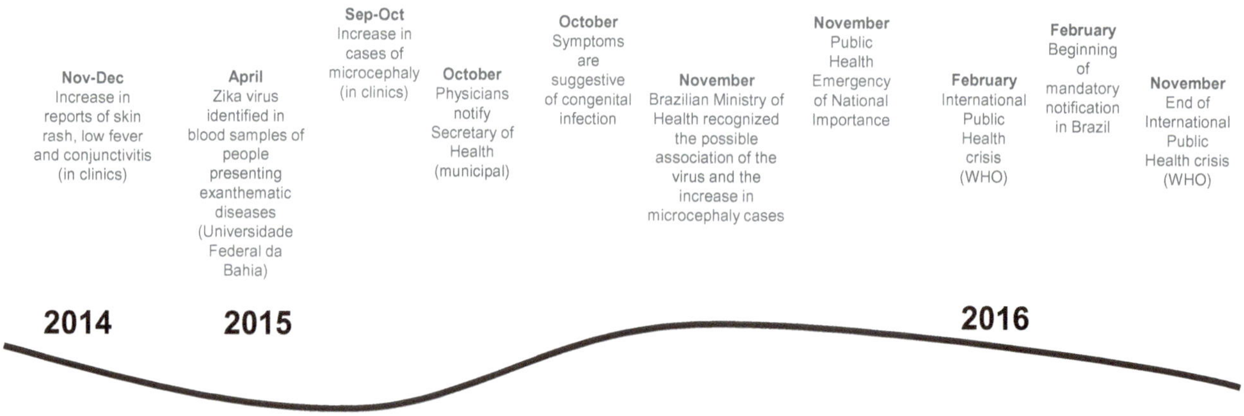

FIG. 2 Timeline of Zika virus epidemics in Brazil. *(Source: Authors own elaboration.)*

country alone (Brasil, Ministério da Saúde, Secretaria de Atenção à Saúde, 2019). In fact the surge in microcephaly in the northeast, which began in August 2015, was preceded by a rise in the incidence of possible cases of Zika virus about half a year earlier (Oliveira et al., 2017). Women of low socioeconomic status were overly represented among the mothers of babies who were born with the congenital Zika syndrome (Diniz, 2016; Vélez & Diniz, 2016), raising the alarm that there could be important social explanations for the phenomenon.

From the moment the link between Zika virus and congenital syndrome was established, two perspectives grounded the immediate responses to the epidemics in Brazil: vector control and advice to delay pregnancy (Ferguson et al., 2016).

The vector control strategy focused on the fact that the *A. aegypti* is overwhelmingly found in the Brazilian territory and is responsible for keeping the incidence of dengue high yearlong. Brazil has been implementing mosquito reduction, elimination, and eradication initiatives in the last 20 years with disappointing results. In 2019 alone more than 1.5 million cases of dengue and over 130,000 cases of chikungunya were recorded (Brasil, Ministério da Saúde, Secretaria de Atenção à Saúde, 2020). Dengue still accounted for the majority of the diseases of mandatory notification when the Zika epidemics peaked between 2015 and 2016, reaching more than 85% of the suspected cases of human arboviruses in the country (Oliveira et al., 2017) (Table 1).

Literature brings several reasons why the dengue epidemic is so pervasive and why Zika virus could rapidly disseminate in Brazil. One reason is that increasing population size with disordered urban growth leads to increases in population density without assuring that basic living conditions, such as sanitation network, are met. Lack of sanitation forces water storage inside the households, which may favor mosquito's breeding. Other reasons are internal migration, geographic displacements, and lack of political continuity, which are important to sustain long-term public health policies and programs (Bueno et al., 2017). Also, growing vector resistance to the pesticides, health professional household visits that do not cover the whole territory, and poor sanitary conditions contribute to mosquito proliferation (Augusto et al., 2016; Santos et al., 2016). According to tropical diseases experts, only vaccination or other new technologies, such as biological mosquito control, might ever be able to counteract the spread of such tropical diseases in Brazil (Bueno et al., 2017).

Public health response on vector control: Communicational strategies and perceptions of risk

At the uncovering of the Zika virus in Brazil, mass campaigns utilized to fight dengue were rapidly reprinted to fight the Zika virus. Pamphlets, banners, and media advertisements remembering the population to check every possible source of standing water (from flower vases and empty bottles to water tanks) were massively distributed by community health agents in households and broadcast on TV, radio, and the Internet. The Brazilian government increased actions for destroying mosquito reservoirs; some included having the military manually helping the population clear their backyards; others had pesticides being sprayed; and others started to test biological mosquito control strategies. In every case, dengue strategies that had been used before were applied again or conjointly (Augusto et al., 2016).

As mosquito eradication was doom to fail given the fragility of the strategies, the general knowledge about Zika virus and its prevention strategies seem to have been built on what the health services understood as dengue prevention, which was to prevent mosquito bites at all cost. Due to the congenital Zika virus syndrome concern, pregnant women were advised to avoid mosquito bites via wearing protective clothing (long-sleeved clothing, pants, and socks), mosquito nets on the bed, mesh screens (window nets), and insect repellent (Melo, Silva, & La Corte, 2019; Rasanathan, MacCarthy, Diniz, Torreele, & Gruskin, 2017) as recommended by the WHO, the Pan-American Health Organization (PAHO), and the Brazilian Ministry of Health.

Melo et al. (2019) investigated the frequency of use of measures of self-protection employed by pregnant women in a Brazilian northeastern state, comparing urban and rural locations (in 2016 and 2017), and found that repellents were the most frequent measure of protection (57%), followed by mosquito nets (47%), long-sleeved clothing (44%), and window

TABLE 1 Number of confirmed cases of dengue, chikungunya, and Zika virus disease in Brazil, 2015–19.

	Year				
Disease	2015	2016	2017	2018	2019
Dengue	1,688,688	1,500,535	239,389	265,934	1,544,987
Chikungunya	38,499	271,824	185,593	87,687	132,205
Zika	37,011[a]	215,319	17,593	8680	10,768

[a]*First case was detected in April 2015.*

Source: Authors own elaboration based on the Epidemiological Surveillance Bulletin, Ministry of Health, Brazil (https://www.saude.gov.br/boletins-epidemiologicos).

nets (6%). While mosquito nets were the most used among low-educated pregnant women (58%) and those living in rural areas (85%), repellents were most used by high-educated ones (83%). Only 2% of women used all of the measures. However, such measures to control mosquito bites are considered merely palliative, mainly because they do not act on the underlying causes of vector proliferation, such as the lack of clean water, poor sanitation and housing, and other causes of climate change (Rasanathan et al., 2017).

Using the Free Words Association Technique (FWAT), an instrument to capture collective mental representations, Guedes, Coutinho, Marteleto, Pereira, and Duarte (2018) researched the inhabitants of a Brazilian midsize southeastern municipality, highly impacted by Zika. The FWAT consists of asking participants what were the first five words that came to their mind when they heard the expression "Zika virus." Classifying the meanings of Zika virus into clusters of thought yielded five different subnetworks of meanings: lack of social and personal responsibility, symptoms, media coverage terror, public health, and microcephaly. These clusters may show information incorporated from the everyday messages spread out by the media, with a focus on vector control and prevention of microcephaly. The survey also inquired participants about the history of infection (if self or family member has been infected) and with what disease (Zika virus, dengue or chikungunya, or none). They found that the map of meanings of those never infected by any disease transmitted by the *A. aegypti* is more diverse and formed by messages spread by the media (like chaos and standing water). Those who were infected by the Zika virus had a more defined network with meanings, mainly formed by the most common symptoms, such as pain, rash, and itching. In conclusion, personal experience seems to have shaped the social representation of the disease.

The way the Zika virus outbreak was reported in two major newspapers in Brazil was analyzed by Ribeiro, Hartley, Nerlich, and Jaspa (2018). They considered the news published between December 2015 and May 2016 and found two main subframes: mosquito eradication and microcephaly control. While 72% of the content were regarding epidemiological or scientific facts, much voiced by scientists and political authorities, only 12% of the material concerned political and socioeconomic discussions by people with less authority, especially poor pregnant women. This absence of women's voice was classified as a "marginal frame," defining the problem of Zika virus as a matter of socioeconomic and political inequalities. In line with that, "avoid and destroy mosquitoes" was the message in 64% of the news, followed by "avoid pregnancy" in 20% of the cases; microcephaly was anchoring the news in 45% of the cases, and women were the interested/target audience of the articles (to whom the article was speaking) in 44% of the times. The authors' main conclusion was that women were to assume the protagonist role of Zika virus prevention strategies both by avoiding the mosquitos' bite and by delaying pregnancy.

Advice to delay pregnancy as a public health strategy

Besides vector control the other strategy adopted as immediate response to the Zika virus epidemics in Brazil was based on the advice by WHO that women in Zika virus affected areas should consider delaying pregnancy. Assuming that fertility control is perfectly effective, a woman who was trying to conceive when the Zika outbreak started may have delayed pregnancy, foregone pregnancy, or decided to get pregnant in spite of the recommendation believing that she was not at risk or could control her exposure to the risk of getting infected (Fig. 3). This means that family planning was considered part of the measures to prevent Zika virus transmission and its worst consequences, the congenital Zika virus syndrome, which may cause microcephaly among other health problems (Burke & Moreau, 2016). Apart from the international agencies, some local governments also voiced recommendations about pregnancy postponement, which contributed to turn the Zika virus epidemics into a reproductive health issue. However, recommendations to delaying pregnancy from local governments are troublesome and incautious since Zika was causing microcephaly in contexts with large proportions of unintended pregnancies, where the contraceptive landscape was based on short-acting contraceptive methods, like the pill and the condom, and where legislation may include legal restrictions toward abortion. In the Brazilian case, such recommendation sounded deceptively simple, considering that 55% of the pregnancies were classified as unintended (Theme-Filha et al., 2016) and even intended pregnancies could turn into unintended on the face of a threat. As Fig. 3 summarizes, a myriad of social determinants may affect reproductive choices and act on a woman's personal vulnerability to unintended pregnancies and to Zika infection.

In Brazil the recommendation to postpone pregnancy had not been made official, yet the director of the Surveillance Department at the Ministry of Health stated in media interviews that women living in high-risk areas who could wait to conceive should do so[b] (Rasanathan et al., 2017). Nevertheless, the country did not focus on including reproductive rights on the top of the agenda. Even though the Brazilian Ministry of Health published guidelines for public services to address

b. Samarasekera U, Triunfol M. Concern over Zika virus grips the world. Lancet 2016 Feb 3. PubMed PMID:26852261.

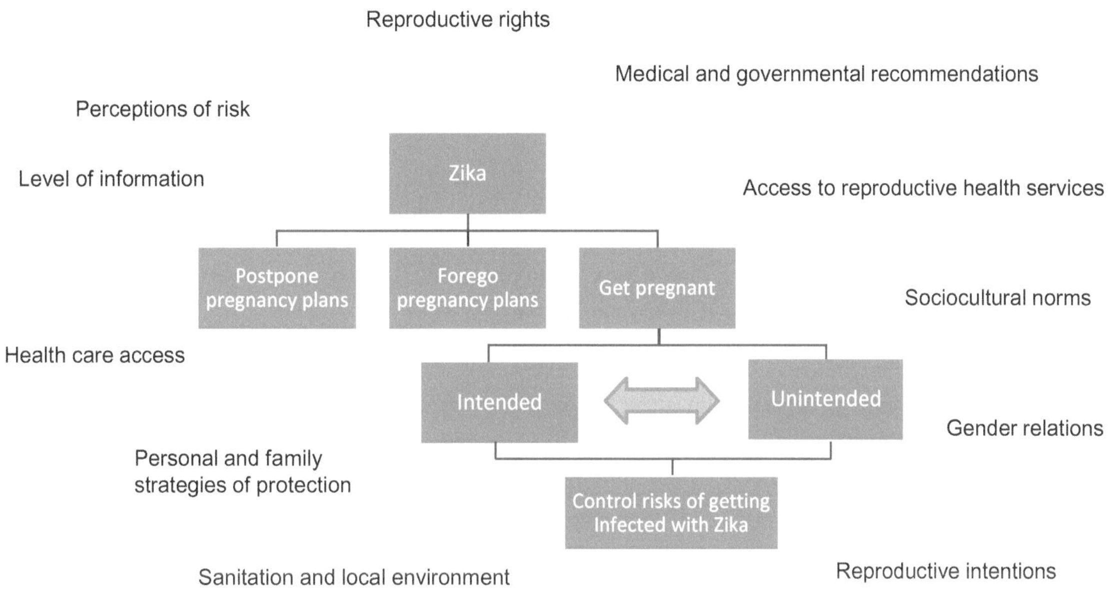

FIG. 3 Social determinants of reproductive outcomes and of risks of infection by Zika. *(Source: Authors own elaboration.)*

the Zika epidemics in March 2016 (Brasil, Ministério da Saúde, Secretaria de Atenção à Saúde, 2016), recognizing that contraceptive methods were important for countering the Zika epidemics, the document ignored the barriers to contraceptive access faced in the country (Baum et al., 2016). Actually the guidelines did not specify any new initiatives to change the contraceptive landscape or to improve access to contraceptive counseling and supplies, like the provision of LARC (Bahamondes et al., 2018), so no changes whatsoever on services and supplies actually took place (Baum et al., 2016; Pitanguy, 2016).

The guidelines to address Zika epidemics (Brasil, Ministério da Saúde, Secretaria de Atenção à Saúde, 2016) and its fetal consequences were also extensively criticized because it failed to address unsafe abortions in Brazil. Although legally restricted, it is estimated that 1 out of 5 women will perform an abortion by the end of their reproductive life in the country (Diniz & Medeiros, 2010).

While the media was concerned about informing general population how to control the mosquito and how to prevent microcephaly cases, with an emphasis on strategies such as pregnancy postponement, Borges, Moreau, Burke, Santos, and Chofakian (2018) showed sparse reproductive health counseling related to the Zika virus outbreak in Brazil in a study conducted in the city of Aracaju, located in the northeastern region. In 2017 the authors observed that only 9% of women reported they were asked about their reproductive intention during the Zika virus outbreak, 11% reported being advised by health professionals to avoid becoming pregnant because of Zika virus, and 14% were recommended to use condoms. Pregnant women were more likely to access counseling about family planning, condom, and pregnancy postponement due to Zika virus than nonpregnant women, which may suggest that health responses followed pregnancy occurrence.

These results reflect missed opportunities for prevention of perinatal transmission of Zika virus through behavioral change (contraception to prevent pregnancy and condoms to prevent perinatal transmission once a woman is pregnant) as a complement to vector control (Borges et al., 2018).

Awareness and knowledge about Zika virus in Brazil

Even though Brazil's public health response to the Zika virus outbreak focused mostly on epidemiological surveillance and vector control, the general population understanding of the novel disease was shaped by the powerful force of the social media and the Internet. Knowledge and speculations, evidences, and counterevidence were presented to the population every single day during 2016, which may have also helped to form the awareness and knowledge about Zika virus in Brazil.

Studies about how women—and general population—absorbed information about the Zika virus in Brazil are scarce. One study that can give us a clue is the one by Borges et al. (2018). The authors showed that awareness about congenital Zika syndrome among Brazilian women living in Aracaju (municipality in the northeast) was high (99%), while knowledge about sexual transmission was low (50%). The knowledge of sexual transmission was similar despite women's age, educational level, socioeconomic status (SES), and parity. While these data concern one municipality in the region most affected by the epidemics, data concerning women living in another municipality in the center-west region of the country (Cuiabá) showed very similar results (98% of women were aware of congenital Zika syndrome, but only 41% knew about sexual transmission—data not published). Such findings suggest that the media highlighted the congenital syndrome when disseminating information about the epidemics to the detriment of other very important topics, like the fact that Zika was sexually transmitted.

Another study (Marteleto, Weitzman, Coutinho, & Alves, 2017) used focus group data with reproductive age women in two major Brazilian cities (Recife, a hotspot for Zika and microcephaly, and Belo Horizonte, a hotspot for dengue, but that was not affected by Zika as hard). Investigating the perceptions about the epidemics and how they varied with education level, the authors found that the concern about Zika virus was mainly for women who were trying to conceive at the surge of the epidemics. The ones that were not trying to conceive reported not being worried about the epidemics despite their previous history of unintended pregnancies (many due to inconsistent contraceptive use and low power of negotiation when it comes to enforcing the use of condom with their sexual partners). High SES women, on the other hand, reported feeling confident about enforcing their reproductive plan successfully, which means they felt confident to prevent a pregnancy and confident they could protect themselves from Zika virus in case they got pregnant during the epidemics. It was also important to notice the high level of fatalism among the poorest and the low level of trust in public services (fear of crowded hospitals and the stigma when looking for contraceptives in the public services).

Lastly, Diniz, Medeiros, and Madeiro (2017) conducted a nationally representative survey of Brazilian urban residents (age 18–39, women) to investigate, in 2016, how reproductive behavior was changing due to the epidemics. They found that 56% of women reported that they had avoided (or tried to avoid) pregnancy because of the Zika virus outbreak.

In the next section, we will show how it has also affected the response of the health services and women's reproductive behavior.

Consequences of the Zika virus epidemics on health services response and on reproductive behaviors

Although knowledge about Zika virus and its reproductive consequences was high, little is known about the intentions to delay or avert a live birth. It seems that contraceptive use dynamics did not change much due to the epidemics. Looking at data on contraceptives sales between September 2014 and August 2016 obtained from pharmaceutical companies based in Brazil (oral contraceptives, vaginal and transdermal contraceptives, injectable contraceptives, IUDs, and emergency contraceptive pills), Bahamondes, Ali, Monteiro, and Fernandes (2017) found that contraceptive sales in the 24 months comprising the Zika virus outbreak showed little variation and no significant change.

When it comes to abortion, the Zika virus outbreak seems to have affected the search for abortion in Brazil despite it being prohibited. To analyze whether abortion requests increased due to the Zika epidemics, Aiken et al. (2016) compared three groups of Latin American countries with legally restricted abortion laws: those with autochthonous Zika virus transmission and a national advisory against pregnancy; those with transmission and no advisory, in which Brazil is included; and those with no transmission. In seven out of the eight countries with Zika transmission and legally restricted abortion, online requests for abortion pills increased significantly during the Zika outbreak, ranging from 36% to 108%, this last percentage concerning Brazil. Their findings may confirm that Brazilian women were trying to avert live births during the Zika virus epidemics irrespective of any governmental initiative or prohibition.

But the outbreak had reproductive consequences yet to be fully understood. Castro, Han, Carvalho, Victora, and França (2018) used time-series analysis to follow the number of live births during the epidemics, fetal deaths, and hospital admissions of women due to abortion complications and compared with projected number of live births, under normal circumstances, to assess the magnitude of the decline in live births. From September 2015 to December 2016, they estimated that 119,095 fewer births occurred in Brazil compared with what was expected, particularly after April 2016, which is when WHO announced the link between the Zika virus epidemic and microcephaly. Fetal death rates remained unchanged. The authors also found declining rates of hospitalization due to postabortion care during the period. This could be a sign of possible increases in medicine-induced abortion at an early gestation age (which usually do not require hospitalization and thus are not captured by routine administrative data collection).

To investigate if Zika virus affected the number of live births in Brazil, other researchers modeled the deficit in live births in a certain week of 2016 based on the incidence of Zika virus in the preceding weeks. To control for seasonality, the deficit in live births in 2016 was estimated based on comparison with the average value of the years 2012–15 for the same week. For a state in the southeast region, for example, they estimated 7484 fewer births occurred (Coelho, Armstrong, Saraceni, & Lemos, 2017).

In the same way, Rangel, Nobles, and Hamoudi (2020) analyzed monthly count of live births by region in Brazil, exploring month-to-month changes in birth cohort sizes (in a log scale) between locations. Drawing on different data sources, they controlled for poverty levels, Human Development Index (HDI), urbanization rates, population size, population age composition, labor market conditions, and inflation and also by level of mosquito infestation using municipal information on the Larval Index Rapid Assay (LIRA). They found a sizeable and significant reduction in birth rates in Recife from September 2016 (15.3%) to February 2017 (12.0%), with peak in November 2016 (19.9%), corresponding to pregnancies that would have happened 1–5 months after the announcement of the link.

The possible effects of Zika on live births were also observed nationwide. Using monthly microdata on live births from the Brazilian Information System on Live Births (SINASC), Marteleto, Guedes, Coutinho, and Weitzman (2020) found that the decline in live births varied according to education level and geographic location, beginning approximately 9 months after the link between Zika virus and microcephaly was announced (see Fig. 4 for aggregate results). Declines were more accentuated in the northeast region (more affected by the epidemics) compared with the southern. Declines in fertility rates were sharper among younger women. However, high-educated women observed the largest percentage decline in fertility rates compared with their low-educated counterparts. While the decline experienced by low-educated women is spread along all the reproductive ages, the decline among high-educated women is concentrated among the younger age groups, those who have time to catch up on their desired family sizes.

Conclusion

It is clear that the Zika virus epidemics had reproductive behavior consequences for Brazilian women. How this was mediated by knowledge around the Zika virus in Brazil cannot be determined, but the awareness about the health risks associated with Zika virus, including the risks of congenital syndrome, seemed to be high among women of reproductive age due to the widespread media descriptions of microcephaly cases, followed by public health informal advisories to delay pregnancy. This situation may have led to a desire to delay pregnancy, but at the same time, outdated contraceptive guidelines and family planning services were not able to respond to this new public health challenge. Neither were gender norms questioned in regard to contraceptive responsibilities. As a result of this structured inequality, women of low SES, especially in the northeast, are the ones who suffered more the impacts of the congenital Zika syndrome.

Almost 4 years after the Zika epidemics in Brazil, little has happened in the field of reproductive rights. The country still has the contraceptive landscape focused on short-acting contraceptive methods as well as an extreme restrictive abortion policy, threated to be made worse by conservative politicians in power, and abortion is treated more as a criminal issue than a public health one; women who perform abortion or people who help women get an abortion may be persecuted and criminalized. It is clear from our perspective that the government not only needs to address problems of vector control, such as

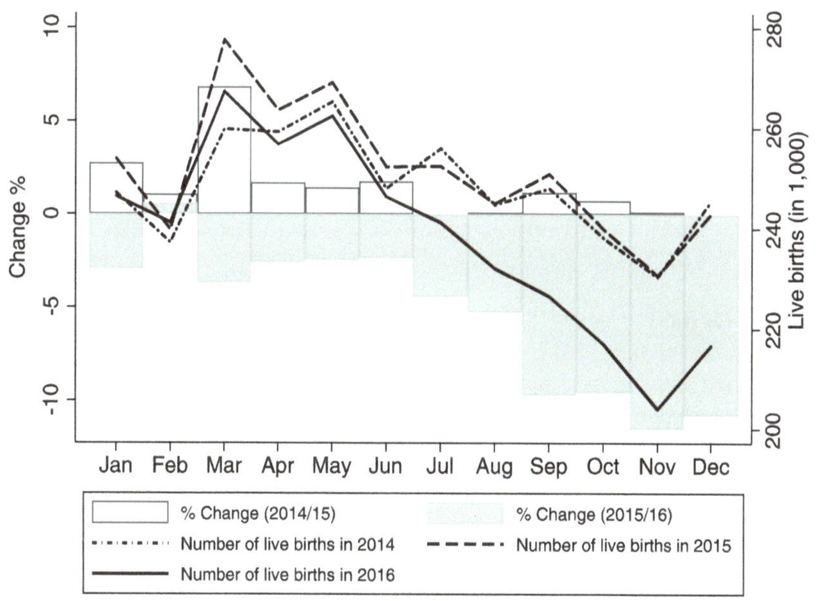

FIG. 4 Live births and year percentage change by month: Brazil, 2014–16. *(Source: Marteleto, L. J., Guedes, G. R., Coutinho, R. Z., & Weitzman, A. (2020). Live births and fertility amid the Zika epidemic in Brazil.* Demography, *57(3), 843–872.)*

reliability and speed of the diagnoses, vaccine production, improve sanitation, housing, and protection measures against mosquito bites, but also provides women with reproductive health services, which includes the right to meet their reproductive goals.

Policy and procedures

To understand how population living in other Zika virus affected areas build their perspectives from the communication and dissemination of information about managing vector control and preventing the severe consequences of being infected during pregnancy, it is suggested to access some studies produced by researchers that report findings from different parts of Latin America, among which we highlight:

Argüelles-Nava, V.G., Alvarez-Bañuelos, M.T., Córdoba-Suárez, D., Sampieri, C.L., Ortiz-León, M.C., Riande-Juárez, G., et al. (2018). Knowledge, attitudes, and practices about Zika among a university community located in an endemic zone in Mexico. *International Journal of Environmental Research and Public Health*, 5, 11, E2548. doi: https://doi.org/10.3390/ijerph15112548.

Brittain, A.W., August, E.M., Romero, L., Sheahan, M., Krashin, J., Ntansah, C., et al. (2019). Community perspectives on contraception in the context of the Zika Virus in the U.S. Virgin Islands: implications for communication and messaging. *Womens Health Issues*, 29, 3, 245–251. doi: https://doi.org/10.1016/j.whi.2019.01.007.

Forero-Martínez, L.J., Murad, R., Calderón-Jaramillo, M., Rivillas-García, J.C. (2020). Zika and women's sexual and reproductive health: Critical first steps to understand the role of gender in the Colombian epidemic. *International Journal of Gynaecology and Obstetrics*, 148, Suppl 2, 15–19. doi: https://doi.org/10.1002/ijgo.13043.

Nelson, E.J., Luetke, M.C., Kianersi, S., Willis, E., Rosenberg, M. (2019). Knowledge and perceptions of Zika virus transmission in the community of Puerto Plata, Dominican Republic. *BMC Infectious Diseases*, 19, 339. doi: https://doi.org/10.1186/s12879-019-3952-0.

Mini-dictionary of terms

Aedes aegypti. This is the scientific name of the mosquito that can transmit dengue, chikungunya, Zika, and yellow fever viruses. It can be recognized by the white markings on its legs.

Dengue. This is a mosquito-borne disease caused by dengue viruses (four types). The most common symptoms are fever, headache, muscle and joint pain, and skin rash.

Chikungunya. This is also a mosquito-borne disease caused by chikungunya virus. The most common symptoms are fever, muscle and joint pain, and skin rash.

Microcephaly. Condition where a newborn's head is smaller than expected.

Congenital Zika virus syndrome. A birth defect related to Zika virus infection during pregnancy that causes, among other negative consequences, a severe microcephaly in which the newborns' skull is partially collapsed.

Autochthonous transmission. Disease transmitted from one individual to another when both are in the same place/location.

Long-acting reversible contraceptives. Known as LARC, they are birth control methods that provide highly effective contraception for an extended period without requiring user action, like injectable, intrauterine devices, and subdermal contraceptive implants.

Key facts of Zika virus knowledge in Brazil

- Awareness about congenital Zika syndrome reached 99% among Brazilian women living in the northeastern region, the most affected by the Zika virus epidemics; knowledge about sexual transmission was reported by 50% of the interviewees.
- The high level of awareness about congenital Zika syndrome may have affected reproductive behaviors.
- Concern about Zika virus was mainly observed among women who were trying to conceive at the surge of the epidemics.
- More than half of reproductive age women (56%) reported that they had avoided (or tried to avoid) pregnancy because of the Zika virus outbreak.
- Fewer births occurred in Brazil compared with what was expected, particularly after the WHO announced the link between the Zika virus epidemic and microcephaly.

Summary points

- Strategies to control the Zika virus outbreak in Brazil focused on vector control and advices to delay pregnancy.
- Vector control strategies were based on what the country had been implementing to prevent dengue and chikungunya—other diseases transmitted by *A. aegypti*—with disappointing results.
- Advices to delay pregnancy among women living in infected areas were not formally performed by the Brazilian government, but the media coverage stressed the need to delay pregnancy due to the consequences of Zika virus in pregnancy, like the congenital syndrome.
- Studies on awareness and knowledge about Zika virus outbreak in Brazil are scarce.
- Data suggest that much of the awareness and knowledge about Zika virus prevention, transmission, and health consequences was derived from the media.
- Brazilian government did not implement any relevant change either in reproductive health services, like an increase in the offer of long-acting reversible contraceptive (LARC) methods, or in abortion or postabortion health services.
- A decrease in the number of births in the months following the onset of the link between the virus and the cases of microcephaly may have been a result of the effects of the Zika virus outbreak on reproductive behaviors.
- Some women were able to reduce fertility in spite of the lack of improvement in reproductive health services.
- Decline in live births was sharper for younger women and those with high levels of education.
- Women with low SES are overrepresented among the mothers of babies with microcephaly.

References

Aiken, A. R., Scott, J. G., Gomperts, R., Trussell, J., Worrell, M., & Aiken, C. E. (2016). Requests for abortion in Latin America related to concern about Zika virus exposure. *New England Journal of Medicine*, *375*, 396–398.

Augusto, L. G. S., Gurgel, A. M., Diderichsen, F., Lacaz, F. A., Parra-Henao, G., Rigotto, R. M., et al. (2016). *Aedes aegypti* control in Brazil. *Lancet*, *387*, 1052–1053.

Bahamondes, L., Ali, M., Monteiro, I., & Fernandes, A. (2017). Contraceptive sales in the setting of the Zika virus epidemic. *Human Reproduction*, *32*(1), 88–93.

Bahamondes, L., Villarroel, C., Guzmán, N. F., Oizerovich, S., Velázquez-Ramírez, N., & Monteiro, N. (2018). The use of long-acting reversible contraceptives in Latin America and the Caribbean: Current landscape and recommendations. *Human Reproduction Open*, *1*. hox030.

Baum, P., Fiastro, A., Kunselman, S., Vega, C., Ricardo, C., Galli, B., et al. (2016). Ensuring a rights-based health sector response to women affected by Zika. *Cadernos Saúde Pública*, *32*(5), e00064416.

Borges, A. L. V., Moreau, C., Burke, A., Santos, O. A., & Chofakian, C. B. N. (2018). Women's reproductive health knowledge, attitudes and practices in relation to the Zika virus outbreak in northeast Brazil. *PLoS One*, *13*(1), e0190024.

Brasil, Ministério da Saúde, Secretaria de Atenção à Saúde. (2016). *Protocolo de atenção à saúde e resposta à ocorrência de microcefalia relacionada à infecção pelo vírus Zika*. Brasília: Ministério da Saúde. http://bvsms.saude.gov.br/bvs/publicacoes/protocolo_resposta_microcefalia_relacionada_infeccao_virus_zika.pdf.

Brasil, Ministério da Saúde, Secretaria de Atenção à Saúde. (2019). *Síndrome congênita associada à infecção pelo vírus Zika. Situação Epidemiológica, Ações Desenvolvidas e Desafios de 2015 a 2019*. Brasília: Ministério da Saúde. Available from: http://portalarquivos2.saude.gov.br/images/pdf/2019/dezembro/05/be-sindrome-congenita-vfinal.pdf.

Brasil, Ministério da Saúde, Secretaria de Atenção à Saúde. (2020). *Boletim Epidemiológico 51, 02. Monitoramento dos casos de arboviroses urbanas transmitidas pelo Aedes (dengue, chikungunya e Zika), Semanas Epidemiológicas 01 a 52*. Brasília: Ministério da Saúde. https://antigo.saude.gov.br/images/pdf/2020/janeiro/20/Boletim-epidemiologico-SVS-02-1-.pdf.

Bueno, F. T. C., García, M., Moya, J., Lowy, I., Benchimol, J. L., Cerqueira, R. C., et al. (2017). Zika and *Aedes aegypti*: New and old challenges. *História, Ciências, Saúde – Manguinhos*, *24*(4), 1–20.

Burke, A., & Moreau, C. (2016). Family planning and Zika virus: The power of prevention. *Seminar in Reproductive Medicine*, *34*(5), 305–312.

Calvet, G., Aguiar, R. S., Melo, A. S. O., Sampaio, S. A., Filippis, S., Fabri, A., et al. (2016). Detection and sequencing of Zika virus from amniotic fluid of fetuses with microcephaly in Brazil: A case study. *Lancet Infectious Diseases*, *16*, 653–660.

Castro, M. C., Han, Q. C., Carvalho, L. R., Victora, C. S., & França, G. A. (2018). Implications of Zika virus and congenital Zika syndrome for the number of live births in Brazil. *Proceedings of the National Academy of Sciences of the United States of America*, *115*(24), 6177–6182.

Coelho, F. C., Armstrong, M., Saraceni, V., & Lemos, C. (2017). Can Zika account for the missing babies? *Frontiers in Public Health*, *5*, 317.

Diniz, S. G. (2016). Zika virus and pregnancy: A perspective from Brazil. *Midwifery*, *35*, 22.

Diniz, D., & Medeiros, M. (2010). Aborto no Brasil: uma pesquisa domiciliar com técnica de urna. *Ciência & Saúde Coletiva*, *15*(Suppl 1), 959–966.

Diniz, D., Medeiros, M., & Madeiro, A. (2017). Brazilian women avoiding pregnancy during Zika epidemic. *Journal of Family Planning and Reproductive Health Care*, *43*, 80. jfprhc-2016-101678.

Ferguson, N. M., Cucunubá, Z. M., Dorigatti, I. D., Nedjati-Gilani, G. L., Donnelly, C. A., Basáñez, M., et al. (2016). Countering the Zika epidemic in Latin America. *Science*, *353*(6297), 353–354.

França, G. V. A., Schuler-Faccini, L., Oliveira, W. K., Henriques, C. M. P., Carmo, E. H., Pedi, V. D., et al. (2016). Congenital Zika virus syndrome in Brazil: A case series of the first 1501 livebirths with complete investigation. *Lancet, 388*, 891–897.

Guedes, G. R., Coutinho, R. Z., Marteleto, L., Pereira, W. H. S., & Duarte, D. (2018). Signifying Zika: Heterogeneity in the representations of the virus by history of infection. *Cadernos de Saúde Pública, 34*(5), e00003217.

Marteleto, L. J., Weitzman, A., Coutinho, R. Z., & Alves, S. V. (2017). Women's reproductive intentions and behaviors during the Zika epidemic in Brazil. *Population and Development Review, 43*(2), 199–227.

Marteleto, L. J., Guedes, G. R., Coutinho, R. Z., & Weitzman, A. (2020). Live births and fertility amid the Zika epidemic in Brazil. *Demography, 57*(3), 843–872.

Melo, V. A. D., Silva, J. R. S., & La Corte, R. (2019). Medidas de proteção individual de gestantes contra a infecção pelo Zika vírus. *Revista de Saúde Pública, 53*(72), 1–10.

Mlakar, J., Korva, M., Tul, N., Popović, M., Poljšak-Prijatelj, M., Mraz, J., et al. (2016). Zika virus associated with microcephaly. *New England Journal of Medicine, 374*, 951–958.

Oliveira, W. K., França, G. V. A., Carmo, E. H., Duncan, B. B., Kuchenbecker, R. S., & Schmidt, M. I. (2017). Infection-related microcephaly after the 2015 and 2016 Zika virus outbreaks in Brazil: A surveillance-based analysis. *Lancet, 390*(10097), 861–870.

Pitanguy, J. (2016). Women's reproductive rights and the Zika virus epidemic. *Cadernos de Saúde Pública, 32*(5), e00066016.

Rangel, M., Nobles, J., & Hamoudi, A. (2020). Brazil's missing infants: Zika risk changes reproductive behavior. *Demography, 57*(5), 1647–1680. https://doi.org/10.1007/s13524-020-00900-9.

Rasanathan, J. J. K., MacCarthy, S., Diniz, D., Torreele, E., & Gruskin, S. (2017). Engaging human rights in the response to the evolving Zika virus epidemic. *American Journal of Public Health, 107*(4), 525–531.

Ribeiro, B., Hartley, S., Nerlich, B., & Jaspa, R. (2018). Media coverage of the Zika crisis in Brazil: The construction of a 'war' frame that masked social and gender inequalities. *Social Science & Medicine, 200*, 137–144.

Santos, D. N., Aquino, E. M. L., Menezes, G. M. S., Paim, J. S., Silva, L. M. V., Souza, L. E. P. F., et al. (2016). *Documento de posição sobre a tríplice epidemia de Zika-Dengue-Chikungunya. 2016*. Salvador: Observatório de Análise Política.

Theme-Filha, M. M., Baldisserotto, M. L., Fraga, A. C. S. A., Ayers, S., Gama, S. G. N., & Leal, M. C. (2016). Factors associated with unintended pregnancy in Brazil: cross-sectional results from the Birth in Brazil National Survey, 2011/2012. *Reproductive Health, 13*(Suppl 1), 118.

Vélez, A. C. G., & Diniz, S. G. (2016). Inequality, Zika epidemics, and the lack of reproductive rights in Latin America. *Reproductive Health Matters, 24* (48), 57–61.

Chapter 5

Zika virus infection and replication organelle biogenesis

Masashi Arakawa and Eiji Morita
Department of Biochemistry and Molecular Biology, Faculty of Agriculture and Life Science, Hirosaki University, Hirosaki-shi, Aomori, Japan

Abbreviations

AMPK	adenosine monophosphate activated protein kinase
ATG	autophagy-related gene
ATL	atlastin
BMV	brome mosaic virus
CHMP	charged multivesicular body protein
Cho2p	choline requiring 2
CM	convoluted membrane
DENV	dengue virus
dsRNA	double-stranded RNA
ER	endoplasmic reticulum
ESCRT	endosomal sorting complex required for transport
FASN	fatty acid synthase
HCV	hepatitis C virus
HMGCR	3-hydroxy-3-methylglutaryl-coenzyme A reductase
HRS	hepatocyte growth factor–regulated tyrosine kinase substrate
JEV	Japanese encephalitis virus
mTOR	mammalian target of rapamycin
NS	nonstructural
nSMase2	neutral SMase 2
OSBP	oxysterol-binding protein
PI4K	phosphatidylinositol-4-phosphate kinase
PI4P	phosphatidylinositol-4-phosphate
prM	precursor membrane
RTN	reticulon
siRNA	small interfering RNA
ssRNA(+)	positive-sense single-stranded RNA
TBSV	Tomato bushy stunt virus
TSG101	tumor susceptibility gene 101
VPs	vesicle packets
WNV	West Nile virus
zER	zippered ER
ZIKV	Zika virus

Introduction

Zika virus (ZIKV) belongs to the *Flaviviridae* family, which also comprises several other human pathogenic viruses, such as dengue virus (DENV), yellow fever virus, Japanese encephalitis virus (JEV), and West Nile virus (WNV) (Neufeldt, Cortese, Acosta, & Bartenschlager, 2018). ZIKV exhibits characteristic features, such as neurotropism and the ability

Zika Virus Impact, Diagnosis, Control, and Models. https://doi.org/10.1016/B978-0-12-820267-8.00005-4

to cause microcephaly. The basic replication mechanism of ZIKV in the cell is reported to be similar to that of other flaviviruses (Neufeldt et al., 2018). Thus previous studies on other flaviviruses can aid in understanding the replication mechanism of ZIKV. ZIKV, an enveloped virus with a diameter of about 50–60nm (Sirohi et al., 2016), has a positive-sense single-stranded RNA genome of 11kb size. The polyprotein translated from the genome is cleaved into 10 viral proteins by the host or viral proteases. The N-terminus of the polyprotein contains the following three structural proteins: capsid, precursor membrane (prM), and envelope proteins. The C-terminus of the polypeptide contains the following seven nonstructural proteins: NS1, NS2A, NS2B, NS3, NS4A, NS4B, and NS5 (Neufeldt et al., 2018). The capsid protein binds to the viral genomic RNA to form a nucleocapsid complex (Ishida et al., 2019). The prM and envelope proteins form the viral particles, which are cleaved by the host proteinase in the Golgi apparatus to generate a mature virion (Kuhn et al., 2002). NS1 is an endoplasmic reticulum (ER) luminal protein that is involved in the virus genome replication, virus particle assembly, and suppression of host immune response (Brown et al., 2016; Ci et al., 2020). NS3 exhibits RNA helicase and protease activities, which are required for viral RNA genome replication. NS3 anchors into the membrane by forming a complex with NS2B, which is a viral transmembrane protein (Rodriguez, Muñoz, Segura, Rangel, & Bello, 2019). NS2A, NS4A, and NS4B are also membrane-integrated proteins, which are reported to be involved in anchoring the viral replication complex to the ER membrane (Chambers, McCourt, & Rice, 1989). Some studies have revealed that NS2A, NS4A, and NS4B are also required for the production of the virus particle and for inhibiting interferon α and β response. The structural and nonstructural proteins have essential roles in the propagation of ZIKV (Chatel-Chaix & Bartenschlager, 2014).

Zika virus infection induces intracellular membrane remodeling and viral replication organelle formation

Generally, the positive-sense single-stranded RNA (ssRNA(+)) viruses induce the formation of a unique organelle-like structure in the cell upon infection (Arakawa & Morita, 2019; Romero-Brey & Bartenschlager, 2016), which is called "viral replication organelle." The viral proteins and double-stranded RNA (dsRNA) are detected within the viral replication organelle (Arakawa & Morita, 2019). This indicated that the viral replication organelle may play a crucial role in the viral genome replication, viral protein synthesis, and viral particle assembly. Several ER marker proteins are localized at this compartment, which suggests that the membrane enclosing the viral replication organelle is derived from the ER. The imaging analysis using fluorescent-labeled concanavalin A and wheat germ agglutinin, which are probes for the respective high-mannose glycans in the ER and Golgi apparatus, revealed that the viral replication organelle could potentially originate from the pre-Golgi zone of ER (Gillespie, Hoenen, Morgan, & Mackenzie, 2010). The electron microscopy analysis revealed that the viral replication organelles are composed of the following two distinct compartments: vesicle packet (VP) and convoluted membrane (CM). The electron tomography analysis of DENV-infected human hepatoma cells revealed that the CM is an unstructured membrane with high electron density, while VP is a small part that budded out from the ER membrane into the ER lumen with a diameter of approximately 90nm. The luminal vesicles in the VP are open toward the cytosol, and the inside of the vesicle area is connected to the cytosol through small pores, which are almost round with an average width of approximately 11nm (Welsch et al., 2009). The progeny positive-stranded RNAs are synthesized in the VP, which contains viral replicase protein complexes and dsRNA. Newly synthesized genomic RNAs are exported to the outside of the vesicles and assembled into the infectious particles, which bud into the ER lumen in an area proximal to the vesicle pore (Chatel-Chaix & Bartenschlager, 2014). Although CM does not contain dsRNA, it contains abundant viral nonstructural membrane proteins. Therefore it has been considered that viral proteins are synthesized and stored in the CM. CM is also reported to function as a lipid storage site. The formation of CM alters the shape of the surrounding mitochondria and disrupts the mitochondrion-associated membranes, which serve as a platform for innate immune signaling. The formation of CM impairs the recruitment of innate immune pattern recognition receptors, which results in the suppression of the innate immune signaling (Chatel-Chaix et al., 2016). However, sometimes, CM is not detected in the flavivirus-infected cells, such as DENV-infected mosquito cells (Junjhon et al., 2014) and ZIKV-infected human neural progenitor cells (Cortese et al., 2017). Currently, there is limited information on the importance of this structure.

In addition to VP and CM, a closely aligned ER cisterna called a zippered ER (zER) membrane is often observed in the ZIKV-infected hepatoma cells (Cortese et al., 2017). Although the role of this structure is unknown, zER and CM may be involved in the neuropathogenesis of ZIKV.

Viral proteins required for viral replication organelle biogenesis

The formation of viral replication organelle is triggered by the production of viral proteins from the first ssRNA(+) genome that enters the host cells. The precise mechanisms underlying the initiation of replication organelle formation are not elucidated. However, the nonstructural NS2A, NS4A, and NS4B viral proteins are reported to be involved in the formation of replication organelles. These nonstructural proteins do not constitute virus particles and do not exhibit enzymatic activities. The nonstructural viral proteins have multiple transmembrane domains that are embedded into the ER membrane like a wedge from the cytoplasm to the ER lumen. The insertion of viral proteins deforms the ER shape, which induces negative membrane curvature (Neufeldt et al., 2018; Stern et al., 2013). The interaction between NS4A and NS4B and their oligomerization are important for this process. However, the ER membrane is deformed in cells expressing NS4A or NS4B or coexpressing NS4A and NS4B (Zou et al., 2015). The membrane structures observed in the NS4A or NS4B protein-expressing cells are different from those observed in the virus-infected cells. This suggested that additional viral proteins and host factors are required for the formation of membrane structure in the infected cells (Neufeldt et al., 2018). One study using the NS1-4A polyprotein of WNV tagged with green fluorescent protein (GFP) at the N-terminus and red fluorescent protein (RFP) at the C-terminus demonstrated that NS1-4A induces the formation of vesicle-like structures at the ER (Yu, Takeda, & Gao, 2017). The electron microscopy analysis revealed that the size of the vesicles is about 90–140nm, which is similar or slightly larger than the diameter of vesicles obtained from the cells infected with viruses. These vesicles have a two-layered membrane enclosing the cytoplasmic contents (Yu et al., 2017). Additionally, polyprotein lacking the N-terminal of NS1 or internal NS2A, NS2B, or NS3 did not induce the formation of a two-layered membrane structure, which indicated that NS2A, NS2B, and NS3 also have essential roles in the formation of replication organelle (Yu et al., 2017).

The polyprotein of DENV contains NS1 to NS5 proteins, which induce the formation of ER membrane invaginations. These structures are very similar to VPs observed in the DENV-infected cells, which indicated that these NS proteins are sufficient to generate viral replication organelle (Płaszczyca et al., 2019). The polyprotein lacking the NS1 coding region does not induce VP structure formation, which indicates the essential role of NS1 in this process. DENV NS1 is reported to interact with the luminal side of NS4A-2K-NS4B precursor. The NS1 mutants do not interact with the NS4A-2K-NS4B precursor, which does not affect the size and number of vesicles. This suggested that the interaction between NS1 and NS4A-2K-NS4B is not associated with this function (Płaszczyca et al., 2019).

ER membrane proteins involved in the formation of viral replication organelle

In addition to the viral proteins, some host factors are also involved in the membrane deformation during the viral replication organelle formation. Reticulon (RTN) and atlastin (ATL) family proteins, which are known as ER-shaping proteins, are involved in the formation of viral replication organelle. Reticulon family proteins are highly conserved in eukaryotes with two short hairpin-like transmembrane domains. These domains are believed to cause local membrane deformation by forming a wedge-like structure that preferentially moves lipids of the outer leaflet of the lipid bilayer. There are four known RTN family proteins (RTN 1–4) in mammalian cells. However, the functional difference among the RTN family proteins is unclear (Wang & Rapoport, 2019). The RTN genes are expressed as multiple isoforms. RTN 3.1A, an RTN3 isoform, is required for flavivirus genome replication. However, the effect of RTN 3.1A silencing varies depending on the flavivirus species. The depletion of RTN 3.1A in the ZIKV-infected cells results in decreased membrane curvatures and replication vesicles in the viral replication organelle. Similarly the depletion of RTN3.1A in the WNV-infected cells markedly decreases the number of replication vesicles in VP. However, RTN3.1A depletion in the DENV-infected cells does not reduce the number of vesicle formation but resulted in the accumulation of elongated vesicles in the VPs of viral replication organelle. RTN3.1A physically interacts with WNV NS4A, which mediates membrane remodeling to promote the ER membrane remodeling characteristic of flavivirus infection. The phenotypes associated with RTN3.1A depletion vary depending on the flavivirus species and may contribute to the differential affinity of these proteins (Aktepe, Liebscher, Prier, Simmons, & Mackenzie, 2017).

In addition to RTN, ATL family proteins are also involved in the formation of flavivirus replication organelle (Neufeldt et al., 2019). ATL family proteins belong to the dynamin superfamily and are composed of a cytoplasmic GTPase domain, a hairpin loop membrane domain, and an amphipathic helix (Hu & Rapoport, 2016; Rismanchi, Soderblom, Stadler, Zhu, & Blackstone, 2008). There are three ATLs (ATL1–3) in mammals, which exhibit different expression patterns. ATL1 is predominately expressed in the neurons, while ATL2 and ATL3 are ubiquitously expressed. ATLs reside at the ER and are reported to be involved in the fusion of ER tubules to maintain the tubular networks (Hu & Rapoport, 2016) (Fig. 1).

Neufeldt et al. reported that the depletion of ATL2 and ATL3 but not that of ATL1 markedly decreases the production of infectious DENV and ZIKV particles (Neufeldt et al., 2019). However, ATL2 and ATL3 have different roles in the DENV and ZIKV life cycle. The intracellular viral RNA levels are decreased only in the ATL2-depleted cells but not in the ATL3-depleted cells. Neufeldt et al. suggested that ATL2 may be required for the viral genome replication during the formation of viral replication organelle because the vesicles were condensed to a small perinuclear region and the vesicle size and shape were distorted in the ATL2-depleted cells. Additionally, ATL3 may be involved in the maturation of infectious viral

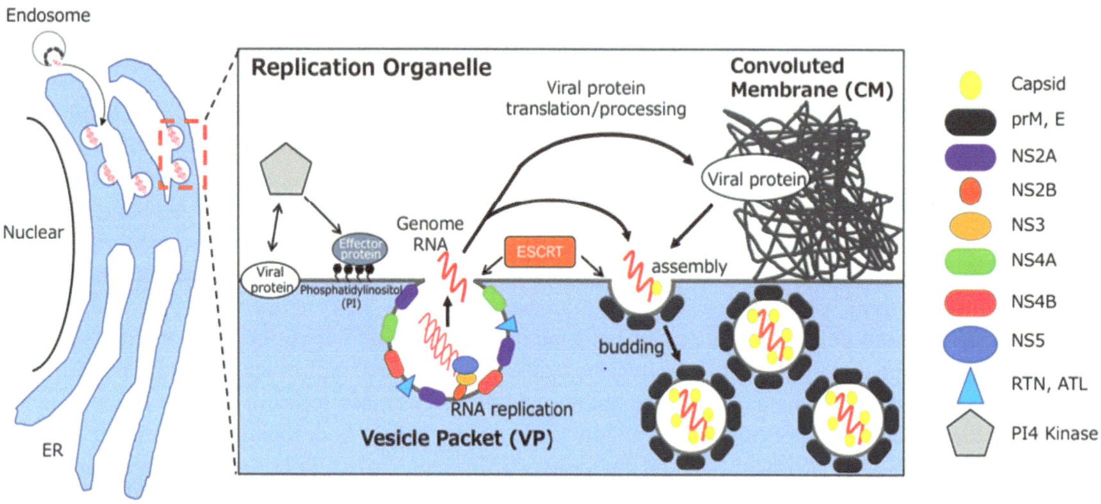

FIG. 1 Model of flavivirus replication organelle biogenesis. Flaviviruses enter in cell via endocytosis pathway and release genome RNA to cytoplasm by membrane fusion of viral particle and endosome. Then, viral nonstructural (NS) proteins are translated on the ER membrane. Initially synthesized viral proteins may a function in generating vesicle packets (VPs), which is the place for viral genome replication. Phosphatidylinositol-4-kinase (PI4K), the viral proteins recruiting factor, and other host factors such as reticulon (RTN) and atlastin (ATL) family proteins play vital roles in the membrane invagination. Newly synthesized viral genome RNAs are exported form VPs, and then they form a nucleocapsid (NC) complex with capsid proteins. NC complex is packaged in virus particles to form an infectious particle. In general, endosomal sorting complex required for transport (ESCRT) proteins are considered to be involved in both VP formation and virus assembly in positive-stranded RNA virus–infected cells. Convoluted membranes (CM) is another typical structure that can be seen in the viral replication organelle, but its physiological roles are unclear.

particles (Neufeldt et al., 2019). Monel et al. also reported that the simultaneous depletion of all ATL family proteins in the ZIKV-infected cells markedly decreased the viral genome replication. Further, ATL3 is recruited to the viral replication organelle along with the viral NS2A and NS2B3 proteins. These results suggest that the ATL-mediated ER tubular network efficiently promotes the formation of the ZIKV replication organelle (Monel et al., 2019).

Endosomal sorting complex required for transport-mediated membrane invagination and fission

Endosomal sorting complex required for transport (ESCRT) was identified from the proteomic analysis of proteins accumulated in the ER of JEV-infected cells (Tabata et al., 2016). ESCRT factors are reported to be involved in the membrane budding toward the luminal side or outside of the cells from the cytoplasm (Hurley, 2015). These factors are involved in the process of enveloped virus budding, intraluminal vesicle formation during multivesicular endosome biogenesis, final membrane fission stage of cytokinesis, nuclear membrane reassembly, and membrane repair of the damaged plasma membrane/organelle, which are commonly required for membrane sealing from the cytoplasmic phase (Hurley, 2015). ESCRT factors are recruited to the viral replication site on the ER, which suggests that similar events are required during the biogenesis of viral replication organelle. In the JEV- and DENV-infected cells, the depletion of ESCRT factors impairs single-round infectious particle production, which suggests that ESCRT factors are necessary for viral particle formation (Tabata et al., 2016). Similarly, ESCRT factors were also reported to be involved in the formation of hepatitis C virus (HCV) particle, which belongs to *Hepacivirus* genus in the *Flaviviridae* family. The depletion of ESCRT factors suppresses HCV budding but not HCV genome replication. These results suggest that ESCRT-mediated membrane sealing events are important for the viral budding process, rather than VP formation. However, other studies using different ssRNA(+) viruses that infect *Saccharomyces cerevisiae* suggest the opposite model. The yeast genetic analyses often serve as a useful reference to identify the host factors and their role in the formation of viral replication organelle. Brome mosaic virus (BMV) and Tomato bushy stunt virus (TBSV) can infect *S. cerevisiae*. BMV and TBSV are commonly used as a model for ssRNA (+) viruses. Nonessential gene screening revealed that several ESCRT factors are also required for the viral RNA genome replication (Barajas, Jiang, & Nagy, 2009; Diaz, Zhang, Ollwerther, Wang, & Ahlquist, 2015). ESCRT factors may be involved in the formation of spherules, a structure similar to the vesicles in VPs, in the viral replication organelle.

There are more than 30 ESCRT factors in mammals. However, only a small proportion of ESCRT factors are involved in each ESCRT-mediated biological process (Morita, 2012). The siRNA knockdown screening revealed that TSG101, CHMP2, CHMP3, and CHMP4 subfamily proteins are preferentially recruited to the viral replication site and have essential roles in the formation of infectious viral particle (Tabata et al., 2016). A study on the HCV reported that hepatocyte growth factor–regulated tyrosine kinase substrate (HRS), a subunit of early-acting ESCRT complex, is required during viral particle formation (Barouch-Bentov et al., 2016). Ubiquitin-interacting motif of HRS interacts with K63 ubiquitinated HCV NS2 protein, which is critical for virus particle formation. K63 ubiquitination of HCV NS2 is reported to be catalyzed by MARCH8, a RING-finger E3 ligase (Kumar et al., 2019). The depletion of MARCH8 also impairs the production of other flaviviruses, such as DENV and ZIKV. However, the molecular mechanisms underlying this process have not been elucidated.

Vps23p, a mammalian homolog of TSG101 subunits of ESCRT-I, and Bro1, a mammalian homolog of ALIX, are important for TBSV genome replication in both plant and yeast cells (Barajas et al., 2009). They interact with TBSV p33 replication protein, which may be ubiquitinated. This is followed by the recruitment of Snf7 and Vps24p, which are ESCRT-III subunits, for the formation of the replication complex. These interactions promote the oligomerization of p33 and the formation of viral spherules with membrane invagination. Several reports have revealed that some specific ESCRT factors are involved in the membrane deformation process during VP formation and viral genome replication. However, the conservation of ESCRT-mediated VP formation in ZIKV infection is not known.

Autophagic machinery is activated in the ZIKV-infected cells

Autophagy is one of the major bulk-degradation pathways for the cytosolic components. Macroautophagy is a degradation process that involves sequestering the target with the isolation membrane and delivering it to the lysosome, which is an acidic organelle in the cell. Autophagy requires a series of autophagy gene (ATG) proteins, which have been identified by yeast genetic screening (Mizushima, 2018). Initially, autophagy was considered to be a process that supplied nutrition by degrading the cells under starvation conditions. However, autophagy is now recognized to be an important process even under nutrient-rich conditions. Autophagy is needed for the elimination of various cellular components, such as damaged organelle, protein aggregates, and pathogens (Mizushima, 2018). This process is known as "selective autophagy," which is mediated by autophagic receptors (Mizushima, 2018). The autophagic receptors are recruited in each target. The accumulation of these receptors on the target can trigger specific sequestering by the isolation membranes. The specificity of the selective autophagy is reported to be mediated by the direct interaction of LC3-II, which is a lipid conjugated form of LC3 and a major surface ATG protein on the isolation membrane, and LC3-interacting region domain of receptors accumulated on the targets (Mizushima, 2018).

ZIKV infection is reported to activate the autophagic machinery. The accumulation of LC3-II form was observed in the ZIKV-infected cells (Chiramel & Best, 2018; Peng et al., 2018), which is also observed in cells infected with other flaviviruses (Metz et al., 2015). Autophagy is inhibited upon treatment with bafilomycin A, which inhibits vacuolar H+ ATPase activities and suppresses ZIKV propagation. This suggested that autophagic machinery positively regulates ZIKV propagation in the cells (Peng et al., 2018; Sabino et al., 2019). Similarly, enhanced autophagy observed after treatment with Torin1 or rapamycin, which are inhibitors of energy-sensing kinase mTOR, subsequently increases ZIKV propagation (Hamel et al., 2015; Liang et al., 2016). The precise relationships of autophagy activation and virus infection are controversial. However, several models have been proposed for the role of virus-induced autophagy activation. The autophagic isolation membrane is derived from the ER membrane, and some viral proteins colocalize with LC3-II expressed on the isolation membrane. Thus one model has proposed that the virus induces autophagy to utilize the autophagic isolation membrane for the materials needed for viral replication organelle (Wang & Ou, 2018). Other reports suggested that viruses induce autophagy due to the induction of autophagic degradation of lipid droplets (Heaton & Randall, 2010). This model suggested that viruses obtain energy by activating lipid metabolism, which enhances viral genome replication. Another model suggested that viruses induce autophagy to suppress innate immune responses. Mitophagy is a type of selective autophagy that involves the elimination of damaged mitochondria (Mizushima, 2018). The accumulation of damaged mitochondria caused by impaired autophagy activates the mitochondrion-mediated antiviral innate immune responses. Suppression of innate immune signaling by mitophagy may also contribute to viral propagation by autophagic activation (Tal et al., 2009).

In contrast to these reports, some studies have suggested that the cellular autophagic machinery negatively regulates virus propagation. In the Drosophila cells, innate immune response activated by ZIKV induces autophagy to negatively regulate virus propagation (Liu et al., 2018). ER-phagy is a type of selective autophagic degradation of the ER membrane that is reported to be involved in the elimination of ER-derived viral replication organelle. The depletion of FAM134B, which is one of the autophagic receptors required for ER-phagy, impairs ER-phagy and increases ZIKV genome replication and production of progeny infectious virus particles (Lennemann & Coyne, 2017). Furthermore, FAM134B is cleaved and inactivated by the DENV proteinase NS2B-NS3 complexes (Lennemann & Coyne, 2017). The virus induces the dysfunction of ER-phagy to prevent cellular intrinsic system degrading the viral replication organelle and subsequently facilitate their own genome replication.

Membrane lipids involved in the formation of viral replication organelle

Flavivirus infection markedly affects the intracellular membrane lipid composition (Martín-Acebes et al., 2014; Melo et al., 2016). The flavivirus-infected cells exhibit enhanced levels of sphingolipid and phosphatidylcholine. Sphingolipid is synthesized in the ER. The changes in the lipid composition influence the curvature and thickness of the local membranes (Hannun & Obeid, 2018). Sphingomyelin, a major sphingolipid in the cell, is hydrolyzed into ceramide and phosphorylcholine by neutral SMase 2 (nSMase2) (Hofmann, Tomiuk, Wolff, & Stoffel, 2000). Treatment with nSMase2 inhibitors, such as GW4869, spiroepoxide, or glutathione decreases the production of WNV recombinant subviral particles (Martín-Acebes et al., 2014). Additionally, siRNA-mediated depletion of nSMase2 leads to the accumulation of viral structural proteins in the cells, which indicates that nSMase is required for the viral particle secretion pathway (Martín-Acebes et al., 2014).

The genetic screening of yeast cells identified choline requiring 2 (Cho2p), a phosphatidylcholine synthase, as a host factor that is required for BMV replication (Zhang et al., 2016). Phosphatidylcholine is enriched in the viral replication organelle, which is associated with the recruitment of Cho2p by interacting with viral protein 1a. Phosphatidylcholine accumulation is also detected in the flavivirus replication organelle. Phosphatidylcholine and ceramide are enriched in the ER compartments around the virus replication site. Maintaining the lipid homeostasis by lipid synthesis plays an important role in the membrane deformation during viral replication organelle biogenesis (Zhang et al., 2016).

Phosphatidylinositol is also involved in the viral replication organelle biogenesis. Phosphatidylinositol-4-phosphate (PI4P) kinase (PI4K) was identified as the host factor that is required for virus propagation based on siRNA screening (Reiss et al., 2011). PI4K phosphorylates the D-4 position of phosphatidylinositol to biosynthesize PI4P. PI4P is present in the ER and Golgi apparatus, as well as the viral replication organelle of other ssRNA(+) viruses, such as coxsackievirus B3 (Hsu et al., 2010). The biosynthesis of PI4K at the viral replication site is reported to serve as a signal to recruit the downstream effector proteins. The host factors, such as oxysterol-binding protein (OSBP) that have a PH domain, a PI4P binding motif, are recruited to the viral replication organelle (Amako, Sarkeshik, Hotta, Yates, & Siddiqui, 2009; Meutiawati et al., 2018). In the normal cells, OSBP is localized at the ER and Golgi apparatus. OSBP is involved in the lipid transfer that regulates cholesterol/PI4P exchange. In the HCV-infected cells, viral NS5A proteins can interact with OSBP to induce the accumulation of cholesterol, which is involved in the formation of viral replication organelle and viral genome replication (Amako et al., 2009). Treatment with methyl-β-cyclodextrin, a reagent that alters the distribution of membrane cholesterol, significantly reduces viral genome replication (Wang et al., 2014). The cholesterol levels are also reported to be important for flavivirus replication. ER membrane deformation induced by flaviviral NS proteins is correlated with the distribution of fatty acid synthase (FASN) and 3-hydroxy-3-methylglutaryl-coenzyme A reductase (HMGCR), a cholesterol-synthesizing enzyme required for the production of the cholesterol precursor, mevalonate (Soto-Acosta, Bautista-Carbajal, Cervantes-Salazar, Angel-Ambrocio, & Del Angel, 2017). DENV NS3 recruits FASN to the ER membrane via direct interactions and facilitates the upregulation of fatty acid biosynthesis at the viral replication site (Heaton et al., 2010). Furthermore the distribution of HMGCR is affected in the DENV-infected cells. The activities of HMGCR, which are regulated by phosphorylating 5′ adenosine monophosphate–activated protein kinase (AMPK), are also increased at the viral replication sites. These events alter the local membrane composition and are required for the formation of viral specific membrane structures (Soto-Acosta et al., 2017).

Concluding remarks

ZIKV forms unique organelle-like structures by markedly remodeling the ER membrane. This event is essential for viral genome replication and viral particle assembly. The precise molecular mechanisms underlying the formation of viral replication organelle are not elucidated. However, several viral proteins and host factors are reported to be involved in the biogenesis of this organelle. Inhibition of organelle formation impairs the viral genome replication. Therefore identifying the host factors and developing specific small compounds that can manipulate this process can be a potential therapeutic strategy for preventing clinical exacerbation associated with ZIKV infection and propagation.

Policy and procedures

There are several analytical methods that can be used for the understanding of the functions of cellular and virus-inducing organelles. In general, they are classified into two categories, which are biochemical analysis and genetic analysis. In biochemical analysis, organelles and proteins are extracted from cells, and their properties are examined in vitro. For instance, identification of unknown factors by mass spectrometry (Tabata et al., 2016) and confirmation by detections with their specific probes. The results obtained by using the previous methods not only reveal the molecular function of the organelle but also clarify the molecular mechanisms for the biogenesis and maintenance of the organelle. In genetic analysis, factors that have been expected to be involved in the function of certain organelles are depleted by siRNA silencing or genetic manipulation, and their phenotypic alterations are investigated by using several types of microscope, such as fluorescent microscope, electron microscope, and electron tomography. These techniques can help understand not only the organelle generation but also virus particle formations and virion transport in infected cells (Welsch et al., 2009).

Mini-dictionary of terms

Vesicle packets. Membrane structure containing luminal vesicles that are open toward the cytosol, and the inside of the vesicle area is connected to the cytosol through small pores. This is a place for the viral genome replication.
Convoluted membrane. An unstructured membrane with high electron density, which is observed in the viral replication organelle, but the function of this structure is still unknown.
Reticulon. Membrane-integrated proteins family involved in the ER membrane shaping.
ESCRT. ESCRT carries out several important functions in membrane fission during intraluminal vesicle formation.
ATG proteins. ATG proteins are factors that are involved in the formation of autophagosome.

Key facts of virus replication organelle biogenesis

- Viral replication organelles are composed of the two distinct compartments: vesicle packets and convoluted membrane.
- Reticulon and atlastin family proteins are involved in the formation of viral replication organelle.
- Some specific subset of ESCRT proteins are recruited to the viral replication site and have essential roles in the viral propagation.
- Depletion of autophagic machinery affects viral propagation both positively and negatively.
- Zika virus infection markedly affects the intracellular membrane lipid composition.

Summary points

- Flaviviruses, including the Zika virus, induce intracellular membrane remodeling, which generates a specific compartment called "viral replication organelle."
- Viral replication organelle is derived from the host endoplasmic reticulum and considered to be the site for efficient viral genome replication and assembly.
- ER membrane–shaping proteins are involved in the formation of replication organelles for membrane remodeling.
- ESCRT-mediated membrane invagination and fission are involved in the viral particle and vesicle packet formation.
- Autophagic machinery is activated in the ZIKV-infected cells.
- Membrane lipid composition is important for the formation of viral replication organelle.

References

Aktepe, T. E., Liebscher, S., Prier, J. E., Simmons, C. P., & Mackenzie, J. M. (2017). The host protein reticulon 3.1A is utilized by flaviviruses to facilitate membrane remodelling. *Cell Reports*, *21*(6), 1639–1654. https://doi.org/10.1016/j.celrep.2017.10.055.

Amako, Y., Sarkeshik, A., Hotta, H., Yates, J., & Siddiqui, A. (2009). Role of oxysterol binding protein in hepatitis C virus infection. *Journal of Virology*, *83*(18), 9237–9246. https://doi.org/10.1128/JVI.00958-09.

Arakawa, M., & Morita, E. (2019). Flavivirus replication organelle biogenesis in the endoplasmic reticulum: Comparison with other single-stranded positive-sense RNA viruses. *International Journal of Molecular Sciences*, *20*(9), 2336. https://doi.org/10.3390/ijms20092336.

Barajas, D., Jiang, Y., & Nagy, P. D. (2009). A unique role for the host ESCRT proteins in replication of Tomato bushy stunt virus. *PLoS Pathogens*, *5*(12). https://doi.org/10.1371/journal.ppat.1000705, e1000705.

Barouch-Bentov, R., Neveu, G., Xiao, F., Beer, M., Bekerman, E., Schor, S., et al. (2016). Hepatitis C virus proteins interact with the endosomal sorting complex required for transport (ESCRT) machinery via ubiquitination to facilitate viral envelopment. *mBio*, *7*(6). https://doi.org/10.1128/mBio.01456-16. e01456–16.

Brown, W. C., Akey, D. L., Konwerski, J. R., Tarrasch, J. T., Skiniotis, G., Kuhn, R. J., et al. (2016). Extended surface for membrane association in Zika virus NS1 structure. *Nature Structural & Molecular Biology*, *23*(9), 865–867. https://doi.org/10.1038/nsmb.3268.

Chambers, T. J., McCourt, D. W., & Rice, C. M. (1989). Yellow fever virus proteins NS2A, NS2B, and NS4B: Identification and partial N-terminal amino acid sequence analysis. *Virology*, *169*(1), 100–109. https://doi.org/10.1016/0042-6822(89)90045-7.

Chatel-Chaix, L., & Bartenschlager, R. (2014). Dengue virus- and hepatitis C virus-induced replication and assembly compartments: The enemy inside—Caught in the web. *Journal of Virology*, *88*(11), 5907–5911. https://doi.org/10.1128/JVI.03404-13.

Chatel-Chaix, L., Cortese, M., Romero-Brey, I., Bender, S., Neufeldt, C. J., Fischl, W., et al. (2016). Dengue virus perturbs mitochondrial morphodynamics to dampen innate immune responses. *Cell Host & Microbe*, *20*(3), 342–356. https://doi.org/10.1016/j.chom.2016.07.008.

Chiramel, A. I., & Best, S. M. (2018). Role of autophagy in Zika virus infection and pathogenesis. *Virus Research*, *254*, 34–40. https://doi.org/10.1016/j.virusres.2017.09.006.

Ci, Y., Liu, Z.-Y., Zhang, N.-N., Niu, Y., Yang, Y., Xu, C., et al. (2020). Zika NS1-induced ER remodeling is essential for viral replication. *Journal of Cell Biology*, *219*(2), 881. https://doi.org/10.1083/jcb.201903062.

Cortese, M., Goellner, S., Acosta, E. G., Neufeldt, C. J., Oleksiuk, O., Lampe, M., et al. (2017). Ultrastructural characterization of Zika virus replication factories. *Cell Reports*, *18*(9), 2113–2123. https://doi.org/10.1016/j.celrep.2017.02.014.

Diaz, A., Zhang, J., Ollwerther, A., Wang, X., & Ahlquist, P. (2015). Host ESCRT proteins are required for bromovirus RNA replication compartment assembly and function. *PLoS Pathogens*, *11*(3). https://doi.org/10.1371/journal.ppat.1004742, e1004742.

Gillespie, L. K., Hoenen, A., Morgan, G., & Mackenzie, J. M. (2010). The endoplasmic reticulum provides the membrane platform for biogenesis of the flavivirus replication complex. *Journal of Virology*, *84*(20), 10438–10447. https://doi.org/10.1128/JVI.00986-10.

Hamel, R., Dejarnac, O., Wichit, S., Ekchariyawat, P., Neyret, A., Luplertlop, N., et al. (2015). Biology of Zika virus infection in human skin cells. *Journal of Virology*, *89*(17), 8880–8896. https://doi.org/10.1128/JVI.00354-15.

Hannun, Y. A., & Obeid, L. M. (2018). Sphingolipids and their metabolism in physiology and disease. *Nature Reviews. Molecular Cell Biology*, *19*(3), 175–191. https://doi.org/10.1038/nrm.2017.107.

Heaton, N. S., Perera, R., Berger, K. L., Khadka, S., Lacount, D. J., Kuhn, R. J., et al. (2010). Dengue virus nonstructural protein 3 redistributes fatty acid synthase to sites of viral replication and increases cellular fatty acid synthesis. *Proceedings of the National Academy of Sciences of the United States of America*, *107*(40), 17345–17350. https://doi.org/10.1073/pnas.1010811107.

Heaton, N. S., & Randall, G. (2010). Dengue virus-induced autophagy regulates lipid metabolism. *Cell Host & Microbe*, *8*(5), 422–432. https://doi.org/10.1016/j.chom.2010.10.006.

Hofmann, K., Tomiuk, S., Wolff, G., & Stoffel, W. (2000). Cloning and characterization of the mammalian brain-specific, Mg^{2+}-dependent neutral sphingomyelinase. *Proceedings of the National Academy of Sciences of the United States of America*, *97*(11), 5895–5900. https://doi.org/10.1073/pnas.97.11.5895.

Hsu, N.-Y., Ilnytska, O., Belov, G., Santiana, M., Chen, Y.-H., Takvorian, P. M., et al. (2010). Viral reorganization of the secretory pathway generates distinct organelles for RNA replication. *Cell*, *141*(5), 799–811. https://doi.org/10.1016/j.cell.2010.03.050.

Hu, J., & Rapoport, T. A. (2016). Fusion of the endoplasmic reticulum by membrane-bound GTPases. *Seminars in Cell & Developmental Biology*, *60*, 105–111. https://doi.org/10.1016/j.semcdb.2016.06.001.

Hurley, J. H. (2015). ESCRTs are everywhere. *EMBO Journal*, *34*(19), 2398–2407. https://doi.org/10.15252/embj.201592484.

Ishida, K., Goto, S., Ishimura, M., Amanuma, M., Hara, Y., Suzuki, R., et al. (2019). Functional correlation between subcellular localizations of Japanese encephalitis virus capsid protein and virus production. *Journal of Virology*, *93*(19), 500. https://doi.org/10.1128/JVI.00612-19.

Junjhon, J., Pennington, J. G., Edwards, T. J., Perera, R., Lanman, J., & Kuhn, R. J. (2014). Ultrastructural characterization and three-dimensional architecture of replication sites in dengue virus-infected mosquito cells. *Journal of Virology*, *88*(9), 4687–4697. https://doi.org/10.1128/JVI.00118-14.

Kuhn, R. J., Zhang, W., Rossmann, M. G., Pletnev, S. V., Corver, J., Lenches, E., et al. (2002). Structure of dengue virus: Implications for flavivirus organization, maturation, and fusion. *Cell*, *108*(5), 717–725.

Kumar, S., Barouch-Bentov, R., Xiao, F., Schor, S., Pu, S., Biquand, E., et al. (2019). MARCH8 ubiquitinates the hepatitis C virus nonstructural 2 protein and mediates viral envelopment. *Cell Reports*, *26*(7), 1800–1814. e5 https://doi.org/10.1016/j.celrep.2019.01.075.

Lennemann, N. J., & Coyne, C. B. (2017). Dengue and Zika viruses subvert reticulophagy by NS2B3-mediated cleavage of FAM134B. *Autophagy*, *13*(2), 322–332. https://doi.org/10.1080/15548627.2016.1265192.

Liang, Q., Luo, Z., Zeng, J., Chen, W., Foo, S.-S., Lee, S.-A., et al. (2016). Zika virus NS4A and NS4B proteins deregulate Akt-mTOR signaling in human fetal neural stem cells to inhibit neurogenesis and induce autophagy. *Cell Stem Cell*, *19*(5), 663–671. https://doi.org/10.1016/j.stem.2016.07.019.

Liu, Y., Gordesky-Gold, B., Leney-Greene, M., Weinbren, N. L., Tudor, M., & Cherry, S. (2018). Inflammation-induced, STING-dependent autophagy restricts Zika virus infection in the Drosophila brain. *Cell Host & Microbe*, *24*(1), 57–68.e3. https://doi.org/10.1016/j.chom.2018.05.022.

Martín-Acebes, M. A., Merino-Ramos, T., Blázquez, A.-B., Casas, J., Escribano-Romero, E., Sobrino, F., et al. (2014). The composition of West Nile virus lipid envelope unveils a role of sphingolipid metabolism in flavivirus biogenesis. *Journal of Virology*, *88*(20), 12041–12054. https://doi.org/10.1128/JVI.02061-14.

Melo, C. F. O. R., de Oliveira, D. N., Lima, E.d. O., Guerreiro, T. M., Esteves, C. Z., Beck, R. M., et al. (2016). A lipidomics approach in the characterization of Zika-infected mosquito cells: Potential targets for breaking the transmission cycle. *PLoS One*, *11*(10). https://doi.org/10.1371/journal.pone.0164377, e0164377.

Metz, P., Chiramel, A., Chatel-Chaix, L., Alvisi, G., Bankhead, P., Mora-Rodriguez, R., et al. (2015). Dengue virus inhibition of autophagic flux and dependency of viral replication on proteasomal degradation of the autophagy receptor p62. *Journal of Virology*, *89*(15), 8026–8041. https://doi.org/10.1128/JVI.00787-15.

Meutiawati, F., Bezemer, B., Strating, J. R. P. M., Overheul, G. J., Žusinaite, E., van Kuppeveld, F. J. M., et al. (2018). Posaconazole inhibits dengue virus replication by targeting oxysterol-binding protein. *Antiviral Research*, *157*, 68–79. https://doi.org/10.1016/j.antiviral.2018.06.017.

Mizushima, N. (2018). A brief history of autophagy from cell biology to physiology and disease. *Nature Cell Biology*, *20*(5), 521–527. https://doi.org/10.1038/s41556-018-0092-5.

Monel, B., Rajah, M. M., Hafirassou, M. L., Sid Ahmed, S., Burlaud-Gaillard, J., Zhu, P.-P., et al. (2019). Atlastin endoplasmic reticulum-shaping proteins facilitate Zika virus replication. *Journal of Virology*, *93*(23), 69. https://doi.org/10.1128/JVI.01047-19.

Morita, E. (2012). Differential requirements of mammalian ESCRTs in multivesicular body formation, virus budding and cell division. *FEBS Journal*, *279* (8), 1399–1406. https://doi.org/10.1111/j.1742-4658.2012.08534.x.

Neufeldt, C. J., Cortese, M., Acosta, E. G., & Bartenschlager, R. (2018). Rewiring cellular networks by members of the Flaviviridae family. *Nature Reviews. Microbiology*, *16*(3), 125–142. https://doi.org/10.1038/nrmicro.2017.170.

Neufeldt, C. J., Cortese, M., Scaturro, P., Cerikan, B., Wideman, J. G., Tabata, K., et al. (2019). ER-shaping atlastin proteins act as central hubs to promote flavivirus replication and virion assembly. *Nature Microbiology*, *4*(12), 2416–2429. https://doi.org/10.1038/s41564-019-0586-3.

Peng, H., Liu, B., Yves, T. D., He, Y., Wang, S., Tang, H., et al. (2018). Zika virus induces autophagy in human umbilical vein endothelial cells. *Viruses*, *10* (5), 259. https://doi.org/10.3390/v10050259.

Płaszczyca, A., Scaturro, P., Neufeldt, C. J., Cortese, M., Cerikan, B., Ferla, S., et al. (2019). A novel interaction between dengue virus nonstructural protein 1 and the NS4A-2K-4B precursor is required for viral RNA replication but not for formation of the membranous replication organelle. *PLoS Pathogens*, *15*(5). https://doi.org/10.1371/journal.ppat.1007736, e1007736.

Reiss, S., Rebhan, I., Backes, P., Romero-Brey, I., Erfle, H., Matula, P., et al. (2011). Recruitment and activation of a lipid kinase by hepatitis C virus NS5A is essential for integrity of the membranous replication compartment. *Cell Host & Microbe*, *9*(1), 32–45. https://doi.org/10.1016/j.chom.2010.12.002.

Rismanchi, N., Soderblom, C., Stadler, J., Zhu, P.-P., & Blackstone, C. (2008). Atlastin GTPases are required for Golgi apparatus and ER morphogenesis. *Human Molecular Genetics*, *17*(11), 1591–1604. https://doi.org/10.1093/hmg/ddn046.

Rodriguez, A. K., Muñoz, A. L., Segura, N. A., Rangel, H. R., & Bello, F. (2019). Molecular characteristics and replication mechanism of dengue, zika and chikungunya arboviruses, and their treatments with natural extracts from plants: An updated review. *EXCLI Journal*, *18*, 988–1006. https://doi.org/10.17179/excli2019-1825.

Romero-Brey, I., & Bartenschlager, R. (2016). Endoplasmic reticulum: The favorite intracellular niche for viral replication and assembly. *Viruses*, *8*(6), 160. https://doi.org/10.3390/v8060160.

Sabino, C., Basic, M., Bender, D., Elgner, F., Himmelsbach, K., & Hildt, E. (2019). Bafilomycin A1 and U18666A efficiently impair ZIKV infection. *Viruses*, *11*(6), 524. https://doi.org/10.3390/v11060524.

Sirohi, D., Chen, Z., Sun, L., Klose, T., Pierson, T. C., Rossmann, M. G., et al. (2016). The 3.8 Å resolution cryo-EM structure of Zika virus. *Science (New York, NY)*, *352*(6284), 467–470. https://doi.org/10.1126/science.aaf5316.

Soto-Acosta, R., Bautista-Carbajal, P., Cervantes-Salazar, M., Angel-Ambrocio, A. H., & Del Angel, R. M. (2017). DENV up-regulates the HMG-CoA reductase activity through the impairment of AMPK phosphorylation: A potential antiviral target. *PLoS Pathogens*, *13*(4). https://doi.org/10.1371/journal.ppat.1006257, e1006257.

Stern, O., Hung, Y.-F., Valdau, O., Yaffe, Y., Harris, E., Hoffmann, S., et al. (2013). An N-terminal amphipathic helix in dengue virus nonstructural protein 4A mediates oligomerization and is essential for replication. *Journal of Virology*, *87*(7), 4080–4085. https://doi.org/10.1128/JVI.01900-12.

Tabata, K., Arimoto, M., Arakawa, M., Nara, A., Saito, K., Omori, H., et al. (2016). Unique requirement for ESCRT factors in flavivirus particle formation on the endoplasmic reticulum. *Cell Reports*, *16*(9), 2339–2347. https://doi.org/10.1016/j.celrep.2016.07.068.

Tal, M. C., Sasai, M., Lee, H. K., Yordy, B., Shadel, G. S., & Iwasaki, A. (2009). Absence of autophagy results in reactive oxygen species-dependent amplification of RLR signaling. *Proceedings of the National Academy of Sciences of the United States of America*, *106*(8), 2770–2775. https://doi.org/10.1073/pnas.0807694106.

Wang, L., & Ou, J.-H. J. (2018). Regulation of autophagy by hepatitis C virus for its replication. *DNA and Cell Biology*, *37*(4), 287–290. https://doi.org/10.1089/dna.2017.4115.

Wang, H., Perry, J. W., Lauring, A. S., Neddermann, P., De Francesco, R., & Tai, A. W. (2014). Oxysterol-binding protein is a phosphatidylinositol 4-kinase effector required for HCV replication membrane integrity and cholesterol trafficking. *Gastroenterology*, *146*(5), 1373–1385. e1–11 https://doi.org/10.1053/j.gastro.2014.02.002.

Wang, N., & Rapoport, T. A. (2019). Reconstituting the reticular ER network – mechanistic implications and open questions. *Journal of Cell Science*, *132* (4). https://doi.org/10.1242/jcs.227611, jcs227611.

Welsch, S., Miller, S., Romero-Brey, I., Merz, A., Bleck, C. K. E., Walther, P., et al. (2009). Composition and three-dimensional architecture of the dengue virus replication and assembly sites. *Cell Host & Microbe*, *5*(4), 365–375. https://doi.org/10.1016/j.chom.2009.03.007.

Yu, L., Takeda, K., & Gao, Y. (2017). Characterization of virus-specific vesicles assembled by West Nile virus non-structural proteins. *Virology*, *506*, 130–140. https://doi.org/10.1016/j.virol.2017.03.016.

Zhang, J., Zhang, Z., Chukkapalli, V., Nchoutmboube, J. A., Li, J., Randall, G., et al. (2016). Positive-strand RNA viruses stimulate host phosphatidylcholine synthesis at viral replication sites. *Proceedings of the National Academy of Sciences of the United States of America*, *113*(8), E1064–E1073. https://doi.org/10.1073/pnas.1519730113.

Zou, J., Xie, X., Wang, Q.-Y., Dong, H., Lee, M. Y., Kang, C., et al. (2015). Characterization of dengue virus NS4A and NS4B protein interaction. *Journal of Virology*, *89*(7), 3455–3470. https://doi.org/10.1128/JVI.03453-14.

Section B

Microcephaly and congenital syndromes

Chapter 6

Microcephaly: Zika and other congenital infections

Lawrence Frenkel[a,b] and Fernando Gomez[c]
[a]Departments of Pediatrics and Microbiology, University of Illinois College of Medicine, Rockford, IL, United States, [b]Department of Pediatrics, Division of Immunology, Allergy, and Infectious Disease, The Children's Hospital at Saint Peter's University Hospital, New Brunswick, NJ, United States, [c]Department of Specialty Medicine, Rocky Vista University, Parker, CO, United States

Abbreviations

CMV	cytomegalovirus
CNS	central nervous system
CRS	congenital rubella syndrome
dNK	decidual natural killer cells
HLA	human leukocyte antigen
IFN	interferon
IL-2	interleukin 2
IUGR	intrauterine growth retardation
NK	natural killer cells
Th1	T helper type 1 cells
Th2	T helper type 2 cells

Microcephaly and TORCH agents

There are numerous causes of microcephaly including genetic and metabolic disorders and environmental insults, such as infectious agents and toxins (von der Hagen et al., 2014). The impact of maternal Zika virus infection on the developing fetus was first appreciated in 2016 after an epidemic of microcephaly had been reported in the Brazilian press in 2015 (Oliveira Melo et al., 2016). Epidemiologic and clinical evidence indicates that Zika virus infection during gestation does cause an appreciable incidence of congenital defects including microcephaly (Rasmussen et al., 2016). The congenital Zika virus syndrome is similar to that described after maternal infection by the original "TORCH" agents including cytomegalovirus (CMV), rubella virus, and *Toxoplasma gondii* (Frenkel, Gomez, & Sabahi, 2018). The clinical syndromes associated with these four pathogens (including Zika virus) are similar (see Table 1 and Fig. 1) and, in addition to microcephaly, include intracranial calcifications, sensorineural hearing loss, intrauterine growth retardation (IUGR), developmental retardation, cataracts, ocular abnormalities, and arthrogryposis, which occurs from decreased fetal movement in utero and appears to occur with severe destruction of the central nervous system (CNS) from the infection (Hall, 2014; Perlman & Argyle, 1992; van der Linden et al., 2016; Wasserman et al., 1956).

Development of microcephaly in congenital infections

There are multiple neurotropic intracellular infectious agents that can lead to development of microcephaly during or after gestation (Table 2). These agents can infect both adult and fetal neural cells, but infection of replicating neural precursor cells in the developing fetus leads to much more CNS damage than would occur with infection of nonreplicating adult neural cells (Frenkel et al., 2018). In addition to the organisms discussed in this chapter, microcephaly has also been described in congenital infections with herpes simplex, varicella zoster, lymphocytic choriomeningitis, chikungunya, and West Nile viruses and other infectious agents (Strausbaugh, Barton, & Mets, 2001; Gérardin et al., 2014; O'Leary et al., 2006; von der Hagen et al., 2014). There are some minor variations in the pathologic findings with individual pathogens, but there are numerous similarities. The infection and subsequent fetal inflammatory response leads to neuronal cell

Zika Virus Impact, Diagnosis, Control, and Models. https://doi.org/10.1016/B978-0-12-820267-8.00006-6

TABLE 1 Major stigmata of congenital infection.

Pathogen	Zika virus	Rubella virus	*Toxoplasma gondii*	Cytomegalovirus
Microcephaly	+	+	+	+
CNS calcifications	+	+	+	+
Hydrocephalus and other CNS abnormalities	+	+	+	+
Developmental delay	+	+	+	+
Contractures	+	+	+	+
Intrauterine growth retardation	+	+	+	+
Sensorineural hearing loss	+	+	+	+
Ophthalmologic abnormalities	+	+	+	+
Heart defects		+		

Teratogenesis by TORCH agents leads to numerous common clinical findings and, occasionally, to unique findings (e.g., heart defects with rubella virus) (Frenkel et al., 2018; Hall, 2014; Perlman & Argyle, 1992; van der Linden et al., 2016; Wasserman et al., 1956).

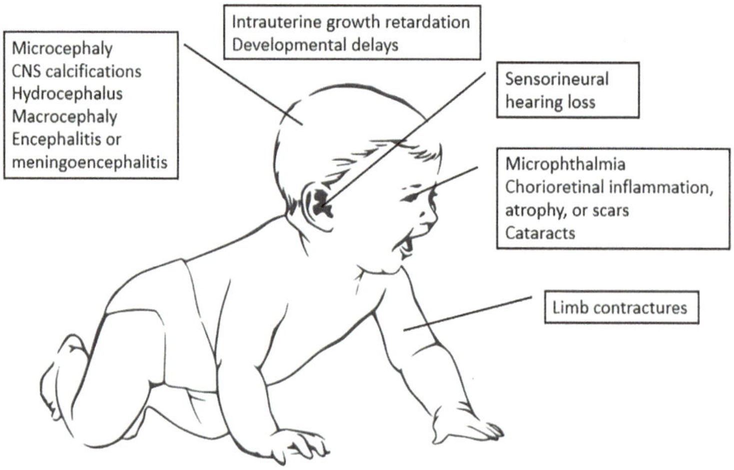

FIG. 1 Shared features. Features that can be found in infants with congenital infections by the pathogens discussed.

death (apoptosis and lysis), decreased replication, and abnormal migration of neural cells. The destruction of CNS tissue leads to calcifications (dystrophic calcification of necrotic tissue) and to decreased brain volume, which leads to microcephaly (Frenkel et al., 2018). Microcephaly often appears to be preceded by normocephaly or even macrocephaly in some instances, followed by prominent destruction of brain tissue and/or intracranial fluid absorption leading to cranial collapse, resulting in overriding of the cranial bones, flattening of the cranium, and redundant scalp skin, which has been previously been described as fetal brain disruption sequence (Russell et al., 1984).

Pathogen characteristics

Zika virus

Zika virus is a congenital microcephaly-associated pathogen that is transmitted by *Aedes* mosquitos (Musso & Gubler, 2016; Vogel, 2016). As is the case with CMV, Zika virus infections can be transmitted sexually (D'Ortenzio et al., 2016; Oster et al., 2016). Zika virus–associated congenital infection seems to occur only with primary maternal infection during gestation. Maternal Zika virus infection is believed to be asymptomatic in most women. Rates of congenital infection following maternal infection range from 6% (Shapiro-Mendoza et al., 2017) to 29% (Brasil et al., 2016). The

TABLE 2 Characteristics of infections.

	Zika virus	Rubella virus	Toxoplasma	CMV
Mode of pathogen transmission to mother	Insect, sexual	Respiratory	Oral	Sexual, oral
Congenital infection with symptomatic or asymptomatic maternal infection	Both	Symptomatic	Asymptomatic	Both
Is primary infection necessary for maternal-fetal transmission	Yes	Yes	No	No
Rate of congenital infection from infected mothers	6%–29%	54%–80%	>65%	30%–40%
Rate of symptomatic congenital infection after maternal infection	1%–29%	>90%	15%, 44%, 71%[a]	10%–20%
Rate in relation to trimester	6%, 5%, 4%[a]	54%, 25%, <1%[a]	57%, 31%, 9.5%[a]	Highest in first and second trimester
Delayed or progressive symptoms in the infected infant	Yes	Yes	Yes	Yes
Pathogen excretion from the congenitally infected infant	Yes	Yes	No	Yes
Vaccine available to prevent congenital infection	No	Yes	No	No

Overview of routes of maternal infection, maternal symptomatology, infection from maternal infection reactivation, likelihood of fetal infection (by pregnancy trimester for some agents), delayed expression of fetal malformations, and if vaccines are available in the United States for prevention (Boppana et al., 2001; Boppana et al., 2013; Brasil et al., 2016; Britt, 2017; Carlson et al., 2010; D'Ortenzio et al., 2016; Dudgeon, 1970; Embil et al., 1970; Frenkel, 2016; Frenkel & Bellanti, 1982; Ho, 2008; Jones & Holland, 2010; Kenneson & Cannon, 2007; Kluge et al., 1960; Lindsay & Dubey, 2011; Musso & Gubler, 2016; Oliveira et al., 2016; Oster et al., 2016; Pass et al., 2006; Peyron et al., 2016; Shapiro-Mendoza et al., 2017; Stagno et al., 1977; Vogel, 2016; Weiss & Dubey, 2009; Wheeler, 2018).
[a]*Data by trimester: first, second, and third.*

higher percentage represents infants identified through brain ultrasound studies. Congenital infection with Zika virus is associated with both symptomatic and asymptomatic maternal infection (61% and 38%, respectively) (Shapiro-Mendoza et al., 2017). This pattern is also true for CMV and toxoplasmosis but unlike the case with rubella virus where almost all maternal infections resulting in fetal infection occur in symptomatic mothers. Among the pregnancies with laboratory evidence of Zika virus infection, birth defects were reported in infants with maternal infection during the first, second, and third trimesters: 6%, 5%, and 4%, respectively (Shapiro-Mendoza et al., 2017). More severe and frequent congenital abnormalities are associated with first-trimester maternal infection (Honein et al., 2017), but severe congenital Zika virus syndrome has been reported with maternal infection in the second and third trimesters (Soares de Souza et al., 2016). A clinically relevant issue is the lack of available vaccines for these pathogens, except rubella virus (Frenkel, 2016). The most common findings in congenital Zika syndrome include severe microcephaly (Adachi & Nielsen-Saines, 2018; Moore et al., 2017), other CNS abnormalities (subcortical calcifications, ventriculomegaly, gyral pattern anomalies, corpus callosum abnormalities, and cerebellar hypoplasia) (Adachi & Nielsen-Saines, 2018), ocular findings (microphthalmia, retinal pigment mottling, chorioretinal atrophy, and optic nerve abnormalities) (Moore et al., 2017; Yepez et al., 2017), arthrogryposis (de Paula Freitas et al., 2016), neurodevelopmental impairment (Wheeler, 2018), and most recently, seizures (van der Linden et al., 2018). It should be noted that manifestations of congenital Zika may be delayed or may become more severe over time (Wheeler, 2018). This may be partly the result of continued virus replication in neural tissue with resultant inflammation and cellular destruction. Prolonged viral shedding from afflicted infants has been reported (Oliveira et al., 2016).

Rubella virus

Rubella virus is a highly contagious pathogen transmitted via respiratory secretions. The congenital rubella syndrome (CRS) first described by Gregg (1942) focused on the association of congenital cataracts and microphthalmia (both often unilateral), congenital heart defects, and the fact that the mothers of these infants all had rubella virus infection early in their pregnancies. CRS is also associated with IUGR, intracranial calcifications, a variety of other CNS structural abnormalities,

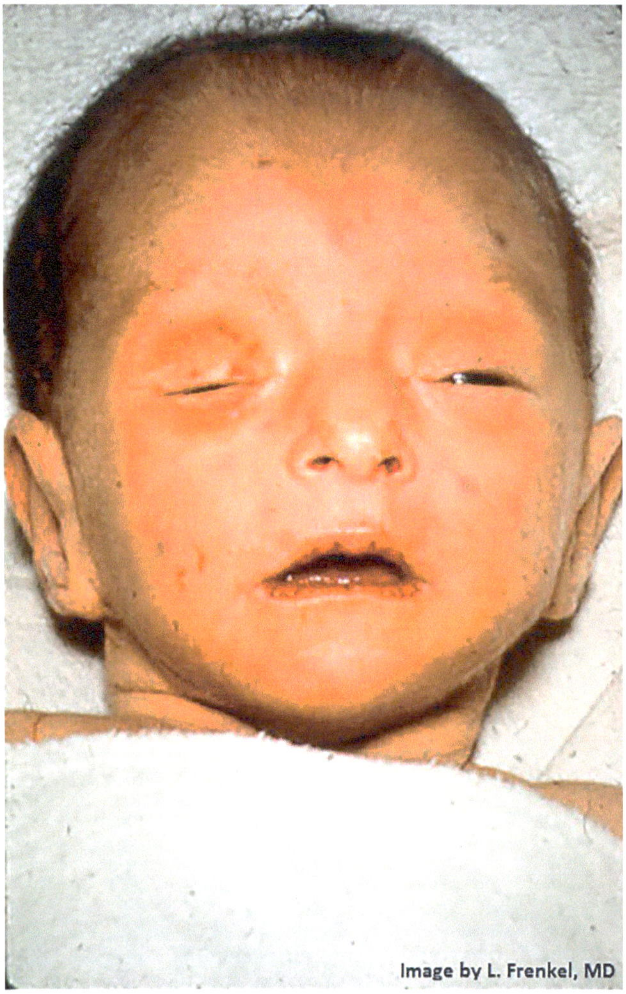

FIG. 2 Infant with congenital rubella. Infant demonstrating microcephaly, petechiae, and distress.

sensorineural hearing loss, chorioretinal inflammation, and other ophthalmologic abnormalities (Dudgeon, 1970; Mets & Chhabra, 2008; Miller, Cradock-Watson, & Pollock, 1982) (see Fig. 2). Dudgeon reported several important and characteristic findings. The first is that congenital infection is the result of primary symptomatic maternal infection; initial estimates of congenital infection following maternal infection range from 80% of women with a rash during the first 12 weeks of pregnancy, 54% during weeks 13 and 14, and 25% at the end of the second trimester (Dudgeon, 1970); 10%–20% of these congenitally infected infants demonstrating major abnormalities (Reef & Plotkin, 2016) and many congenitally infected infants are stillborn (Plotkin, 2001). Congenital infection does occur later in gestation but is not generally associated with CRS (Dudgeon, 1970). Prior maternal rubella virus infection and/or immunization is not likely to be associated with manifestations of severe congenital infection. Rubella virus can be isolated from congenitally infected infants for a year or more after birth (Dudgeon, 1970; Frenkel & Bellanti, 1982). In congenital rubella infection, as with CMV, seemingly uninfected infants may demonstrate delayed findings (e.g., hearing loss and developmental delay) after several years, which is thought to be a manifestation of persistent infection and inflammation (Frenkel & Bellanti, 1982).

Toxoplasma gondii

T. gondii is one of the most common parasitic infections worldwide (Robert-Gangneux & Dardé, 2012). It was not until 1949 that Sabin connected the syndrome of congenital infection with the pathogen *T. gondii* (Sabin & Feldman, 1949). This parasite is transmitted orally, often from exposure to cat feces (McAuley, 2014) or eating undercooked infected meat. Congenital infection is usually a result of primary maternal *T. gondii* infection during or immediately preceding pregnancy (McAuley, 2014; Weiss & Dubey, 2009). However, although uncommon, reactivation infection in immunocompromised women and infection with a second serotype manifesting as congenital toxoplasmosis have been reported (Lindsay &

Dubey, 2011; Peyron et al., 2016). Most congenital toxoplasmosis infections occur in asymptomatic women (Weiss & Dubey, 2009) although congenital anomalies are reported after symptomatic maternal disease. Evidence suggests that congenital infection with *T. gondii* progresses through maternal parasitemia leading to infection of the placenta and fetus (McAuley, 2014). There is a dichotomy between the trimester of primary maternal infection and symptomatic fetal infection, thought to reflect differences in placental cell susceptibility to *T. gondii* infection (Robbins et al., 2011). While 65%–71% of *primary maternal infections* occur during the third trimester, 44%–54% during the second, and 15%–25% during the first, more frequent *severe congenital anomalies* occur during the first (57%) and second trimesters (31%) compared with the third (9.5%) (Peyron et al., 2016). Severe congenital infections manifest with microcephaly (5%–15%), developmental delay, intracranial calcifications, seizures, hydrocephalus, and ophthalmologic abnormalities including chorioretinitis and microphthalmia in 1%–2% of congenitally infected infants (Frenkel, 1974; McAuley, 2014; Peyron et al., 2016; Weiss & Dubey, 2009; Wilson et al., 1981) (see Fig. 3). A unique phenomenon of acute chorioretinitis first manifesting in individuals 15–20 years of age after mild or asymptomatic congenital infection is described (Jones & Holland, 2010; Peyron et al., 2016). Ocular toxoplasmosis is more often bilateral in congenital disease than in acquired disease (Peyron et al., 2016). Subclinically, congenitally infected individuals who first manifest acute chorioretinitis during adolescence may have subtle changes in their cell-mediated immune status, which may allow ocular *T. gondii* cysts to reactivate and rupture (Peyron et al., 2016). Microcephaly from overt or subclinical congenital toxoplasmosis may also not occur until 12–24 months of age.

Cytomegalovirus

Congenital CMV disease was first described as cytomegalic inclusion disease in the 1950s (Fetterman, 1952; Margileth, 1955; Mercer, Luse, & Guyton, 1953) prior to the identification of CMV as the causal agent in 1956 (Weller et al., 1957). The virus was renamed cytomegalovirus, and the congenital syndrome further described in 1962 (Weller, Hanshaw, & Scott, 1962). It is both sexually transmitted (Ho, 2008) and transmitted via salivary secretions. This latter mode of transmission is especially important in situations when an infected child infects a seronegative second child, which then passes the virus to a seronegative pregnant mother (Carlson, Norwitz, & Stiller, 2010). Maternal infection with CMV is generally asymptomatic or mild (Boppana, Ross, & Fowler, 2013; Carlson et al., 2010; Kenneson & Cannon, 2007). The majority (85%–90%) of congenitally infected infants are asymptomatic at birth; the remainder exhibit abnormalities at birth (Carlson et al., 2010; Kenneson & Cannon, 2007), which include microcephaly, IUGR, chorioretinitis, sensorineural hearing loss, anophthalmia, developmental retardation, and seizures (Frenkel et al., 1980; Kenneson & Cannon, 2007) (see Fig. 4). When primary maternal infection occurs during the first or second trimesters of pregnancy, there appears to be a greater chance of more severe congenital infection (Britt, 2017; Ho, 2008; Kenneson & Cannon, 2007; Pass et al., 2006). It is now generally appreciated that congenital CMV infection can occur in mothers with preconception CMV seropositivity (Boppana et al., 2001; Britt, 2017; Embil, Ozere, & Haldane, 1970; Ho, 2008; Kenneson & Cannon, 2007; Pass et al., 2006; Stagno et al., 1977). It is not yet clear if the severity of congenital infection is greater when associated with primary versus recurrent or reactivation maternal infection. Maternal CMV infection is associated with a 30%–40% chance of congenital fetal infection (Carlson et al., 2010) and as much as a 10%–20% chance of clinically apparent manifestations in the infant at birth (symptomatic congenital CMV), with up to 50% manifesting microcephaly (Boppana et al., 1992; Britt, 2017; Picone et al., 2013). CMV is the major cause of congenital infection in the United States occurring in 0.5% of pregnancies (Britt, 2017) and is a

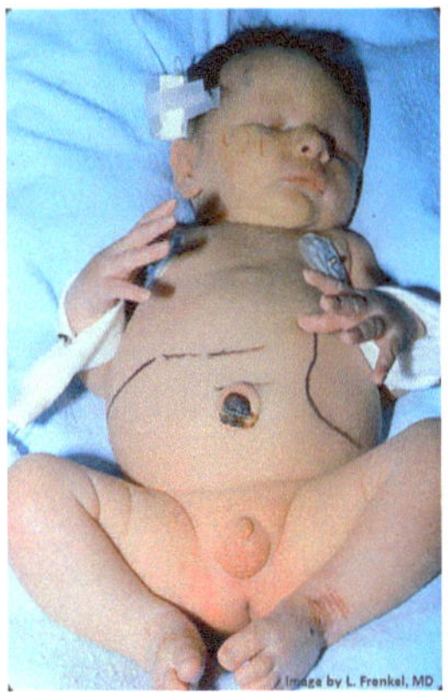

FIG. 3 Infant with congenital toxoplasmosis. Infant demonstrating hepatosplenomegaly and microcephaly. Infant with congenital cytomegalovirus. Infant demonstrating microphthalmia and microcephaly.

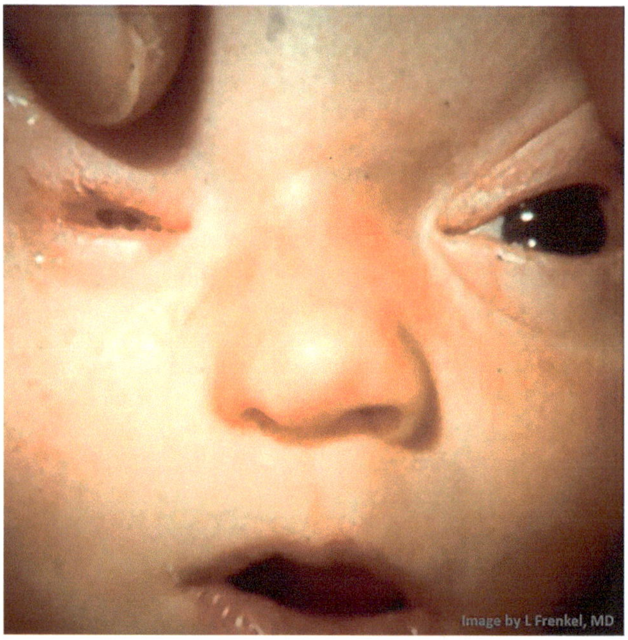

FIG. 4 Infant with congenital cytomegalovirus. Infant demonstrating microphthalmia and microcephaly.

leading cause of sensorineural hearing loss and mental retardation (Boppana et al., 2013; Britt, 2017). Infants with congenital CMV could shed the virus in their urine for years after birth, and manifestation of microcephaly and hearing loss can be delayed for months or even years after birth (Kluge, Wicksman, & Weller, 1960). In the immunocompetent nonpregnant host, the vast majority of CMV infections are subclinical. In spite of a robust host immune response, once a normal host is infected by CMV, the virus is never eliminated, causing a persistent asymptomatic infection (La Rosa & Diamond, 2012). In the immune competent fetus and infant, the delayed or deficient immune responses can result in both persistent viruria and delayed manifestations of affliction.

The pathogenesis of congenital infections

The major factors associated with congenital infection, with or without abnormalities at birth, including microcephaly, apply with greater or lesser extent for all the pathogens discussed in this chapter. These factors are determined by:

1. Characteristics of the maternal infections
2. Invasion of the maternal circulation
3. Invasion of the placenta
4. Fetal cellular tropism, and
5. Fetal tissue response to both microbial invasion and the subsequent fetal inflammatory and immune response

Since the fetus is essentially a graft of foreign tissue, several immune tolerance mechanisms, including downregulation of innate and acquired maternal immune responses, are involved in protecting the developing fetus (Sabahi et al., 1995). Unfortunately the downregulation of maternal immune responses during pregnancy is known to result in increased severity of a variety of microbial pathogens during gestation, not only those associated with congenital infections (Harris, 1919; Jamieson, Theiler, & Rasmussen, 2006; Sappenfield, Jamieson, & Kourtis, 2013; Siston, 2010). The immune status of each pregnant individual and pathogen factors seem to determine pathogen load in both the maternal circulation and at the maternal-fetal placental interface; higher pathogen loads facilitate invasion of the fetal circulation and dissemination to fetal organs.

It is of interest that primary maternal infection with these pathogens is generally asymptomatic or mild. This suggests that early maternal immune and inflammatory responses to these agents are insufficient to prevent invasion of the pathogen into the maternal circulation. On the other hand, the recognition that primary maternal infection is generally necessary for the development of Zika virus, rubella virus, and *T. gondii* congenital infection suggests that specific humoral and/or cell-mediated immunity prior to subsequent infection of the pregnant woman is generally protective.

Another consideration is that endogenous pathogen factors such as "strain" differences may determine host susceptibility to infection by pathogens associated with congenital infections (Klase et al., 2016). Alternatively, genetically determined individual variations in immune mechanisms may contribute to the severity of these infections in the mother and the outcome of the infection in the fetus. For example, more severe depression of maternal cell-mediated immunity in the individual pregnant host may be a deciding factor in the genesis of congenital infection and determine its severity. Additionally, the most important events in embryogenesis, such as the development of the central nervous system, take place early in gestation, and thus first-trimester and second-trimester maternal infections more commonly lead to congenital anomalies. It has been suggested that some pathogens have a tropism for specific fetal cells and tissues, such as neuronal cells in the CNS.

Zika virus

Among the proteins coded by the Zika virus genome, the envelope (E) protein is responsible for host cell binding and membrane fusion (Hussain et al., 2018); this could facilitate initial infection of fibroblasts and macrophages (which subsequently can infiltrate the placenta) (Hussain et al., 2018).

In vitro studies have shown that human uterine fibroblasts are susceptible to Zika virus invasion and replication (Robbiani et al., 2019). Primary Zika virus infection stimulates the production of interferons and various interferon stimulatory genes (Bayer et al., 2016; Hamel et al., 2015; Quicke et al., 2016) as well as other inflammatory mediators, which seem to be especially active at the placental maternal-fetal interface. Zika virus also manipulates the host cell intrinsic autophagy response that promotes pathogen replication, viral load, and end-organ disease establishment (Olagnier et al., 2016).

Zika virus has been shown to replicate in first-trimester human maternal-decidual tissues in ex vivo organ cultures (Weisblum et al., 2016). In this model, efficient viral replication and spread to the adjacent cells were associated with a rapid increase in Zika virus load and high titers of free virus infectious particles (Weisblum et al., 2016). This would be expected to lead to high virus load in the placenta and subsequently in fetal tissues. This may explain, in part, the rapid and severe degree of CNS tissue destruction. Midgestation maternal-decidual tissues also remained susceptible to the virus. Fetal chorionic villi cells were most susceptible to Zika virus in the first trimester, with decreased viral titers as the fetal age increased (Weisblum et al., 2016).

A genome-wide transcriptome analysis demonstrated an increase in components of innate immune responses in the decidual tissues of the placenta (Klase et al., 2016). When these results were compared with those of CMV, Zika virus was shown not to increase immune cell activation in the maternal-fetal interface; instead, Zika virus upregulated placental cell death and apoptosis. In contrast to CMV, type I and III interferons are stimulated by Zika virus (Klase et al., 2016). This points to the importance of interferons in Zika virus infection and perhaps to fetal tissue destruction and severity of the congenital infection.

The kinetics of replication and spread of Zika virus in the fetal host compared with cytomegalovirus may be explained, in part, by the efficient cell-free mode of Zika virus spread in comparison to the cell-associated mode of CMV spread (Klase et al., 2016). In addition, kinetic differences may explain the severity and character of microcephaly associated with congenital Zika virus infection (Weisblum et al., 2016). An interesting recent finding, whose implications are not yet fully understood, is that higher maternal Zika virus enhancement antibodies were highly correlated with congenital microcephaly (Robbiani et al., 2019). The mechanism proposed is that an antibody-mediated mechanism may be associated with fetal brain injury.

Rubella virus

Generally, congenital rubella is a result of acute mildly symptomatic viremic infection with high titers of virus in the maternal circulation and rapid spread to the placenta. The factors, which determine viral load in the maternal circulation, remain unclear but seem to be unrelated to "strain" differences. Thirteen serotypes of rubella virus forming two clades are recognized (Abernathy et al., 2013). Phylogenetic analysis of rubella virus strains from an outbreak in Spain, from 2004 to 2005, demonstrated that only one genotype (strain) was isolated. All the samples were from unvaccinated individuals; two had symptomatic congenital rubella (Martinez-Torres et al., 2009). The last major epidemic of rubella in the United States occurred in 1964 and almost certainly was caused by one strain of rubella virus, with an estimated 30,000 cases of congenital rubella syndrome (Reef & Plotkin, 2016). Two rubella vaccines are generally used in the United States and have been very successful in terminating rubella epidemics and thus congenital rubella infection. This, along with the lack of strain differences in rubella viruses within a geographic region, suggests that strain differences are not associated with the development of congenital rubella.

Rubella virus spread from the placenta to the developing fetus is rapid and affects many fetal tissues, including the CNS, before the fetus can mount an immune response. In spite of the acute and self-limited nature of rubella infection in the normal nonpregnant host, a hallmark of congenital rubella is chronic infection of the fetal organs and the ability to isolate virus from the congenitally infected infant for months after birth (Reef & Plotkin, 2016). The persistence of rubella virus infection and the ongoing, albeit delayed and weak, fetal immune responses may account for some of the tissue damage of congenital infection. Persistent fetal infection is also seen in congenital Zika virus and CMV infections. Finally, the presence of vasculitis in the tissues of the congenitally infected infants is well described but apparently not well understood (Reef & Plotkin, 2016).

Toxoplasma gondii

The pathogenesis of congenital *T. gondii* is believed to be related to several, as yet unknown, genetic factors determined by serotypes of the parasite (Peyron et al., 2016). Human studies have suggested that clinical manifestations and severity of congenital toxoplasmosis are determined, in part, by parasitic genotype, especially atypical strains (Ajzenberg et al., 2002).

Host genetic susceptibility and host immune status in the infected mother and in the fetus are also important. One of the most important factors that determine both the transmission rate of parasites from the mother to the fetus and thereafter the severity of fetal abnormalities is the gestational status at the time of maternal infection. Transmission rate is directly related to the week of gestation at which maternal infection occurs, while the severity of congenital infection is inversely related to the week of gestation at which maternal infection occurs (Peyron et al., 2016). NK-cell function is known to limit acute toxoplasma infection, but T helper cell and CD8+ cytotoxic T-cell responses are necessary for resolution of active infection (Peyron et al., 2016). A variety of regulatory cells and cytokines are believed to protect against untoward tissue injury. However, IFN-γ and a host of innate immune responses also enhance the transplacental transmission of *T. gondii* via the upregulation of the intracellular adhesion molecule ICAM-1. *T. gondii* infected macrophages and dendritic cells and may thereafter infect placental trophoblasts, followed by transmission of the parasites to developing fetal organs.

IFN-γ seems to have an important role in the recruitment of T cells into the brain after congenital infection. T-cell hyporesponsiveness to *T. gondii* antigens has been observed in congenitally infected infants, followed by resolution of immune responses later in childhood (Guglietta et al., 2007), as is the case with congenital CMV. On the other hand, upregulated inflammatory responses also amplify the chronic tissue destruction. As is the case with CMV, reinfection or reactivation of *T. gondii* infection can give rise to a severe congenitally infected infant in an otherwise normal mother, albeit an apparently rare event (Valdès et al., 2011).

Cytomegalovirus

Human CMV is an ancient and ubiquitous pathogen that elicits a robust immune response in the immunocompetent host and has evolved effective immune defense strategies to circumvent innate, humoral, and cell-mediated host immune defenses. Specific genes that determine the cellular tropism and that modify the immune responses to cytomegalovirus have been identified (Britt, 2016). Innate and adaptive immune responses can limit, but not prevent, the spread and latency of CMV (Britt, 2016; Rola-Pleszczynski et al., 1977). It is accepted that maternal viremia, placental infection, and subsequent dissemination to fetal organs, including the CNS, are all involved in the pathogenesis of congenital CMV. However, the complex role of immunologic factors remains largely unresolved (Burny et al., 2004). CD8+ T lymphocytes play a central role in the control of CMV replication (Burny et al., 2004). Both asymptomatic and symptomatic congenitally infected infants showed similar CD8+ responses, suggesting that they were not sufficient to prevent congenital CMV infection.

Cytomegalovirus infection is clearly different than rubella virus infection, and probably Zika virus infection, in that it does occur in women who were seropositive prior to their current pregnancy (nonprimary) (Boppana et al., 2001; Britt, 2016). This reactivation or reinfection is unpredictable in terms of which women can be so affected and in terms of which trimester it occurs. Restriction enzyme analysis of CMV isolated from pregnancies, in which pairs (conceived and born at different times) of congenitally infected siblings were studied demonstrated that two of the pairs reflected reactivation infection—the viruses were identical—but in one pair, the CMV strains were different, reflecting CMV infection with a new strain (Stagno et al., 1982). The infants born to two of these subsequent pregnancies (reactivation) were asymptomatic, whereas the reinfection pregnancy produced a severe congenitally infected infant. In general, however, the impression is that pregnancies in seropositive women produce less severely or even subclinically infected infants must be confirmed (Britt, 2016).

Some seropositive pregnant women have positive CMV viral cultures intermittently during their pregnancies; this finding does not seem to be correlated with fetal infection (Britt, 2016). Similarly, seropositive nonpregnant women also demonstrate intermittent CMV shedding. Over 40% of normal postpartum seropositive women with normal pregnancies and normal infants also demonstrate CMV shedding into their breast milk (Britt, 2016). CMV congenitally infected infants have positive urine cultures for months or even years after birth (Frenkel et al., 1980).

Several, but not all, investigators have reported that CMV-seropositive women demonstrate depressed CMV-specific T-cell-mediated immunity during gestation (Sabahi et al., 1995; Sabahi, Frenkel, & O'Connell, 1993). CMV-specific lymphocyte proliferation, IL-2, and IFN-γ production are significantly reduced in the first and third trimesters of normal pregnancy (Frenkel et al., 2018). Apparently, in CMV-seropositive subjects, most of the memory T cells are specific for CMV antigens, and these CMV-specific memory T cells increase as these seropositive subjects age (Sylwester et al., 2005).

Specific impairment of cell-mediated immunity occurs in mothers of infants with congenital CMV infection (Rola-Pleszczynski et al., 1977), but not all investigators have been able to reproduce this finding (Britt, 2016). Suppressed CMV-specific lymphocyte proliferation and interferon secretion in severely symptomatic congenitally infected infants and their mothers have been reported (Starr et al., 1979). A significantly delayed lymphoproliferative response has been reported in mothers who transmitted CMV to their babies compared with nontransmitting mothers (Revello et al., 2006). Furthermore, the transmitting mothers (i.e., those who delivered CMV-infected infants) had lower CD4+ and CD8+ T cells compared with nontransmitting women (Fornara et al., 2011; Lilleri et al., 2007). Taken together, various studies indicate that defective function and delayed maternal CD4+ T-cell production may be important contributors to gestational CMV transmission from mother to fetus and defective fetal CD4+ T-cell response may contribute to CMV congenital infection and affliction.

In CMV-seropositive pregnant women, naive CD8+ effector T cells are reduced by 50%. In pregnant seronegative women, no such change is observed (Lissauer et al., 2011). During pregnancy, levels of maternal CD8+ T-cell activation are normal. CMV-infected fetuses showed a significant increase in activated CD8+ T cells as early as 22 weeks' gestation. However, fewer CD8+ T cells from these infected fetuses were capable of releasing IFN-γ upon stimulation with CMV antigen (Britt, 2016; Burny et al., 2004). The definitive role of CD8+ T cells, with regard to congenital infection, remains unclear. However, induction of proinflammatory cytokines associated with fetal CMV infection is manifested in a shift toward the proinflammatory Th1 profile and away from the normal antiinflammatory Th2 pattern. In combination with direct cytopathic effects, this results in damage to, and impairment of, critical organ functions in the fetus (Adams Waldorf & McAdams, 2013).

CMV can downregulate HLA-A, HLA-B, HLA-C, and even HLA-E and HLA-G molecules on infected cells. CMV can escape immune control by inhibiting the optimum presentation of their peptides to CD8+ T lymphocytes. The loss of HLA molecules on infected cells can make them susceptible to attack by NK cells (Jun et al., 2000), facilitating cellular and tissue destruction. NK cells play a role in specifically controlling CMV infection (Burny et al., 2004). Infants with symptomatic congenital CMV have lower NK activity than asymptomatic infants (Cauda et al., 1987). On the other hand, inadequate NK-cell function has been found to be associated with severe CMV infections in children (Biron, Byron, & Sullivan, 1989). NK cells may be helpful in the prevention of CMV transmission from mother to fetus during early pregnancy. Another factor in the pathogenesis of CMV infection of the CNS is that CMV initiates a robust inflammatory response characterized by recruitment of activated resident and peripheral immune cells as well as expression of a large number of proinflammatory cytokines and IFN-regulated genes (Slavuljica et al., 2014), which facilitates the destruction of CNS cells and tissues.

Policy and procedures

A review of pertinent literature was performed for intrauterine infections due to Zika virus, cytomegalovirus, rubella virus, and *T. gondii*. Facts about the individual infectious agents were analyzed and summarized for this report highlighting the similarities and differences of the syndromes caused by these agents, highlighting the common finding of microcephaly.

Mini-dictionary of terms

Apoptosis. A genetically programmed sequence of events that leads to death of unneeded or defective cells.
Arthrogryposis. Congenital limb contractures.
Congenital. Existing at the time of birth; the condition may be clinically identified at birth or later in life.
Developmental retardation. Delay in physical development for age.

Macrocephaly. Head circumference above two standard deviations from the expected mean for age (or gestational age) and sex.

Microcephaly. Head circumference below two or three standard deviations (varies by study) from the expected mean for age (or gestational age) and sex.

Sensorineural hearing loss. Hearing loss from abnormalities in the vestibulocochlear nerve (cranial nerve VIII) or inner ear structures.

TORCH syndrome. Infection of a developing fetus by group of infectious agents that lead to similar clinical findings in infected infants.

Key facts on microcephaly secondary to intrauterine infections

- There are numerous causes of microcephaly, including maternal infection by Zika virus during gestation.
- Primary, and generally asymptomatic, maternal infection by Zika virus, CMV, and *T. gondii* leads to congenital infection. Maternal infection by rubella virus is usually symptomatic. Both primary and reactivation/recurrent toxoplasma and CMV may lead to congenital infections.
- Maternal infections that occur during the first and second trimesters are associated with more severe stigmata of congenital infection.
- Strain differences are particularly important in congenital toxoplasma disease, less so with CMV and Zika virus but not so with rubella virus.
- Clinical manifestations of congenital infections may be delayed or may become more severe over time, and the pathogens may be isolated from infected infants for months to years.

Summary points

- Congenital Zika virus syndrome is similar to that described after maternal infection by cytomegalovirus (CMV), rubella virus, and *T. gondii*.
- Multiple neurotropic intracellular infectious agents can lead to development of microcephaly during or after gestation.
- Downregulation of maternal immunity during pregnancy facilitates maternal infection.
- Microorganism-specific factors such as receptors on cells, replication kinetics, and strain differences contribute to both the incidence and severity of congenital infection.
- Maternal immune and inflammatory responses determine placental pathogen load and the outcome of fetal infection.
- Fetal immune and inflammatory responses to the microbial invasion facilitate ongoing tissue destruction.
- Early fetal infection increases the likelihood of significant fetal tissue destruction and congenital defects, including microcephaly.

References

Abernathy, E., et al. (2013). Analysis of whole genome sequences of 16 strains of rubella virus from the United States, 1961–2009. *Virology Journal, 10*(1), 32. https://doi.org/10.1186/1743-422X-10-32.

Adachi, K., & Nielsen-Saines, K. (2018). Zika clinical updates: Implications for pediatrics. *Current Opinion in Pediatrics, 30*(1), 105–116. https://doi.org/10.1097/MOP.0000000000000582.

Adams Waldorf, K. M., & McAdams, R. M. (2013). Influence of infection during pregnancy on fetal development. *Reproduction, 146*(5), R151–R162. https://doi.org/10.1530/REP-13-0232. BioScientifica.

Ajzenberg, D., et al. (2002). Genotype of 86 *Toxoplasma gondii* isolates associated with human congenital toxoplasmosis, and correlation with clinical findings. *The Journal of Infectious Diseases, 186*(5), 684–689. https://doi.org/10.1086/342663.

Bayer, A., et al. (2016). Type III interferons produced by human placental trophoblasts confer protection against Zika virus infection. *Cell Host & Microbe, 19*(5), 705–712. https://doi.org/10.1016/j.chom.2016.03.008. Elsevier BV.

Biron, C. A., Byron, K. S., & Sullivan, J. L. (1989). Severe herpesvirus infections in an adolescent without natural killer cells. *New England Journal of Medicine, 320*(26), 1731–1735. https://doi.org/10.1056/NEJM198906293202605.

Boppana, S. B., Ross, S. A., & Fowler, K. B. (2013). Congenital cytomegalovirus infection: Clinical outcome. *Clinical Infectious Diseases, 57*(suppl 4), S178–S181. https://doi.org/10.1093/cid/cit629.

Boppana, S. B., et al. (1992). Symptomatic congenital cytomegalovirus infection: Neonatal morbidity and mortality. *The Pediatric Infectious Disease Journal, 11*(2), 93–98. Ovid Technologies (Wolters Kluwer Health).

Boppana, S. B., et al. (2001). Intrauterine transmission of cytomegalovirus to infants of women with preconceptional immunity. *New England Journal of Medicine*, *344*(18), 1366–1371. https://doi.org/10.1056/NEJM200105033441804.

Brasil, P., et al. (2016). Zika virus infection in pregnant women in Rio de Janeiro. *New England Journal of Medicine*, *375*(24), 2321–2334. https://doi.org/10.1056/NEJMoa1602412.

Britt, W. J. (2016). Cytomegalovirus. In C. B. Wilson, et al. (Eds.), *Remington and Klein's infectious diseases of the fetus and newborn infant* (8th ed., pp. 724–781). Philadelphia: Saunders.

Britt, W. J. (2017). Congenital human cytomegalovirus infection and the enigma of maternal immunity. *Journal of Virology*, *91*(15). https://doi.org/10.1128/JVI.02392-16, e02392-16. Edited by F. Goodrum.

Burny, W., et al. (2004). Epidemiology, pathogenesis and prevention of congenital cytomegalovirus infection. *Expert Review of Anti-Infective Therapy*, *2*(6), 881–894. https://doi.org/10.1586/14789072.2.6.881. Available from http://www.ncbi.nlm.nih.gov/pubmed/15566332.

Carlson, A., Norwitz, E. R., & Stiller, R. J. (2010). Cytomegalovirus infection in pregnancy: Should all women be screened? *Reviews in Obstetrics and Gynecology*, *3*(4), 172–179. https://doi.org/10.3909/riog0131.

Cauda, R., et al. (1987). Congenital cytomegalovirus: Immunological alterations. *Journal of Medical Virology*, *23*(1), 41–49. https://doi.org/10.1002/jmv.1890230106.

D'Ortenzio, E., et al. (2016). Evidence of sexual transmission of Zika virus. *New England Journal of Medicine*, *374*(22), 2195–2198. https://doi.org/10.1056/NEJMc1604449.

de Paula Freitas, B., et al. (2016). Ocular findings in infants with microcephaly associated with presumed Zika virus congenital infection in Salvador, Brazil. *JAMA Ophthalmology*, *134*(5), 529–535. https://doi.org/10.1001/jamaophthalmol.2016.0267. American Medical Association (AMA).

Dudgeon, J. A. (1970). Teratogenic effect of rubella virus. *Proceedings of the Royal Society of Medicine*, *63*(12), 1254–1257. Available from http://www.ncbi.nlm.nih.gov/pubmed/4992488.

Embil, J. A., Ozere, R. L., & Haldane, E. V. (1970). Congenital cytomegalovirus infection in two siblings from consecutive pregnancies. *The Journal of Pediatrics*, *77*(3), 417–421. https://doi.org/10.1016/S0022-3476(70)80009-9.

Fetterman, G. H. (1952). A new laboratory aid in the clinical diagnosis of inclusion disease of infancy. *American Journal of Clinical Pathology*, *22*(5), 424–425. https://doi.org/10.1093/ajcp/22.5.424.

Fornara, C., et al. (2011). Kinetics of effector functions and phenotype of virus-specific and γδ T lymphocytes in primary human cytomegalovirus infection during pregnancy. *Journal of Clinical Immunology*, *31*(6), 1054–1064. https://doi.org/10.1007/s10875-011-9577-8.

Frenkel, J. K. (1974). Pathology and pathogenesis of congenital toxoplasmosis. *Bulletin of the New York Academy of Medicine*, *50*(2), 182–191. Available from https://www.ncbi.nlm.nih.gov/pmc/articles/PMC1749352/.

Frenkel, L. D. (2016). Comparing congenital Zika and cytomegalovirus affliction. *The Pediatric Infectious Disease Journal*, *35*(12), 1371–1372. https://doi.org/10.1097/INF.0000000000001273.

Frenkel, L. D., & Bellanti, J. A. (1982). Immunology of measles, mumps and rubella viruses. In A. J. Nahmias, & R. J. O'Reilly (Eds.), *Immunology of human infection (comprehensive immunology)* (pp. 135–163). Boston: Springer.

Frenkel, L. D., Gomez, F., & Sabahi, F. (2018). The pathogenesis of microcephaly resulting from congenital infections: Why is my baby's head so small? *European Journal of Clinical Microbiology & Infectious Diseases*, *37*(2), 209–226. https://doi.org/10.1007/s10096-017-3111-8.

Frenkel, L. D., et al. (1980). Unusual eye abnormalities associated with congenital cytomegalovirus infection. *Pediatrics*, *66*(5), 763–766. Available from http://www.ncbi.nlm.nih.gov/pubmed/6253868.

Gérardin, P., et al. (2014). Neurocognitive outcome of children exposed to perinatal mother-to-child chikungunya virus infection: The CHIMERE cohort study on Reunion Island. *PLoS Neglected Tropical Diseases*, *8*(7), e2996. https://doi.org/10.1371/journal.pntd.0002996. Edited by A. M. Powers.

Gregg, N. M. (1942). Congenital cataract following German measles in the mother. Sydney: Australasian Medical Publishing Company.

Guglietta, S., et al. (2007). Age-dependent impairment of functional helper T cell responses to immunodominant epitopes of *Toxoplasma gondii* antigens in congenitally infected individuals. *Microbes and Infection*, *9*(2), 127–133. https://doi.org/10.1016/j.micinf.2006.10.017.

Hall, J. G. (2014). Arthrogryposis (multiple congenital contractures): Diagnostic approach to etiology, classification, genetics, and general principles. *European Journal of Medical Genetics*, *57*(8), 464–472. https://doi.org/10.1016/j.ejmg.2014.03.008. Elsevier Masson SAS.

Hamel, R., et al. (2015). Biology of Zika virus infection in human skin cells. *Journal of Virology*, *89*(17), 8880–8896. https://doi.org/10.1128/jvi.00354-15. Edited by M. S. Diamond. American Society for Microbiology.

Harris, J. W. (1919). Influenza occurring in pregnant women. *Journal of the American Medical Association*, *72*(14), 978–980. https://doi.org/10.1001/jama.1919.02610140008002. American Medical Association (AMA).

Ho, M. (2008). The history of cytomegalovirus and its diseases. *Medical Microbiology and Immunology*, *197*(2), 65–73. https://doi.org/10.1007/s00430-007-0066-x.

Honein, M. A., et al. (2017). Birth defects among fetuses and infants of US women with evidence of possible Zika virus infection during pregnancy. *JAMA*, *317*(1), 59–68. https://doi.org/10.1001/jama.2016.19006. American Medical Association (AMA).

Hussain, A., et al. (2018). A comprehensive review of the manifestations and pathogenesis of Zika virus in neonates and adults. *Cureus*, *10*(9). https://doi.org/10.7759/cureus.3290.

Jamieson, D. J., Theiler, R. N., & Rasmussen, S. A. (2006). Emerging infections and pregnancy. *Emerging Infectious Diseases*, *12*(11), 1638–1643. https://doi.org/10.3201/eid1211.060152.

Jones, J. L., & Holland, G. N. (2010). Annual burden of ocular toxoplasmosis in the United States. *American Journal of Tropical Medicine and Hygiene*, *82*(3), 464–465. https://doi.org/10.4269/ajtmh.2010.09-0664.

Jun, Y., et al. (2000). Human cytomegalovirus gene products US3 and US6 down-regulate trophoblast class I MHC molecules. *The Journal of Immunology*, *164*(2), 805–811. https://doi.org/10.4049/jimmunol.164.2.805.

Kenneson, A., & Cannon, M. J. (2007). Review and meta-analysis of the epidemiology of congenital cytomegalovirus (CMV) infection. *Reviews in Medical Virology*, *17*(4), 253–276. https://doi.org/10.1002/rmv.535.

Klase, Z. A., et al. (2016). Zika fetal neuropathogenesis: Etiology of a viral syndrome. *PLoS Neglected Tropical Diseases*, *10*(8). https://doi.org/10.1371/journal.pntd.0004877.

Kluge, R. C., Wicksman, R. S., & Weller, T. H. (1960). Cytomegalic inclusion disease of the newborn: Report of case with persistent viruria. *Pediatrics*, *25* (1), 35–39. Available from http://www.ncbi.nlm.nih.gov/pubmed/14410006.

La Rosa, C., & Diamond, D. J. (2012). The immune response to human CMV. *Future Virology*, *7*(3), 279–293. https://doi.org/10.2217/fvl.12.8.

Lilleri, D., et al. (2007). Development of human cytomegalovirus–specific T cell immunity during primary infection of pregnant women and its correlation with virus transmission to the fetus. *The Journal of Infectious Diseases*, *195*(7), 1062–1070. https://doi.org/10.1086/512245.

Lindsay, D. S., & Dubey, J. P. (2011). Toxoplasma gondii: The changing paradigm of congenital toxoplasmosis. *Parasitology*, *138*(14), 1829–1831. https://doi.org/10.1017/S0031182011001478.

Lissauer, D., et al. (2011). Cytomegalovirus sero positivity dramatically alters the maternal CD8 + T cell repertoire and leads to the accumulation of highly differentiated memory cells during human pregnancy. *Human Reproduction*, *26*(12), 3355–3365. https://doi.org/10.1093/humrep/der327.

Margileth, A. M. (1955). The diagnosis and treatment of generalized cytomegalic inclusion disease of the newborn. *Pediatrics*, *15*(3), 270–283. Available from: https://pubmed.ncbi.nlm.nih.gov/14356799/.

Martinez-Torres, A. O., et al. (2009). Phylogenetic analysis of rubella virus strains from an outbreak in Madrid, Spain, from 2004 to 2005. *Journal of Clinical Microbiology*, *47*(1), 158–163. https://doi.org/10.1128/JCM.00469-08.

McAuley, J. B. (2014). Congenital toxoplasmosis. *Journal of the Pediatric Infectious Diseases Society*, *3*(Suppl 1), S30–S35. https://doi.org/10.1093/jpids/piu077.

Mercer, R. D., Luse, S., & Guyton, D. H. (1953). Clinical diagnosis of generalized cytomegalic inclusion disease. *Pediatrics*, *11*(5), 502–514. Available from http://www.ncbi.nlm.nih.gov/pubmed/13055362.

Mets, M. B., & Chhabra, M. S. (2008). Eye manifestations of intrauterine infections and their impact on childhood blindness. *Survey of Ophthalmology*, *53* (2), 95–111. https://doi.org/10.1016/j.survophthal.2007.12.003.

Miller, E., Cradock-Watson, J. E., & Pollock, T. M. (1982). Consequences of confirmed maternal rubella at successive stages of pregnancy. *The Lancet*, *320*(8302), 781–784. Available from http://www.ncbi.nlm.nih.gov/pubmed/6126663.

Moore, C. A., et al. (2017). Congenital Zika syndrome: Characterizing the pattern of anomalies for pediatric healthcare providers. *JAMA Pediatrics*, *171* (3), 288–295. https://doi.org/10.1001/jamapediatrics.2016.3982.

Musso, D., & Gubler, D. J. (2016). Zika virus. *Clinical Microbiology Reviews*, *29*(3), 487–524. https://doi.org/10.1128/CMR.00072-15.

O'Leary, D. R., et al. (2006). Birth outcomes following West Nile virus infection of pregnant women in the United States: 2003–2004. *Pediatrics*, *117*(3), e537–e545. https://doi.org/10.1542/peds.2005-2024.

Olagnier, D., et al. (2016). Dengue virus immunopathogenesis: Lessons applicable to the emergence of Zika virus. *Journal of Molecular Biology*, *428*(17), 3429–3448. https://doi.org/10.1016/j.jmb.2016.04.024. Elsevier BV.

Oliveira Melo, A. S., et al. (2016). Zika virus intrauterine infection causes fetal brain abnormality and microcephaly: Tip of the iceberg? *Ultrasound in Obstetrics & Gynecology*, *47*(1), 6–7. https://doi.org/10.1002/uog.15831.

Oliveira, D. B. L., et al. (2016). Prolonged shedding of Zika virus associated with congenital infection. *New England Journal of Medicine*, *375*(12), 1202–1204. https://doi.org/10.1056/NEJMc1607583.

Oster, A. M., et al. (2016). Interim guidelines for prevention of sexual transmission of Zika virus — United States, 2016. *MMWR Morbidity and Mortality Weekly Report*, *65*(5), 120–121. https://doi.org/10.15585/mmwr.mm6505e1.

Pass, R. F., et al. (2006). Congenital cytomegalovirus infection following first trimester maternal infection: Symptoms at birth and outcome. *Journal of Clinical Virology*, *35*(2), 216–220. https://doi.org/10.1016/j.jcv.2005.09.015.

Perlman, J. M., & Argyle, C. (1992). Lethal cytomegalovirus infection in preterm infants: Clinical, radiological, and neuropathological findings. *Annals of Neurology*, *31*(1), 64–68. https://doi.org/10.1002/ana.410310112.

Peyron, F., et al. (2016). Toxoplasmosis. In C. B. Wilson, et al. (Eds.), *Remington and Klein's infectious diseases of the fetus and newborn infant* (8th ed., pp. 949–1042). Philadelphia: Saunders.

Picone, O., et al. (2013). A series of 238 cytomegalovirus primary infections during pregnancy: Description and outcome. *Prenatal Diagnosis*, *33*(8), 751–758. https://doi.org/10.1002/pd.4118.

Plotkin, S. A. (2001). Rubella eradication. *Vaccine*, *19*(25–26), 3311–3319. https://doi.org/10.1016/S0264-410X(01)00073-1.

Quicke, K. M., et al. (2016). Zika virus infects human placental macrophages. *Cell Host & Microbe*, *20*(1), 83–90. https://doi.org/10.1016/j.chom.2016.05.015. Elsevier BV.

Rasmussen, S. A., et al. (2016). Zika virus and birth defects—Reviewing the evidence for causality. *New England Journal of Medicine*, *374*(20), 1981–1987. https://doi.org/10.1056/NEJMsr1604338.

Reef, S. E., & Plotkin, S. A. (2016). Rubella. In C. B. Wilson, et al. (Eds.), *Remington and Klein's infectious diseases of the fetus and newborn infant* (8th ed., pp. 894–932). Philadelphia: Saunders.

Revello, M. G., et al. (2006). Lymphoproliferative response in primary human cytomegalovirus (HCMV) infection is delayed in HCMV transmitter mothers. *The Journal of Infectious Diseases*, *193*(2), 269–276. https://doi.org/10.1086/498872.

Robbiani, D. F., et al. (2019). Risk of Zika microcephaly correlates with features of maternal antibodies. *The Journal of Experimental Medicine*, *216*(10), 2302–2315. https://doi.org/10.1084/jem.20191061.

Robbins, J. R., et al. (2011). Tissue barriers of the human placenta to infection with *Toxoplasma gondii*. *Infection and Immunity*, *80*(1), 418–428. https://doi.org/10.1128/IAI.05899-11.

Robert-Gangneux, F., & Dardé, M.-L. (2012). Epidemiology of and diagnostic strategies for toxoplasmosis. *Clinical Microbiology Reviews*, *25*(2), 264–296. https://doi.org/10.1128/CMR.05013-11.

Rola-Pleszczynski, M., et al. (1977). Specific impairment of cell-mediated immunity in mothers of infants with congenital infection due to cytomegalovirus. *Journal of Infectious Diseases*, *135*(3), 386–391. https://doi.org/10.1093/infdis/135.3.386. Oxford University Press (OUP).

Russell, L. J., et al. (1984). In utero brain destruction resulting in collapse of the fetal skull, microcephaly, scalp rugae, and neurologic impairment: The fetal brain disruption sequence. *American Journal of Medical Genetics*, *17*(2), 509–521. https://doi.org/10.1002/ajmg.1320170213.

Sabahi, F., Frenkel, L. D., & O'Connell, S. (1993). Depression of cell-mediated-immunity in pregnancy. *Journal of Immunology*, *150*(8), A101.

Sabahi, F., et al. (1995). Qualitative and quantitative analysis of T lymphocytes during normal human pregnancy. *American Journal of Reproductive Immunology*, *33*(5), 381–393. https://doi.org/10.1111/j.1600-0897.1995.tb00907.x. Wiley-Blackwell.

Sabin, A. B., & Feldman, H. A. (1949). Persistence of placentally transmitted toxoplasmic antibodies in normal children in relation to diagnosis of congenital toxoplasmosis. *Pediatrics*, *4*(5), 660–664. Available from http://www.ncbi.nlm.nih.gov/pubmed/15391048.

Sappenfield, E., Jamieson, D. J., & Kourtis, A. P. (2013). Pregnancy and susceptibility to infectious diseases. *Infectious Diseases in Obstetrics and Gynecology*, *2013*, 1–8. https://doi.org/10.1155/2013/752852. Hindawi Publishing.

Shapiro-Mendoza, C. K., et al. (2017). Pregnancy outcomes after maternal Zika virus infection during pregnancy—U.S. territories, January 1, 2016–April 25, 2017. *MMWR Morbidity and Mortality Weekly Report*, *66*(23), 615–621. https://doi.org/10.15585/mmwr.mm6623e1.

Siston, A. M. (2010). Pandemic 2009 influenza A(H1N1) virus illness among pregnant women in the United States. *Journal of the American Medical Association*, *303*(15), 1517–1525. https://doi.org/10.1001/jama.2010.479. American Medical Association (AMA).

Slavuljica, I., et al. (2014). Immunobiology of congenital cytomegalovirus infection of the central nervous system—The murine cytomegalovirus model. *Cellular and Molecular Immunology*, *12*(2), 180–191. https://doi.org/10.1038/cmi.2014.51. Springer Nature.

Soares de Souza, A., et al. (2016). Fetal infection by Zika virus in the third trimester: Report of 2 cases. *Clinical Infectious Diseases*, *63*(12), 1622–1625. https://doi.org/10.1093/cid/ciw613. Oxford University Press (OUP).

Stagno, S., et al. (1977). Congenital cytomegalovirus infection – occurrence in an immune population. *New England Journal of Medicine*, *296*(22), 1254–1258. https://doi.org/10.1097/00003081-198209000-00014.

Stagno, S., et al. (1982). Maternal cytomegalovirus infection and perinatal transmission. *Clinical Obstetrics and Gynecology*, *25*(3), 563–576.

Starr, S. E., et al. (1979). Impaired cellular immunity to cytomegalovirus in congenitally infected children and their mothers. *Journal of Infectious Diseases*, *140*(4), 500–505. https://doi.org/10.1093/infdis/140.4.500.

Strausbaugh, L. J., Barton, L. L., & Mets, M. B. (2001). Congenital lymphocytic choriomeningitis virus infection: Decade of rediscovery. *Clinical Infectious Diseases*, *33*(3), 370–374. https://doi.org/10.1086/321897.

Sylwester, A. W., et al. (2005). Broadly targeted human cytomegalovirus-specific CD4 + and CD8 + T cells dominate the memory compartments of exposed subjects. *The Journal of Experimental Medicine*, *202*(5), 673–685. https://doi.org/10.1084/jem.20050882. Rockefeller University Press.

Valdès, V., et al. (2011). Toxoplasmose congénitale secondaire à une réinfection maternelle pendant la grossesse. *Archives de Pédiatrie*, *18*(7), 761–763. https://doi.org/10.1016/j.arcped.2011.04.011.

van der Linden, V., et al. (2016). Congenital Zika syndrome with arthrogryposis: Retrospective case series study. *BMJ (Online)*, *354*. https://doi.org/10.1136/bmj.i3899.

van der Linden, H., et al. (2018). Epilepsy profile in infants with congenital Zika virus infection. *New England Journal of Medicine*, *379*(9), 891–892. https://doi.org/10.1056/NEJMc1716070.

Vogel, G. (2016). Top mosquito suspect found infected with Zika. *Science*. https://doi.org/10.1126/science.aaf5743.

von der Hagen, M., et al. (2014). Diagnostic approach to microcephaly in childhood: A two-center study and review of the literature. *Developmental Medicine and Child Neurology*, *56*(8), 732–741. https://doi.org/10.1111/dmcn.12425.

Wasserman, L., et al. (1956). Contributii anatomo-clinice în toxoplasmoza congenitala umana. *Studii si Cercetari Injramicrowbiol., Microbiol. si Parazitol.*, *7*(1/2), 203–219.

Weisblum, Y., et al. (2016). Zika virus infects early- and midgestation human maternal decidual tissues, inducing distinct innate tissue responses in the maternal-fetal interface. *Journal of Virology*, *91*(4), e01905–e01916. https://doi.org/10.1128/jvi.01905-16. Edited by S. Perlman. American Society for Microbiology.

Weiss, L. M., & Dubey, J. P. (2009). Toxoplasmosis: A history of clinical observations. *International Journal for Parasitology*, *39*(8), 895–901. https://doi.org/10.1016/j.ijpara.2009.02.004.

Weller, T. H., Hanshaw, J. B., & Scott, D. E. (1962). Virologic and clinical observations on cytomegalic inclusion disease. *New England Journal of Medicine*, *266*(24), 1233–1244. https://doi.org/10.1056/NEJM196206142662401.

Weller, T. H., et al. (1957). Isolation of intranuclear inclusion producing agents from infants with illnesses resembling cytomegalic inclusion disease. *Experimental Biology and Medicine*, *94*(1), 4–12. https://doi.org/10.3181/00379727-94-22841.

Wheeler, A. C. (2018). Development of infants with congenital zika syndrome: What do we know and what can we expect? *Pediatrics*, *141*(suppl 2), S154–S160. https://doi.org/10.1542/peds.2017-2038D.

Wilson, C. B., et al. (1981). Development of adverse sequelae in children born with subclinical congenital toxoplasma infection. *Obstetrical & Gynecological Survey*, *36*(8), 436–437.

Yepez, J. B., et al. (2017). Ophthalmic manifestations of congenital Zika syndrome in Colombia and Venezuela. *JAMA Ophthalmology*, *135*(5), 440. https://doi.org/10.1001/jamaophthalmol.2017.0561.

Chapter 7

Zika and impact on the nervous system in children

Aline Almeida Bentes[a] and Erna Geessien Kroon[b]
[a]*Departamento de Pediatria, Faculdade de Medicina, Universidade Federal de Minas Gerais, Belo Horizonte, MG, Brazil,* [b]*Laboratório de Vírus, Departamento de Microbiologia, Instituto de Ciências Biológicas, Universidade Federal de Minas Gerais, Belo Horizonte, MG, Brazil*

List of abbreviations

AXL	receptor tyrosine kinase type AXL
DNA	deoxyribonucleic acid
Cas9	associated protein 9
CRISPR	clustered regularly interspaced short palindromic repeats
CSF	cerebrospinal fluid
CZS	congenital Zika syndrome
DEET	N,N-diethyl-m-toluamide
GBS	Guillain-Barré syndrome
MAC-ELISA	antibody capture enzyme-linked immunosorbent assay
MER	receptor tyrosine kinase type MER
NCR	noncoding protein
NPC	neuronal progenitor cells
NS	nonstructural protein
IgM	immunoglobulin M
IRF	interferon receptor
PAHO	Pan-American Health Organization
p53	protein 53
PRNT	plaque neutralization test
RNA	ribonucleic acid
TAM	receptor tyrosine kinase TAM family
TLR3	receptor Toll-like 3
TYRO3	receptor tyrosine kinase
TIM	mucin domain
ZIKV	Zika virus

Introduction

Zika virus (ZIKV) belonging to the genus *Flavivirus* and *Flaviviridae* family is a single-stranded RNA virus with two noncoding regions (5′NCR and 3′NCR) and one open reading frame that encodes for a polyprotein that by proteolytic cleavage generates three structural proteins: capsid (C), premembrane/membrane (PrM), envelope (E), and seven nonstructural proteins: NS1, NS2A, NS2B, NS3, NS4, NS4B, and NS5 (Musso & Gublier, 2016).

The first phylogenetic study of ZIKV was based on NS5 region and revealed the circulation of three different ZIKV lineages: East African, West African, and Asian (Lanciotti et al., 2008). Viruses isolated in American countries were 99.8% similar to the Asian lineages, making it likely that the infection reached America from Asia (Zanluca et al., 2015).

The virus is mainly transmitted by domestic mosquitoes *Aedes aegypti* and *Aedes albopictus*. In tropical urban areas with a large population of *Aedes* sp., there is a high potential for Zika emergence. But since there are also areas of circulation of other arboviruses, the actual seroprevalence of ZIKV is difficult to determine (Ali et al., 2017).

In addition to transmission by vector mosquitoes, ZIKV can be transmitted transplacentally (Pomar et al., 2018), by blood transfusion, by bone marrow or organ transplantation (Musso & Gublier, 2016), and sexually (Musso et al., 2015).

Zika Virus Impact, Diagnosis, Control, and Models. https://doi.org/10.1016/B978-0-12-820267-8.00007-8

TABLE 1 Zika symptoms and differential diagnosis with other arboviruses.

Main symptoms	Zika	Dengue	Chikungunya
Fever	Low fever or no fever	High fever 4–7 days	High fever 2–5 days
Arthralgia	Mild pain +	Moderate pain ++	Severe pain +++
Rash	Appears in the first 24 h of illness	Appears on the third or fourth day of illness	Appears on the first or second day of illness
Itch	Mild	Mild to intense	Mild
Conjunctivitis	Often occurs	Rarely occurs	Rarely occurs

ZIKV infection may show no symptoms or nonspecific symptoms of fever, myalgia, and rash similar to other arboviruses (Table 1). However, it was the considerable increase in neurological manifestations following ZIKV infection that made this virus a public health problem around the world.

Epidemiology

ZIKV was first identified in monkeys in Uganda in 1947 through a network that monitored yellow fever (Dick, Kitchen, & Haddow, 1952). In a few years, there was reported seroprevalence of ZIKV antibodies from the human population in most of the African countries including Central Africa, Egypt, Gabon, Nigeria, Sierra Leone, and Tanzania (Moore et al., 1975; Smithburn, 1954; Weaver et al., 2016). First reports of ZIKV outside Africa were from Malaysia and Indonesia (Olson, Ksiazek, Suhandiman, & Triwibowo, 1981; Weaver et al., 2016). ZIKV remained confined to Africa and Asia causing sporadic outbreaks for six decades since its discovery.

The virus later spread over the Pacific and caused outbreak in Yap Islands, Federated States of Micronesia in Western Pacific, causing infection in approximately 75% of the total population in 2007 (Duffy et al., 2009) and French Polynesia in 2013–14 that resulted in over 30,000 cases of illness (Cao-Lormeau et al., 2014).

Between 2013 and 2015 the virus arrived to Americas most likely from the Pacific. In Brazil the first infection cases were reported from Rio Grande do Norte in early 2015 (Zanluca et al., 2015). First case of autochthonous transmission was reported from Bahia in May 15, and in the following months, 18 states of Brazil have reported autochthonous ZIKV transmission. This outbreak accounted for 440,000–1,300,000 suspected infection cases (Hennessey, Fischer, & Staples, 2016; Musso, Nilles, & Cao-Lormeau, 2014). In Brazil the considerable increase in children born with microcephaly during the Zika outbreak led researchers to investigate and confirm this association (de Araújo et al., 2016; Pomar et al., 2018).

ZIKV continues to develop and spread silently throughout the world in the form of asymptomatic infections. In 2018 the Pan-American Health Organization reported 1.003.509 suspected or confirmed cases of Zika in 52 countries or territories in the Americas ((PAHO) Pan American Health Organization, 2018).

Clinical manifestations

ZIKV most often causes an oligosymptomatic febrile exanthematic disease, similar to dengue fever (Table 1). The patient has low fever, headache, myalgia, conjunctive hyperemia, and micropapular rash that most commonly appear on the first day of fever (Table 1). The symptoms last from 3 to 5 days (Musso & Gublier, 2016).

However, if ZIKV infection occurs during pregnancy, it can cause congenital infection in 2%–26% of babies (Pomar et al., 2018). The risk of congenital infection and the severity of manifestations depend on the gestational period in which the mother contracted the infection (Marques et al., 2019; Ximenes et al., 2019). ZIKV infection has also been associated with Guillain-Barré syndrome (GBS), meningoencephalitis, myelitis, and ophthalmic changes in children and adults. Neurological manifestations often do not occur associated with classic symptoms of a ZIKV infection (Fig. 1) (Bentes, Kroon, & de Romanelli, 2019; Marinho et al., 2019).

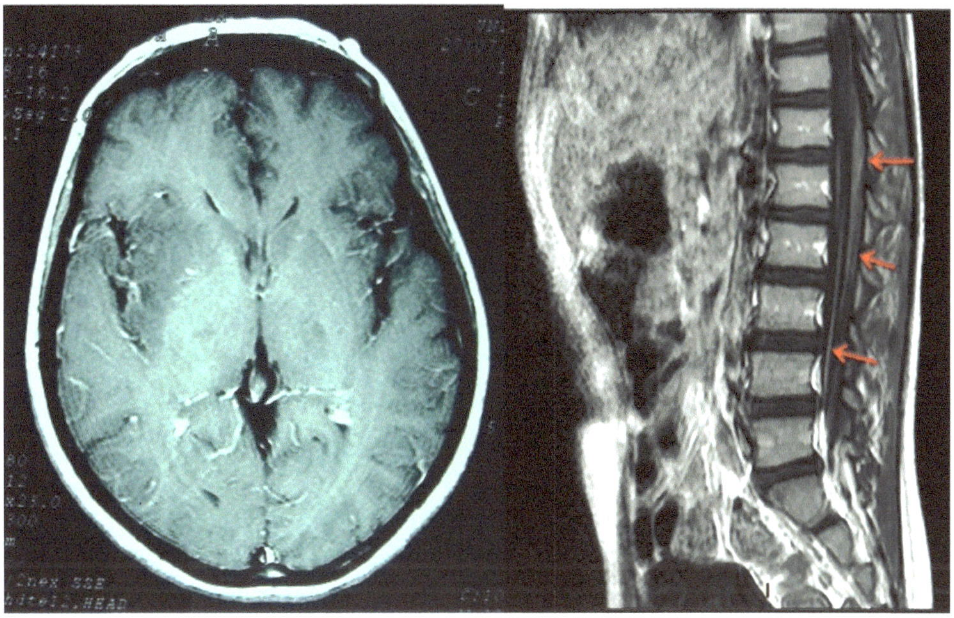

FIG. 1 (A) Skull tomography with edema, alteration commonly seen in viral encephalitis (UNICAMP website). (B) Inflammation and narrowing of the spinal cord roots in a child with myelitis and detection of CSF Zika virus. *(Image kindly provided by a patient from Dr. Aline Bentes.)*

Congenital Zika virus infection

Congenital Zika syndrome (CZS) is characterized by severe microcephaly, ocular lesions, and congenital contractures. CZS causes intrauterine growth restriction and several fetal malformations such as arthrogryposis, craniosynostosis, craniofacial malformations, and pulmonary hypoplasia (de Araújo et al., 2016).

Fetal loss occurs in approximately 5%–14% of pregnancies with documented Zika virus infection (Brasil et al., 2016; Pomar et al., 2018).

The stage of pregnancy at which the infection occurs is an important risk factor for the development of CZS. Severe destruction of brain occurs if ZIKV is acquired during the first trimester (Chimelli & Avvad-Portari, 2018), although the chance of mother's infection transmission to her baby is much lower in the first trimester (Ximenes et al., 2019).

Prospective study conducted during a Zika outbreak in French Guiana, in 2016, with a cohort of pregnant women with proven ZIKV infection, detected maternal-fetal transmission in 26% of infants. Among those infected, 45% had no symptoms, 20% developed moderate CZS symptoms, 21% with severe complications, and 14% with fetal death (Pomar et al., 2018).

Ventriculomegaly commonly seen in congenital Zika infection may be caused by mesencephalic atresia, aqueduct stenosis, or cerebral cortex hypoplasia (ex vacuo). In this syndrome, frequently important brain structures do not develop such as corpus callosum, hippocampus, thalamus, and even the brainstem, which can be seen on fetal ultrasounds (Marques et al., 2019). However, when the infection occurs in the last trimester of pregnancy, the brain has been macroscopically formed, but white matter calcifications are common (Marques et al., 2019).

The baby's brain with CZS has impaired at some point during pregnancy disrupting proper development. Studies evaluating autopsies of infants with CZS and life-threatening severe hypoplastic brain lesions detected apoptosis and degeneration of nerve cells with disordered migration and absence of descending fibers in the cerebral hemispheres, cerebellum, and brainstem (Chimelli et al., 2017; Chimelli & Avvad-Portari, 2018).

Children with congenital infection have lifelong neurological complications such as seizures, developmental delay, intellectual disability, hearing loss, and vision problems (Marques et al., 2019).

Neurological complications associated with Zika infection

GBS is an autoimmune condition that destroys myelinated neural cells and leads to gradual muscle weakness, reduced nerve function, or even paralysis (Sebastián et al., 2017). During the outbreak of ZIKV in 2013–14 in French Polynesia, the incidence rate of GBS was increased 20-fold (Cao-Lormeau et al., 2014). A study in seven different countries found an increase in the number of GBS following Zika outbreaks ranging from 2% to 9.8% (Counotte et al., 2018).

TABLE 2 Neurological manifestations in children and adolescents associated with zika virus infection.

Neurological manifestations	Diagnostic method	Outcomes/complication	Reference/country
Recent memory loss, slowed speech	Urine and CSF PCR+	Symptoms persisted for 15 weeks	Zucker et al., 2017, United States
Myelitis, hemiparesis	CSF, blood, and urine PCR+	Weakness in the legs for 1 month	Mécharles et al., 2016, Guadalupe Island
Cortical infarction with hemiparesis	Serum PCR+	Arm weakness for 3 months	Landais et al., 2017, Guadalupe Island
Guillain-Barré	Serum IgM and PRNT CSF+	So sequelae after 4 months	Raboni et al., 2017, Brazil
Hemiparesis with encephalitis	CSF PCR+	Walking disability for 1 year	Marinho et al., 2019, Brazil

Neurological complications reported in children after nonneonatal ZIKV infection have been associated with encephalitis and myeloradiculopathy (Marinho et al., 2019), hemiparesis, myelitis (Mécharles et al., 2016), GBS (Raboni et al., 2017), cortical infarction (Landais et al., 2017), and behavioral changes (Zucker et al., 2017) (Table 2).

A prospective study that we made evaluating 278 children with neurological manifestations between 2015 and 2018 found 13 children (4.7%) infected with ZIKV as confirmed by CSF PCR. The neurological manifestations presented by these children in this cohort was encephalitis (54%), seizures (38.5%), meningitis (23%), bulging fontanel (15.4%); motor disorders such as hypotonia, walking disability, and neuropathic pain (15.4%); and muscle weakness, paresis, or ataxia (7.7%).

In clinical follow-up of these children, 30% had some type of neurological complications, 23% were discharged with anticonvulsants due to electroencephalogram abnormalities or neuroimaging examination, and 15.4% evolved with learning disorders and memorization difficulties.

Neuropathogenesis

ZIKV enter host cells after interaction between viral envelope glycoproteins and cell surface receptors. Some receptors were already described as facilitating entry of ZIKV into human cells as TAM family tyrosine kinase receptors (TYRO3, AXL, and MER), T-cell immunoglobulin, C-type lecithin receptors, phosphatidylserine receptors, and mucin domain proteins (TIM) (Li, Saucedo-Cuevas, Regla-Nava, et al., 2016; Li, Saucedo-Cuevas, Shresta, & Gleeson, 2016).

The viral genome transduction experiments using CRISPR/Cas9 identified that the ZIKV nonstructural proteins NS4A and NS4B can suppress neurogenesis and induce neuronal cell autophagy (Liang et al., 2016). Neurobiological studies have identified that neuronal progenitor cells (NPCs) infected with ZIKV activate genes associated with innate antiviral response, such as TLR3, p53, and IRF, which consequently induce these cells to apoptosis which causes neuronal death (Grant et al., 2016; Li, Saucedo-Cuevas, Regla-Nava, et al., 2016; Li, Saucedo-Cuevas, Shresta, et al., 2016; Tang et al., 2016). As NPCs differentiate into mature brain cells and this process of differentiation and maturation continues throughout childhood and adulthood, this is possibly one of the mechanisms for neurological complications following Zika infection.

Studies with ZIKV-infected mice observed neurological manifestations in the animals similar to meningoencephalitis symptoms in human patients with reduced proliferation and death in stem cell niches, demonstrating that not only embryonic neuronal cells are susceptible to ZIKV infection. The brain areas of mice most affected by ZIKV infection were cellular niches responsible for brain plasticity, learning, and memory (Li, Saucedo-Cuevas, Regla-Nava, et al., 2016; Li, Saucedo-Cuevas, Shresta, et al., 2016). Prospective studies of patients with neurological manifestations after ZIKV infection are still needed for assessment of psychomotor development, memory, and learning.

Diagnosis

The specific laboratory diagnosis of ZIKV infection is mainly based on the detection of viral RNA present in blood from the first day of clinical onset to around 5 days (Fig. 2) (Landry & St George, 2017). But viremia was detected until day 10 (Musso & Gublier, 2016). In children with neurological manifestations after ZIKV infection, viral RNA was detected

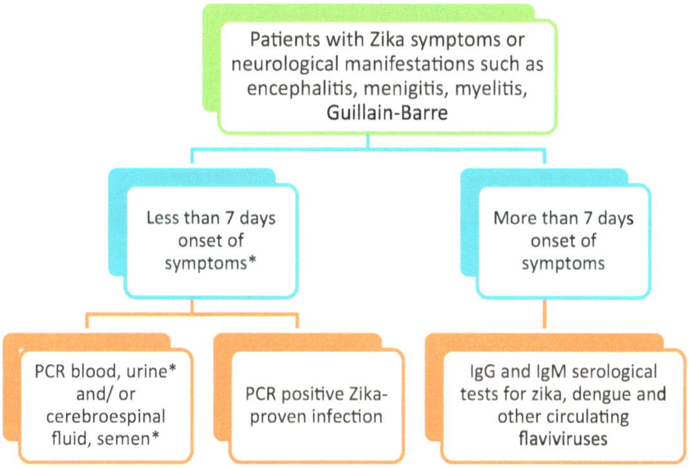

FIG. 2 Diagnostic flowchart of Zika virus infection. *In semen and urine, ZIKV was detected until 30 days. In CSF, ZIKV was detected until 11 days.

in cerebrospinal fluid (CSF) between 1 and 11 days from the onset of symptoms (Musso & Gublier, 2016). Viral RNA can also be detected in urine for more than 10 days and in semen for several weeks (Musso et al., 2015).

However, many ZIKV infections are asymptomatic or oligosymptomatic, and thus, for diagnosis of autoimmune post-infection manifestations or pregnancy, diagnostic tests based on antibody detection are required. Some commercial anti-Zika IgM antibody detection kits are currently available, such as ZIKV MAC-ELISA and ZIKV Detect IgM capture ELISA. All these kits are highly sensitive but lack specificity due to cross-reactivity with other flaviviruses such as dengue virus and yellow fever virus (Landry & St George, 2017).

Rapid tests using recombinant monoclonal antibodies to NS1 nonstructural protein sequences of the four dengue virus serotypes and ZIKV are in the final phase of validation for commercialization. These rapid tests did not show cross-reaction with a sensitivity and specificity of 0.76 and 0.89 for DENV1; 0.89 and 0.98 for DENV2; 1 and 1 for DENV3; 1 and 0.96 for DENV4; 0.88 and 1 for the pan-DENV; and 0.81 and 0.86 for ZIKV (Bosch et al., 2019).

The plaque reduction neutralization test (PRNT) can be used to confirm the presence of specific anti-ZIKV IgM antibodies. PRNT assays are not generally performed in routine clinical laboratory tests but may be available through state and national health departments for confirmatory diagnosis (Rabe et al., 2016). The PRNT test takes about 14 weeks, so it is not very useful in clinical practice, only in research. It is very important that rapid and specific diagnostic tests be developed to assist clinicians in detecting ZIKV infections (Landry & St George, 2017).

Nonspecific alterations may occur during ZIKV infection such as leukopenia, thrombocytopenia, and mild elevation of transaminases, lactate dehydrogenase, gamma-glutamyl transferase, and markers of inflammatory activity such as C-reactive protein, fibrinogen, and ferritin (Musso & Gublier, 2016).

Management

There are no specific medications to treat ZIKV infection yet, but for infants with confirmed or probable CZS, important aspects of follow-up include monitoring growth parameters and development, providing routine immunizations, and psychosocial support. All the babies should be referred to a neurologic, ophthalmologic, and audiology examination. Referrals for early intervention services should be made. Infant feedings should be monitored, and if there are signs of swallowing dysfunction, an assessment for dysphagia should be performed (Brasil et al., 2016; Pomar et al., 2018).

Children with postnatal ZIKV infection and neurological manifestations should be accompained by neurologists with assessments of development, memory, and learning.

Prevention

For all the mosquito-borne diseases, integrated vector management is the most effective approach to prevent and control arboviruses disease like ZIKV (Fig. 3). In ZIKV-endemic areas, systematic mosquito control programs should be carried out by public health organizations such as community sanitation programs (Ali et al., 2017; Hotez, 2016).

FIG. 3 Prevention *Aedes aegypti* proliferation. Basic sanitation and piped water are the main programs to control the proliferation of *Aedes aegypti*. *(Personal archive images Aline Bentes).*

Insecticides and environment management practices like destroying their breeding sites, removing stagnant water bodies, and spraying larvicides and insecticides are necessary to control the mosquito population (Weaver et al., 2016).

Mosquito bites should be avoided by using mosquito repellent, using permethrin cream for treating cloths or skin, or using bed nets and window screens. Mosquito repellents like N,N-diethyl-m-toluamide (DEET) have been reported as safe for use in pregnant and nursing women and in the children above the age of two months (Hennessey et al., 2016; Weaver et al., 2016) (Fig. 4).

Sexual transmission of ZIKV infections has been reported, and safe sex practice should be followed (Musso et al., 2015). Healthcare workers and laboratory technicians should wear protective equipment when handling biological materials from Zika patients (Rather, Kumar, Bajpai, Lim, & Park, 2017).

There are different anti-Zika vaccines being researched and developed including DNA-based, live-attenuated, inactivated, and subunit vaccines but are still in the process of evaluating effectiveness (Garg, Sedano, Plata, Punke, & Joshi, 2017; Shan et al., 2017).

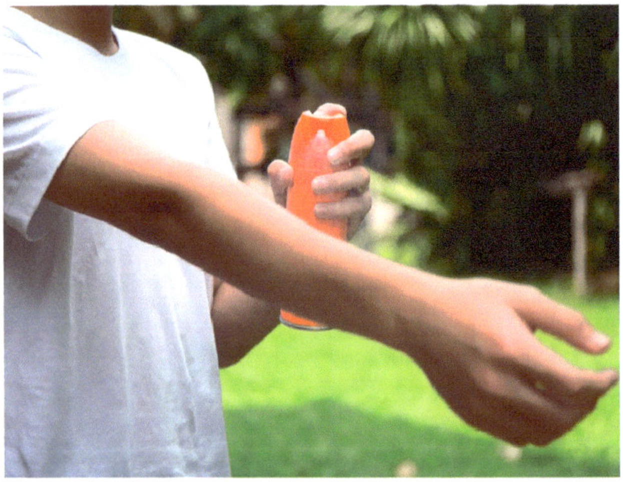

FIG. 4 Repellents should be applied to clothing and exposed areas of the body when playing and walking in open areas. Never apply the product in the hands of children, as they can contaminate eyes and mouth. *(Image from: idec.org.br).*

Mini-dictionary of terms

1. **Arboviruses**: Arthropod-borne viruses.
2. **Arthrogryposis**: Severe congenital deformity of joints and limbs characterized by marked stiffness.
3. **Ataxia**: Change in balance or walk caused by inflammation or injury to the cerebellum.
4. **Craniosynostosis**: Early closure of the cranial sutures that does not allow proper brain growth.
5. **Encephalitis**: Inflammation of the cerebral parenchyma that causes neurological manifestations that persist for at least 24 h.
6. **Meningitis**: Inflammatory process of the meningeal membranes lining the brain. This inflammation may be aseptic or caused by an infectious pathogen such as virus, bacteria, or fungus.
7. **Microcephaly**: Term used to describe an abnormally small head, an indicator of reduced brain growth. To characterize microcephaly the head circumference must be less than two standard deviations from the normal mean for age and sex.
8. **Myelitis**: Inflammation of the spinal cord and nerve roots.
9. **Ventriculomegaly**: Enlargement of the cerebral ventricles that can be caused by hypoplasia or atresia, degeneration, ischemia, or infection.

Key facts

- There has been a considerable increase in ZIKV transmission since 2007, reaching more than 70 countries, according to the World Health Organization.
- To control Zika outbreaks, it is critical to implement public policies to reduce the vector population.
- Rapid and specific diagnostic tests can greatly assist the care of children with neurological manifestations following ZIKV infection.

Summary points

- Transplacental transmission of Zika virus from the infected mother to the baby may cause different malformations in the fetus according to the gestational period in which the infection occurred.
- Zika virus infections outside the neonatal period can also cause neurological manifestations such as Guillain-Barré syndrome, myelitis, meningoencephalitis, and behavioral change.
- There is no specific treatment for ZIKV infection, but all children should be followed for a long time to evaluate and rehabilitate possible neurological complications.

References

Ali, S., Gugliemini, O., Harber, S., Harrison, A., Houle, L., Ivory, J., … Mordecai, E. A. (2017). Environmental and social change drive the explosive emergence of Zika virus in the Americas. *PLoS Neglected Tropical Diseases*, *11*, 1–16. https://doi.org/10.1371/journal.pntd.0005135.

Bentes, A. A., Kroon, E. G., & de Romanelli, R. M. C. (2019). Neurological manifestations of pediatric arboviral infections in the Americas. *Journal of Clinical Virology*, *116*, 49–57. https://doi.org/10.1016/j.jcv.2019.04.006.

Bosch, I., Puig, H., Hiley, M., Carré-Camps, M., Pordomi-Celis, F., … Gehrke, L. (2019). Rapid antigen tests for dengue virus serotypes and Zika virus in patient serum. HHS Public Access. *Science Translational Medicine*, *176*, 139–148. https://doi.org/10.1016/j.physbeh.2017.03.040.

Brasil, P., Pereira, J. P., Moreira, M. E., Jr., Ribeiro Nogueira, R. M., Damasceno, L., Wakimoto, M., … Nielsen-Saines, K. (2016). Zika virus infection in pregnant women in Rio de Janeiro. *The New England Journal of Medicine*, *375*, 2321–2334. https://doi.org/10.1056/NEJMoa1602412.

Cao-Lormeau, V. M., Roche, C., Teisser, E. R., Berry, A.-L., Mallet, H.-P., … Musso, D. (2014). Zika virus, French Polynesia, South Pacific, 2013. *Emerging Infectious Diseases*, *20*, 1085–1086. https://doi.org/10.3201/eid2011.141380.

Chimelli, L., & Avvad-Portari, E. (2018). Congenital Zika virus infection: A neuropathological review. *Child's Nervous System*, *34*, 95–99. https://doi.org/10.1007/s00381-017-3651-3.

Chimelli, L., Melo, A. S. O., Avvad-Portari, E., Wiley, C. A., Camacho, A. H. S., Lopes, V. S., … Tanuri, A. (2017). The spectrum of neuropathological changes associated with congenital Zika virus infection. *Acta Neuropathologica*, *133*, 983–999. https://doi.org/10.1007/s00401-017-1699-5.

Counotte, M. J., Egli-Gany, D., Riesen, M., Abraha, M., Porgo, T. V., Wang, J., & Low, N. (2018). Zika virus infection as a cause of congenital brain abnormalities and Guillain-Barré syndrome: From systematic review to living systematic review [version 1; referees: 2 approved, 1 approved with reservations]. *F1000Research*, *7*, 1–20. https://doi.org/10.12688/f1000research.13704.1.

de Araújo, T. V. B., Rodrigues, L. C., de Alencar Ximenes, R. A., de Barros Miranda-Filho, D., Montarroyos, U. R., de Melo, A. P. L., … Martelli, C. M. T. (2016). Association between Zika virus infection and microcephaly in Brazil, January to May, 2016: preliminary report of a case-control study. *The Lancet Infectious Diseases*, *16*, 1356–1363. https://doi.org/10.1016/S1473-3099(16)30318-8.

Dick, G. W. A., Kitchen, S. F., & Haddow, A. J. (1952). Zika virus. Isolation and serological specificity. *Transactions of the Royal Society of Tropical Medicine and Hygiene*, *s7-V*(116), 509–520. https://doi.org/10.1093/nq/s7-v.116.219c.

Duffy, M. R., Chen, T. H., Hancock, W. T., Powers, A. M., Kool, J. L., Lanciotti, R. S., ... Hayes, E. B. (2009). Zika virus outbreak on Yap Island, Federated States of Micronesia. *New England Journal of Medicine*, *360*, 2536–2543. https://doi.org/10.1056/NEJMoa0805715.

Garg, H., Sedano, M., Plata, G., Punke, E. B., & Joshi, A. (2017). Development of virus-like-particle vaccine and reporter assay for Zika virus. *Journal of Virology*, *91*(20). https://doi.org/10.1128/jvi.00834-17.

Grant, A., Ponia, S. S., Tripathi, S., Balasubramaniam, V., Miorin, L., Sourisseau, M., ... García-Sastre, A. (2016). Zika virus targets human STAT2 to inhibit type i interferon signaling. *Cell Host and Microbe*, *19*, 882–890. https://doi.org/10.1016/j.chom.2016.05.009.

Hennessey, M., Fischer, M., & Staples, J. E. (2016). Zika virus spreads to new areas—Region of the Americas, May 2015-January 2016. *American Journal of Transplantation*, *16*, 1031–1034. https://doi.org/10.1111/ajt.13743.

Hotez, P. J. (2016). What does Zika virus mean for the children of the Americas? *JAMA Pediatrics*, *170*, 787–789. https://doi.org/10.1001/jamapediatrics.2016.1465.

Lanciotti, R. S., Kosoy, O. L., Laven, J. J., Velez, J. O., Lambert, A. J., Johnson, A. J., ... Duffy, M. R. (2008). Genetic and serologic properties of Zika virus associated with an epidemic, Yap State, Micronesia, 2007. *Emerging Infectious Diseases*, *14*, 1232–1239. https://doi.org/10.3201/eid1408.080287.

Landais, A., Césaire, A., Fernandez, M., Breurec, S., Herrmann, C., Delion, F., & Desprez, P. (2017). Zika vasculitis: A new cause of stroke in children? *Journal of the Neurological Sciences*, *383*, 211–213. https://doi.org/10.1016/j.jns.2017.10.045.

Landry, M. L., & St George, K. (2017). Laboratory diagnosis of Zika virus infection. *Archives of Pathology and Laboratory Medicine*, *141*, 60–67. https://doi.org/10.5858/arpa.2016-0406-SA.

Li, H., Saucedo-Cuevas, L., Regla-Nava, J. A., Chai, G., Sheets, N., Tang, W., ... Gleeson, J. G. (2016). Zika virus infects neural progenitors in the adult mouse brain and alters proliferation. *Cell Stem Cell*, *19*, 593–598. https://doi.org/10.1016/j.stem.2016.08.005.

Li, H., Saucedo-Cuevas, L., Shresta, S., & Gleeson, J. G. (2016). The neurobiology of Zika virus. *Neuron*, *92*, 949–958. https://doi.org/10.1016/j.neuron.2016.11.031.

Liang, Q., Luo, Z., Zeng, J., Chen, W., Foo, S. S., Lee, S. A., ... Jung, J. U. (2016). Zika virus NS4A and NS4B proteins deregulate Akt-mTOR signaling in human fetal neural stem cells to inhibit neurogenesis and induce autophagy. *Cell Stem Cell*, *19*, 663–671. https://doi.org/10.1016/j.stem.2016.07.019.

Marinho, P. E. S., Alvarenga, P. P. M., Lima, M. T., de Souza Andrade, A., Candiani, T. M. S., Crispim, A. P. C., ... Kroon, E. G. (2019). Central and peripheral nervous system involvement in Zika virus infection in a child. *Journal of NeuroVirology*, *25*, 893–896. https://doi.org/10.1007/s13365-019-00770-x.

Marques, V.d. M., Santos, C. S., Santiago, I. G., Marques, S. M., Nunes Brasil, M.d. G., Lima, T. T., & Costa, P. S. (2019). Neurological complications of congenital Zika virus infection. *Pediatric Neurology*, *91*, 3–10. https://doi.org/10.1016/j.pediatrneurol.2018.11.003.

Mécharles, S., Herrmann, C., Poullain, P., Tran, T. H., Deschamps, N., Mathon, G., ... Lannuzel, A. (2016). Acute myelitis due to Zika virus infection. *The Lancet*, *387*, 1481. https://doi.org/10.1016/S0140-6736(16)00644-9.

Moore, D. L., Causey, O. R., Carey, D. E., Reddy, S., Cooke, A. R., Akinkugbe, F. M., ... Kemp, G. E. (1975). Arthropod-borne viral infections of man in nigeria, 1964–1970. *Annals of Tropical Medicine and Parasitology*, *69*, 49–64. https://doi.org/10.1080/00034983.1975.11686983.

Musso, D., Nilles, E. J., & Cao-Lormeau, V. M. (2014). Rapid spread of emerging Zika virus in the Pacific area. *Clinical Microbiology and Infection*, *20*, 595–596. https://doi.org/10.1111/1469-0691.12707.

Musso, D., Roche, C., Robin, E., Nhan, T., Teissier, A., & Cao-Lormeau, V. M. (2015). Potential sexual transmission of Zika virus. *Emerging Infectious Diseases*, *21*, 509–520. https://doi.org/10.3201/eid2102.141363.

Musso, D., & Gublier, D. J. (2016). Zika virus. *Clinical Microbiology Reviews*, *29*, 487–524. https://doi.org/10.1128/CMR.00072-15. Address.

Olson, J. G., Ksiazek, T. G., Suhandiman, G., & Triwibowo, V. (1981). Zika virus, a cause of fever in central java, indonesia. *Transactions of the Royal Society of Tropical Medicine and Hygiene*, *75*, 389–393. https://doi.org/10.1016/0035-9203(81)90100-0.

(PAHO) Pan American Health Organization. (2018). *Zika cases and congenital syndrome associated with Zika virus reported by countries and territories in the Americas, 2015–2018. Cumulative cases.* Pan American Health Organization.

Pomar, L., Vouga, M., Lambert, V., Pomar, C., Hcini, N., Jolivet, A., ... Baud, D. (2018). Maternal-fetal transmission and adverse perinatal outcomes in pregnant women infected with Zika virus: Prospective cohort study in French Guiana. *BMJ (Online)*, 363. https://doi.org/10.1136/bmj.k4431.

Rabe, I. B., Staples, J. E., Villanueva, J., Hummel, K. B., Johnson, J. A., Rose, L., ... Powers, A. M. (2016). Interim guidance for interpretation of Zika virus antibody test results. *Morbidity and Mortality Weekly Report*, *65*, 543–546. https://doi.org/10.15585/mmwr.mm6521e1.

Raboni, S. M., Bonfim, C., Almeida, B. M., Zanluca, C., Koishi, A. C., Rodrigues, P. R. V. P., ... Duarte dos Santos, C. N. (2017). Flavivirus cross-reactivity in serological tests and Guillain-Barré syndrome in a hematopoietic stem cell transplant patient: A case report. *Transplant Infectious Disease*, *19*(4). https://doi.org/10.1111/tid.12700.

Rather, I. A., Kumar, S., Bajpai, V. K., Lim, J., & Park, Y. H. (2017). Prevention and control strategies to counter Zika epidemic. *Frontiers in Microbiology*, *8*, 1–8. https://doi.org/10.3389/fmicb.2017.00305.

Sebastián, U. U., Ricardo, A. V. A., Alvarez, B. C., Cubides, A., Luna, A. F., Arroyo-Parejo, M., ... Marshall, J. C. (2017). Zika virus-induced neurological critical illness in Latin America: Severe Guillain-Barre syndrome and encephalitis. *Journal of Critical Care*, *42*, 275–281. https://doi.org/10.1016/j.jcrc.2017.07.038.

Shan, C., Muruato, A. E., Jagger, B. W., Richner, J., Nunes, B. T. D., Medeiros, D. B. A., ... Shi, P. Y. (2017). A single-dose live-attenuated vaccine prevents Zika virus pregnancy transmission and testis damage. *Nature Communications*, *8*(1). https://doi.org/10.1038/s41467-017-00737-8.

Smithburn, K. C. (1954). Neutralizing antibodies against arthropod-borne viruses in the sera of long-time residents of Malaya and borneo. *American Journal of Epidemiology*, *59*, 157–163. https://doi.org/10.1093/oxfordjournals.aje.a119630.

Tang, H., Hammack, C., Ogden, S. C., Wen, Z., Qian, X., Li, Y., … Ming, G. L. (2016). Zika virus infects human cortical neural progenitors and attenuates their growth. *Cell Stem Cell*, *18*, 587–590. https://doi.org/10.1016/j.stem.2016.02.016.

Weaver, S. C., Costa, F., Garcia-blanco, M. A., Ko, A. I., Ribeiro, G. S., Saade, G., … Vasilakis, N. (2016). Zika virus: History, emergence, biology, and prospects for control. *Antiviral Research*, *130*, 69–80. https://doi.org/10.1016/j.antiviral.2016.03.010. Zika.

Ximenes, R., Ramsay, L. C., Miranda, R. N., Morris, S. K., Murphy, K., & Sander, B. (2019). Health outcomes associated with Zika virus infection in humans: A systematic review of systematic reviews. *BMJ Open*, *9*, 1–11. https://doi.org/10.1136/bmjopen-2019-032275.

Zanluca, C., De Melo, V. C. A., Mosimann, A. L. P., Dos Santos, G. I. V., dos Santos, C. N. D., & Luz, K. (2015). First report of autochthonous transmission of Zika virus in Brazil. *Memorias Do Instituto Oswaldo Cruz*, *110*, 569–572. https://doi.org/10.1590/0074-02760150192.

Zucker, J., Neu, N., Chiriboga, C. A., Hinton, V. J., Leonardo, M., Sheikh, A., & Thakur, K. (2017). Zika virus-associated cognitive impairment in adolescent, 2016. *Emerging Infectious Diseases*, *23*, 1047–1048. https://doi.org/10.3201/eid2306.162029.

Chapter 8

Use of induced pluripotent stem cells and cerebral organoids to profile Zika virus infection: Features and findings

Fernanda Majolo[a,b], Daniel Rodrigo Marinowic[a], Pamella Nunes Azevedo[a], Guilherme Liberato da Silva[c], Denise Cantarelli Machado[a], and Jaderson Costa DaCosta[a]

[a] *Brain Institute of Rio Grande do Sul (BraIns), Medical School, Pontifical Catholic University of Rio Grande do Sul, Rio Grande do Sul, Brazil,*
[b] *Post-Graduate Program in Biotechnology, Universidade do Vale do Taquari—Univates, Lajeado, Rio Grande do Sul, Brazil,*
[c] *Medical Science Center, Universidade do Vale do Taquari—Univates, Lajeado, Rio Grande do Sul, Brazil*

Introduction

This chapter aims to address two main types of disease modeling tools, namely, induced pluripotent stem cells (iPSCs) and cerebral organoids, associated with the several findings applied to Zika virus infection. We will highlight the progress made so far, as well as the prospects in this area of research.

The ZIKV flavivirus transmitted by dipterous mosquitoes has spread across the globe, comprising more than 70 countries (CDC, 2016; Heukelbach, Alencar, Kelvin, de Oliveira, & de Góes Cavalcanti, 2016; Qian, Nguyen, Jacob, Song, & Ming, 2017). ZIKV infection is widely known to be related with the increased incidence of newborns with microcephaly, a condition in which babies are born with an abnormally small head, with about 80% of infections being asymptomatic (Ming, Tang, & Song, 2016; Qian et al., 2017). After the release of the World Health Organization (WHO), ranking ZIKV a global emergency, several microcephalic fetal tissues were evaluated, showing, especially in infected women during pregnancy, brain damage in those fetuses (Driggers et al., 2016; Heymann et al., 2016; Mlakar et al., 2016; Ventura et al., 2016). The Centers for Disease Control and Prevention (CDC), in recent years, confirms the real evidence of ZIKV as causing microcephaly and severe fetal brain damage (CDC, 2016; Ming et al., 2016; Qian et al., 2017).

Researchers had several questions about some patterns of occurrence, causes, long-term consequences, and the mechanism of ZIKV transplacental invasion. Such questions fostered epidemiological and animal model studies (primates and nonprimates) (Platt & Miner, 2017). However, the difficulty in working with human living tissues and the complications due to variability in *postmortem* tissue quality have directed their efforts in new study models, such as production of cerebral organoids from stem cells enabling studies on cell tropism and pathogenic mechanisms related to ZIKV (Ming et al., 2016; Qian et al., 2017).

But what would such stem cells be? They are characterized as specialized cells that have the ability to differentiate into other cell types (Ming et al., 2016; Weissman, 2000). When used in certain organs, the stem cells produce descendants that have the ability to maintain tissue homeostasis and also cells with the same functions as those that were not generated from this differentiation (Bond, Ming, & Song, 2015; Ming et al., 2016). Such class of cells represents a revolution within this study area, enabling its use in patients with various disorders.

A major milestone for science was the development of iPSCs. These cells are cultivated in monolayer (Tao & Zhang, 2016) and will be discussed later in text. However, the big step has been taken with the discovery of organ-like tissues, called organoids, derived from these iPSCs (Clevers, 2016; Ming et al., 2016). They are three-dimensional (3D) tissue structures derived from self-assembling stem cells, thus producing a morphology similar to the developing human brain, in the case of the cerebral organoids (Jo et al., 2016; Kadoshima et al., 2013a, 2013b; Lancaster et al., 2013a, 2013b; Mariani et al., 2015; Paşca et al., 2015; Qian et al., 2016, 2017) (Fig. 1).

Cerebral organoids can be infected by different strains of ZIKV, allowing deepening of the pathogenic mechanisms studies. In these cases, it was proved that the virus targeted the cortical progenitor, inducing apoptosis and autophagy, mainly from V/VI layers, thus triggering microcephaly (Mansour et al., 2018; Setia & Muotri, 2019). This type of discovery

Zika Virus Impact, Diagnosis, Control, and Models. https://doi.org/10.1016/B978-0-12-820267-8.00008-X

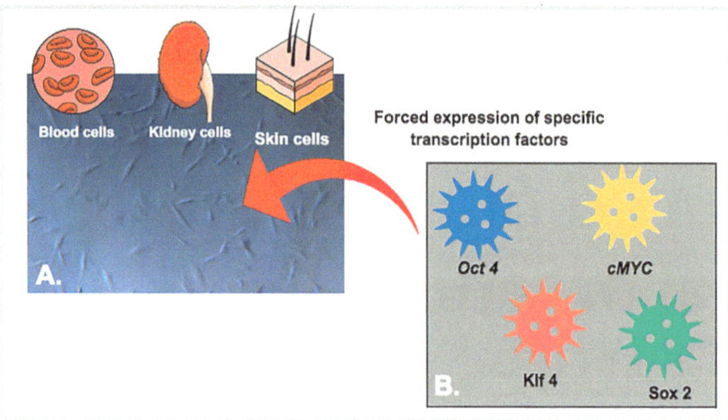

FIG. 1 Representation of the cell reprogramming technique discovered in 2006 by scientists Takahashi and Yamanaka. (A) Fibroblast culture with approximately 70% confluence before viral transfection (increase 200 ×). (B) The reprogramming of differentiated cells in iPSCs occurs through the introduction and forced expression of specific transcription factors, for example, *Oct3/4*, *Klf4*, *Sox2*, and *c-Myc*, in somatic cells, a process called reprogramming. Dermal fibroblasts are the primary source of somatic cells most used, although it is also feasible to use less invasive sources such as hematopoietic cells or urine-derived kidney epithelial cells. *(Image by authors/https://mindthegraph.com/.)*

would be impossible if it were tested in rodents because they have a different gestational period from humans. Thus this study model allows us to deepen studies on the efficacy of future treatments against ZIKV (Mesci et al., 2018; Setia & Muotri, 2019).

Those new study models substitute mouse experiments, facilitating the modeling of various human diseases. For example, diseases with high genetic diversity when studied in animals would be practically impossible to model (Chen, Song, & Ming, 2019; Lizarraga et al., 2010; Pulvers et al., 2010; Qian et al., 2017; Setia & Muotri, 2019; Takahashi et al., 2007; Wen, Christian, Song, & Ming, 2016).

With the advent of the application of multitechnologies in stem cells, there are several potentialities in health, such as disease modeling, regenerative therapy, development of new therapies, toxicity, and new drug tests. Personalized cell therapies pave the way for greater understanding of tissues and organs from patients suffering the most diverse disorders and would be possible to reconstruct regions from human brain damaged by disorders (Platt & Miner, 2017; Wiegand & Banerjee, 2019). They provide specific results that assist to verify drug efficacy and other therapeutic interventions, making it a potentially transformative substrate for brain circuit reconstruction after injury. With this advance in the therapeutic line, it will be possible that in the coming decades doctors will use brain organoids from their patients serving as a diagnostic tool or even to test various medications and different doses before prescribing a prescription (Chen et al., 2019; Platt & Miner, 2017).

Experimental models to study neurological diseases

Induced pluripotent stem cells

Initially, it was believed that cell survival orientation could only occur in one direction, from a state of cellular immaturity to a mature, differentiated state. However, a number of important experiments presented changes in the concept related to the fate of cells. This mature final outcome can be flexible and reversible, and indeed, cells can move from a differentiated state to a pluripotent state through techniques that promote cell reprogramming.

In 2006 researchers Kazutoshi Takahashi and Shinya Yamanaka accomplished an unprecedented feat, breaking a paradigm of cell biology by promoting an undifferentiated state in adult somatic cells by transfecting genes that are expressed only during the embryonic period (Figs. 1 and 2). The cells produced by this method were called iPSCs, being extremely similar to embryonic stem cells, presenting the same self-renewal characteristics and differentiation potential in ectoderm, mesoderm, and endoderm cells (Fig. 3). For this, 24 transcription factors selected for maintaining the "stem state" present in all embryonic tissues were selected. These transcription factors are "turned off" in adult cells. Through the gene transfection technique, different combinations of the 24 candidate genes were inserted into the somatic cells, resulting in a combination of four genes, being simultaneously expressed in the cells. They are able to promote the cellular undifferentiating state with plasticity and behaviors again, similar to embryonic stem cells (Takahashi et al., 2007; Takahashi & Yamanaka, 2006).

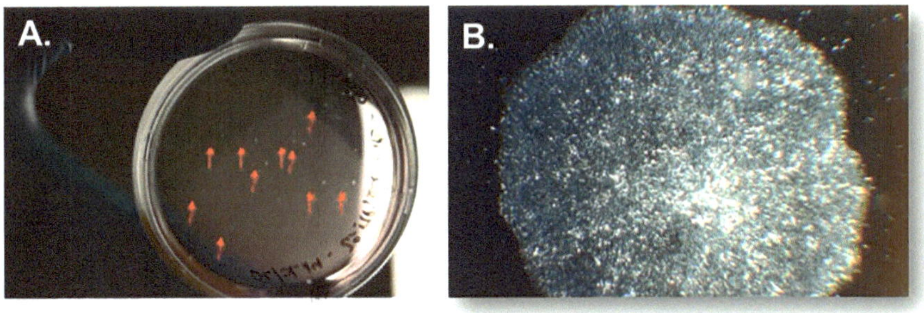

FIG. 2 Induced pluripotent stem cell culture. (A) Image of the macroscopic view of the generated clones (iPSCs) grown in a 60 mm^2 plate on Matrigel. (A) Arrows indicate clones of iPSCs. (B) Clone without differentiation areas (40× magnification). *(Image by authors.)*

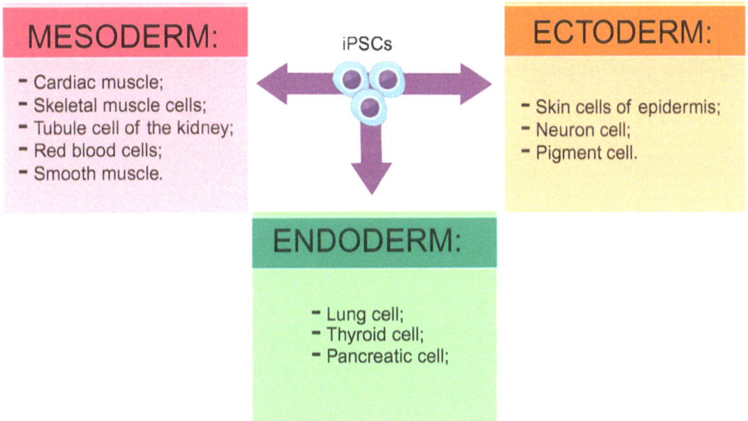

FIG. 3 Illustrative scheme of the iPSC differentiation potential. They are similar to embryonic stem cells, presenting the same characteristics of self-renewal and differentiation potential into the three embryonic leaflets. *(Image by authors/https://mindthegraph.com/.)*

The use of iPSC technology is underpinned by one major premise: one can now have an unlimited supply of reprogrammed autologous pluripotent cells in a culture dish and transform them into any type of adult cell. This new ability has profoundly changed the approach to the study of human development biology. The use of iPSCs for the study of pathogenic changes has been emphasized with great importance over the years, since the study of several of these diseases can still be considered a major challenge for researchers. The iPSCs provided a unique platform for better understanding of the pathophysiology, mechanisms of action, identification of therapeutic targets, testing of new drugs, and new therapeutic alternatives for the treatment of various pathologies (Dolmetsch & Geschwind, 2011; Dragunow, 2008) (Fig. 4).

iPSCs X Zika virus

Congenital Zika virus infection during pregnancy has been associated with distinct fetal anomalies. The greatest risk of severe fetal sequelae occurs within first and second gestational trimester infection and may occur among pregnant women who have symptoms of the disease and are asymptomatic (Dasgupta et al., 2016).

iPSCs generated from human fibroblasts can simulate the embryonic neurodevelopment environment, thus providing the possibility of investigating this process from a pluripotent cell with embryonic characteristics to the formation of an adult neuron, passing through all possible stages of maturation and differentiation during brain formation. These cells can be exposed to environmental factors potentially causing some kind of change during brain formation, thus allowing a temporal analysis of cytotoxicity, gene expression, aberrant cell behavior, and protein synthesis mediated by the candidate agent to cause a phenotypic change, simulating, for example, a viral infection during pregnancy (Dolmetsch & Geschwind, 2011).

The use of iPSC as a model for the study of congenital infections and its consequent neuropathologies has increased significantly; however, there is still a shortage, and many new studies need to be produced (Majolo, Marinowic, Moura,

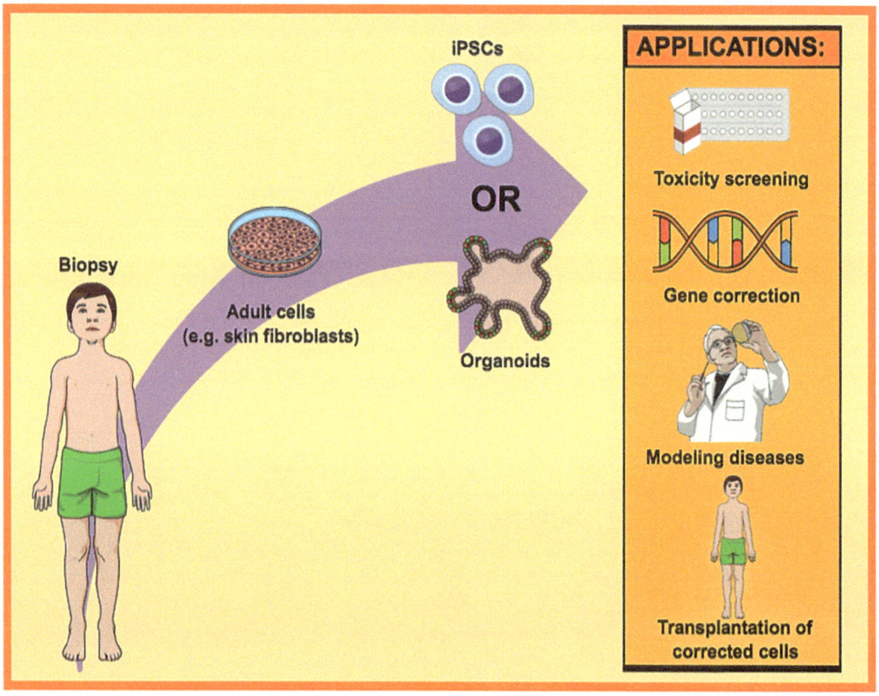

FIG. 4 Potential applications of iPSCs and organoids for cell therapy and disease modeling. It is possible to generate iPSCs and organoids from fibroblasts from an individual affected by a pathology, for example, and from the generated cells; perform toxicity screening and gene correction; model the disease; and transplant the corrected cells.

Machado, & da Costa, 2019). Due to the possibility of simulating embryonic neurogenesis, the iPSC model can be an excellent platform to study the mechanisms related to ZIKV infection and changes at the cellular and molecular levels in different periods of nerve cell formation.

Some studies involving iPSCs and ZIKV have been able to demonstrate important mechanisms that certainly could not be described in clinical studies or animal models. Exposure of iPSCs to ZIKV has shown molecular changes in signaling pathways that are related not only to microcephaly but to other central nervous system disorders such as epilepsy (Rolfe et al., 2016).

Zika virus tropism for nervous cells and its action on microglial cell activation were evidenced in a study in which iPSCs were exposed to ZIKV strains. In this study the researchers concluded that microglial cells initiate an inflammatory process and are able to invade adjacent tissue when infected with Zika (Majolo et al., 2019). In addition, a study using iPSCs and ZIKV showed the high susceptibility of neural precursor cells to virus infection, resulting in a higher rate of cell death, reduced proliferation, structural changes, and autophagy (Souza et al., 2016; Tang et al., 2016).

Gabriel et al. showed that ZIKV-infected iPSCs present rupture of centrosomes and consequent alteration in the plane of cell division (Gabriel et al., 2017). Changes in the plane of cell division during cerebral cortex formation may irreversibly compromise the formation of cortical layers that depend directly on the distribution of specific proteins in the cytoplasm when precursor cells initiate the phase of cell migration mediated by symmetrical (generation of new precursor cells) and asymmetric (generation of migratory neurons) divisions. This polarization after cell division directly depends on the integrity of the centrosome and the position of this precursor cell in the germinal matrix (Uzquiano et al., 2018).

Inflammatory mechanisms have also been associated with ZIKV infection through studies using iPSCs. After ZIKV infection, iPSC-derived neural precursor cells were able to signal proinflammatory mediators, presenting alterations in the expression of genes related to inflammatory cytokines, suggesting new therapeutic possibilities to attenuate the damage caused by infection through immunomodulation (Rolfe et al., 2016).

Thus it can be considered that the use of iPSCs to study the infection of neural precursor cells by Zika virus is highly relevant, providing excellent conditions to understand the mechanisms. It is a powerful tool for high-performance experiments, including the construction of platforms for screening new treatments and elucidating cellular and molecular mechanisms associated with infection and neurological damage caused by it.

Cerebral organoids

After the iPSCs, another notable advance in in vitro modeling of human brain disease was the generation of three-dimensional (3D) brain culture, named cerebral organoids or "minibrain" (Fig. 5). Technology for 3D organoid cultures was first introduced in 2009 using adult intestinal stem cells to model an organoid that had characteristics similar to those of the human gut (Lancaster & Knoblich, 2014; Sato et al., 2009). Lancaster and Knoblich (2014) defined organoids as 3D structures generated from iPSCs or organ-specific stem cells that must have multiple organ-specific cell types. Then, they must be capable of self-organization reproducing some organ function, and lastly, they must be spatially organized according to the cell lineage, similar to the organ in vivo (Lancaster & Knoblich, 2014). Since then, organoids from a variety of human organs have been widely used in the experimental investigation of human diseases, such as retina, thyroid, liver, kidney, inner ear, and specific brain regions (Lancaster & Knoblich, 2014) (Figs. 6 and 7).

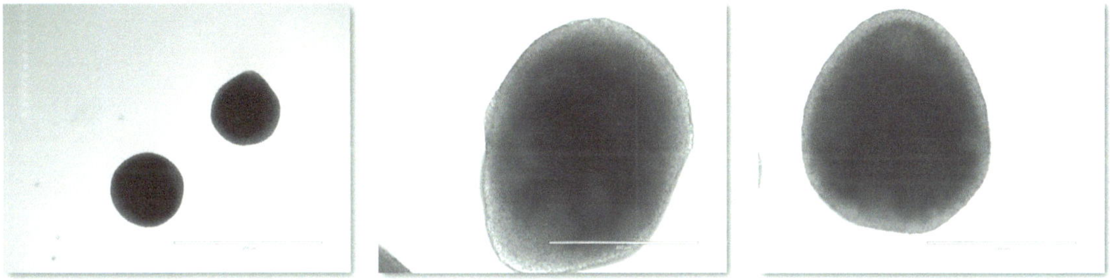

FIG. 5 Cerebral organoid culture. Macroscopic image of cerebral organoids under cultivation for 39 days in suspension. *(Image by authors.)*

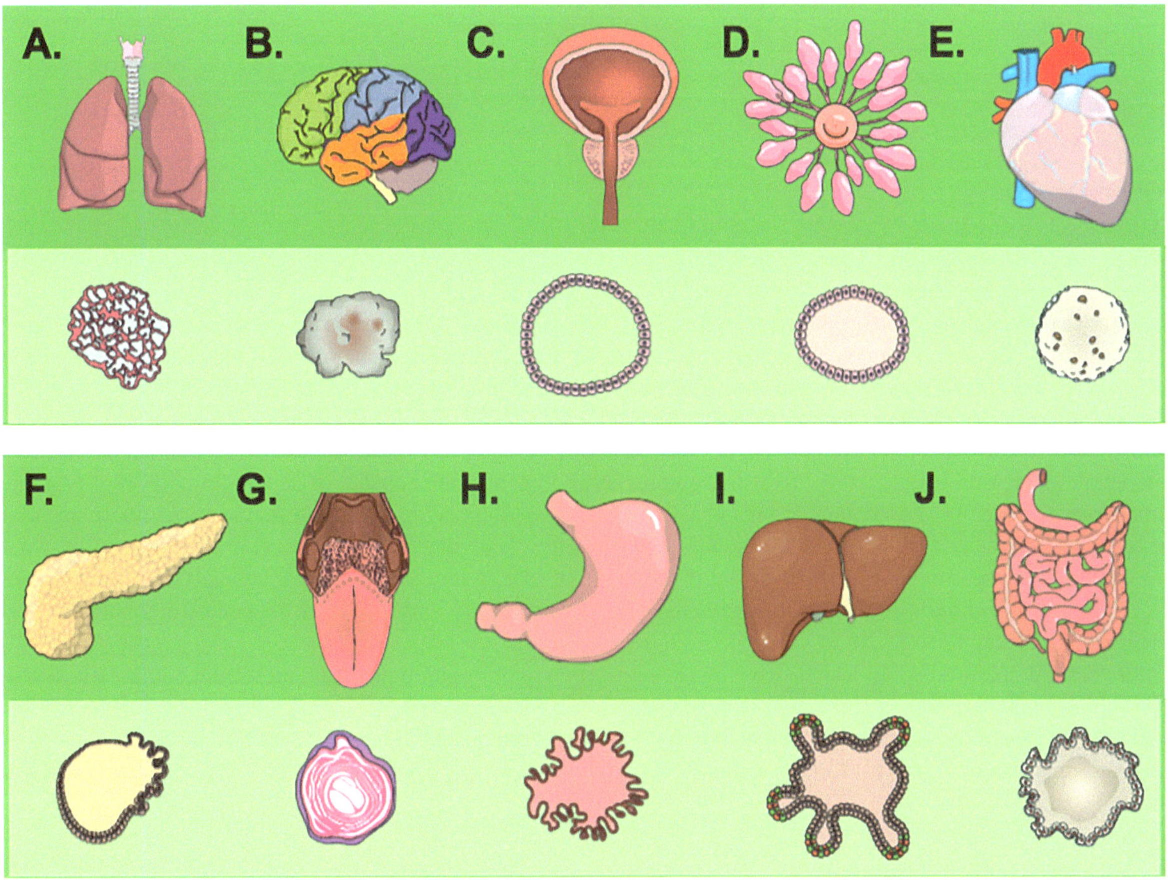

FIG. 6 Illustration of some organoids established from different organs. Organoids from (A) lung, (B) brain, (C) prostate, (D) mammary gland, (E) cardiac, (F) pancreas, (G) tongue, (H) stomach, (I) intestine, and (J) liver. *(Image by authors/https://mindthegraph.com/.)*

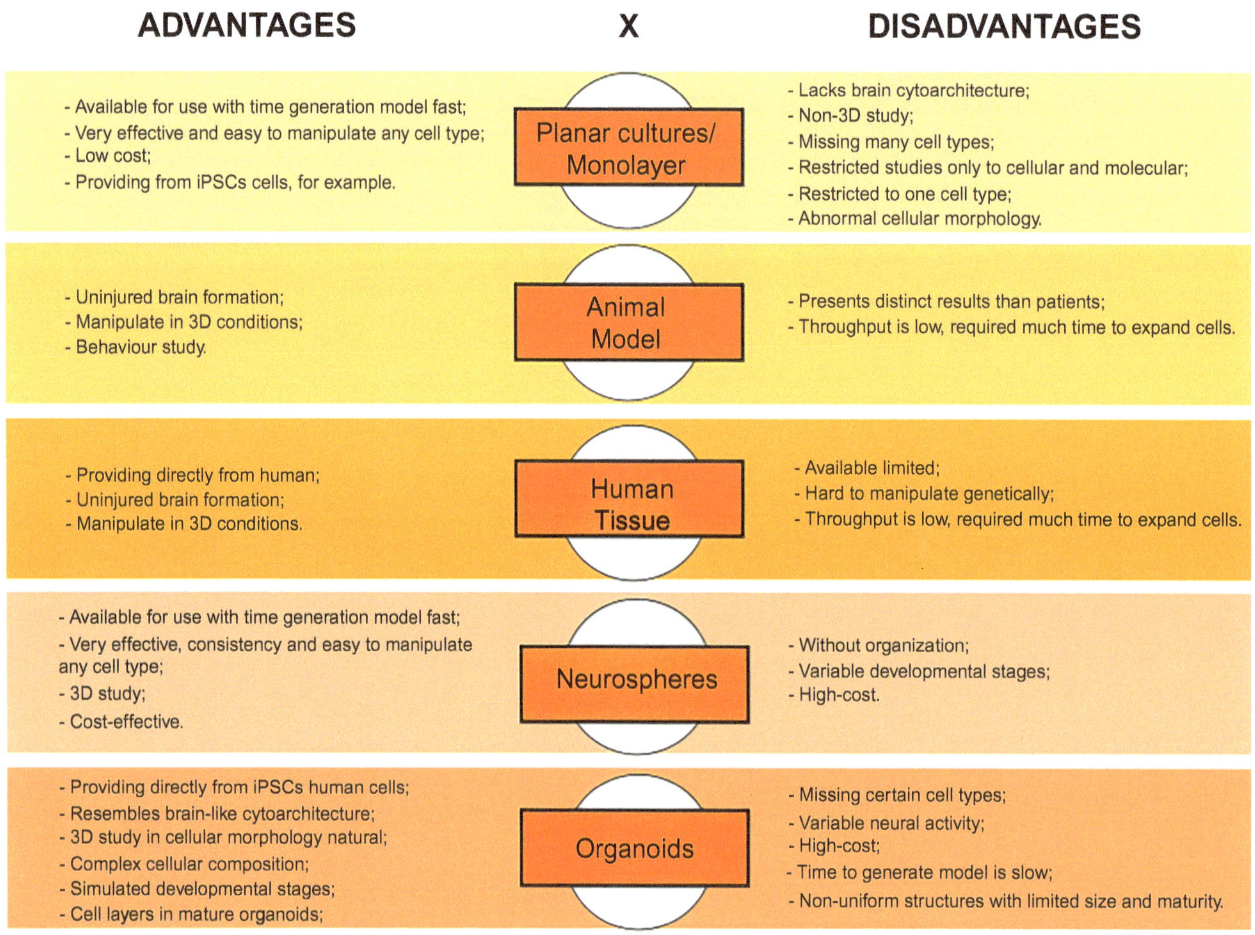

FIG. 7 Advantages and disadvantages from different systems of cultures. *(Adapted from: Chen et al., 2018; Qian et al., 2017. Image by authors/https://mindthegraph.com/.)*

Animal models, especially rodent models, are widely used in research on neurodevelopment and corticogenesis. However, due to the evolutionary distance of rodents from humans and differences in brain physiology, the relevance of this model is uncertain. To close this gap, brain organoids generated from the iPSCs are able to accurately reproduce early human brain development process in vivo, making them more reliable models for translating findings into the human spectrum. Generated under proper suspension culture conditions, brain organoids can reach a few millimeters and can be visible by human eye (Kelava & Lancaster, 2016;Lancaster et al., 2013a, 2013b). They presented intrinsic capacity of self-organization in the absence of growth factors or other signaling, which brings them closer to mammalian embryonic brain development (Lancaster et al., 2013a, 2013b). These cerebral organoids are capable of developing different brain regions with functional intercellular communication and also originated cortical layers with different types of neural cells such as in vivo corticogenesis and well-defined ventricular and subventricular zones (Kadoshima et al., 2013a, 2013b; Kelava & Lancaster, 2016; Lancaster et al., 2013a, 2013b; Paşca et al., 2015). In addition, comparative analyses of gene expression between organoids and brain development in vivo have shown a strong resemblance to what occurs in the human first gestational trimester (Paşca et al., 2015).

Thus cerebral organoids show a significant increase in structural, morphological, and molecular complexity when compared with monolayer or neurosphere cultures even surpassing animal models in the modeling of human central nervous system development. They are a versatile platform for modeling human central nervous system diseases with a high investigative and therapeutic property (Kelava & Lancaster, 2016; Paşca et al., 2015).

Cerebral organoids X Zika virus

The possible tropism of Zika virus by neural precursors and immature neurons could explain the contribution of infection in triggering microcephaly, a congenital condition characterized by abnormal cell division or proliferation that causes changes in cortical layer thickness (Pang, Atefy, & Sheen, 2008; Cugola et al., 2016; Gaburro et al., 2018; Li et al., 2016). In this context, cerebral organoids have become an indispensable tool in the investigation of the pathophysiological mechanisms by which ZIKV acts in the central nervous system, since these organoids are now the model that present the most reliable human developmental processes of recapitulation, consequently providing the best modeling to study of cortical malformations, including microcephaly (Lancaster et al., 2013a, 2013b).

The use of cerebral organoids, compared with investigations of monolayer neural cell cultures, allowed a better understanding of the interaction of ZIKV between different cell types and the impact of infection during early neurodevelopment (Cugola et al., 2016; Qian et al., 2016; Salick, Wells, Eggan, & Kaykas, 2017). Studies with various approaches have shown that cerebral organoids exposed to ZIKV showed a significant reduction in total organoid growth (Cugola et al., 2016; Garcez et al., 2016; Majolo et al., 2019; Qian et al., 2016). Consistent with this result the ZIKVBR strain also reduced the neuronal population of cortex layers of cerebral organoids leading to reduced thickness in cortical lamination, as well as increasing apoptotic cell death and reducing growth in an in vitro model of neurospheres (Cugola et al., 2016). In addition, alterations in the ventricular zone cell proliferation of the organoids with reduction of progenitor cells were observed after ZIKVBR infection (Cugola et al., 2016). Despite being detected in immature neurons, intermediate progenitors, and astrocytes, ZIKV exhibits greater tropism for neural precursor cells as demonstrated in a human forebrain-specific brain organoid model (Qian et al., 2016). Qian and colleagues also showed that transient exposure of early-stage brain organoids triggers, in addition to decreased organoid size and dilution of cortical layer thickness, increased ventricular lumen size, which is also characteristic of microcephaly in vivo (Qian et al., 2016). Thus Salik and colleagues described a methodology for modeling ZIKV in brain organoids that exhibited high levels of virus in their neural progenitors, leading to the development of microcephaly over time (Salick et al., 2017).

In addition to the apparent higher affinity of the Brazilian ZIKV strain to the central nervous system, when compared with the African strain, the infections with other flaviviruses such as dengue virus or yellow fever virus in brain organoids also caused no morphophysiological changes, indicating that this characteristic seems to be unique to the ZIKVBR (Cugola et al., 2016; Garcez et al., 2016). However, ZIKVBR failed to influence the size of organoids generated from nonhuman primates. This finding suggests a potential adaptation of the virus resulting in species-specific affinity of ZIKVBR (Cugola et al., 2016).

Despite the recent association of ZIKV with neurological changes and brain malformations from the exponential increase in microcephaly cases during the 2015 virus epidemic in Brazil, the emergency efforts of the world scientific community are reflected in the literature (Majolo et al., 2019). In vitro 3D modeling of brain development has proved to be a valuable tool to investigate the impact of ZIKV infection during the neurodevelopment. This observation provided important evidence of the cellular and molecular mechanisms of infection, as well as demonstrating easy access to the placenta during pregnancy by the virus and the active participation of ZIKV in triggering microcephaly in infants who suffered contamination by vertical transmission.

Perspectives and conclusions

Given the different study models for ZIKV, most of them still focus on characterizing the pathological impact of virus infection. The most profound advances came from the cerebral organoids, due to their ease of manipulation, allowing valuable genetic and biochemical studies. Therefore organoids to validate therapeutic agents and to evaluate molecular and cellular mechanisms associated with ZIKV damage become potential tools for future studies.

Gene manipulation techniques such as genome editing and infection with other viruses are important to be done in organoid models (Qian et al., 2017). Liang et al. (2016) point out that the development of organoids may be affected by the protein chain synthesized or derived from noncoding ZIKV RNAs. Although the technique of "minibrains" is fantastic, it has some disadvantages as it has high heterogeneity due to different cell types even among simultaneously grown organoids under the same conditions (Chukwurah, Osmundsen, Davis, & Lizarraga, 2019; Quadrato et al., 2017). As the growth of organoids increases in complexity, there is a need to provide nutrients, adequate oxygenation, and vascular structures. Current methods are known to have been able to grow "minibrains" for up to 510 days without a vascular system (Chukwurah et al., 2019; Sloan et al., 2017; Sloan, Andersen, Paşca, Birey, & Paşca, 2018). However, many studies are searching to improve the vascularized culture systems because it is more suited to the functional and developmental aspect of the brain (Chukwurah et al., 2019; Paredes, Himmels, & de Almodovar, 2018).

Efficiency in mutation correction and disease modeling is intrinsically related to recent technologies like zinc-finger nucleases (ZFN) (Hockemeyer et al., 2009; Schlaeger et al., 2015). Transcription activator such as effector nucleases and the CRISPR/Cas9 system-specific nucleases may induce double-stranded DNA breaks at the site of genetic modification (Christian et al., 2010; Cong et al., 2013; Hockemeyer et al., 2011; Shalem et al., 2014). The success of this technique is also associated with the successful generation of mature functional organs (Wiegand & Banerjee, 2019). Specific modeling through the addition of distinct morphogenetic proteins assists the development of brain region-specific organoids (Paşca, 2018), thus allowing to discuss brain region-specific issues (Birey et al., 2017; Chukwurah et al., 2019).

Another challenge is the search for a greater similarity with the complexity of human neural circuits; some success already has been made in transplanting human organoid grafts reproducing functional neuronal networks in their host (Chukwurah et al., 2019; Mansour et al., 2018). As mentioned earlier, in vitro organoid vascularization still has some failures, because not all blood vessels mature in a transplant. Bioengineering would be a way of solving some of these in vitro system failures, such as using 3D blood vessels or "organ on a chip" (Chukwurah et al., 2019; Lin, Guo, & Zhang, 2018; Trujillo & Muotri, 2018).

Diseases characterized as structural malformations such as microcephaly are visible conditions that allow to verify the action of a treatment in organoids. Thus, with the construction of organoids with sophisticated electrical activity, it is possible to compare with the patients' electroencephalograms (EEGs) (Platt & Miner, 2017). Zika epidemics pose a major challenge; however, with the incorporation of stem cell-based bioengineering techniques from ZIKV exposure, organoids would represent this futuristic scenario with vascularized fluid systems that would facilitate drug research and development and even to study other models of brain disease (Ming et al., 2016; Platt & Miner, 2017).

Policy and procedures

In our laboratory placed at Brain Institute of Rio Grande do Sul (BraIns), it established, in 2017, an applied cellular model to study focal cortical dysplasia based on iPSCs from fibroblasts of patients affected (Marinowic et al., 2017). Following are the protocols with emphasis on cell reprogramming.

Materials and methods

Human fibroblasts were obtained from skin biopsies. The skin fragments were cut into pieces of approximately 5 mm^2 and placed in a 60-mm Petri dish with the dermis face facing the plate. IPSCs were generated by exposing fibroblasts to viral vectors containing the genes OCT4, KLF4, SOX2, and c-MYC utilizing CytoTune-iPS 2.0 Sendai Reprogramming Kit (Thermo Fisher Scientific) according to the guidelines of manufacturer. Fibroblasts were cultured in culture plates, and for transfection, those with a confluence of 70% of cultivation area were selected. The number of viral particles used was calculated using the MOI equation. The determined amount of virus was diluted in 1 mL of DMEM/F12 culture medium (Life Technologies) adding 20% of Knockout Serum Replacement (KSR) (Life Technologies), Non-Essential Amino Acids of DMEM (1 ×), GlutaMAX (1 ×) (Life Technologies), and antibiotics. The fibroblasts were exposed to the medium containing the viruses and incubated at 37°C—5% of CO_2 for 24 h. The cells were washed with DPBS (Dulbecco's Phosphate Buffered Saline) and cultured for six days under the same conditions.

On the seventh day after transfection, the cultures were treated with Trypsin-EDTA and transferred to a culture plate prepared with Matrigel (BD Corning). After 24 h the culture medium was aspirated, and cells started to be cultured in embryonic cells medium culture mTeSR (STEMCELL Technologies). After about 20 days the clones were removed manually and cultured in new plates containing Matrigel preparation. Clones were characterized using the antibodies anti-Nanog, anti-Sox2, anti-Oct4, anti-TRA1-60, and anti-TRA1-81 (Merck). For this the clones were cultured in plates, the culture medium was removed, and cells were fixed with 4% PFA (Sigma-Aldrich). Blocking buffer was added and after antibodies conjugated to fluorophores diluted (1:100) in dilution buffer. The cells were incubated for 2 h at room temperature. Two washes were performed with DPBS buffer and nuclear labeling using DAPI. Upon reaching the time limit manipulation of a plate containing Matrigel (about 15 days), manual pickings were made by cutting the clone with the aid of a glass Pasteur pipette for recultivation or freezing.

- The article can be found by the following reference: Marinowic, D. R., Majolo, F., Sebben, A. D., da Silva, V. D., Lopes, T. G., Paglioli, E., Palmini, A., Machado, D. C., & da Costa, J. C. (2017). Induced pluripotent stem cells from patients with focal cortical dysplasia and refractory epilepsy. *Molecular Medicine Reports, 15*, 2049–2056. https://doi.org/10.3892/mmr.2017.6264.
- The complete protocol can be found by the link: https://www.thermofisher.com/br/en/home/life-science/stem-cell-research/induced-pluripotent-stem-cells/sendai-virus-reprogramming.html.

Mini-dictionary of terms

Cerebral organoids or "minibrains": Three-dimensional (3D) tissue structures that mimic the brain.

Corticogenesis: It's considered a formation process of cerebral cortex during the development of the nervous system. This process can be promoted from iPSCs in cerebral organoids.
Embryonic stem cells: Pluripotent cells able to differentiate into all derivatives cells of the ectoderm, endoderm, and mesoderm layers (primary germ layers).
Induced pluripotent stem cells (iPSCs): They are similar to embryonic stem cells, presenting the same characteristics of self-renewal and differentiation potential into the three embryonic leaflets.
Neurospheres: Aggregate of neural progenitor cells that do not tend to generate complex structures.

Key facts

- Influence of ZIKV against iPSCs: Exposure of iPSCs to ZIKV has shown molecular changes in signaling pathways that are related not only to microcephaly but to other central nervous system disorders such as epilepsy.
- iPSCs as a future tool: The use of iPSCs to study the infection of neural precursor cells by Zika virus is highly relevant, providing excellent conditions to understand the mechanisms.
- Plasticity of "minibrains": Cerebral organoids are capable of developing different brain regions with functional intercellular communication and originate cortical layers with different types of neural cells such as in vivo corticogenesis and well-defined ventricular and subventricular zones.
- Effects of ZIKV on cerebral organoids: Cerebral organoids exposed to ZIKV showed a significant reduction in total organoid growth leading to reduced thickness in cortical lamination.
- The future of organoids: Gene manipulation techniques such as genome editing and infection with other viruses are important to be done in organoid models, although the technique of "minibrains" has some disadvantages, which still need to be resolved, for example, the high heterogeneity due to different cell types even among simultaneously grown organoids under the same conditions.

Summary points

- iPSCs may be considered extremely similar to embryonic stem cells, presenting the same self-renewal characteristics and differentiation potential in ectoderm, mesoderm, and endoderm cells.
- Studies involving iPSCs and ZIKV have been able to demonstrate important mechanisms that certainly could not be described in clinical studies or animal models.
- After ZIKV infection, iPSC-derived neural precursor cells were able to signal proinflammatory mediators, presenting alterations in the expression of genes related to inflammatory cytokines, suggesting new therapeutic possibilities to attenuate the damage caused by infection through immunomodulation.
- The use of cerebral organoids, compared with investigations of monolayer neural cell cultures, allowed a better understanding of the interaction of ZIKV between different cell types and the impact of infection during early neurodevelopment.
- Cerebral organoids have become an indispensable tool in the investigation of the pathophysiological mechanisms by which ZIKV acts in the central nervous system.

Acknowledgments

The first author thanks Coordenação de Aperfeiçoamento de Pessoal de Nível Superior, Brasil (CAPES), Finance Code 001, for the postdoctoral scholarship. DaCosta is 1A researcher from Conselho Nacional de Desenvolvimento Científico e Tecnológico (CNPq).

References

Birey, F., Andersen, J., Makinson, C. D., Islam, S., Wei, W., Huber, N., … O'Rourke, N. A. (2017). Assembly of functionally integrated human forebrain spheroids. *Nature*, *545*(7652), 54.

Bond, A. M., Ming, G. L., & Song, H. (2015). Adult mammalian neural stem cells and neurogenesis: Five decades later. *Cell Stem Cell*, *17*(4), 385–395.

CDC. (2016). *Zika virus*. Atlanta, GA: US Department of Health and Human Services, United States Centers for Disease Control and Prevention.

Chen, H. I., Song, H., & Ming, G. L. (2019). Applications of human brain organoids to clinical problems. *Developmental Dynamics*, *248*(1), 53–64.

Christian, M., Cermak, T., Doyle, E. L., Schmidt, C., Zhang, F., Hummel, A., … Voytas, D. F. (2010). Targeting DNA double-strand breaks with TAL effector nucleases. *Genetics*, *186*(2), 757–761.

Chukwurah, E., Osmundsen, A., Davis, S. W., & Lizarraga, S. B. (2019). All together now: Modeling the interaction of neural with non-neural systems using organoid models. *Frontiers in Neuroscience*, *13*.

Clevers, H. (2016). Modeling development and disease with organoids. *Cell*, *165*(7), 1586–1597.

Cong, L., Ran, F. A., Cox, D., Lin, S., Barretto, R., Habib, N., … Zhang, F. (2013). Multiplex genome engineering using CRISPR/Cas systems. *Science*, *339*(6121), 819–823.

Cugola, F. R., Fernandes, I. R., Russo, F. B., Freitas, B. C., Dias, J. L. M., Guimarães, K. P., et al. (2016). The Brazilian Zika virus strain causes birth defects in experimental models. *Nature*, *534*(7606), 267–271.

Dasgupta, S., Reagan-Steiner, S., Goodenough, D., Russell, K., Tanner, M., Lewis, L., et al. (2016). Patterns in Zika virus testing and infection, by report of symptoms and pregnancy status—United States, January 3-March 5, 2016. *MMWR. Morbidity and Mortality Weekly Report*, *65*(15), 395–399.

Dolmetsch, R., & Geschwind, D. H. (2011). The human brain in a dish: The promise of iPSC-derived neurons. *Cell*, *145*(6), 831–834.

Dragunow, M. (2008). The adult human brain in preclinical drug development. *Nature Reviews Drug Discovery*, *7*(8), 659–666.

Driggers, R. W., Ho, C. Y., Korhonen, E. M., Kuivanen, S., Jääskeläinen, A. J., Smura, T., … Timofeev, J. (2016). Zika virus infection with prolonged maternal viremia and fetal brain abnormalities. *New England Journal of Medicine*, *374*(22), 2142–2151.

Gabriel, E., Ramani, A., Karow, U., Gottardo, M., Natarajan, K., Gooi, L. M., et al. (2017). Recent Zika virus isolates induce premature differentiation of neural progenitors in human brain organoids. *Cell Stem Cell*, *20*(3), 397–406.

Gaburro, J., Bhatti, A., Harper, J., Jeanne, I., Dearnley, M., Green, D., et al. (2018). Neurotropism and behavioral changes associated with Zika infection in the vector *Aedes aegypti* article. *Emerging Microbes and Infections*, *7*(1), 68.

Garcez, P. P., Loiola, E. C., Da Costa, R. M., Higa, L. M., Trindade, P., Delvecchio, R., et al. (2016). Zika virus: Zika virus impairs growth in human neurospheres and brain organoids. *Science*, *352*(6287), 816–818.

Heukelbach, J., Alencar, C. H., Kelvin, A. A., de Oliveira, W. K., & de Góes Cavalcanti, L. P. (2016). Zika virus outbreak in Brazil. *The Journal of Infection in Developing Countries*, *10*(02), 116–120.

Heymann, D. L., Hodgson, A., Freedman, D. O., Staples, J. E., Althabe, F., Baruah, K., … Menon, K. U. (2016). Zika virus and microcephaly: Why is this situation a PHEIC? *The Lancet*, *387*(10020), 719–721.

Hockemeyer, D., Soldner, F., Beard, C., Gao, Q., Mitalipova, M., DeKelver, R. C., … Meng, X. (2009). Efficient targeting of expressed and silent genes in human ESCs and iPSCs using zinc-finger nucleases. *Nature Biotechnology*, *27*(9), 851.

Hockemeyer, D., Wang, H., Kiani, S., Lai, C. S., Gao, Q., Cassady, J. P., … Zeitler, B. (2011). Genetic engineering of human pluripotent cells using TALE nucleases. *Nature Biotechnology*, *29*(8), 731.

Jo, J., Xiao, Y., Sun, A. X., Cukuroglu, E., Tran, H. D., Göke, J., … Lee, Y. (2016). Midbrain-like organoids from human pluripotent stem cells contain functional dopaminergic and neuromelanin-producing neurons. *Cell Stem Cell*, *19*(2), 248–257.

Kadoshima, T., Sakaguchi, H., Nakano, T., Soen, M., Ando, S., Eiraku, M., et al. (2013b). Self-organization of axial polarity, inside-out layer pattern, and species-specific progenitor dynamics in human ES cell-derived neocortex. *Proceedings of the National Academy of Sciences of the United States of America*, *110*(50), 20284–20289.

Kadoshima, T., Sakaguchi, H., Nakano, T., Soen, M., Ando, S., Eiraku, M., & Sasai, Y. (2013a). Self-organization of axial polarity, inside-out layer pattern, and species-specific progenitor dynamics in human ES cell-derived neocortex. *Proceedings of the National Academy of Sciences*, *110*(50), 20284–20289.

Kelava, I., & Lancaster, M. A. (2016). *Dishing out mini-brains: Current progress and future prospects in brain organoid research. Developmental biology. vol. 420* (pp. 199–209). Academic Press Inc.

Lancaster, M. A., & Knoblich, J. A. (2014). *Organogenesisin a dish: Modeling development and disease using organoid technologies*. Science American Association for the Advancement of Science.

Lancaster, M. A., Renner, M., Martin, C. A., Wenzel, D., Bicknell, L. S., Hurles, M. E., et al. (2013a). Cerebral organoids model human brain development and microcephaly. *Nature*, *501*(7467), 373–379.

Lancaster, M. A., Renner, M., Martin, C. A., Wenzel, D., Bicknell, L. S., Hurles, M. E., … Knoblich, J. A. (2013b). Cerebral organoids model human brain development and microcephaly. *Nature*, *501*(7467), 373.

Li, C., Xu, D., Ye, Q., Hong, S., Jiang, Y., Liu, X., … Xu, Z. (2016). Zika virus disrupts neural progenitor development and leads to microcephaly in mice. *Cell Stem Cell*, *19*(1), 120–126.

Liang, Q., Luo, Z., Zeng, J., Chen, W., Foo, S. S., Lee, S. A., … Zhao, Z. (2016). Zika virus NS4A and NS4B proteins deregulate Akt-mTOR signaling in human fetal neural stem cells to inhibit neurogenesis and induce autophagy. *Cell Stem Cell*, *19*(5), 663–671.

Lin, D. S., Guo, F., & Zhang, B. (2018). Modeling organ-specific vasculature with organ-on-a-chip devices. *Nanotechnology*, *30*(2), 024002.

Lizarraga, S. B., Margossian, S. P., Harris, M. H., Campagna, D. R., Han, A. P., Blevins, S., … Fleming, M. D. (2010). Cdk5rap2 regulates centrosome function and chromosome segregation in neuronal progenitors. *Development*, *137*(11), 1907–1917.

Majolo, F., Marinowic, D. R., Moura, A. A., Machado, D. C., & da Costa, J. C. (2019). Use of induced pluripotent stem cells (iPSCs) and cerebral organoids in modeling the congenital infection and neuropathogenesis induced by Zika virus. *Journal of Medical Virology*, *91*(4), 525–532.

Mansour, A. A., Gonçalves, J. T., Bloyd, C. W., Li, H., Fernandes, S., Quang, D., … Gage, F. H. (2018). An in vivo model of functional and vascularized human brain organoids. *Nature Biotechnology*, *36*(5), 432–441.

Mariani, J., Coppola, G., Zhang, P., Abyzov, A., Provini, L., Tomasini, L., … Gerstein, M. (2015). FOXG1-dependent dysregulation of GABA/glutamate neuron differentiation in autism spectrum disorders. *Cell*, *162*(2), 375–390.

Mesci, P., et al. (2018). Author correction: blocking Zika virus vertical transmission. *Scientific Reports*, *8*(1), 8794.

Ming, G. L., Tang, H., & Song, H. (2016). Advances in Zika virus research: Stem cell models, challenges, and opportunities. *Cell Stem Cell*, *19*(6), 690–702.

Mlakar, J., Korva, M., Tul, N., Popović, M., Poljšak-Prijatelj, M., Mraz, J., … Vizjak, A. (2016). Zika virus associated with microcephaly. *New England Journal of Medicine*, *374*(10), 951–958.

Pang, T., Atefy, R., & Sheen, V. (2008). Malformations of cortical development. *The Neurologist*, *14*, 181–191.

Paredes, I., Himmels, P., & de Almodovar, C. R. (2018). Neurovascular communication during CNS development. *Developmental Cell*, *45*(1), 10–32.

Paşca, A. M., Sloan, S. A., Clarke, L. E., Tian, Y., Makinson, C. D., Huber, N., … Smith, S. J. (2015). Functional cortical neurons and astrocytes from human pluripotent stem cells in 3D culture. *Nature Methods*, *12*(7), 671.

Paşca, S. P. (2018). The rise of three-dimensional human brain cultures. *Nature*, *553*(7689), 437–445.

Platt, D. J., & Miner, J. J. (2017). Consequences of congenital Zika virus infection. *Current Opinion in Virology*, *27*, 1–7.

Pulvers, J. N., Bryk, J., Fish, J. L., Wilsch-Bräuninger, M., Arai, Y., Schreier, D., … Nitsch, R. (2010). Mutations in mouse Aspm (abnormal spindle-like microcephaly associated) cause not only microcephaly but also major defects in the germline. *Proceedings of the National Academy of Sciences*, *107*(38), 16595–16600.

Qian, X., Nguyen, H. N., Jacob, F., Song, H., & Ming, G. L. (2017). Using brain organoids to understand Zika virus-induced microcephaly. *Development*, *144*(6), 952–957.

Qian, X., Nguyen, H. N., Song, M. M., Hadiono, C., Ogden, S. C., Hammack, C., … Yoon, K. J. (2016). Brain-region-specific organoids using mini-bioreactors for modeling ZIKV exposure. *Cell*, *165*(5), 1238–1254.

Quadrato, G., Nguyen, T., Macosko, E. Z., Sherwood, J. L., Yang, S. M., Berger, D. R., … Boyden, E. S. (2017). Cell diversity and network dynamics in photosensitive human brain organoids. *Nature*, *545*(7652), 48.

Rolfe, A. J., Bosco, D. B., Wang, J., Nowakowski, R. S., Fan, J., & Ren, Y. (2016). Bioinformatic analysis reveals the expression of unique transcriptomic signatures in Zika virus infected human neural stem cells. *Cell & Bioscience*, *6*, 42.

Salick, M. R., Wells, M. F., Eggan, K., & Kaykas, A. (2017). Modelling zika virus infection of the developing human brain in vitro using stem cell derived cerebral organoids. *Journal of Visualized Experiments*, *127*, 56404.

Sato, T., Vries, R. G., Snippert, H. J., Van De Wetering, M., Barker, N., Stange, D. E., et al. (2009). Single Lgr5 stem cells build crypt-villus structures in vitro without a mesenchymal niche. *Nature*, *459*(7244), 262–265.

Schlaeger, T. M., Daheron, L., Brickler, T. R., Entwisle, S., Chan, K., Cianci, A., … Gupta, D. (2015). A comparison of non-integrating reprogramming methods. *Nature Biotechnology*, *33*(1), 58.

Setia, H., & Muotri, A. R. (2019). Brain organoids as a model system for human neurodevelopment and disease. In *Seminars in cell & developmental biology* Academic Press.

Shalem, O., Sanjana, N. E., Hartenian, E., Shi, X., Scott, D. A., Mikkelsen, T. S., … Zhang, F. (2014). Genome-scale CRISPR-Cas9 knockout screening in human cells. *Science*, *343*(6166), 84–87.

Sloan, S. A., Andersen, J., Paşca, A. M., Birey, F., & Paşca, S. P. (2018). Generation and assembly of human brain region-specific three-dimensional cultures. *Nature Protocols*, *13*(9), 2062.

Sloan, S. A., Darmanis, S., Huber, N., Khan, T. A., Birey, F., Caneda, C., … Paşca, S. P. (2017). Human astrocyte maturation captured in 3D cerebral cortical spheroids derived from pluripotent stem cells. *Neuron*, *95*(4), 779–790.

Souza, B. S., Sampaio, G. L., Pereira, C. S., Campos, G. S., Sardi, S. I., Freitas, L. A., et al. (2016). Zika virus infection induces mitosis abnormalities and apoptotic cell death of human neural progenitor cells. *Scientific Reports*, *6*, 39775.

Takahashi, K., & Yamanaka, S. (2006). Induction of pluripotent stem cells from mouse embryonic and adult fibroblast cultures by defined factors. *Cell*, *126*(4), 663–676.

Takahashi, K., Tanabe, K., Ohnuki, M., Narita, M., Ichisaka, T., Tomoda, K., & Yamanaka, S. (2007). Induction of pluripotent stem cells from adult human fibroblasts by defined factors. *Cell*, *131*(5), 861–872.

Tang, H., Hammack, C., Ogden, S. C., Wen, Z., Qian, X., Li, Y., et al. (2016). Zika virus infects human cortical neural progenitors and attenuates their growth. *Cell Stem Cell*, *18*(5), 587–590.

Tao, Y., & Zhang, S. C. (2016). Neural subtype specification from human pluripotent stem cells. *Cell Stem Cell*, *19*(5), 573–586.

Trujillo, C. A., & Muotri, A. R. (2018). Brain organoids and the study of neurodevelopment. *Trends in Molecular Medicine*, *24*(12), 982–990.

Uzquiano, A., Gladwyn-Ng, I., Nguyen, L., Reiner, O., Gotz, M., Matsuzaki, F., et al. (2018). Cortical progenitor biology: key features mediating proliferation versus differentiation. *Journal of Neurochemistry*, *146*(5), 500–525.

Ventura, C. V., Maia, M., Ventura, B. V., Linden, V. V. D., Araújo, E. B., Ramos, R. C., … Ventura, L. O. (2016). Ophthalmological findings in infants with microcephaly and presumable intra-uterus Zika virus infection. *Arquivos Brasileiros de Oftalmologia*, *79*(1), 1–3.

Weissman, I. L. (2000). Stem cells: Units of development, units of regeneration, and units in evolution. *Cell*, *100*(1), 157–168.

Wen, Z., Christian, K. M., Song, H., & Ming, G. L. (2016). Modeling psychiatric disorders with patient-derived iPSCs. *Current Opinion in Neurobiology*, *36*, 118–127.

Wiegand, C., & Banerjee, I. (2019). Recent advances in the applications of iPSC technology. *Current Opinion in Biotechnology*, *60*, 250–258.

Chapter 9

Zika, miRNAs, and microcephaly genes

Omar Bagasra[a] and Ewen McLean[b]

[a]*Department of Biology, South Carolina Center for Biotechnology, School of Natural Sciences and Mathematics, Claflin University, Orangeburg, SC, United States,* [b]*Aqua Cognoscenti, West Columbia, SC, United States*

List of abbreviations

ASPM	abnormal spindle-like microcephaly-associated protein
MCPH	autosomal recessive primary microcephaly (MCPH)
BACE1	beta-secretase 1
BDNF	brain-derived neurotrophic factor
CRS	congenital rubella syndrome
DYRK1A	dual-specificity tyrosine-phosphorylation-regulated kinase 1A
EFNB1	Ephrin B1
GDNF	glial cell-derived neurotrophic factor
NSC	human neuronal stem cells
MCPH1	microcephalin
miR	microRNA
NGF	nerve growth factor
MCPH11	PCH1
RAS/MAPK	RAS/mitogen-activated protein kinase
RV	Rubella virus
MCPH7	STIL
SUCLG2	succinyl-CoA ligase subunit beta
TFRC	transferrin receptor
VEGF	vascular endothelial growth factor
WDR62	WD repeat domain 62
WES	whole-exome sequencing
ZIKV	Zika virus

Introduction

Zika virus (ZIKV) is a positive-sense, single-stranded small enveloped RNA member of the Spondweni serocomplex within the genus *Flavivirus*, family *Flaviviridae* (Hayes, 2009). Originally isolated in 1947 from a platform-caged sentinel rhesus macaque posted in the Zika forest, Uganda (Dick, 1953; Dick, Kitchen, & Haddow, 1952), the first report of ZIKV in humans was from Nigeria in 1954 (Simpson, 1964; Fig. 1). Since that time, ZIKV infections have been recorded only infrequently with cases occurring in west and central Africa and Asia (Hayes, 2009; Henderson, Hewitt, & Lule, 1968; Olson, Ksiazek, & Suhandiman, 1981; Rasmussen, Jamieson, Honein, & Petersen, 2016). In 2007 the first recorded case of ZIKV, beyond its apparent natural range, was recorded in Yap (Duffy et al., 2009; Fig. 1). This outbreak was coupled with infected individuals exhibiting neurological symptoms associated with Guillain-Barré syndrome (GBS). Early in 2015 a ZIKV outbreak was recorded in Camaçari, Brazil, and, like the Yap incident, patients presented with disorders matching those of GBS. From 2015, ZIKV marched onward, infecting other countries in Central and South America and the Caribbean (Fig. 1). One of the most disturbing attributes of ZIKV is its capacity to cause severe microcephaly in preterm infants, which were reported throughout the region. Microcephaly, together with other brain defects, was first noted in Brazilian newborns during 2015. By January 2016 over 3500 cases had been reported for infants whose mothers lived in, or had visited, ZIKV-infected areas (Kleber de Oliveira et al., 2016). Similar outcomes were documented, retrospectively, in

Zika Virus Impact, Diagnosis, Control, and Models. https://doi.org/10.1016/B978-0-12-820267-8.00009-1

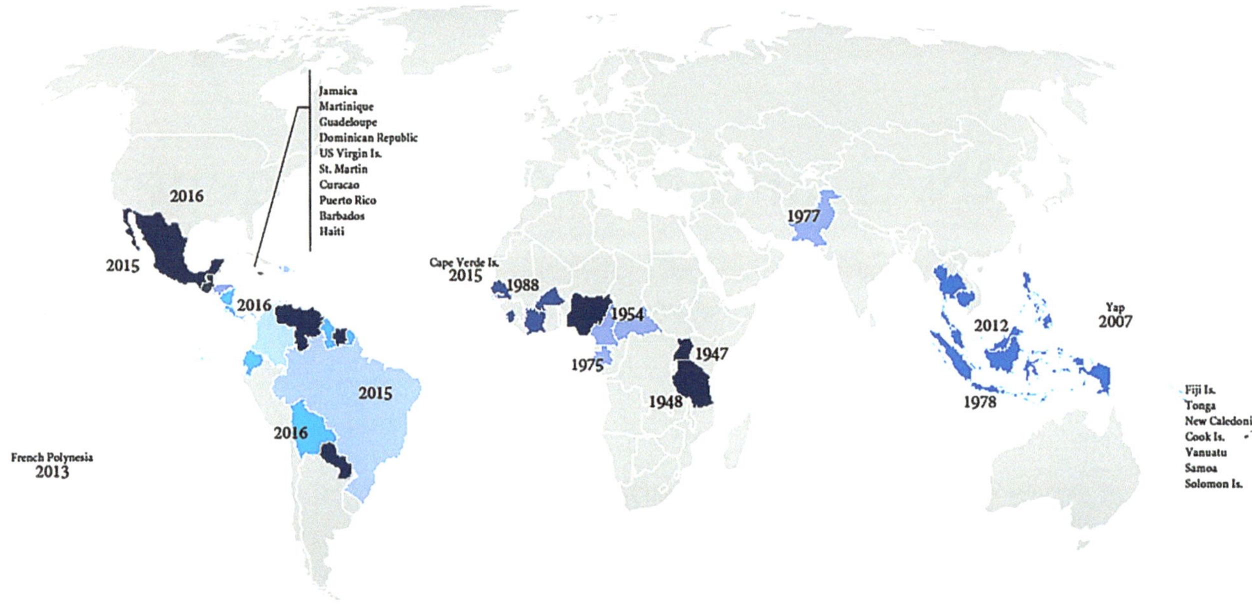

FIG. 1 Summary of major timeline components describing discovery and dispersal of Zika virus from 1947 until 2016. It should be noted that reports of microcephaly in newborns only surfaced in 2015, and it is relatively a new phenomenon. It remains a mystery as to why no reports of ZIKV-induced microcephaly in newborns were published prior to 2015. It may be due to ZIKV mutation or lack of recognition of association.

infants born in French Polynesia (Cauchemez et al., 2016). In the first instance, however, the medical community took a precautionary approach and did not declare a link between ZIKV and microcephaly. This reticence likely stemmed from the fact that no flavivirus had previously been shown to conclusively cause birth defects. A review of the evidence for causality using various assessment criteria, however, ultimately led to declaring a categorical link between ZIKV and neurological complications (Mlakar et al., 2016; Rasmussen et al., 2016; Salgado et al., 2020; Štrafela et al., 2017; Tolosa et al., 2017).

ZIKV-induced cranial morphology

The appearance of ZIKV-impacted crania is diverse. They include disruptions and reorganization of the lambdoid and coronal sutures and occipital plate prominence (Culjat et al., 2016; Moore et al., 2017). These changes are often associated with calvaria collapse, reduced cranial vault volume, craniofacial asymmetry, disturbance in occipitofrontal head circumference, and nuchal skin rumpling (Culjat et al., 2016; Horovitz, da Silva Pone, Pone, Salles, & Boechat, 2016; Moura da Silva et al., 2016). It is significant that while 7 of 13 infants with congenital Zika had regular occipitofrontal head circumference at birth, head growth decelerated in 12 individuals, and all expressed malformations in cortical development and calcification of the subcortical and/or basal ganglia 5–9 months after birth (van der Linden, Pessoa, Dobyns, Barkovich, et al., 2016). These observations, therefore, determined that ZIKV-induced microcephaly could elaborate postnatally. The brain typically expresses subcortical calcifications that are believed to result due to cell death (Moore et al., 2017). Other features include cortical thinning, absence of ridges or convolutions (lissencephaly), incomplete or absent *corpus callosum*, underdevelopment of the *cerebellar vermis*, enlargement of the *cisterna magna*, and ocular anomalies (Besnard et al., 2016; Chimelli & Avvad-Portari, 2017; de Oliveira-Szejnfeld et al., 2016; de Oliveira Melo et al., 2016; Ventura et al., 2016). A disruption of myelination and changes to the *arachnoid* and *pia mater* also occur, with the latter suggesting disrupted cortical neuron generation (Nakagomi & Matsuyama, 2017). It is noteworthy that cerebral calcifications at the cortical white matter junction diminish or disappear one year post birth (de Lima Petribu et al., 2017), thus indicating that these manifestations should not be used as criteria for late diagnosis.

Molecular pathogenesis

Recent research has started to clarify the molecular mechanisms associated with microcephaly and ZIKV (Chavali et al., 2017; Rosa et al., 2019; Zhang et al., 2016, 2019), and it has been hypothesized that miRNAs may influence ZIKV-host interactions. First identified in the early 1990s, miRNAs interfere with the expression of complementary mRNA and regulation of gene expression (Ambros, 2004). miRNAs function by binding to the three untranslated regions (3′UTR) of target mRNA, blocking translation, or degrading the molecule. Host cell miRNAs can inhibit replication and dissemination of viruses, which themselves may encode miRNAs that assist their replication and control dormancy. An evolving warehouse of evidence has established the importance of miRNAs to normal brain development, function, and pathologies (Barry, 2014; Dang, Tiwari, Qin, & Rana, 2019; Fineberg, Kosik, & Davidson, 2009; Follert, Cremer, & Béclin, 2014). Since miRNAs appear to be comprehensively engaged in the hippocampus, frontal cortex, cerebellum, midbrain, and other nervous tissue genesis (Chen & Qin, 2015; Petri, Malmevik, Fasching, Åkerblom, & Jakobsson, 2014), it follows that their dysfunction likely leads to nerve-related pathologies (Ma, Zhang, & Pearce, 2019; Pan, Sun, & Feng, 2017). Deciphering the roles of miRNAs in response to ZIKV-induced microcephaly provides an approach for identifying key genes and pathways that are connected to the pathology. This includes genes that are silenced, repressed, activated, and/or enhanced. Evidence to support a significant role for miRNAs in the elaboration of ZIKV-induced microcephaly has been derived using in silico, in vitro, and model animal and, among others, human organoid studies (Azouz, Arora, Krause, Nerurkar, & Kumar, 2019; Castro et al., 2019; Dang et al., 2019; Iannolo et al., 2019; Islam, Khan, Murad, Karim, & Islam, 2019; Kozak et al., 2017; Li et al., 2016; McLean, Bhattarai, Hughes, Mahalingam, & Bagasra, 2016; Onorati et al., 2016; Pylro et al., 2016; Xu et al., 2019; Zhang et al., 2016, 2019). These investigations have determined that miRNAs are engaged in disrupting normal cell cycle progression, apoptosis, DNA replication and repair, mRNA transcription, synaptic transmission, signal transduction, and other molecular mechanics that leads to the morphological characteristics of the condition.

Microcephaly genes

To decipher the role of ZIKV in the pathogenesis of microcephaly, we first examined the potential role of miRNAs that share mutual homologies to autosomal recessive primary microcephaly (MCPH). MCPH is a sporadic neurodevelopmental disease characterized by small head circumference >3–4 standard deviations with a significant decrease in cerebral cortex size when compared with age- and sex-matched controls (Fig. 2: 29–30). MCPH is a neurogenic mitotic disorder present at birth and is characterized by nonprogressive mental retardation (26–28). MCPH patients demonstrate abnormal neuronal migration, neuronal apoptosis, and neural function. Generally, MCPH is classified into two forms. Primary microcephaly is a prenatal developmental neurogenic disorder, whereas secondary microcephaly is associated with progressive neurodegenerative disease with or without other syndromic features (Cowie, 1960; Naveed et al., 2018). The incidence of MCPH ranges from 1:30,000 to 1:250,000 per live birth depending on the population (Zaqout, Morris-Rosendahl, & Kaindl, 2017). Therefore, in the regions of the world where consanguineous marriages are permitted (i.e., within Asian and Arabian families), MCPH is more common than in areas where marriage to first cousins is rare or prohibited (Muhammad et al., 2019). Worldwide, MCPH has been reported in ~300 families with patients generally presenting IQs of 30–80, hyperactivity, and a tapered slanting forehead (Bhat et al., 2011; Kaindl et al., 2010).

So far, 18 loci have been mapped for MCPH (MCPH1–MCPH18), and the locations of each of the mutated genes are summarized in Fig. 2 and Table 1 (Faheem, Naseer, & Rasool, 2015; Zaqout et al., 2017). The genes linked to MCPH play a pivotal role in the early stages of neurogenesis, particularly in determining brain size. The proteins produced from these genes play an important role in cell growth, differentiation, and neuronal migration. Mutation in any of the 18 genes is associated with the MCPH phenotype. The most frequent triggers of MCPH are biallelic mutations in ASPM (abnormal spindle-like microcephaly-associated protein) > WDR62 (WD repeat domain 62) gene > MCPH1 (microcephalin) gene, at 68.6%, 14.1%, and 8% incidence, respectively. ASPM is located on the long arm of chromosome 1 as 1q31.3; WDR62 is located on chromosome 19 as 19q13.12, and MCPH1 is found on chromosome 8 as 8p23.1. We believe more genetic mutations will be discovered in the future (Hussain et al., 2012; Passemard, Kaindl, & Verloes, 2013). The following examines both newly identified and previously recognized genes associated with MCPH, twelve of which we have described previously with regard to MCPH and miRNAs (McLean, Bhattarai, Hughes, Mahalingam, & Bagasra, 2017). Each miRNA described shared 100% uniformity at the seed sequences (3′-2-8 base pairs) and between 75% and 100% resemblance at the genetic level, implying that the genetic targets we found most likely retain a significant role in triggering MCPH.

ZIKV and miRNAs

ZIKV has been demonstrated to dysregulate a wide variety of pathways by altering host miRNA profiles (Ferreira et al., 2018). A variety of resources are now available (Gupta et al., 2016; Maracaja-Coutinho et al., 2019; Pylro et al., 2016) that assist in linking and integrating research and clinical findings with ZIKV infection, miRNA, and gene expression profiles. These ZIKV-specific and general databases have increased our understanding not only of the relationships between

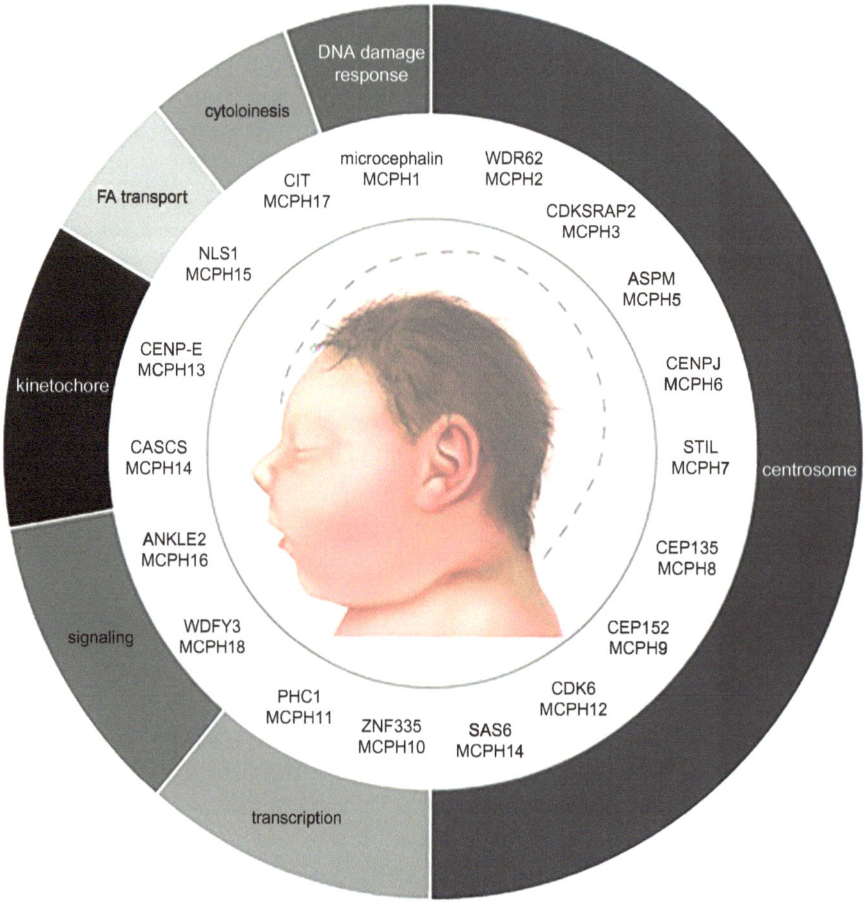

FIG. 2 Side view of microcephalic infant (Wikimedia, CDC) illustrating cranial size reduction from normal (dashed outline), noting MCPH-associated genes and their subcellular targets (Jayaraman, Bae, & Walsh, 2018; www.genecards.org). MCPH, autosomal recessive primary microcephaly; WDR62, WD repeat-containing protein 62; CDK5RAP2, cyclin-dependent kinase-5 regulatory subunit associated protein; CASC5, cancer susceptibility candidate-5; ASPM, abnormal spindle-like primary microcephaly; CENPJ, centromere-associated protein J; STIL, SCL/TAL1 interrupting locus; CEP135, centrosomal protein 135; CEP152, centrosomal protein 152; ZNF335, zinc finger protein 335; PHC1, polyhomeotic-like protein 1; CDK6, cyclin-dependent kinase 6; CENE-P, centromere-associated protein E; SAS6, spindle assembly abnormal protein 6 homolog; NLS1, nuclear localization sequence; ANKLE2, ankyrin repeat and LEM domain-containing protein 2; CI, citron rho-interacting serine/threonine kinase; WDFY3, WD repeat and FYVE domain-containing 3.

TABLE 1 Locations and functions of the 18 MCPH genes.

Gene	Location	Function
ASPM	1q31.3	The most common cause of MCPH (~69%). It is present on spindle poles and centrosomes during mitosis. ASPM plays a fundamental role in organizing microtubules and focusing spindle poles during mitosis. It is involved in orientation of mitotic spindles during embryonic neurogenesis, centriole duplication, spindle organization, and kinetochore association. Mutations in this gene can decrease the size of the brain by influencing the orientation of the mitotic spindle
WDR62	19q13.12	The second most common cause of MCPH (~14%). It encodes the "WD repeat-containing protein 62′," which plays a significant task in the formation of several cellular layers in the cerebral cortex during embryogenesis. It plays a compelling role in the proliferation and migration of neurons and in the duplication of centrioles that are dependent on maternal centrioles
Microcephalin	8p23.1	The third most common mutation in MCPH (~1%) children. It is a BRAT repeats inhibitor of hTERT expression and engaged in chromosomal condensation, cell cycle checkpoint, and DNA damage response. Mutations in this gene affects neurogenesis
CDKRAP2	9q33.2	This gene is engaged in centriole duplication and regulates microtubule function, spindle organization, and kinetochore association. Mutation in CDK5RAP2 reduces the progenitor pool/ decreases the number of neurons and reduces cell survival

TABLE 1 Locations and functions of the 18 MCPH genes—cont'd

Gene	Location	Function
CASC5	15q15.1	The protein encoded by this gene is an important part of the multiprotein assembly that is used for the accurate development of kinetochore–microtubule coupling and chromosome separation. It is vital for the spindle checkpoint of the mitotic cycle. CASC5 underscores the role of kinetochore integrity in the precise volumetric development of the human brain
CENPJ	13q12.12-q12.13	This gene carries out microtubule assembly in the centrosome by causing nucleation and depolymerization of microtubules. It controls centriole length/microtubule function, spindle organization, and kinetochore association. Its deletion causes an increased incidence of multiple spindle poles, apoptosis, and mitosis arrest
STIL	STIL	Mutation in STIL is a rare cause of microcephaly. This gene regulates apoptosis/cell cycle progression, centriole duplication, and ciliogenesis. Its mutation in zebrafish causes an embryonic lethal defect, and STIL knockout mice (Stil-/-) exhibit numerous developmental abnormalities including a decreased size/defective midline neural tube. The expression study demonstrated that STIL is critically involved in early forebrain development that may be associated with the Shh signaling pathway
CEP135	4q12	This gene maintains organization/structure of the centrosome and centriole duplication. CEP135 knockdown has shown decreased growth rate/disorganized microtubules. The centrosomal protein CEP135 plays a key role in the biosynthesis of centrosomes
CEP152	5q21.1	CEP152 gene is localized in centrosomes, and in the brain, it is localized in the cerebral cortex. Centrosomal protein 152 organizes the microtubules of cells and has critical roles in shaping the cell, polarity, receptivity, and cellular division. It is involved in centriole duplication/shape and cell/polarity/motility; conversion of glutamine into proline disturbs potential coiled-coiled protein domain (CCD)/reduced head size
ZNF335	20q13.12	This gene, localized in the cerebral cortex, plays a vital role in neural progenitor cell division/differentiation, neurogenesis, neuronal migration, and spindle orientation. Mutated ZNF335 causes degeneration of neurons, and knockdown of ZNF335 caused a small brain size with an absent cortex/disrupted proliferation and differentiation of neuronal cells
PHC1	12p13.31	PHC1 takes part in the regulation of the cell cycle, and its mutation highlights the role of chromatin remodeling in the pathogenesis of primary microcephaly. It regulates cell cycle and controls DNA damage response
CDK6	7q21.2	The CDK6 protein controls the mechanism of the cell cycle and plays a very significant role in differentiation of different cell types. A mutated CDK6 gene or knockdown CDK6 causes distorted nuclei, disorganization of the spindles, and microtubule and supernumerary centrosomes. It controls cell cycle/organizes microtubules, spindle organization, and kinetochore association. CDK6 mutation affects apical neuronal precursor cell proliferation, reduces the progenitor pool, decreases neuronal production, and causes primary microcephaly
CENPE	4q24	This gene encodes a protein that has signal transducer activity and sequence-specific DNA binding activity. This gene in involved in microtubule, spindle organization, and kinetochore association. Mutation leads to the dysregulation of a large number of genes involved in cell cycle and causes defective DNA repair and cell cycle with reduced proliferative activity
SASS6	1p21.2	The gene is found within the cytoplasm, cytoskeleton, microtubule-organizing center, centrosome, and centrioles. The most critical and fundamental role of the SAS-6 gene is to help in centriole formation by producing a procentriole. Centrioles play a crucial role in cilia and flagella and so are of much importance, and their formation is strongly linked to cell duplication and centrosome replication. Centriole formation through the procentriole complex
MFSD2A	1p34.2	Protein encoded by MFSD2A gene is transmembrane; the encoded protein plays important role in blood–brain barrier and is also important for normal brain function and growth. MFSD2A is mainly localized to the cytosol, plasma membrane, and cytoplasmic bodies in the human brain and other cells. The MFSD2A gene is also present in the cerebral cortex (endothelial cells, neuronal cells, etc.). Homozygous mutation causes problems in the activity of the blood–brain barrier, which leads to severe primary microcephaly
ANKLE2	12q24.33	The protein from this gene is highly involved in proliferation of neuroprogenitor cells during initial brain development and biogenesis of CNS. ANKLE2-mutated genes lead to impaired brain development that reduces brain size as a result of decreased number of neuroblasts, reduced mitosis, less cell proliferation, and increased apoptosis. This improper development of brain contributes to MCPH16 due to recessive mutation of this gene

Continued

TABLE 1 Locations and functions of the 18 MCPH genes—cont'd

Gene	Location	Function
CIT	12q24.23	CIT is critical for the development of normal brain size, and it is also essential for the successful completion of cytokinesis. Homozygous mutation in the kinase domain of CIT causes MCPH by the loss or activation of protein citron kinase
WDFY3	4q21.23	This gene is required for the organization of misfolded proteins that are degraded by autophagy. It acts as a ligand for some molecules involved in the selective macroautophagy in the degradation of the midbody ring. Mutation in this gene leads to abnormal cerebral cortex size and other neurodevelopmental disorders

miRNAs and gene expression but also the numerous pathways implicated in Zika pathogenesis. The following provides examples of some of the dozens of miRNAs understood to be actively engaged in changing gene expression and the possible implications of such associations to development of microcephaly. The variety of studies undertaken so far appear to expose differences between viral strains in terms of pathogenicity and cellular responses of the host (Lee et al., 2018) while also suggesting tight connections, virtually a network, between miRNAs and control over specific pathways.

Kozak et al. (2017) determined that human fetal astrocytes were infectable by ZIKV and supported viral replication. Moreover, ZIKV-infected cells displayed both up- and downregulation in miRNA expression with changes in expression levels over a 72-h timeframe. During the first 24 h postinfection, *miR-17-5p*, *miR-146b-5p*, and *miR-30e-3p* were upregulated $\geq$3-fold. *miR-17-5p* acts on the G1/S-phase cell cycle through its action on more than 20 genes (Cloonan et al., 2008), *miR-146b-5p* is engaged in apoptosis, and *miR-30e* may, through its inhibition of the cytokine TGF-ß, impact cell differentiation, proliferation, and apoptosis. The *miR-30* family is also intimately engaged in endothelial cell growth and angiogenesis and may be partially responsible for vascular remodeling of the microcephalic brain. During the first 24 h of cellular ZIKV infection, therefore, miRNAs work to impact genes that impinge on viral processing and life cycle dynamics while increasing apoptosis and decreasing cell viability as initial phases of the pathology. Even after 72 h, *miR-9-5p*, which was enhanced 7+-fold, resulted in the downregulation of 6 genes involved in mRNA processing, gene expression, and viral processes.

Zhang et al. (2019) demonstrated a role for *miR-9* in mouse cortical development. Overexpression of the microRNA led to animals with smaller cortices, with glial cell-derived neurotrophic factor (GDNF) being an explicit target; GDNF promotes neuronal differentiation and survival, whereas its obstruction caused reduced numbers of neural progenitors and newborn neurons. The same authors reported a significantly reduced number of SOX2-expressing neural progenitors 5 days after infecting E13.5 mice. SOX2 is indispensable for normal myelinization of the brain (Zhang et al., 2018) and development of the eye (Kondoh, Uchikawa, & Ishii, 2016). *miR-9* also plays roles in astrogliosis, axon outgrowth and guidance, and communication between neurons, and microglia *miR-9* and *miR-1273* likewise play a role in the quiescence, renewal, and differentiation of neural stem cells via targeting of the PAX3 and NOTCH2 genes (Bhagat et al., 2018). NOTCH2 is engaged also in vascular remodeling (Peterson et al., 2018). Individual miRNAs, therefore, play different roles at different stages of neuronal development and in different parts of the brain (Liu, Githinji, Wilczzek, & Nolta, 2012; Volvert, Rogister, Moonen, Malgrange, & Nguyen, 2012).

Using a neuroblastoma cell line, Castro et al. (2019) revealed downregulation of eight miRNAs and upregulation of *miR-145*. Furthermore, they were able to confirm the upregulation of *miR-145 and miR-148a* in two stillborn ZIKV-infected brains. Overexpression of *miR-145* suppresses cell proliferation but is also involved in the amelioration of astrocyte damage during ischemia. In fact, *miR-145* appears to be promiscuous and regulates multiple cellular pathways including embryonic angiogenesis via its action on the signal protein VEGF (vascular endothelial growth factor). Ricketson and Lyons-Weiler (2016a, b) identified the human homolog *miR-147a* by BLASTn query of a ZIKV miRNA sequence that targeted 251 predicted human transcripts. Of these neurofascins, which are involved in synapse formation during neural development, synaptic vesicle glycoprotein 2A, a synaptic vesicle protein, and neurofibromin, which relates to cell growth and neural development, were identified as targets. Similar to *miR-145*, *miR-147a* is known to target the VEGF gene, the activin A receptor gene, and, like *miR-17-5p*, impacts the G1 phase progression and G1/S transition through its action on the CDK6 gene.

Among 599 miRNAs examined by array, 62 were found to be downregulated, but only 5 upregulated at least twofold 24 and 48 h following ZIKV infection of mouse cortical neurons (Azouz et al., 2019). Twenty-six of the former and three of the latter (*miR-29a*, *miR-29b*, and *miR-155*) were common to both timepoints, with *miR-124-3p*, *miR-883a-3p*, and *miR-2173* expressing highest levels of downregulation. *miR-124-3p* targets SP1 that is involved with neuronal differentiation, the methylation enzymes DNMT3A/B, the angiogenesis inhibitors COL4A1/2, apoptosis regulators BCL2 and MCL1, the cell cycle regulator PMP22, the astrocytosis and axonal proliferation and neurotrophic factor S100B, and many others. Indeed, 16 of the significantly modulated miRNAs were associated with neurological development, function, and disease. Upregulated miRNAs were apparently involved in viral immunity and inflammatory responses.

In undertaking a bioinformatics analysis of Argonaute-bound RNAs in ZIKV-infected human neuronal stem cells (NSC), Dang et al. (2019) identified upregulated *miR-124-3p* as disrupting their normal maintenance by repression of the transferrin receptor (TFRC). *miR-124-3p* also impacts expression of the brain-derived neurotrophic factor (BDNF) gene. BDNF is active in the cortex and basal forebrain and is engaged in encouraging growth and differentiation of new neurons. The microRNA also targets the ephrin B1 (EFNB1) gene, the encoded protein of which participates in the maintenance of the nervous system. Interestingly, mutations of this gene are causative of craniofrontonasal syndrome that, like ZIKV-induced microcephaly, is characterized by reorganization of the coronal sutures and facial asymmetry. *miR-124-3p* additionally targets beta-secretase 1 (BACE1), which is crucial to myelinization, and the succinyl-CoA ligase subunit beta (SUCLG2) mitochondrial gene, which is expressed in actively biosynthetic tissues and is reported to be involved in supporting microvasculature growth in the brain. Dang et al. (2019) also determined that ZIKV infection of NSC upregulated *hsa-let-7c*. Target genes for *hsa-let-7c* included HGMA2, which is accountable for gene transcription during early development (Dröge & Davey, 2008); the transcription factor MYC; BCL2L1 involved in apoptosis; and, among others, NRAS, which acts as a G-like regulatory protein involved in normal cell growth. *miR-1246* and *miR-335* were also found to be upregulated in ZIKV-infected NSC. The former microRNA is known to impact the dual-specificity tyrosine-phosphorylation-regulated kinase 1A (DYRK1A) gene, mutations of which have been implicated in ASD; the gene may also be involved in learning disorders related with Down syndrome. *miR-1246* has been associated with the chromodomain-helicase-DNA-binding protein 7 (CHD7) gene, the absence of which results in a lack of *corpus callosum*. Targets for *miR-335* are broad and include PTPRN2, involved with cell growth, differentiation, and the mitotic cycle, and BCL2L2, which diminishes apoptosis and may assist in the persistence of nerve growth factor (NGF) and BDNF. The BIRC5 (baculoviral inhibitor of apoptosis repeat-containing 5) or survivin gene, which is an inhibitor of apoptosis, is also targeted at by *miR-335*.

Microcephaly genes

To decipher the role of ZIKA in pathogenesis of microcephaly, we first looked at the potential role of miRs that share mutually homologies to autosomal recessive primary microcephaly (MCPH). MCPH is sporadic neurodevelopmental disease with a small head circumference that is >3–4 standard deviations with a significant decrease in cerebral cortex size as compared with age- and sex-matched controls (Fig. 2: Faheem et al., 2015). MCPH is a neurogenic mitotic disorder present at birth and is characterized by nonprogressive mental retardation (Woods, 2004; Zhu, Chan, Tee, et al., 2016). MCPH patients demonstrate normal neuronal migration, neuronal apoptosis, and neural function. Generally, MCPH is classified into two forms: primary microcephaly (at birth) or secondary microcephaly (onset after birth) with and without other syndromic features (Cowie, 1960). Primary microcephaly is a prenatal developmental neurogenic disorder, whereas secondary microcephaly is associated with progressive neurodegenerative disease (Naveed et al., 2018). The incidence of MCPH ranges from 1:30,000 to 1:250,000 per live birth depending on the population (Zaqout et al., 2017). Therefore, in the regions of the world where consanguineous marriages are permitted (i.e., among Asian and Arab populations), MCPH is more common than in areas of the world where marrying first cousins is rare (Muhammad et al., 2019). Worldwide, MCPH has been reported in ~300 families. Usually, MCPH patients present with IQs of between 30 and 70–80, hyperactivity, and a tapered slanting forehead (Bhat et al., 2011; Kaindl et al., 2010).

The role of miRs in gene regulation is well established and well documented. Currently, there are numerous clinical trials that are underway to treat certain diseases utilizing miRs that target specific genes (Rupaimoole & Slack, 2017; Shen & Corey, 2018). The therapeutic potential of miRs, either as monotherapies or adjuvants, has been recognized for various types of cancer including small cell lung cancer, nonclear cell renal carcinoma, and breast and pancreatic cancers (Barth et al., 2019; Ding et al., 2019; Petrek & Yu, 2019). miRs are also being investigated to treat diabetes, Parkinson's disease, multiple sclerosis, and various cardiopathies (Chen et al., 2019; Goh, Chao, Dheen, Tan, & Tay, 2019; Pang, Phua, & Soh, 2019). Other areas of research include modulation of cell signaling pathways (Chen et al., 2019), and the utility of miRs as diagnostic or prognostic biomarkers represents a major area for discovery (Dai, Kaushik, & Zhang, 2019; Silantyev et al., 2019). In this arena, use of exosomes from blood or other bodily fluids has taken center stage in molecular diagnostics with miRs being investigated to diagnose various types of cancer, allergic diseases, neurologic maladies, neurodevelopmental

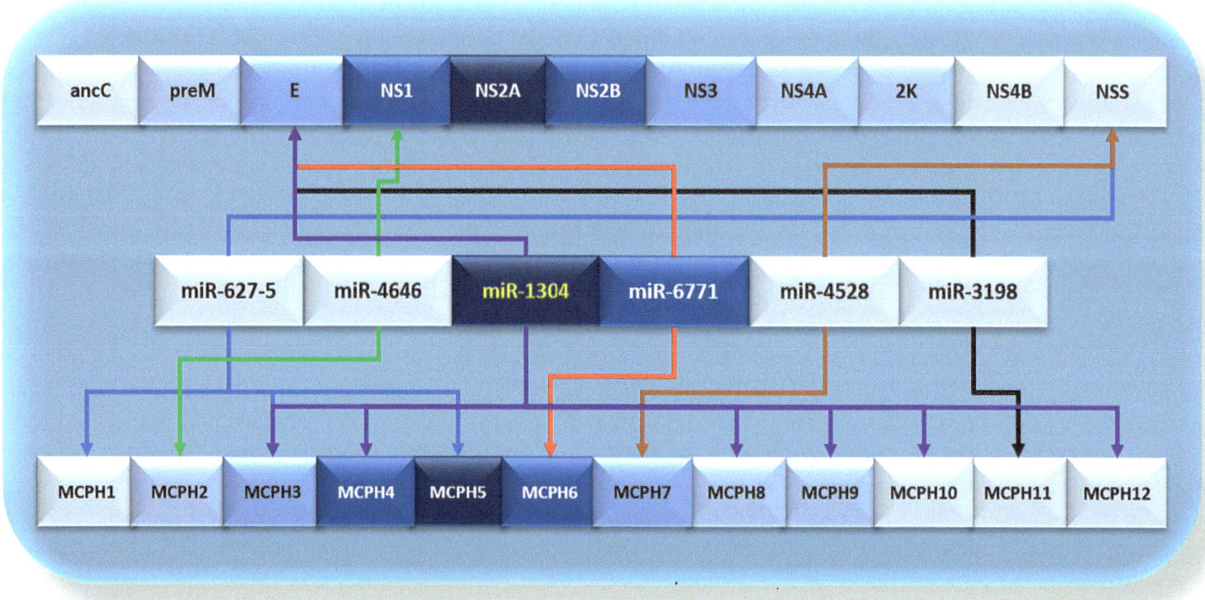

FIG. 3 Locations of miRNA to ZIKV and MCPH. The upper panel illustrates the 10.79-KB genomic map of ZIKV, while the lower panel presents the 12 MCPH genes involved in microcephaly primary heredity (MCPH) or autosomal recessive primary microcephaly. Six miRNAs aligned perfectly at the seed sequences (shown in the middle panel) with the ZIKV and 12 MCPH genomes. Of note, *miR-1304* targets six out of 12 MCPH genes.

disorders, and almost every other type of malady. Research on targeted delivery for miRs has intensified, and use of nanoparticles to convey miRs to precise areas of organs is being investigated (Lee et al., 2019).

In the present study, we have carried out a bioinformatics search for miR binding sites in the prototypic ZIKV MR766 genome to determine their precise locations. We discovered six miRs that share significant mutual homology with the 18 MCPH genetic sequences, which may play a role in inducing MCPH in the developing human fetal brain. The six miRs, namely, *miR-627-5*, *miR-4646*, *miR-1304*, *miR-6771*, *miR-4528*, and *miR-3198*, shared high degrees of mutual homology to the ZIKV genome and the 18 MCPH genes (Table 1 and Fig. 3). In all cases, these miRs shared 100% identity at the seed sequences (3′-2-8 base pairs) and between 75% and 100% homologies at the genetic level, suggesting that the genetic targets we identified impart some degree of influence in regulating MCPH induction (Nunes, Carlini, Marinowic, et al., 2016). Table 1 summarizes the functions of each of the 18 MCPH genes. It appears that three MCPH genes-ASPM and WDR62 genes impart a major effect on neurogenesis during early fetal brain development. It is still unknown as to how each of the 18 MCPH genes precisely cause microcephaly and during which stages of fetal brain development. The dysregulations in the temporal expressions of each of the genes, alone or in combination with MCPH associated genes, due to mutations or ZIKV-induced interference via miRs may be a key factor in development of microcephaly.

MCPH genes and the miRs connection

We analyzed the alignment of various miRs with various MCPH genes. We found one particular microRNA, *hsa-miR-1304*, which showed 100% identify at the seed sequences of six different MCPH genes, namely, microcephalin, CASC5, CEP135, CEP152, ZNF335, and CDK6, and the ZIKV genome (Fig. 3). Therefore it would be logical to hypothesize that this particular miRs plays a pivotal role in development of the cerebral cortex of a developing fetus and dysregulation in these miRs may interfere in fetal neurodevelopment. Of note, any mutations in any of the six MCPH genes, namely, microcephalin, CASC5, CEP135, CEP152, ZNF335, and CDK6, most likely would perturb the normal functioning of associated miRs and likely dislodge their binding sites and perhaps bring about microcephaly among other neurologic abnormalities in the newborn infected with ZIKV. As summarized in Table 1, these genes regulate chromosomal condensation, cell cycle check points, centriole duplication, spindle organization kinetochore-microtubule coupling and chromosome separation, biosynthesis of centrosomes, neurogenesis, neuronal migration and spindle orientation, apical neuronal precursor cell proliferation, and DNA damage response. More recent investigations have shown that *hsa-miR-1304* also plays a role in development of RASopathies. RASopathies are a group of rare genetic diseases resulting from germline mutations in genes involved in the RAS/mitogen-activated protein kinase (RAS/MAPK) pathway. Carvalho, Oliveira, Argolo, and

Ferragut (2017) carried out whole-exome sequencing (WES) in five patients with clinical diagnosis of RASopathies. They determined that variations in nine miRs (i.e., *hsa-miR-1304*, *hsa-miR-146a*, *hsa-miR-196a2*, *hsa-miR-499a/b*, *hsa-miR-449b*, *hsa-miR-548l*, *hsa-miR-575*, and *hsa-miR-593*) can result in dysregulation in RAS, MAPK, RAP1, and PI3K/Akt signaling pathways, neuronal differentiation, neurogenesis and nervous system development, congenital cardiac defects (hypertrophic cardiomyopathy, dilated cardiomyopathy, and arrhythmogenic right ventricular cardiomyopathy), and the phenotypes and syndromes of RASopathies (Noonan syndrome, Legius syndrome, Costello syndrome, multiple café au lait spots, subaortic stenosis, pulmonary valve stenosis, and LEOPARD syndrome).

hsa-miR-627-5 also shares 100% identity to three different MCPH genes—microcephalin, ASPM and CDKRAP2—as well as to ZIKV, HIV-1, and several flaviviruses (Fig. 3: Li et al., 2019), suggesting an important role in neurodevelopment and neurogenesis. Of note, ASPM plays a fundamental role in organizing microtubules and focusing spindle poles during mitosis. It is involved in orientation of mitotic spindles during embryonic neurogenesis, centriole duplication, spindle organization, and kinetochore association. Mutations in this gene can decrease the size of the brain by influencing the orientation of the mitotic spindle. In addition, *hsa-miR-4646* showed 100% identity for MCPH2. MCPH 2 (WDR62), the second most common cause of microcephaly, is responsible for producing multiple cellular layers in the cerebral cortex during early stages of fetal neurogenesis and a key factor in the proliferation and migration of neurons and in the duplication of centrioles that are dependent on mother centrioles.

hsa-miR-4528 and STIL (MCPH7); STIL is a rare cause of microcephaly and in the zebrafish model mutation in this gene causes apoptosis/cell cycle progression, centriole duplication, and ciliogenesis. It plays a critical role in forebrain development. Recent investigations have shown that *miR-4528* is differentially expressed in conjunctival melanoma (Mikkelsen et al., 2019); however, its role in forebrain development still awaits further research. *hsa-miR-3198* with PCH1 (MCPH11) and the ZIKV genome (Tanman, Yangın, & Cansaran-Duman, 2019); PCH1 is one of the genes that regulates the cell cycle—one of the most important processes in fetal development and differentiation. Mutations in this gene dysregulate chromatin remodeling and exhibits its role in the pathogenesis of primary microcephaly. Recent studies have shown that *miR-3198* is differentially expressed in breast cancer cells treated with usnic acid and in primary and recurrent epithelial ovarian cancer (Chong et al., 2015).

MCPH mutations and the miR link

Microcephaly is a rare developmental disorder associated with an abnormally small brain. In our attempt to decipher the underlying mechanisms involved in ZIKV-associated microcephaly, we hypothesized that the 18 microcephaly genes that are associated with the malady might have been dysregulated due to infection during fetal development. Since, in ZIKV-infected fetuses, MCPH genes are not mutated in the newborn, it would be reasonable to assume that ZIKV may be exerting its adverse effects via miRs. Therefore we utilized gene alignment tools to discover the putative miRs that may be dysregulated by ZIKV infection in utero. It should be noted that in microcephalic brains there is an altered cleavage plane during progenitor cell division reducing vertical, symmetric cell division and promoting horizontal, asymmetric cell division. This results in decreased number of progenitor neurons and significantly reduces cortical volume. In mild cases the decrease in the total number of neurons is not accompanied by a commensurate loss of cortical folding. On the other hand, in more severe cases, an additional increase in programmed cell death results in thinned cortices and abnormally small convolutions (Budday, Steinmann, & Kuhl, 2015; Wakabayashi, Hidaka, Fujimaki, et al., 2014). Before the recent ZIKV-associated microcephaly reports, in utero rubella virus (RV)-induced microcephaly was well documented (Lee & Bowden, 2000). Congenital rubella syndrome (CRS) exhibited meningoencephalitis, retinopathy, cataract, deafness, and various other non-neuropathic manifestations (Lee & Bowden, 2000). It is interesting to note that there is an overlap between CRS and the ZIKV-induced neuropathies as exemplified by prominent ocular abnormities in ZIKV-infected newborns (Lee & Bowden, 2000). In addition, there are reports of seizures, low range of motion, and disturbed hearing and vision that overlap with the symptoms of CRS, suggesting that, like RV, an early in utero infection may be the underlying mechanism in both types of infection. This may also explain why the fetus appears to be somewhat sheltered from ZIKV-induced neuropathy (Cerbino-Neto, Mesquita, Souza, et al., 2016; Plourde & Bloch, 2016; Simpson, 1964).

The mechanisms involved in brain development have to wait until appropriate animal models are available, and even if rodent models are developed, such investigations would be difficult to directly applied to *Homo sapiens* fetal brain development since, in an adult *H. sapiens*, there are over 100 billion neurons and 0.15 quadrillion synapses. If one is looking for genetic mutations and comparing research on humans to various rodent animal models, then one must be aware that these comparisons are far from precise. For example, the average adult human weighs 62 kg, is made of 3.7 trillion cells (3.7×10^{13} cells), and lives to about 70 years of average age. On the other hand, a mouse weighs ~20 g, has roughly 1.2×10^{10} cells, and has life span of 3 years. The human genome has about 3 billion base pairs on its 23 pairs of chromosomes, whereas a mouse genome contains approximately 2.7 billion base pairs on 20 pairs of chromosomes. The human brain is astronomically more complex than that of the small rodent, and the inference of findings from studies with mice to the human situation is, at best, perilous. This is not to imply that animal models lack usefulness in suggesting broader implications; it is to say that in the case of microcephaly research, rodent or other small animal models have limitations (Bagasra et al., 2017).

Summary points

(1) microRNAs (miRNAs) are expansively engaged in typical brain development, function, and dysfunction.
(2) They are comprehensively involved in the genesis of the hippocampus, frontal cortex, cerebellum, midbrain, and other nervous tissues.
(3) miRNA dysfunction has been demonstrated to lead to nerve-related pathologies.
(4) The putative role of human miRs is described for the molecular pathogenesis of microcephaly.
(5) There are 12 genes involved in autosomal recessive microcephaly (MCPH).
(6) The role of ZIKV-induced miR dysregulation is described and related to congenital microcephaly.

References

Ambros, V. (2004). The functions of animal microRNAs. *Nature*, *431*, 350–355.

Azouz, F., Arora, K., Krause, K., Nerurkar, V. R., & Kumar, M. (2019). Integrated microRNA and mRNA profiling in Zika virus-infected neurons. *Viruses*, 11. https://doi.org/10.3390/v11020162.

Bagasra, O, Addanki, KC, Goodwin, GR, Hughes, BW, Pandey, P, & McLean, E. (2017). Cellular targets and receptor of sexual transmission of Zika virus. *Applied Immunohistochemistry & Molecular Morphology*, *25*(10), 679–686. https://doi.org/10.1097/PAI.0000000000000580.

Barry, G. (2014). Integrating the roles of long and small non-coding RNA in brain function and disease. *Molecular Psychiatry*, *19*, 410–416.

Barth, D. A., Slaby, O., Klec, C., Juracek, J., Drula, R., Calin, G. A., & Pichler, M. (2019). Current concepts of non-coding RNAs in the pathogenesis of non-clear cell renal cell carcinoma. *Cancers (Basel)*, *11*(10). https://doi.org/10.3390/cancers11101580. pii: E1580.

Besnard, M., Eyrolle-Guignot, D., Guillemette-Artur, P., Lastère, S., Bost-Bezeaud, F., Marcelis, L., et al. (2016). Congenital cerebral malformations and dysfunction in fetuses and newborns following the 2013 to 2014 Zika virus epidemic in French Polynesia. *Euro Surveillance*, *21*, 30181. https://doi.org/10.2807/1560-7917.ES.2016.21.13.30181.

Bhagat, R., Prajapati, B., Narwal, S., Agnihotri, N., Adlakha, Y. K., Sen, J., et al. (2018). Zika virus E protein alters the properties of human fetal neural stem cells by modulating microRNA circuitry. *Cell Death and Differentiation*, *25*, 1837–1854.

Bhat, V, Girimaji, SC, Mohan, G, Arvinda, HR, Singhmar, P, Duvvari, MR, & Kumar, A. (2011). Mutations in WDR62, encoding a centrosomal and nuclear protein, in Indian primary microcephaly families with cortical malformations. *Clinical Genetics*, *80*(6), 532–540. https://doi.org/10.1111/j.1399-0004.2011.01686.x.

Budday, S., Steinmann, P., & Kuhl, E. (2015). Physical biology of human brain development. *Frontiers in Cellular Neuroscience*, *9*, 257. https://doi.org/10.3389/fncel.2015.00257.

Carvalho, A. N., Oliveira, A. R., Argolo, P. S., & Ferragut, F. J. (2017). Three new species of phytoseiid mites (Acari: Phytoseiidae) from Bahia State, Brazil. *Zootaxa*, *4272*(1), 142–150. https://doi.org/10.11646/zootaxa.4272.1.8.

Castro, F. L., Geddes, V. E. V., Monteiro, F. L. L., Gonçalves, R. M. D. T., Campanati, L., Pezzuto, P., et al. (2019). MicroRNAs 145 and 148a are upregulated during congenital Zika virus infection. *ASN Neuro*, *11*. https://doi.org/10.1177/1759091419850983.

Cauchemez, S., Besnard, M., Bompard, P., Dub, T., Guillemette-Artur, P., Eyrolle-Guigno, D., et al. (2016). Association between Zika virus and microcephaly in French Polynesia, 2013–15: a retrospective study. *Lancet*, *387*, 2125–2132.

Cerbino-Neto, J., Mesquita, E. C., Souza, T. M., et al. (2016). Clinical manifestations of zika virus infection, Rio de Janeiro, Brazil. *Emerging Infectious Diseases*, *22*(6), 124–126.

Chavali, P. L., Stojic, L., Meredith, L. W., Joseph, N., Nahorski, M. S., Sanford, T. J., ... Gergely, F. (2017). Neurodevelopmental protein Musashi-1 interacts with the Zika genome and promotes viral replication. *Science*, *357*(6346), 83–88. https://doi.org/10.1126/science.aam9243.

Chen, W., & Qin, C. (2015). General hallmarks of microRNAs in brain evolution and development. *RNA Biology*, *12*, 701–708.

Chen, B., Li, J., Chi, D., Sahnoune, I., Calin, S., Girnita, L., & Calin, G. A. (2019). Non-coding RNAs in IGF-1R signaling regulation: The underlying pathophysiological link between diabetes and cancer. *Cells*, *8*(12). https://doi.org/10.3390/cells8121638. pii: E1638.

Chimelli, L., & Avvad-Portari, E. (2017). Congenital Zika virus infection: A neuropathological review. *Child's Nervous System*. https://doi.org/10.1007/s00381-017-3651-3.

Chong, G. O., Jeon, H. S., Han, H. S., Son, J. W., Lee, Y. H., Hong, D. G., ... Cho, Y. L. (2015). Differential MicroRNA expression profiles in primary and recurrent epithelial ovarian cancer. *Anticancer Research*, *35*(5), 2611–2617.

Cloonan, N., Brown, M. K., Steptoe, A. L., Wani, S., Chan, W.-L., Forrest, A. R. R., et al. (2008). The miR-17-5p microRNA is a key regulator of the G1/S phase cell cycle transition. *Genome Biology*, *9*(8), R127.

Cowie, V. (1960). The genetics and sub-classification of microcephaly. *Journal of Mental Deficiency Research*, *4*, 42–47. https://doi.org/10.1111/j.1365-2788.1960.tb00751.x.

Culjat, M., Darling, S. E., Nerurkar, V. R., Ching, N., Kumar, M., et al. (2016). Clinical and imaging findings in an infant with Zika embryopathy. *Clinical Infectious Diseases*, *63*, 805–811. https://doi.org/10.1093/cid/ciw324.

Dai, X., Kaushik, A. C., & Zhang, J. (2019). The emerging role of major regulatory RNAs in cancer control. *Frontiers in Oncology*, *9*, 920. https://doi.org/10.3389/fonc.2019.00920.

Dang, J. W., Tiwari, S. K., Qin, Y., & Rana, T. M. (2019). Genome-wide integrative analysis of Zika-virus-infected neuronal stem cells reveals roles for microRNAs in cell cycle and stemness. *Cell Reports*, *27*, 3618–3628. https://doi.org/10.1016/j.celrep.2019.05.059.

de Lima Petribu, N. C., de Fatima Vasco Aragao, M., van der Linden, V., Parizel, P., Jungmann, P., Araújo, L., et al. (2017). Follow-up brain imaging of 37 children with congenital Zika syndrome: Case series study. *BMJ*, *359*, j4188. https://doi.org/10.1136/bmj.j4188.

de Oliveira Melo, A. S., Malinger, G., Ximenes, R., Szejnfeld, P. O., Alves Sampaio, S., & Bispo de Filippis, A. M. (2016). Zika virus intrauterine infection causes fetal brain abnormality and microcephaly: Tip of the iceberg? *Ultrasound in Obstetrics & Gynecology*, *47*, 6–7. https://doi.org/10.1002/uog.15831.

de Oliveira-Szejnfeld, P. S., Levine, D., de Oliveira Melo, A. S., Amorim, M. M. R., Batista, A. G. M., Chimelli, L., et al. (2016). Congenital brain abnormalities and Zika virus: What the radiologist can expect to see prenatally and postnatally. *Radiology*, *281*. https://doi.org/10.1148/radiol.2016161584.

Dick, G. W. A. (1953). Epidemiological notes on some viruses isolated in Uganda. Yellow fever, Rift Valley fever, Bwamba fever, West Nile, Semliki Forest, Bunyamwera, Ntaya, Uganda S. and Zika viruses. *Transactions of The Royal Society of Tropical Medicine and Hygiene*, *47*, 13–43. https://doi.org/10.1016/0035-9203(53)90021-2.

Dick, G. W. A., Kitchen, S. F., & Haddow, A. J. (1952). Zika virus (I). Isolations and serological specificity. *Transactions of the Royal Society of Tropical Medicine and Hygiene*, *46*, 509–534. https://doi.org/10.1016/0035-9203(52)90042-4.

Ding, L., Gu, H., Xiong, X., Ao, H., Cao, J., Lin, W., … Cui, Q. (2019). MicroRNAs involved in carcinogenesis, prognosis, therapeutic resistance and applications in human triple-negative breast cancer. *Cells*, *8*(12). https://doi.org/10.3390/cells8121492. pii: E1492.

Dröge, P., & Davey, C. A. (2008). Do cells let-7 determine stemness? *Cell Stem Cell*, *2*, 8–9.

Duffy, M. R., Chen, T. H., Hancock, W. T., Powers, A. M., Kool, J. L., Lanciotti, R. S., et al. (2009). Zika virus outbreak on Yap Island, federated states of Micronesia. *The New England Journal of Medicine*, *360*, 2536–2543. https://doi.org/10.1056/NEJMoa0805715.

Faheem, M., Naseer, M. I., Rasool, M., et al. (2015). Molecular genetics of human primary microcephaly: An overview. *BMC Medical Genomics*, *8*, S4.

Ferreira, R. N., Holanda, G. M., Pinto Silva, E. V., Casseb, S. M. M., Melo, K. F. L., Carvalho, C. A. M., et al. (2018). Zika virus alters the expression profile of microRNA-related genes in liver, lung, and kidney cell lineages. *Viral Immunology*, *31*, 583–588.

Fineberg, S. K., Kosik, K. S., & Davidson, B. L. (2009). MicroRNAs potentiate neural development. *Neuron*, *64*, 303–309. https://doi.org/10.1016/j.neuron:2009.10.020.

Follert, P., Cremer, H., & Béclin, C. (2014). MicroRNAs in brain development and function: A matter of flexibility and stability. *Frontiers in Molecular Neuroscience*, *7*, 5. https://doi.org/10.3389/fnmol.2014.00005.

Goh, S. Y., Chao, Y. X., Dheen, S. T., Tan, E. K., & Tay, S. S. (2019). Role of MicroRNAs in Parkinson's disease. *International Journal of Molecular Sciences*, *20*(22). https://doi.org/10.3390/ijms20225649. pii: E5649.

Gupta, A. K., Kaur, K., Rajput, A., Dhanda, S. K., Sehgal, M., … Khan, Md. S. (2016). ZikaVR: an integrated Zika virus resource for genomics, proteomics, phylogenetic and therapeutic analysis. *Scientific Reports*, *6*, 32713. https://doi.org/10.1038/srep32713.

Hayes, E. B. (2009). Zika virus outside Africa. *Emerging Infectious Diseases*, *15*, 1347–1350.

Henderson, B. E., Hewitt, L. E., & Lule, M. (1968). *Serology of wild mammals*. Virus Research Institute Annual Report (pp. 48–51). Nairobi, Kenya: East African Printer.

Horovitz, D. D. G., da Silva Pone, M. V., Pone, S. M., Salles, T. R. D. S., & Boechat, M. C. B. (2016). Cranial bone collapse in microcephalic infants prenatally exposed to Zika virus infection. *Neurology*, *87*, 118–119.

Hussain, MS, Baig, SM, Neumann, S, Nürnberg, G, Farooq, M, Ahmad, I, … Nürnberg, P. (2012). A truncating mutation of CEP135 causes primary microcephaly and disturbed centrosomal function. *American Journal of Human Genetics*, *90*(5), 871–878. https://doi.org/10.1016/j.ajhg.2012.03.016.

Iannolo, G., Sciuto, M. R., Cuscino, N., Pallini, R., Douradinha, B., Vitiani, L. R., et al. (2019). Zika virus infection induces MiR34c expression in glioblastoma stem cells: New perspectives for brain tumor treatments. *Cell Death & Disease*, *10*, 263.

Islam, M. S., Khan, M. A., Murad, M. W., Karim, M., & Islam, A. B. M. M. K. (2019). In silico analysis revealed Zika virus miRNAs associated with viral pathogenesis through alteration of host genes involved in immune response and neurological functions. *Journal of Medical Virology*, *91*(9), 1584–1594. https://doi.org/10.1002/jmv.25505.

Jayaraman, D., Bae, B.-I., & Walsh, C. A. (2018). The genetics of primary microcephaly. *Annual Review of Genomics and Human Genetics*, *19*, 177–200. https://doi.org/10.1146/annurev-genom-083117-021441.

Kaindl, AM, Passemard, S, Kumar, P, Kraemer, N, Issa, L, Zwirner, A, … Gressens, P. (2010). Many roads lead to primary autosomal recessive microcephaly. *Progress in Neurobiology*, *90*(3), 363–383. https://doi.org/10.1016/j.pneurobio.2009.11.002.

Kleber de Oliveira, W., Cortez-Escalante, J., De Oliveira, W. T., Ikeda do Carmo, G. M., Henriques, C. M. P., Coelho, G. E., & De Franco, G. V. A. (2016). Increase in reported prevalence of microcephaly in infants born to women living in areas with confirmed Zika virus transmission during the first trimester of pregnancy—Brazil, 2015. *MMWR Morbidity Mortality Weekly Reports*, *65*, 242–247. https://doi.org/10.15585/mmwr.mm6509e2.

Kondoh, H., Uchikawa, H., & Ishii, Y. (2016). Multiple roles for SOX2 in eye development. In H. Kondoh, & R. Lovell-Badge (Eds.), *Sox2 biology and role in development and disease* (pp. 217–233). London, UK: Academic Press. 344 pp.

Kozak, R. A., Majer, A., Biondi, M. J., Medina, S. J., Goneau, L. W., Sajesh, B. V., et al. (2017). MicroRNA and mRNA dysregulation in astrocytes infected with Zika virus. *Viruses*, *9*, 297. https://doi.org/10.3390/v9100297.

Lee, J. Y., & Bowden, D. S. (2000). Rubella virus replication and links to teratogenicity. *Clinical Microbiology Reviews*, *13*(4), 571–587.

Lee, I., Bos, S., Li, G., Wang, S., Gadea, G., Despres, P., & Zhao, R. Y. (2018). Probing molecular insights into Zika-host interactions. *Viruses*, *10*. https://doi.org/10.3390/v10050233. pii: E233.

Lee, S. W. L., Paoletti, C., Campisi, M., Osaki, T., Adriani, G., Kamm, R. D., … Chiono, V. (2019). MicroRNA delivery through nanoparticles. *Journal of Controlled Release*, *313*, 80–95. https://doi.org/10.1016/j.jconrel.2019.10.007.

Li, C., Xu, D., Ye, Q., Hong, S., Jiang, Y., Liu, X., et al. (2016). Zika virus disrupts neural progenitor development and leads to microcephaly in mice. *Cell Stem Cell*, *19*, 120–126.

Li, J., Kong, X., Jiang, S., Liao, W., Zhang, Z., Song, J., … Zhang, W. (2019). miR-627/HMGB1/NF-κB regulatory loop modulates TGF-β1-induced pulmonary fibrosis. *Journal of Cellular Biochemistry*, *120*(3), 2983–2993. https://doi.org/10.1002/jcb.27038.

Liu, J., Githinji, M. L. B., Wilczzek, K., & Nolta, J. (2012). Role of miRNAs in neuronal differentiation from human embryonic stem cell-derived neural stem cells. *Stem Cell Reviews*, *8*, 1129–1137.

Ma, Q., Zhang, L., & Pearce, W. J. (2019). MicroRNAs in brain development and cerebrovascular pathophysiology. *American Journal of Physiology. Cell Physiology*, *317*, C3–C19.

Maracaja-Coutinho, V., Paschoal, A. R., Caris-Maldonado, J. C., Borges, P. V., Ferreira, A. J., & Durham, A. M. (2019). Noncoding RNAs databases: Current status and trends. In X. Lai, S. Gupta, & J. Vera (Eds.), *Vol. 1912*. *Computational biology of non-coding RNA*. *Methods in molecular biology*. New York, NY: Humana Press. 451 pp.

McLean, E., Bhattarai, R., Hughes, B. W., Mahalingam, K., & Bagasra, O. (2017). Computational identification of mutually homologous Zika virus miRNAs that target microcephaly genes. *Libyan Journal of Medicine*, *12*(1), 1304505. https://doi.org/10.1080/19932820.2017.1304505. PMID: 28385119.

McLean, E., Bhattarai, R., Hughes, B. W., Mahalingam, K., & Bagasra, O. (2016). Computational identification of mutually homologous Zika virus miRNAs that target microcephaly genes. *Libyan Journal of Medicine*, *V12*. https://doi.org/10.1080/19932820.2017.1304505.

Mikkelsen, L. H., Andersen, M. K., Andreasen, S., Larsen, A. C., Tan, Q., Toft, P. B., … Heegaard, S. (2019). Global microRNA profiling of metastatic conjunctival melanoma. *Melanoma Research*, *29*(5), 465–473. https://doi.org/10.1097/CMR.0000000000000606.

Mlakar, J., Korva, M., Tul, N., Popović, M., Poljšak-Prijatelj, M., Mraz, J., et al. (2016). Zika virus associated with microcephaly. *The New England Journal of Medicine*, *374*, 951–958.

Moore, C. A., Staples, J. E., Dobyns, W. B., Pessoa, A., Ventura, C. V., et al. (2017). Characterizing the pattern of anomalies in congenital Zika syndrome for pediatric clinicians. *JAMA Pediatrics*, *171*, 288–295.

Moura da Silva, A., Ganz, J., Sousa, P., Doriqui, M., Ribeiro, M., Branco, M., et al. (2016). Early growth and neurologic outcomes of infants with probable congenital Zika virus syndrome. *Emerging Infectious Diseases*, *22*, 1953–1956.

Muhammad, S., Tang, Q., Wei, L., Zhang, Q., Wang, G., Muhammad, B. U., … Wang, X. (2019). miRNA-30d serves a critical function in colorectal cancer initiation, progression and invasion via directly targeting the GNA13 gene. *Experimental and Therapeutic Medicine*, *17*(1), 260–272. https://doi.org/10.3892/etm.2018.6902.

Nakagomi, T., & Matsuyama, T. (2017). Leptomeninges: A novel stem cell niche with neurogenic potential. *Stem Cell Investigation*, *4*(22). https://doi.org/10.21037/sci.2017.03.09.

Naveed, M., Kazmi, S. K., Amin, M., Asif, Z., Islam, U., Shahid, K., & Tehreem, S. (2018). Comprehensive review on the molecular genetics of autosomal recessive primary microcephaly (MCPH). *Genetics Research*, *100*, e7. https://doi.org/10.1017/S0016672318000046.

Nunes, M. L., Carlini, C. R., Marinowic, D., et al. (2016). Microcephaly and Zika virus: A clinical and epidemiological analysis of the current outbreak in Brazil. *Pediatrics*, *S0021-7557*(16), 30001–30008.

Olson, J. G., Ksiazek, T. G., & Suhandiman, T. (1981). Zika virus, a cause of fever in Central Java, Indonesia. *Transactions Royal Society Tropical Medicine Hygiene.*, *75*, 389–393.

Onorati, M., Li, Z., Liu, F., Sousa, A. M., Nakagawa, N., Li, M., et al. (2016). Zika virus disrupts phospho-TBK1 localization and mitosis in human neuroepithelial stem cells and radial glia. *Cell Report*, *16*, 2576–2592. https://doi.org/10.1016/j.celrep.2016.08.038.

Pan, Y.-B., Sun, Z.-L., & Feng, D.-F. (2017). The role of microRNA in traumatic brain injury. *Neuroscience*, *367*, 189–199.

Pang, J. K. S., Phua, Q. H., & Soh, B. S. (2019). Applications of miRNAs in cardiac development, disease progression and regeneration. *Stem Cell Research & Therapy*, *10*(1), 336. https://doi.org/10.1186/s13287-019-1451-2.

Passemard, S., Kaindl, A. M., & Verloes, A. (2013). Microcephaly. *Handbook of Clinical Neurology*, *111*, 129–141. https://doi.org/10.1016/B978-0-444-52891-9.00013-0.

Peterson, S. M., Turner, J. E., Harrington, A., Davis-Knowlton, J., Linder, V., Grindley, T., et al. (2018). Notch2 and proteomic signatures in mouse neointimal lesion formation. *Arteriosclerosis, Thrombosis, and Vascular Biology*, *38*, 1576–1593.

Petrek, H., & Yu, A. M. (2019). MicroRNAs in non-small cell lung cancer: Gene regulation, impact on cancer cellular processes, and therapeutic potential. *Pharmacology Research & Perspectives*, *7*(6). https://doi.org/10.1002/prp2.528, e00528.

Petri, R., Malmevik, J., Fasching, L., Åkerblom, M., & Jakobsson, J. (2014). miRNAs in brain development. *Experimental Cell Research*, *321*, 84–89. https://doi.org/10.1016/j.yexcr.2013.09.022.

Plourde, A. R., & Bloch, E. M. (2016). A literature review of zika virus. *Emerging Infectious Diseases*, *22*(7), 1185–1192.

Pylro, V. S., Oliveira, F. S., Morais, D. K., Cuadros-Orellana, S., Pais, F. S.-M., Medeiros, J. D., et al. (2016). ZIKV—CDB: A collaborative database to guide research linking SncRNAs and ZIKA virus disease symptoms. *PLoS Neglected Tropical Diseases*, *10*. https://doi.org/10.1371/journal.pntd.0004817, e0004817.

Rasmussen, S. A., Jamieson, D. J., Honein, M. A., & Petersen, L. R. (2016). Zika virus and birth defects—reviewing the evidence for causality. *The New England Journal of Medicine*, *374*, 1981–1987.

Ricketson, R., & Lyons-Weiler, J. (2016a). The Zika virus sfRNA secondary structure reveals a miR-147a homologue that targets neurofascin as a potential cause of its neurologic syndromes. *Virology*, *7*, WMC005076.

Ricketson, R., & Lyons-Weiler, J. (2016b). Virus induced neurotropic brain injury: Lessons for the study of disease etiology and vaccine development against pathogens. *Virology*, *7*, WMC005082.

Rosa, F. L., Santi, L., Berger, M., Tureta, E. F., Quincozes-Santos, A., Souza, D. O., … Beys-da-Silva, W. O. (2019). ZIKAVID—Zika virus infection database: a new platform to analyze the molecular impact of Zika virus infection. *Journal of Neurovirology*. https://doi.org/10.1007/s13365-019-00799-y.

Rupaimoole, R., & Slack, F. J. (2017). MicroRNA therapeutics: towards a new era for the management of cancer and other diseases. *Nature Reviews. Drug Discovery*, *16*(3), 203–222. https://doi.org/10.1038/nrd.2016.246.

Salgado, D. M., Vega, R., Rodríguez, J. A., Niño, Á., Rodríguez, R., Ortiz, Á., et al. (2020). Clinical, laboratory and immune aspects of Zika virus-associated encephalitis in children. *International Journal of Infectious Diseases*, *90*, 104–110.

Shen, X., & Corey, D. R. (2018). Chemistry, mechanism and clinical status of antisense oligonucleotides and duplex RNAs. *Nucleic Acids Research*, *46*(4), 1584–1600. https://doi.org/10.1093/nar/gkx1239.

Silantyev, A. S., Falzone, L., Libra, M., Gurina, O. I., Kardashova, K. S., Nikolouzakis, T. K., … Tsatsakis, A. (2019). Current and future trends on diagnosis and prognosis of glioblastoma: From molecular biology to proteomics. *Cells*, *8*(8). https://doi.org/10.3390/cells8080863. pii: E863.

Simpson, D. I. (1964). Zika virus infection in man. *Transactions of the Royal Society of Tropical Medicine and Hygiene*, *58*, 335–338.

Tanman, Ü., Yangın, S., & Cansaran-Duman, D. (2019). Determination of dysregulated miRNA expression levels by qRT-PCR after the application of usnic acid to breast cancer. *Anti-Cancer Agents in Medicinal Chemistry*. https://doi.org/10.2174/1871520619666190923163552.

Tolosa, N., Tinker, S. C., Pacheco, O., Valencia, D., Botero, D. S., Tong, V. T., et al. (2017). Zika virus disease in children in Colombia, August 2015 to May 2016. *Paediatric and Perinatal Epidemiology*, *31*, 537–545.

van der Linden, V., Pessoa, A., Dobyns, W., Barkovich, J. A., van der Linden H Jr, Filho ELR, et al. (2016). Description of 13 infants born during October 2015-January 2016 with congenital Zika virus infection without microcephaly at birth—Brazil. *MMWR Morbidity Mortality Weekly Reports*. https://doi.org/10.15585/mmwr.mm6547e2.

Ventura, C. V., Maia, M., Travassos, S. B., Martins, T. T., Patriota, F., Nunes, M. E., et al. (2016). Risk factors associated with the ophthalmoscopic findings identified in infants with presumed Zika virus congenital infection. *JAMA Ophthalmology*, *134*, 912–918.

Volvert, M.-L., Rogister, F., Moonen, G., Malgrange, B., & Nguyen, L. (2012). MicroRNAs tune cerebral cortical neurogenesis. *Cell Death and Differentiation*, *19*, 1573–1581.

Wakabayashi, T., Hidaka, R., Fujimaki, S., et al. (2014). MicroRNAs and epigenetics in adult neurogenesis. *Advances in Genetics*, *86*, 27–44.

Woods, C. G. (2004). Human microcephaly. *Current Opinion in Neurobiology*, *14*, 112–117.

Xu, Y.-P., Qiu, Y., Zhang, B., Chen, G., Chen, Q., Wang, M., et al. (2019). Zika virus infection induces RNAi-mediated antiviral immunity in human neural progenitors and brain organoids. *Cell Research*, *29*, 265–273.

Zaqout, S., Morris-Rosendahl, D., & Kaindl, A. M. (2017). Autosomal recessive primary microcephaly (MCPH): An update. *Neuropediatrics*, *48*(3), 135–142. https://doi.org/10.1055/s-0037-1601448.

Zhang, F., Hammack, C., Ogden, S. C., Cheng, Y., Lee, E. M., Wen, Z., et al. (2016). Molecular signatures associated with ZIKV exposure in human cortical neural progenitors. *Nucleic Acids Research*, *44*, 8610–8620.

Zhang, S., Zhu, K., Gui, K., Croteau, C., Song, L., Xu, J., et al. (2018). Sox2 is essential for oligodendroglial proliferation and differentiation during postnatal brain myelination and CNS remyelination. *The Journal of Neuroscience*, *38*, 1802–1820.

Zhang, H., Chang, Y., Zhang, L., Kim, S.-N., Otaegi, G., Zhang, Z., et al. (2019). Upregulation of microRNA miR-9 is associated with microcephaly and Zika virus infection in mice. *Molecular Neurobiology*, *56*, 4072–4085.

Zhu, Z., Chan, J. F., Tee, K. M., et al. (2016). Comparative genomic analysis of pre-epidemic and epidemic Zika virus strains for virological factors potentially associated with the rapidly expanding epidemic. *Emerging Microbes & Infections*, *5*, e22.

Štrafela, P., Vizjak, A., Mraz, J., Mlakar, J., Pižem, J., Tul, N., … Popović, M. (2017). Zika virus-associated micrencephaly: A thorough description of neuropathologic findings in the fetal central nervous system. *Archives of Pathology & Laboratory Medicine*, *141*, 78–81.

Further reading

Kalsum, A., Kanak, M., & Bagasra, O. (2018). Intercellular communications of microRNAs are mediated by Triplex-Forming Complexes. *Trends in Cell & Molecular Biology*, *13*, 1–17.

O'Connell, R. M., Rao, D. S., & Baltimore, D. (2012). microRNA regulation of inflammatory responses. *Annual Review of Immunology*, *30*, 295–312.

O'Leary, D. R., Kuhn, S., Kniss, K. L., Hinckley, A. F., Rasmussen, S. A., Pape, W. J., et al. (2006). Birth outcomes following West Nile Virus infection of pregnant women in the United States: 2003–2004. *Pediatrics*, *117*, e537–e545.

Specjalski, K., & Jassem, E. (2019). MicroRNAs: Potential biomarkers and targets of therapy in allergic diseases? *Archivum Immunologiae et Therapiae Experimentalis (Warsz)*, *67*(4), 213–223. https://doi.org/10.1007/s00005-019-00547-4.

Tang, H., Hammack, C., Ogden, S. C., Wen, Z., Qian, X., Li, Y., … et al. (2016). Zika virus infects human cortical neural progenitors and attenuates their growth. *Cell Stem Cell*, *18*, 587–590.

Thounaojam, M. C., Kaushik, D. K., Kundu, K., & Basu, A. (2014). MicroRNA-29b modulates Japanese encephalitis virus-induced microglia activation by targeting tumor necrosis factor alpha-induced protein 3. *Journal of Neurochemistry*, *129*, 143–154.

Yang, D., Li, N. L., Wei, D., Liu, B., Guo, F., Elbahesh, H., … Li, K. (2019). The E3 ligase TRIM56 is a host restriction factor of Zika virus and depends on its RNA-binding activity but not miRNA regulation, for antiviral function. *PLOS Neglected Tropical Diseases*, *13*(6), e0007537. https://doi.org/10.1371/journal.pntd.0007537.

Chapter 10

Adherens junctions and cell polarity: What they are and how they relate to congenital Zika virus syndrome

Felipe A. Bustamante-Barrientos[b,c], Roberto Henzi[d], and Luis Federico Bátiz[a,d]

[a]*School of Medicine, Facultad de Medicina, Universidad de los Andes, Santiago, Chile,* [b]*Cells for Cells, Santiago, Chile,* [c]*Immunology Program, Center for Biomedical Research and Innovation (Centro de Investigación e Innovación Biomédica [CiiB]), Universidad de los Andes, Santiago, Chile,* [d]*Neuroscience Program, Center for Biomedical Research and Innovation (Centro de Investigación e Innovación Biomédica [CiiB]), Universidad de los Andes, Santiago, Chile*

Abbreviations

AA	amino acid
AJs	adherens junctions
aPKCλ	atypical protein kinase C-lambda
cKO	conditional knockout
Cyfip1	cytoplasmic FMR1-interacting protein 1
CZS	congenital Zika syndrome
DENV	dengue virus
EC	extracellular cadherin domains or repeats
ECT2	nucleotide exchange factor 2
GJs	GAP junctions
hScrib	human ortholog of Scribble
IPCs	intermediate progenitor cells
JEV	Japanese encephalitis virus
KO	knockout
LAP	leucine-rich repeats and PSD/Dlg/ZO-1 (PDZ)
Lgl1	mammalian homolog of Drosophila lethal giant larvae-1
NFκB	nuclear factor kappa B
NMII-b	nonmuscle myosin II-B
NS2A	nonstructural protein 2A
NS5	nonstructural protein 5
NSPCs	neural stem/progenitor cells
NUMBL	Numb-like protein
RGCs	radial glial cells
RTKs	receptor tyrosine kinases
SMAD7	mothers against decapentaplegic homolog protein
SVZ	subventricular zone
TBEV	tick-borne encephalitis virus
TBK1	TANK binding kinase 1
TJs	tight junctions
TLR3	Toll-like receptor 3
VZ	ventricular zone
Wnt	wingless-related integration site
WNV	West Nile virus

Zika Virus Impact, Diagnosis, Control, and Models. https://doi.org/10.1016/B978-0-12-820267-8.00010-8

ZIKV	Zika virus
ZO-1	zonula occludens-1
ZO-2	zonula occludens-2

Cadherin-based adherens junctions and neural development

Cadherins are calcium-dependent transmembrane glycoproteins that function as anchor points for mediating cell-cell interaction. Cadherin superfamily is a heterogeneous group of proteins that include (i) classic cadherins, (ii) desmosome cadherins, (iii) protocadherins, and (iv) unconventional cadherins (Angst, Marcozzi, & Magee, 2001; Hulpiau & Van Roy, 2009). All cadherins share structural features at the extracellular interaction domain: the extracellular cadherin (EC) repeats, which are similar to the Ig-like domains. Classic (type I) cadherins are characterized by having five ECs (EC1 to EC5). It has been shown that these domains function in *trans*-homophilic adhesion; that is, cells expressing a particular cadherin will preferentially interact with cells expressing the same cadherin. Extracellular calcium concentration is crucial for inducing a structural configuration of ECs that allows the interaction with the opposite EC (Hulpiau & Van Roy, 2009; Tamura, Shan, Hendrickson, Colman, & Shapiro, 1998). Once ECs interact, cadherins function as a bridge through which neighboring cells transmitted mechanical forces connecting actin filaments. It is well established that β-catenin binds to the cytoplasmic tail of the cadherin and also interact with α-catenin, which binds to actin filaments (Jou, Stewart, Stappert, Nelson, & Marrs, 1995). The arrival and stabilization of cadherins at the plasma membrane are finely and dynamically regulated by exocytosis, endocytosis, and recycling (Fig. 1) (Le, Yap, & Stow, 1999). Besides the dimerization and stabilization of cadherins at plasma membrane are regulated by p120ctn, a catenin that interacts with their juxtamembrane region (Yap, Niessen, & Gumbiner, 1998).

Cadherins also function as receptors that transduce cell-cell interaction into intracellular signals (Fig. 1). They modulate various cellular processes and signaling pathways, such as those mediated by Wnt, RTKs, NFκB, and Notch (Klezovitch & Vasioukhin, 2015), pathways with a well-recognized importance on NSPC biology (Navarro Quiroz et al., 2018). Also, cell-cell adhesion plays an essential role in the maintenance of cell survival. Irreversible disruption of cadherin signaling has been proposed as an underlying mechanism for differential susceptibility to death in diverse cell types (Li et al., 2010). Thus normal (nontumoral) cells undergo apoptosis upon loss of cell-cell and/or cell-extracellular matrix attachment, a process known as anoikis (Fouquet et al., 2004; Oliver et al., 2013). On the other hand, loss of apical-lateral interactions triggers abnormal proliferation and differentiation of diverse populations of stem cells, including NSPCs (Kadowaki et al., 2007; Li et al., 2010).

Two major isoforms of classic cadherins are expressed along with brain formation, including epithelial (E)-cadherin and neural (N)-cadherin. E-cadherin is broadly expressed at the ectoderm during early brain formation. With the onset of neurulation, N-cadherin increases its expression and virtually replace E-cadherin in most of the AJs at the ventricular lining, being the main isoform expressed at late stages of embryonic life (Hatta, Takagi, Fujisawa, & Takeichi, 1987). After the neural tube is formed, neuroepithelial cells are organized as a specialized epithelium—the neuroepithelium—that lines the lumen of the ventricular system. These cells preserve typical epithelial features and apical-basal polarity given by the clustering of specific transmembrane proteins that subdivides the plasma membrane into structurally and functionally different domains: apical and basolateral domains. As neurogenesis begins, the gradual differentiation of the neuroepithelium into radial glial cells (RGCs) implies the downregulation of the majority of the molecular constituents of tight junctions (TJs); however, some of them remain clustered at the subapical region of RGCs, becoming part of AJ complexes (Fig. 1) (Aaku-Saraste, Hellwig, & Huttner, 1996). In this context, N-cadherin-based AJs become the primary determinant of the apical-lateral polarity of RGCs, and its functionality is crucial for maintaining the integrity of the ventricular/subventricular zone (VZ/SVZ). Furthermore, we have shown that blocking N-cadherin function in mature ependymal cells with specific peptides that interfere with ECs provokes disruption of AJs and denudation of brain ventricular walls (Oliver et al., 2013).

Since the genetic deletion of N-cadherin causes embryonic lethality (Radice et al., 1997), various animal models have been employed for studying its roles during brain development. A genetic screening performed in zebrafish embryos revealed a series of loci involved in the maintenance of the neural retinal morphology, including the glass onion locus that encodes N-cadherin. In this regard, mutations in the gene encoding N-cadherin induce a marked loss of polarity at the retinal neuroepithelial sheet and several defects in the retinal cytoarchitecture (Malicki, Jo, & Pujic, 2003). On the other hand, experiments performed in chicken embryos reveal that the functional blockade of N-cadherin induces severe morphological defects at specific brain areas, including the presence of neuroepithelial rosettes at the tectum and dorsal thalamus areas. These embryos showed patch-like multifocal lesions characterized by the loss of the apical-basal polarity of NSPCs, as well as disruption of the ventricular lining and the exposure and/or protrusion of the neuropil

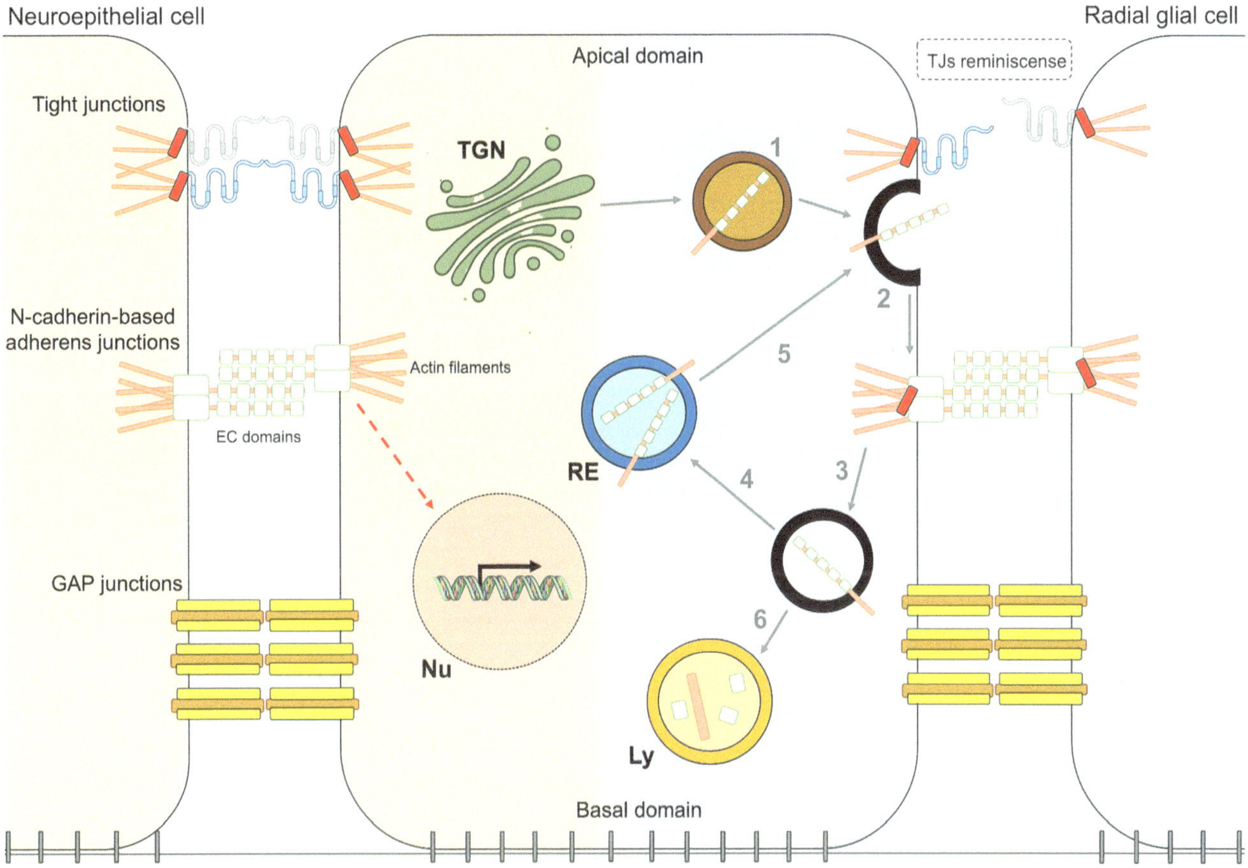

FIG. 1 Junctional complexes between neuroepithelial and radial glial cells (RGCs) and N-cadherin trafficking and cycling. Neuroepithelial cells are joined together by tight junctions (TJs), adherens junctions (AJs), and gap junctions (GJs). RGCs downregulate most of TJ components, making N-cadherin-based AJs (zonula adherens) the junctional complexes responsible for apical-basal cell polarity. ZO-1 proteins *(red squares)* are associated with AJs as neuroepithelial cells become RGCs, and at TJ reminiscences. Note that cadherins not only link actin cytoskeleton between neighboring cells but also function as receptors that transduce cell-cell interaction into intracellular signals *(red dotted arrow)* that can regulate gene expression in the nucleus (Nu). Newly synthesized N-cadherin molecules follow the secretory pathway from the trans-Golgi network (TGN) to the lateral plasma membrane (1), carrier vesicles fuse with plasma membrane, and cadherin molecules laterally diffuse and stabilize at adherens junction (AJ) complexes (2). Extracellular cadherin (EC) domains mediate homophilic interaction with opposite cadherins, and intracellular domains interact with actin filaments (through catenins, not represented). Surface N-cadherin can be endocytosed (3), moved to recycling endosomes (RE) (4), and delivered back to the plasma membrane (5) or targeted for degradation in the lysosomes (Ly) (6). The complexity of the endosomal system and degradative pathways are not fully depicted.

(Ganzler-Odenthal & Redies, 1998). Likewise the development of an N-cadherin conditional knockout (cKO) mouse by inducing specific Cre-recombinase activity at E10.5 in the dorsal telencephalon, which maintains its expression restricted to the cortex and hippocampus, showed a set of molecular and cellular changes that lead to a phenotype characterized by (i) severe disruption of the VZ/SVZ, (ii) protrusion of nonpolarized cells into the ventricles, (iii) defects in neuron positioning, (iv) formation of periventricular rosette-like structures, and (v) ventriculomegaly and/or hydrocephalus (Kadowaki et al., 2007). In summary, these studies and others have demonstrated that the formation and stabilization of N-cadherin-based AJs are fundamental not only to preserve the integrity of the ventricular lining but also to maintain RGC polarity. Thus a failure in N-cadherin-based AJs generates a wide range of molecular and cellular changes, including the relocation of several transmembrane proteins and the disorganization of their basal processes, also known as radial fibers. These phenomena reflect the loss of apical-basal polarity and suppose a poorly favorable scenario to asymmetric cell divisions and cortical lamination mechanisms (Bustamante et al., 2019).

Cell polarity and symmetric versus asymmetric cell division

Apical-basal cell polarity of both neuroepithelial and RGCs comprises a fundamental property for brain development. These cells show (i) an apical domain that points to the ventricular system and conserves specific proteins such as

CD133 (or prominin-1) (Corti et al., 2007), (ii) a lateral domain characterized by a series of adhesive molecules involved in the formation and stabilization of both TJs and AJs (Aaku-Saraste et al., 1996), and (iii) a basal domain that functions as an anchoring structure by retaining some lamina-interacting proteins such as integrin α6 (Haubst, Georges-Labouesse, De Arcangelis, Mayer, & Gotz, 2006).

Why does this polarized organization represent a crucial determinant for brain formation? RGCs give rise to most of the cell types contained at the central nervous system, including neurons, astrocytes, oligodendrocytes, and ependymocytes (Gotz & Huttner, 2005). The genesis of these cells is the outcome of a transition from proliferative (self-renewal) cell divisions to differentiative (neuro/gliogenic) divisions, given by the commitment of multipotent NSPCs with some particular lineage, a process that is known as cell fate decision (Gotz & Huttner, 2005).

Transition to neuro/gliogenic divisions implies asymmetric cell divisions. Symmetric and asymmetric cell divisions are defined as equal versus unequal distribution of cellular components or determinants between two daughter cells, respectively (Huttner & Brand, 1997). In this sense, symmetric divisions imply the inheritance of virtually the same amount and type of apical and basal cellular constituents. In contrast, asymmetric divisions imply an unproportionate inheritance of these determinants and, therefore, the generation of two daughter cells with a different identity and degree of potency (Huttner & Brand, 1997; Kosodo et al., 2004). In this context, apical-basal polarity and the orientation of the cleavage plane may be considered as the basis for determining the ability of NSPCs to divide asymmetrically.

It is well known that asymmetric divisions of founder (apical) NSPCs can generate diverse types of nonmitotic differentiated cells; however, asymmetric divisions can also generate proliferative cellular amplifiers, the so-called basal or intermediate progenitor cells (IPCs). These cells represent a growing population of lineage-committed precursors that reside in the developing subventricular zone (SVZ) and denote an evolutionary strategy to optimize the morphogenesis of diverse tissues, including the brain (Huttner & Brand, 1997; Huttner & Kosodo, 2005). Interestingly, some evidence suggests that defects in the generation of IPCs are a common cause of neurodevelopmental disorders, such as microcephaly (Fish, Kosodo, Enard, Paabo, & Huttner, 2006; Thornton & Woods, 2009).

In addition to be the basis of asymmetric cell divisions, apical-basal polarity of RGCS is also relevant for neuroblast migration and brain morphogenesis. The radial unit hypothesis of neocortical expansion proposes that the long basal processes of RGCs serve as columns connecting the apical (ventricular) and basal (pial) surfaces, through which newly generated neurons are guided and assembled in developing neural circuits (Rakic, 2000). Defects in RGCs and/or their basal processes can provoke defects in neuronal migration and periventricular heterotopias (Ferland et al., 2009).

As stated earlier, adherens junctions play a pivotal role in the polarized architecture of RGCs. Several animal models have emphasized the implication of proteins involved in the formation and/or stabilization of AJs on apical-basal cell polarity of RGCs and brain development. Animal models with a reduced or abolished function of proteins involved in the formation, stabilization, and/or turnover of AJ components, such as (1) neogenin (O'Leary et al., 2017), (2) afadin (Yamamoto et al., 2013), (3) mammalian homolog of *Drosophila lethal giant larvae* (Lgl1) (Klezovitch, Fernandez, Tapscott, & Vasioukhin, 2004), (4) atypical protein kinase C-lambda (aPKCλ) (Imai et al., 2006), (5) Rho-GTPase Cdc42 (Cappello et al., 2006), (6) soluble NSF (*N*-ethylmaleimide-sensitive factor) attachment protein alpha (α-SNAP) (Ferland et al., 2009; Wagner et al., 2003), and (7) cytoplasmic FMR1-interacting protein 1 (Cyfip1) (Yoon et al., 2014), present dysfunctional AJs and loss of cell polarity of NSPCs during brain development. Interestingly, most of these models show similar neurodevelopmental defects, characterized by (i) disruption of the ventricular lining, (ii) disorganization of radial glial fibers, (iii) formation of rosette-like structures, (iv) cortical defects, and (v) ventricular enlargement/hydrocephalus, among others (Table 1). Interestingly, most of the neuropathological changes observed in congenital Zika virus (ZIKV) syndrome (CZS) are surprisingly similar to those observed in these animal models, which highlight the possibility of a pathogenic link between ZIKV and AJ pathology (Bustamante et al., 2019). Table 1 summarizes the neuropathological alterations observed in CZS patients and in different animal models with dysfunctional AJs.

Flaviviruses and their impact on the expression and distribution of cell adhesion molecules

The massive widespread of ZIKV in the northeast of Brazil led to the World Health Organization to declare a public emergency of international concern since it carried a sudden increase in the number of newborns with microcephaly and many other cytoarchitectural brain defects (De Fatima Vasco Aragao et al., 2016; Rasmussen, Jamieson, Honein, & Petersen, 2016). On the other hand, experimental approaches both in vitro and in vivo, as well as neuroimaging and histopathological studies performed in humans, have provided evidence on the pathogenic mechanisms underlying congenital ZIKV infection (Dang et al., 2016; Liang et al., 2016; Onorati et al., 2016). Interestingly, defects in NSPC biology might explain the entire spectrum of brain cytoarchitectural changes in congenital ZIKV disease (Chimelli et al., 2017).

TABLE 1 Comparison of neuropathological alterations in congenital ZIKV syndrome (CZS) patients and animal models with reduced/abolished function of proteins that participate in the maintenance and/or dynamic turnover of AJs.

		Alterations in NSPCs				Developmental neuropathology			
ZIKV or protein (gene)	Patients or animal models	Disruption of AJs	Loss of cell polarity	Defects in cell cycle/ divisions (cell fate)	Disruption of VZ	Protrusion of nonpolarized cells into the ventricles and/or formation of rosette-like structures	Abnormal cortical development	Ventricular enlargement/ hydrocephalus	References
ZIKV	CZS patients	?	?	✓	✓	✓	✓	✓	Melo et al. (2016), De Fatima Vasco Aragao et al. (2016), Chimelli et al. (2017)
Lgl1 (Llgl1)	Llgl1 KO mice	✓	✓	✓	✓	✓	✓	✓	Klezovitch et al. (2004)
NMII-b (Myh10)	NMII-b (Myh10) KO and mutant NMII-b (Myh10) KI mice	✓	N/D	N/D	✓	✓	N/D	✓	Ma, Bao, and Adelstein (2007)
aPKCλ (Prkci)	aPKCλ (Prkci) cKO mice	✓	✓	✓	✓	✓	N/D	✓	Imai et al. (2006)
Cdc42 (Cdc42)	Cdc42 cKO mice	✓	✓	✓	✓	N/D	✓	N/D	Cappello et al. (2006)
Afadin (Afdn)	Afadin (Afdn) cKO mice	✓	✓	✓	✓	✓	✓	✓	Ooshio et al. (2007), Yamamoto et al. (2013)
Neogenin (Neo1)	Neogenin KO mice	✓	✓	N/D	✓	✓	✓	✓	O'Leary et al. (2017)
CYFIP1 (Cyfip1)	CYFIP1 downregulation in developing cortex by in utero electroporation in mice	✓	✓	✓	✓	N/D	✓	N/D	Yoon et al. (2017)

Continued

TABLE 1 Comparison of neuropathological alterations in congenital ZIKV syndrome (CZS) patients and animal models with reduced/abolished function of proteins that participate in the maintenance and/or dynamic turnover of AJs—cont'd

		Alterations in NSPCs			Developmental neuropathology				
ZIKV or protein (gene)	Patients or animal models	Disruption of AJs	Loss of cell polarity	Defects in cell cycle/ divisions (cell fate)	Disruption of VZ	Protrusion of nonpolarized cells into the ventricles and/or formation of rosette-like structures	Abnormal cortical development	Ventricular enlargement/ hydrocephalus	References
KAP3 (Kap3)	Kap3 cKO mice	✓	N/D	✓	✓	✓	✓	✓	Teng et al. (2005)
Numb/ Numb-like (Numb/ Numbl)	Numb/Numbl cKO mice	✓	✓	N/D	✓	✓	✓	✓	Rasin et al. (2007)
Dlg5 (Dlg5)	Dlg5 KO mice	✓	✓	N/D	✓	✓	N/D	✓	Nechiporuk, Fernandez, and Vasioukhin (2007)
α-SNAP (Napa)	hyh (hydrocephalus with hop gait) spontaneous mutant mice.	✓	✓	✓(*)	✓	✓	✓	✓	Wagner et al. (2003), Ferland et al. (2009)

N/D, not determined in the study. (?) Unknown. (*) Unpublished data.
Adapted from Bustamante, F. A., Miro, M. P., Velasquez, Z. D., Molina, L., Ehrenfeld, P., Rivera, F. J., & Batiz, L. F. (2019). Role of adherens junctions and apical-basal polarity of neural stem/progenitor cells in the pathogenesis of neurodevelopmental disorders: A novel perspective on congenital Zika syndrome. *Translational Research*.

Flaviviridae comprises a family of 53 virus species, whose mode of transmission is given by the bite of mosquitoes, ticks, and nonarthropod vectors (Lindenbach, Thiel, & Rice, 2007). Several members of this family, including (i) dengue virus (DENV) (Puerta-Guardo et al., 2013), (ii) Japanese encephalitis virus (JEV) (Agrawal, Sharvani, Nair, & Medigeshi, 2013), (iii) Western Nile virus (WNV) (Xu, Waeckerlin, Urbanowski, Van Marle, & Hobman, 2012), and (iv) tick-borne encephalitis virus (TBEV) (Werme, Wigerius, & Johansson, 2008), can provoke dysfunction of intercellular complexes by inducing abnormal expression and/or distribution of junctional proteins.

An interesting high-throughput mass spectrometry study performed by Chiu et al. revealed a broad panel of nuclear and cytoplasmic molecules that participate in the human response to DENV (Chiu, Hannemann, Heesom, Matthews, & Davidson, 2014). The proteome of DENV-infected A549 cells showed a significant alteration in several proteins involved in the formation and/or stabilization of AJs, such as Ras-related protein Rab-8B, nucleotide exchange factor ECT2, contactin-1, and Dlg5 (Chiu et al., 2014). Similar observations were performed by Puerta-Guardo et al., which demonstrated that flaviviruses are capable of inducing the disorganization of AJs in polarized cells. They showed that DENV-infected U937 human monocyte–derived conditioned medium induces a redistribution of diverse junctional proteins in MDCK cells, such as ZO-2, E-cadherin, and F-actin. Furthermore the epithelial barrier becomes permeable in an irreversible manner (Puerta-Guardo et al., 2013).

Likewise, JEV has proven to be enriched at the points of cell-to-cell contact during early phases of viral replication, generating a severe dysregulation in the expression and distribution of several junctional proteins (Agrawal et al., 2013). In epithelial and hepatocyte-derived cell lines (huh7, Caco-2, and HUVEC), there is a significant reduction in claudin-5, occludin, α-catenin, β-catenin, and E-cadherin levels, 48 h after infection, while claudin-1 levels are reduced through lysosomal degradation during the first 24 h. These effects lead to severe disorganization and dysfunction of junctional complexes, as well as increased epithelial barrier permeability. Interestingly, this function is efficiently rescued by blocking viral replication (Agrawal et al., 2013). On the other hand, the authors described that WNV-infected polarized epithelial and endothelial cells showed increased endocytosis of a specific subset of proteins involved in the stabilization of TJ, such as claudin-1 and JAM-1. Although both of them are destined for lysosomal degradation (Xu et al., 2012), the authors did not observe reduced levels nor mislocalization of TJ membrane proteins, including ZO-1 and occludin. However, these junctional changes were enough to induce loss of the epithelial barrier function (Agrawal et al., 2013).

Lastly, it has been demonstrated that TBEV-derived nonstructural protein 5 (TBEV-NS5) can act as a multifunctional RNA-dependent RNA polymerase with an essential task in viral replication and has been well described by its capability to generate disruption of junctional complexes and loss of cell polarity (Werme et al., 2008). Protein-protein interaction assays reveal that TBEV-NS5 can directly bind to the endogenous human Scribble (hScrib) for maintaining a membranous distribution pattern (Dow et al., 2003). hScrib comprises a scaffold protein that belongs to the LAP (leucine-rich repeats and PSD/Dlg/ZO-1 (PDZ)) family containing 16 leucine-rich repeats and four nonidentical PDZ domains (Santoni, Pontarotti, Birnbaum, & Borg, 2002); both hScrib knockdown and NS5 site-specific mutation lead to the intracellular relocalization of NS5, showing that TBVE subcellular distribution depends on NS5 activity (Werme et al., 2008). Indeed, immunostaining and confocal laser scanning microscopy analyses show that NS5 and hScrib colocalize at the points of cell-to-cell contact and their recruitment occurs depending on the enrichment of cadherins at the apical-lateral domain (Werme et al., 2008).

The current evidence supports that various members of the *Flaviviridae* family can induce loss of cell polarity through abnormal expression and distribution of a broad spectrum of junctional proteins, such as those contained in TJs and AJs. In this regard, Yoon et al. pioneer experiments have ratified that ZIKV-encoded proteins can deregulate the molecular composition and functionality of N-cadherin-based AJs (Yoon et al., 2017). Research studies using human-derived NSPCs, animal models, and in vitro approaches have revealed that ZIKV has a marked tropism by neural cells; however, it is particularly elevated in RGCs (Brault et al., 2016; De Fatima Vasco Aragao et al., 2016; Gabriel et al., 2017; Souza et al., 2016). Considering that N-cadherin-based AJs mainly stabilize the apical-lateral domain of RGCs, and knowing that their disruption has been extensively described as an underlying pathogenic mechanism for various neurodevelopmental defects, an issue arises: How critical are AJs and cell polarity of RGCs during ZIKV infection?

Current knowledge on the impact of N-cadherin-based AJs and apical-basal polarity of RGCs in ZIKV-induced neuropathology

Current research on congenital ZIKV neuropathology has delved into molecular and cellular etiopathogenic mechanisms, including (1) hyperactivation of TLR3, an innate immune receptor (Dang et al., 2016); (2) relocation of phospho-TBK1, a serine/threonine kinase involved in mitotic spindle organization (Onorati et al., 2016); and (3) increased autophagy, via downregulation of Akt/mTOR pathway (Liang et al., 2016). These mechanisms have been associated with decreased survival and abnormal proliferation and differentiation of NSPCs, as well as viral replication.

On the other hand, given that N-cadherin-based AJs become the main type of adhesive complex at the most apical-lateral domain in RGCs (Fig. 1), and their disruption implies a clear loss of cell polarity that precede various brain defects, most of which are also induced by ZIKV (Table 1), a hypothetical interaction between AJs and ZIKV could explain most of the pathological changes so far linked to the previously described mechanisms. As stated before, members of the *Flaviviridae* family conserve the capability to interfere and disorganize AJs, leading to a disruption of the apical-basal polarity of infected cells. These changes seem to be a conserved viral mechanism to favor their replication and enhance their infectivity.

Recent experiments developed by Yoon et al. comprise the first evidence regarding the interaction of ZIKV-encoded proteins and N-cadherin-based AJs. The authors systematically introduced individual proteins encoded by ZIKV and DENV at the mouse developing cortex. Using interactome mapping and biochemical assays to evaluate the interaction of viral proteins with AJ constituents, these authors revealed that ZIKV-NS2A, but not DENV-NS2A, can dysregulate the turnover of junctional proteins at the apical-lateral domain of RGCs, such as (i) β-catenin, (ii) SMAD7, (iii) NUMBL, (iv) PKCλ, and (v) ZO-1 (Yoon et al., 2017). Interestingly, mouse-derived NSPCs showed a tendency to increase N-cadherin expression after being exposed to ZIKV-NS2A and DENV-NS2A, suggesting that these proteins may be pivotal in the adaptive response to the infection by diverse flaviviruses, including ZIKV (Yoon et al., 2017).

In this regard, immunostaining and confocal microscopy approaches demonstrate that ZIKV can induce severe disorganization of N-cadherin-based AJs between telencephalon-residing RGCs, leading to protrusion of nonpolarized cells positioned below the ventricular lining, through patch-like disruption foci (Yoon et al., 2017). On the other hand, in utero electroporation experiments disclose that ZIKV-NS2A expression resulted in a dramatic decrease in the proliferation index, measured as the ratio between EdU+ cells and GFP+Pax+ RGCs (or GFP+ total cells). Interestingly the percentage of Edu+ cells among GFP+Tbr2+ neuronal progenitor cells—a type of IPCs already committed with the neuronal lineage (Gotz & Huttner, 2005)—was also reduced (Yoon et al., 2017). Additionally the number of GFP+Tuj1+ immature neurons among GFP+ cells suggests that RGCs are prematurely differentiating into neurons, which explain the reduced number of GFP+Tbr2+ neuronal progenitor cells (Yoon et al., 2017). Finally, this study shows that ZIKV-NS2A, but not DENV-NS2A, induces marked disorganization of radial fiber scaffolding, which denotes an evident loss of cell polarity (Yoon et al., 2017). Given that newly generated neurons are guided and assembled through radial fibers during brain formation (Gotz & Huttner, 2005), it is expected that this might contribute to aberrant corticogenesis. In this context the authors prove that diverse groups of neurons showed aberrant cortical localization at E19.5, measured as cell body position of GFP+SATB2+ and GFP-CTIP2+ neurons (Yoon et al., 2017).

These results are coherent with previously data concerning the effects of ZIKV infection (Brault et al., 2016; Gabriel et al., 2017) and show that disruption of N-cadherin-based AJs and loss of cell polarity can explain some of the pathogenic hallmarks that characterize ZIKV-infected brains either at the molecular, cellular, and neuropathological level. Therefore disruption of AJs and loss of cell polarity of NSPCs should be considered as a pathogenic pathway in congenital ZIKV syndrome. Furthermore, future research should focus on AJs as a novel therapeutic target to treat this devastating disease successfully.

Policy and procedures

In utero electroporation (IUE). Technique developed in 2001 to introduce plasmid DNA into embryonic mouse (or rat) brains while the embryos are still alive in the uterus (Saito & Nakatsuji, 2001). Thus IUE allows to perform in vivo genetic manipulation of specific cell types, mainly NSPCs located in the ventricular zone, at different brain locations. This procedure, described in detail by Wang and Mei (2013), is an excellent tool for studying genes involved in cell proliferation, fate determination, and migration in the developing mouse/rat brain.

Depending on the embryonic stage, this results in spatially and temporally restricted cell populations carrying the gene of interest. In this context, several research groups have used IUE to study the biology of NSPCs and/or their daughter cells and show how the overexpression or knockdown of specific genes can disrupt normal processes. Furthermore, electroporated embryos can be characterized at any developmental stage, including both prenatal and postnatal (adult) stages, using biochemical, structural, and/or functional studies (Gómez-Molina et al., 2019; Yoon et al., 2017).

Mini-dictionary of terms

Radial glia. Radial glial cells are direct descendant of neuroepithelial cells that preserve their ability to self-renew and differentiate into neural cells; therefore they are considered as neural stem/progenitor cells. Their long basal processes act as scaffolds that guide the migration of newborn neurons.

Cell polarity. The tendency of cells to adopt a particular shape with two or more poles, which differ both structurally and functionally. It implies the asymmetric organization (segregation and maintenance) of several components of the cell, including plasma membrane proteins/lipids and organelles.

N-Cadherin-based adherens junctions. Junctional complexes located at the apical-lateral plasma membrane, which stabilizes cell-to-cell interaction through the homophilic binding with opposite N-cadherin (located in neighboring cells).

Flaviviruses. A single-stranded membrane enveloped RNA virus's family mainly transmitted through the bite of mosquitoes, ticks, and nonarthropod vectors. They are recognized by their harmful effects on human health.

Plasma membrane protein turnover. Referred to as the net result of protein synthesis/insertion at cell surface (exocytosis) and sequestration (endocytosis) for degradation or recycling. Under steady-state conditions, most membrane proteins are continuously turned over.

Key facts of neurodevelopmental disorders

- Neurodevelopmental disorders are a group of pathologies in which the development of the central nervous system is disrupted. This includes structural and functional alterations originated during brain development that leads to brain dysfunction.
- The prevalence of neurodevelopmental disorders varies worldwide; however, financial costs, mortality, and morbidity quantified as quality-adjusted life years (QALYs) or disability-adjusted life years (DALYs) are significant, yet underestimated.
- Even though most classifications of neurodevelopmental disorders include genetic disabilities and disorders due to neurotoxicants, environmental factors such as vertical transmission of viruses from mother to fetus can also induce diverse pathological changes in the developing brain and should be considered.
- Neurodevelopmental disorders are originated at embryologic stages, but assessment and follow-up of these patients are relevant to figure the consequences of these anomalies on child's outcome.
- Increased knowledge of molecular and cellular pathogenic mechanisms common to different neurodevelopmental disorders (AJs pathologies) can pave the way for novel diagnosis and therapeutic strategies.

Summary points

- Adherens junctions (AJs) comprise cell-cell adhesion complexes between radial glial cells (RGCs) that preserve the cytoarchitecture of the ventricular zone and also act as receptors that modulate intracellular signaling pathways.
- Studies performed in both human and animal models have shown that loss of function of proteins involved in the formation and/or stabilization of AJ components gives rise to a neuropathological phenotype characterized by disruption of the ventricular zone, increased cell death, and defective corticogenesis.
- Zika virus (ZIKV) is a flavivirus with a high tropism by neural stem cells.
- A growing body of evidence has revealed that ZIKV replicate a wide spectrum of defects also expressed by disruption of AJs in RGCs.
- Diverse flaviviruses can interact with junctional proteins, inducing their mislocalization and/or abnormal turnover.
- ZIKV proteins can interact with N-cadherin-based AJ components and triggers their disruption, leading to a phenotype that shares several neuropathological changes previously associated to disruption of AJs and loss of cell polarity in RGCs.

Acknowledgments

The authors acknowledge the support of Úrsula Wyneken (Facultad de Medicina, Universidad de los Andes), Nicolás Muñoz (CiiB, Universidad de los Andes), and Jean Carol Arce (Doctorado en Biomedicina, Facultad de Medicina, Universidad de los Andes). This work was supported by Chilean FONDECYT Grants 1141015 and 1211384 (LFB), 3190646 (RH), Chilean CONICYT Doctoral Scholarship 21160084 (FAB-B), and FAI-UANDES Postdoctoral Fellowship (FAB-B).

References

Aaku-Saraste, E., Hellwig, A., & Huttner, W. B. (1996). Loss of occludin and functional tight junctions, but not ZO-1, during neural tube closure—Remodeling of the neuroepithelium prior to neurogenesis. *Developmental Biology*, *180*, 664–679.

Agrawal, T., Sharvani, V., Nair, D., & Medigeshi, G. R. (2013). Japanese encephalitis virus disrupts cell-cell junctions and affects the epithelial permeability barrier functions. *PLoS One*, *8*, e69465.

Angst, B. D., Marcozzi, C., & Magee, A. I. (2001). The cadherin superfamily: Diversity in form and function. *Journal of Cell Science*, *114*, 629–641.

Brault, J. B., Khou, C., Basset, J., Coquand, L., Fraisier, V., Frenkiel, M. P., ... Baffet, A. D. (2016). Comparative analysis between flaviviruses reveals specific neural stem cell tropism for Zika virus in the mouse developing neocortex. *eBioMedicine*, *10*, 71–76.

Bustamante, F. A., Miro, M. P., Velasquez, Z. D., Molina, L., Ehrenfeld, P., Rivera, F. J., & Batiz, L. F. (2019). Role of adherens junctions and apical-basal polarity of neural stem/progenitor cells in the pathogenesis of neurodevelopmental disorders: A novel perspective on congenital Zika syndrome. *Translational Research*, *210*, 57–79.

Cappello, S., Attardo, A., Wu, X. W., Iwasato, T., Itohara, S., Wilsch-Brauninger, M., ... Gotz, M. (2006). The Rho-GTPase cdc42 regulates neural progenitor fate at the apical surface. *Nature Neuroscience*, *9*, 1099–1107.

Chimelli, L., Melo, A. S. O., Avvad-Portari, E., Wiley, C. A., Camacho, A. H. S., Lopes, V. S., ... Tanuri, A. (2017). The spectrum of neuropathological changes associated with congenital Zika virus infection. *Acta Neuropathologica*, *133*, 983–999.

Chiu, H. C., Hannemann, H., Heesom, K. J., Matthews, D. A., & Davidson, A. D. (2014). High-throughput quantitative proteomic analysis of dengue virus type 2 infected A549 cells. *PLoS One*, *9*, e93305.

Corti, S., Nizzardo, M., Nardini, M., Donadoni, C., Locatelli, F., Papadimitriou, D., ... Comi, G. P. (2007). Isolation and characterization of murine neural stem/progenitor cells based on Prominin-1 expression. *Experimental Neurology*, *205*, 547–562.

Dang, J., Tiwari, S. K., Lichinchi, G., Qin, Y., Patil, V. S., Eroshkin, A. M., & Rana, T. M. (2016). Zika virus depletes neural progenitors in human cerebral organoids through activation of the innate immune receptor TLR3. *Cell Stem Cell*, *19*, 258–265.

De Fatima Vasco Aragao, M., Van Der Linden, V., Brainer-Lima, A. M., Coeli, R. R., Rocha, M. A., Sobral Da Silva, P., ... Valenca, M. M. (2016). Clinical features and neuroimaging (CT and MRI) findings in presumed Zika virus related congenital infection and microcephaly: Retrospective case series study. *BMJ*, *353*, i1901.

Dow, L. E., Brumby, A. M., Muratore, R., Coombe, M. L., Sedelies, K. A., Trapani, J. A., ... Humbert, P. O. (2003). hScrib is a functional homologue of the Drosophila tumour suppressor Scribble. *Oncogene*, *22*, 9225–9230.

Ferland, R. J., Batiz, L. F., Neal, J., Lian, G., Bundock, E., Lu, J., ... Sheen, V. L. (2009). Disruption of neural progenitors along the ventricular and subventricular zones in periventricular heterotopia. *Human Molecular Genetics*, *18*, 497–516.

Fish, J. L., Kosodo, Y., Enard, W., Paabo, S., & Huttner, W. B. (2006). Aspm specifically maintains symmetric proliferative divisions of neuroepithelial cells. *Proceedings of the National Academy of Sciences of the United States of America*, *103*, 10438–10443.

Fouquet, S., Lugo-Martinez, V. H., Faussat, A. M., Renaud, F., Cardot, P., Chambaz, J., ... Thenet, S. (2004). Early loss of E-cadherin from cell-cell contacts is involved in the onset of Anoikis in enterocytes. *Journal of Biological Chemistry*, *279*, 43061–43069.

Gabriel, E., Ramani, A., Karow, U., Gottardo, M., Natarajan, K., Gooi, L. M., ... Gopalakrishnan, J. (2017). Recent Zika virus isolates induce premature differentiation of neural progenitors in human brain organoids. *Cell Stem Cell*, *20*, 397–406. e5.

Ganzler-Odenthal, S. I., & Redies, C. (1998). Blocking N-cadherin function disrupts the epithelial structure of differentiating neural tissue in the embryonic chicken brain. *Journal of Neuroscience*, *18*, 5415–5425.

Gómez-Molina, C., Sandoval, M., Henzi, R., Ramírez, J. P., Varas-Godoy, M., Luarte, A., ... Wyneken, U. (2019). Small extracellular vesicles in rat serum contain astrocyte-derived protein biomarkers of repetitive stress. *International Journal of Neuropsychopharmacology*, *22*(3), 232–246. https://doi.org/10.1093/ijnp/pyy098.

Gotz, M., & Huttner, W. B. (2005). The cell biology of neurogenesis. *Nature Reviews Molecular Cell Biology*, *6*, 777–788.

Hatta, K., Takagi, S., Fujisawa, H., & Takeichi, M. (1987). Spatial and temporal expression pattern of N-cadherin cell-adhesion molecules correlated with morphogenetic processes of chicken embryos. *Developmental Biology*, *120*, 215–227.

Haubst, N., Georges-Labouesse, E., De Arcangelis, A., Mayer, U., & Gotz, M. (2006). Basement membrane attachment is dispensable for radial glial cell fate and for proliferation, but affects positioning of neuronal subtypes. *Development*, *133*, 3245–3254.

Hulpiau, P., & Van Roy, F. (2009). Molecular evolution of the cadherin superfamily. *International Journal of Biochemistry & Cell Biology*, *41*, 349–369.

Huttner, W. B., & Brand, M. (1997). Asymmetric division and polarity of neuroepithelial cells. *Current Opinion in Neurobiology*, *7*, 29–39.

Huttner, W. B., & Kosodo, Y. (2005). Symmetric versus asymmetric cell division during neurogenesis in the developing vertebrate central nervous system. *Current Opinion in Cell Biology*, *17*, 648–657.

Imai, F., Hirai, S., Akimoto, K., Koyama, H., Miyata, T., Ogawa, M., ... Ohno, S. (2006). Inactivation of aPKC lambda results in the loss of adherens junctions in neuroepithelial cells without affecting neurogenesis in mouse neocortex. *Development*, *133*, 1735.

Jou, T. S., Stewart, D. B., Stappert, J., Nelson, W. J., & Marrs, J. A. (1995). Genetic and biochemical dissection of protein linkages in the cadherin-catenin complex. *Proceedings of the National Academy of Sciences of the United States of America*, *92*, 5067–5071.

Kadowaki, M., Nakamura, S., Machon, O., Krauss, S., Radice, G. L., & Takeichi, M. (2007). N-Cadherin mediates cortical organization in the mouse brain. *Developmental Biology*, *304*, 22–33.

Klezovitch, O., Fernandez, T. E., Tapscott, S. J., & Vasioukhin, V. (2004). Loss of cell polarity causes severe brain dysplasia in Lgl1 knockout mice. *Genes & Development*, *18*, 559–571.

Klezovitch, O., & Vasioukhin, V. (2015). Cadherin signaling: Keeping cells in touch. *F1000Res*, *4*, 550.

Kosodo, Y., Roper, K., Haubensak, W., Marzesco, A. M., Corbeil, D., & Huttner, W. B. (2004). Asymmetric distribution of the apical plasma membrane during neurogenic divisions of mammalian neuroepithelial cells. *EMBO Journal*, *23*, 2314–2324.

Le, T. L., Yap, A. S., & Stow, J. L. (1999). Recycling of E-cadherin: A potential mechanism for regulating cadherin dynamics. *Journal of Cell Biology*, *146*, 219–232.

Li, L., Wang, B. H., Wang, S., Moalim-Nour, L., Mohib, K., Lohnes, D., & Wang, L. (2010). Individual cell movement, asymmetric colony expansion, rho-associated kinase, and E-cadherin impact the clonogenicity of human embryonic stem cells. *Biophysical Journal*, *98*, 2442–2451.

Liang, Q., Luo, Z., Zeng, J., Chen, W., Foo, S. S., Lee, S. A., … Jung, J. U. (2016). Zika virus NS4A and NS4B proteins deregulate Akt-mTOR signaling in human fetal neural stem cells to inhibit neurogenesis and induce autophagy. *Cell Stem Cell*, *19*, 663–671.

Lindenbach, B. D., Thiel, H. J., & Rice, C. M. (2007). *Flaviviridae: The viruses and their replication*. Philadelphia, Pennsylvania: Lippincott, Williams & Wilkins.

Ma, X., Bao, J., & Adelstein, R. S. (2007). Loss of cell adhesion causes hydrocephalus in nonmuscle myosin II-B-ablated and mutated mice. *Molecular Biology of the Cell*, *18*, 2305–2312.

Malicki, J., Jo, H., & Pujic, Z. (2003). Zebrafish N-cadherin, encoded by the glass onion locus, plays an essential role in retinal patterning. *Developmental Biology*, *259*, 95–108.

Melo, A. S., Aguiar, R. S., Amorim, M. M., Arruda, M. B., Melo, F. O., Ribeiro, S. T., … Tanuri, A. (2016). Congenital Zika virus infection: Beyond neonatal microcephaly. *JAMA Neurology*, *73*, 1407–1416.

Navarro Quiroz, E., Navarro Quiroz, R., Ahmad, M., Gomez Escorcia, L., Villarreal, J. L., Fernandez Ponce, C., & Aroca Martinez, G. (2018). Cell signaling in neuronal stem cells. *Cells*, *7*, 7–75.

Nechiporuk, T., Fernandez, T. E., & Vasioukhin, V. (2007). Failure of epithelial tube maintenance causes hydrocephalus and renal cysts in Dlg5−/− mice. *Developmental Cell*, *13*, 338–350.

O'Leary, C. J., Nourse, C. C., Lee, N. K., White, A., Langford, M., Sempert, K., … Cooper, H. M. (2017). Neogenin recruitment of the WAVE regulatory complex to ependymal and radial progenitor adherens junctions prevents hydrocephalus. *Cell Reports*, *20*, 370–383.

Oliver, C., Gonzalez, C. A., Alvial, G., Flores, C. A., Rodriguez, E. M., & Batiz, L. F. (2013). Disruption of CDH2/N-cadherin-based adherens junctions leads to apoptosis of ependymal cells and denudation of brain ventricular walls. *Journal of Neuropathology and Experimental Neurology*, *72*, 846–860.

Onorati, M., Li, Z., Liu, F. C., Sousa, A. M. M., Nakagawa, N., Li, M. F., … Sestan, N. (2016). Zika virus disrupts phospho-TBK1 localization and mitosis in human neuroepithelial stem cells and radial glia. *Cell Reports*, *16*, 2576–2592.

Ooshio, T., Fujita, N., Yamada, A., Sato, T., Kitagawa, Y., Okamoto, R., … Takai, Y. (2007). Cooperative roles of Par-3 and afadin in the formation of adherens and tight junctions. *Journal of Cell Science*, *120*, 2352–2365.

Puerta-Guardo, H., Raya-Sandino, A., Gonzalez-Mariscal, L., Rosales, V. H., Ayala-Davila, J., Chavez-Mungia, B., … Del Angel, R. M. (2013). The cytokine response of U937-derived macrophages infected through antibody-dependent enhancement of dengue virus disrupts cell apical-junction complexes and increases vascular permeability. *Journal of Virology*, *87*, 7486–7501.

Radice, G. L., Rayburn, H., Matsunami, H., Knudsen, K. A., Takeichi, M., & Hynes, R. O. (1997). Developmental defects in mouse embryos lacking N-cadherin. *Developmental Biology*, *181*, 64–78.

Rakic, P. (2000). Radial unit hypothesis of neocortical expansion. *Novartis Foundation Symposium*, *228*, 30–42. (discussion 42–52.

Rasin, M. R., Gazula, V. R., Breunig, J. J., Kwan, K. Y., Johnson, M. B., Liu-Chen, S., … Sestan, N. (2007). Numb and Numbl are required for maintenance of cadherin-based adhesion and polarity of neural progenitors. *Nature Neuroscience*, *10*, 819–827.

Rasmussen, S. A., Jamieson, D. J., Honein, M. A., & Petersen, L. R. (2016). Zika virus and birth defects—Reviewing the evidence for causality. *New England Journal of Medicine*, *374*, 1981–1987.

Saito, T., & Nakatsuji, N. (2001). Efficient gene transfer into the embryonic mouse brain using in vivo electroporation. *Developmental Biology*, *240*(1), 237–246. https://doi.org/10.1006/dbio.2001.0439.

Santoni, M. J., Pontarotti, P., Birnbaum, D., & Borg, J. P. (2002). The LAP family: A phylogenetic point of view. *Trends in Genetics*, *18*, 494–497.

Souza, B. S. F., Sampaio, G. L. A., Pereira, C. S., Campos, G. S., Sardi, S. I., Freitas, L. A. R., … Soares, M. B. P. (2016). Zika virus infection induces mitosis abnormalities and apoptotic cell death of human neural progenitor cells. *Scientific Reports*, *6*, 39775.

Tamura, K., Shan, W. S., Hendrickson, W. A., Colman, D. R., & Shapiro, L. (1998). Structure-function analysis of cell adhesion by neural (N-) cadherin. *Neuron*, *20*, 1153–1163.

Teng, J., Rai, T., Tanaka, Y., Takei, Y., Nakata, T., Hirasawa, M., … Hirokawa, N. (2005). The KIF3 motor transports N-cadherin and organizes the developing neuroepithelium. *Nature Cell Biology*, *7*, 474–482.

Thornton, G. K., & Woods, C. G. (2009). Primary microcephaly: Do all roads lead to Rome? *Trends in Genetics*, *25*, 501–510.

Wagner, C., Batiz, L. F., Rodriguez, S., Jimenez, A. J., Paez, P., Tome, M., … Rodriguez, E. M. (2003). Cellular mechanisms involved in the stenosis and obliteration of the cerebral aqueduct of hyh mutant mice developing congenital hydrocephalus. *Journal of Neuropathology and Experimental Neurology*, *62*, 1019–1040.

Wang, C., & Mei, L. (2013). In utero electroporation in mice. *Methods in Molecular Biology*, *1018*, 151–163. https://doi.org/10.1007/978-1-62,703-444-9_15.

Werme, K., Wigerius, M., & Johansson, M. (2008). Tick-borne encephalitis virus NS5 associates with membrane protein scribble and impairs interferon-stimulated JAK-STAT signalling. *Cellular Microbiology*, *10*, 696–712.

Xu, Z., Waeckerlin, R., Urbanowski, M. D., Van Marle, G., & Hobman, T. C. (2012). West Nile virus infection causes endocytosis of a specific subset of tight junction membrane proteins. *PLoS One*, *7*, e37886.

Yamamoto, H., Maruo, T., Majima, T., Ishizaki, H., Tanaka-Okamoto, M., Miyoshi, J., … Takai, Y. (2013). Genetic deletion of afadin causes hydrocephalus by destruction of adherens junctions in radial glial and ependymal cells in the midbrain. *PLoS One*, *8*, e80356.

Yap, A. S., Niessen, C. M., & Gumbiner, B. M. (1998). The juxtamembrane region of the cadherin cytoplasmic tail supports lateral clustering, adhesive strengthening, and interaction with p120ctn. *Journal of Cell Biology*, *141*, 779–789.

Yoon, K. J., Nguyen, H. N., Ursini, G., Zhang, F., Kim, N. S., Wen, Z., ... Ming, G. L. (2014). Modeling a genetic risk for schizophrenia in iPSCs and mice reveals neural stem cell deficits associated with adherens junctions and polarity. *Cell Stem Cell*, *15*, 79–91.

Yoon, K. J., Song, G., Qian, X., Pan, J., Xu, D., Rho, H. S., ... Ming, G. L. (2017). Zika-virus-encoded NS2A disrupts mammalian cortical neurogenesis by degrading adherens junction proteins. *Cell Stem Cell*, *21*, 349–358. e6 https://doi.org/10.1016/j.stem.2017.07.014.

Section C

Guillain-Barré syndrome

Chapter 11

Severe Guillain-Barré syndrome

Selman Kesici and Benan Bayrakci

Department of Pediatrics, Division of Pediatric Critical Care Medicine, Hacettepe University, İhsan Doğramacı Children's Hospital, Altındağ/Ankara, Turkey

Abbreviations

AIDP acute inflammatory demyelinating polyneuropathy
AMAN acute motor axonal neuropathy
CSF cerebrospinal fluid
EMG electromyography
GBS Guillain-Barré syndrome
ICU intensive care unit
IVIG intravenous immunoglobulin
LOS lipooligosaccharides
MAC membrane attack complex
MRI magnetic resonance imaging
PE plasma exchange

Introduction

Guillain-Barré syndrome (GBS) is an acute, usually postinfectious neuropathy of common occurrence with a yearly incidence rate between 1.1 and 1.8 per100,000 (McGrogan, Madle, Seaman, & de Vries, 2009). GBS incidence increases exponentially with age, and male subjects are more commonly affected (Sejvar, Baughman, Wise, & Morgan, 2011). The disorder is a polyradiculoneuropathy and involves mainly motor occasionally sensory and autonomic nerves presenting as acute, usually symmetrical, typically ascending paralysis. It may involve bulbar weakness, respiratory failure, and autonomic nervous system dysfunction. GBS usually follows a preceding infection or rarely vaccination. The most common preceding infection causing GBS is *Campylobacter jejuni* enteritis. Other infectious agents preceding GBS include cytomegalovirus, Epstein-Barr virus, *Mycoplasma pneumoniae*, and *Haemophilus influenzae* (Parra et al., 2016; van Doorn, Ruts, & Jacobs, 2008). Molecular mimicry between microbial and nerve antigens is clearly underlying pathophysiological mechanism behind the development of the disorder. Less than 1:1000 patients with a *C. jejuni* infection develop GBS. Host factors may influence this susceptibility to develop GBS or the extent of nerve damage and outcome (van Doorn et al., 2008). Rapidly progressive weakness is the main clinical feature of GBS. The acute progression of limb weakness, often with sensory and cranial nerve involvement 1–2 weeks after immune stimulation, proceeds to its peak clinical deficit in 2–4 weeks (Fokke et al., 2014). Maximal weakness is reached within 4 weeks, but most patients have reached their maximal weakness within 2 weeks (Asbury & Cornblath, 1990). Patients then have a plateau phase of variable duration (days to weeks); this phase is followed by a recovery phase of variable duration (van Doorn, 2013). GBS can be classified into two major subtypes, acute inflammatory demyelinating polyneuropathy (AIDP) and acute motor axonal neuropathy (AMAN), depending on whether the myelin or the axonal components of the peripheral nerves are affected (El-Bayoumi et al., 2011). Although demyelinating form is more common, motor symptoms and weakness are more prominent in axonal form (Hicks, Kay, Worley, & Moodley, 2010). Treatment of GBS comprises intravenous immunoglobulin (IVIG) or plasma exchange (PE). But both IVIG and PE are well-accepted and proven treatments for GBS (van der Meche & Schmitz, 1992). The prognosis for recovery in patients with GBS can be favorable, with only minor residual symptoms; however, the mortality rate ranges from 2% to 12%. Approximately one-third of GBS patients are admitted to an intensive care unit (ICU), and many require mechanical ventilation (Sharshar et al., 2003). There are clinical and electrophysiological findings that related with poor prognosis (Table 1). Despite the demonstrated efficacy of PE and IVIG, GBS however remains a disabling disease in a significant proportion of patients, and these treatments have not improved mortality (Rajabally & Uncini, 2012).

Zika Virus Impact, Diagnosis, Control, and Models. https://doi.org/10.1016/B978-0-12-820267-8.00011-X

TABLE 1 Predictors of poor prognosis in GBS.

Clinical	Age >40 or 50 years
	Preceding diarrhea (*C. jejuni*)
	Short interval between weakness onset and admission
	Reduced vital capacity
	Need for mechanical ventilation
	Low MRC sum score at admission
	Low MRC sum score at day 7 postadmission
	Facial and/or bulbar weakness
Electrophysiological	Inexcitable nerves
	Low summated distal CMAP (<20% of normal)
	Motor axonal form

MRC, Medical Research Council; *CMAP*, compound muscle action potential.

Pathophysiology of Guillain-Barré syndrome

Accumulated evidence supports the likelihood that GBS is an autoimmune disorder, caused by self-antibodies targeting various components of peripheral nerve myelin and sometimes the axon. Antecedent infections appear to play an important role in the development of GBS. Both cellular and humoral parts of the immune system are involved in GBS pathogenesis (Rosen, 2012) (Table 2). Antiganglioside antibodies, most commonly GM1, occur frequently in GBS. Some appear to target the myelin sheath, others the nodes of Ranvier and the axon (Rosen, 2012). *C. jejuni* isolates from GBS patients express lipooligosaccharides (LOS) that mimic the carbohydrates of gangliosides. Antibodies in these patients recognize LOS as well as gangliosides or ganglioside complexes. GBS after *Campylobacter* infection in anti-GM1/GD1a/GQ1b antibody-related cases is considered to be true example of molecular mimicry-related disease (van Doorn et al., 2010). The carbohydrate moieties of gangliosides elicit a T-cell-independent humoral response, and antiganglioside antibodies exist as low-affinity IgM isotypes in normal subjects. The level and affinity of these antibodies are controlled by tolerance in normal subjects, and 99% of humans infected with ganglioside-mimicking strains of *C. jejuni* do not develop significant anti-LOS/ganglioside antibodies or GBS (Willison, 2005b). In GBS the antibody response has class-switched from IgM to complement-fixing IgG1 and IgG3 isotypes (Burns, 2008) (Table 2). Nerve terminal damage follows antibody binding and

TABLE 2 Chain reaction playing role in the pathophysiology of GBS.

Preceding infection (*C. jejuni*, Zika virus)
Emergence of antibodies in the form of Ig G1-G3 subtype
Binding of autoantibodies to Fc receptors
Leukocyte degranulation
Activation of the complement pathway
Membrane attack complex formation
Macrophage activation
Hypercytokinemia
Phagocytosis

The pathophysiological mechanism in each step triggers the next step, and also each step stimulates the pathophysiological mechanisms in the previous steps.

complement fixation. Activation of the complement pathway leads to membrane attack complex (MAC) formation with degradation of the terminal axonal cytoskeleton and mitochondrial injury (Burns, 2008). The macrophages dissect into the internodal periaxonal spaces, displacing axons from inner Schwann cell plasmalemma, and appear to cause axonal degeneration (Griffin et al., 1996). Macrophages are considered crucial effector cells in the pathogenesis of GBS, and myelin-loaded macrophages have been detected in most demyelinating diseases (Griffin, Stoll, Li, Tyor, & Cornblath, 1990). Macrophages are potent immunomodulatory phagocytes capable of such diverse effector functions as cytokine production, cytotoxicity, and phagocytosis (Table 2). FcgRIIIa is expressed by macrophages, which are important effector cells in the pathogenesis of GBS (Griffin et al., 1990). FcgR provides a link between the cellular and humoral immune system. FcgRIIIa and FcgRIIIb genotypes were associated with severity of GBS (van Sorge et al., 2005). GBS has two major subtypes: acute inflammatory demyelinating neuropathy (AIDP) and acute motor axonal neuropathy (AMAN). The pathological processes underlying AIDP and AMAN are different, but the final pathway is common. In AIDP, complement activation and membrane attack complex formation on Schwann cell surface mainly cause vesicular demyelination (Hafer-Macko, Sheikh, et al., 1996). In AMAN, antibodies against motor axons lead to complement-mediated membrane attack complex formation at the nodal axolemma, and, in severe cases, axonal degeneration of the motor axons develops (Hafer-Macko, Hsieh, et al., 1996). Patients who develop AMAN following *C. jejuni* enteritis have IgG antibodies against GM1 (Yuki, Watanabe, Nakajima, & Spath, 2011). But patients who have *C. jejuni* enteritis with no neurological involvement do not have similar autoantibodies. Autopsy studies show axolemmal deposition of IgG on the anterior spinal root in AMAN patients (Hafer-Macko, Hsieh, et al., 1996), indicating that IgG, which binds effectively with complement, is an important factor in the development of disease. These findings suggest that GM1 is an autoantigen for IgG antibodies in patients with AMAN related to *C. jejuni* enteritis. IgG anti-GD1a antibodies are also associated with AMAN subsequent to *C. jejuni* infection (Ho et al., 1999). Patients with severe, rapidly progressive GBS most likely have an intense autoantibody production with a high percentage of these antibodies already bound to nerves upon development of respiratory failure.

Clinical presentation

GBS should be considered as a diagnosis in patients who have rapidly progressive bilateral weakness of the legs and/or arms. Patients with the classic sensorimotor form of GBS present with distal paresthesias or sensory loss, followed by weakness that starts in the legs and progresses to the arms and cranial muscles (Leonhard et al., 2019). Rapidly progressive bilateral weakness is the key presenting symptom in most patients (Asbury & Cornblath, 1990). Weakness is classically described as ascending and usually starts in the distal lower extremities. Symmetric weak muscles are accompanied by depressed or absent deep tendon reflexes. Facial and oropharyngeal muscles are affected in half of the cases, and weakness of these muscle groups may be the initial manifestation (Wijdicks & Klein, 2017). A small number of patients present with paraparesis, which can remain during the course of the disease (van den Berg, Fokke, Drenthen, van Doorn, & Jacobs, 2014). Some of the patients present with cranial nerve involvement resulting in facial, oculomotor, or bulbar weakness, as in Miller Fisher syndrome (Willison, Jacobs, & van Doorn, 2016). Patients might initially have sensory signs, ataxia, and features of autonomic dysfunction. Patients can have progression of weakness within 4 weeks. Most patients, however, reach the nadir within 2 weeks (Fokke et al., 2014). Respiratory failure develops in 20%–30% of the patients, and they need mechanical ventilation (Fokke et al., 2014). During the all phases of the disease, patients might have signs or symptoms of autonomic dysfunction like cardiac arrhythmia, excessive sweating, blood pressure instability, and ileus and bladder dysfunction (van den Berg, Walgaard, et al., 2014). GBS has a monophasic clinical course, but in few patients treatment-related fluctuations and relapses may occur in a minority of patients (Ruts, Drenthen, Jacobs, van Doorn, & Dutch, 2010).

Severe GBS

GBS disability scale (Hughes score) is used to determine the severity of disease (Hughes, Newsom-Davis, Perkin, & Pierce, 1978; Plasma Exchange/Sandoglobulin Guillain-Barre Syndrome Trial Group, 1997) (Table 3). Severity of symptoms is assessed using Hughes grades and patients with Hughes grade greater or equal to 3 defined as severe GBS (Galldiks, Dohmen, Neveling, Fink, & Haupt, 2009). Severity of the illness is associated with respiratory failure requiring mechanical ventilation. Approximately one-third of hospitalized GBS patients require mechanical ventilation because of respiratory muscle or oropharyngeal weakness (Burns, 2008). Rapid disease progression, bulbar dysfunction, bilateral facial weakness, or inability to stand, inability to lift the elbows or head, elevated liver enzymes, abnormal pulmonary function test, electrophysiological evidence of demyelination, and autonomic disturbance are previously reported indicators for severity of illness (Durand et al., 2003; Sharshar et al., 2003; Walgaard et al., 2010). The epidemiological association established

TABLE 3 Guillain-Barre syndrome disability scale. Patients with grade greater or equal to 3 defined as severe GBS.

0.	Healthy
1.	Minor symptoms or signs of neuropathy but capable of manual work/capable of running
2.	Able to walk without support of a stick (5 m across an open space) but incapable of manual work/running
3.	Able to walk with a stick, appliance, or support (5 m across an open space)
4.	Confined to bed or chair bound
5.	Requiring assisted ventilation (for any part of the day or night)
6.	Death

between *C. jejuni* infection and GBS noted that patients with an antecedent *C. jejuni* infection had a more severe form of GBS with axonal degeneration (Rees, Soudain, Gregson, & Hughes, 1995).

In patients with Zika virus–associated GBS, ICU admission was needed for 15%–59%, with 4% mortality (Nascimento & da Silva, 2017). Mechanical ventilation was required in 29%–31% of the patients with Zika virus–associated GBS (Nascimento & da Silva, 2017).

Laboratory tests

Cerebrospinal fluid analysis

Cerebrospinal fluid (CSF) analysis is typical and reveals high protein level with normal cell count (the classic albumino-cytologic dissociation). Protein levels may be normal in 30%–50% of patients in the first week after disease onset and 10%–30% of patients in the second week (Doets et al., 2018; Fokke et al., 2014). Normal CSF protein level in early phase of the disease does not rule out the GBS. In the case of pleocytosis, >50 cells/μL suggests the diseases other than GBS.

Electrodiagnostic studies

Electrodiagnostic studies are helpful in supporting the diagnosis, particularly in patients with an atypical presentation. Sensorimotor polyradiculoneuropathy or polyneuropathy indicated by reduced conduction velocities, reduced sensory and motor evoked amplitudes, abnormal temporal dispersion, and/or partial motor conduction blocks are found in electromyography (EMG) of the patients with GBS (Vucic, Cairns, Black, Chong, & Cros, 2004; Willison et al., 2016). This is a typical where the sural sensory nerve action potential is normal while the median and ulnar sensory nerve action potentials are abnormal or even absent (Vucic et al., 2004; Willison et al., 2016). Electrodiagnostic studies can also differentiate between the electrophysiological subtypes of classical GBS (Hadden et al., 1998).

Antiganglioside antibodies and preceding infections

Antiganglioside antibodies can be found in a proportion of GBS patients, especially in patients with AMAN (especially IgG anti-GM1 antibodies), and in patients with MFS (anti-GQ1b antibodies) (Willison, 2005a; Yuki & Hartung, 2012). The presence of anti-GQ1b antibodies can be helpful in the diagnosis of patients with MFS.

Testing for preceding infections can provide important epidemiological information during outbreaks of infectious diseases, as was seen in previous outbreaks of Zika virus and *C. jejuni* infection (Ho et al., 1995; Parra et al., 2016).

Imaging

Magnetic resonance imaging (MRI) is not routinely used in diagnosis of GBS, but can be helpful for excluding differential diagnoses such as brainstem infection, stroke, spinal cord or anterior horn cell inflammation, nerve root compression, or leptomeningeal malignancy. The presence of nerve root enhancement on gadolinium-enhanced MRI is a nonspecific but sensitive for GBS (Gorson, Ropper, Muriello, & Blair, 1996). Clinical presentation of acute flaccid myelitis and GBS can mimic each other especially in children, and MRI also can be used to distinguish between these two diagnoses (Hopkins, Elrick, & Messacar, 2019).

Treatment

Guillain-Barré syndrome is a potentially life-threatening disease. Both general medical care and immunomodulatory treatment are crucial. Supportive care is needed to prevent or to manage complications (Hughes et al., 2005; van den Berg, Walgaard, et al., 2014). Measures include monitoring of respiratory function by frequent measurement of vital capacity and other clinical outcomes and timely transfer to ICU when needed.

IVIG (0.4 g/kg body weight daily for 5 days) and plasma exchange (200–250 mL plasma/kg body weight in five sessions) are equally effective treatments for GBS (Hughes, Swan, & van Doorn, 2014; Verboon, van Doorn, & Jacobs, 2017).

It was shown by randomized controlled studies that both IVIG and PE are effective in GBS treatment (Hughes, Raphael, Swan, & van Doorn, 2006; Raphael, Chevret, Hughes, & Annane, 2002). Both of the treatment modalities shorten recovery time and improve disability scores when administered within the early stages of disease (Hughes et al., 2006; Raphael et al., 2002; Raphael, Chevret, Hughes, & Annane, 2012). A systematic review showed there was no evidence for a better outcome with either of the two modalities of treatment (Hughes et al., 2007). But a more recent randomized controlled study favors PE for the subset of children with severe rapidly progressive GBS regarding the shorter duration of mechanical ventilation (El-Bayoumi et al., 2011). Dada et al. also suggested PE to be superior to IVIG as a treatment modality in patients with the severe form of GBS and evidence of axonal involvement on EMG (Dada & Kaplan, 2004). In another pediatric case series, the outcome of patients treated with plasma exchange was found to be better (Hicks et al., 2010).

PE is presumed to act by removal of circulating autoantibodies (Creange et al., 1998), while IVIG, besides other functions related to complement system, is thought to work through blocking antibody production (Yuki et al., 2011) (Table 4). For resistant cases to either treatment regiment, combination modalities aroused. The combination in which PE course is followed by IVIG course was studied, but this treatment modality did not confer a significant advantage (Plasma Exchange/Sandoglobulin Guillain-Barre Syndrome Trial Group, 1997). Also, in other study, IVIG treatment followed by PE was compared with IVIG alone, and there was no significant benefit of PE treatment applied after IVIG treatment (Oczko-Walker, Manousakis, Wang, Malter, & Waclawik, 2010).

Although these therapies have been in use for many years, particularly the prognosis of patients with rapid progression and severe involvement has not changed, and the rate of mortality has remained around 10% (Merkies, Schmitz, Samijn, van der Meche, & van Doorn, 1999).

No significant differences were found between IVIG and plasma exchange therapy in previous randomized controlled trials (van der Meche & Schmitz, 1992). This is because neither of them can control all components of the cellular and humoral immune system as a single treatment.

IVIG

IVIG product is derived from the isolation of immunoglobulin from thousands of human donor sources. The primary component of IVIG is IgG that is responsible for its therapeutic effects. The mechanism of action of IVIG is not fully known, but it is thought to modulate cellular and humoral immune system processes by antibody neutralization, complement attenuation, and the regulation of cytokine, B-cell, and T-cell activity (Sederholm, 2010).

TABLE 4 **Impact points of the treatment modalities.**

Pathophysiological mechanisms	IVIG	Plasma exchange
Emergence of antibodies in the form of Ig G1–G3 subtype	[a]	[a]
Binding of autoantibodies to Fc receptors	[a]	
Leukocyte degranulation	[a]	[a]
Activation of the complement pathway	[a]	[a]
Membrane attack complex formation	[a]	
Macrophage activation		[a]
Hypercytokinemia	[a]	[a]
Phagocytosis		[a]

[a] *Treatment modality blocks the stage of pathophysiological mechanism.*

The mechanism of action of IVIG includes the suppression of antibody production and the provision of antiidiotypic antibodies that can neutralize autoantibodies; inhibition of complement activation and membrane attack complex formation; modulation of the expression and function of Fc receptors on macrophages and other effector cells; and suppression of cytokine, chemokine, and adhesion molecules (Dalakas, 2004) (Table 4). IVIG is preferable as it can be administered immediately and has fewer complications and contraindications than PE.

Plasma exchange

In 1985 PE was established as a treatment for GBS following a large trial that compared PE with conventional therapy (Dyck & Kurtzke, 1985). PE was particularly effective in patients who received the treatment within 7 days of onset and in patients who had a more severe disease, requiring mechanical ventilation.

The basic mechanism for PE includes elimination of offending toxins or pathologic autoimmune antibodies from the bloodstream through extracorporeal separation from blood components. During PE a patient's plasma and cells are separated through centrifugation or a filtration process (Lehmann, Hartung, Meyer Zu Horste, & Kieseier, 2008) (Table 4).

Combination treatment options

In the randomized trials, neither PE nor IVIG alone improved mortality, and the mortality in hospital and population-based studies has not changed, ranging from 5% to 15% (Hughes & Cornblath, 2005). It is well recognized that clinicians quite commonly face the therapeutic dilemma of a patient with GBS who is initially treated with IVIG but continues to deteriorate during the course of treatment, frequently necessitating intubation, mechanical ventilation, or artificial support of feeding. Many of those patients have been treated with subsequent PE, the latter being viewed as a possible "rescue" intervention (Oczko-Walker et al., 2010). In previous studies, combined treatment was defined as five PE sessions following 5 days of IVIG or 5 days of IVIG following PE sessions (Oczko-Walker et al., 2010). Some researchers also studied the sequential treatment with PE followed by IVIG, but this treatment did not appear to confer additional benefit to the patients as compared with PE alone (Hughes et al., 2003; Plasma Exchange/Sandoglobulin Guillain-Barre Syndrome Trial Group, 1997).

Zipper method

Recently a novel treatment modality named as zipper method was defined for severe GBS (Kesici, Tanyildiz, Yetimakman, & Bayrakci, 2019). In zipper method, following diagnosis of GBS, plasma exchange was started immediately. In the first session of PE, one and a half volume of patients' plasma was removed by using 5% albumin as replacement solution. At the end of the PE session, 0.4 g/kg IVIG infusion was started immediately. Second PE session was applied with one volume change after 24 h from the end of the IVIG infusion. Each PE session was followed by IVIG infusions. This PE-IVIG cycle was repeated for five times. It was aimed to finish the whole course in about 7 days. Because of hypofibrinogenemia risk, fresh frozen plasma was used as replacement fluid in the third PE session (Kesici et al., 2019) (Fig. 1).

No significant differences were found between IVIG and plasma exchange therapy in previous randomized controlled trials (van der Meche & Schmitz, 1992). This is because neither of them can control all components of the cellular and humoral immune system as a single treatment. In the zipper method, PE removes the antibodies intensively from the plasma, controls hypercytokinemia, decreases leukocyte degranulation, and inhibits macrophage activation and phagocytosis when initiated immediately. Besides, IVIG infusion started at the end of each PE course blocks Fc receptors and prevents formation of membrane attack complex by inhibiting complement activation (Hughes, Swan, & van Doorn, 2012). Since the plasma is cleared from the autoantibodies after the plasma exchange, an autoantibody transition from the tissue to the plasma starts (Vriesendorp, Mayer, & Koski, 1991). Also, it was postulated in a study that removal of autoantibodies by PE may stimulate rebound synthesis of autoantibodies (Plasma Exchange/Sandoglobulin Guillain-Barre Syndrome Trial Group, 1997). IVIG given immediately after PE neutralizes these regenerated autoantibodies. Antibody movement from tissue to plasma completes in 24 h (Vriesendorp et al., 1991). This is the reason why we preferred to initiate the next PE session 24 h after the IVIG infusion in the zipper method. This is the logic underlying the initiation of the next PE session to be 24 h after the IVIG infusions. The following PE session is expected to sweep away the soiled antibodies. This way of ranking leads to a consistency of both regiments augmenting each other's effect. Considering the mentioned pathophysiological diversity of AMAN, zipper method is currently suggested for severe cases. Key point of success underlines in prompt administration of the zipper method prior to irreversible damage.

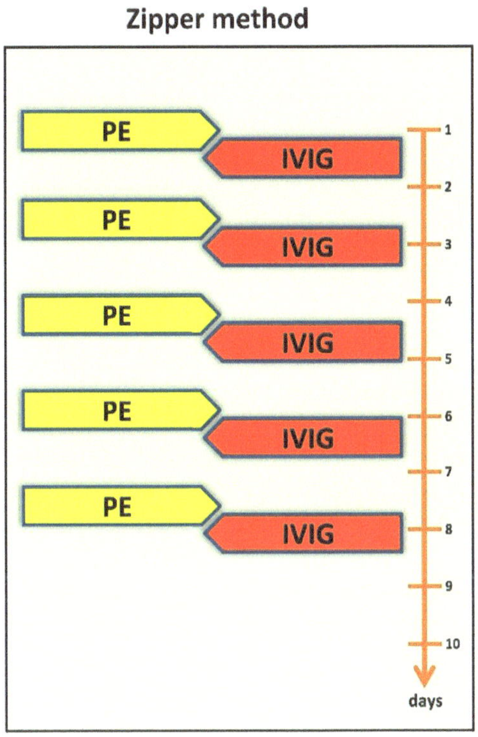

FIG. 1 Zipper method. In the first session of PE, one and a half volume of patients' plasma was removed by using 5% albumin as replacement solution. At the end of the PE session, 0.4 g/kg IVIG infusion was started immediately. Second PE session was applied with one volume change after 24 h from the end of the IVIG infusion. Each PE session was followed by IVIG infusions. This PE-IVIG cycle was repeated for five times.

The zipper method as a novel treatment modality seems to reduce mortality, speed up weaning from mechanical ventilation, and shorten hospital stay with an excellent outcome in severe GBS patients who require intensive care (Kesici et al., 2019).

Policy and procedures: Diagnosis and treatment of Guillain-Barré syndrome

Guillain-Barré syndrome can be difficult to diagnose in its earliest stages. Its signs and symptoms are similar to those of other neurological disorders and may vary from person to person.

Diagnostic process usually starts with a medical history and thorough physical examination. The following diagnostic procedures will be planned:

- **Spinal tap (lumbar puncture).** A small amount of fluid is withdrawn from the spinal canal in lower back. The fluid is tested for a type of change that commonly occurs in people who have Guillain-Barré syndrome.
- **Electromyography.** Thin-needle electrodes are inserted into the muscles that are affected. The electrodes measure nerve activity in the muscles.
- **Nerve conduction studies.** Electrodes are taped to the skin above nerves. A small shock is passed through the nerve to measure the speed of nerve signals.

Treatment

There's no cure for Guillain-Barré syndrome. But two types of treatments can speed recovery and reduce the severity of the illness:

- **Plasma exchange (plasmapheresis).** The liquid portion of part of blood (plasma) is removed and separated from blood cells. The blood cells are then put back into body, which manufactures more plasma to make up for what was removed. Plasmapheresis may work by ridding plasma of certain antibodies that contribute to the immune system's attack on the peripheral nerves.

- **Immunoglobulin therapy.** Immunoglobulin containing healthy antibodies from blood donors is given through a vein (intravenously). High doses of immunoglobulin can block the damaging antibodies that may contribute to Guillain-Barré syndrome.

These treatments are equally effective. Mixing them or administering one after the other is no more effective than using either method alone.

Mayo Clinic (2019): Patient Care & Health Information/Diseases & Conditions/Guillain-Barré syndrome https://www.mayoclinic.org/diseases-conditions/guillain-barre-syndrome/diagnosis-treatment/drc-20363006.

Mini-dictionary of terms

Polyradiculoneuropathy. Inflammatory disorder affecting peripheral nerves and the nerve roots of the spinal nerves.
Molecular mimicry. Theoretical possibility that sequence similarities between foreign and self-peptides.
Autoimmune disease. Illness that causes the immune system to produce antibodies that attack normal body tissues.
Myelin. The fatty substance that covers and protects nerves.
Axon. The appendage of the neuron that transmits impulses away from the cell body.

Key facts of plasma exchange

- During plasma exchange (PE), a patient's plasma and cells are separated through centrifugation or a filtration process.
- Patient's plasma is replaced by albumin or fresh frozen plasma.
- The basic mechanism for PE is the elimination of offending toxins or pathologic autoimmune antibodies from the bloodstream.
- Removal of autoantibodies, by PE, creates a concentration gradient between the lowered blood level and the extravascular space forcing antibody movement from the extra to the intravascular space to be removed during the subsequent session.
- PE was introduced as a possible treatment for GBS in 1978.

Summary points

- Guillain-Barré syndrome (GBS) is currently the most common cause of acute flaccid paralysis.
- Rapidly progressive weakness is the main clinical feature of GBS. It may involve bulbar weakness, respiratory failure, and autonomic nervous system dysfunction.
- In countries where Zika virus infection is common, an increase in GBS cases has been reported.
- Zika virus infection seems to be a trigger of GBS.
- Intravenous immunoglobulin and plasma exchange are equally effective in treatment of GBS.
- Despite the proven treatment modalities, prognosis of severe GBS is still not satisfactory.
- The best treatment option for patients with GBS that continue to deteriorate despite initial therapy with IVIG or PE alone is unclear.
- The zipper method as a novel treatment modality seems to be effective in the treatment of severe GBS.

References

Asbury, A. K., & Cornblath, D. R. (1990). Assessment of current diagnostic criteria for Guillain-Barre syndrome. *Annals of Neurology*, *27*(Suppl), S21–S24.

Burns, T. M. (2008). Guillain-Barre syndrome. *Seminars in Neurology*, *28*(2), 152–167.

Creange, A., Belec, L., Clair, B., Degos, J. D., Raphael, J. C., & Gherardi, R. K. (1998). Circulating transforming growth factor beta 1 (TGF-beta1) in Guillain-Barre syndrome: Decreased concentrations in the early course and increase with motor function. *Journal of Neurology, Neurosurgery, and Psychiatry*, *64*(2), 162–165.

Dada, M. A., & Kaplan, A. A. (2004). Plasmapheresis treatment in Guillain-Barre syndrome: Potential benefit over IVIg in patients with axonal involvement. *Therapeutic Apheresis and Dialysis*, *8*(5), 409–412.

Dalakas, M. C. (2004). Intravenous immunoglobulin in autoimmune neuromuscular diseases. *JAMA*, *291*(19), 2367–2375.

Doets, A. Y., Verboon, C., van den Berg, B., Harbo, T., Cornblath, D. R., Willison, H. J., … Consortium, I. (2018). Regional variation of Guillain-Barre syndrome. *Brain*, *141*(10), 2866–2877.

Durand, M. C., Lofaso, F., Lefaucheur, J. P., Chevret, S., Gajdos, P., Raphael, J. C., & Sharshar, T. (2003). Electrophysiology to predict mechanical ventilation in Guillain-Barre syndrome. *European Journal of Neurology*, *10*(1), 39–44.
Dyck, P. J., & Kurtzke, J. F. (1985). Plasmapheresis in Guillain-Barre syndrome. *Neurology*, *35*(8), 1105–1107.
El-Bayoumi, M. A., El-Refaey, A. M., Abdelkader, A. M., El-Assmy, M. M., Alwakeel, A. A., & El-Tahan, H. M. (2011). Comparison of intravenous immunoglobulin and plasma exchange in treatment of mechanically ventilated children with Guillain Barre syndrome: A randomized study. *Critical Care*, *15*(4), R164.
Fokke, C., van den Berg, B., Drenthen, J., Walgaard, C., van Doorn, P. A., & Jacobs, B. C. (2014). Diagnosis of Guillain-Barre syndrome and validation of Brighton criteria. *Brain*, *137*(Pt 1), 33–43.
Galldiks, N., Dohmen, C., Neveling, M., Fink, G. R., & Haupt, W. F. (2009). Selective immune adsorption treatment of severe Guillain Barre syndrome in the intensive care unit. *Neurocritical Care*, *11*(3), 317–321.
Gorson, K. C., Ropper, A. H., Muriello, M. A., & Blair, R. (1996). Prospective evaluation of MRI lumbosacral nerve root enhancement in acute Guillain-Barre syndrome. *Neurology*, *47*(3), 813–817.
Griffin, J. W., Li, C. Y., Ho, T. W., Tian, M., Gao, C. Y., Xue, P., … Asbury, A. K. (1996). Pathology of the motor-sensory axonal Guillain-Barre syndrome. *Annals of Neurology*, *39*(1), 17–28.
Griffin, J. W., Stoll, G., Li, C. Y., Tyor, W., & Cornblath, D. R. (1990). Macrophage responses in inflammatory demyelinating neuropathies. *Annals of Neurology*, *27*(Supplement), S64–S68.
Hadden, R. D., Cornblath, D. R., Hughes, R. A., Zielasek, J., Hartung, H. P., Toyka, K. V., & Swan, A. V. (1998). Electrophysiological classification of Guillain-Barre syndrome: Clinical associations and outcome. Plasma Exchange/Sandoglobulin Guillain-Barre Syndrome Trial Group. *Annals of Neurology*, *44*(5), 780–788.
Hafer-Macko, C., Hsieh, S. T., Li, C. Y., Ho, T. W., Sheikh, K., Cornblath, D. R., … Griffin, J. W. (1996). Acute motor axonal neuropathy: An antibody-mediated attack on axolemma. *Annals of Neurology*, *40*(4), 635–644.
Hafer-Macko, C. E., Sheikh, K. A., Li, C. Y., Ho, T. W., Cornblath, D. R., McKhann, G. M., … Griffin, J. W. (1996). Immune attack on the Schwann cell surface in acute inflammatory demyelinating polyneuropathy. *Annals of Neurology*, *39*(5), 625–635.
Hicks, C. W., Kay, B., Worley, S. E., & Moodley, M. (2010). A clinical picture of Guillain-Barre syndrome in children in the United States. *Journal of Child Neurology*, *25*(12), 1504–1510.
Ho, T. W., Mishu, B., Li, C. Y., Gao, C. Y., Cornblath, D. R., Griffin, J. W., … McKhann, G. M. (1995). Guillain-Barre syndrome in northern China. Relationship to Campylobacter jejuni infection and anti-glycolipid antibodies. *Brain*, *118*(Pt 3), 597–605.
Ho, T. W., Willison, H. J., Nachamkin, I., Li, C. Y., Veitch, J., Ung, H., … McKhann, G. M. (1999). Anti-GD1a antibody is associated with axonal but not demyelinating forms of Guillain-Barre syndrome. *Annals of Neurology*, *45*(2), 168–173.
Hopkins, S. E., Elrick, M. J., & Messacar, K. (2019). Acute flaccid myelitis-keys to diagnosis, questions about treatment, and future directions. *JAMA Pediatrics*, *173*(2), 117–118.
Hughes, R. A., & Cornblath, D. R. (2005). Guillain-Barre syndrome. *Lancet*, *366*(9497), 1653–1666.
Hughes, R. A., Newsom-Davis, J. M., Perkin, G. D., & Pierce, J. M. (1978). Controlled trial prednisolone in acute polyneuropathy. *Lancet*, *2*(8093), 750–753.
Hughes, R. A., Raphael, J. C., Swan, A. V., & van Doorn, P. A. (2006). Intravenous immunoglobulin for Guillain Barre syndrome. *Cochrane Database of Systematic Reviews*, *1*, CD002063.
Hughes, R. A., Swan, A. V., Raphael, J. C., Annane, D., van Koningsveld, R., & van Doorn, P. A. (2007). Immunotherapy for Guillain-Barre syndrome: A systematic review. *Brain*, *130*(Pt 9), 2245–2257.
Hughes, R. A., Swan, A. V., & van Doorn, P. A. (2012). Intravenous immunoglobulin for Guillain-Barre syndrome. *Cochrane Database of Systematic Reviews*, *7*, CD002063.
Hughes, R. A., Swan, A. V., & van Doorn, P. A. (2014). Intravenous immunoglobulin for Guillain-Barre syndrome. *Cochrane Database of Systematic Reviews*, *9*, CD002063.
Hughes, R. A., Wijdicks, E. F., Barohn, R., Benson, E., Cornblath, D. R., Hahn, A. F., … Quality Standards Subcommittee of the American Academy of, N. (2003). Practice parameter: Immunotherapy for Guillain-Barre syndrome: report of the Quality Standards Subcommittee of the American Academy of Neurology. *Neurology*, *61*(6), 736–740.
Hughes, R. A., Wijdicks, E. F., Benson, E., Cornblath, D. R., Hahn, A. F., Meythaler, J. M., … Multidisciplinary Consensus, G. (2005). Supportive care for patients with Guillain-Barre syndrome. *Archives of Neurology*, *62*(8), 1194–1198.
Kesici, S., Tanyildiz, M., Yetimakman, F., & Bayrakci, B. (2019). A novel treatment strategy for severe Guillain-Barre syndrome: Zipper method. *Journal of Child Neurology*, *34*(5), 277–283.
Lehmann, H. C., Hartung, H. P., Meyer Zu Horste, G., & Kieseier, B. C. (2008). Plasma exchange in immune-mediated neuropathies. *Current Opinion in Neurology*, *21*(5), 547–554.
Leonhard, S. E., Mandarakas, M. R., Gondim, F. A. A., Bateman, K., Ferreira, M. L. B., Cornblath, D. R., … Jacobs, B. C. (2019). Diagnosis and management of Guillain-Barre syndrome in ten steps. *Nature Reviews. Neurology*, *15*(11), 671–683.
McGrogan, A., Madle, G. C., Seaman, H. E., & de Vries, C. S. (2009). The epidemiology of Guillain-Barre syndrome worldwide. A systematic literature review. *Neuroepidemiology*, *32*(2), 150–163.
Merkies, I. S., Schmitz, P. I., Samijn, J. P., van der Meche, F. G., & van Doorn, P. A. (1999). Fatigue in immune-mediated polyneuropathies. European Inflammatory Neuropathy Cause and Treatment (INCAT) Group. *Neurology*, *53*(8), 1648–1654.
Nascimento, O. J. M., & da Silva, I. R. F. (2017). Guillain-Barre syndrome and Zika virus outbreaks. *Current Opinion in Neurology*, *30*(5), 500–507.

Oczko-Walker, M., Manousakis, G., Wang, S., Malter, J. S., & Waclawik, A. J. (2010). Plasma exchange after initial intravenous immunoglobulin treatment in Guillain-Barre syndrome: Critical reassessment of effectiveness and cost-efficiency. *Journal of Clinical Neuromuscular Disease*, *12* (2), 55–61.

Parra, B., Lizarazo, J., Jimenez-Arango, J. A., Zea-Vera, A. F., Gonzalez-Manrique, G., Vargas, J., … Pardo, C. A. (2016). Guillain-Barre syndrome associated with Zika virus infection in Colombia. *The New England Journal of Medicine*, *375*(16), 1513–1523.

Plasma Exchange/Sandoglobulin Guillain-Barre Syndrome Trial Group. (1997). Randomised trial of plasma exchange, intravenous immunoglobulin, and combined treatments in Guillain-Barre syndrome. Plasma Exchange/Sandoglobulin Guillain-Barre Syndrome Trial Group. *Lancet*, *349*(9047), 225–230.

Rajabally, Y. A., & Uncini, A. (2012). Outcome and its predictors in Guillain-Barre syndrome. *Journal of Neurology, Neurosurgery, and Psychiatry*, *83*(7), 711–718.

Raphael, J. C., Chevret, S., Hughes, R. A., & Annane, D. (2002). Plasma exchange for Guillain-Barre syndrome. *Cochrane Database of Systematic Reviews*, *2*, CD001798.

Raphael, J. C., Chevret, S., Hughes, R. A., & Annane, D. (2012). Plasma exchange for Guillain-Barre syndrome. *Cochrane Database of Systematic Reviews*, *7*, CD001798.

Rees, J. H., Soudain, S. E., Gregson, N. A., & Hughes, R. A. (1995). Campylobacter jejuni infection and Guillain-Barre syndrome. *The New England Journal of Medicine*, *333*(21), 1374–1379.

Rosen, B. A. (2012). Guillain-Barre syndrome. *Pediatrics in Review*, *33*(4), 164–170 (quiz 170-161).

Ruts, L., Drenthen, J., Jacobs, B. C., van Doorn, P. A., & Dutch, G. B. S. S. G. (2010). Distinguishing acute-onset CIDP from fluctuating Guillain-Barre syndrome: A prospective study. *Neurology*, *74*(21), 1680–1686.

Sederholm, B. H. (2010). Treatment of acute immune-mediated neuropathies: Guillain-Barre syndrome and clinical variants. *Seminars in Neurology*, *30* (4), 365–372.

Sejvar, J. J., Baughman, A. L., Wise, M., & Morgan, O. W. (2011). Population incidence of Guillain-Barre syndrome: A systematic review and meta-analysis. *Neuroepidemiology*, *36*(2), 123–133.

Sharshar, T., Chevret, S., Bourdain, F., Raphael, J. C., & French Cooperative Group on Plasma Exchange in Guillain-Barre, S. (2003). Early predictors of mechanical ventilation in Guillain-Barre syndrome. *Critical Care Medicine*, *31*(1), 278–283.

van den Berg, B., Fokke, C., Drenthen, J., van Doorn, P. A., & Jacobs, B. C. (2014). Paraparetic Guillain-Barre syndrome. *Neurology*, *82*(22), 1984–1989.

van den Berg, B., Walgaard, C., Drenthen, J., Fokke, C., Jacobs, B. C., & van Doorn, P. A. (2014). Guillain-Barre syndrome: Pathogenesis, diagnosis, treatment and prognosis. *Nature Reviews. Neurology*, *10*(8), 469–482.

van der Meche, F. G., & Schmitz, P. I. (1992). A randomized trial comparing intravenous immune globulin and plasma exchange in Guillain-Barre syndrome. Dutch Guillain-Barre Study Group. *The New England Journal of Medicine*, *326*(17), 1123–1129.

van Doorn, P. A. (2013). Diagnosis, treatment and prognosis of Guillain-Barre syndrome (GBS). *Presse Médicale*, *42*(6 Pt 2), e193–e201.

van Doorn, P. A., Kuitwaard, K., Walgaard, C., van Koningsveld, R., Ruts, L., & Jacobs, B. C. (2010). IVIG treatment and prognosis in Guillain-Barre syndrome. *Journal of Clinical Immunology*, *30*(Supplement 1), S74–S78.

van Doorn, P. A., Ruts, L., & Jacobs, B. C. (2008). Clinical features, pathogenesis, and treatment of Guillain-Barre syndrome. *Lancet Neurology*, *7*(10), 939–950.

van Sorge, N. M., van der Pol, W. L., Jansen, M. D., Geleijns, K. P., Kalmijn, S., Hughes, R. A., … van den Berg, L. H. (2005). Severity of Guillain-Barre syndrome is associated with Fc gamma receptor III polymorphisms. *Journal of Neuroimmunology*, *162*(1–2), 157–164.

Verboon, C., van Doorn, P. A., & Jacobs, B. C. (2017). Treatment dilemmas in Guillain-Barre syndrome. *Journal of Neurology, Neurosurgery, and Psychiatry*, *88*(4), 346–352.

Vriesendorp, F. J., Mayer, R. F., & Koski, C. L. (1991). Kinetics of anti-peripheral nerve myelin antibody in patients with Guillain-Barre syndrome treated and not treated with plasmapheresis. *Archives of Neurology*, *48*(8), 858–861.

Vucic, S., Cairns, K. D., Black, K. R., Chong, P. S., & Cros, D. (2004). Neurophysiologic findings in early acute inflammatory demyelinating polyradiculoneuropathy. *Clinical Neurophysiology*, *115*(10), 2329–2335.

Walgaard, C., Lingsma, H. F., Ruts, L., Drenthen, J., van Koningsveld, R., Garssen, M. J., … Jacobs, B. C. (2010). Prediction of respiratory insufficiency in Guillain-Barre syndrome. *Annals of Neurology*, *67*(6), 781–787.

Wijdicks, E. F., & Klein, C. J. (2017). Guillain-Barre syndrome. *Mayo Clinic Proceedings*, *92*(3), 467–479.

Willison, H. J. (2005a). Ganglioside complexes: New autoantibody targets in Guillain-Barre syndromes. *Nature Clinical Practice. Neurology*, *1*(1), 2–3.

Willison, H. J. (2005b). The immunobiology of Guillain-Barre syndromes. *Journal of the Peripheral Nervous System*, *10*(2), 94–112.

Willison, H. J., Jacobs, B. C., & van Doorn, P. A. (2016). Guillain-Barre syndrome. *Lancet*, *388*(10045), 717–727.

Yuki, N., & Hartung, H. P. (2012). Guillain-Barre syndrome. *The New England Journal of Medicine*, *366*(24), 2294–2304.

Yuki, N., Watanabe, H., Nakajima, T., & Spath, P. J. (2011). IVIG blocks complement deposition mediated by anti-GM1 antibodies in multifocal motor neuropathy. *Journal of Neurology, Neurosurgery, and Psychiatry*, *82*(1), 87–91.

Chapter 12

Oxidative stress in Guillain-Barré syndrome and linkage with neurology

Serdar Baraklı, Şadiye Gümüşyayla, Gönül Vural, and Orhan Deniz
Department of Neurology, Faculty of Medicine, Ankara Yildirim Beyazit University, Ankara, Turkey

Abbreviations

AIDP	acute inflammatory demyelinating polyneuropathy
AOA	antioxidant activity
CEHC	carboxyethyl-hydroxychromans
CNS	central nervous system
COX-2	cyclooxygenase 2
EPR	electron paramagnetic resonance
fMLF	*N*-formyl-Met-Leu-Phe
GBS	Guillain-Barré syndrome
GSH	glutathione
GST	glutathione-*S*-transferase
HbNO	nonhemic iron
Hp	haptoglobin
MAC	membrane attack complex
MDA	malondialdehyde
MEPP	miniature end-plate potential
MG	myasthenia gravis
MS	multiple sclerosis
NMO	neuromyelitis optica
NO	nitric oxide
PN	peroxynitrite
ROS	reactive oxygen species
SOD	superoxide dismutase
UA	uric acid
WKYMVM	Trp-Lys-Tyr-Met-Val-Met-NH2

Introduction

Acute inflammatory demyelinating polyneuropathy (AIDP) or Guillain-Barré syndrome (GBS) is an autoimmune process that usually presents with progressive areflexic symmetrical ascending weakness. It may be associated with autonomic dysfunction and about 20% of patients end up with respiratory failure. Sensory systems may be involved and present with paresthesias or numbness. Sensory symptoms may precede motor weakness (Daroff, Jankovic, Mazziotta, & Pomeroy, 2016).

Axonal degeneration and demyelination may be involved to varying degrees. If the patient's symptoms continue to present chronically and progress beyond 8 weeks, it is termed chronic inflammatory demyelinating polyneuropathy (CIDP). Some resources identify this period as 4 weeks for CIDP (Lewis, 2015).

AIDP has been associated with many infectious diseases, including *Campylobacter jejuni*, mycoplasma pneumonia, the herpes virus, and many other viral and bacterial infections, along with vaccinations. These infections have been reported in

Zika Virus Impact, Diagnosis, Control, and Models. https://doi.org/10.1016/B978-0-12-820267-8.00012-1

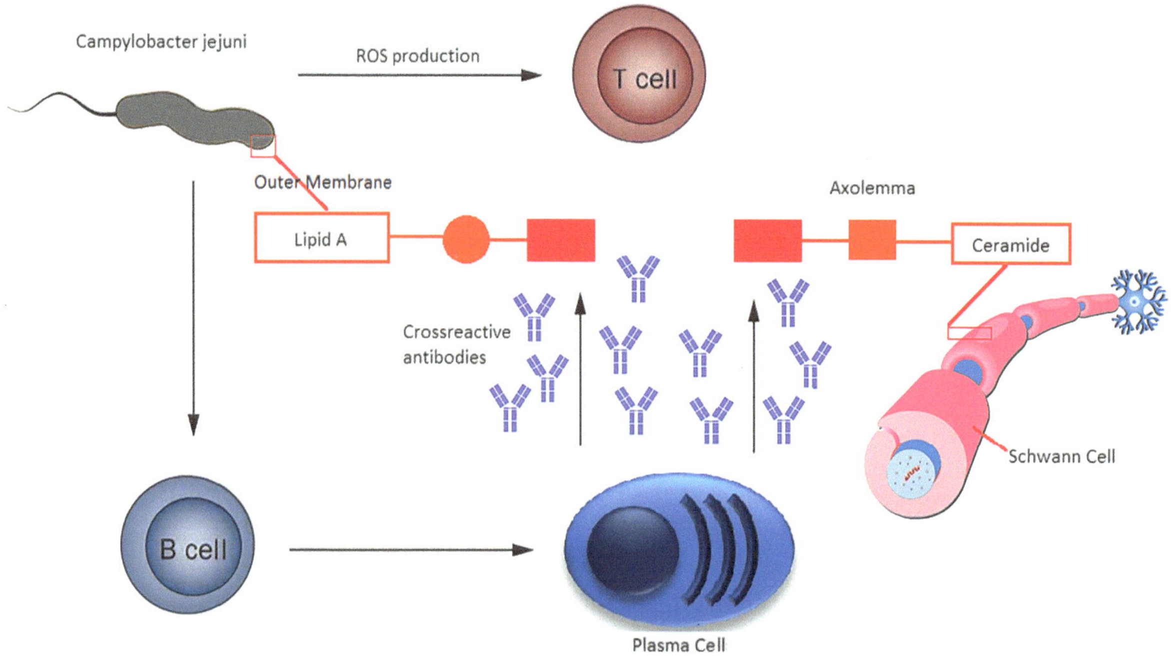

FIG. 1 Acute polyneuropathy caused by *Campylobacter jejuni*. Guillain-Barré syndrome has an association with precedent infection with *Campylobacter jejuni*.

up to 90% of the previous 30 days of GBS. One of the most identifiable microbial agents is *C. jejuni* (Fig. 1) (Daroff et al., 2016). Surgery and pregnancy can also trigger an abnormal immunological response that leads to the degeneration of myelin sheaths and/or axons (Meyer zu Hörste, Hartung, & Kieseier, 2007).

Peripheral nerve damage in GBS is histopathologically divided into two types: demyelinative forms and forms with axonal degeneration (Kieseier et al., 2004). The most common disease subset is acute inflammatory demyelinating polyradiculoneuropathy; other subgroups include acute motor axonal neuropathy, acute motor and sensory axonal neuropathy, and Miller-Fisher syndrome, which are more prevalent as axonal forms (Mitsui et al., 2005).

Oxidative stress in demyelinating diseases

AIDP or GBS is characterized pathologically by endoneurial perivascular mononuclear cell infiltration including lymphocytes, especially T lymphocytes, class II-positive monocytes, and macrophages together with multifocal demyelination. It has been shown that macropages play a major role in demyelination by stripping off myelin lamellae from the axon. Some autoantibodies to myelin glycolipids, including GM1, GD1a, and GD1b, have been identified. It has been shown in pathology specimens that complement fixation induces antibody-mediated demyelination. In some cases, axonal damage can be seen, and it is believed to be a result of bystander damage (Daroff et al., 2016).

It is known that calpain-mediated proteolysis is a feature of many neurodegenerative diseases. It has also been shown that calpain activation may enhance axonal degeneration in animal models. Antibody-mediated complement activation leads to irregular Ca^{2+} influx through the nonspecific pores formed by the membrane attack complex (MAC) at the motor nerve terminal. By this way, increased intracellular [Ca^{2+}] would be a factor that may trigger an uncontrolled increase in miniature end-plate potential (MEPP) frequency and calpain activation. Therefore, this may lead to the degradation of substrates, resulting in the structural disorganization of the nerve terminal (O'Hanlon et al., 2003).

Inflammatory and autoimmune-mediated tissue damage can lead to the increased production of reactive oxygen molecules, which can disrupt the cellular metabolism and lead to cell death. Antioxidant protection systems aim to limit the production of reactive oxygen species (ROS) to ensure cellular equilibrium (Peng et al., 2012). Oxidative stress is an impaired balance between ROS and antioxidant molecules (Fig. 2).

The central nervous system (CNS) is highly susceptible to oxidative stress due to its high oxygen consumption, and has low antioxidant activity due to high iron and lipid content. Therefore, it is sensitive to damage caused by ROS. As for that, there is slightly poor evidence about the effect of oxidative stress in the demyelinating diseases of the peripheral nervous system. Oxidative stress plays an important role in the pathogenesis of many autoimmune and inflammatory diseases of the

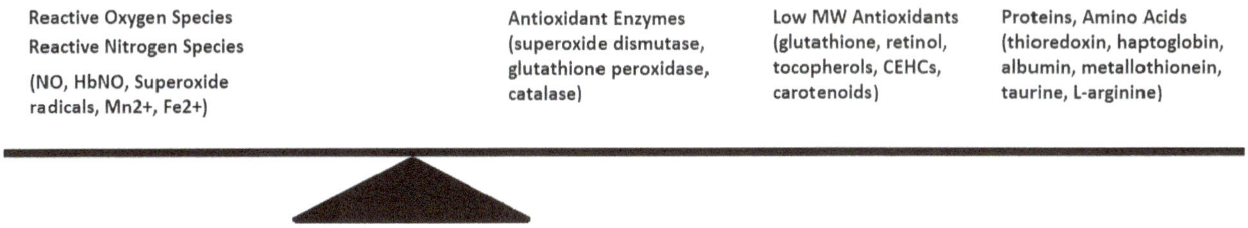

FIG. 2 Redox balance and oxidative stress. Oxidative stress is characterized by the balance between the oxidative and antioxidative mechanisms of living organisms. It is caused by an imbalance between the production and detoxification of reactive oxygen species.

nervous system, such as multiple sclerosis (MS), neuromyelitis optica (NMO), and myasthenia gravis (MG) (Fuhua et al., 2012; Ghafourifar et al., 2008). GBS, MS, and NMO are inflammatory, autoimmune, and demyelinating diseases of the nervous system, respectively (Davison, Tibbits, Shi, & Moon, 1988; Ghafourifar et al., 2008; Su et al., 2017). Common pathogenic features of these diseases include demyelination and axonal damage. MS and NMO affect the central nervous system, whereas GBS affects the peripheral nervous system. Several studies demonstrated that oxidative stress plays a role in the pathogenesis of MS (Karlik et al., 2015; Miller, Walczak, Saluk, Ponczek, & Majsterek, 2012; Tasset et al., 2012). In addition, active oxygen molecules have been linked to many neuropathies and myopathies (Davison et al., 1988). Myelin, which is affected by the most common form of GBS, is vulnerable to oxidative stress, especially during inflammatory processes, due to its high lipid content (Peng et al., 2012). In addition, molecular mimicry is held responsible for GBS pathogenesis and mainly phagocytic cells play a role in this pathogenetic mechanism. ROS is produced by phagocytic cells and forms an important part of the immune deficiency initiation against microorganisms (Babior, 2000; Klebanoff & Coombs, 1992). Therefore, the oxidative stress of the organism is likely to be affected by GBS (Ghabaee, Jabedari, Al-E-Eshagh, Ghaffarpour, & Asadi, 2010; Kumar et al., 2004).

Free radicals and enzymatic antioxidation in GBS

Axonal degeneration accompanying its demyelination is the main reason of neurological insufficiency of GBS. Although all the evidence was mentioned before, the mechanism of axonal degeneration is not established yet. There are lots of studies researching the destructive influence of free radicals, which are atoms, molecules, or ions that contain at least one unpaired electron. By using the electron paramagnetic resonance (EPR) method that aims to study materials with unpaired electrons, a study has shown the increase of intensive EPR signals of nitric oxide (NO), complexes of NO with nonhemic iron (HbNO), and lypo- and superoxide radical content in the blood and cerebrospinal fluid. It has also been shown that the signals of free Mn^{+2}, Fe^{+2}, the activity of blood antioxidant enzymes, ceruloplasmin, and katalasa increases. The evidence has also shown a reduction in the activity of superoxide dismutase and glutathione reductase. Afterward, the oxidative stress that reactive oxygen species lead to damages the intracellular pathways of products, possibly causing axon demyelination and degeneration (Dogonadze, Ninua, Gordeziani, Kavlashvili, & Sanikidze, 2006).

There are many parameters that show the redox status of the organism, and these parameters are examined in a few studies in patients with GBS. A previous study that examined the serum samples of 10 GBS patients differed from these results without the increase of glutathione peroxidase, myeloperoxidase, and lactoferrin levels (Sakai, Inoue, Koh, & Ikeda, 2000).

Another study has demonstrated that erythrocyte glutathione (GSH) levels in blood samples were significantly decreased in the cases of GBS when compared to the control group. Erythrocyte superoxide dismutase (SOD) levels were not markedly altered. Plasma vitamin C levels also did not exhibit any significant changes in contrast to increased vitamin E and serum total glutathione-*S*-transferase (GST) levels. Elevated levels of vitamin E may be a compensatory mechanism for the lack of SOD and vitamin C. It has been seen that erythrocyte malondialdehyde (MDA) levels significantly decreased in the pretreatment cases of GBS. An effective purification of lipid peroxidation substrates by increased GST levels may have an important role in the decreases in MDA levels. The total effect of all these enzymatic and nonenzymatic antioxidants may cause an initial decrease of MDA (Kumar et al., 2004).

A later study that included 15 Iranian patients with GBS showed an increase in the MDA values in the CSF but a decrease in the serum MDA levels. When compared to the control group, the serum antioxidant activity (AOA) has been seen to decrease. However, a significant increase of AOA was shown in the CSF. According to this study's results, the CSF and serum did not show the same changes for the parameters of oxidative stress in GBS patients (Ghabaee et al., 2010).

GBS and lipophilic antioxidants

Only 80% of GBS patients have increased protein concentrations in CSF, and CSF protein content is generally normal during disease onset. So, this has prompted researchers to find new biomarkers. Some of the lipophilic antioxidants such as tocopherols, retinol, and carotenoids protect unsaturated fatty acid chains of phospholipids against reactive oxygen species (Farias et al., 2012; Schweigert, 2001; Vijayaraghavan et al., 2005).

A study that included 12 GBS patients showed no significant changes in MDA concentrations between GBS patients and healthy controls. Concentrations of γ-tocopherol and δ-tocopherol markedly decreased, but α-tocopherol did not. No significant difference between GBS patients and healthy controls was found in tocopherol metabolites, including γ-CEHC, α-CEHC, and δ-CEHC. The α-caroten concentration was higher than in GBS patients in contrast to healthy controls. Other lipophilic antioxidants such as retinol, lutein, β-carotene, and Q10 were decreased, but this did not indicate a statistical significance. The concentration of substrates including γ-tocopherol, δ-tocopherol, α-tocopherol, lutein, Q10, γ-CEHC, and δ-CEHC correlated positively. This evidence suggested that the lipophilic antioxidant defense mechanism has fallen down in GBS patients. In addition, this study showed that decreased γ-tocopherol is a good predictor for GBS.

γ-Tocopherol inhibits cyclooxygenase 2 (COX-2) in macrophages more than α-tocopherol (Jiang, Elson-Schwab, Courtemanche, & Ames, 2000). γ-Tocopherol and δ-tocopherol offer better protection against oxidative stress and have more potent antiinflammatory activity than α-tocopherol (Jiang & Ames, 2003; Jiang, Christen, Shigenaga, & Ames, 2001; Li et al., 2011; Liu, Wallmon, Olsson-Mortlock, Wallin, & Saldeen, 2003). It is also known that γ-tocopherol and δ-tocopherol prevent disease progression related to inflammation and oxidative stress (Chang, Chen, Ke, & Hu, 2005; Jiang et al., 2000).

CEHCs (carboxyethyl-hydroxychromans) that is metabolites of tocopherols show antioxidant activities like inhibition of inflammation (Bruno, Leonard, Li, Bray, & Traber, 2005; Chang et al., 2005; Christen et al., 1997; Hensley et al., 2004; Jiang et al., 2000; Tang et al., 2017).

Uric acid, albumin, and haptoglobin as the antioxidant substrate in patients with GBS

We discussed enzymatic processes and metabolites above. However, it is also known that some proteins are on duty in the antioxidant metabolism. Human plasma haptoglobin (Hp) is a protein produced by the liver. It clears free hemoglobin outside of erythrocytes from circulation. It has been previously demonstrated that Hp is a potent antioxidant (Tseng, Lin, Huang, Liu, & Mao, 2004). Former research proved that the levels of serum haptoglobin increased in the pretreatment of GBS patients compared to controls (Gutowski, Pinkham, Akanmu, Chirico, & Murphy, 1998).

Uric acid is a substrate produced in the body as a final product of the purine nucleotide metabolism. It is thought to be a naturally occurring antioxidant in human blood that cleans peroxynitrite (PN) (Fig. 3) (Hooper et al., 1997; So & Thorens, 2010). Many studies have also shown that serum albumin has potent antioxidant properties that inhibit the production of free hydroxyl radicals and have the ability to scavenge peroxy radicals (Halliwell, 1988; Marx & Chevion, 1986; Wayner, Burton, Ingold, & Locke, 1985). In light of this information, in a study that included 88 GBS patients in China, the serum uric acid (UA) levels of patients with GBS were significantly lower than those in the healthy control group. It was also discovered that the UA levels in GBS patients were significantly lower in male and female subgroups. In this study, it

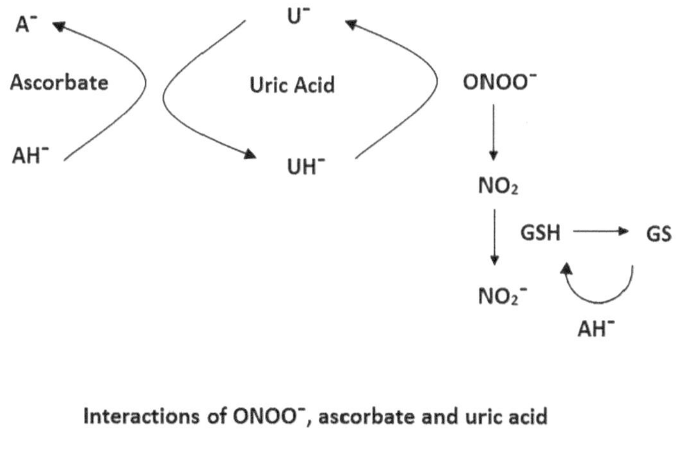

FIG. 3 Metobolic pathways of uric acid formation. Uric acid is a potent hydrophilic antioxidant that is modified by diet and drugs. Several lines of evidence suggest that plasma uric acid may modulate outcomes in neurologic disease.

was proven that the serum levels of albumin in GBS patients were significantly lower than the control group. When patients were divided into two subgroups of women and men, a similar phenomenon to the uric acid mentioned above was found. Furthermore, it has been demonstrated that both high UA and high albumin were protective factors for patients with GBS (Su et al., 2017).

T cell activity in GBS

GBS is usually considered to be monophasic, but recurrences are reported in 2%–5% of patients (Das, Kalita, & Misra, 2004; Grand'Maison, Feasby, Hahn, & Koopman, 1992). The recurrent GBS (RGBS) is characterized by episodes of relapsing and remitting symptoms with complete or near complete functional recovery between episodes. A study determined that the severity of the disease in patients with RGBS was associated with superoxide anion production by leukocytes in response to NADPH oxidase-activating peptides such as *N*-formyl-Met-Leu-Phe (fMLF) and Trp-Lys-Tyr-Met-Val-Met-NH2 (WKYMVM). It has also been shown that the superoxide anion response to fMLF and WKYMVM was significantly lower in GBS patients with incomplete recovery, higher residual motor deficit, and longer duration of the plateau phase during attacks (Mossberg et al., 2010).

In addition, some studies have shown that changes in ROS production increase T cell-dependent autoimmunity (Gelderman, Hultqvist, Holmberg, Olofsson, & Holmdahl, 2006; Hultqvist, Backlund, Bauer, Gelderman, & Holmdahl, 2007). Therefore, the T cell activity, which plays a dominant role in GBS patients, becomes activated due to the increase in ROS production (Fig. 4). Individuals with defective antioxidant defense mechanisms are susceptible to GBS. Thus, the innate system can be protective by clearing the pathogens that trigger or exacerbate the disease, or by regulating the

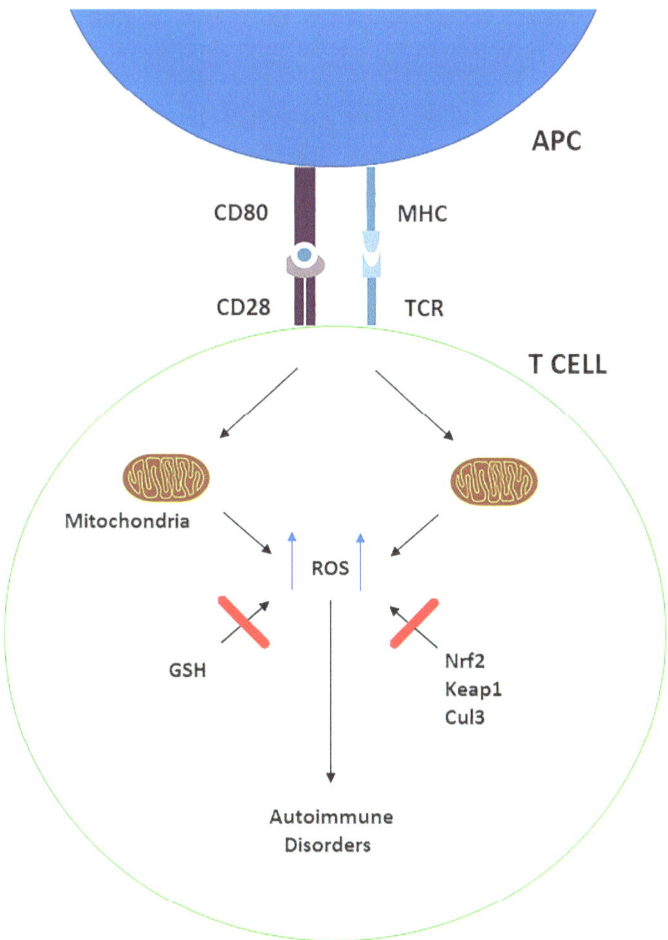

FIG. 4 Reactive oxygen species and T cell activity. Reactive oxygen species changes T cell activity.

presentation of antigens to T lymphocytes. Also, the defective eradication of microorganisms with a persisting antigen stimulation or failure in the clearance of self-reactive lymphocytes can result in organ-specific autoimmune diseases (Carroll, 2001). Because ROS changes T cell activity, it can be speculated that oxidative stress in GBS is a cause of demyelination.

In conclusion, the role of oxidative stress is still unknown in GBS but this is an important topic for understanding that exogenous antioxidant support may facilitate disease improvement. Monitoring the antioxidant profiles with clinical and laboratory findings will be certainly useful for assessing the prognosis, and this should be considered in future studies.

Policy and procedures

GBS is a neurological disease that can lead to serious morbidity and mortality. It is usually caused by an autoimmune reaction caused by an antecedent infection. It causes various degrees of demyelination and axonal degeneration. Oxidative stress is the disturbance of the balance between reactive oxygen species and antioxidant molecules. Oxidative stress is now held responsible in many degenerative and demyelinating neurological diseases.

Mini-dictionary of terms

Polyneuropathy. A disease or damage that affects peripheral nerves because of various reasons including hereditary, toxic, metabolic or nutritional insufficiency; the patients end up with numbness, weakness, and burning pain.

Campylobacter jejuni. It is a Gram-negative, rod-shaped, nonspore forming, microaerophilic bacterium that is the most common bacterial cause of foodborne infection in the United States.

Myelin sheath. An insulating layer and modified plasma membrane around nerves that originates from and is a part of the oligodendroglial cells in the central nervous system and the Schwann cells in the peripheral nervous system.

Miller-Fisher syndrome. A rare variant of Guillain-Barré syndrome characterized by external ophthalmoplegia, areflexia, and ataxia.

Membrane attack complex. A potentially cytolytic complex formed on the surface of pathogen cell membranes that is as a result of the activation of the host's complement system.

Miniature end-plate potential (MEPP). A slight (~0.4 mV) depolarization of the postsynaptic terminal caused by the release of a single vesicle into the synaptic space.

Reactive oxygen species. A chemical species that easily reacts with another molecule due to it containing oxygen.

Tocopherol. Major forms of vitamin E that have lipophilic features.

Key facts of oxidative stress in GBS

- There is an increase in the intensive EPR signals of nitric oxide (NO), complexes of NO with nonhemic iron (HbNO), and lypo- and superoxide radical content in the blood and cerebrospinal fluid in GBS.
- The reduction of the activity of superoxide dismutase and glutathione reductase in GBS has been shown.
- CEHCs show antioxidant activities like inhibition of inflammation.
- Increased levels of serum haptoglobin in the pretreatment of GBS patients compared to controls has been previously demonstrated.
- Uric acid is thought to be a naturally occurring antioxidant in human blood that cleans peroxynitrite, and it is significantly lower in GBS patients.
- High albumin levels are protective factors for patients with GBS.

Summary points

- Acute inflammatory demyelinating polyneuropathy (AIDP) or Guillain-Barré syndrome (GBS) is an autoimmune process that usually presents with progressive areflexic symmetrical ascending weakness.
- AIDP has been associated with many infectious diseases, including *C. jejuni*, mycoplasma pneumonia, the herpes virus, and many other viral and bacterial infections, along with vaccinations, pregnancy, and surgery.
- AIDP or GBS is characterized pathologically by immune system cells together with multifocal demyelination. Some autoantibodies to myelin glycolipids, including GM1, GD1a, and GD1b, have been identified.
- It has been demonstrated that the erythrocyte glutathione (GSH) levels in blood samples are significantly decreased in cases of GBS when compared to the control group.

- The total effect of all the enzymatic and nonenzymatic antioxidants may cause an initial decrease of MDA in patients with GBS.
- The concentrations of γ-tocopherol and δ-tocopherol markedly decrease in GBS patients. Decreased γ-tocopherol is a good predictor for GBS.
- No significant difference between GBS patients and healthy controls is found in tocopherol metabolites, including γ-CEHC, α-CEHC, and δ-CEHC.
- α-Caroten concentration is higher than in GBS patients in contrast to healthy controls.
- Other lipophilic antioxidants such as retinol, lutein, β-carotene, and Q10 decrease in patients with GBS.

References

Babior, B. M. (2000). Phagocytes and oxidative stress. *American Journal of Medicine*, *109*, 33–44.

Bruno, R. S., Leonard, S. W., Li, J., Bray, T. M., & Traber, M. G. (2005). Lower plasma alpha-carboxyethyl-hydroxychroman after deuterium-labeled alpha-tocopherol supplementation suggests decreased vitamin E metabolism in smokers. *American Journal of Clinical Nutrition*, *81*(5), 1052–1059.

Carroll, M. (2001). Innate immunity in the etiopathology of autoimmunity. *Nature Immunology*, *2*, 1089–1090.

Chang, C. Y., Chen, J. Y., Ke, D., & Hu, M. L. (2005). Plasma levels of lipophilic antioxidant vitamins in acute ischemic stroke patients: Correlation to inflammation markers and neurological deficits. *Nutrition*, *21*(10), 987–993.

Christen, S., Woodall, A. A., Shigenaga, M. K., Southwell-Keely, P. T., Duncan, M. W., & Ames, B. N. (1997). Gamma-tocopherol traps mutagenic electrophiles such as NO(X) and complements alpha-tocopherol: Physiological implications. *Proceedings of the National Academy of Sciences*, *94*(7), 3217–3222.

Daroff, R. B., Jankovic, J., Mazziotta, J. C., & Pomeroy, S. L. (2016). Neuroimmunology. In *Bradley's neurology in clinical practice* (7th ed.). London, New York, Oxford, Philadelphia, St Louis, Sydney, Toronto: Elsevier (Chapter 51).

Das, A., Kalita, J., & Misra, U. K. (2004). Recurrent Guillain–Barré syndrome. *Electromyography and Clinical Neurophysiology*, *44*, 95–102.

Davison, A., Tibbits, G., Shi, Z. G., & Moon, J. (1988). Active oxygen in neuromuscular disorders. *Molecular and Cellular Biochemistry*, *84*, 199–216.

Dogonadze, S. I., Ninua, N. G., Gordeziani, M. G., Kavlashvili, M. S., & Sanikidze, T. V. (2006). The role of oxidative stress in pathogenesis of GBS. *Georgian Medical News*, (140), 43–47. Nov.

Farias, M. S., Budni, P., Ribeiro, C. M., Parisotto, E. B., Santos, C. E., Dias, J. F., et al. (2012). Antioxidant supplementation attenuates oxidative stress in chronic hepatitis C patients. *Gastroenterologia y Hepatologia*, *35*(6), 386–394.

Fuhua, P., Xuhui, D., Zhiyang, Z., Ying, J., Yu, Y., Feng, T., et al. (2012). Antioxidant status of bilirubin and uric acid in patients with myasthenia gravis. *Neuroimmunomodulation*, *19*, 43–49.

Gelderman, K. A., Hultqvist, M., Holmberg, J., Olofsson, P., & Holmdahl, R. (2006). T cell surface redox levels determine T cell reactivity and arthritis susceptibility. *Proceedings of the National Academy of Sciences*, *103*, 12831–12836.

Ghabaee, M., Jabedari, B., Al-E-Eshagh, N., Ghaffarpour, M., & Asadi, F. (2010). Serum and cerebrospinal fluid antioxidant activity and lipid peroxidation in Guillain-Barre syndrome and multiple sclerosis patients. *International Journal of Neuroscience*, *120*(4), 301–304.

Ghafourifar, P., Mousavizadeh, K., Parihar, M. S., Nazarewicz, R. R., Parihar, A., & Zenebe, W. J. (2008). Mitochondria in multiple sclerosis. *Frontiers in Bioscience*, *13*, 3116–3126.

Grand'Maison, F., Feasby, T. E., Hahn, A. F., & Koopman, W. J. (1992). Recurrent Guillain–Barré syndrome. Clinical and laboratory features. *Brain*, *115*, 1093–1106.

Gutowski, N. J., Pinkham, J. M., Akanmu, D., Chirico, S., & Murphy, R. P. (1998). Free radicals in inflammatory neurological disease: Increased lipid peroxidation and haptoglobin levels in Guillain Barre syndrome. *Irish Journal of Medical Science*, *167*(1), 43–46.

Halliwell, B. (1988). Albumin—An important extracellular antioxidant? *Biochemical Pharmacology*, *37*, 569–571.

Hensley, K., Benaksas, E. J., Bolli, R., Comp, P., Grammas, P., Hamdheydari, L., et al. (2004). New perspectives on vitamin E: Gamma-tocopherol and carboxyelthyl-hydroxychroman metabolites in biology and medicine. *Free Radical Biology and Medicine*, *36*(1), 1–15.

Hooper, D. C., Bagasra, O., Marini, J. C., Zborek, A., Ohnishi, S. T., Kean, R., et al. (1997). Prevention of experiment alallergic encephalomyelitis by targeting nitric oxide and peroxynitrite: Implications for the treatment of multiple sclerosis. *Proceedings of the National Academy of Sciences*, *94*, 2528–2533.

Hultqvist, M., Backlund, J., Bauer, K., Gelderman, K. A., & Holmdahl, R. (2007). Lack of reactive oxygen species breaks T cell tolerance to collagen type II and allows development of arthritis in mice. *Journal of Immunology*, *179*, 1431–1437.

Jiang, Q., & Ames, B. N. (2003). γ-Tocopherol, but not α-tocopherol, decreases proinflammatory eicosanoids and inflammation damage in rats. *FASEB Journal*, *17*(8), 816–822.

Jiang, Q., Christen, S., Shigenaga, M. K., & Ames, B. N. (2001). γ-Tocopherol, the major form of vitamin E in the US diet, deserves more attention. *American Journal of Clinical Nutrition*, *74*(6), 714–722.

Jiang, Q., Elson-Schwab, I., Courtemanche, C., & Ames, B. N. (2000). γ-Tocopherol and its major metabolite, in contrast to α-tocopherol, inhibit cyclooxygenase activity in macrophages and epithelial cells. *Proceedings of the National Academy of Sciences*, *97*(21), 11494–11499.

Karlik, M., Valkovic, P., Hancinova, V., Krizova, L., Tothova, L., & Celec, P. (2015). Markers of oxidative stress in plasma and saliva in patients with multiple sclerosis. *Clinical Biochemistry*, *48*, 24–28.

Kieseier, B. C., Kiefer, R., Gold, R., Hemmer, B., Willison, H. J., & Hartung, H. P. (2004). Advances in understanding and treatment of immune-mediated disorders of the peripheral nervous system. *Muscle and Nerve*, *30*, 131–156.

Klebanoff, S. J., & Coombs, R. W. (1992). Viricidal effect of polymorphonuclear leukocytes on human immunodeficiency virus-1. Role of the myeloperoxidase system. *Journal of Clinical Investigation*, *89*, 2014–2017.

Kumar, K. T., Chandrika, A., Sumanth, K. N., Sireesha, P., Rao, S., & Rao, A. (2004). Free radical toxicity and antioxidants in Guillain-Barre syndrome, a preliminary study. *Clinica Chimica Acta*, *346*(2), 205–209.

Lewis, R. A. (2015). *Chronic inflammatory demyelinating polyneuropathy*. https://rarediseases.org/rare-diseases/chronic-inflammatory-demyelinating-polyneuropathy/. (Accessed 26 November 2019).

Li, G. X., Lee, M. J., Liu, A. B., Yang, Z., Lin, Y., Shih, W. J., et al. (2011). δ-Tocopherol is more active than α- or γ-tocopherol in inhibiting lung tumorigenesis in vivo. *Cancer Prevention Research*, *4*(3), 404–413.

Liu, M., Wallmon, A., Olsson-Mortlock, C., Wallin, R., & Saldeen, T. (2003). Mixed tocopherols inhibit platelet aggregation in humans: Potential mechanisms. *American Journal of Clinical Nutrition*, *77*(3), 700–706.

Marx, G., & Chevion, M. (1986). Site-specific modification of albumin by free radicals. Reaction with copper(II) and ascorbate. *Biochemical Journal*, *236*, 397–400.

Meyer zu Hörste, G., Hartung, H. P., & Kieseier, B. C. (2007). From bench to bedside-experimental rationale for immune-specific therapies in the inflamed peripheral nerve. *Nature Clinical Practice Neurology*, *3*, 198–211.

Miller, E., Walczak, A., Saluk, J., Ponczek, M. B., & Majsterek, I. (2012). Oxidative modification of patient's plasma proteins and its role in pathogenesis of multiple sclerosis. *Clinical Biochemistry*, *45*, 26–30.

Mitsui, Y., Kusunoki, S., Arimura, K., Kaji, R., Kanda, T., Kuwabara, S., et al. (2005). A multicentre prospective study of Guillain-Barre syndrome in Japan: A focus on the incidence of subtypes. *Journal of Neurology, Neurosurgery and Psychiatry*, *86*(1), 110–114.

Mossberg, N., Andersen, O., Nordin, M., Nilsson, S., Svedhem, A., Bergström, T., et al. (2010). Leukocyte oxygen radical production determines disease severity in the recurrent Guillain-Barré syndrome. *Journal of Inflammation*, *7*(1), 40.

O'Hanlon, G. M., Humphreys, P. D., Goldman, R. S., Halstead, S. K., Bullens, R. W., Plomp, J. J., et al. (2003). Calpain inhibitors protect against axonal degeneration in a model of anti-ganglioside antibody-mediated motor nerve terminal injury. *Brain*, *126*, 2497–2509.

Peng, F., Yang, Y., Liu, J., Jiang, Y., Zhu, C., Deng, X., et al. (2012). Low antioxidant status of serum uric acid, bilirubin and albumin in patients with neuromyelitis optica. *European Journal of Neurology*, *19*, 277–283.

Sakai, T., Inoue, A., Koh, C. S., & Ikeda, S. (2000). A study of free radical defense and oxidative stress in the sera of patients with neuroimmunological disorders. *Arerugi*, *49*(1), 12–18.

Schweigert, F. J. (2001). Inflammation-induced changes in the nutritional biomarkers serum retinol and carotenoids. *Current Opinion in Clinical Nutrition and Metabolic Care*, *4*(6), 477–481.

So, A., & Thorens, B. (2010). Uric acid transport and disease. *Journal of Clinical Investigation*, *120*, 1791–1799.

Su, Z., Chen, Z., Xiang, Y., Wang, B., Huang, Y., Yang, D., et al. (2017). Low serum levels of uric acid and albumin in patients with Guillain-Barre syndrome. *Medicine (Baltimore)*, *96*(15), e6618.

Tang, H. Y., Ho, H. Y., Chiu, D. T., Huang, C. Y., Cheng, M. L., & Chen, C. M. (2017). Alterations of plasma concentrations of lipophilic antioxidants are associated with Guillain-Barre syndrome. *Clinica Chimica Acta*, *470*, 75–80.

Tasset, I., Agüera, E., Sanchez-Lopez, F., Feijoo, M., Giraldo, A. I., Cruz, A. H., et al. (2012). Peripheral oxidative stress in relapsing-remitting multiple sclerosis. *Clinical Biochemistry*, *45*, 440–444.

Tseng, C. F., Lin, C. C., Huang, H. Y., Liu, H. C., & Mao, S. J. (2004). Antioxidant role of human haptoglobin. *Proteomics*, *4*(8), 2221–2228.

Vijayaraghavan, R., Suribabu, C. S., Sekar, B., Oommen, P. K., Kavithalakshmi, S. N., Madhusudhanan, N., et al. (2005). Protective role of vitamin E on the oxidative stress in Hansen's disease (Leprosy) patients. *European Journal of Clinical Nutrition*, *59*(10), 1121–1128.

Wayner, D. D., Burton, G. W., Ingold, K. U., & Locke, S. (1985). Quantitative measurement of the total, peroxyl radical-trapping antioxidant capability of human blood plasma by controlled peroxidation. The important contribution made by plasma proteins. *FEBS Letters*, *187*, 33–37.

Chapter 13

Neuromuscular effects and rehabilitation in Guillain-Barré syndrome

Thomas Harbo and Henning Andersen
Department of Neurology, Aarhus University Hospital, Aarhus N, Denmark

Abbreviations

FIM	Functional Independence Measure
GBS	Guillain-Barré syndrome
ICU	intensive care unit
IVIG	intravenous immunoglobulin
MFS	muscle fiber stimulation
NMES	neuromuscular electrical stimulation
PE	plasma exchange
VMR	virtual motor rehabilitation

Introduction

Guillain-Barré syndrome (GBS) is an acute inflammatory disease affecting peripheral nerves and nerve roots (Leonhard et al., 2019). Most commonly, GBS is preceded by an infection a few weeks prior to the occurrence of neuropathic symptoms (Sipila & Soilu-Hanninen, 2015). Therefore the incidence of GBS can increase during outbreaks of infectious diseases, which most recently was observed during the 2015–16 Zika virus epidemics in the French Polynesia and Latin America with severalfold increased incidence of GBS in several countries (Cao-Lormeau et al., 2016; Parra et al., 2016). GBS typically presents with muscle weakness and sensory symptoms combined with loss of tendon reflexes and presents initially in the lower extremities progressing to the arms and the respiratory and cranial muscles (Willison, Jacobs, & van Doorn, 2016). The progressive phase usually last for days to weeks with most patients reaching nadir within 4 weeks of symptoms followed by a plateau phase and slowly recovery. Besides the typical presentation of sensory and motor neuropathy, patients may have clinical variants like the triad of ophthalmoplegia, ataxia, and areflexia known as Miller Fischer syndrome, pure motor, or paraparetic or pharyngeal-cervical-brachial variant (Wakerley & Yuki, 2015). Furthermore the neuropathy can be classified as demyelinating or axonal according to the electrophysiological examination (Hadden et al., 1998).

The prognosis of GBS is very heterogeneous. Some patients are mildly affected with fast recovery and no disabilities irrespective of any treatment given. On the other side, 20%–30% of patients develop complete paralysis, severe respiratory, or autonomic failure and remain treated at intensive care unit (ICU) for months (Al-Hakem et al., 2019). In a group of prolonged mechanical ventilated patients, 31% was able to walk after 1 year and 58% after maximum time of follow-up (van den Berg, Storm, Garssen, Blomkwist-Markens, & Jacobs, 2018). Despite the lack of evidence, multidisciplinary supportive care and rehabilitation are important in GBS. In the acute phase, consensus-based recommendations include (1) monitoring of respiratory and autonomic function in a setting with available artificial ventilation and neurointensive care, (2) prophylactic antithrombotic treatment for deep vein thrombosis, (3) treatment of pain, (4) management of nutrition and bladder and bowel dysfunction, and (5) physiotherapy to prevent muscle shortening and joint contractures (Hughes et al., 2005). All this should be followed by a rehabilitation and exercise program to regain physical abilities as soon as possible. Recovery can take months to years and end up with significant chronic disabilities despite immunomodulatory treatment. As shown in the largest prospective cohort of GBS patients studied to date, a large proportion of patients have long-term motor dysfunction with 17% of patients from Europe and America being unable to walk unaided after 12 months (Doets et al., 2018), emphasizing the importance of identifying more effective neuromuscular rehabilitation. Motor

Zika Virus Impact, Diagnosis, Control, and Models. https://doi.org/10.1016/B978-0-12-820267-8.00013-3

dysfunctions like weakness, wasting, and contractures are major long-term complications of severely affected patients. In this review, we present an overview of existing evidence of treatment to prevent muscle weakness and disabilities after GBS with special emphasis on the effect of neuromuscular rehabilitation in the acute and chronic phase of the disease.

Pharmacological treatment

In several large randomized controlled clinical trials, treatment with plasma exchange (PE) or intravenous immunoglobulin (IVIG) initiated in the acute phase of GBS has been proven effective. Compared with placebo, treatment with PE or IVIG result in reduced need for respiratory support and an increased chance to regain mobility and muscle strength after 1 month and 12 months (Chevret, Hughes, & Annane, 2017; Hughes, Swan, & van Doorn, 2014). Despite immunomodulatory treatment a group of GBS patients remains with a very poor prognosis. In a combined cohort study of 526 patients and a cross-sectional study including 63 ventilated patients (van den Berg et al., 2018), it was shown that 6% of GBS patients required mechanical ventilation for more than 2 months. The prolonged mechanical ventilated patients had a median (range) stay at ICU of 101 (97–126) days and at hospital of 129 (104–162) days, followed by 252 (177–403) days of clinical rehabilitation and 198 (183–502) days of outpatient rehabilitation. At an 11-year follow-up, only 58% had regained ambulation, and the median time to regain ambulation was 548 (270–730) days. This emphasizes the need for more effective treatment in GBS. Recently, small clinical studies have indicated that monoclonal antibodies against complement proteins given early during disease course could have some benefit in GBS, but larger studies are needed to prove it (Davidson et al., 2017; Misawa et al., 2018). Eventually, it is important to underscore that as of today no evidence-based pharmacological treatments are available to prevent muscle atrophy or muscle weakness in GBS (Pritchard, Hughes, Hadden, & Brassington, 2016).

Multidisciplinary rehabilitation

Most patients with moderate to severe GBS are offered multidisciplinary rehabilitation, which means two or more coordinated interventions under medical supervision by a neurologist or rehabilitation physician. Multidisciplinary rehabilitation aims at regaining autonomy with the ability to perform all activities of daily living. This may include not only physio- or occupational therapy and exercise programs but also nursing, dietitian advice, psychotherapy, speech therapy, and social rehabilitation depending on the individual patient's need. Since multidisciplinary rehabilitation varies considerably between countries and local facilities and requires an individually adjusted plan to the individual patient, it has been difficult to design research trials to assess the efficacy of the intervention. In a systematic review of rehabilitation interventions in patients with GBS (Khan & Amatya, 2012), only five original studies could be identified evaluating the effectiveness of multidisciplinary rehabilitation. This includes only one good quality randomized controlled study comparing high- and low-intensity rehabilitation in patients with disability more than 1 year after GBS (Khan et al., 2011). In this study, 79 adult patients were included 1–12 years after GBS diagnosis and randomized to receive either individualized outpatient center-based high-intensive rehabilitation (intervention, $n = 40$) or a lower-intensity home-based program (control, $n = 39$). The intervention comprised three one-hour individualized sessions weekly for 12 weeks. It included physical and occupational therapy for strengthening, endurance, and gait training and task-specific rehabilitation to improve every day, community, and work function. The control group did 30-min maintenance training program twice weekly and was also allowed to perform other rehabilitation activities if needed. Outcome was assessed 1 year after the intervention and included measurements of activity, participation, and perceived impact of their problems. Based on the total and the motor scales of the Functional Independence Measure (FIM), in an intention-to-treat analysis, there was a small but statistically significant improvement in the high-intensity rehabilitation group compared with the controls. Furthermore, 80% of the patients complying with the high-intensity protocol had a clinical meaningful improvement in the FIM motor score (at least three points) compared with only 8% of the controls. Adverse effects were not reported; however, only 22 (55%) of the 40 patients assigned to high-intensity rehabilitation completed the study due to loss to follow-up or inability or unwillingness to comply with the protocol. This low number of follow-up reduces the applicability and external validity of the study as it suggests that applicability of the intervention is challenging. Other original studies have included (1) one case-control study ($n = 34$) of inpatient rehabilitation with a control group of healthy subjects (Demir & Koseoglu, 2008), (2) one prospective case series ($n = 35$) of inpatient rehabilitation followed by a home-based training program (Gupta, Taly, Srivastava, & Murali, 2010), and (3) two retrospective case series ($n = 39$ and 24) of inpatient rehabilitation (Meythaler, DeVivo, & Braswell, 1997; Nicholas, Playford, & Thompson, 2000). In these studies, GBS patients improved during multidisciplinary rehabilitation, but they were not designed to distinguish between spontaneous recovery and the effect of the rehabilitation intervention.

Despite several limitations the authors of the review concluded that there is good evidence (grade level II) to support ambulatory, outpatient multidisciplinary rehabilitation in obtaining long-term improvements at the level of activity and participation in patients with GBS in the later stages of recovery. Further, they concluded that there is satisfactory (grade level III) evidence to support (1) inpatient rehabilitation followed by outpatient rehabilitation, thereby inducing functional recovery, and (2) physical therapy and exercise to reduce joint contractures and muscle weakness. In another more recent case series of 51 GBS patients, motor recovery, following the acute pharmacological treatment response, is assessed during the acute inpatient care and after outpatient and home-based rehabilitation (Prada et al., 2019). A description of the intervention was not provided, but it included physical therapy for 61 ± 58 (mean $\pm$ SD) days for inpatient, 96 ± 70 days for outpatients, and 75 ± 15 days during home rehabilitation. Again the natural history with spontaneous improvement after GBS and the lack of a control group impair the possibility to draw any final conclusions regarding the effectiveness of rehabilitation in the study. However, it was shown that muscle strength measured with a MRC sum score (Medical Research Council, 1976) and ambulation assessed with the GBS disability score (Hughes, Newsom-Davis, Perkin, & Pierce, 1978) continue to improve beyond the first 6 months of rehabilitation.

Exercise

In a systematic review, Simatos and colleagues have evaluated the available literature of exercise as an intervention of rehabilitation for adult patients with GBS (Simatos, Vincent, Yu, Bastien, & Sweeney, 2016). Papers from 1951 to 2016 were identified in PubMed searches, and the quality of the studies was assessed and classified according to a modified version of the Centre for Evidence-Based Medicine level of evidence. Seven papers with exercise as the main intervention were identified, including four uncontrolled single cases with minimal evidence level, one trial that included multidisciplinary rehabilitation (reviewed in the previous section) (Khan et al., 2011), and two Dutch papers of a case series in an open-label standardized exercise protocol (evidence level 5) (Bussmann, Garssen, van Doorn, & Stam, 2007; Garssen et al., 2004). In the Dutch study, 16 patients were included 0.5–15 years after their GBS as well as four patients with stable chronic inflammatory demyelinating polyradiculoneuropathy. All patients were ambulatory and reported fatigue as a major complaint. The exercise intervention consisted of three 45-min cycle-training sessions every week for 12 weeks. During the 12-week period, training intensity was gradually increased. The target heart rate increasing from 65% to a maximum of 90% of maximal heart rate, and an increasing workload was applied on the home trainer also. The intervention resulted in lower fatigue levels, increased isokinetic muscle strength, and a higher peak oxygen uptake. Further, patients improved on a handicap scale and on the physical components score of the SF-36 quality of life scale. Two patients did not complete the study for nonstudy-related reasons, and 25% reported mild and transient muscle cramps, paresthesias, or pain. Overall the conclusion is that exercise as intervention for late disabilities and fatigue in GBS is feasible and can be of benefit for some patients.

Neuromuscular electrical stimulation

In the acute phase of severe GBS, rehabilitation training is challenged by limited patient participation due to severe weakness or even paralysis. For practical reasons, training may also be challenged if patients are in the ICU, intubated, and with respiratory support by a ventilator. Inactive and denervated muscles will indisputable and fast undergo degeneration, and muscle atrophy develops (Griffiths, Palmer, Helliwell, MacLennan, & MacMillan, 1995; Reid, Campbell, & Little, 2004). Neuromuscular electrical stimulation (NMES) is a method to induce muscle contractions without patient participation. This may be an alternative therapeutic approach in the acute setting of GBS, which can minimize inactivation and denervation wasting until patients have recovered to a level where a multidisciplinary rehabilitation effort can be initiated (Maddocks, Gao, Higginson, & Wilcock, 2013; Maffiuletti, Roig, Karatzanos, & Nanas, 2013). In a small proof-of-concept study, this has proven feasible with satisfactory safety. There was also a trend for an effect of NMES on muscle wasting as add-on to established standard of care in the acute and subacute phase of GBS (Harbo et al., 2019). Seventeen patients with moderate to severe GBS were randomized to receive an hour of NMES every weekday on the right or left quadriceps femoris muscle with the nonstimulated muscle serving as control. Stimulation was initiated within 2 weeks after the first sign of weakness and was continued through the acute hospital admission and the following inpatient rehabilitation. The median (range) time of participation was 27 days (10–95) and included 17 (4–53) stimulation sessions. During the study, each patient had a mean loss of lean body mass (muscle) of 3.4 kg, pointing out that GBS patients with weakness experience substantial muscle wasting. NMES was found to be safe and feasible as add-on to standard supportive therapy and rehabilitation in the acute and subacute phase of GBS. There was a trend toward a preventive effect of NMES on muscle atrophy, but the study was not designed to explore effect on disability of the patients.

Virtual motor rehabilitation system

Virtual motor rehabilitation (VMR) is a new technology that combines novel rehabilitation software with low-cost commercially available devices like the Nintendo Wii platform. To be effective, multidisciplinary rehabilitation in GBS is very time demanding including several daily sessions for as long as 6, 12, and 18 months (El Mhandi, Calmels, Camdessanche, Gautheron, & Feasson, 2007). Often the rehabilitation offered is limited due to lack of time and resources, and patients may find the training tedious and monotonous, resulting in lack of compliance. Therefore VMR could be an attractive supplement to the established rehabilitation. The method is still under development, and so far, only one paper has been published, describing VMR applied 4 and 5 months after admission in two severely affected GBS patients as an add-on to conventional multidisciplinary rehabilitation (Albiol-Perez et al., 2015). In this study the Nintendo Wii Balance Board and a virtual environmental tool were applied in 20 rehabilitation sessions consisting of 30-min traditional therapy and 30-min VMR. Compliance was good and the patients improved. VMR could be developed further to include more aspects of the rehabilitation process in the future.

Safety

In anecdotal case reports and experimental animal studies, it has been indicated that overexercising during rehabilitation after GBS can damage motor units and cause paradoxical weakening, which have led to hesitation with intensive and strenuous exercise in many recommendations (Hughes et al., 2005). The clinical data to support this concern are negligible, and overall, it is reasonable to believe that the benefit of exercising weakened muscles after GBS excess the risk of harm. However, systematic registration of safety and complications should always be included in future studies.

Neuromuscular rehabilitation after GBS is important for the functional outcome of each individual patient. However, the quality of the present evidence of rehabilitation efficacy is low; the rehabilitation is both complex, time consuming and expensive; and there is currently no standardized care for patients with neuromuscular disabilities after GBS. Therefore the rehabilitation effort can suffer from limited available resources and expertise. Because the monophasic course and spontaneous recovery in GBS challenge the interpretation of noncontrolled studies, to improve the interpretation of neuromuscular rehabilitation trials in GBS in the future, large controlled studies and standardized sensitive efficacy outcome measures are needed.

Policy and procedures

Neuromuscular electrical stimulation protocol

Stimulation of the quadriceps muscle was performed using a STIWELL med4 stimulation unit (https://www.ottobock.co.th/neurorehabilitation/solutions/solutions-with-functional-electrical-stimulation/stiwell-med-4/) (Otto Bock, Königsee, Germany) and two large stimulation pads (6 × 8 cm). The intensity of electrical stimulation was titrated individually at entry and weekly during the study to the point of maximal contraction or the highest tolerable intensity. During the first session of stimulation, the skin under the pads was inspected every 5 min for redness or other signs of tissue damage. Trained physical therapists attached the equipment and titrated the stimulation intensity, but after being attached to the patient, the individualized stimulation protocol ran automatically.

Direct muscle fiber stimulation (MFS). With MFS, contraction is induced directly through the muscle fiber membrane independent of the neuromuscular junction, which means that complete distally denervated muscle fibers can be activated. The disadvantage is that higher-intensity stimulation, especially in atrophic muscle, is needed, which can lead to discomfort and irritation of the skin. MFS was applied by two pads placed over the proximal and distal part of the muscle with triangular dual-phase stimulation pulses. The initiation protocol was 1 Hz frequency, 250 ms pulse width, and 3/6 on/off ratio. The lowest pulse width with maximal contraction was chosen, and then the frequency was increased to the highest tolerated level.

Neuromuscular electrical stimulation (NMES). With NMES the muscle is activated through the muscle spindle and neuromuscular endplate. As a result the contraction is more physiological, and less electrical stimulation needs to be applied. NMES was applied by one pad placed on the middle of the muscle bulk, where the neuromuscular transmission is located, with rectangular dual-phase stimulation pulses. The protocol included four phases of 5, 15, 15, and 5 min, with frequencies of 10, 40, 60, and 3 Hz, with a pulse width of 0.3 ms. The intensity could be adjusted from 0 to 100 mA and was increased to the highest tolerated level.

The intention was to stimulate 5–7 days a week including 20 min of MFS followed by 40 min of NMES. Also the NMES was applied to patients where no visible contraction could be observed.

Mini-dictionary of terms

Neuromuscular electrical stimulation. A method to induce muscle contraction by applying an electrical impulse to the neuromuscular endplate by an electronic device.

Multidisciplinary rehabilitation. Two or more coordinated interventions for disabled patients to regain autonomy and functions of daily living. Usually, this is performed by physical and occupational therapists but may include other specialties.

Motor dysfunction. Several methods are used to describe the motor dysfunction in GBS. Muscle weakness is a main feature of GBS that develops quickly in the acute phase. Weakness can be assessed manually with the MRC score on a scale from 0 to 5 (0, paralysis with no visible contraction; 1, visible contraction but no limb movement; 2, limb movement only with gravity eliminated; 3, active movement against gravity; 4, active movement against gravity and resistance but reduced strength; 5, normal strength). Weakness may quantified on a linear scale using a dynamometer. In addition to weakness, chronic muscle dysfunction can result in muscle wasting and muscle and joint contracture and shortening, which is very disabling.

Impairment and disability. Impairment is the direct damage caused by the disease, for example, weakness of leg muscles (as described earlier) or loss of sensation, while disability is the loss of function that impairment causes, for example, loss of ambulation. Often the GBS disability score is used in GBS to describe the severity of the disease on the level of disability (0, healthy; 1, minor symptoms and capable of running; 2, able to walk 10m without assistance but unable to run; 3, able to walk 10m across an open space with help; 4, bedridden or chair bound; 5, requiring assisted ventilation for at least part of the day; 6, dead).

Key facts of neuromuscular rehabilitation in GBS

Neuromuscular rehabilitation in Guillain-Barré syndrome can include:

- Physical therapy to prevent muscle and joint shortening and contractures.
- Multidisciplinary rehabilitation with two or more coordinated interventions for disabled patients to regain autonomy and functions of daily living.
- Exercise and training to improve or maintain function.
- Neuromuscular electrical stimulation to prevent muscle wasting.

The prognosis of Guillain-Barré syndrome:

- Guillain-Barré syndrome is a heterogenous disorder with a monophasic course.
- Clinical severity ranges from mild impairment to complete paralysis combined with respiratory and autonomic failure.
- In 20%–30% of patients, mechanical ventilation is required at nadir of GBS.
- The most severely affected patients have a long recovery phase and poor prognosis.
- More than half of all mechanically ventilated patient remain unable to walk unassisted at 1-year follow-up (Fig. 1).

Summary points of neuromuscular rehabilitation in GBS

- Most commonly, GBS is preceded by an infection; therefore the incidence of GBS can increase during outbreaks of infectious diseases, which most recently was observed during the Zika virus outbreak in the French Polynesia and Latin America with severalfold increased incidence of GBS in several countries.
- Despite optimal evidence-based treatment with immunoglobulin and plasma exchange, a large proportion of GBS patients will have substantial neuromuscular disabilities more than 1 year after disease onset. Among patients receiving mechanical ventilation, more than half will be able to walk unassisted.
- In the acute phase of GBS, physical therapy is important to prevent muscle shortening and joint contractures.
- Patients may still improve their physical function several years after onset of GBS.
- There is evidence to support high-intensity multidisciplinary rehabilitation and exercise, which improves level of activity and participation in the late and chronic stages of GBS.
- New approaches like neuromuscular electrical stimulation and virtual motor rehabilitation seem to be feasible methods in the acute and late-stage recovery of GBS, but the efficacy needs to be explored in future studies.

Acute phase (one to eight weeks) **Patients are at risk of deteriorating and to suffer complications**	**Rehabilitation focus: Preserve joint mobility (no grade 1 or 2 evidence)**
Patients are at hospital for acute care including respiratory and autonomic monitoring and support, thromboprophylaxis, pain treatment etc.	* Physiotherapy to prevent muscle shortening and contractures. * Neuromuscular electrical stimulation. Safe and feasible, efficacy still to be explored.
Intermediate phase (weeks to several months) **Patients improve gradually**	**Rehabilitation focus: Regain autonomy and ability to perform activities of daily living.** **(no grade 1 or 2 evidence)**
Patients are at rehabilitation centers or outpatient rehabilitation facilities.	* Multidisciplinary rehabilitation * Task specific training * Neuromuscular electrical stimulation. Safe and feasible, efficacy still to be determined. * Cardiovascular and strengthening exercises
Chronic phase (months to years) **Patients are stable with residual disabilities**	**Rehabilitation focus: Further improve or maintain function**
Patients are exercising and performing rehabilitation training at home, at outpatient facilities or fitness centers.	* High intensity multidisciplinary rehabilitation improves functional independence **(grade level 2)**. * Multidisciplinary rehabilitation * Task specific training * Exercise cardiovascular and strength * Virtual Motor Rehabilitation

FIG. 1 Neuromuscular rehabilitation in three phases of Guillain-Barré syndrome. Rehabilitation-focused and studied interventions in three phases of Guillain-Barré syndrome, the acute, subacute/intermediate, and chronic/long-term phase. Level of evidence is indicated using the following grade system: level 1, meta-analysis of multiple well-designed randomized controlled trials; level 2, at least one randomized controlled trial; level 3–5, nonrandomized controlled trials, descriptive studies, or case series.

References

Albiol-Perez, S., Forcano-Garcia, M., Munoz-Tomas, M. T., Manzano-Fernandez, P., Solsona-Hernandez, S., Mashat, M. A., et al. (2015). A novel virtual motor rehabilitation system for Guillain-Barre syndrome. Two single case studies. *Methods of Information in Medicine*, *54*(2), 127–134. Available from: PM:25609504.

Al-Hakem, H., Sindrup, S. H., Andersen, H., de la Cour, C. D., Lassen, L. L., van den Berg, B., et al. (2019). Guillain-Barre syndrome in Denmark: A population-based study on epidemiology, diagnosis and clinical severity. *Journal of Neurology*, *266*(2), 440–449. Available from: PM:30536111.

Bussmann, J. B., Garssen, M. P., van Doorn, P. A., & Stam, H. J. (2007). Analysing the favourable effects of physical exercise: Relationships between physical fitness, fatigue and functioning in Guillain-Barre syndrome and chronic inflammatory demyelinating polyneuropathy. *Journal of Rehabilitation Medicine*, *39*(2), 121–125. Available from: PM:17351693.

Cao-Lormeau, V. M., Blake, A., Mons, S., Lastere, S., Roche, C., Vanhomwegen, J., et al. (2016). Guillain-Barre syndrome outbreak associated with Zika virus infection in French Polynesia: A case-control study. *Lancet*, *387*(10027), 1531–1539. Available from: PM:26948433.

Chevret, S., Hughes, R. A., & Annane, D. (2017). Plasma exchange for Guillain-Barre syndrome. *Cochrane Database of Systematic Reviews*, *2*, CD001798. Available from: PM:28241090.

Davidson, A. I., Halstead, S. K., Goodfellow, J. A., Chavada, G., Mallik, A., Overell, J., et al. (2017). Inhibition of complement in Guillain-Barre syndrome: The ICA-GBS study. *Journal of the Peripheral Nervous System*, *22*(1), 4–12. Available from: PM:27801990.

Demir, S. O., & Koseoglu, F. (2008). Factors associated with health-related quality of life in patients with severe Guillain-Barre syndrome. *Disability and Rehabilitation*, *30*(8), 593–599. Available from: PM:17852306.

Doets, A. Y., Verboon, C., van den Berg, B., Harbo, T., Cornblath, D. R., Willison, H. J., et al. (2018). Regional variation of Guillain-Barre syndrome. *Brain*, *141*(10), 2866–2877. Available from: PM:30247567.

El Mhandi, M. L., Calmels, P., Camdessanche, J. P., Gautheron, V., & Feasson, L. (2007). Muscle strength recovery in treated Guillain-Barre syndrome: A prospective study for the first 18 months after onset. *American Journal of Physical Medicine and Rehabilitation*, *86*(9), 716–724. Available from: PM:17709995.

Garssen, M. P., Bussmann, J. B., Schmitz, P. I., Zandbergen, A., Welter, T. G., Merkies, I. S., et al. (2004). Physical training and fatigue, fitness, and quality of life in Guillain-Barre syndrome and CIDP. *Neurology*, *63*(12), 2393–2395. Available from: PM:15623709.

Griffiths, R. D., Palmer, T. E., Helliwell, T., MacLennan, P., & MacMillan, R. R. (1995). Effect of passive stretching on the wasting of muscle in the critically ill. *Nutrition*, *11*(5), 428–432. Available from: PM:8748193.

Gupta, A., Taly, A. B., Srivastava, A., & Murali, T. (2010). Guillain-Barre Syndrome—rehabilitation outcome, residual deficits and requirement of lower limb orthosis for locomotion at 1 year follow-up. *Disability and Rehabilitation*, *32*(23), 1897–1902. Available from: PM:20331413.

Hadden, R. D., Cornblath, D. R., Hughes, R. A., Zielasek, J., Hartung, H. P., Toyka, K. V., et al. (1998). Electrophysiological classification of Guillain-Barre syndrome: Clinical associations and outcome. Plasma Exchange/Sandoglobulin Guillain-Barre Syndrome Trial Group. *Annals of Neurology*, *44* (5), 780–788. Available from: PM:9818934.

Harbo, T., Markvardsen, L. K., Hellfritzsch, M. B., Severinsen, K., Nielsen, J. F., & Andersen, H. (2019). Neuromuscular electrical stimulation in early rehabilitation of Guillain-Barre syndrome: A pilot study. *Muscle and Nerve*, *59*(4), 481–484. Available from: PM:30549053.

Hughes, R. A., Newsom-Davis, J. M., Perkin, G. D., & Pierce, J. M. (1978). Controlled trial prednisolone in acute polyneuropathy. *Lancet*, *2*(8093), 750–753. Available from: PM:80682.

Hughes, R. A., Swan, A. V., & van Doorn, P. A. (2014). Intravenous immunoglobulin for Guillain-Barre syndrome. *Cochrane Database of Systematic Reviews*, *9*, CD002063. Available from: PM:25238327.

Hughes, R. A., Wijdicks, E. F., Benson, E., Cornblath, D. R., Hahn, A. F., Meythaler, J. M., et al. (2005). Supportive care for patients with Guillain-Barre syndrome. *Archives of Neurology*, *62*(8), 1194–1198. Available from: PM:16087757.

Khan, F., & Amatya, B. (2012). Rehabilitation interventions in patients with acute demyelinating inflammatory polyneuropathy: A systematic review. *European Journal of Physical and Rehabilitation Medicine*, *48*(3), 507–522. Available from: PM:22820829.

Khan, F., Pallant, J. F., Amatya, B., Ng, L., Gorelik, A., & Brand, C. (2011). Outcomes of high- and low-intensity rehabilitation programme for persons in chronic phase after Guillain-Barre syndrome: A randomized controlled trial. *Journal of Rehabilitation Medicine*, *43*(7), 638–646. Available from: PM:21667009.

Leonhard, S. E., Mandarakas, M. R., Gondim, F. A. A., Bateman, K., Ferreira, M. L. B., Cornblath, D. R., et al. (2019). Diagnosis and management of Guillain-Barre syndrome in ten steps. *Nature Reviews. Neurology*, *15*(11), 671–683. Available from: PM:31541214.

Maddocks, M., Gao, W., Higginson, I. J., & Wilcock, A. (2013). Neuromuscular electrical stimulation for muscle weakness in adults with advanced disease. *Cochrane Database of Systematic Reviews*, (1), CD009419. Available from: PM:23440837.

Maffiuletti, N. A., Roig, M., Karatzanos, E., & Nanas, S. (2013). Neuromuscular electrical stimulation for preventing skeletal-muscle weakness and wasting in critically ill patients: A systematic review. *BMC Medicine*, *11*, 137. Available from: PM:23701811.

Medical Research Council. (1976). *Aids to examination of the peripheral nervous system*. London: HMSO. Memorandum No. 45.

Meythaler, J. M., DeVivo, M. J., & Braswell, W. C. (1997). Rehabilitation outcomes of patients who have developed Guillain-Barre syndrome. *American Journal of Physical Medicine and Rehabilitation*, *76*(5), 411–419. Available from: PM:9354496.

Misawa, S., Kuwabara, S., Sato, Y., Yamaguchi, N., Nagashima, K., Katayama, K., et al. (2018). Safety and efficacy of eculizumab in Guillain-Barre syndrome: A multicentre, double-blind, randomised phase 2 trial. *Lancet Neurology*, *17*(6), 519–529. Available from: PM:29685815.

Nicholas, R., Playford, E. D., & Thompson, A. J. (2000). A retrospective analysis of outcome in severe Guillain-Barre syndrome following combined neurological and rehabilitation management. *Disability and Rehabilitation*, *22*(10), 451–455. Available from: PM:10950498.

Parra, B., Lizarazo, J., Jimenez-Arango, J. A., Zea-Vera, A. F., Gonzalez-Manrique, G., Vargas, J., et al. (2016). Guillain-Barre syndrome associated with Zika virus infection in Colombia. *The New England Journal of Medicine*, *375*(16), 1513–1523. Available from: PM:27705091.

Prada, V., Massa, F., Salerno, A., Fregosi, D., Beronio, A., Serrati, C., et al. (2019). Importance of intensive and prolonged rehabilitative treatment on the Guillain-Barre syndrome long-term outcome: A retrospective study. *Neurological Sciences*. Available from: PM:31586288.

Pritchard, J., Hughes, R. A., Hadden, R. D., & Brassington, R. (2016). Pharmacological treatment other than corticosteroids, intravenous immunoglobulin and plasma exchange for Guillain-Barre syndrome. *Cochrane Database of Systematic Reviews*, *11*, CD008630. Available from: PM:27846348.

Reid, C. L., Campbell, I. T., & Little, R. A. (2004). Muscle wasting and energy balance in critical illness. *Clinical Nutrition*, *23*(2), 273–280. Available from: PM:15030968.

Simatos, A. N., Vincent, P. O., Yu, B. H., Bastien, R., & Sweeney, A. (2016). Influence of exercise on patients with Guillain-Barre syndrome: A systematic review. *Physiotherapy Canada*, *68*(4), 367–376. Available from: PM:27904236.

Sipila, J. O., & Soilu-Hanninen, M. (2015). The incidence and triggers of adult-onset Guillain-Barre syndrome in southwestern Finland 2004–2013. *European Journal of Neurology*, *22*(2), 292–298. Available from PM:25196425.

van den Berg, B., Storm, E. F., Garssen, M. J. P., Blomkwist-Markens, P. H., & Jacobs, B. C. (2018). Clinical outcome of Guillain-Barre syndrome after prolonged mechanical ventilation. *Journal of Neurology, Neurosurgery, and Psychiatry*, *89*(9), 949–954. Available from: PM:29627773.

Wakerley, B. R., & Yuki, N. (2015). Mimics and chameleons in Guillain-Barre and Miller Fisher syndromes. *Practical Neurology*, *15*(2), 90–99. Available from: PM:25239628.

Willison, H. J., Jacobs, B. C., & van Doorn, P. A. (2016). Guillain-Barre syndrome. *Lancet*, *388*(10045), 717–727. Available from: PM:26948435.

Chapter 14

Postinfectious demyelinating diseases: Guillain-Barré syndrome and beyond

T. Foiadelli, C. Trabatti, G.L. Marseglia, and S. Savasta
Pediatric Clinic, Fondazione IRCCS Policlinico "San Matteo", University of Pavia, Pavia (PV), Italy

Abbreviations

A-CIDP acute-onset CIDP
ADEM acute disseminated encephalomyelitis
AIDP acute inflammatory demyelinating polyneuropathy
AMAN acute motor axonal neuropathy
AMSAN acute motor and sensory axonal neuropathy
ATM acute transverse myelitis
BBE Bickerstaff's brainstem encephalitis
CDC Centers for Disease Control and Prevention
CHIKV chikungunya virus
CIDP chronic inflammatory demyelinating polyneuropathy
CIS clinically isolated syndrome
CNS central nervous system
CSF cerebrospinal fluid
DENV dengue virus
EAE experimental allergic encephalomyelitis
EAN experimental allergic neuritis
GBS Guillain-Barré syndrome
IVIG intravenous immunoglobulin
MAC-ELISA antibody capture enzyme-linked immunosorbent assay
MFS Miller-Fisher syndrome
MRI magnetic resonance imaging
PNS peripheral nervous system
WHO World Health Organization
ZIKV Zika virus

Introduction

The first evidence of a link between viral infections and demyelinating diseases was reported by Van Bogaert (1950). Focusing on inflammatory and demyelinating diseases of the central nervous system (CNS) (e.g., acute disseminated encephalomyelitis (ADEM) and multiple sclerosis), scientists initially speculated that as a consequence of a previous or concomitant infection, antigen-antibody complexes could deposit in the cerebral blood vessels and cause damage (Behan, 1978). Soon, animal disease models such as "allergic encephalomyelitis" (EAE) and "allergic neuritis" (EAN) (Lisak & Behan, 1975) helped to understand that central and peripheral nervous system (PNS) demyelination was more likely to result from cell-mediated immune reactions against myelin antigenic proteins (Soliven, 2014).

Indeed the reasons leading immune response to cause autoreactivity are nor fully clarified, and many genetic and environmental factors may play a role. Nevertheless, "molecular mimicry" between the antigens of the infectious agent and those presented on the neuronal surface is a widely accepted hypothesis (Pinto-Díaz et al., 2017). Hence, in predisposed subjects, demyelination seems to be caused by immunological processes, rather than by the direct viral invasion and damage of the nervous system.

Zika Virus Impact, Diagnosis, Control, and Models. https://doi.org/10.1016/B978-0-12-820267-8.00014-5

Different viruses have been associated with the onset of demyelinating diseases: measles, mumps, varicella zoster, rubella, and less frequently also influenza, coxsackie, cytomegalovirus, and Epstein-Barr virus (Bennetto & Scolding, 2004). The first neurological complications of Zika virus (ZIKV) infection were described during the large Polynesian outbreak in the early 2010s when a net increase in the incidence of Guillain-Barré syndrome (GBS) was recorded (Leonhard, Lant, et al., 2018; Sudre et al., 2014). The following outbreak that occurred in Brazil in 2015–16 definitely paved the way for further studies on the neurological complications of ZIKV infection. Since then the virus has spread in several countries, leading the World Health Organization to declare ZIKV a Public Health Emergency of International Concern (WHO, 2016). There is increasing interest in the relationship between ZIKV and demyelinating diseases of the central and peripheral nervous system, and the literature on this topic is rapidly growing. Herein, we will try to offer the readers a complete and concise overview of the most recent evidences, data, and controversies.

Zika virus and peripheral nervous system inflammatory and demyelinating diseases

Guillain-Barré syndrome (GBS) was the first acquired demyelinating disease related to ZIKV infection in South America (Cao-Lormeau et al., 2016; Reid, Rimmer, & Thakur, 2018; Sebastian et al., 2017). It occurs more frequently in young adults (from the third to fifth decade) and is slightly more prevalent in males (male to female ratio = 1.8:1) (Nascimento & Da Silva, 2017). The most common ZIKV-related GBS subtype is *acute inflammatory demyelinating polyneuropathy (AIDP)*. AIDP is usually a monophasic disease presenting with progressive ascending paraparesis and areflexia/hyporeflexia, followed by lower limb pain, distal paresthesia, autonomic dysfunction, and eventually cranial nerve involvement (i.e., facial palsy) or bulbar signs.

Weakness roughly develops 6–10 days after the initial symptoms of ZIKV infection (Cao-Lormeau et al., 2016; Parra et al., 2016), which is shorter than what usually occurs in the classical form of postinfectious GBS (typically developing 2–4 weeks after an infection).

Cranial nerve deficits are observed in nearly 50% of the cases and are associated with a higher risk of respiratory failure and major disability after hospital discharge (Muñoz, Parra, & Pardo, 2017). Autonomic dysfunction in ZIKV-related GBS is also frequently reported (75%) and contributes to a poorer prognosis: it usually occurs as blood pressure instability (i.e., hypertension and hypotension) and cardiac arrhythmias (Anaya et al., 2017).

The Brighton Collaboration criteria are the basis for the diagnosis of GBS (Sejvar et al., 2011). Regarding cerebrospinal fluid (CSF) analysis, patients usually show cytoalbuminologic dissociation and slight increase of CSF white cell count (5–7 cells/μL). Brain and spine MRI, although useless for the diagnosis of GBS, might reveal cranial nerve enhancement and cauda equina or nerve root contrast enhancement. These findings do not differ from what is occasionally observed in non-ZIKV-related GBS patients (Nascimento & Da Silva, 2017).

Electrophysiological data confirm that ZIKV-related GBS has predominantly demyelinating features, with significant distal nerve involvement. These terminal areas, where the blood-nerve barrier is anatomically deficient, are mainly affected. Prolonged distal latency and reduced terminal latency index are the earliest and most frequent nerve conduction abnormalities (Uncini, Shahrizaila, & Kuwabara, 2017). Depending on the geographical area and on the study design, the prevalence of the AIDP subtype in ZIKV-related GBS varies between 60% and 80%. However, data are limited by heterogeneous clinical criteria and neurophysiological techniques (Anaya et al., 2016; da Silva, Frontera, Bispo de Filippis, & Nascimento, 2017; Parra et al., 2016). Latency between the first GBS signs and neurophysiological testing is crucial, because axonal characteristics may be present early in the disease even in demyelinating subtypes (Muñoz et al., 2017). A single electrophysiological study performed in early stages of the disease may therefore not be reliable for an accurate GBS subtyping. Table 1 summarizes the main clinical, biochemical, neurophysiologic, and imaging characteristics of GBS.

Intravenous immunoglobulins (IVIG) are the most commonly used treatment for all GBS variants, and they are usually administered within 7 days after the onset of symptoms. The prognosis is generally favorable, with almost complete recovery after 3 months in about 80% of patients (Anaya et al., 2017), and does not differ from what has been observed in patients with GBS secondary to other causes (Nascimento & Da Silva, 2017). However, motor sequelae in lower limbs and chronic pain have been observed in up to 40% of ZIKV-positive patients (da Silva et al., 2017; Ferreira et al., 2017), with a significant impact on quality of life (Barbi, Coelho, de Alencar, & Crovella, 2018).

Other less frequent ZIKV-related ***GBS subtypes*** have been described according to the "nodoparanodopathy" and the GBS Classification Group approach (Uncini, Susuki, & Yuki, 2013; Wakerley, Uncini, & Yuki, 2014).

Acute motor axonal neuropathy (AMAN), described in studies from Polynesia and Colombia (Cao-Lormeau et al., 2016), is characterized by the absence of sensory or autonomic symptoms and a rarer cranial nerve involvement compared with AIDP. Clinical outcome is generally favorable, despite the rapid onset and the short "plateau" phase. Neurophysiological

TABLE 1 Common findings in Guillain-Barré syndrome.

Clinical features

- Progressive ascending paraparesis
- Areflexia/hyporeflexia
- Ataxia
- Lower limb pain
- Distal paresthesia
- Autonomic dysfunction: blood pressure instability (i.e., hypertension or hypotension) and cardiac arrhythmias
- Possible cranial nerve involvement (e.g., facial palsy) or bulbar signs (respiratory failure)

Laboratory findings

- Cytoalbuminologic dissociation in the CSF
- Slight increase of CSF white cell count (5–7 cells/μL)
- Possible detection of specific antiganglioside antibodies in serum

Neurophysiologic and imaging data

- Nerve conduction studies: demyelinating features predominate in AIDP, with significant distal nerve involvement (prolonged distal latency and reduced terminal latency index)

  In AMAN, prolonged distal motor latencies and reduction of the distal compound muscle action potential, without significant decrease in the motor conduction velocity, are usually seen

- Brain and spine MRI: possible cranial nerve enhancement and cauda equina or nerve root contrast enhancement

Summary of clinical, biochemical, neurophysiologic, and imaging data in Guillain-Barré syndrome (GBS). The most common ZIKV-related GBS subtype is acute inflammatory demyelinating polyneuropathy (AIDP), which usually has a monophasic course. Other less frequent GBS subtypes can manifest without sensory or autonomic symptoms. Electrophysiological patterns can help the physician to distinguish between the various GBS subtypes, but alterations can be absent in the first 7 days after the onset of symptoms. The Brighton Collaboration GBS Working Group (Sejvar et al., 2011) has developed a standardized definition for GBS, dividing into three levels of diagnostic certainty on the basis of the number of clinical, laboratory, and electrophysiologic criteria which are met. *CFS*, cerebrospinal fluid; *AIDP*, acute inflammatory demyelinating polyneuropathy; *AMAN*, acute motor axonal neuropathy.

changes consist of prolonged distal motor latencies and reduction of the distal compound muscle action potential, without significant decrease in the motor conduction velocity (Cao-Lormeau et al., 2016). Besides axonal degeneration, at a first examination, conduction block and slowing can mimic demyelination. However, these features rapidly resolve within a few weeks, and in the absence of abnormal temporal dispersion, they can be referred to as "reversible conduction failure." These pathological changes can be explained by antibody-mediated and complement-mediated dysfunction, causing microstructural damage in the axolemma at the nodes of Ranvier (Uncini et al., 2017). The typical association of AMAN to IgG antibodies against GM1 and GD1a, though, is rare in patients with ZIKV-related GBS.

Rare cases of *acute motor and sensory axonal neuropathy (AMSAN)*, *pure sensory GBS*, and *pharyngeal-cervical-brachial weakness* were reported in Colombia following ZIKV infection, but clinical and laboratory characteristics were similar to those described in ZIKV-negative patients (Anaya et al., 2017; Parra et al., 2016).

Few cases of *Miller-Fisher syndrome (MFS)* have also been described in Haiti and Brazil (da Silva et al., 2017; Kassavetis, Joseph, Francois, Perloff, & Berkowitz, 2016; Parra et al., 2016). This entity is characterized by bilateral complete ophthalmoplegia, areflexia, ataxia, and distal sensory alterations. Diagnosis can be challenging, as typical symptoms may be clearly identified several days after the onset. IVIG treatment seems to be effective in ZIKV-positive patients, even though evidence is based on few case reports and data over long-term prognosis are lacking.

A single case of *Bickerstaff's brainstem encephalitis (BBE)*, which is characterized by ataxia, ophthalmoplegia, hyperreflexia, and altered mental status, was reported in Colombia (Anaya et al., 2017).

As already described in ZIKV-negative demyelinating disorders (Michev et al., 2019), overlapping forms of GBS and CNS demyelination are possible. In a large Brazilian cohort, two cases of GBS were associated with encephalitis (da Silva et al., 2017), while an association of GBS with ADEM has been reported in a Colombian patient (Mancera-Páez, Román, Pardo-Turriago, Rodríguez, & Anaya, 2018). In the latter case the authors speculate that neuroinflammation could have been induced by direct ZIKV neurotropism and subsequently boosted by active immunity against other arboviruses. These cases support the hypothesis that ZIKV may have a role in determining both PNS and CNS inflammatory diseases. Brain neuroimaging should therefore be performed in GBS patients with clinical CNS involvement to identify cerebral lesions and support differential diagnosis (e.g., bilateral lesions in the cerebellar peduncles may mimic West Nile Virus encephalitis) (Nascimento & Da Silva, 2017).

TABLE 2 Possible pathogenetic mechanisms for ZIKV-related GBS.

Pathogenetic mechanism	Description
Molecular mimicry	ZIKV antigen determinants can resemble host nerve components; ZIKV-induced neutralizing antibodies therefore can cross-react with peripheral nerve proteins, causing immune-mediated damage of nodal and paranodal structures (Muñoz, Barreras, & Pardo, 2016)
Cross-reactive response and antibody-dependent enhancement	ZIKV can trigger a memory immune response in subjects who have been previously exposed to other flavivirus (e.g., dengue). Cross-reactive antibodies can drive enhancement of ZIKV infection, leading to more severe manifestations and complications (Dejnirattisai et al., 2016; Muñoz et al., 2016; Solomon et al., 1998)
Autoreactive T-cell response	Antigen-presenting cells show ZIKV-derived peptides to naive T $CD4^{+}$ lymphocytes, driving activation of these cells and inflammation. Cytokines and inflammatory response can mediate peripheral nerve damage (Muñoz et al., 2017)
Direct viral damage	Direct ZIKV infection of Schwann cells and neurons can induce cytopathic changes and can cause cytotoxicity mediated by T $CD8^{+}$ lymphocyte activation (Muñoz et al., 2016; Muñoz et al., 2017)

Several studies have tried to define the pathogenetic mechanisms for ZIKV-related GBS, but there is still much uncertainty on this topic. The high structural overlapping between ZIKV polyprotein and human proteins related to myelin can support the hypothesis of "molecular mimicry," while the hypothesis of "cross-reactive antibody enhancement" could explain the short time between acute infection and the development of neurological symptoms in some patients. These and other presumed pathogenetic mechanisms need further study to be confirmed.

The ***pathogenetic mechanisms underlying ZIKV-related GBS*** are not yet clarified. On the basis of clinical and serological evidence, several possibilities can be listed, as shown in Table 2.

In some cases the time course between the onset of ZIKV-like symptoms and the development of neurological complications can be sufficiently long for antibody production and would be compatible with the role of an adaptive immune response in this process (Lynch et al., 2019). Instead the hypothesis of cross-reactive antibody enhancement could explain the relatively short latency between acute infection and neurological symptoms that is seen in several ZIKV-related GBS patients, which would be too rapid to allow an autoimmune reaction to a primary viral exposure (Nascimento & Da Silva, 2017). This suggests a parainfectious temporal profile, in contrast to the postinfectious profile described in other GBS cases (Muñoz et al., 2017).

Furthermore, antigens involved in the ZIKV-related "molecular mimicry" might be different from the ones usually involved in GBS secondary to other causes. Analysis of blood samples from the French Polynesian outbreak revealed that less than 50% of the patient's sera had an autoimmune response against glycolipids (including gangliosides and glycolipid complexes) (Parra et al., 2016). A high structural overlapping between the ZIKV polyprotein and human proteins related to myelin, (de)myelination, and axonal neuropathy has been demonstrated. When altered, these proteins can be specifically related to the ZIKV-induced pathologic sequelae, suggesting that the spectrum of neurological disease may be broader than initially thought (Lucchese & Kanduc, 2016).

The emergence of neurological complications associated with ZIKV infection may also be explained by intrinsic host factors (e.g., genetic susceptibility) or by the magnitude of the outbreaks. In fact, low-risk neurological events may be noticed in epidemics with large number of cases, such as those occurring in French Polynesia, Brazil, and Colombia, but may go unrecognized in smaller outbreaks (Muñoz et al., 2017).

Besides GBS and its variants, other inflammatory demyelinating diseases of the PNS have been recently associated with ZIKV infection. ***Acute transient polyneuritis***, which occurs rapidly after a febrile illness, may represent a milder form of GBS. Affected patients usually do not require or seek medical care. This raises the possibility of underreporting of these minor cases, as none of the patients meet the Brighton criteria for GBS (Nascimento, Frontera, Amitrano, Bispo de Filippis, & Da Silva, 2017). Distal pain, stocking-glove hypoesthesia, mild distal weakness, and hyporeflexia are the predominant symptoms, with onset within 1–2 days from ZIKV systemic symptom debut. No alterations of mental status are reported, nor cerebellar or cranial nerve dysfunction. Electrophysiological, neuroradiological, and CSF testing are in most cases unrevealing, and rapid spontaneous clinical improvement is observed within the first 7–10 days. Etiopathological mechanisms may involve segmental nerve swelling and inflammation, resulting in edema and nerve dysfunction.

Chronic inflammatory demyelinating polyneuropathy (CIDP) after Zika virus infection is reported approximately in 5% of the patients initially diagnosed as GBS. These patients may show initial recovery after IVIG administration but experience subsequent disease progression with increased limb weakness. This particular form, called "acute-onset CIDP"

(A-CIDP), rarely causes cranial nerve palsy or respiratory failure at onset and is usually diagnosed during the follow-up period, when relapses or deterioration beyond 8 weeks from onset occurs (da Silva et al., 2017; Leonhard, Munts, Van Der Eijk, & Jacobs, 2018). The presence of pain or sensory symptoms before the start of weakness may be an early indicator of a chronic form. All patients initially diagnosed with ZIKV-related GBS need a careful follow-up to eventually identify a clinical progression to chronic forms, which may require a periodical maintenance treatment with IVIG (approximately every 2 weeks) to prevent progression and relapses.

Zika virus and central nervous system inflammatory and demyelinating diseases

The CNS is considerably more susceptible to infection than the PNS. In particular, CNS neuronal axons and myelinating oligodendrocytes are especially vulnerable to injury. As for GBS a certain overlapping between immune-mediated and direct viral mechanisms of inflammation and injury is plausible, but the lack of neuropathological evidence from human studies hinders any conclusion on such hypothesis (Muñoz et al., 2016). The developing of ex vivo models of ZIKV infection would be essential for biological studies on viral neuroinvasion and its clinical and therapeutic consequences (Cumberworth et al., 2017). Host and environmental factors also need to be further investigated in the future (Muñoz et al., 2016).

Several studies have highlighted the possible relationship between ZIKV infection and a heterogeneous spectrum of CNS inflammatory and demyelinating diseases (Acosta-Ampudia et al., 2018; Medina & Medina-Montoya, 2017).

Acute transverse myelitis (ATM) is characterized by inflammatory and demyelinating lesions that involve an entire medullar segment. Affected patients are prevalently young males, and the neurological symptoms occur after a median lag of 18 days after ZIKV infection. Clinical presentation commonly comprises pelvic/back pain, flaccid paraplegia, and urinary retention. All patients show hyperreflexia, suggesting a central white matter dysfunction (Anaya et al., 2017; Mécharles et al., 2016; Palacios, Clavijo-Prado, Ruiz, Arias Antun, & Julian Duran, 2019). Cervicothoracic spinal segments are the most commonly affected, and medullar MRI can sometimes show longitudinally extensive lesions ($\geq$3 metamers), while brain MRI is negative. CSF analysis usually shows lymphocytic pleocytosis and elevated protein levels. In addition to serological evidence of ZIKV infection, most of the patients show immunity also against dengue virus (DENV) and chikungunya virus (CHIKV), which may play an additive effect in the disease process. Interestingly, ATM has also been reported in subjects who are hyperresponders to hepatitis B (Mancera-Páez et al., 2018). High-dosage steroid therapy has proven to be effective in these patients, while complementary plasma exchange can be required in more severe cases.

Single cases of ***encephalitis/encephalomyelitis*** have been reported in association with ZIKV infection (Araujo, Ferreira, & Nascimento, 2016; Ferreira et al., 2017; Galliez et al., 2016; Roth, Tyshkov, Thakur, & Vargas, 2017). When both brain and spinal cord are involved by demyelinating lesions, along with an acute deterioration in consciousness, ***acute disseminated encephalomyelitis (ADEM)*** can be diagnosed (Niemeyer, Niemeyer, Borges, & Marchiori, 2017). The disease usually affects young adults between the second and the third decade. The latency period between ZIKV systemic symptoms and neurological onset ranges from 1 to 21 days. Clinical presentation can include motor and sensory deficits in the lower limbs (which may gradually extend proximally to the trunk) and pyramidal signs. Tetraplegia, urinary retention, and respiratory failure can be observed at the "plateau" stage. Fever at onset is not uncommon (Mancera-Páez et al., 2018). Neuroimaging findings of ADEM typically include multiple asymmetrically distributed and poorly defined lesions, involving the white matter and the deep gray matter nuclei. More rarely, multifocal lesions can be found in the subcortical and callosal white matter (Fig. 1). However, no specific imaging finding for ZIKV-related ADEM can distinguish it from other causes of ADEM. Open-ring postcontrast enhancement is common, but complete-ring or punctate enhancement and a complete lack of enhancement have also been reported (Zare Mehrjardi, Keshavarz, Poretti, & Hazin, 2016). An accurate differential diagnosis is required, as these MRI features can mimic other nondemyelinating conditions (e.g., primary CNS lymphoma) (da Silva et al., 2017).

The first monophasic event compatible with a demyelinating process of the CNS, in the absence of an alteration of mental status, is called ***clinically isolated syndrome***. If other similar episodes follow (polyphasic course), fulfilling imaging 2010 McDonald criteria, the diagnosis of ***multiple sclerosis*** can be done. All these forms show a good clinical response to high-dose corticosteroid treatment (Trabatti et al., 2016).

In some cases of CNS demyelination, genetic material from ZIKV was detected in brain biopsies. This finding keeps with viral latency rather than with virus-associated encephalitis and thus poses the possibility of subsequent reactivation (Benjamin, 2019) and clinical relapse. Electroencephalography can reveal abnormal slowing in these patients, and in some cases, ZIKV has been linked to the development of seizures (Asadi-Pooya, 2016), particularly in patients with subcortical lesions. ZIKV-associated CNS demyelinating encephalitis appears to have a predominant neuroinflammatory profile, as CSF analysis depicts lymphocytic pleocytosis and elevated protein levels (Muñoz et al., 2017).

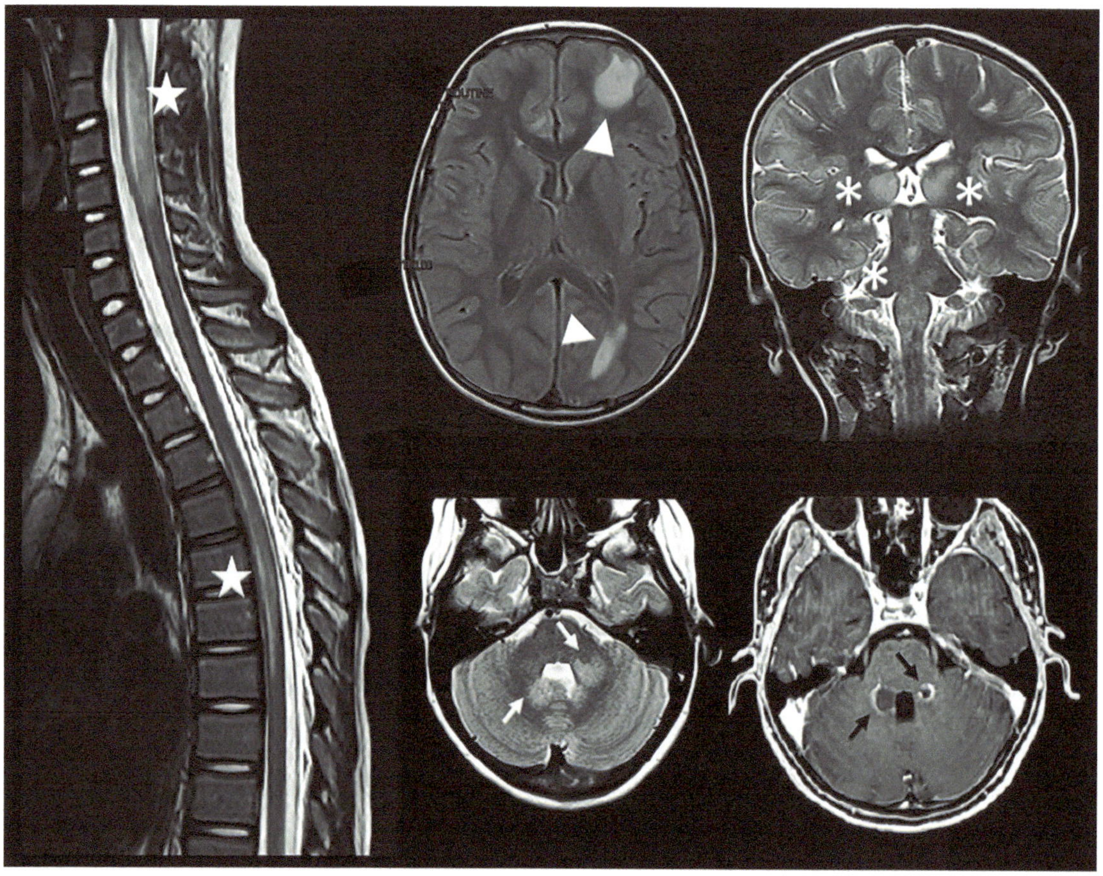

FIG. 1 Typical MRI findings in acute disseminated encephalomyelitis (ADEM). Multiple T2-hyperintense lesions affecting the spinal cord, with spinal swelling *(white stars)*. Additional central nervous system involvement includes multiple large asymmetrical demyelinating lesions that appear hyperintense on T2 or fluid-attenuated inversion recovery (FLAIR) sequences, variably localized in the subcortical and deep white matter *(white triangles)*, in the deep gray matter and brainstem *(asterisks)*, and in the cerebellum *(white arrows)*, with or without T1 contrast enhancement *(black arrows)*. The diagnosis is confirmed when the typical MRI findings are combined with encephalopathic symptoms and an acute onset of polysymptomatic neurological deficits.

Only few reports have described the ocular complications of acute ZIKV infection in adults, including conjunctivitis, iridocyclitis, and chorioretinitis (de Paula Freitas, Ventura, Maia, & Belfort, 2017). ***Optic neuritis*** is rare and has been reported in only two patients so far: one with a form of multifocal encephalomyelitis (Ferreira et al., 2017) and one with a AQP4-negative form of ***neuromyelitis optica*** (Aspahan et al., 2019).

Challenges in the diagnosis of ZIKV and the role of other flavivirus infection

There are several clinical and technical difficulties in confirming ZIKV infection and ZIKV-related neurological complications. In fact, by the time of neurological symptom presentation, ZIKV has cleared from blood in most cases. Furthermore the duration and kinetics of ZIKV-RNA detection in the CSF are yet unknown (partly because CSF sampling is not routinely performed for the diagnosis of flavivirus infection). Finally, serological diagnosis is limited by the extensive cross-reactivity of ZIKV with other flaviviruses, particularly when a previous exposure to DENV has occurred (Muñoz et al., 2017). Both ZIKV and DENV neutralizing antibodies are boosted following ZIKV infection, but such antibodies derive from distinct B-cell populations, and the anamnestic dengue response occurs first, followed by a de novo ZIKV response (Lynch et al., 2019).

To date, real-time polymerase chain reaction (RT-PCR) for ZIKV is the gold standard test to confirm acute ZIKV infection. However, its positivity in blood or CSF samples from patients with GBS or other demyelinating diseases may be limited by the timing of sample collection and eventually by the low or transient viremia that usually lasts no longer than 5–7 days. For this reason, concomitant sampling of other biological materials is often recommended, as ZIKV may persist in reservoir organs for longer periods. For instance, ZIKV-RNA positivity can persist for a few weeks in urine samples. These reservoirs may also allow viral replication and represent the antigenic source of an antiviral inflammatory response, which can trigger inflammatory neurological syndromes (CDC, 2017; Muñoz et al., 2017).

It is important to highlight that in many studies that we have referred to in the present chapter (i.e., during the French Polynesian outbreak), the association of neurological demyelinating diseases with ZIKV infection was exclusively based on serological investigations. In an epidemic context, and in regions with a high prevalence of non-ZIKV flavivirus infections, this poses a significant limitation. To standardize methodologies the Centers for Disease Control and Prevention (CDC) published in 2017 a fact sheet on the correct use and interpretation of serological tests (CDC, 2017), encouraging the use of antibody capture enzyme-linked immunosorbent assay (MAC-ELISA) to support serological diagnoses. Anti-ZIKV IgM are typically detectable soon after the onset of symptoms, and they remain detectable for approximately 12 weeks following infection. Although more specific, however, MAC-ELISA still may produce false positivity in patients with a recent history of flavivirus infection such as DENV. Finally, in patients who have been vaccinated for yellow fever or Japanese encephalitis, the presence of cross-reactive antibodies may lead to further misinterpretation.

During the 2015–16 ZIKV outbreak, specific surveillance protocols were established, and all patients who presented with neurological symptoms were investigated for arbovirus positivity. In these patients, fever and myalgia were the most frequent initial findings associated with either DENV, ZIKV, or CHIKV infection (Vieira et al., 2018). As expected, early symptoms had low specificity, and there was an extensive overlap between the classic arboviral clinical manifestations and infection by other neurotropic viruses. Of notice, coinfections with more than one virus (identified by RT-PCR) were common. Particular attention must be paid to persistent flavivirus infections, which can be misleading in endemic areas. The intrinsic neurotropism of flavivirus belonging to the same family also plays a role and could at least partly explain the different clinical severity and the development of epidemics. DENV, as well as other flaviviruses like Japanese encephalitis and West Nile virus, can cause neurological syndromes similar to those described during ZIKV infection, involving both the CNS and PNS (Savasta et al., 2018). Also, CHIKV, which belongs to a different genus (*Alphavirus*, *Togaviridae* family) has been related in few reports with GBS, encephalitis, and myelitis, even if its role has not been completely assessed (Mehta et al., 2018). No serological cross-reactivity with flaviviruses occurs in these cases, making diagnosis more straightforward.

Conclusions

ZIKV is associated with a wide range of neurological manifestations, including both CNS and PNS diseases (Table 3). Neurological involvement can comprise brain parenchyma, spinal cord, and peripheral nerves in various combinations, as evidenced by clinical, imaging, and neurophysiological findings (Mehta et al., 2018). However, the causative evidence of ZIKV has been limited by the lack of well-designed case-control studies, the lack of reliable epidemiological and historical data on the baseline incidence of demyelinating diseases in endemic and epidemic regions (Muñoz et al., 2017), and some major technical limitations for the diagnosis of ZIKV infection itself. The absence of neuropathological studies and the existence of a significant immunological and serological cross-reactivity between flaviviruses further hamper definitive conclusions.

This lack of evidence was highlighted in a recent review that showed no statistically significant association between ZIKV infection and GBS, despite the increased number of hospitalizations since ZIKV started circulating and caught the attention of academic communities, health services, and media (Wachira, Peixoto, & de Oliveira, 2019). Further studies are needed to better characterize the pathogenic role of ZIKV in inflammatory and demyelinating diseases of the nervous system, supporting evidence for the development of specific therapies and preventive strategies (Lynch et al., 2019).

Policy and procedures

International classification criteria for inflammatory demyelinating diseases of the central and peripheral nervous system

The classification of inflammatory demyelinating diseases of the central and peripheral nervous system is recent, and frequently evolving due to new evidence, particularly due to the detection of novel autoantibodies. Specific diagnostic criteria for GBS have been published by the Brighton Collaboration GBS Working Group in 2005 and periodically reviewed. However, on the basis of the detection of various antiganglioside antibodies, a new classification of "nodoparanodopathies" has been proposed in 2013, which is now widely accepted. On this basis the various subtypes and variants of GBS have been reclassified. Concerning CNS demyelinating diseases, MS is the best known and described entity. The latest criteria for the diagnosis of MS are the 2017 revised McDonald criteria, which are also suitable for the pediatric age. The emerging entity of clinically isolated syndromes (CIS) was well defined in adult patients by Miller in 2012; in the same year the

TABLE 3 Demyelinating diseases related to ZIKV infection.

Demyelinating diseases of the peripheral nervous system
• Guillain-Barré syndrome (GBS) and its variants – Acute inflammatory demyelinating polyneuropathy (AIDP) – Acute motor axonal neuropathy (AMAN) – Acute motor and sensory axonal neuropathy (AMSAN) – Pure sensory GBS – Pharyngeal-cervical-brachial weakness – Miller-Fisher syndrome (MFS) – Bickerstaff's brainstem encephalitis (BBE) • Acute transient polyneuritis • Chronic inflammatory demyelinating polyneuropathy (CIDP)
Demyelinating diseases of the central nervous system
• Acute transverse myelitis (ATM) • Encephalitis/encephalomyelitis • Acute disseminated encephalomyelitis (ADEM) • Clinically isolated syndrome (CIS) • Multiple sclerosis (MS) • Optic neuritis (ON) • Neuromyelitis optica (NMO) spectrum disorder
Combined central and peripheral nervous system demyelination
• GBS+encephalitis • GBS+ADEM

Several inflammatory and demyelinating diseases of both central and peripheral nervous systems have been reported in relation to ZIKV infection. However, lack of well-designed studies and some technical limitations in the diagnosis still prevent from drawing definitive conclusions.

International Pediatric Multiple Sclerosis Study Group provided a systematic revised classification and definition for all pediatric CNS demyelinating diseases, especially acute disseminated encephalomyelitis (ADEM), which typically manifests in childhood. Finally, in 2015, the International Consensus Diagnostic Criteria for Neuromyelitis Optica Spectrum Disorders were made available.

Zika antibody and nucleic acid amplification testing

The Centers for Disease Control and Prevention (CDC) published a fact sheet in 2017 on the correct use and interpretation of ZIKV serological tests, encouraging the use of antibody capture enzyme-linked immunosorbent assay (MAC-ELISA) in all reference laboratories. This assay provides qualitative in vitro detection of human IgM antibodies to Zika virus. The Zika MAC-ELISA is intended for use in serum of individuals meeting CDC Zika clinical and/or epidemiological criteria for testing. It can also be used in cerebrospinal fluid (CSF) when submitted with a patient-matched serum sample.

An adequate CDC's algorithm for Zika and dengue virus testing is provided to guide the diagnosis (updated in June 2019), suggesting the different role of nucleic acid amplification tests and antibody tests. Particular attention is given to the timing of sample collection, the type of biological samples, and the correct screening in pregnant women.

References (related to policy and procedures)

CDC. (2017). *Fact sheet for healthcare providers: Interpreting Zika MAC-ELISA test results* (pp. 1–4). Retrieved from https://www.cdc.gov/%0Azika/pdfs/interpreting-zika-mac-elisa-results.pdf.

CDC. (2019). Dengue and Zika virus diagnostic testing for patients with a clinically compatible illness and risk for infection with both viruses. *Morbidity and Mortality Weekly Report*, *68*(1). https://www.cdc.gov/mmwr/volumes/68/rr/pdfs/rr6801a1-H.pdf.

Krupp, L. B., Tardieu, M., Amato, M. P., Banwell, B., Chitnis, T., Dale, R. C., …, Wassmer, E. (2013). International Pediatric Multiple Sclerosis Study Group criteria for pediatric multiple sclerosis and immune-mediated central nervous

system demyelinating disorders: Revisions to the 2007 definitions. *Multiple Sclerosis Journal*, *19*(10), 1261–1267. https://doi.org/10.1177/1352458513484547.
Miller, D. H., Chard, D. T., & Ciccarelli, O. (2012). Clinically isolated syndromes. *The Lancet Neurology*, *11*(2), 157–169. https://doi.org/10.1016/S1474-4422(11)70274-5.
Sejvar, J. J., Kohl, K. S., Gidudu, J., Amato, A., Bakshi, N., Baxter, R., …, Wiznitzer, M. (2011). Guillain-Barre syndrome and Fisher syndrome: Case definitions and guidelines for collection, analysis, and presentation of immunization safety data. *Vaccine*, *29*(3), 599–612. https://doi.org/10.1016/j.vaccine.2010.06.003.
Thompson, A. J., Banwell, B. L., Barkhof, F., Carroll, W. M., Coetzee, T., Comi, G., …, Cohen, J. A. (2018). Diagnosis of multiple sclerosis: 2017 revisions of the McDonald criteria. *The Lancet Neurology*, *17*(2), 162–173. https://doi.org/10.1016/S1474-4422(17)30470-2.
Uncini, A., Susuki, K., & Yuki, N. (2013). Nodo-paranodopathy: Beyond the demyelinating and axonal classification in anti-ganglioside antibody-mediated neuropathies. *Clinical Neurophysiology: Official Journal of the International Federation of Clinical Neurophysiology*, *124*(10), 1928–1934. https://doi.org/10.1016/j.clinph.2013.03.025.
Wingerchuk, D. M., Banwell, B., Bennett, J. L., Cabre, P., Carroll, W., Chitnis, T., …, Weinshenker, B. G. (2015). International consensus diagnostic criteria for neuromyelitis optica spectrum disorders. *Neurology*, *85*(2), 177–189. https://doi.org/10.1212/WNL.0000000000001729.

Mini-dictionary of terms

Arbovirus. Heterogeneous group of viruses transmitted by an arthropod vector. They belong to the following families: *Togaviridae*, *Flaviviridae*, and *Bunyaviridae*. The carriers are ticks, sandflies, or mosquitoes. Arboviruses are widespread in tropical countries, and they all have a marked tropism for small vessels and the nervous system.

Autoreactivity. Abnormal immune response produced by an organism, acting against its own cells or tissues.

Blood-nerve barrier (BNB). Anatomical and functional unit formed by endoneurial microvascular endothelial cells, which regulate the movement of solutes, macromolecules, and inflammatory cells from the circulating blood to peripheral nerves' axons and Schwann cells. A structure with similar functions can be identified in the central nervous system, called *blood-brain barrier (BBB)*. It is formed by capillary endothelial cells and astrocyte end feet, which build a sheath around the capillaries.

Demyelinating diseases. Group of disorders characterized by inflammatory alteration of the myelin, which wraps most of the nerve fibers in both the central and peripheral nervous system. The function of myelin sheaths is to isolate the axons and allow optimal conduction of nerve impulses.

Experimental allergic encephalitis (EAE). Animal model in which an inflammatory and demyelinating response in the brain is elicited by injection of antigenic protein components. This reaction, observed in many animal species (mostly in rodents and primates), was first hypothesized by Rivers et al. in 1933 and is still considered a model for investigating the mechanisms of cellular immune reactions against brain tissue, leading to demyelinating diseases in humans.

Intravenous immunoglobulins (IVIG). Therapeutic administration of antibodies (mainly polyvalent IgG) obtained from healthy multidonor plasma. They are widely used as a replacement therapy for primitive immunodeficiencies, but they are also considered the first-line treatment in some inflammatory and demyelinating diseases of the nervous system, such as Guillain-Barré syndrome.

Molecular mimicry. Process determined by sequence similarities between foreign (e.g., infectious) and self-peptides/epitopes, resulting in triggering autoreactive immune response and inflammation.

Neurophysiological testing. Methods that allow investigation of the bioelectrical activity of the central and peripheral nervous system by positioning of different kinds of electrodes able to detect both spontaneous and stimulated signals. These techniques can provide a precise definition of the site and entity of a lesion. Electroencephalography, evoked potentials, electromyography, and nerve conduction studies are particularly useful in clinical practice and can help the physician in differential diagnosis and in the choice of the correct therapeutic approach.

Nodoparanodopathy. Term applicable to a group of diseases with a common pathogenic mechanism of dysfunction or disruption at the node of Ranvier, as nodes and paranodes can be the target of antiganglioside antibodies. The first categorization of this entity was proposed by Uncini et al. (2013) in the attempt to bypass the traditional classification into axonal or demyelinating neuropathies, which could be misleading in the diagnosis of GBS subtypes.

Key facts

Key facts of Zika virus-related neurological disorders

- The first neurological complications linked to acute Zika virus (ZIKV) infection were observed during the viral outbreak in Polynesia in the early 2010s, where a net increase in the incidence of Guillain-Barré syndrome (GBS) was noticed. The following outbreak in Brazil in 2015–16 definitely set the stage for further studies on this topic.
- In 2016 ZIKV was declared a Public Health Emergency of International Concern by the World Health Organization (WHO).
- Both central and peripheral inflammatory and demyelinating diseases have been linked to ZIKV infection. These include GBS and its variants, acute transverse myelitis, meningoencephalitis, and acute disseminated encephalomyelitis. Generally the clinical and laboratory features of such diseases do not differ from the corresponding ZIKV-negative forms.
- The main problem in confirming ZIKV-related neurological disorders is that, in most cases, ZIKV is no longer replicating (thus detectable) at the time the first neurological symptoms appear. Serological diagnosis is limited by the extensive cross-reactivity of ZIKV with other flaviviruses.
- ZIKV detection in multiple biological samples (including the cerebrospinal fluid) using RT-PCR is the gold standard diagnostic test to confirm acute ZIKV infection in clinical practice.

Key facts of demyelinating diseases

- Myelin is a lipid-rich structure that covers the nerve cells axons, allowing fast conduction of the electrical impulses (action potentials). Demyelinating diseases cause myelin damage and sometimes axonal and neuronal loss due to inflammation. In turn the reduction in conduction ability causes deficiency in sensation, movement, cognition, or other functions depending on which nerves are involved.
- The myelin structure differs between the central (CNS) and peripheral nervous system (PNS). Specific epitopes can be targeted by the immune attack, leading to different neurological syndromes.
- While the main postinfectious demyelinating diseases involving the PNS are Guillain-Barré syndrome and its variants, the principal CNS demyelinating disorders include acute disseminate encephalomyelitis, transverse myelitis, optic neuritis, neuromyelitis optica (and its spectrum), and multiple sclerosis.
- The mechanisms involved in shifting immune response to autoreactivity are not fully clarified, and many genetic and environmental factors could play a role.
- "Molecular mimicry" between the antigens of some infectious agents and those present on neuronal and myelinic surfaces is a generally accepted hypothesis, but direct viral neuroinvasion and damage cannot be excluded.
- Some autoantibodies targeting surface myelin epitopes have recently been identified, including antiaquaporin-4 (AQP4) and antimyelin oligodendrocyte glycoprotein (MOG) antibodies.

Summary points

- ZIKV infection has been associated with many neurological inflammatory and demyelinating diseases, involving both the central (CNS) and the peripheral (PNS) nervous system.
- The pathogenic mechanisms underlying ZIKV-related demyelinating diseases are yet to be clarified: both molecular mimicry, cross-reactive immune reactivity, T-cell autoreactivity, and direct viral cytotoxicity have been proposed.
- Guillain-Barré syndrome (GBS) is the most frequent acquired demyelinating disease related to ZIKV infection, mostly classified as acute inflammatory demyelinating polyneuropathy (AIDP).
- Among ZIKV-related CNS demyelinating diseases, acute transverse myelitis and acute disseminated encephalomyelitis have been the most commonly described.
- Molecular diagnosis of ZIKV infection is difficult in patients with demyelinating diseases, because serological tests lack specificity in patients with past or recent infection by other flaviviruses and RT-PCR (which is considered the gold standard test for acute infection) is often negative at this stage.
- Most of the available evidence of an association between ZIKV and neurological disorders has been limited by the lack of reliable epidemiological data and of well-designed prospective studies; therefore further studies are needed to better characterize the pathogenic role of ZIKV in inflammatory and demyelinating diseases of the nervous system, paving the way for the development of specific therapies and preventive strategies.

References

Acosta-Ampudia, Y., Monsalve, D. M., Castillo-Medina, L. F., Rodríguez, Y., Pacheco, Y., Halstead, S., … Ramírez-Santana, C. (2018). Autoimmune neurological conditions associated with Zika virus infection. *Frontiers in Molecular Neuroscience*, *11*(April). https://doi.org/10.3389/fnmol.2018.00116.

Anaya, J. M., Ramirez-Santana, C., Salgado-Castaneda, I., Chang, C., Ansari, A., & Gershwin, M. E. (2016). Zika virus and neurologic autoimmunity: The putative role of gangliosides. *BMC Medicine*, *14*(1), 1–3. https://doi.org/10.1186/s12916-016-0601-y.

Anaya, J.-M., Rodriguez, Y., Monsalve, D. M., Vega, D., Ojeda, E., Gonzalez-Bravo, D., … Acosta-Ampudia, Y. (2017). A comprehensive analysis and immunobiology of autoimmune neurological syndromes during the Zika virus outbreak in Cucuta, Colombia. *Journal of Autoimmunity*, *77*, 123–138. https://doi.org/10.1016/j.jaut.2016.12.007.

Araujo, L. M., Ferreira, M. L. B., & Nascimento, O. J. M. (2016). Síndrome de Guillain-Barré associada ao surto de infecção por vírus Zika no Brasil. *Arquivos de Neuro-Psiquiatria*, *74*(3), 253–255. https://doi.org/10.1590/0004-282X20160035.

Asadi-Pooya, A. A. (2016). Zika virus-associated seizures. *Seizure*, *43*, 13. https://doi.org/10.1016/j.seizure.2016.10.011.

Aspahan, M. C., Leonhard, S. E., Gomez, R. S., Rocha, E.d. S., Vilela, M. R.d. S., Alvarenga, P. P. M., … Meira, F. (2019). Neuromyelitis optica spectrum disorder associated with Zika virus infection. *Neurology. Clinical Practice*, *9*, e1–e3. https://doi.org/10.1212/CPJ.0000000000000546.

Barbi, L., Coelho, A. V. C., de Alencar, L. C. A., & Crovella, S. (2018). Prevalence of Guillain-Barre syndrome among Zika virus infected cases: A systematic review and meta-analysis. *The Brazilian Journal of Infectious Diseases: An Official Publication of the Brazilian Society of Infectious Diseases*, *22*(2), 137–141. https://doi.org/10.1016/j.bjid.2018.02.005.

Behan, P. O. (1978). Post-infectious encephalomyelitis: Some aetiological mechanisms. *Postgraduate Medical Journal*, *54*(637), 755–759. https://doi.org/10.1136/pgmj.54.637.755.

Benjamin, L. A. (2019). A tale of two demyelinating diseases and Zika virus. *Multiple Sclerosis Journal*, *25*(3), 430–431. https://doi.org/10.1177/1352458518799584.

Bennetto, L., & Scolding, N. (2004). Inflammatory/post-infectious encephalomyelitis. *Journal of Neurology, Neurosurgery, and Psychiatry*, *75*(Suppl 1), i22–i28.

Cao-Lormeau, V.-M., Blake, A., Mons, S., Lastere, S., Roche, C., Vanhomwegen, J., … Ghawche, F. (2016). Guillain-Barre syndrome outbreak associated with Zika virus infection in French Polynesia: A case-control study. *Lancet (London, England)*, *387*(10027), 1531–1539. https://doi.org/10.1016/S0140-6736(16)00562-6.

CDC. (2017). *Fact sheet for healthcare providers: Interpreting Zika MAC-ELISA test results* (pp. 1–4). Retrieved from https://www.cdc.gov/%0Azika/pdfs/interpreting-zika-mac-elisa-results.pdf.

Cumberworth, S. L., Barrie, J. A., Cunningham, M. E., de Figueiredo, D. P. G., Schultz, V., Wilder-Smith, A. J., … Edgar, J. M. (2017). Zika virus tropism and interactions in myelinating neural cell cultures: CNS cells and myelin are preferentially affected. *Acta Neuropathologica Communications*, *5*(1), 50. https://doi.org/10.1186/s40478-017-0450-8.

da Silva, I. R. F., Frontera, J. A., Bispo de Filippis, A. M., & Nascimento, O. J. M.d. (2017). Neurologic complications associated with the Zika virus in Brazilian adults. *JAMA Neurology*, *74*(10), 1190–1198. https://doi.org/10.1001/jamaneurol.2017.1703.

de Paula Freitas, B., Ventura, C. V., Maia, M., & Belfort, R. J. (2017). Zika virus and the eye. *Current Opinion in Ophthalmology*, *28*(6), 595–599. https://doi.org/10.1097/ICU.0000000000000420.

Dejnirattisai, W., Supasa, P., Wongwiwat, W., Rouvinski, A., Barba-Spaeth, G., Duangchinda, T., … Screaton, G. R. (2016). Dengue virus sero-cross-reactivity drives antibody-dependent enhancement of infection with Zika virus. *Nature Immunology*, *17*(9), 1102–1108. https://doi.org/10.1038/ni.3515.

Ferreira, M. L. B., De Brito, C. A. A., Moreira, Á. J. P., Machado, M.Í.d. M., Henriques-Souza, A., Cordeiro, M. T., … Pena, L. J. (2017). Guillain–Barré syndrome, acute disseminated encephalomyelitis and encephalitis associated with Zika virus infection in Brazil: Detection of viral RNA and isolation of virus during late infection. *American Journal of Tropical Medicine and Hygiene*, *97*(5), 1405–1409. https://doi.org/10.4269/ajtmh.17-0106.

Galliez, R. M., Spitz, M., Rafful, P. P., Cagy, M., Escosteguy, C., Germano, C. S. B., … Moll, F. T. (2016). Zika virus causing encephalomyelitis associated With immunoactivation. *Open Forum Infectious Diseases*, *3*(4). https://doi.org/10.1093/ofid/ofw203, ofw203.

Kassavetis, P., Joseph, J.-M. B., Francois, R., Perloff, M. D., & Berkowitz, A. L. (2016). Zika virus-associated Guillain-Barre syndrome variant in Haiti. *Neurology*, *87*(3), 336–337. https://doi.org/10.1212/WNL.0000000000002759.

Leonhard, S. E., Lant, S., Jacobs, B. C., Wilder-Smith, A., Ferreira, M. L. B., Solomon, T., & Willison, H. J. (2018). Zika virus infection in the returning traveller: What every neurologist should know. *Practical Neurology*, *18*(4), 271–277. https://doi.org/10.1136/practneurol-2017-001789.

Leonhard, S. E., Munts, A. G., Van Der Eijk, A. A., & Jacobs, B. C. (2018). Acute-onset chronic inflammatory demyelinating polyneuropathy after Zika virus infection. *Journal of Neurology, Neurosurgery and Psychiatry*, *89*(10), 1118–1119. https://doi.org/10.1136/jnnp-2017-317346.

Lisak, R. P., & Behan, P. O. (1975). Experimental autoimmune demyelinating diseases: Experimental allergic encephalomyelitis and experimental allergic neuritis. *Biomedicine/[Publiee Pour l'A.A.I.C.I.G.]*, *22*(2), 81–87.

Lucchese, G., & Kanduc, D. (2016). Zika virus and autoimmunity: From microcephaly to Guillain-Barré syndrome, and beyond. *Autoimmunity Reviews*, *15*(8), 801–808. https://doi.org/10.1016/j.autrev.2016.03.020.

Lynch, R. M., Mantus, G., Encinales, L., Pacheco, N., Li, G., Porras, A., … Chang, A. Y. (2019). Augmented Zika and dengue neutralizing antibodies are associated with Guillain-Barré syndrome. *Journal of Infectious Diseases*, *219*(1), 26–30. https://doi.org/10.1093/infdis/jiy466.

Mancera-Páez, O., Román, G. C., Pardo-Turriago, R., Rodríguez, Y., & Anaya, J. M. (2018). Concurrent Guillain-Barré syndrome, transverse myelitis and encephalitis post-Zika: A case report and review of the pathogenic role of multiple arboviral immunity. *Journal of the Neurological Sciences*, *395*(February), 47–53. https://doi.org/10.1016/j.jns.2018.09.028.

Mécharles, S., Herrmann, C., Poullain, P., Tran, T. H., Deschamps, N., Mathon, G., … Lannuzel, A. (2016). Acute myelitis due to Zika virus infection. *The Lancet*, *387*(10026), 1481. https://doi.org/10.1016/S0140-6736(16)00644-9.

Medina, M. T., & Medina-Montoya, M. (2017, December). New spectrum of the neurologic consequences of Zika. *Journal of the Neurological Sciences*, *383*, 214–215. https://doi.org/10.1016/j.jns.2017.10.046.

Mehta, R., Soares, C. N., Medialdea-Carrera, R., Ellul, M., da Silva, M. T. T., Rosala-Hallas, A., … Solomon, T. (2018). The spectrum of neurological disease associated with Zika and chikungunya viruses in adults in Rio de Janeiro, Brazil: A case series. *PLoS Neglected Tropical Diseases*, *12* (2). https://doi.org/10.1371/journal.pntd.0006212.

Michev, A., Musso, P., Foiadelli, T., Trabatti, C., Lozza, A., Franciotta, D., … Savasta, S. (2019). Bickerstaff brainstem encephalitis and overlapping Guillain-Barré syndrome in children: Report of two cases and review of the literature. *European Journal of Paediatric Neurology*, *23*(1), 43–52. https://doi.org/10.1016/j.ejpn.2018.11.008.

Muñoz, L. S., Barreras, P., & Pardo, C. A. (2016). Zika virus-associated neurological disease in the adult: Guillain-Barré syndrome, encephalitis, and myelitis. *Seminars in Reproductive Medicine*, *345*, 273–279. https://doi.org/10.1055/s-0036-1592066.

Muñoz, L. S., Parra, B., & Pardo, C. A. (2017). Neurological implications of Zika virus infection in adults. *Journal of Infectious Diseases*, *216*(Suppl_10), S897–S905. https://doi.org/10.1093/infdis/jix511.

Nascimento, O. J. M., & Da Silva, I. R. F. (2017). Guillain-Barré syndrome and Zika virus outbreaks. *Current Opinion in Neurology*, *30*(5), 500–507. https://doi.org/10.1097/WCO.0000000000000471.

Nascimento, O. J. M., Frontera, J. A., Amitrano, D. A., Bispo de Filippis, A. M., & Da Silva, I. R. F. (2017). Zika virus infection-associated acute transient polyneuritis. *Neurology*, *88*(24), 2330–2332. https://doi.org/10.1212/WNL.0000000000004026.

Niemeyer, B., Niemeyer, R., Borges, R., & Marchiori, E. (2017). Acute disseminated encephalomyelitis following Zika virus infection. *European Neurology*, *77*(1–2), 45–46. https://doi.org/10.1159/000453396.

Palacios, E., Clavijo-Prado, C., Ruiz, A., Arias Antun, A., & Julian Duran, E. (2019). Longitudinal extensive transverse myelitis and Zika virus: A diagnostic challenge in a hospital in Colombia. *Neurologia (Barcelona, Spain)*, *34*, 204–206. https://doi.org/10.1016/j.nrl.2016.08.006.

Parra, B., Lizarazo, J., Jiménez-Arango, J. A., Zea-Vera, A. F., González-Manrique, G., Vargas, J., … Pardo, C. A. (2016). Guillain-Barré syndrome associated with Zika virus infection in Colombia. *New England Journal of Medicine*, *375*(16), 1513–1523. https://doi.org/10.1056/NEJMoa1605564.

Pinto-Díaz, C. A., Rodríguez, Y., Monsalve, D. M., Acosta-Ampudia, Y., Molano-González, N., Anaya, J. M., & Ramírez-Santana, C. (2017). Autoimmunity in Guillain-Barré syndrome associated with Zika virus infection and beyond. *Autoimmunity Reviews*, *16*(4), 327–334. https://doi.org/10.1016/j.autrev.2017.02.002.

Reid, S., Rimmer, K., & Thakur, K. (2018). Zika virus and neurologic disease. *Neurologic Clinics*, *36*(4), 767–787. https://doi.org/10.1016/j.ncl.2018.06.003.

Roth, W., Tyshkov, C., Thakur, K., & Vargas, W. (2017). Clinical/scientific notes: Encephalomyelitis following definitive zika virus infection. *Neurology: Neuroimmunology and Neuroinflammation*, *4*(4), 2–5. https://doi.org/10.1212/NXI.0000000000000349.

Savasta, S., Rovida, F., Foiadelli, T., Campana, A. M., Percivalle, E., Marseglia, G. L., & Baldanti, F. (2018). West-Nile virus encephalitis in an immunocompetent pediatric patient: Successful recovery. *Italian Journal of Pediatrics*, *44*(1), 140. https://doi.org/10.1186/s13052-018-0574-x.

Sebastian, U. U., Ricardo, A. V. A., Alvarez, B. C., Cubides, A., Luna, A. F., Arroyo-Parejo, M., … Marshall, J. C. (2017). Zika virus-induced neurological critical illness in Latin America: Severe Guillain-Barre syndrome and encephalitis. *Journal of Critical Care*, *42*, 275–281. https://doi.org/10.1016/j.jcrc.2017.07.038.

Sejvar, J. J., Kohl, K. S., Gidudu, J., Amato, A., Bakshi, N., Baxter, R., … Wiznitzer, M. (2011). Guillain-Barre syndrome and Fisher syndrome: Case definitions and guidelines for collection, analysis, and presentation of immunization safety data. *Vaccine*, *29*(3), 599–612. https://doi.org/10.1016/j.vaccine.2010.06.003.

Soliven, B. (2014). Animal models of autoimmune neuropathy. *ILAR Journal*, *54*(3). https://doi.org/10.1093/ilar/ilt054.

Solomon, T., Kneen, R., Dung, N. M., Khanh, V. C., Thuy, T. T. N., Ha, D. Q., … White, N. J. (1998). Poliomyelitis-like illness due to Japanese encephalitis virus. *Lancet*, *351*(9109), 1094–1097. https://doi.org/10.1016/S0140-6736(97)07509-0.

Sudre, B., Danielsson, N., Rakotoarivony, L. M., Van Bortel, W., Zeller, H., & Jansa, J. (2014). *Zika virus infection outbreak French Polynesia. Rapid risk assessment* (pp. 1–12). ECDC. (February). Retrieved from http://ecdc.europa.eu/en/publications/Publications/Zika-virus-French-Polynesia-rapid-risk-assessment.pdf?time=0.

Trabatti, C., Foiadelli, T., Spartà, M. V., Gagliardone, C., Rinaldi, B., Delmonte, M., … Savasta, S. (2016). Paediatric clinically isolated syndromes: Report of seven cases, differential diagnosis and literature review. *Child's Nervous System*, *32*(1), 69–77. https://doi.org/10.1007/s00381-015-2959-0.

Uncini, A., Shahrizaila, N., & Kuwabara, S. (2017). Zika virus infection and Guillain-Barré syndrome: A review focused on clinical and electrophysiological subtypes. *Journal of Neurology, Neurosurgery and Psychiatry*, *88*(3), 266–271. https://doi.org/10.1136/jnnp-2016-314310.

Uncini, A., Susuki, K., & Yuki, N. (2013). Nodo-paranodopathy: Beyond the demyelinating and axonal classification in anti-ganglioside antibody-mediated neuropathies. *Clinical Neurophysiology: Official Journal of the International Federation of Clinical Neurophysiology*, *124*(10), 1928–1934. https://doi.org/10.1016/j.clinph.2013.03.025.

Van Bogaert, L. (1950). Post-infectious encephalomyelitis and multiple sclerosis; the significance of perivenous encephalomyelitis. *Journal of Neuropathology and Experimental Neurology*, *9*(3), 219–249. https://doi.org/10.1097/00005072-195007000-00001.

Vieira, M. A.d. C. E. S., Costa, C. H. N., Linhares, A.d. C., Borba, A.d. S., Henriques, D. F., Silva, E. V. P.d., … Vasconcelos, P. F.d. C. (2018). Potential role of dengue virus, chikungunya virus and Zika virus in neurological diseases. *Memorias Do Instituto Oswaldo Cruz*, *113*(11). https://doi.org/10.1590/0074-02760170538, e170538.

Wachira, V. K., Peixoto, H. M., & de Oliveira, M. R. F. (2019). Systematic review of factors associated with the development of Guillain–Barré syndrome 2007–2017: What has changed? *Tropical Medicine and International Health*, *24*(2), 132–142. https://doi.org/10.1111/tmi.13181.

Wakerley, B. R., Uncini, A., & Yuki, N. (2014). Guillain-Barre and Miller Fisher syndromes—New diagnostic classification. *Nature Reviews. Neurology*, *10*(9), 537–544. https://doi.org/10.1038/nrneurol.2014.138.

WHO. (2016). *Zika virus microcephaly and Guillain-Barré syndrome*. Retrieved from https://apps.who.int/iris/bitstream/handle/10665/204454/zikasitrep_19Feb2016_eng.pdf?sequence=1.

Zare Mehrjardi, M., Keshavarz, E., Poretti, A., & Hazin, A. N. (2016). Neuroimaging findings of Zika virus infection: A review article. *Japanese Journal of Radiology*, *34*(12), 765–770. https://doi.org/10.1007/s11604-016-0588-5.

Chapter 15

Linking in placental alterations, Zika virus, and Guillain-Barré syndrome

Kíssila Rabelo[a], Natália Gedeão Salomão[b], and Marciano Viana Paes[b,†]

[a]*Laboratory of Ultrastructure and Tissue Biology, Rio de Janeiro State University, Rio de Janeiro, Brazil,* [b]*Laboratory of Interdisciplinary Medical Research, Oswaldo Cruz Institute, Rio de Janeiro, Brazil*

Abbreviations

ADE	antibody-dependent enhancement
AIDP	acute inflammatory demyelinating polyradiculoneuropathy
AMAN	acute motor axonal neuropathy
AMSAN	acute motor sensory axonal neuropathy
CCL	chemokine (C-C motif) ligand
CMV	cytomegalovirus
CSF	cerebrospinal fluid
CTB	cytotrophoblasts
CXCL	chemokine (C-X-C motif) ligand
CZS	congenital Zika syndrome
DC-SIGN	dendritic cell-specific intercellular adhesion molecule-3-grabbing nonintegrin
FcR	Fc receptor
GBS	Guillain-Barré syndrome
GM-CSF	granulocyte-macrophage colony-stimulating factor
HIV	human immunodeficiency virus
IFN	interferon
IL	interleukin
IP	protein from interferon (gamma)-induced
MCP	monocyte chemoattractant protein
Mertk	tyrosine-protein kinase Mer
NFKB1	nuclear factor NF-kappa-B p105 subunit
NINDS	National Institute of Neurological Disorders and Stroke
NK	natural killer cells
NLRP	nucleotide-binding oligomerization domain, leucine-rich repeat and pyrin domain containing
PNS	peripheral nervous system
STBs	syncytiotrophoblasts
TGF-β	transforming growth factor beta
TIM	T-cell immunoglobulin and mucin domain
TLR	Toll-like receptor
TNF	tumor necrosis factor
TORCH	toxoplasmosis, other—including syphilis and Zika, rubella, cytomegalovirus, and herpes simplex virus
Tyro	tyrosine-protein kinase receptor
ZIKV	Zika virus

[†]After finishing this chapter, Dr. Marciano Viana Paes passed away, victim of Covid-19 infection. For friends, students, and science, a huge loss. The collaborating authors are grateful for their time with Dr. Paes and dedicate this chapter to his memory.

Zika Virus Impact, Diagnosis, Control, and Models. https://doi.org/10.1016/B978-0-12-820267-8.00015-7

Introduction

Zika fever is an important arbovirus, transmitted mainly by mosquitoes of the *Aedes* genus (subgenus *Stegomyia*) but also by other routes including sexual contact, blood transfusion, and vertical transmission. The etiological agent of this illness, Zika virus (ZIKV), was first isolated in 1947 from the blood of sentinel rhesus monkeys during a study on yellow fever transmission in the Zika Forest of Uganda, which gave rise to its name. Up through 2006 the occurrence of ZIKV infections in humans had been detected sporadically through serology in African and Asian continents. The Zika fever epidemic occurred in Latin America in 2015–17 and quickly became a worldwide public health concern (Musso, Ko, & Baud, 2019).

This massive outbreak highlighted possible correlations between infection and dangerous complications such as Guillain-Barré syndrome (GBS) (Araujo, Silva, & Araujo, 2016; Malkki, 2016). Despite one of the biggest concerns about ZIKV infection being in the pregnant women, the fact that the Zika epidemic was very unexpected made the follow-up of pregnant women flawed. There is a large gap in the literature and even in clinical records about GBS associated with Zika virus infection in pregnancy. Our group, other Brazilian scientists and many other researchers around the world have started a search to elucidate what happens in pregnant women and specifically in the placenta during Zika infection. We do know is that, in fact, there is the installation of a proinflammatory environment in this temporary organ, capable of causing great variations in the needs of the fetus (El Costa et al., 2016; Rabelo et al., 2017, 2018; Rosenberg, Yu, Hill, Reyes, & Schwartz, 2017).

These are two important points in isolation—the placental aspects and the Guillain-Barré syndrome—which in the Zika outbreak, specifically, brought us a new concern: the cases of GBS in pregnant patients and the consequences for the mother and the maintenance of the pregnancy and the fetus, with the placental changes. In this chapter, we will approach the aspects that the placental tissue presents facing the infection by ZIKV, such as inflammation and histopathological damage. In addition, we will deliberate the increase in cases of GBS resulting from the ZIKV outbreak. Finally, we want to associate the conditions in which pregnant patients presented GBS. For this, we will begin by describing the basic characteristics of the placenta and GBS, and then, we will discuss the histopathological and immunological alterations of the ZIKV-infected placentas, as well as the occurrences of GBS associated with this infection.

The placenta and vertical transmission

The placenta is characterized as a temporary and chimerical organ, formed during pregnancy from maternal and fetal tissue. This organ is responsible for nutrition, gas exchange, and removal of toxic residues, as well as providing endocrine and immunological support to the fetus (Simmons, Fortier, & Cross, 2007). Problems in placenta formation can cause gestational complications that result in morbidity and even mortality of the fetus or mother (Cross, 2005; Rossant & Cross, 2001). The placenta has a very complex structure formed by maternal tissue and several types of trophoblastic cells derived from the zygote. Trophoblastic cells are epithelial cells specialized for various functions during pregnancy (Simmons et al., 2007). The functional structure of the placenta consists of the chorionic villus and has three layers: the syncytiotrophoblast (STBs) layer, which is on the surface; the cytotrophoblast (CTB) layer; and the mesoderm with the endothelium of the fetal vessels (which separates the maternal blood from the fetal blood). The basal decidua is the maternal component of the placenta, formed from the endometrium, composed of decidual cells and entirely covered by the amnion.

During pregnancy the deciduous basal contains a rich immunological infiltrate, which affects and may be affected according to the dynamics of the maternal-fetal interface. It has already been observed that a proinflammatory environment is necessary before and not as a result of successful implantation. Decisive leukocytes, especially dendritic and natural killer (NK) cells, assist in the apposition, adhesion and invasion of the blastocyst (Gnainsky et al., 2015; Hanna et al., 2006). As the blastocyst invades the endometrium and the decidua is formed, trophoblastic cells increase the population of decidual leukocytes by chemotaxis, via secretion of cytokines such as CXCL12, CXCL8, TGF-β, and CCL2, and induce the differentiation of NK cells, macrophages, and CD4+ cells through secretion of IL-15, GM-CSF, and TGF-β, respectively (Mor & Cardenas, 2010; Ramhorst et al., 2012). Approximately 70% of decidual leukocytes are NK cells, 20%–25% are macrophages, 1.7% are dendritic cells, and approximately 3%–10% are T cells (Mor & Cardenas, 2010).

Vertical transmission through placenta is one of the most worrisome issues related to ZIKV considering the cases of microcephaly and the other characteristic alterations of the congenital Zika syndrome (CZS). The first report of this kind of transmission occurred during the first outbreak in French Polynesia and was later confirmed in the epidemic in Brazil in 2015 (Baud, Gubler, Schaub, Lanteri, & Musso, 2017; Besnard, Lastere, Teissier, Cao-Lormeau, & Musso, 2014). Despite other flaviviruses not having this type of transmission as a characteristic, it is known that the placenta is a target organ for "TORCH"-type infections (World Health Organization (WHO), 2015). ZIKV and symptomatic congenital infection are not observed in all exposed fetuses, similar to cytomegalovirus and toxoplasmosis (Baud et al., 2017).

Prevention of maternal-fetal transmission is carried out in the context of avoiding as much as possible the exposure of pregnant women to possible mosquito bites, with the use of clothing that covers the skin and repellents, as well as safe sexual practice (Baud et al., 2017).

Placental permissiveness to ZIKV

The mechanism by which the virus crosses the placenta to establish the infection of the developing fetus has not been elucidated, so many groups have tried to clarify the mechanism of placental transposition. Some studies have identified the ZIKV genome in the amniotic fluid, in the intervillary, decidual, and chorionic villi of the placenta, in addition to the fetal tissues, including the brain (Calvet et al., 2016; Driggers et al., 2016; Martines et al., 2016; Rabelo et al., 2018). An in vitro study in cells explanted from placentas and cell lines showed that several cell types such as CTBs, fibroblasts, and decidual, immune, and endothelial cells had different surface receptors: Tyro3, Axl, Mertk, TIM1, an immunoglobulin produced by T cells, and the DC-SIGN proteins, to which the flaviviruses are able to bind and mediate their entry into the cell by endocytosis (Tabata, Petitt, Puerta-Guardo, et al., 2016). Another possible pathway for cell-to-cell infection may be through the paracellular pathway, since in ZIKV-infected placentas there is a modification in the expression of the junction protein claudin-4, thus altering the permeability of STBs (Miranda et al., 2019).

Additionally, other in vitro study with isolated cells of mature placenta showed that placental macrophages (Hofbauer cells) and CTBs are permissive to ZIKV infection. Hofbauer cells are stimulated to present a postinfection response, with increased secretion of IFN-α and proinflammatory cytokines IL-6, IP-10, and MCP-1. Both macrophages and infected CTBs showed an increase in antiviral transcript levels after ZIKV infection. These results suggest that ZIKV is able to access the fetus after infection and proliferation in placental cells (Quicke et al., 2016). It has been observed by Costa et al. that ZIKV replicates in a wide range of placental cells, including decidual fibroblasts and macrophages, trophoblasts, Hofbauer cells, and mesenchymal cord blood stem cells (El Costa et al., 2016). A histopathological analysis carried out in premature placenta demonstrated the proliferation of Hofbauer cells and an increase in chorionic villus. The presence of ZIKV was detected in stromal and chorionic villus cells by in situ hybridization, showing that placental cells were not only permissive to infection but also to viral replication (Rosenberg et al., 2016). Although there is evidence that Hofbauer cells are susceptible to ZIKV infection, the mechanisms by which the virus reaches these cells in the fetal portion of the placenta and how it is subsequently transmitted to the fetus are still uncertain. Until now the phenotypic and functional changes of Hofbauer cells after ZIKV infection have not yet been fully elucidated (Zulu, Martinez, Gordon, & Gray, 2019).

Histopathological aspects and changes in ZIKV-infected placentas

Despite recent efforts to elucidate the consequences of ZIKV infection in the maternal-fetal context and of vertical transmission, few studies investigating the infected placenta have been published, and there is still much to be studied. In the earliest studies following the major outbreaks in 2015, authors developed two hypotheses on the role of the placenta. The first is that the placenta could directly transmit the ZIKV to the embryo or fetus. Alternatively the placenta itself could be setting up a response to exposure, and this response could contribute to the neurological defect of the fetus (Adibi, Marques, Cartus, & Beigi, 2016). With the knowledge generated after these outbreaks, today we know that ZIKV is capable of infecting placental cells, which would then be transmitted to the fetus (Miranda et al., 2019; Tabata, Petitt, Puerta-guardo, et al., 2016).

The histological analysis of explanted cultures infected ex vivo with ZIKV showed drastic alterations in the architecture of placental villus with rupture of trophoblastic layers and changes in the morphology of nuclei, such as karyorrhexis and karyolysis (El Costa et al., 2016). Beaufrère and colleagues also described the characteristics of inflammatory changes in the histopathological analysis of three placentas of patients diagnosed with Zika. The placentas had chorioamnionitis, villitis, and intervillitis (inflammation in the membranes, chorionic villus, and intervillus, respectively), with inflammatory infiltrate, hyperplasia of Hofbauer cells, and endothelial thickening (Beaufrère et al., 2018). The analysis of placentas infected in the three different trimesters showed stromal edema in the chorionic villus, with irregular dilatation and prominent aggregates of vacuolated cells. The fetal membranes also showed several reactive changes, including an increase in the number of vacuolated mononuclear cells in the subamniotic connective tissue, according to the increased infiltration of macrophages (Lum et al., 2019). Similarly, our group evaluated two cases of ZIKV-infected placentas, in which we observed the presence of large and diffuse areas of fibrinoid necrosis in the maternal decidua, such as edema, fibrosis, vascular endothelial thickening, degeneration, vascular congestion, calcification, extracellular matrix degradation, and focal areas of mononuclear or perivascular inflammatory infiltrates (Fig. 1D–I), unlike control cases, which presented normal aspects of placental tissue (Fig. 1A–C) (Rabelo et al., 2017, 2018).

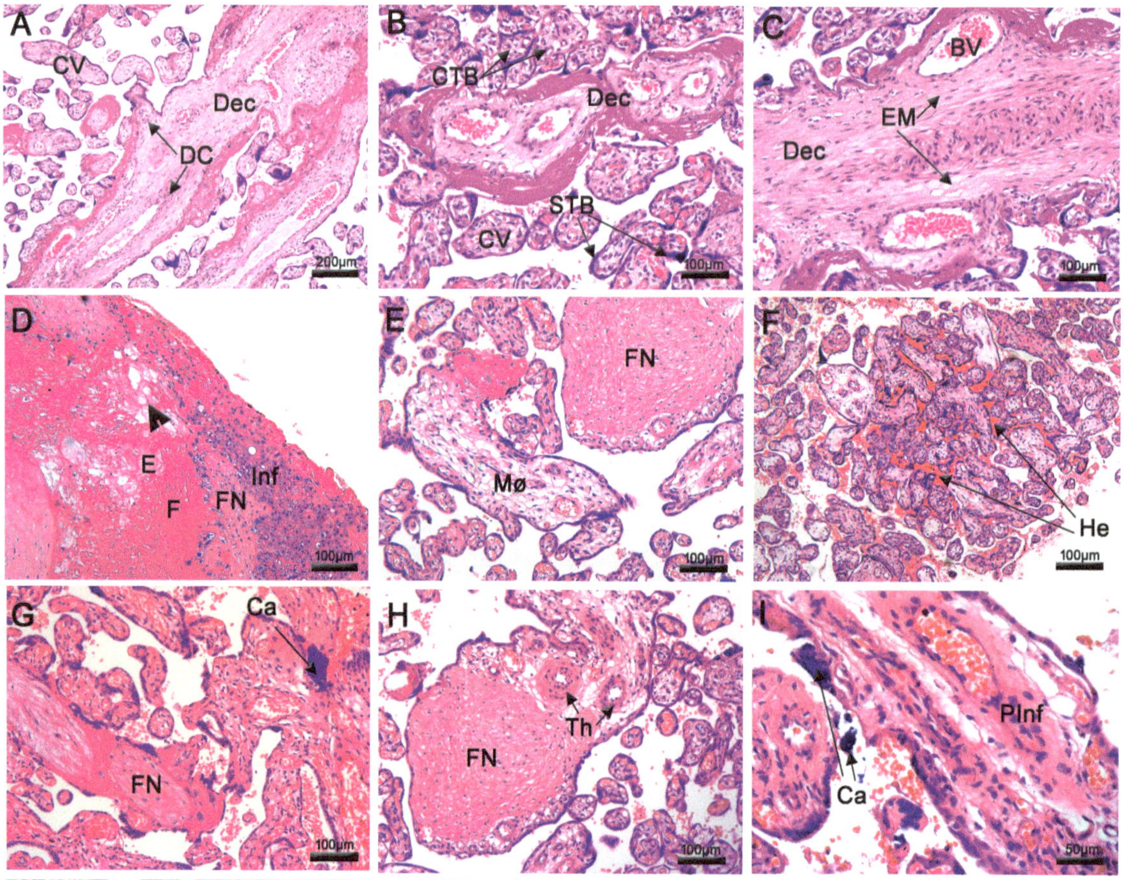

FIG. 1 Histopathological analysis of the placenta. (A–C) Placenta of a non-ZIKV patient stained with H&E and presenting normal features: maternal decidua (Dec), decidual cells (DC), chorionic villi (CV), syncytiotrophoblasts (STB), cytotrophoblasts (CTB), extracellular matrix (EM), and blood vessels (BV). (D–I) Sections of ZIKV-infected placental tissue stained with H&E, showing abnormalities in the decidua, including edema (e), fibrosis (F), fibrinoid necrosis (FN), mononuclear inflammatory infiltrate (Inf), macrophages (Mø), endothelial thickening (Th), cellular degeneration (*arrowhead* in panel (D)), calcification (Ca), and other pathological features in CV, such as perivascular inflammatory infiltrate (PInf) and hemorrhage (He). *(Modified from Rabelo, K., de Souza Campos Fernandes, R. C., de Souza, L. J., de Souza, T. L., dos Santos, F. B., Nunes, P. C. G., … Paes, M. V. (2017). Placental histopathology and clinical presentation of severe congenital Zika syndrome in a human immunodeficiency virus-exposed uninfected infant.* Frontiers in Immunology, *8(DEC), 1–8. doi:10.3389/fimmu.2017.01704.)*

Cell subpopulations and cytokines in ZIKV-infected placentas

The inflammatory infiltrate is recurrent in the infection, and its profile has been investigated. These cells were characterized and quantified in a case studied by our group, which a number of CD68+ and CD8+ cells were sixfold and fourfold higher in the infected placenta, respectively. Furthermore, proinflammatory cytokines such as TNF-α and IFN-γ and mediators related to vascular permeability such as VEGFR-2 and RANTES were also significantly increased (Rabelo et al., 2018). Some studies show that the ZIKV antigen located in Hofbauer cells and other placental cells persisted until birth, regardless of the period of infection (Lum et al., 2019; Rabelo et al., 2017, 2018). A recent study describes Hofbauer cell hyperplasia and numerous histiocytic cells in the maternal portion and deciduitis with lymphocyte prevalence. In this case, there was an increase in CD3+ T lymphocytes, as well as CD4+ and CD8+ cells (Santos et al., 2019).

The placental membrane region also presented molecular changes in infected patients in three different quarters, with transcripts indicating inflammation, genes, and response pathways were increased in abundance, including NFKB1, IL-1B, TNF, TLR2, TLR4, CXCL8, NLRP3, and NLRP6 (Lum et al., 2019).

Because of the major damage caused to the placenta and the loss of the product of conception documented with the occurrence of miscarriages and retained fetuses (Rabelo et al., 2018; Rivadeneyra-Espinar, Venegas-Esquivel, & Díaz-Espinoza, 2019), ZIKV infection should be suspected during pregnancy to make a diagnosis and provide appropriate care.

The Guillain-Barré syndrome

Guillain-Barré syndrome (GBS) is the most common acute paralytic neuropathy of the peripheral nervous system (PNS) in the world, which is often resulted from an infection. Several infectious agents are strongly associated with GBS, such as *Campylobacter jejuni*, *Cytomegalovirus*, *Haemophilus influenzae*, Epstein-Barr virus, and *Mycoplasma pneumoniae* (Jacobs et al., 1998). In addition to infectious agents, GBS may occur due to surgeries, vaccines, and injuries. Its incidence worldwide ranges from 0.81 to 1.89 per 100,000 person-years (Sejvar, Baughman, Wise, & Morgan, 2011). GBS frequency is higher in man than in woman, and the incidence increases with age (Sejvar, Baughman, et al., 2011).

One of the proposed mechanisms to explain the pathogenesis of this disease is "mimicry molecular." According to this hypothesis, antibodies are produced against an agent that is structurally similar to gangliosides. Thus these antibodies attack the gangliosides of the host, leading to demyelination and axonal degeneration (Nyati & Nyati, 2013). Due to the absence of sufficiently sensitive and specific disease biomarkers, to conclude the GBS diagnosis, the patient history associated with neurological, electrophysiological, and cerebrospinal fluid (CSF) examinations is necessary (Leonhard et al., 2019). To exclude other causes of acute flaccid paralysis, laboratory tests are necessary, since GBS patients have complete blood counts and blood tests for glucose, electrolytes, kidney function, and liver enzymes. CSF often presents increased protein level and normal cell count (albuminocytological dissociation). The most commonly used diagnostic criteria for GBS were developed by the National Institute of Neurological Disorders and Stroke (NINDS) in 1978 (revised in 1990) and the Brighton Collaboration in 2011 (Leonhard et al., 2019).

GBS divides into subtypes: acute inflammatory demyelinating polyradiculoneuropathy (AIDP), acute motor axonal neuropathy (AMAN), and acute motor sensory axonal neuropathy (AMSAN) (Sejvar, Kohl, et al., 2011). AIDP is the most common type and is characterized by focal demyelination of motor and sensory peripheral nerves and roots, more predominant in North America and Europe. AMAN has a more rapid progression of weakness and is characterized by axonal damage, mainly motor nerve. AMSAN is less common, and the clinical presentation is more severe, associated with a slow and incomplete recovery (Sejvar, Kohl, et al., 2011).

Until now the only effective treatments available are intravenous immunoglobulin and plasma exchange (Sejvar, Baughman, et al., 2011). Meanwhile, prognosis varies depending on GBS type and care availability. Fortunately, 62%–92% of patients recover completely or almost completely its motor functions in 1 year since the onset of the neurological signs (Bernsen, De Jager, Van Der Meché, & Suurmeijer, 2005).

The clinical of GBS

GBS has a monophasic clinical course. The patient presents alterations in sensation or pain along with muscle weakness. First is in the feet and hands, and quickly progressive bilateral weakness of the legs/arms and ascend to cranial muscles, without CNS involvement. The patient may report distal paresthesias or sensory loss, decrease/absence of reflexes, and dysautonomia, including cardiac arrhythmias, blood pressure instability, pupillary dysfunction, and bladder dysfunction and pain. After comes the plateau phase, which can last from days to months. And then comes the recovery phase, which can last several years or never end, even with treatment (Dyck, 2005).

In addition to the symptoms already cited, the patient may also experience other complications such as bladder (urinary retention) and gastrointestinal dysfunctions, such as constipation, ileus, gastric distension, diarrhea, and fecal incontinence (Leonhard et al., 2019). Approximately 20% of patients evolve to neuromuscular respiratory failure, requiring mechanical ventilation (Willison, Jacobs, & van Doorn, 2016). Whether the patient reports involvement of the autonomic nervous system, it may contribute to death (3%–10% of cases) (Van Den Berg, Bunschoten, Van Doorn, & Jacobs, 2013).

The ZIKV infection is already one of the most related diseases associated with SGB. The incidence of GBS as a consequence of ZIKV infection is 2.5 cases per 10,000 infections, which is comparable with the risk associated with *Campylobacter* infection (Mier-Y-Teran-Romero, Delorey, Sejvar, & Johansson, 2018; Musso et al., 2019). All GBS subtypes have been associated with ZIKV infection, but the proportions of these subtypes vary between studies and regions (dos Santos, Rodriguez, Almiron, & Sanhueza, 2016; Musso et al., 2019). The interval between the previous infection and the onset of GBS is 5–10 days to 9 weeks, which led to speculation about a contributory parainfectious process (Muñoz, Parra, & Pardo, 2017; Paploski et al., 2016). In patients affected by ZIKV infection, areflexia, limb weakness, and paresthesia were the most commonly reported symptoms. In addition, especially in Zika infection, cases of Miller Fisher syndrome, characterized by areflexia, ataxia, and ophthalmoplegia, were observed (da Silva, Frontera, de Filippis, & Nascimento, 2017; Parra et al., 2016).

The GBS associated with ZIKV infection

The association between ZIKV and GBS started in the epidemic that occurred in French Polynesia, as there was an unusual increase in neurological disease in adults, turning around 42 cases, which massively exceeded the expected annual average of 3–8 cases (Watrin et al., 2016). In 2015 there was the outbreak of Zika in Brazil, which quickly spread throughout the Americas, increasing not only the cases of microcephaly but also GBS, with almost 2000 reported cases (Counotte et al., 2018; Faria et al., 2017; Ferreira et al., 2017; Muñoz et al., 2017; Soto-Hernández, Ponce, & Rosales, 2019; WHO, 2017).

Little is known about the pathogenesis of GBS, but according to the clinical and pathological findings associated with ZIKV, it seems that the neuronal involvement occurs directly by tropism of the virus by these cells, as well as the permissiveness of immune cells (Muñoz et al., 2017; Rabelo et al., 2017, 2018). One of the facilitators of this process would be the phenomenon of antibody-dependent enhancement (ADE), which antibodies produced in a primary infection. Other flaviviruses form complexes with ZIKV without neutralizing, and these virus-antibody complexes are endocitated by cells that express Fc receptors, thus increasing viral replication (Muñoz et al., 2017) (Fig. 2A), a phenomenon that has already been confirmed for damage specifically to the placenta (Brown et al., 2019). Based on immunological information and the entry of ZIKV into cells, Muñoz et al. developed a theory about the pathogenesis of ZIKV-GBS. Neuronal damage could occur

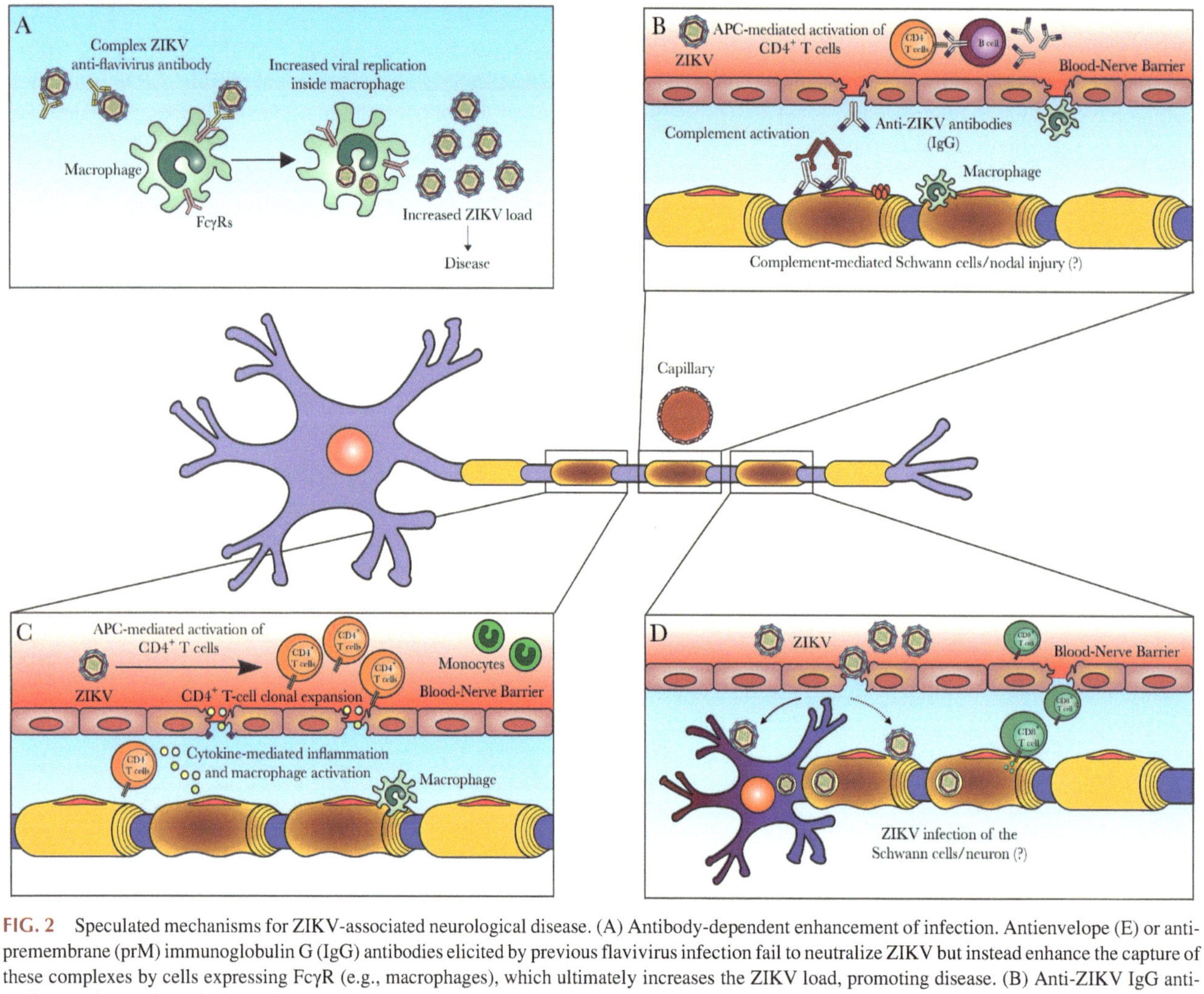

FIG. 2 Speculated mechanisms for ZIKV-associated neurological disease. (A) Antibody-dependent enhancement of infection. Antienvelope (E) or antipremembrane (prM) immunoglobulin G (IgG) antibodies elicited by previous flavivirus infection fail to neutralize ZIKV but instead enhance the capture of these complexes by cells expressing FcγR (e.g., macrophages), which ultimately increases the ZIKV load, promoting disease. (B) Anti-ZIKV IgG antibodies attach to antigens in the peripheral nerves (i.e., cross-reactivity), which activates the complement system and triggers local inflammation driven by macrophages. (C) Antigen-presenting cells (APCs) present ZIKV-derived peptides on class II major histocompatibility complex (MHC) molecules to naive CD4+ T lymphocytes, driving activation of these cells and inflammation. Cytokines and inflammatory cells are attracted to the peripheral nervous system, mediating peripheral nerve injury. (D) ZIKV infects Schwann cells/neurons and causes cytopathic effects; viral infection activates CD8+ T lymphocytes, causing cytotoxicity and peripheral nerve injury. *(Extracted from Muñoz, L. S., Parra, B., & Pardo, C. (2017). Neurological implications of Zika virus infection in adults.* The Journal of Infectious Diseases, 216, *897–905. doi:10.1093/infdis/jix512.)*

through activation of the complement system, taking the features developed in AMAN (Fig. 2B), while demyelination of AIDP would be induced by cell-mediated inflammation, complement, and activation of macrophages (Fig. 2C). The latest hypothesis is that ZIKV would directly infect Schwann cells causing axonal damage (Fig. 2D) (Muñoz et al., 2017).

Although the investigations show that men are more susceptible to GBS, in the case of association with ZIKV, the concern with pregnant patients is always very relevant. Studies describing GBS cases in pregnancy are rare, and little is known about immunological changes and other risk factors that may make pregnant women more prone to neurological disorders. At present, there is only one study on histopathological and immunological alterations in the placenta and retained fetus tissues in a patient infected with ZIKV who evolved to GBS. This work was performed by our group, which described a case occurred in 2016, in Brazil outbreak. In this case the patient developed to ascending and symmetrical flaccid tetraparesis, paresthesia, areflexia, hands with pendular movement, dysautonomia (resting tachycardia and hypertension), and signs of respiratory insufficiency, characterizing the GBS. The histological analysis of the patient's placenta and fetal organs revealed different tissue defects: inflammation, hemorrhage, edema, and necrosis in placenta and tissue disorganization in the fetus. In the particular case of the observed placental alterations, one fact that drew our attention was the local increase of numbers of Hofbauer cells in association with the viral infection and proinflammatory cytokine production (Fig. 3) (Rabelo et al., 2018).

Conclusion

The detection of Zika has already been added to the battery of essential tests for pregnant women, the TORCH, but an attentive monitoring during pregnancy is still necessary. In fact, studies related to placenta and stillbirths or even babies born to mothers who had Zika during pregnancy are essential to clarify the doubts about the pathogenesis of this disease still so mysterious. Babies born to mothers who had Zika, even at the end of gestation, should have a close monitoring of their development, repeating neurological, ophthalmological, auditory, and even intellectual development tests. Similarly, cases of GBS should be monitored, and a multidisciplinary treatment should be performed to minimize the possible sequels of the disease.

Policy and procedures

- Placental samples collected in studies carried out by our group—At delivery, samples from the cases placenta were collected and fixed following various histopathological techniques. Samples were collected at the Hospital Plantadores de Cana, Campos dos Goytacazes, RJ, Brazil. As a reference control, samples of term placenta from healthy donors were included.
- Diagnosis of Guillain-Barré syndrome—The diagnosis is made initially by a doctor's evaluation based on the pattern of symptoms. However, tests are performed to confirm the diagnosis, such as electromyography and nerve conduction studies, magnetic resonance imaging, blood tests, and lumbar puncture. A combination of high protein levels and few or no white blood cells in the cerebrospinal fluid, as well as specific indications observed in electromyography, suggests GBS.

Mini-dictionary of terms

Areflexia. Condition in which the muscles are not responsive to stimulation.
Ataxia. Lack of voluntary coordination of muscle movements.
Karyorrhexis. Fragmentation of the cell nucleus.
Fibrinoid necrosis. Descriptive term for necrosis with amorphous necrotic debris.
Karyolysis. Dissolution of the chromatin due to the enzymatic degradation.
Paresthesia. Abnormal dermal sensation.
Tetraparesis. Condition in which all four limbs are weak.

Key facts of ZIKV-infected placentas and GBS associated to ZIKV

- The ZIKV caused a major epidemic in 2015–17 in Brazil, leading to a series and cases of congenital Zika syndrome.
- Congenital Zika syndrome is a unique pattern of birth defects and disabilities found among fetuses and babies infected with Zika virus during pregnancy.

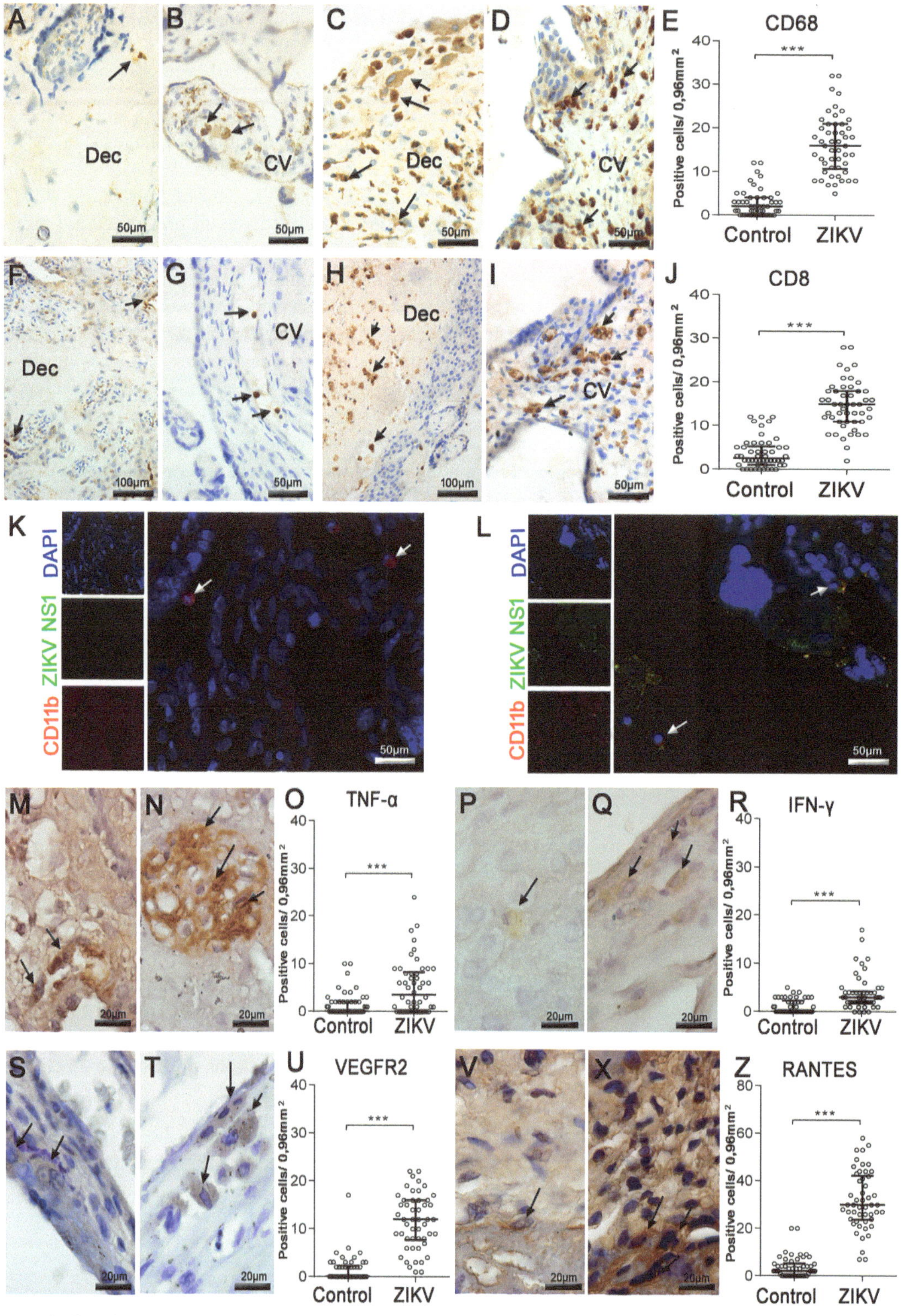

FIG. 3 Characterization of mononuclear cell subpopulations in ZIKV-infected placental tissue, colocalization with virus and cytokine-producing cell profile. (A, B) Detection of CD68+ cells (Hofbauer cells) by immunohistochemistry in decidua and chorionic villi of control placenta, respectively. (C, D) Hofbauer cells in decidua and chorionic villi of ZIKV-infected placental tissue, respectively. (F, G) Detection of CD8+ cells by immunohistochemistry in decidua and chorionic villi of control placenta, respectively. (H, I) CD8+ cells immunostained in decidua and chorionic villi of ZIKV-infected placental tissue, respectively. (E, J) Quantification of CD68+ and CD8+ cells in ZIKV case and control, respectively. (K, L) Colocalization by immunofluorescence of the NS1 protein *(fluorescent green)* and CD11b for identification of leukocytes *(fluorescent red)*. Nuclei were stained using DAPI *(fluorescent blue)*. (K) ZIKV NS1 antigen was not detected in the control placenta. (L) Cells presenting dual staining *(green and red)* were observed in the ZIKV-infected placenta. (M, N). Detection of TNF-α in cells of chorionic villi of control and ZIKV-infected placenta by immunohistochemistry, respectively. (P, Q) Production of IFN-γ in cells of membrane of control and ZIKV-infected placenta, respectively. (S, T) VEGFR2-expressing cells of decidua in control and ZIKV-infected placenta, respectively. (V, X) Detection of RANTES in cells of chorionic villi of control and ZIKV-infected placenta, respectively. (O, R, U, Z) Quantification of the number of cells expressing cytokines, in ZIKV case and control, respectively. *Asterisks* indicate differences that are statistically significant between groups (***$P < 0.001$). *(Extracted from Rabelo, K., Souza, L. J., Salomão, N. G., Oliveira, E. R. A., Sentinelli, L. de P., Lacerda, M. S., … Paes, M. V. (2018). Placental inflammation and fetal injury in a rare Zika case associated with Guillain-Barré Syndrome and abortion.* Frontiers in Microbiology, 9*(MAY), 1–10. doi:10.3389/fmicb.2018.01018.)*

- The ZIKV accesses the fetus causing damage from the placental infection.
- Guillain-Barré syndrome is an acute paralytic neuropathy of the peripheral nervous system and can be caused by ZIKV infection.
- Guillain-Barré syndrome associated with ZIKV infection causes severe disease, and the infection can lead to miscarriage.

Summary points

- ZIKV causes an epidemic in Latin America in 2015–17 and became a worldwide concern.
- The ZIKV epidemic increases the cases of congenital Zika syndrome in babies and Guillain-Barré syndrome in adults.
- Placental cells are permissive to ZIKV infection, leading to vertical transmission.
- Placental tissue showed several alterations and injury in the infection by ZIKV.
- ZIKV led to increased Guillain-Barré syndrome and was reported in a pregnant patient.

References

Adibi, J. J., Marques, E. T. A., Cartus, A., & Beigi, R. H. (2016). Teratogenic effects of the Zika virus and the role of the placenta. *The Lancet*, *387*(10027), 1587–1590. https://doi.org/10.1016/S0140-6736(16)00650-4.

Araujo, A. Q., Silva, M. T., & Araujo, A. P. (2016). Zika virus-associated neurological disorders: A review. *Brain*, *139*(8), 2122–2130. https://doi.org/10.1093/brain/aww158.

Baud, D., Gubler, D. J., Schaub, B., Lanteri, M. C., & Musso, D. (2017). An update on Zika virus infection. *The Lancet*, *6736*(17). https://doi.org/10.1016/S0140-6736(17)31450-2.

Beaufrère, A., Bessières, B., Bonnière, M., Driessen, M., Alfano, C., Couderc, T., … Encha-Razavi, F. (2018). A clinical and histopathological study of malformations observed in fetuses infected by the Zika virus. *Brain Pathology (Zurich, Switzerland)*. https://doi.org/10.1111/bpa.12644.

Bernsen, R. A. J. A. M., De Jager, A. E. J., Van Der Meché, F. G. A., & Suurmeijer, T. P. B. M. (2005). How Guillain-Barré patients experience their functioning after 1 year. *Acta Neurologica Scandinavica*. https://doi.org/10.1111/j.1600-0404.2005.00429.x.

Besnard, M., Lastere, S., Teissier, A., Cao-Lormeau, V. M., & Musso, D. (2014). Evidence of perinatal transmission of zika virus, French Polynesia, December 2013 and February 2014. *Eurosurveillance*, *19*(13). Retrieved from http://www.eurosurveillance.org/ViewArticle.aspx?ArticleId=20751%5Cnhttp://ovidsp.ovid.com/ovidweb.cgi?T=JS&PAGE=reference&D=emed16&NEWS=N&AN=372788392.

Brown, J. A., Singh, G., Acklin, J. A., Krammer, F., Lim, J. K., Brown, J. A., … Chokola, A. N. (2019). Dengue virus immunity increases Zika virus-induced damage during pregnancy article dengue virus immunity increases Zika virus-induced damage during pregnancy. *Immunity*, *50*, 751–762. https://doi.org/10.1016/j.immuni.2019.01.005.

Calvet, G., Aguiar, R. S., Melo, A. S. O., Sampaio, S. A., De Filippis, I., Fabri, A., … De Sequeira, P. C. (2016). Detection and sequencing of Zika virus from amniotic fluid of fetuses with microcephaly in Brazil: a case study. *The Lancet*, *16*, 653–660. https://doi.org/10.1016/S1473-3099(16)00095-5.

Counotte, M. J., Egli-Gany, D., Riesen, M., Wang, J., Low, N., Porgo, T. V., & Rodriguez-Morales, A. (2018). Zika virus infection as a cause of congenital brain abnormalities and Guillain-Barré syndrome: from systematic review to living systematic review [vers]. *F100Research*, *7*, 1–20.

Cross, J. C. (2005). How to make a placenta: Mechanisms of trophoblast cell differentiation in mice – A review. *Placenta*, *26*(Suppl). https://doi.org/10.1016/j.placenta.2005.01.015.

da Silva, I. R. F., Frontera, J. A., de Filippis, A. M. B., & Nascimento, O. J. M.d. (2017). Neurologic complications associated with the Zika virus in Brazilian adults. *JAMA Neurology*, *74*(10), 1190–1198. https://doi.org/10.1001/jamaneurol.2017.1703.

dos Santos, T., Rodriguez, A., Almiron, M., & Sanhueza, A. (2016). Zika virus and the Guillain – Barré syndrome—Case series from seven countries. *New England Journal of Medicine*, 1–4. https://doi.org/10.1056/NEJMc1609015.

Driggers, R. W., Ho, C. Y., Korhonen, E. M., Kuivanen, S., Jaaskelainen, A. J., Smura, A., … Vapalahti, O. (2016). Zika virus infection with prolonged maternal viremia and fetal brain abnormalities. *The New England Journal of Medicine*, 1–10. https://doi.org/10.1056/NEJMoa1601824.

Dyck, P. (2005). *Peripheral neuropathy*. Elsevier Inc.

El Costa, H., Gouilly, J., Mansuy, J., Chen, Q., Levy, C., Cartron, G., … Izopet, J. (2016). ZIKA virus reveals broad tissue and cell tropism during the first trimester of pregnancy. *Scientific Reports*, (October), 1–9. https://doi.org/10.1038/srep35296.

Faria, N. R., Quick, J., Morales, I., Theze, J., de Jesus, J. G., Giovanetti, M., … Pybus, O. G. (2017). Epidemic establishment and cryptic transmission of Zika virus in Brazil and the Americas. *Nature*, *546*, 406–424. https://doi.org/10.1038/nature22401.

Ferreira, M. L. B., de Brito, C. A. A., Moreira, Á. J. P., Machado, M.Í.d. M., Henriques-Souza, A., Cordeiro, M. T., … Pena, L. J. (2017). Guillain–Barré syndrome, acute disseminated encephalomyelitis and encephalitis associated with Zika virus infection in Brazil: Detection of viral RNA and isolation of virus during late infection. *The American Journal of Tropical Medicine and Hygiene*, *97*(5), 1405–1419.

Gnainsky, Y., Granot, I., Aldo, P., Barash, A., Or, Y., Mor, G., & Dekel, N. (2015). Biopsy-induced inflammatory conditions improve endometrial receptivity: The mechanism of action. *Reproduction*, *149*(1), 75–85. https://doi.org/10.1530/REP-14-0395.

Hanna, J., Goldman-Wohl, D., Hamani, Y., Avraham, I., Greenfield, C., Natanson-Yaron, S., … Mandelboim, O. (2006). Decidual NK cells regulate key developmental processes at the human fetal-maternal interface. *Nature Medicine*, *12*(9), 1065–1074. https://doi.org/10.1038/nm1452.

Jacobs, B. C., Rothbarth, P. H., Van Der Meché, F. G. A., Herbrink, P., Schmitz, P. I. M., De Klerk, M. A., & Van Doorn, P. A. (1998). The spectrum of antecedent infections in Guillain-Barre syndrome: A case-control study. *Neurology*. https://doi.org/10.1212/WNL.51.4.1110.

Leonhard, S. E., Mandarakas, M. R., Gondim, F. A. A., Bateman, K., Ferreira, M. L. B., Cornblath, D. R., … Jacobs, B. C. (2019). Diagnosis and management of Guillain–Barré syndrome in ten steps. *Nature Reviews Neurology*. https://doi.org/10.1038/s41582-019-0250-9.

Lum, F., Narang, V., Hue, S., Chen, J., Mcgovern, N., Rajarethinam, R., … Liu, X. (2019). Immunological observations and transcriptomic analysis of trimester-specific full-term placentas from three Zika virus-infected women. *Clinical & Translational Immunology*, *8*, 1–15. https://doi.org/10.1002/cti2.1082.

Malkki, H. (2016). CNS infections: Zika virus infection could trigger Guillain-Barré syndrome. *Nature Reviews Neurology*, *12*(4), 187. https://doi.org/10.1038/nrneurol.2016.30.

Martines, R. B., Bhatnagar, J., Keating, M. K., Silva-Flannery, L., Gary, J., Goldsmith, C., … Zaki, S. (2016). Evidence of Zika virus infection in brain and placental tissues from two congenitally infected newborns and two fetal losses-Brazil 2015. *Morbidity and Mortality Weekly Report*, *65*(6), 2015–2016.

Mier-Y-Teran-Romero, L., Delorey, M. J., Sejvar, J. J., & Johansson, M. A. (2018). Guillain–Barré syndrome risk among individuals infected with Zika virus: A multi-country assessment. *BMC Medicine*, *16*(67), 1–8.

Miranda, J., Martín-Tapia, D., Valdespino-Vázquez, Y., Alarcón, L., Espejel-Nuñez, A., Guzmán-Huerta, M., … González-Mariscal, L. (2019). Syncytiotrophoblast of placentae from women with Zika virus infection has altered tight junction protein expression and increased paracellular permeability. *Cell*, *8*(10), 1174. https://doi.org/10.3390/cells8101174.

Mor, G., & Cardenas, I. (2010). The immune system in pregnancy: A unique complexity. *American Journal of Reproductive Immunology*, *63*(6), 425–433.

Muñoz, L. S., Parra, B., & Pardo, C. (2017). Neurological implications of Zika virus infection in adults. *The Journal of Infectious Diseases*, *216*, 897–905. https://doi.org/10.1093/infdis/jix512.

Musso, D., Ko, A. I., & Baud, D. (2019). Zika virus infection—After the pandemic. *The New England Journal of Medicine*, *381*(15), 1444–1457. https://doi.org/10.1056/NEJMra1808246.

Nyati, K. K., & Nyati, R. (2013). Role of campylobacter jejuni infection in the pathogenesis of Guillain-Barré syndrome: An update. *BioMed Research International*. https://doi.org/10.1155/2013/852195.

Paploski, I. A. D., Prates, A. P. P. B., Cardoso, C. W., Kikuti, M., Silva, M. M. O., Waller, L. A., … Ribeiro, G. S. (2016). Time lags between exanthematous illness attributed to Zika virus, Guillain-Barré syndrome, and microcephaly, Salvador, Brazil. *Emerging Infectious Diseases*, *22*(8), 1438–1444.

Parra, B., Lizarazo, J., Jiménez-Arango, J. A., Zea-Vera, A. F., González-Manrique, G., Vargas, J., … Pardo, C. A. (2016). Guillain–Barré syndrome associated with Zika virus infection in Colombia. *New England Journal of Medicine*, *375*(16), 1513–1523. https://doi.org/10.1056/NEJMoa1605564.

Quicke, K. M., Bowen, J. R., Johnson, E. L., Schinazi, R. F., Chakraborty, R., Suthar, M. S., … Neal, J. T. O. (2016). Zika virus infects human placental macrophages. *Cell Host and Microbe*, 1–8. https://doi.org/10.1016/j.chom.2016.05.015.

Rabelo, K., de Souza Campos Fernandes, R. C., de Souza, L. J., de Souza, T. L., dos Santos, F. B., Nunes, P. C. G., … Paes, M. V. (2017). Placental histopathology and clinical presentation of severe congenital Zika syndrome in a human immunodeficiency virus-exposed uninfected infant. *Frontiers in Immunology*, *8*(DEC), 1–8. https://doi.org/10.3389/fimmu.2017.01704.

Rabelo, K., Souza, L. J., Salomão, N. G., Oliveira, E. R. A., Sentinelli, L.d. P., Lacerda, M. S., … Paes, M. V. (2018). Placental inflammation and fetal injury in a rare Zika case associated with Guillain-Barré syndrome and abortion. *Frontiers in Microbiology*, *9*(MAY), 1–10. https://doi.org/10.3389/fmicb.2018.01018.

Ramhorst, R., Fraccaroli, L., Aldo, P., Alvero, A. B., Cardenas, I., Leirós, C. P., & Mor, G. (2012). Modulation and recruitment of inducible regulatory T cells by first trimester trophoblast cells. *American Journal of Reproductive Immunology*, *67*(1), 17–27. https://doi.org/10.1111/j.1600-0897.2011.01056.x.

Rivadeneyra-Espinar, P. G., Venegas-Esquivel, G. A., & Díaz-Espinoza, C. M. (2019). Zika como causa de aborto espontáneo en zonas endémicas. *Boletín Médico del Hospital Infantil de México*, *76*, 193–197. https://doi.org/10.24875/BMHIM.19000116.

Rosenberg, A., Weiying, Y., Hill, A., Reyes, C. A., Schwartz, D., & Hyg, M. S. (2016). Placental pathology of Zika virus. *College of American Pathologists*, *1*, 1–6. https://doi.org/10.5858/arpa.2016-0401-OA.

Rosenberg, A. Z., Yu, W., Hill, D. A., Reyes, C. A., & Schwartz, D. A. (2017). Placental pathology of Zika virus: Viral infection of the placenta induces villous stromal macrophage (Hofbauer cell) proliferation and hyperplasia. *Archives of Pathology & Laboratory Medicine*, *141*(1), 43–48.

Rossant, J., & Cross, J. C. (2001). Placental development: Lessons from mouse mutants. *Nature Reviews. Genetics*, *2*(7), 538–548. https://doi.org/10.1038/35080570.

Santos, G. R., Pinto, C. A. L., Prudente, R. C. S., Bevilacqua, E. M. A. F., … Study, C. (2019). Case report: Histopathologic changes in placental tissue associated with vertical transmission of Zika virus. *International Journal of Gynecological Pathology*, *1*, 1–6. https://doi.org/10.1097/PGP.0000000000000586.

Sejvar, J. J., Baughman, A. L., Wise, M., & Morgan, O. W. (2011). Population incidence of Guillain-Barré syndrome: A systematic review and meta-analysis. *Neuroepidemiology*. https://doi.org/10.1159/000324710.

Sejvar, J. J., Kohl, K. S., Gidudu, J., Amato, A., Bakshi, N., Baxter, R., … Patrick, Z. (2011). Guillain-Barré syndrome and Fisher syndrome: Case definitions and guidelines for collection, analysis, and presentation of immunization safety data. *Vaccine*. https://doi.org/10.1016/j.vaccine.2010.06.003.

Simmons, D. G., Fortier, A. L., & Cross, J. C. (2007). Diverse subtypes and developmental origins of trophoblast giant cells in the mouse placenta. *Developmental Biology*, *304*(2), 567–578. https://doi.org/10.1016/j.ydbio.2007.01.009.

Soto-Hernández, J. L., Ponce, S., & Rosales, D. L. (2019). Guillain – Barré syndrome associated with Zika virus infection: A prospective case series from. *Frontiers in Neurology*, *10*(April), 1–6. https://doi.org/10.3389/fneur.2019.00435.

Tabata, T., Petitt, M., Puerta-guardo, H., Fang-Hoover, J., Harris, E., Tabata, T., … Fang-Hoover, J. (2016). Zika virus targets different primary human placental cells, suggesting two routes for vertical article Zika virus targets different primary human placental cells, suggesting two routes for vertical transmission. *Cell Host and Microbe*, 1–12. https://doi.org/10.1016/j.chom.2016.07.002.

Tabata, T., Petitt, M., Puerta-Guardo, H., Michlmayr, D., Wang, C., Fang-Hoover, J., … Pereira, L. (2016). Zika virus targets different primary human placental cells, suggesting two routes for vertical transmission. *Cell Host and Microbe*, *20*(2), 155–166.

Van Den Berg, B., Bunschoten, C., Van Doorn, P. A., & Jacobs, B. C. (2013). Mortality in Guillain-Barré syndrome. *Neurology*. https://doi.org/10.1212/WNL.0b013e3182904fcc.

Watrin, L., Ghawche, F., Larre, P., Neau, J.-P., Mathis, S., & Fournier, E. (2016). Guillain–Barre´ syndrome (42 cases) occurring during a Zika virus outbreak in French Polynesia. *Medicine*, *95*(14), 1–8. https://doi.org/10.1097/MD.0000000000003257.

WHO. (2017). *Guillain-Barré syndrome—Brazil.*

Willison, H. J., Jacobs, B. C., & van Doorn, P. A. (2016). Guillain-Barré syndrome. *The Lancet*. https://doi.org/10.1016/S0140-6736(16)00339-1.

World Health Organization (WHO). (2015). *Congenital anomalies*. Retrieved from http://www.who.int/mediacentre/%0Afactsheets/fs370/en.

Zulu, M. Z., Martinez, O., Gordon, S., & Gray, M. (2019). The elusive role of placental macrophages: The Hofbauer cell. *Journal of Innate Immunity*, *11*, 447–456. https://doi.org/10.1159/000497416.

Section D

Case studies and short reports

Chapter 16

Case study: Neuroimaging of adults and Zika virus

Diogo Goulart Corrêa and Luiz Celso Hygino da Cruz Júnior
Department of Radiology, Clinic of Diagnostic Imaging/DASA, Rio de Janeiro, Rio de Janeiro, Brazil

Abbreviations

ADEM	acute disseminated encephalomyelitis
CNS	central nervous system
CZS	congenital Zika syndrome
FLAIR	fluid-attenuated inversion recovery
MRI	magnetic resonance imaging
PCR	polymerase chain reaction
STIR	short tau inversion recovery
ZIKV	Zika virus

Introduction

In addition to congenital Zika syndrome (CZS), Zika virus (ZIKV) infection is associated with a number of neurological complications, when the infection occurs in adults and older children. These neurologic complications, associated with ZIKV infection, prompted the declaration of the public health emergency of international concern by the World Health Organization (Reid, Rimmer, & Thakur, 2018).

Among the neurologic complications associated with ZIKV infection in adolescents and adults, the main ones are Guillain-Barré syndrome, encephalitis/meningoencephalitis, myelitis, and acute disseminated encephalomyelitis (ADEM).

The first reported cases of neurological complications related to ZIKV infection in Brazil included four Guillain-Barré syndrome cases, one patient with encephalitis, and one case of ADEM (Brito Ferreira, Antunes de Brito, Moreira, et al., 2017).

ADEM is an immune-mediated inflammatory demyelinating disorder of the central nervous system (CNS) that predominantly affects the white matter of the brain and spinal cord. Clinically, it is characterized by new-onset multifocal neurologic symptoms, including encephalopathy, with neuroimaging evidence of multifocal white matter inflammatory demyelinating lesions. In the absence of specific biologic markers, the diagnosis of ADEM is based on clinical and radiologic features (Pohl et al., 2016).

In this chapter, we present a case study of ADEM related to ZIKV infection, with emphasis on the neuroimaging presentation.

Case study

A 48-year-old woman presented initially with fever (39.5°C), associated with disseminated skin rash, during the 2016 Zika epidemic in the city of Rio de Janeiro. ZIKV infection was confirmed by polymerase chain reaction (PCR) in the peripheral blood. Supportive treatment was instituted, and the patient's symptoms began to improve. However, 10 days after the initial symptoms, the patient started to present reduced level of conscience, loss of upper and lower limbs coordination, and paresis in the lower limbs.

Brain and spine magnetic resonance imaging (MRI) showed T2-weighted image hyperintense lesions in the middle cerebellar peduncles and a hyperintense lesion in short tau inversion recovery (STIR) in the upper portion of the dorsal spine, without gadolinium enhancement in the postcontrast T1-weighted image (Fig. 1). The possibility of encephalitis and myelitis secondary to ZIKV infection was considered.

Zika Virus Impact, Diagnosis, Control, and Models. https://doi.org/10.1016/B978-0-12-820267-8.00016-9

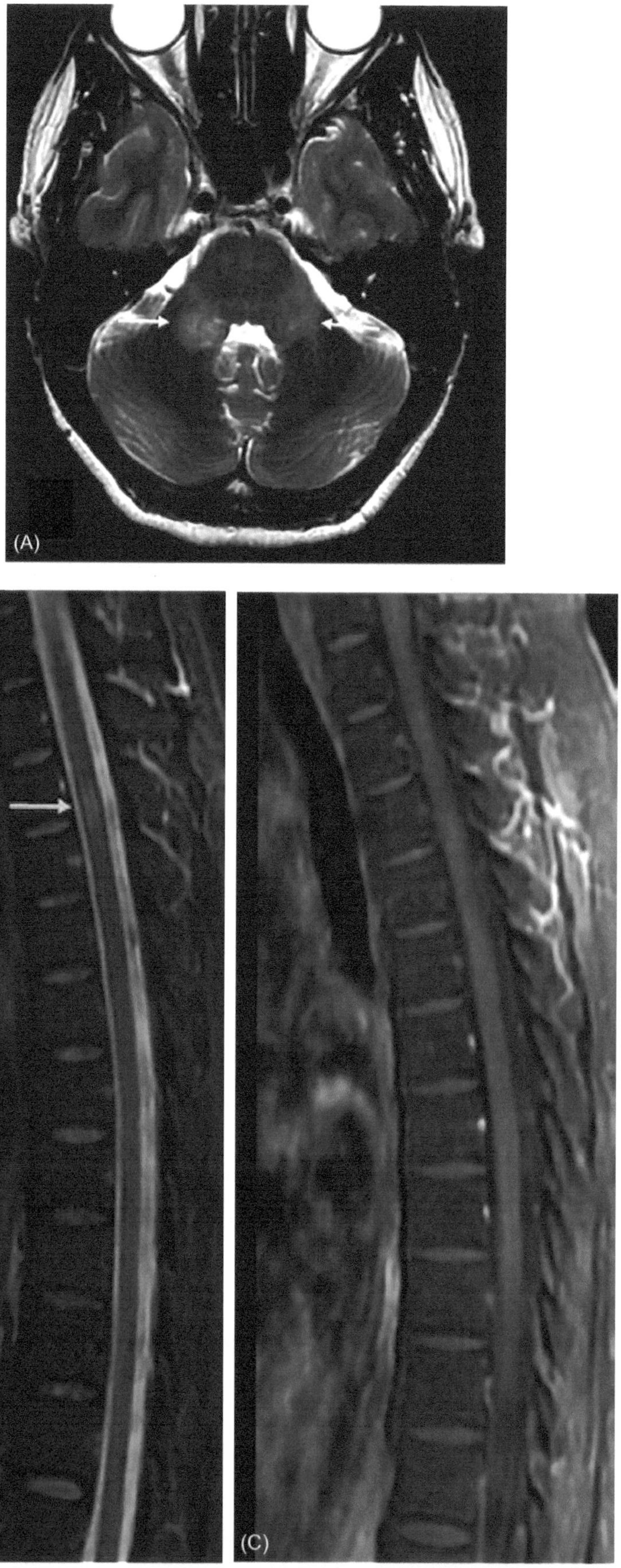

FIG. 1 Acute disseminated encephalomyelitis secondary to Zika virus infection. Brain MRI showing bilateral hyperintense lesions in the middle cerebellar peduncles in the axial T2-weighted image (*arrows* in A). Spine MRI showing a hyperintense lesion in the upper dorsal portion of the spine in the sagittal STIR (*arrow* in B). The lesion did not present gadolinium enhancement in the sagittal postcontrast T1-weighted image (C). *(From Hygino da Cruz, L. C., Jr., Nascimento, O. J. M., Lopes, F. P. P. L., da Silva, I. R. F. (2018). Neuroimaging findings of Zika virus-associated neurologic complications in adults.* AJNR. American Journal of Neuroradiology, 39, *1967–1974, with permission.)*

However, due to the worsening of the lower limb symptoms, the MRI was repeated 2 weeks later. Follow-up MRI of the brain and spine showed improvement of the cerebellar peduncle lesions and an enlargement of the spine lesion, which at that moment came to present contrast enhancement (Fig. 2). Cerebrospinal fluid analysis revealed just a pleocytosis and elevated protein. At this moment the possibility of ADEM was considered, and the patient was treated with 1 g intravenous methylprednisolone for 5 days, followed by an oral taper over with prednisone. The patient's symptoms have improved, and at this moment, she has only mild weakness in the lower limbs.

Acute disseminated encephalomyelitis

ADEM is an immune-mediated demyelinating disorder of the CNS, characterized, clinically, by new-onset multifocal neurologic symptoms including encephalopathy, coupled with neuroimaging evidence of multifocal demyelination. Classically, ADEM is considered a monophasic illness, with highest incidence in early childhood (Pohl et al., 2016). Most cases are associated with an antigenic challenge, which may be a vaccination or an infection that occurs just days before neurological symptoms (Marin & Callen, 2013).

ADEM is a diagnosis of exclusion, necessitating thorough consideration of alternate diagnoses, based on clinical and imaging findings. The diagnostic criteria are (all are required) (1) a multifocal clinical event in the CNS, with a presumed inflammatory demyelinating cause; (2) encephalopathy (alteration in consciousness or behavior unexplained by fever, systemic illness, or postictal symptoms); (3) brain MRI abnormalities consistent with demyelination during the acute phase (3 months); and (4) no new clinical or MRI findings 3 months or more after the clinical onset (Krupp, Tardieu, Amato, et al., 2013; Pohl et al., 2016).

It is important to remember that ADEM is a heterogeneous entity and is best viewed as a syndrome rather than a specific disorder. Also the clinical symptoms and MRI findings of ADEM can fluctuate in severity and evolve in the first 3 months following disease onset. The presence of encephalopathy is obligatory to its diagnosis (Krupp et al., 2013).

Imaging findings of acute disseminated encephalomyelitis

Computed tomography is not a sensitive technique to evaluate suspected cases of ADEM. The most commonly reported abnormalities are discrete hypodense areas within cerebral white matter and juxtacortical areas (Marin & Callen, 2013).

MRI of the brain and spine is the main imaging technique available in the diagnosis of ADEM and to distinguish the clinical presentation from other inflammatory and noninflammatory neurologic diseases (Marin & Callen, 2013).

The typical MRI findings described in ADEM are widespread, bilateral, asymmetric irregular lesions with hyperintense signal on T2-weighted imaging and fluid-attenuated inversion recovery (FLAIR), within the white matter and spinal cord. The most commonly affected portion of the white matter is juxtacortical. But the periventricular white matter and corpus callosum are rarely affected, in contrast to multiple sclerosis, in which these portions of the brain are commonly affected. Gray matter, particularly the basal ganglia, also can be involved (Zare Mehrjardi, Carteaux, Poretti, et al., 2017). The lesions vary from small and round to large, amorphous, and irregular. Contrast enhancement is variable, ranging from 30% to 100% of the patients, and occur in nonspecific patterns (nodular, diffuse, gyral, complete, or incomplete ring). Diffusion restriction may be present in the acute stage of the disease (Marin & Callen, 2013; Zare Mehrjardi et al., 2017).

Infratentorial lesions are common in ADEM, as in our case, including the brainstem and cerebellar white matter. Unenhanced T1-weighted images reveal that lesions are typically inconspicuous unless the lesions are large, in which case a faint hypodensity is seen within the affected areas (Hynson, Kornberg, Coleman, et al., 2001; Khong, Ho, Cheng, et al., 2002; Rossi, 2008).

A longitudinal extensive transverse myelitis has been reported in 92% of ADEM cases, in general, with a 50% rate of contrast enhancement. MRI typically shows contiguous lesions, with hyperintense signal on T2-weighted imaging, spanning 3 or more vertebral bodies, variably enhancing (Goh, Phal, & Desmond, 2011; Mirbagheri et al., 2016).

Typically, the lesions appear simultaneously with clinical presentation. However, delayed appearance of abnormalities up to 1 month after clinical onset has been described. In the case presented in this chapter, the spinal lesion worsened and began to show contrast enhancement 2 weeks after the initial presentation (Marin & Callen, 2013).

After a short period of treatment, usually occurs a decrease in the size and number of lesions and a change in signal intensity of lesions, detected by MRI, together with clinical improvement. Complete resolution has been noted in up to 70% of patients within months of presentation; however, residual deficits may persist in up to one-third of patients 2 years after disease onset (Callen, Shroff, Branson, et al., 2009). Our patient has shown a strikingly improvement of her symptoms, but 1 year after the symptom onset, she still has discrete lower limb paresis.

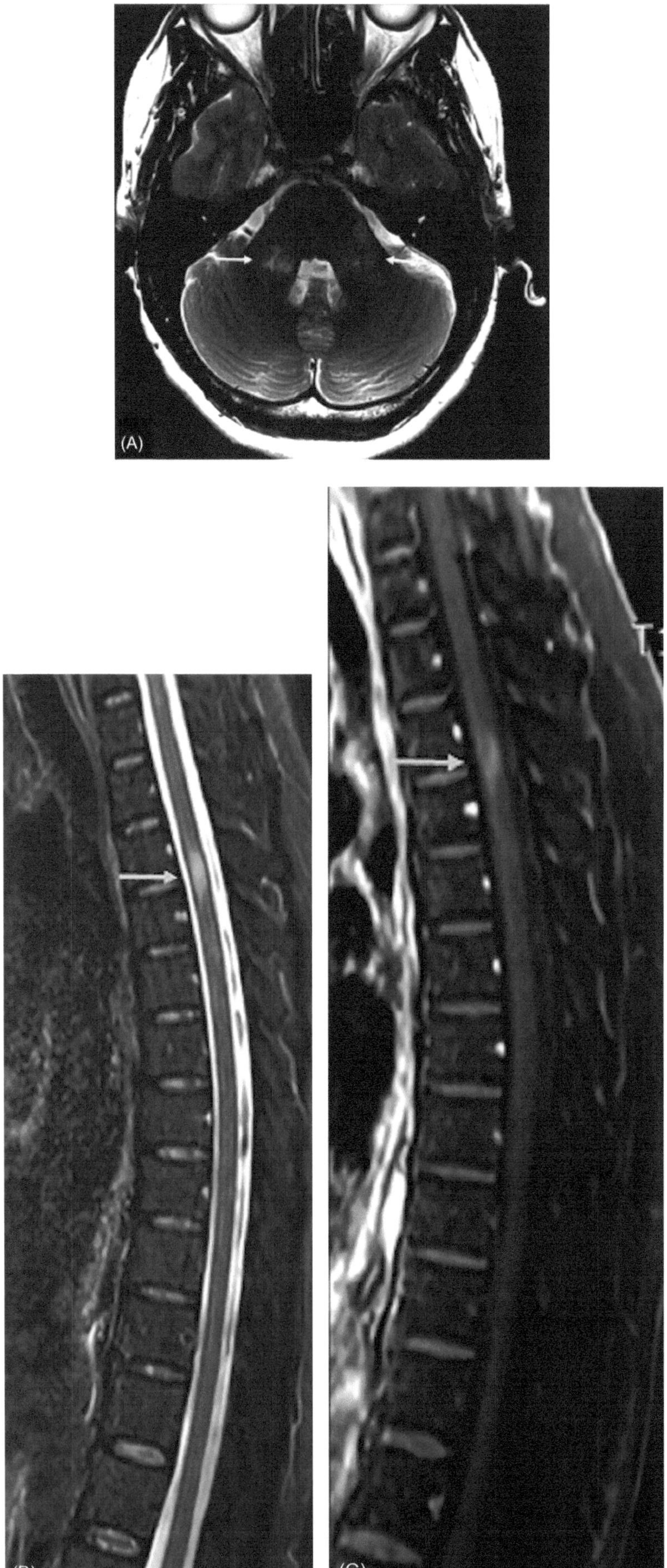

FIG. 2 Acute disseminated encephalomyelitis secondary to Zika virus infection. Follow-up brain and spine MRI, 2 weeks after the first neurological symptoms, showing improvement of the cerebellar peduncles lesions (*arrows* in A). However, the spinal cord lesion is enlarged (*arrow* in B) and started to present contrast enhancement (*arrow* in C). *(From Hygino da Cruz, L. C., Jr., Nascimento, O. J. M., Lopes, F. P. P. L., da Silva, I. R. F. (2018). Neuroimaging findings of Zika virus-associated neurologic complications in adults.* AJNR. American Journal of Neuroradiology, 39, *1967–1974, with permission.)*

Acute disseminated encephalomyelitis and Zika virus

The association between ZIKV infection and ADEM has been suggested by few reports (Brito Ferreira et al., 2017; Niemeyer, Niemeyer, Borges, et al., 2017).

It is important to remember that the clinical features and neuroimaging findings in ADEM secondary to ZIKV infection do not differ from those caused by other etiologies (Zare Mehrjardi et al., 2017). The clinical course of ADEM secondary to ZIKV is very similar to that of the most typical ADEM cases, characterized as a self-limited and, typically, monophasic course, appearing weeks after the viral infection. There is no specific imaging finding for ZIKV-related ADEM distinguishing it from other causes of ADEM (Hygino da Cruz, Nascimento, Lopes, & da Silva, 2018).

Conclusion

ADEM is a rare neurological complication of ZIKV infection. Although there are no clinical and/or imaging findings that differentiate ADEM secondary to ZIKV from the other causes, the knowledge of its imaging characteristic is important, and in an epidemic scenario, it may suggest the correct diagnosis.

Policy and procedures

Computed tomography is not adequately sensitive to investigate cases suspected of ADEM. MRI is the most important imaging technique available to diagnose cases of ADEM and to distinguish it from other inflammatory and noninflammatory neurologic diseases. In suspected cases of ADEM, it is important to perform the MRI of the entire neuroaxis to assess brain lesions and spinal cord lesions. Intravenous gadolinium-based contrast injection is also important.

The typical imaging findings of ADEM are multiple, asymmetrical, and ill-defined lesions, with hyperintense signal on T2-weighted imaging and FLAIR, involving the white matter and deep gray matter nuclei. Contrast enhancement is variable, ranging from absent enhancement to ring enhancement. A longitudinal extensive transverse myelitis is a common finding, extending over at least three vertebral bodies. MRI demonstrates contiguous lesions, with hyperintense signal on T2-weighted imaging, spanning 3 or more vertebral bodies, variably enhancing.

Typically the lesions appear together with clinical presentation. However, delayed appearance of abnormalities up to 1 month after clinical onset has been described. Then a normal initial MRI does not exclude the diagnosis of ADEM.

After treatment, the lesions decrease in size and number, together with clinical improvement. Complete resolution has been noted in the majority of cases, but residual deficits may persist in up to one-third of patients.

Although there is no specific clinical and imaging finding of ADEM secondary to ZIKV infection capable to distinguish it from other causes of ADEM, knowledge of its imaging findings is important, and in the correct clinical scenario, it may suggest the final diagnosis.

Mini-dictionary of terms

Acute disseminated encephalomyelitis. Immunologically mediated demyelinating disease of the central nervous system, coursing with multifocal lesions in the brain and spinal cord, commonly associated with an antigenic challenge, as a trigger to the inflammatory response.

Encephalopathy. A reduction in the level of consciousness (e.g., stupor and lethargy) or behavioral change, unexplained by fever, systemic illness, or postictal symptoms.

Longitudinal extensive transverse myelitis. Spinal cord lesion, secondary to myelitis, presenting with hyperintense signal on T2-weighted imaging, extending craniocaudally over at least three vertebral bodies on an MRI.

Monophasic ADEM. ADEM that evolve in the first 3 months after symptom onset. It is the most common form of the disease.

Multiphasic ADEM. Two episodes consistent with ADEM, separated by 3 months and not followed by any further events.

Key facts of Zika virus infection complicated by acute disseminated encephalomyelitis

- Acute disseminated encephalomyelitis typically follows a monophasic disease course, although confirmation of monophasic disease is retrospective and requires prolonged observation.

- Although the majority of acute disseminated encephalomyelitis cases are monophasic, multiphasic acute disseminated encephalomyelitis has been defined as recurrent disseminated encephalomyelitis. However, there are no cases of recurrent acute disseminated encephalomyelitis reported complicating a Zika virus infection.
- Typical magnetic resonance imaging findings of acute disseminated encephalomyelitis are widespread, bilateral, asymmetric patchy areas of T2-weighted imaging and fluid-attenuated inversion recovery hyperintensities within the white matter, deep gray nuclei, and spinal cord. Contrast enhancement is variable and has been reported in nonspecific patterns.
- A longitudinal extensive transverse myelitis is a common finding in spinal magnetic resonance imaging of patients with acute disseminated encephalomyelitis.
- Neuroimaging findings of Zika virus-related acute disseminated encephalomyelitis do not differ from those caused by other etiologies.

Summary points

- In addition to congenital Zika syndrome, Zika virus infection is associated with a number of neurological complications in adults and older children.
- Among the main neurologic complications associated with Zika virus infection in adolescents and adults are Guillain-Barré syndrome, encephalitis/meningoencephalitis, myelitis, and acute disseminated encephalomyelitis.
- MRI is the most important paraclinical tool for the diagnosis of acute disseminated encephalomyelitis.
- The clinical course of acute disseminated encephalomyelitis secondary to Zika virus is very similar to that of other causes.
- There is no specific imaging finding for Zika virus-related acute disseminated encephalomyelitis distinguishing it from other causes of acute disseminated encephalomyelitis.

References

Brito Ferreira, M. L., Antunes de Brito, C. A., Moreira, Á. J. P., et al. (2017). Guillain-Barré syndrome, acute disseminated encephalomyelitis and encephalitis associated with Zika virus infection in Brazil: Detection of viral RNA and isolation of virus during late infection. *The American Journal of Tropical Medicine and Hygiene*, *97*, 1405–1409.

Callen, D. J., Shroff, M. M., Branson, H. M., et al. (2009). Role of MRI in the differentiation of ADEM from MS in children. *Neurology*, *72*, 968–973.

Goh, C., Phal, P. M., & Desmond, P. M. (2011). Neuroimaging in acute transverse myelitis. *Neuroimaging Clinics of North America*, *21*, 951–973.

Hygino da Cruz, L. C., Jr., Nascimento, O. J. M., Lopes, F. P. P. L., & da Silva, I. R. F. (2018). Neuroimaging findings of Zika virus-associated neurologic complications in adults. *AJNR. American Journal of Neuroradiology*, *39*, 1967–1974.

Hynson, J. L., Kornberg, A. J., Coleman, L. T., et al. (2001). Clinical and neuroradiologic features of acute disseminated encephalomyelitis in children. *Neurology*, *56*, 1308–1312.

Khong, P. L., Ho, H. K., Cheng, P. W., et al. (2002). Childhood acute disseminated encephalomyelitis: The role of brain and spinal cord MRI. *Pediatric Radiology*, *32*, 59–66.

Krupp, L. B., Tardieu, M., Amato, M. P., et al. (2013). International Pediatric Multiple Sclerosis Study Group criteria for pediatric multiple sclerosis and immune-mediated central nervous system demyelinating disorders: Revisions to the 2007 definitions. *Multiple Sclerosis*, *19*, 1261–1267.

Marin, S. E., & Callen, D. J. (2013). The magnetic resonance imaging appearance of monophasic acute disseminated encephalomyelitis: An update post application of the 2007 consensus criteria. *Neuroimaging Clinics of North America*, *23*, 245–266.

Mirbagheri, S., Eckart Sorte, D., Zamora, C. A., Mossa-Basha, M., Newsome, S. D., & Izbudak, I. (2016). Evaluation and management of longitudinally extensive transverse myelitis: A guide for radiologists. *Clinical Radiology*, *71*, 960–971.

Niemeyer, B., Niemeyer, R., Borges, R., et al. (2017). Acute disseminated encephalomyelitis following Zika virus infection. *European Neurology*, *77*, 45–46.

Pohl, D., Alper, G., Van Haren, K., Kornberg, A. J., Lucchinetti, C. F., Tenembaum, S., et al. (2016). Acute disseminated encephalomyelitis: Updates on an inflammatory CNS syndrome. *Neurology*, *87*(9 Suppl. 2), S38–S45.

Reid, S., Rimmer, K., & Thakur, K. (2018). Zika virus and neurologic disease. *Neurologic Clinics*, *36*, 767–787.

Rossi, A. (2008). Imaging of acute disseminated encephalomyelitis. *Neuroimaging Clinics of North America*, *18*, 149–161.

Zare Mehrjardi, M., Carteaux, G., Poretti, A., et al. (2017). Neuroimaging findings of postnatally acquired Zika virus infection: A pictorial essay. *Japanese Journal of Radiology*, *35*, 341–349.

Chapter 17

Case study: Magnetic resonance imaging and babies with Zika virus infection

Diogo Goulart Corrêa, Heron Werner, and Luiz Celso Hygino da Cruz Júnior
Department of Radiology, Clinic of Diagnostic Imaging/DASA, Rio de Janeiro, Rio de Janeiro, Brazil

Abbreviations

CNS	central nervous system
CT	computed tomography
CMV	cytomegalovirus
CZS	congenital Zika syndrome
MRI	magnetic resonance imaging
NPC	neural progenitor cells
rRT-PCR	real-time reverse transcription polymerase chain reaction
RNA	ribonucleic acid
SD	standard deviations
US	ultrasound
WHO	World Health Organization
ZIKV	Zika virus

Introduction

The term congenital Zika syndrome (CZS) describes the spectrum of findings associated with maternal-fetal Zika virus (ZIKV) transmission (Zorrilla, Rivera-Vinas, De La Vega-Pujols, et al., 2018). Usually, pregnant women infected with ZIKV report no symptoms or only self-limiting flu-like symptoms, whereas a broad range of clinical presentations has been reported in fetuses and newborns of women infected with ZIKV during pregnancy (Liang, Guida, Costa Do Nascimento, & Mysorekar, 2019).

At the beginning of 2015, the first autochthonous transmission of ZIKV was reported in the northeastern region of Brazil (Campos, Bandeira, & Sardi, 2015; Song, Yun, Woolley, & Lee, 2017). In the same year a dramatic increase of fetuses and newborns with microcephaly was reported in Brazil, particularly in the state of Pernambuco. Posteriorly, congenital ZIKV infection was linked to microcephaly. In 2016 the World Health Organization (WHO) declared the spreading of ZIKV infection a global public health emergency (Werner, Fazecas, Guedes, et al., 2016).

Although microcephaly has been the hallmark of fetuses and newborns with CZS, not all affected fetuses demonstrated microcephaly. Additionally, other clinical signs of this syndrome include brain atrophy, malformations of cortical development, hypoplasia of the cerebellum and cerebellar vermis, and eye abnormalities. Arthrogryposis and polyhydramnios, likely due to swallowing impairment because of brain injury, are also common findings (Chibueze, Tirado, Lopes, et al., 2017).

Different imaging modalities, both prenatally and postnatally, are important to evaluate and follow up vertically ZIKV-infected patients. Brain abnormalities related to congenital ZIKV infection are generally observed for the first time with prenatal ultrasound (US), especially in the third trimester of pregnancy. Fetal magnetic resonance imaging (MRI) better identifies brain malformations, such as cerebral atrophy, ventriculomegaly, and microphthalmia, usually being performed when abnormalities are seen at fetal US. Postnatal computed tomography (CT) and MRI can also unveil additional findings with regard to involvement of the central nervous system (CNS), including the occurrence of microcephaly with

Zika Virus Impact, Diagnosis, Control, and Models. https://doi.org/10.1016/B978-0-12-820267-8.00017-0

malformation of cortical development, ventriculomegaly, and multifocal dystrophic calcifications in the cortex (Ribeiro, Werner, Lopes, et al., 2017).

In this chapter, we present a case study of congenital ZIKV infection, with emphasis on the prenatal and postnatal imaging exams.

Case study

A 27-year-old woman, pregnant for the first time, with no history of congenital anomalies or consanguinity in the family, presented with fever, disseminated maculopapular cutaneous rash, and arthralgia at the 12th week of the gestation.

Serology testing for *Toxoplasma gondii*, rubella, cytomegalovirus, syphilis, dengue, and chikungunya were negative. Real-time reverse transcription polymerase chain reaction (rRT-PCR) for ZIKV was positive. First- and second-trimester US, performed at 12 and 21 weeks of gestation, respectively, were normal.

Obstetric US at 37 weeks, performed in a Voluson E8 (GE Healthcare Ultrasound, Milwaukee, WI, USA), equipped with transabdominal (RAB 4–8D) and transvaginal two- and three-dimensional (3D) volumetric probes (RIC 5–9W), revealed microcephaly and diffuse brain calcifications (Fig. 1).

Fetal MRI confirmed the microcephaly and demonstrated a simplified brain gyral pattern, suggesting pachygyria/lissencephaly and asymmetric ventriculomegaly (Fig. 2).

Induction of a vaginal labor with vaginal misoprostol did not succeed. Then a Cesarean section was performed at 38 weeks. A male newborn was delivered, weighting 2210 g, with Apgar scores of 8 and 9 on the first and fifth minutes, respectively. Neonatal head circumference was 29 cm.

Postnatal transfontanellar ultrasound examination identified subcortical and periventricular calcifications with ventricular dilatation. Postnatal CT with 3D postprocessing performed at the 10th day of life demonstrated microcephaly, with brain calcifications, and cortical atrophy. 3D postprocessing showed more clearly the microcephaly, as well as a small anterior fontanel, and exuberant external occipital protuberance (Fig. 3). A 3D virtual physical model was obtained from CT scan data and printed onto thermoplastic acrylonitrile butadiene styrene (Stratasys U-Print, Stratasys, Ltd., Eden Prairie, MN, USA) (Fig. 4).

Postnatal brain MRI of the 1-month-old neonate demonstrated a cephalic circumference of 32.0 cm, brain atrophy, with a reduced and simplified brain gyration, especially in the frontoparietal lobes, a finding consistent with pachygyria (Fig. 5). Dysgenesis of the corpus callosum was also identified, with ventriculomegaly. In addition, axial T1-weighted imaging showed multiple brain subcortical frontoparietal hyperintense foci, suggested of calcifications.

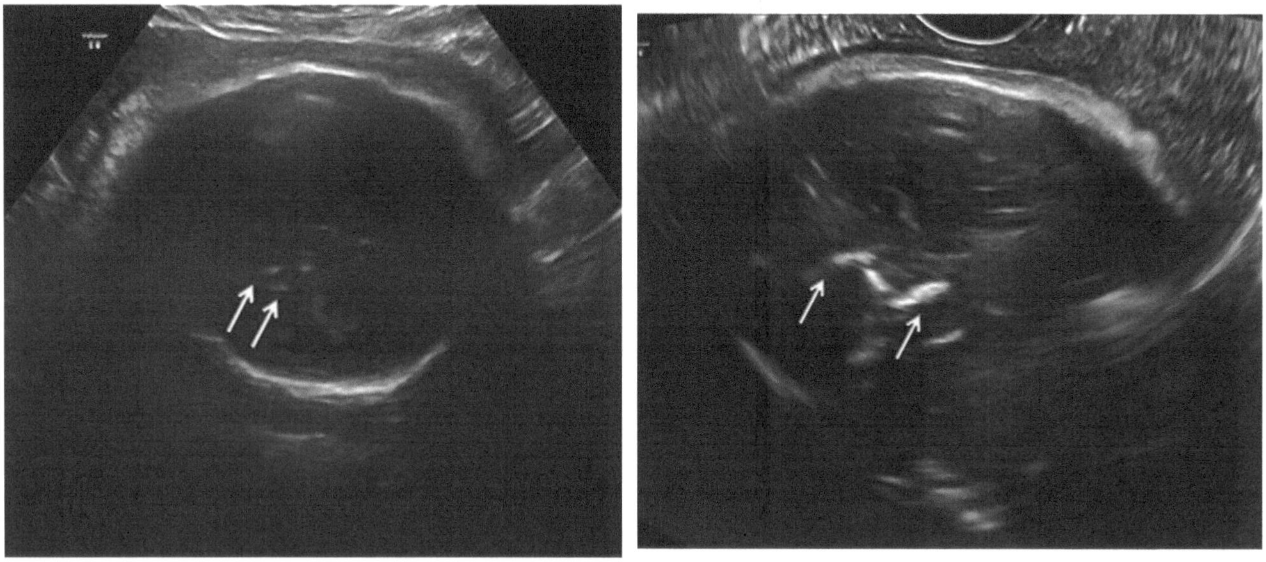

FIG. 1 Obstetric ultrasound of a fetus with congenital Zika syndrome. Transabdominal (A) and transvaginal (B) ultrasound of the 37-week fetus showed microcephaly and brain calcifications *(arrows)*. Note that ultrasound evaluation at a later gestational age limits brain assessment because of ossified skull artifacts. *(From Werner, H., Fazecas, T., Guedes, B., et al. (2016). Intrauterine Zika virus infection and microcephaly: Correlation of perinatal imaging and three-dimensional virtual physical models.* Ultrasound in Obstetrics & Gynecology, 47*, 657–60, with permission.)*

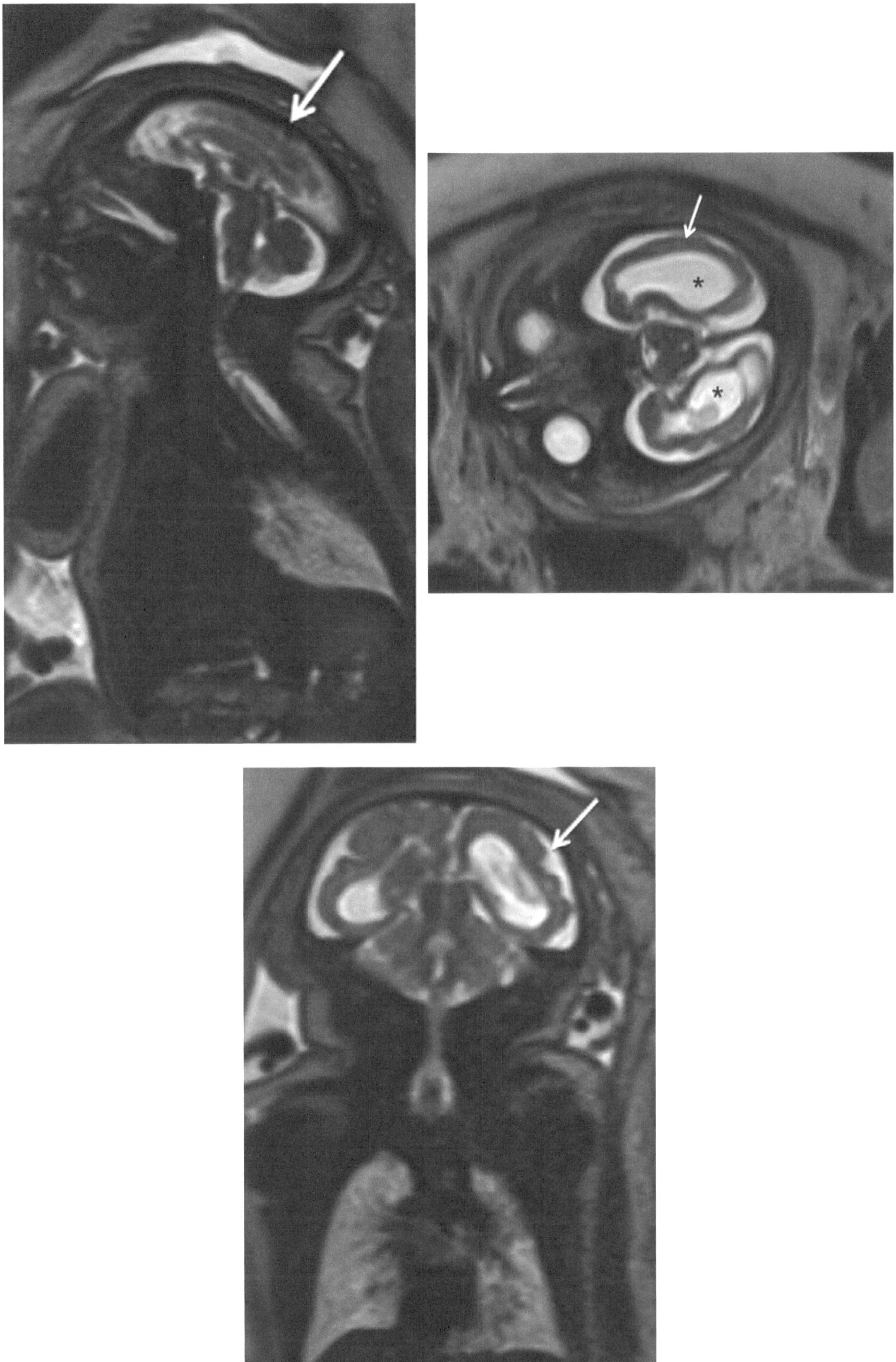

FIG. 2 Prenatal magnetic resonance imaging of a fetus with congenital Zika syndrome. T2-weighted MRI, at 37 weeks of gestation, in the sagittal (A), axial (B), and coronal (C) planes demonstrating microcephaly, ventriculomegaly (*asterisk* in B), and a simplified pattern of brain gyri *(arrows)*. *(From Werner, H., Fazecas, T., Guedes, B., et al. (2016). Intrauterine Zika virus infection and microcephaly: Correlation of perinatal imaging and three-dimensional virtual physical models.* Ultrasound in Obstetrics & Gynecology, 47, *657–60, with permission.)*

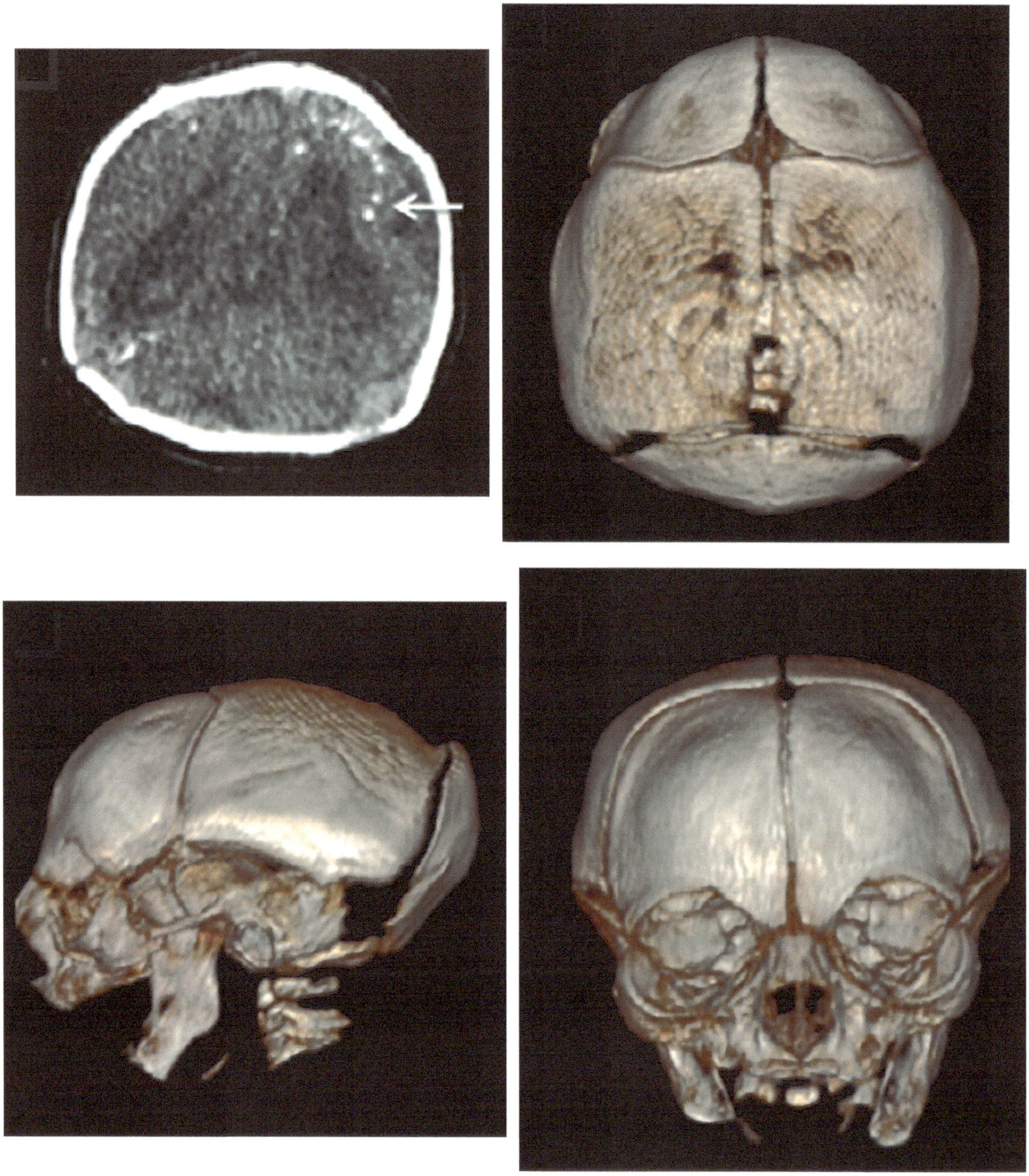

FIG. 3 Postnatal computed tomography of a 10-day-old boy with congenital Zika syndrome. Brain computed tomography in the axial plane (A) showing microcephaly, cortical atrophy, and calcifications in a predominantly subcortical location *(arrows)*. Three-dimensional reconstruction in axial (B), sagittal (C), and coronal (D) views, better demonstrating the microcephaly, and small anterior fontanel, as well as the exuberant external occipital protuberance (caused by collapsed bone and constrained intracranial contents). *(From Werner, H., Fazecas, T., Guedes, B., et al. (2016). Intrauterine Zika virus infection and microcephaly: Correlation of perinatal imaging and three-dimensional virtual physical models.* Ultrasound in Obstetrics & Gynecology, 47, *657–60, with permission.)*

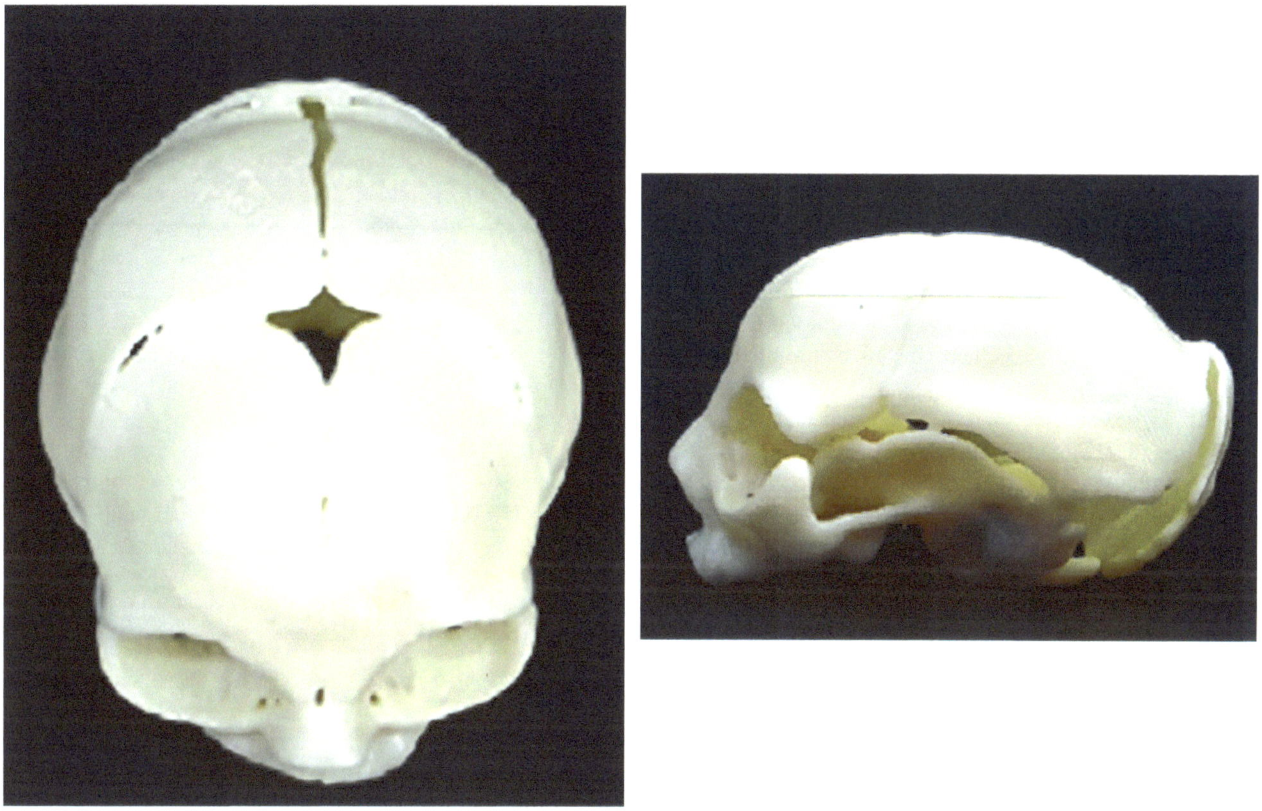

FIG. 4 Pictures of the three-dimensional physical model of the neonate's skull with microcephaly secondary to congenital Zika syndrome, constructed using data from computed tomography. Microcephaly, small anterior fontanel, and the exuberant external occipital protuberance are clearly reconstructed. *(From Werner, H., Fazecas, T., Guedes, B., et al. (2016). Intrauterine Zika virus infection and microcephaly: Correlation of perinatal imaging and three-dimensional virtual physical models.* Ultrasound in Obstetrics & Gynecology, 47, *657–60, with permission.)*

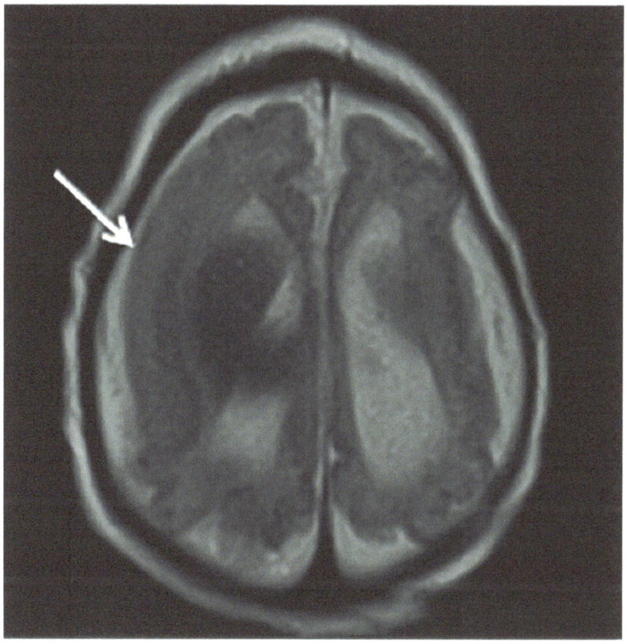

FIG. 5 Postnatal brain MRI of a 1-month-old boy with congenital Zika syndrome. Axial T2-weighted MRI showing a smooth and thickened cortical surface in the frontoparietal regions, compatible with pachygyria *(arrow)*. *(From Werner, H., Fazecas, T., Guedes, B., et al. (2016). Intrauterine Zika virus infection and microcephaly: Correlation of perinatal imaging and three-dimensional virtual physical models.* Ultrasound in Obstetrics & Gynecology, 47, *657–60, with permission.)*

Transmission and pathogenesis

Congenital transplacental transmission may happen at any time during pregnancy, as well as during the delivery. Maternal-fetal transmission can be confirmed by the detection of ZIKV ribonucleic acid (RNA) in the amniotic fluid from symptomatic ZIKV-infected pregnant females (Carod-Artal, 2018).

Although vertical transmission of ZIKV during pregnancy can affect multiple fetal organs, ZIKV has proven to have a particular neurotropism. However, ZIKV vertical transmission does not occur in all ZIKV-infected pregnant women, and symptomatic CZS will not be observed in all exposed fetuses. The exact rate of vertical transmission and congenital infection remains to be identified (Baud, Gubler, Schaub, Lanteri, & Musso, 2017). Besides microcephaly, a wide range of congenital abnormalities are observed in infants prenatally infected by ZIKV, such as decreased brain parenchymal volume, lissencephaly/pachygyria, ventriculomegaly, dysgenesis of the corpus callosum, cerebral calcifications, chorio-meningitis, arthrogryposis, and various ocular abnormalities. An association seems to exist between the time of ZIKV infection during the pregnancy and the severity of congenital malformations. The mother infection during the first or second trimester seems to have more serious consequences for the infant (Sejvar, 2018).

In a pregnant woman, maternal viremia might lead to fetal viremia. It is not well known how the virus is transmitted to developing nervous system following the establishment of fetal viremia. However, while in circulation, ZIKV infects the fetal monocytes, which can carry the infection to the developing CNS (Noorbakhsh, Kamal Abdolmohammadi, Fatahi, et al., 2019). Previous studies have shown that the virus can proliferate inside neural progenitor cells (NPC) in a developing brain (Mlakar, Korva, Tul, et al., 2016).

In the first trimester, the neural specificity of ZIKV may result in cortical damage due to direct invasion and injury of the NPC, leading to abnormal neural proliferation and/or migration. However, the ZIKV infection later in pregnancy is associated with brain abnormalities mainly due to ischemic insults to the corpus callosum, thalami, cerebellar hemispheres, and brainstem, leading to parenchymal atrophy and multiple parenchymal dystrophic calcifications.

ZIKV preferentially infects human NPC and triggers cell apoptosis (Tang, Hammack, Ogden, et al., 2016), damaging the developing brain and affecting the process of multiplication and migration of nervous system cells. As a consequence, apoptosis, modification in the pattern of myelin formation, and disrupted synaptogenesis occur in the CNS (Saad, Pennae Costa, de Góes, et al., 2018).

In the case presented in this chapter, the mother was infected at the end of the first trimester of pregnancy, which led to a severe fetal brain malformation, characterized by microcephaly, pachygyria, and dystrophic subcortical calcifications, which could be seen in prenatal US and MRI and confirmed in postnatal examinations.

Clinical manifestations of congenital Zika syndrome

Microcephaly, defined as head circumference 2 standard deviations (SD) below the mean for the gestational age or under the 2nd percentile for age and sex, is the most striking imaging finding in congenital ZIKV infection. Additionally, it can also be accompanied by other findings consistent with fetal brain disruption sequence (FBDS) (Moore, Staples, Dobyns, et al., 2017; Ribeiro et al., 2017).

FBDS is clinically and imaging characterized by a severe microcephaly, prominent occipital bones associated with redundant occipital scalp skin, overlapping cranial sutures, frontal bone depression, overlapping parietal bones, and extreme craniofacial disproportion (Corona-Rivera et al., 2001; Moore et al., 2017).

The FBDS phenotype is not pathognomonic of CZS. Although rare, it has been reported to be the result of ischemic disease and other viral infections. It is believed to be the result of brain volume loss and decrease in intracranial pressure (Corona-Rivera et al., 2001), secondary to extensive brain destruction, leading to the upper cranial bone collapse over the occiput, pushing it backward (Dain Gandelman Horovitz, da Silva Pone, Moura Pone, Dias Saad Salles, & Bastos Boechat, 2016).

Additionally, imaging methods can identify a spectrum of other brain abnormalities, including, in decreasing frequency, subcortical calcifications; increased fluid spaces (ventricular and extraaxial); abnormal gyral patterns, such as polymicrogyria and pachygyria; dysgenesis of the corpus callosum; decreased myelination; and cerebellar or cerebellar vermis hypoplasia (Cavalheiro, Lopez, & Serra, 2016).

Brain calcifications are likely dystrophic, related to cell death, either by necrosis, apoptosis, or both (Fujita, Yamamoto, & Ogino, 2014). Although brain calcifications secondary to congenital ZIKV infection are typically subcortical, calcifications have been also identified in the basal ganglia, periventricular white matter, and brainstem (Moore et al., 2017).

In general, these brain abnormalities can be detected prenatally with ultrasound or MRI. However, fetal MRI is more sensitive for detecting malformations of cortical development, dysgenesis of the corpus callosum, decreased myelination, and cerebellar hypoplasia.

Postnatally, transfontanellar US can be a challenge because of the small or closed fontanels (Ribeiro et al., 2017). Postnatal CT is the best method to detect brain calcifications, and MRI is better to detect congenital brain malformations.

Eye abnormalities, such as microphthalmia, coloboma, cataracts, and intraocular calcifications, are also reported in CZS. The pathogenesis of the posterior eye lesions might be related to direct cellular damage by ZIKV itself or be an inflammatory squeal. Arthrogryposis or congenital contractures have been reported in fetuses with CZS. It reflects the CNS damage. Neurogenic factors that impact the corticospinal tract, motor neurons, or their synapses can cause fetal motor abnormalities, leading to diminished fetal movements and contractures.

Our patient presented brain atrophy, subcortical brain calcifications, pachygyria, and dysgenesis of the corpus callosum. In addition to the conventional imaging methods, 3D printed model of the skull with microcephaly, constructed using data from CT, allows a better evaluation of the FBDS, such as small fontanels, premature closure of cranial sutures, and occipital protuberance. It can also improve the parents' understanding of the disease.

Conclusion

Different imaging methods play important and complementary roles in the diagnosis of CZS, both before and after birth, with particular advantages and disadvantages of each method. 3D printing is a new tool that can be used to better understand the clinical spectrum of the disease.

Policy and procedures

A baseline obstetric US should be performed in all cases suspected to be of congenital ZIKV infection, and then every 4–6 weeks, in endemic areas. If any abnormality is found, US evaluations should be performed every 2–3 weeks. Fetal brain MRI better detects brain abnormalities and may be performed when fetal US shows abnormalities suggestive of ZIKV infection.

Although microcephaly has been the hallmark of fetuses and newborns with CZS, not all affected fetuses demonstrated microcephaly. In addition, there are other signs of this syndrome, such as brain atrophy, malformations of cortical development, dysgenesis of the corpus callosum, hypoplasia of the cerebellum and cerebellar vermis, and eye abnormalities, as well as arthrogryposis and polyhydramnios.

In general, the brain abnormalities can be detected prenatally, with US or MRI. However, fetal MRI is more sensitive for detecting malformations of cortical development, dysgenesis of the corpus callosum, decreased myelination, and cerebellar hypoplasia. Postnatally, transfontanellar US can be difficult to perform, because of the small or closed fontanels. Postnatal brain CT is the best method to detect brain calcifications, and postnatal MRI is better to detect congenital brain malformations.

Although the FBDS phenotype, characterized by severe microcephaly, prominent occipital bones with redundant occipital scalp skin, overlapping cranial sutures, frontal bone depression, overlapping parietal bones, and extreme craniofacial disproportion, is typical of CZS, it is not pathognomonic. Although rare, it has been reported to be the result of ischemic disease and other viral infections.

Mini-dictionary of terms

Congenital Zika syndrome. Spectrum of clinical findings in the fetus and/or newborn, associated with maternal-fetal Zika virus transmission.

Microcephaly. Abnormally small head size. Head circumference 2 standard deviations below the mean for gestational age or under the 2nd percentile for age and sex.

Dysgenesis of the corpus callosum. Congenital malformation of the corpus callosum, which may be complete (agenesis) or partial (hypoplasia).

Arthrogryposis. Congenital articular contracture.

Three-dimensional printing. Constructing of a three-dimensional object from a digital file.

Key facts of imaging of congenital Zika syndrome

- Prenatal and postnatal ultrasound and magnetic resonance imaging and postnatal computed tomography have complementary roles in the diagnosis of congenital Zika syndrome.
- Although microcephaly is the hallmark of fetuses and newborns with congenital Zika syndrome, not all affected fetuses have microcephaly.

- In addition to microcephaly, other clinical signs of congenital Zika syndrome include brain atrophy, malformations of cortical development, dysgenesis of the corpus callosum, hypoplasia of the cerebellum and cerebellar vermis, and eye abnormalities.
- Arthrogryposis is also a feature of congenital Zika syndrome.
- Three-dimensional printing can be used to better understand the clinical spectrum of congenital Zika syndrome.

Summary points

- Since 2015, congenital Zika virus infection is linked to microcephaly.
- There is a wide range of clinical and imaging signs of congenital Zika syndrome, in addition to microcephaly.
- Transplacental transmission may happen at any time during pregnancy.
- There is an association between the time of Zika virus infection during the pregnancy and the severity of congenital malformations.
- Different imaging modalities, both prenatally and postnatally, are important to evaluate and follow up vertically Zika virus-infected patients.

References

Baud, D., Gubler, D. J., Schaub, B., Lanteri, M. C., & Musso, D. (2017). An update on Zika virus infection. *Lancet, 390*, 2099–2109.

Campos, G. S., Bandeira, A. C., & Sardi, S. I. (2015). Zika virus outbreak, Bahia, Brazil. *Emerging Infectious Diseases, 21*, 1885–1886.

Carod-Artal, F. J. (2018). Neurological complications of Zika virus infection. *Expert Review of Anti-Infective Therapy, 16*, 399–410.

Cavalheiro, S., Lopez, A., Serra, S., et al. (2016). Microcephaly and Zika virus: Neonatal neuroradiological aspects. *Child's Nervous System, 32*, 1057–1060.

Chibueze, E. C., Tirado, V., Lopes, K. D., et al. (2017). Zika virus infection in pregnancy: A systematic review of disease course and complications. *Reproductive Health, 14*, 28.

Corona-Rivera, J. R., Corona-Rivera, E., Romero-Velarde, E., Hernandez-Rocha, J., Bobadilla-Morales, L., & Corona-Rivera, A. (2001). Report and review of the fetal brain disruption sequence. *European Journal of Pediatrics, 160*, 664–667.

Dain Gandelman Horovitz, D., da Silva Pone, M. V., Moura Pone, S., Dias Saad Salles, T. R., & Bastos Boechat, M. C. (2016). Cranial bone collapse in microcephalic infants prenatally exposed to Zika virus infection. *Neurology, 87*, 118–119.

Fujita, H., Yamamoto, M., Ogino, T., et al. (2014). Necrotic and apoptotic cells serve as nuclei for calcification on osteoblastic differentiation of human mesenchymal stem cells in vitro. *Cell Biochemistry and Function, 32*, 77–86.

Liang, B., Guida, J. P., Costa Do Nascimento, M. L., & Mysorekar, I. U. (2019). Host and viral mechanisms of congenital Zika syndrome. *Virulence, 10*, 768–775.

Mlakar, J., Korva, M., Tul, N., et al. (2016). Zika virus associated with microcephaly. *The New England Journal of Medicine, 374*, 951–958.

Moore, C. A., Staples, J. E., Dobyns, W. B., et al. (2017). Congenital Zika syndrome: Characterizing the pattern of anomalies for pediatric healthcare providers. *JAMA Pediatrics, 171*, 288–295.

Noorbakhsh, F., Kamal Abdolmohammadi, K., Fatahi, Y., et al. (2019). Zika virus infection, basic and clinical aspects: A review article. *Iranian Journal of Public Health, 48*, 20–31.

Ribeiro, B. G., Werner, H., Lopes, F. P. P. L., et al. (2017). Central nervous system effects of intrauterine Zika virus infection: A pictorial review. *Radiographics, 37*, 1840–1850.

Saad, T., Pennae Costa, A. A., de Góes, F. V., et al. (2018). Neurological manifestations of congenital Zika virus infection. *Child's Nervous System, 34*, 73–78.

Sejvar, J. J. (2018). Zika virus and other emerging arboviral central nervous system infections. *Continuum (Minneap Minn), 24*, 1512–1534.

Song, B. H., Yun, S. I., Woolley, M., & Lee, Y. M. (2017). Zika virus: History, epidemiology, transmission, and clinical presentation. *Journal of Neuroimmunology, 308*, 50–64.

Tang, H., Hammack, C., Ogden, S. C., et al. (2016). Zika virus infects human cortical neural progenitors and attenuates their growth. *Cell Stem Cell, 18*, 587–590.

Werner, H., Fazecas, T., Guedes, B., et al. (2016). Intrauterine Zika virus infection and microcephaly: Correlation of perinatal imaging and three-dimensional virtual physical models. *Ultrasound in Obstetrics & Gynecology, 47*, 657–660.

Zorrilla, C. D., Rivera-Vinas, J. I., De La Vega-Pujols, A., et al. (2018). The Zika virus infection in pregnancy: Review and implications for research and care of women and infants in affected areas. *Puerto Rico Health Sciences Journal, 37*(Special issue), S66–S72.

Chapter 18

A case study of Guillain-Barre syndrome associated with Zika virus infection

José Luis Soto-Hernández[a], Karina Carrillo Loza[b], and Steven Vargas Cañas[c]
[a]Department of Infectious Diseases, National Institute of Neurology and Neurosurgery Manuel Velasco Suárez, Tlalpan, México City, Mexico, [b]Neurological Emergency Service, National Institute of Neurology and Neurosurgery Manuel Velasco Suarez, Tlalpan, México City, Mexico, [c]Nerve and Muscle Clinic, Department of Neurology, National Institute of Neurology and Neurosurgery Manuel Velasco Suárez, Tlalpan, Mexico City, Mexico

Abbreviations

AMAN	acute motor axonal neuropathy
AMP	amplitude
CSF	cerebrospinal fluid
DML	distal motor latencies
GBS	Guillain-Barré syndrome
ICU	intensive care unit
L	left
N	nerve
NCV	nerve conduction velocities
R	right
RT-PCR	reverse transcription polymerase chain reaction
ZKV	Zika virus

Setting

Guillain-Barré syndrome (GBS) is an autoimmune acute paralytic polyneuropathy with distinct clinical features and heterogeneous clinical presentation. In its severe forms, GBS endangers life due to respiratory failure that occurs in about one-third of the affected patients (Willison et al., 2016). A GBS diagnosis requires clinical experience in neurology. It is necessary to perform other diagnostic procedures such as a lumbar puncture to obtain cerebrospinal fluid (CSF) for chemical tests, cell count and other tests, and electrophysiological studies that complement the diagnosis with a history and neurological examination. This allows medical personnel to distinguish GBS from other diseases. Patients with GBS who develop respiratory failure require intensive care admission and multidisciplinary management. The available treatments with proven efficacy by controlled studies are the intravenous administration of high-dose immunoglobulin or plasmapheresis (Hughes et al., 2014). The prognosis of the disease is variable, with about 80% of patients recovering the ability to walk at 6 months after disease onset (Doets et al., 2018). Since 2015–16, an increase in the number of cases of Zika virus (ZIKV) infection was observed first in Brazil, followed rapidly by reports from other South American, Caribbean, and Central American countries. It reached North America by the end of 2016, along with a parallel increase in cases of GBS (Nascimento and da Silva, 2017). In México, a study performed in 2016 in the state of Veracruz on 34 hospitalized patients with GBS found in two patients a positive reverse transcription polymerase chain reaction (RT-PCR) to detect ZIKV in urine (Del Carpio-Orantes et al., 2018). We reported a prospective case series of hospitalized patients with GBS from Mexico in whom the serum, CSF, urine, and saliva were tested by RT-PCR for ZIKV, dengue virus (DENV), and chikungunya virus in patients referred from states with transmissions of ZIKV infection. Clinical, electrophysiological tests, and the long-term outcome were registered. All five patients had areflexic flaccid weakness and the cranial nerves were affected in three of them. The electrophysiological patterns were axonal in three patients and demyelinating in two; the outcome was favorable. ZIKV RT-PCR was positive in all five patients on at least one sample, and the DENV and chikungunya virus RT-PCR tests were negative (Soto-Hernández et al., 2019).

Zika Virus Impact, Diagnosis, Control, and Models. https://doi.org/10.1016/B978-0-12-820267-8.00018-2

Clinical case study of Guillain-Barre syndrome associated with Zika virus infection

Onset of disease

A 32-year-old male who was born and resided in Guadalajara Jalisco, México, attended a social event outside the city weeks before the onset of his neurological symptoms. It was at an open place near a lake. In the afternoon and at dusk, numerous mosquitoes could be seen, and the patient stated that "They could enter your mouth when you talked." He received numerous mosquito bites. One week after the event, he presented unquantified fever, generalized myalgias, moderate-to-severe generalized headaches, and retro-ocular pain. Three days later, nausea and vomiting appeared, with poor tolerance to the oral route. He denied having presented diarrhea, conjunctivitis, rash, or arthritis. At that moment, he was assessed by an internist who considered an acute gastrointestinal virus infection as a diagnostic possibility. No laboratory studies were performed and only oral hydration and general measures were indicated; all his symptoms were remitted within 48 h. The patient is married and has a university degree. He plays basketball regularly, does not smoke, and only drinks small amounts of alcohol at occasional social events without getting intoxicated. He does not use illicit drugs nor does he have risk factors for sexually transmitted diseases. In his past medical history, he suffered bronchial asthma in childhood. His sister-in-law (neighbor) was diagnosed with DENV fever 1 week prior to the onset of his symptoms. Two weeks after having suffered the mosquito bites, he began to have neurological symptoms that consisted of distal tingling in the feet and hands. The next day, he presented distal weakness of the pelvic limbs with an ascending pattern up to the knees in 24 h. That was accompanied by difficulty in walking, so he decided to go for an evaluation at a third-level hospital.

Hospital course

He was hospitalized and upon admission his vital signs were blood pressure 120/70 mmHg, pulse 86 beats/minute, respiration 20 breaths per minute, and temperature 36.9°C. Upon physical examination, he looked moderately ill, anxious, and worried. Upon mental examination, he centered, directed, and maintained attention and he was also oriented in person, place, time, and circumstance. Both semantic and nondeclarative episodic memory were preserved. His language was fluent and coherent. He was able to nominate, repeat, read, and write normally. Judgment, calculation, and abstraction were preserved. Upon neurological examination after admission, he was found with normal cranial nerves, and symmetrical distal motor weakness, more prominent in lower limbs, with hyporeflexia, and hypotonia. The plantar response was indifferent on both sides. The sensibility to a soft touch and a pin prick in the hands and feet was reduced. The next day after admission, he developed bilateral facial palsy. A lumbar puncture was performed, and the CSF was reported with 4 cells/mm^3, a total protein concentration of 68 mg/dL, and a glucose concentration of 66 mg/dL with a simultaneous blood glucose of 92 mg/dL, meeting the criteria of albuminocytologic dissociation (increase in the concentration of total protein with a normal cell count in the CSF). Treatment with intravenous immunoglobulin at a dose of 0.4 g/kg body weight for 5 days was started. The bacterial, mycobacterial, and fungal cultures of CSF remained negative, as did the Gram stain, India ink preparation, and acid-fast bacilli stains of CSF. A cranial magnetic resonance image was normal, and the nerve conduction study performed on the second day in the hospital was compatible with acute axonal motor polyneuropathy (Table 1).

Transfer to intensive care unit

On the fourth day in the hospital, the patient presented difficulty in swallowing solid foods. He was immediately moved to the intensive care unit (ICU), where he required endotracheal intubation due to acute ventilatory failure. On the fifth hospital day, a reverse transcription polymerase chain reaction (RT-PCR) test in the blood was reported positive for ZIKV; the IgM tests for DENV and ZIKV were reported not reactive (Fig. 1). Serum tests for human immunodeficiency virus infection, hepatitis B, hepatitis C, and cytomegalovirus were also negative. During his stay in the ICU, he presented diarrhea the first 5 days; stool cultures were negative for salmonella, shigella, and campylobacter while no ova or parasites were observed. Several attempts at early extubation were unsuccessful because he developed pulmonary atelectasis, the treatment of which required alternating pronation and alveolar recruitment maneuvers. On the 12th day of the hospital stay, it was decided to perform a tracheostomy. The subsequent evolution was favorable. He initiated rehabilitation sessions in which a progressive increase in the proximal muscular strength and gradually of the inspiratory force was noted. This allowed weaning him off the mechanical ventilator and made it possible to discharge him from the ICU after 20 days of hospitalization. He was discharged from the hospital the next day with a Hughes Guillain-Barré syndrome disability scale score of 4 and a Modified Ranking Scale (mRS) score of 4 (Fig. 1).

TABLE 1 Electrophysiological studies obtained in the patient on his second day of hospitalization.

	DML		AMP		NCV		F wave latency	
Motor nerve	R	L	R	L	R	L	R	L
N. Median	3.7	2.9	1.4	3.2	53	61	NR	42
	8.0	6,4	0.6	1.4				
N. Ulnar	3.0	2.8	1.7	2.5	88	75	52	46
	5.4	5.6	1.1	1.8				
N. Tibial	4.5	5.1	2.1	2.3	43	48	68	NR
	16.3	18.0	1.3	1.3				
N. Peroneal	5.5	5.6	2.1	1.5	40	34	NR	NR
	13.7	14.2	1.3	1.1				

All sensitives potentials were normal

Low distal compound muscle action potentials with preserved sensory nerve action potentials in the same nerve segments strongly suggestive of acute motor axonal neuropathy (AMAN) (Hadden et al., 1988).
DML, distal motor latencies; *AMP*, amplitude; *NCV*, nerve conduction velocities; *N*, nerve; *R*, right; *L*, left.

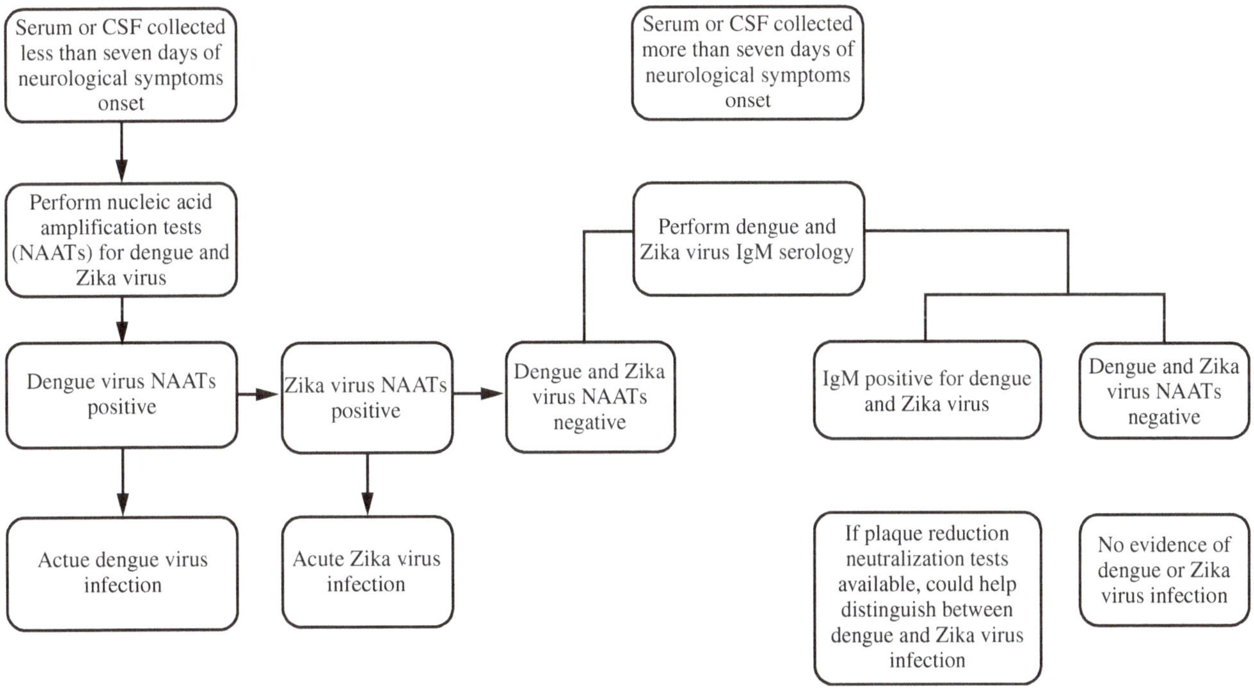

FIG. 1 Testing for dengue and ZIKV infection in patients with Guillain-Barré syndrome living in endemic areas for both viruses. Some centers perform nucleic-acid amplification tests (NAATs) in urine and saliva where the viral RNA may appear for several days after it disappears from serum. Plaque reduction neutralization tests require highly specialized reference laboratories. *(Modified from CDC. https://www.cdc.gov/mmwr/volumes/68/rr/rr6801a1.htm?s_cid=rr6801a1_w#F1_down.)*

Evolution after hospital discharge

The patient was followed in the outpatient clinic, and the tracheostomy cannula was removed 2 weeks after discharge. The patient received early rehabilitation in the ICU, and this continued after discharge three times a week. Twenty-one days after hospital admission, he stood, and 12 days later he was able to walk with a walker. One week later he managed to throw a basketball. Fifty days after hospital admission he walked without help and with normal speed, but was unable to run. In a new neurological exam 68 days after hospitalization revealed that his mental state was preserved, he had normal eye movements, bilateral facial diparesis persisted (right: House Brackmann IV, Left: House Brackmann V) (House and Brackmann, 1985), rest without abnormalities. A generalized muscle strength of 5/5, a conserved tone, generalized hyporeflexia, a bilateral flexor plantar response, and all sensitive modalities were preserved. The cerebellum and sphincters were without alteration. He had a gait with a wide lift base and could walk without help, but was still unable to run. His Hughes Guillain-Barré syndrome disability scale score was 2, and his mRS score was also 2 (Fig. 1).

Policy and procedures

Diagnosis of ZIKV infection in patients with GBS

ZIKV and DENV share antigenic characteristics and occur in the same geographical areas, so the clinical manifestation of the two diseases overlaps. The definitive diagnosis relies on nucleic acid amplification tests (NAATs) and serologic testing (Fig. 2). NAATs should be performed on whole blood or serum samples obtained during the first 7 days of illness. CSF specimens, urine, and saliva in some centers may also be utilized for NAATs. A positive nucleic acid test shows the presence of ZIKV RNA. Serodiagnosis is complicated by false positive results owing to cross-reactivity in persons previously exposed to other flaviviruses. Screening by the detection of ZIKV IgM is performed on serum samples obtained 2–12 weeks after the onset of illness. Confirmation of inconclusive and positive IgM results requires further testing with a plaque reduction neutralization test (PRNT), which can be performed only at highly specialized reference laboratories (Musso et al., 2019).

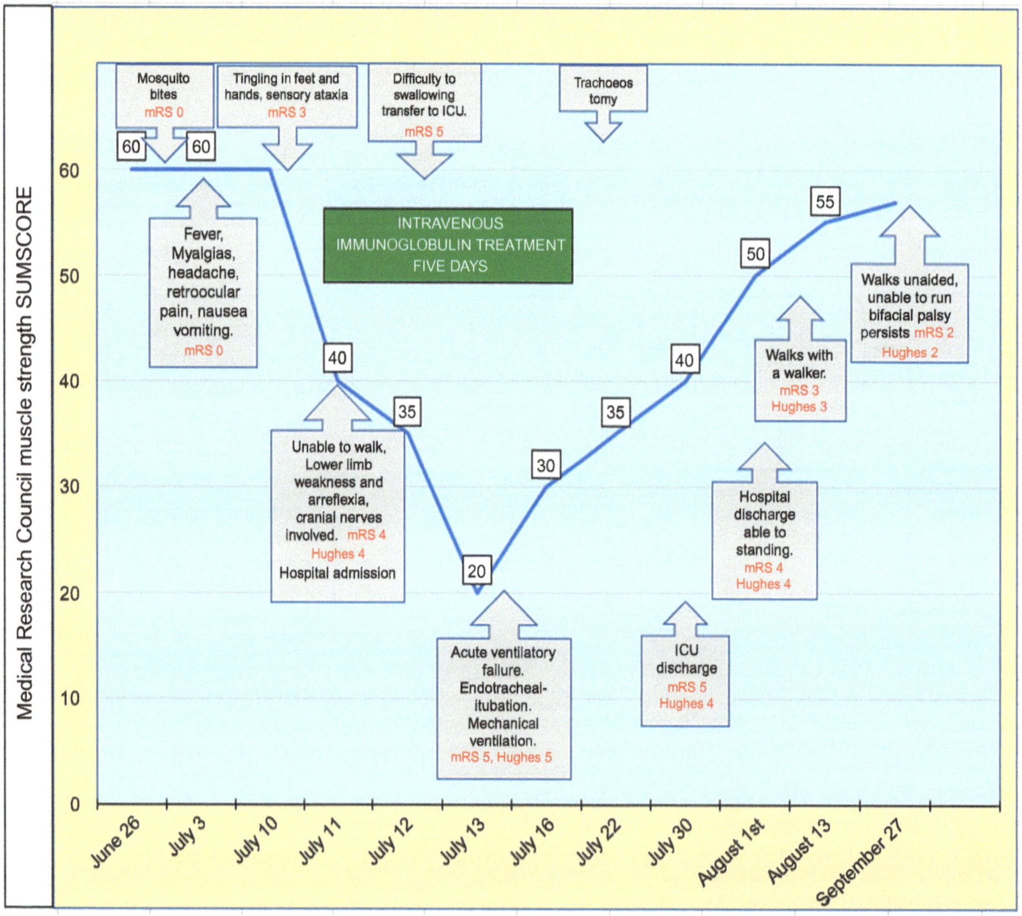

FIG. 2 Clinical course of the patient with GBS associated with ZIKV infection. Summary of the clinical evolution of the patient during hospital stay and after hospital discharge up to day 68 after the onset of neurological symptoms.

Mini-dictionary of terms

Features required for the diagnosis of Guillaín-Barré syndrome. This set of clinical, cerebrospinal fluid, and electrodiagnostic features was recently updated (Leonhard et al., 2019) (Table 2).

Guillain-Barré syndrome disability scale. This scale makes it easy to establish the severity of functional disability caused by Guillaín-Barre syndrome (Hughes et al., 1978) (Table 3).

TABLE 2 Key facts about Guillaín-Barre syndrome and Zika virus infection.

Features required for diagnosis of Guillaín-Barré syndrome (Leonhard et al., 2019)

- Progressive bilateral weakness of arms and legs (initially only legs may be involved)
- Absent or decreased tendon reflexes in affected limbs (at some point in clinical course)

Features that strongly support diagnosis

- Progressive phase lasts from days to 4 weeks (usually <2 weeks)
- Relative symmetry of symptoms and signs
- Relatively mild sensory symptoms and signs (absent in pure motor variant)
- Cranial nerve involvement, especially bilateral facial palsy
- Autonomic dysfunction
- Muscular or radicular back or limb pain
- Increased protein level in cerebrospinal fluid (CSF); normal protein levels do not rule out the diagnosis
- Electrodiagnostic features of motor or sensorimotor neuropathy (normal electrophysiology in the early stages does not rule out the diagnosis)

This set of clinical, CSF, and electrophysiological studies was recently updated by a consensus of experts (Leonhard et al., 2019).

TABLE 3 Disability scale and functional ranking scale commonly used to describe the severity of Guillain-Barre syndrome.

Guillain-Barré syndrome disability scale, adapted from Hughes et al. (1978)

0 A healthy state

1 Minor symptoms, and capable of running

2 Able to walk 10 m or more without assistance but unable to run

3 Able to walk 10 m across an open space with help

4 Bedridden or chair-bound

5 Requiring assisted ventilation for at least part of the day

6 Dead

The Modified Rankin Scale (van Swieten et al., 1988)

Grade description

0 No symptoms at all

1 No significant disability despite symptoms: able to carry out all usual duties and activities

2 Slight disability: unable to carry out all previous activities but able to look after own affairs without assistance

3 Moderate disability: requiring some help, but able to walk without assistance

4 Moderately severe disability: unable to walk without assistance, and unable to attend to own bodily needs without assistance

5 Severe disability: bedridden, incontinent, and requiring constant nursing care and attention

The use of these clinical scales from the initial evaluation of the patient makes it possible to evaluate the progression or recovery from GBS and other neurological diseases.

Modified Ranking Scale (mRS). This scale is widely used to assess handicaps in stroke patients but it is also used as an indicator of functional capacity in other neurological conditions (van Swieten et al., 1988) (Table 3).

Medical Research Council (MRC) muscle scale. The Medical Research Council muscle scale grades muscle power on a scale of 0–5 in relation to the maximum expected for that muscle. https://mrc.ukri.org/research/facilities-and-resources-for-researchers/mrc-scales/mrc-muscle-scale/.

Medical Research Council (MRC) sumscore. This score is a summation of the strength of six muscle groups tested on both sides: Abduction of the arm, flexion of the forearm, extension of the wrist, flexion of the leg, extension of the knee, and dorsal flexion of the foot, according to the Medical Research Council (MRC) scale, ranging from 0 (paralysis) to 60 (normal strength) (Kleyweg et al., 1991).

House-Brackmann classification of facial function. This grading system is used to characterize the degree of facial paralysis. Grade I is assigned to normal function, and grade VI indicates complete paralysis. Intermediate grades vary with function observed at rest and with effort (House & Brackmann 1985).

Summary points

- This chapter presents a case study of Guillaín-Barre syndrome associated with active Zika virus infection.
- Our patient lives in an area that had active circulation of the dengue and Zika viruses throughout 2019, and he was exposed to multiple mosquito bites.
- His clinical picture is similar to other cases reported in Mexico: Dysesthesias, ascending limb weakness, cranial nerves involved, and dysphagia and ventilatory failure, requiring intensive care unit admission.
- Clinical progression, as often happens, was observed despite the onset of early treatment with immunoglobulin.
- We found a positive Zika virus reverse transcription polymerase chain reaction in serum early in hospitalization.
- The clinical evolution, albuminocytologic dissociation in cerebrospinal fluid, acute axonal motor polyneuropathy on electrophysiological testing, and a positive serum reverse transcription polymerase chain reaction test for Zika virus conferred strong support for the diagnosis of Zika virus-associated Guillaín-Barre Syndrome.
- Starting rehabilitation as early as possible during hospitalization is essential for functional recovery.

References

Del Carpio-Orantes, L., Peniche Moguel, K. G., Sánchez Díaz, J. S., Pola-Ramirez, M. D. R., Mata Miranda, M. D. P., García-Méndez, S., ... González-Flores, E. E. (2018). Guillain Barré syndrome associated with Zika virus infection: Analysis of a cohort from the region of northern Veracruz in 2016–2017. *Neurologia*. Jul 30. pii: S0213-4853(18)30173-7.

Doets, A. Y., Verboon, C., van den Berg, B., et al. (2018). Regional variation of Guillain-Barré syndrome. *Brain*, *141*(10), 2866–2877. https://doi.org/10.1093/brain/awy232.

Hadden, R. D., Cornblath, D. R., Hughes, R. A., Zielasek, J., Hartung, H. P., Toyka, K. V., & Swan, A. V. (1988). Electrophysiological classification of Guillain-Barré syndrome: clinical associations and outcome. Plasma Exchange/Sandoglobulin Guillain-Barré Syndrome Trial Group. *Annals of Neurology*, *44*(5), 780–788.

House, J. W., & Brackmann, D. E. (1985). Facial nerve grading system. *Otolaryngology-Head and Neck Surgery*, *93*, 146–147. https://doi.org/10.1177/019459988509300202.

Hughes, R. A., Newsom-Davis, J. M., Perkin, G. D., & Pierce, J. M. (1978). Controlled trial prednisolone in acute polyneuropathy. *Lancet*, *2*(8093), 750–753. https://doi.org/10.1016/s0140-6736(78)92644-2.

Hughes, R. A., Swan, A. V., & van Doorn, P. A. (2014). Intravenous immunoglobulin for Guillain-Barré syndrome. *Cochrane Database of Systematic Reviews*, (9). https://doi.org/10.1002/14651858.CD002063. CD002063.

Kleyweg, R. P., van der Meché, F. G., & Schmitz, P. I. (1991). Interobserver agreement in the assessment of muscle strength and functional abilities in Guillain-Barré syndrome. *Muscle & Nerve*, *14*(11), 1103–1109.

Leonhard, S. E., Mandarakas, M. R., Gondim, F. A. A., Bateman, K., Ferreira, M. L. B., Cornblath, D. R., ... Jacobs, B. C. (2019). Diagnosis and management of Guillain-Barré syndrome in ten steps. *Nature Reviews Neurology*, *15*(11), 671–683. https://doi.org/10.1038/s41582-019-0250-9. Epub 2019.

Musso, D., Ko, A. I., & Baud, D. (2019). Zika virus infection – after the pandemic. *The New England Journal of Medicine*, *381*(15), 1444–1457. https://doi.org/10.1056/NEJMra1808246.

Nascimento, O. J. M., & da Silva, I. R. F. (2017). Guillain-Barré syndrome and Zika virus outbreaks. *Current Opinion in Neurology*, *30*(5), 500–507. https://doi.org/10.1097/WCO.0000000000000471.

Soto-Hernández, J. L., Ponce de León Rosales, S., Vargas Cañas, E. S., Cárdenas, G., Carrillo Loza, K., Díaz-Quiñonez, J. A., … Kuri Morales, P. (2019). Guillain-Barré syndrome associated with Zika virus infection: A prospective case series from Mexico. *Frontiers in Neurology*, *10*, 435. https://doi.org/10.3389/fneur.2019.00435.

van Swieten, J. C., Koudstaal, P. J., Visser, M. C., Schouten, H. J., & van Gijn, J. (1988). Interobserver agreement for the assessment of handicap in stroke patients. *Stroke*, *19*(5), 604–607.

Willison, H. J., Jacobs, B. C., & van Doorn, P. A. (2016). Guillain-Barré syndrome. *The Lancet*, *388*, 717–727.

Chapter 19

Clinical manifestations and outcomes of Guillain-Barre syndrome complicating Zika virus infection

Ayman Alboudi[a] and Andrew Jameson[b]

[a]*Department of Neurology, Hauenstein Neuroscience Center, Mercy Health Saint Mary's Hospital, Grand Rapids, MI, United States,* [b]*Michigan State University—College of Human Medicine, Grand Rapids, MI, United States*

Abbreviations

AMAN	acute motor axonal neuropathy
AIDP	acute inflammatory demyelinating polyneuropathy
CSF	cerebral spinal fluid
DNKV	dengue fever virus
ELISA	enzyme-linked immunosorbent assay
GBS	Guillain-Barré syndrome
IFIT-1	interferon-induced protein with tetratricopeptide repeats 1
IFN-λ	interferon gamma
IL	interleukin
IVIG	intravenous immunoglobulin
HLA	human leukocyte antigen
RNA	ribonucleic acid
SC	Schwann cells
TNFα	tumor necrosis factor α
ZIKV	Zika virus

Introduction

Guillain-Barré syndrome (GBS) is an autoimmune acute neuropathy with a classic presentation of symmetric, ascending paralysis. Involvement of cranial nerves and respiratory muscles may happen, and many patients may have sensory symptoms including pain, numbness, or tingling. Currently, GBS is considered the most common cause of acute flaccid paralysis. There is a known relationship with preceding infections, the most common pathogens being upper respiratory infections and gastrointestinal infections. Molecular mimicry is the most widely accepted theory through which these organisms can precipitate the inflammatory reaction and initiate the autoimmune response (Doorn, Ruts, & Jacobs, 2008).

Zika virus (ZIKV) is an arbovirus of the *Flavivirus* genus of the *Flaviviridae* family, which include dengue fever virus (DENV), yellow fever virus, Japanese encephalitis virus, West Nile virus, and Saint Louis encephalitis virus. Many of these viruses have neurological complications, either through precipitating autoimmune response (Oliveira et al., 2014) or direct invasion of the nervous system (Sejvar, 2016).

The usual interval between symptomatic infection and the onset of neurological signs in GBS is 2–8 weeks. Prognosis is different depending on the precipitating infection with the worst outcomes following *Campylobacter jejuni*. Also affecting prognosis is the time between infection and the onset of GBS. Prolonged recovery and worse residual deficits are common in cases where that interval is short. Other factors known to be associated with poor outcomes include the rapid progression of the symptoms and reaching the nadir over 1 week.

Zika Virus Impact, Diagnosis, Control, and Models. https://doi.org/10.1016/B978-0-12-820267-8.00019-4

An increase in the incidence of GBS was established during the Zika virus pandemic (Cao-Lormeau et al., 2016; Parra et al., 2016; Uncini, Shahrizaila, & Kuwabara, 2017). Other autoimmune neurological complications were also documented including transverse myelitis and ADEM (da Silva, Frontera, Bispo de Filippis, & Nascimento, 2017).

Clinical description

A 42-year-old woman presented to the ER with complaints of numbness and paresthesia in her hands and legs for 1 day prior to presentation. The patient was assessed in the emergency department and diagnosed with a panic attack. She was sent home with reassurance and symptom management. The next day, the patient noted weakness in her feet and worsening in her gait. Upon reassessment in the same emergency department, she was found to have absent reflexes, weakness in both lower limbs, no sensory loss, and mute plantar reflexes.

There was no history of recent trauma or drug abuse. No history of skin rash or tick bites were noted by the patient. She describes no recent gastrointestinal complaints. She never eats undercooked meats or unpasteurized milk products.

The patient does describe flu-like symptoms including fever up to 103°F, malaise, and a diffuse macular skin rash about 1 week ago. Symptoms started when she was on a business trip to Colombia. She works for a large coffee roaster and travels to Colombia multiple times a year to keep the supply chain intact. She describes her entire party encountering mosquitoes as she traveled into the region surrounding Cucuta. She describes this most recent illness to be less severe than a case of DENV she contracted in 2010 during a similar business trip.

The patient was quickly diagnosed with Guillain-Barré syndrome.

Laboratory test

A lumbar puncture was performed revealing a protein of 144, glucose of 65, 5 white blood cells, and no red blood cells. Nerve conduction studies on day 2 of symptom onset revealed a normal study.

Campylobacter serology and stool antigen were negative. Lyme disease IgM was negative in both the serum and the CSF by ELISA followed by no bands identified by Western blot assay. The patient tested positive for both ZIKV and DENV by NS1 antibody testing, but the neutralization titer (NT_{50}) was eightfold higher against ZIKV H/PF/2013 than DENV2 16,681 (212,788 vs 33,485). This assay combined with a positive ZIKV PCR from the urine demonstrated current ZIKV infection with previous DENV2 exposure.

Treatment and recovery

Intravenous immunoglobulin was started with a dose of 0.5 g/kg daily for a planned 5-day course of therapy. On the second day the patient developed left facial weakness of peripheral type with mild difficulty swallowing. She had progressive weakness distally along with weakness of neck flexion throughout the day. Overnight the patient had new-onset dyspnea and respiratory failure, resulting in endotracheal intubation for airway protection.

Unfortunately the patient's response to IVIG was not favorable. On day 14 of following completion of IVIG, she was still quadriparetic and requiring ventilator support. She was able to move her fingers minimally. Plasma exchange was initiated as salvage therapy. The patient had 5 sessions of 30 mL/kg volume exchange. Over the next week the patient started to improve slowly, with spontaneous breathing and improvement in her limb power. Tracheostomy was removed on day 30, and the patient was able to do some remarkable movements. With continuous rehabilitation and physiotherapy, she was discharged to an acute rehabilitation center where she stayed for 3 weeks. She ultimately returned home with a Hughes score of 1.

Discussion

Many reports showed an increase in the incidence of GBS during the Zika virus outbreaks. The electrophysiological studies showed acute motor axonal neuropathy (AMAN) to be the predominant presenting subtype. GBS after Zika virus infection characteristically is rapid progression to nadir within 6 days and being acute motor axonal neuropathy. These are shown to be poor prognostic indicators (Cornblath et al., 1988; Sharshar, Chevret, Bourdain, & Raphael, 2003). In addition, having GBS after Zika virus infection has been shown to be unusually severe (Cao-Lormeau et al., 2016; Lynch et al., 2019).

Patients with AMAN and rapid progression tend to have poor response to treatment (Cornblath et al., 1988; Walgaard et al., 2011). Many of these patients end up with quadriplegia and requiring ventilator support. The hospital course is prolonged with higher Hughes score. Studies have shown both IVIG and plasma exchange are effective in treating GBS (Ye, Li, & Li, 2015). However, other studies demonstrated a better response and prognosis using a second treatment course of either IVIG or plasma exchange (Alboudi et al., 2019). The electrophysiological characteristics of GBS patients post Zika virus infection were different from one study to another. The majority of the patients were AMAN subtype in French Polynesia (Arias et al., 2017; Cao-Lormeau et al., 2016), while it was acute inflammatory demyelinating subtype (AIDP) in other literature (Dirlikov et al., 2016; Parra et al., 2016). The reason behind this difference is unclear. It was suggested that the genetic background in different affected populations may represent different antigens leading to different reacting antibodies on the molecular mimicry theory (Magira et al., 2003; Monos, Papaioakim, Ho, Li, & McKhann, 1997). It has been shown that either AMAN or AIDP predominates depending on the HLA type encountered in a given patient.

The exact pathogenesis of GBS after Zika virus infection is not yet fully understood. One study showed Zika virus can infect Schwann cells (SC) directly andincrease SC expression of IL-6, interferon (IFN) β1, IFN-λ, IFIT-1, TNFα, and IL-23A mRNAs as well as IFN-λ receptors (Dhiman, Abraham, & Griffin, 2019); this does not fully explain the delay between the infection symptoms and the neurological symptoms. In addition, the symptoms compatible with Zika virus infection were reported at a median of 6days before the onset of the neurological symptoms, at which point most ZIKV viremia is significantly waning (Cao-Lormeau et al., 2016). The molecular mimicry model will continue to be the most likely explanation of autoimmune disease, including in GBS. The antiglycolipid IgG antibodies, directed to ganglioside, have been described particularly in the axonal type of GBS (Rinaldi & Willison, 2008; Willison, 2007). Interestingly, this autoantibody was detected in less than 50% of post-Zika virus GBS in one study (Cao-Lormeau et al., 2016). This either indicates the presence of autoantibodies not identified or adds additional weight to the direct invasion argument.

Finally, it is very interesting that not all flaviviruses have a propensity to cause GBS. For instance, DENV is not routinely associated with neurological complications. However, there may be an immune priming that occurs as one large study demonstrated a dramatic association between GBS associated with ZIKV infection in patients that were previously exposed to DENV (Cao-Lormeau et al., 2016; Lynch et al., 2019). Whether this is caused by a higher ZIKV viral load or represents an elevated level of immune activation is still not clear.

Policy and procedures: Diagnostic criteria of GBS

GBS is diagnosed usually by the diagnostic criteria proposed in 1978 from the National Institute of Neurological Disorders and Stroke (NINDS). These criteria were based on expert consensus and have been modified over time to reflect advances in the understanding of GBS. However, the World Health Organization recommends the use of the Brighton criteria for the case definition of GBS in regions affected by Zika virus transmission, which can be found on this link: http://apps.who.int/iris/bitstream/10665/204474/1/WHO_ZIKV_MOC_16.4_eng.pdf?ua=1 (Accessed on October 03, 2016).

Brighton criteria has three levels of certainty, which are different from each other by the presence or absence of cytoalbuminologic dissociation and electrophysiologic findings consistent with GBS (Sejvar et al., 2011).

Mini-dictionary of terms

Flaviviridae. All belong to a single genus, *Flavivirus*, and represent a significant burden of disease to humans. They are positive, single-stranded, enveloped RNA viruses. They are primarily spread through arthropod vectors.

IVIG. It is a mixture of antibodies taken from the plasma of thousands blood donors and used to treat many autoimmune diseases.

Plasma exchange. It is removal of the plasma from the patient blood and replace it with treated plasma.

Hughes scale. It is a scale used to assess disability in Guillain-Barré syndrome. It ranges from 0 as normal to 6 as death.

Cytoalbuminologic dissociation. It is the elevation of the CSF protein with no associated increase in the WBC.

Key facts of Guillain-Barré syndrome

- Overall incidence of 1–2 cases per 100,000 per year worldwide.
- The initial diagnosis of GBS is based upon the clinical presentation.
- Two-thirds of patients report antecedent acute infectious illness.
- Ventilatory support may result from severe respiratory muscles weakness in 10%–30%.
- Mortality form GBS is 3%–5% even in more developed countries.

Summary points

- GBS after ZIKA virus infection is associated with many poor prognostic factors. These include rapid progression to nadir and being axonal-type neuropathy.
- Poor clinical response to initial therapy is not uncommon. Utilizing plasma exchange or IVIG as a secondary treatment modality may result in a positive treatment response.
- The etiology by which ZIKA virus infection is related to GBS is still not clear. Both direct invasion by the virus and the autoimmune mechanism are feasible etiologies.
- Previous infection by dengue virus may act as a primary for a more severe presentation of Zika virus infection, specifically an increased risk of GBS.

References

Alboudi, A. M., Sarathchandran, P., Geblawi, S. S., Kayed, D. M., Inshasi, J., Purayil, S. P., … Katirji, B. (2019). Rescue treatment in patients with poorly responsive Guillain-Barre syndrome. *SAGE Open Medicine*, *7*. https://doi.org/10.1177/2050312119840195, 2050312119840195.

Arias, A., Torres-Tobar, L., Hernandez, G., Paipilla, D., Palacios, E., Torres, Y., … Castellanos, G. (2017). Guillain-Barre syndrome in patients with a recent history of Zika in Cucuta, Colombia: A descriptive case series of 19 patients from December 2015 to March 2016. *Journal of Critical Care*, *37*, 19–23. https://doi.org/10.1016/j.jcrc.2016.08.016.

Cao-Lormeau, V. M., Blake, A., Mons, S., Lastere, S., Roche, C., Vanhomwegen, J., … Ghawche, F. (2016). Guillain-Barre syndrome outbreak associated with Zika virus infection in French Polynesia: A case-control study. *Lancet*, *387*(10027), 1531–1539. https://doi.org/10.1016/s0140-6736(16)00562-6.

Cornblath, D. R., Mellits, E. D., Griffin, J. W., McKhann, G. M., Albers, J. W., Miller, R. G., … Quaskey, S. A. (1988). Motor conduction studies in Guillain-Barre syndrome: Description and prognostic value. *Annals of Neurology*, *23*(4), 354–359. https://doi.org/10.1002/ana.410230407.

da Silva, I. R. F., Frontera, J. A., Bispo de Filippis, A. M., & Nascimento, O. (2017). Neurologic complications associated with the Zika virus in Brazilian adults. *JAMA Neurology*, *74*(10), 1190–1198. https://doi.org/10.1001/jamaneurol.2017.1703.

Dhiman, G., Abraham, R., & Griffin, D. E. (2019). Human Schwann cells are susceptible to infection with Zika and yellow fever viruses, but not dengue virus. *Scientific Reports*, *9*(1), 9951. https://doi.org/10.1038/s41598-019-46389-0.

Dirlikov, E., Major, C. G., Mayshack, M., Medina, N., Matos, D., Ryff, K. R., … Rivera-Garcia, B. (2016). Guillain-Barre syndrome during ongoing Zika virus transmission—Puerto Rico, January 1-July 31, 2016. *MMWR. Morbidity and Mortality Weekly Report*, *65*(34), 910–914. https://doi.org/10.15585/mmwr.mm6534e1.

Doorn, P. A. V., Ruts, L., & Jacobs, B. C. (2008). Clinical features, pathogenesis, and treatment of Guillain-Barré syndrome. *The Lancet Neurology*, *7*(10), 939–950. https://doi.org/10.1016/S1474-4422(08)70215-1.

Lynch, R. M., Mantus, G., Encinales, L., Pacheco, N., Li, G., Porras, A., … Chang, A. Y. (2019). Augmented Zika and dengue neutralizing antibodies are associated with Guillain-Barre syndrome. *The Journal of Infectious Diseases*, *219*(1), 26–30. https://doi.org/10.1093/infdis/jiy466.

Magira, E. E., Papaioakim, M., Nachamkin, I., Asbury, A. K., Li, C. Y., Ho, T. W., … Monos, D. S. (2003). Differential distribution of HLA-DQ beta/DR beta epitopes in the two forms of Guillain-Barre syndrome, acute motor axonal neuropathy and acute inflammatory demyelinating polyneuropathy (AIDP): Identification of DQ beta epitopes associated with susceptibility to and protection from AIDP. *Journal of Immunology*, *170*(6), 3074–3080. https://doi.org/10.4049/jimmunol.170.6.3074.

Monos, D. S., Papaioakim, M., Ho, T. W., Li, C. Y., & McKhann, G. M. (1997). Differential distribution of HLA alleles in two forms of Guillain-Barre syndrome. *The Journal of Infectious Diseases*, *176*(Suppl. 2), S180–S182. https://doi.org/10.1086/513786.

Oliveira, A. C., Maria Henrique da Mota, L., Dos Santos-Neto, L. L., De Carvalho, J. F., Caldas, I. R., Martins Filho, O. A., & Tauil, P. L. (2014). Occurrence of autoimmune diseases related to the vaccine against yellow fever. *Autoimmune Diseases*, *2014*. https://doi.org/10.1155/2014/473170, 473170.

Parra, B., Lizarazo, J., Jimenez-Arango, J. A., Zea-Vera, A. F., Gonzalez-Manrique, G., Vargas, J., … Pardo, C. A. (2016). Guillain-Barre syndrome associated with Zika virus infection in Colombia. *The New England Journal of Medicine*, *375*(16), 1513–1523. https://doi.org/10.1056/NEJMoa1605564.

Rinaldi, S., & Willison, H. J. (2008). Ganglioside antibodies and neuropathies. *Current Opinion in Neurology*, *21*(5), 540–546. https://doi.org/10.1097/WCO.0b013e32830b84b7.

Sejvar, J. J. (2016). West Nile virus infection. *Microbiology Spectrum*, *4*(3). https://doi.org/10.1128/microbiolspec.EI10-0021-2016.

Sejvar, J. J., Kohl, K. S., Gidudu, J., Amato, A., Bakshi, N., Baxter, R., … Wiznitzer, M. (2011). Guillain-Barré syndrome and fisher syndrome: Case definitions and guidelines for collection, analysis, and presentation of immunization safety data. *Vaccine*, *29*(3). https://doi.org/10.1016/j.vaccine.2010.06.003.

Sharshar, T., Chevret, S., Bourdain, F., & Raphael, J. C. (2003). Early predictors of mechanical ventilation in Guillain-Barre syndrome. *Critical Care Medicine*, *31*(1), 278–283. https://doi.org/10.1097/00003246-200301000-00044.

Uncini, A., Shahrizaila, N., & Kuwabara, S. (2017). Zika virus infection and Guillain-Barre syndrome: A review focused on clinical and electrophysiological subtypes. *Journal of Neurology, Neurosurgery, and Psychiatry*, *88*(3), 266–271. https://doi.org/10.1136/jnnp-2016-314310.

Walgaard, C., Lingsma, H. F., Ruts, L., van Doorn, P. A., Steyerberg, E. W., & Jacobs, B. C. (2011). Early recognition of poor prognosis in Guillain-Barre syndrome. *Neurology*, *76*(11), 968–975. https://doi.org/10.1212/WNL.0b013e3182104407.

Willison, H. J. (2007). Gangliosides as targets for autoimmune injury to the nervous system. *Journal of Neurochemistry*, *103*(Suppl. 1), 143–149. https://doi.org/10.1111/j.1471-4159.2007.04718.x.

Ye, Y., Li, S. L., & Li, Y. J. (2015). Comparison on therapeutic effect of plasma exchange and intravenous immunoglobulin for Guillian-Barre syndrome. *Transfusion Medicine*, *25*(2), 79–84. https://doi.org/10.1111/tme.12169.

Chapter 20

Auditory brainstem in Zika virus: Insights about brain development in microcephaly

Dimitri Marques Abramov[a], Tania Regina Saad Salles[a], Cecília Hedin-Pereira[b], Maria Elizabeth Lopes Moreira[c], and Vladimir V. Lazarev[a]

[a]*Laboratory of Neurobiology and Clinical Neurophysiology, National Institute of Women, Children and Adolescents Health Fernandes Figueira (IFF), Oswaldo Cruz Foundation (FIOCRUZ), Rio de Janeiro, Brazil,* [b]*Vice Presidency of Research and Biological Collections (VPPCB), Oswaldo Cruz Foundation (FIOCRUZ), Rio de Janeiro, RJ, Brazil,* [c]*Unit of Clinical Research, National Institute of Women, Children and Adolescents Health Fernandes Figueira (IFF), Oswaldo Cruz Foundation (FIOCRUZ), Rio de Janeiro, RJ, Brazil*

Abbreviations

A1, A2	left and right auricular leads
BAEPs	brainstem auditory evoked potentials
CMV	cytomegalovirus
CT	computed tomography
Cz	central EEG lead (on the head vertex)
dB nHL	decibels from normal hearing level
MC	microcephaly
MRI	magnetic resonance imaging
NPCs	neuroprogenitor cells
SD	standard deviation
ZIKV	Zika virus

Introduction

In the cohort studies of the National Institute of Women, Children and Adolescents Health Fernandes Figueira (IFF) of the Oswaldo Cruz Foundation (FIOCRUZ) in Rio de Janeiro, Brazil, children with microcephaly (MC) due to Zika virus (ZIKV) infection frequently manifest a persistent brainstem syndrome since early life, revealed by discordant swallowing, without compromised respiratory function, also controlled by brainstem (Saad et al., 2018). Brainstem thinning observed in MRI images of many cases may be one of the explanations. As in other infectious congenital syndromes, the ZIKV can alter both visual (Tsui et al., 2018) and auditory (Barbosa et al., 2019) sensory systems. Studying the brainstem auditory evoked potentials (BAEPs) as a way to test for the integrity of brainstem's intrinsic circuitry in MC, we showed that brainstem function is normal in these children, although we confirmed hearing alterations (Marques Abramov et al., 2018).

Our work includes the study on the wave complex derived from the BAEPs in ZIKV congenital microcephalic individuals. The BAEPs explore brainstem circuitry hierarchically from the eighth cranial nerve up to the inferior colliculus (Chiappa, 1997). Our observations contradict the clinical impressions and the imaging information related to the brainstem syndrome since no impairment was detected. In this chapter, we present a series of clinical cases (Marques Abramov et al., 2018) to clarify this paradox.

Brainstem auditory evoked potentials in microcephaly caused by the Zika virus

In a recent integrative review, Barbosa et al. (2019) revised 27 papers with audiological measurements in ZIKV cases. While several studies report neurosensory hearing loss, the review concluded that there was no consensus on the

Zika Virus Impact, Diagnosis, Control, and Models. https://doi.org/10.1016/B978-0-12-820267-8.00020-0

impairments caused by the ZIKV infection, be it in the presence or not of MC. Most recent studies involving BAEP measurements are audiological evaluations rather than brainstem structural investigations.

In 2016 a few months after the the outbreak of the ZIKV epidemics and the birth of the first children with the congenital infection of ZIKV in the hospital of IFF, we started to record signals of auditory processing in this population by BAEPs in response to sound clicks. We found normo- and microcephalic children with neurosensory deficits revealed by the increase in latency of the wave I, sometimes poorly defined (unpublished data).

From August 2016 to February 2017, we gathered the data of brainstem function of microcephalic children with congenital syndrome by ZIKV in a cross-sectional study using BAEPs. A detailed methodological description and results can be found in Marques Abramov et al. (2018). From the 23 tested individuals, 19 (12 girls) were included, aged, on average, 29.57 ± 17.84 weeks (mean $\pm$ SD; ranging from 12 to 62 weeks' postnatal). All were born with a head circumference <32 cm. Four children were excluded due to other malformations and concomitant comorbidities.

Subjects were examined by a standard protocol for the BAEP study (Chiappa, 1997), stimulated using earphones and the "click" sound (composed by a spectrum of frequencies from 500 to 2000 Hz, durations of 200 μs, 70 dB nHL, with 30 dB nHL of contralateral mask, at a rate of 11.1 stim/s), rarefaction, and condensation polarities, during spontaneous sleep or calm wakefulness. The EEG waves were registered at auricular positions, referred to Cz electrode (International 10–20 System), using standard patterns.

We considered only the latency for analysis since it is the most robust and stable parameter well established in the literature, as well as virtually refractory to artifacts. We used Picton et al. (Chiappa, 1997) normative values, related to the latencies of the wave peaks I, III, and V, and interpeak latency I–V (Fig. 1) comprising the ages of 6, 12, 26, 52, and 108 weeks. We performed a weekly table of values by interpolating the normal latencies for the analysis. After ensuring the sample statistical independence, we evaluated the ears as individual subjects (see following text). We normalized the latencies by the number of SDs around normative values, considering any equal or higher than 3 SDs as out of standard (Chiappa, 1997). We employed parametric and nonparametric tests corrected by Bonferroni's FWER method for comparisons and correlations.

Our primary outcome results from the qualitative analysis of all traces from BAEPs (Fig. 1) showed that the exams were practically normal. Only one child evidenced deafness to the left ear with isoelectric trace when stimulated until 90 dB nHL. Neurosensory deficits probably caused that alteration. Another girl exhibited conductive impairment in the left ear. The remaining 36 ears had well-defined waves, consistent with those of typical children. All 19 children showed normal

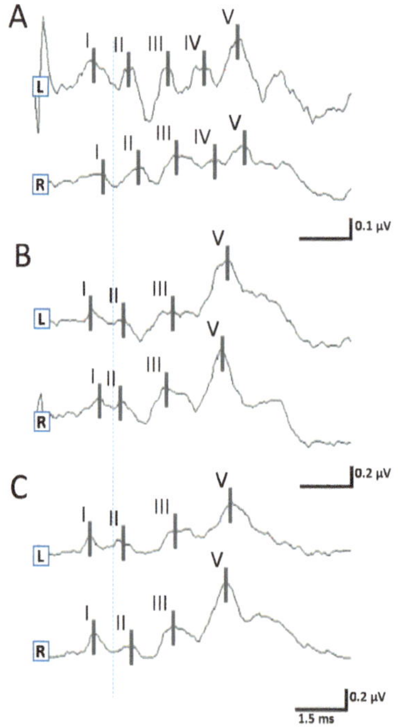

FIG. 1 Selected brainstem auditory evoked potentials of infected microcephalic children (unpublished). All plots show a typical pattern of brainstem activity for left (L) and right (R) ears. I–V, wave numbers. (A) Male, 18 weeks, CP: 29 cm. (B) Male, 59 weeks, CP = 30 cm. (C) Male, 24 weeks, CP not informed. All subjects were infected at first trimester. *Dashed line:* 3SD from normal mean latency. See respective scales for each plot. *L*, A1-Cz, left auricular channel; *R*, A2-Cz, right auricular channel. *CP*, head circumference at birth.

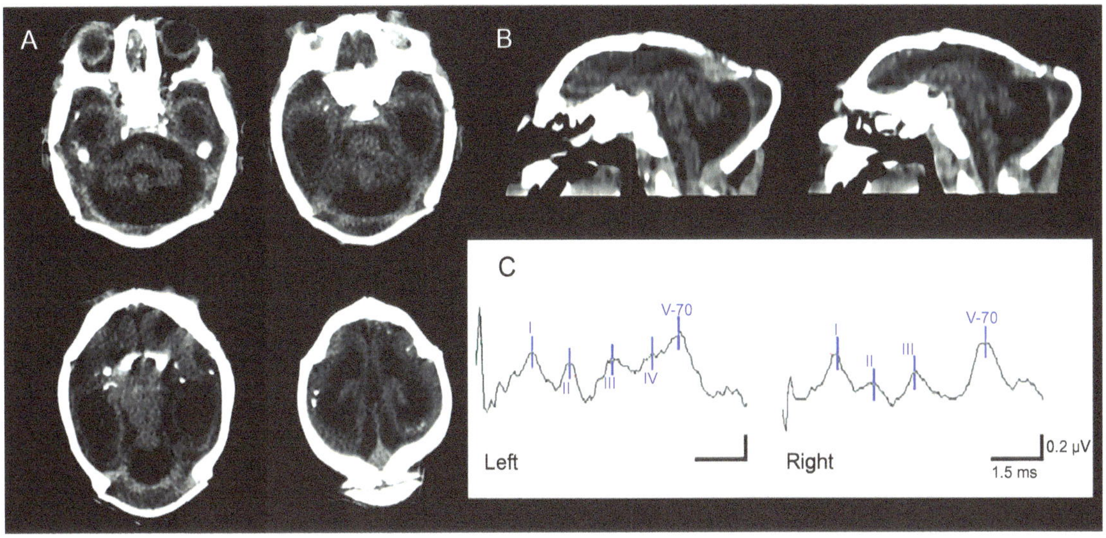

FIG. 2 Severe brain damage with normal brainstem auditory function (unpublished). (A) Axial slices by head computed tomography and (B) sagittal reconstruction, showing severe telencephalic disruption and cerebellum agenesis, telencephalic calcifications, enlarged ventriculi, and apparent brainstem hypoplasia. (C) BAEP pattern for left and right ears evidencing preserved intrinsic brainstem circuitry (see Fig. 1).

auditory brainstem functions, which we believe means that brainstem nuclei and auditory paths may be well organized to express typical traces of BAEPs and no contiguous alterations seemed to be present as well.

The quantitative analysis of latencies concerning SD around normal values revealed only one abnormal case: shorter I–V interval in the right ear of one subject. Despite being in normality interval (3 SD), the wave I latencies in the group were almost all higher than the normative. This may be consistent with subnormal neurosensory impairments due to ZIKV infection in different proportions. The wave I latencies and I–V intervals correlated with the age of the subjects ($r = 0.45$, $P = 0.005$; $e = -0.37$, $P = 0.022$, respectively). Perhaps, somehow, the neurosensory impairment may be progressive, as occurs in the cytomegalovirus (CMV) infection due to late immunity related inflammation (Cohen, Durstenfeld, & Roehm, 2014).

Since short-latency evoked potentials are considered the gold standard to evaluate the brainstem functions, and once we observed small alterations of them in ZIKV congenital syndrome children with microcephaly, we concluded that the brainstem of these individuals is close to normality (Fig. 2).

The brainstem in congenital Zika virus syndrome

Anatomical alterations in CT of children with MC due to ZIKV infection are frequent, but far from being a rule in the syndrome, even less with a similar pattern to telencephalic alterations. Imaging studies have systemically reported brainstem hypoplasia in MC by ZIKV, however, with discordant prevalence, from 16.8% to 70% (Del Campo et al., 2017; Levine, Jani, Castro-Aragonk, & Cannie, 2017; Zare Mehrjardi et al., 2017). Consistently, Cavalheiro et al. (2016) have not found any anatomical alteration in the brainstem neuroimaging of a series of 13 cases of MC by ZIKV. In one of the first anatomopathological studies, Melo et al. (2016) identified calcifications in the brainstem of three stillbirths. In the case series described by de Fatima Vasco Aragao et al. (2016), of the 22 children studied by computed tomography, 8 children showed calcifications in CT, and 10 had hypoplasia, while only three displayed both alterations.

Despite the structural alterations described in several works, the studies of our case series indicated brainstem normality concerning the auditory function. Considering BAEP sensitivity, the functional normality correlates, firstly, with normal neural circuits, which would only be possible in normal fetal brainstem development. Our neurofunctional finding is counterintuitive, forcing us to rethink the primary cause of the brainstem clinical syndrome observed in children with ZIKV congenital syndrome, given the previously described results. Indeed, both brainstem hypoplasia and calcifications might be secondary to telencephalic disorganization. The reduction of motor efferent pathways could be responsible for the volumetric anatomical alteration of the brainstem and the severe swallowing dysphagia. We support that idea emphasizing that the most prominent neurological syndrome observed is pyramidal (Chimelli et al., 2017). Even in normocephalic children the congenital ZIKV infection is linked to motor alterations and developmental delay (Vianna et al., 2019). On the other hand, the calcifications may be due to degenerative inflammatory processes. This issue is discussed later. Unfortunately, we are not aware of any brainstem electrophysiological studies in other microcephalies to discuss if this pattern is specific to the infection.

Zika pathogenesis and brainstem ontogenesis

The pathophysiology of microcephaly in the infection by ZIKV has been extensively covered in other chapters. We aimed to discuss the extent of damage of the brainstem observed in CTs in light of the data from BAEPs and what possible differential mechanisms may be in play.

In the congenital infection by ZIKV, pathological mechanisms interrupt brain growth, affecting the proliferation and differentiation of neuroprogenitor cells (NPCs). The virus appears to infect cerebral cortex neuroprogenitor cells, immature neurons, and mature neurons of lesser volume, inducing massive apoptosis and structural cellular defects (Souza et al., 2016; Wen, Song, & Ming, 2017). The hypothesis is that the decreasing NPCs in proliferative zones (ventricular and subventricular) disturb the generation of neurons, making thinner layers and affecting gyrification of the cortical mantle. In parallel an inflammatory process would occur, causing calcifications, necrosis, gliosis, and also secondary degeneration of NPCs (Wen et al., 2017). Thus two different pathological mechanisms are in play simultaneously. The pathological inflammatory/degenerative process is compatible with the significant heterogeneity of our clinical and image findings, as well as the brainstem and cerebral calcifications.

The formation of the cerebral cortex first begins with symmetric divisions of stem cells (radial glial cells), which increase the progenitor pool until the first neurons are born by asymmetric proliferation in the ventricular zone. These cells migrate in radial glia to generate the typical inside-out organization of the cerebral cortex (Stiles & Jernigan, 2010). While the proliferation of NPCs occurs between the fourth and seventh weeks, neuronal migration occurs between the 12th and 24th weeks in the human brain, and axon growth and connection for many weeks more, a very protracted time window (deAzevedo et al., 2003; Stiles & Jernigan, 2010; ten Donkelaar, Yamada, Shiota, & van der Vliet, 2014).

The formation of the brainstem, in contrast, is quite precocious between 6 and 8 weeks. Neurons are generated, and the cranial nerve nuclei and internal nuclei, as well as their connections, are defined (Yachnis & Rorke, 1999). Although many mothers have been contaminated in the first 8 weeks of gestation, the brainstems of their babies are normal. The pathogenic effects of the ZIKV probably appear after this period of time or affecting specifically the telencephalic NPCs. In both cases the question remains of whether there is specific viral tropism toward telencephalic progenitors or the virus has more access to telencephalon in its invasive trajectories. The fact is that the brainstem remains mysteriously protected in its intrinsic circuitry.

Policy and procedures

Brainstem auditory evoked potentials (BAEPs) are very sensitive exams with very low amplitude waves whose profile is potentially influenced by the topography of the space between the signal source and the body surface. Therefore the recording and signal parameters (filters, sampling rates, click characteristics, and others) and the procedures for performing the exams are stringent. Deviations from these protocols can easily mimic the manifestation of functional changes.

We usually evaluate wave latencies as a clinical parameter, since amplitudes (absolute and relative) are influenced, respectively, by the conductance of the circuit and the intrinsic variability of the vector components of the leads.

Latencies vary in the first year of life, given the maturation of brainstem circuits. After 54 weeks the latencies are virtually the same as in adults. Several studies make normative value tables (some are in Chiappa (1997)), but some argue that each laboratory should develop its normative value tables, examining samples of people without functional changes.

International and regional standards are established as good practice for conducting BAEPs exams, which are frequently reviewed and updated. See, for example, the standards offered by the American Clinical Neurophysiology Society (2008): https://www.acns.org/pdf/guidelines/Guideline-9C.pdf.

Mini-dictionary of terms

Brainstem. The posterior part of the brain, continuous with the spinal cord. Includes the midbrain, the pons, and medulla oblongata. Has integrative functions being involved in cardiovascular system control, respiratory control, pain sensitivity control, alertness, awareness, and consciousness. Contains nuclei of pairs of cranial nerves and auditory and vestibular pathways.

Brainstem auditory evoked potentials (BAEPs). Bioelectrical potentials generated by neuronal populations of the brainstem nuclei, synchronized and related to various auditory stimuli (from broadband clicks to specific frequencies). BAEPs are a method of maximum sensitivity to changes in the brainstem auditory circuits that cause changes in wave latency or profile. The specific relationship between each wave and its generating source is commonly established as follows: wave I, the spiral ganglion and the distal portion of the eighth nerve (vestibulocochlear); wave II (usually poorly defined), the insertion of the eighth nerve in the brainstem; wave III, cochlear nucleus (caudal bridge); wave IV, lateral lemniscus and superior olivary complex; V, inferior colliculus (mainly).
Conductive hearing impairment or loss. Occurs when there is a problem transferring sound waves anywhere along the pathway through the outer ear, tympanic membrane, or middle ear.
Decibel (dB). A logarithmic unit used to measure sound pressure level.
Neural pathways. The connection formed by axons that project from neurons to make synapses onto neurons in another location to enable a signal to be sent from one region of the nervous system to another.
Neurosensory hearing impairment, deficit or loss. A type of hearing problems in which the root cause lies in the inner ear or sensory organ (cochlea and associated structures) or the vestibulocochlear nerve (cranial nerve VIII).
Telencephalon. The uppermost region of the central nervous system, a large part of the brain containing the cerebral cortex and subcortical structures, including the hippocampus, basal ganglia, and olfactory bulb.

Key facts of brainstem auditory evoked potentials

- Brainstem auditory evoked potentials (BAEPs) are a highly sensitive method for detecting functional alterations caused by structural changes in the brainstem.
- BAEP waves have very short latencies (<10 ms) and small amplitudes ($<2\,\mu$V).
- BAEP wave latencies form patterns of low population variability for each age group.
- BAEP waves show early maturation (first year of life).
- Minimal neuronal or axonal changes primarily interfere with wave latencies.
- We can differentiate auditory alterations from brainstem alterations by varying the intensity of the sound stimulus.

Summary points

- MC by ZIKV appears to be associated with structural changes in the brainstem.
- BAEPs of children with ZIKV MC show patterns consistent with brainstem functional normality.
- Identified evidence of brainstem alterations should be secondary to telencephalic damage.
- In the pathophysiology of MC by ZIKV, proliferation, differentiation, and axonal migration of intrinsic brainstem structures appear to be unaffected.

References

Barbosa, M. H. M., Magalhães-Barbosa, M. C., Robaina, J. R., Prata-Barbosa, A., Lima, M. A. M. T., & Cunha, A. J. L. A. D. (2019). Auditory findings associated with Zika virus infection: An integrative review. *Brazilian Journal of Otorhinolaryngology*, *85*, 642–663.

Cavalheiro, S., Lopez, A., Serra, S., Da Cunha, A., da Costa, M. D., Moron, A., et al. (2016). Microcephaly and Zika virus: Neonatal neuroradiological aspects. *Child's Nervous System*, *32*, 1057–1060.

Chiappa, K. H. (1997). *Evoked potentials in clinical medicine* (3rd ed.). Philadelphia: Lippincott Raven (Chapter 6).

Chimelli, L., Mello, A. S. O., Avvad-Portari, E., Wiley, C. A., Camacho, A. H. S., Lopes, V. S., et al. (2017). The spectrum of neuropathological changes associated with congenital Zika virus infection. *Acta Neuropathologica*, *133*, 983–999.

Cohen, B. E., Durstenfeld, A., & Roehm, P. C. (2014). Viral causes of hearing loss: A review for hearing health professionals. *Trends in Hearing*, *18*, 2331216514541361.

deAzevedo, L. C., Fallet, C., Moura-Neto, V., Daumas-Duport, C., Hedin-Pereira, C., & Lent, R. (2003). Cortical radial glial cells in human fetuses: Depth-correlated transformation into astrocytes. *Journal of Neurobiology*, *55*(3), 288–298.

de Fatima Vasco Aragao, M., van der Linden, V., Brainer-Lima, A. M., Coeli, R. R., Rocha, M. A., Sobral da Silva, P., et al. (2016). Clinical features and neuroimaging (CT and MRI) findings in presumed Zika virus related congenital infection and microcephaly: Retrospective case series study. *BMJ*, *353*, i1901.

Del Campo, M., Feitosa, I. M., Ribeiro, E. M., Horovitz, D. D., Pessoa, A. L., França, G. V., et al. (2017). Zika Embryopathy Task Force-Brazilian Society of Medical Genetics ZETF-SBGM. The phenotypic spectrum of congenital Zika syndrome. *American Journal of Medical Genetics. Part A*, *173*, 841–857.

Levine, D., Jani, J. C., Castro-Aragonk, I., & Cannie, M. (2017). How does imaging of congenital Zika compare with imaging of other TORCH infections? *Radiology*, *285*, 744–761.

Marques Abramov, D., Saad, T., Gomes-Junior, S. C., Souza E Silva, D., Araújo, I., Lopes Moreira, M. E., et al. (2018). Auditory brainstem function in microcephaly related to Zika virus infection. *Neurology*, *90*, e606–e614.

Melo, A. S., Aguiar, R. S., Amorim, M. M., Arruda, M. B., Melo, F. O., Ribeiro, S. T., et al. (2016). Congenital Zika virus infection: Beyond neonatal microcephaly. *JAMA Neurology*, *73*, 1407–1416.

Saad, T., PennaeCosta, A., de Goes, F. V., de Freitas, M., de Almeida, J. V., Santa Ignêz, L. J., et al. (2018). Neurological manifestations of congenital Zika virus infection. *Child's Nervous System*, *34*, 73–78.

Souza, B. S. F., Sampaio, G. L. A., Pereira, C. S., Campos, G. S., Sardi, S. I., Freitas, L. A. R., et al. (2016). Zika virus infection induces mitosis abnormalities and apoptotic cell death of human neural progenitor cells. *Scientific Reports*, *6*, 39775.

Stiles, J., & Jernigan, T. L. (2010). The basics of brain development. *Neuropsychology Review*, *20*, 327–348.

ten Donkelaar, H. J., Yamada, S., Shiota, K., & van der Vliet, T. (2014). Overview of the development of the human brain and spinal cord. In H. J. ten Donkelaar, M. Lammens, & A. Hori (Eds.), *Clinical neuroembryology* (pp. 1–52). Berlin, Heidelberg: Springer.

Tsui, I., Moreira, M. E. L., Rossetto, J. D., Vasconcelos, Z., Gaw, S. L., Neves, L. M., et al. (2018). Eye findings in infants with suspected or confirmed antenatal Zika virus exposure. *Pediatrics*, *142*, e20181104.

Vianna, R. A. O., Lovero, K. L., Oliveira, S. A., Fernandes, A. R., Santos, T., Lima, L., et al. (2019). Children born to mothers with rash during Zika virus epidemic in Brazil: First 18 months of life. *Journal of Tropical Pediatrics*, *65*, 592–602.

Wen, Z., Song, H., & Ming, G. L. (2017). How does Zika virus cause microcephaly? *Genes & Development*, *31*, 849–861.

Yachnis, A. T., & Rorke, L. B. (1999). Cerebellar and brainstem development: An overview in relation to Joubert syndrome. *Journal of Child Neurology*, *14*, 570–573.

Zare Mehrjardi, M., Poretti, A., Huisman, T. A., Werner, H., Keshavarz, E., & Araujo Júnior, E. (2017). Neuroimaging findings of congenital Zika virus infection: A pictorial essay. *Japanese Journal of Radiology*, *35*, 89–94.

Section E

Methods, biomarkers, and diagnosis

Chapter 21

Magnetic resonance imaging use in detecting neurological abnormalities in Zika virus infection

Diogo Goulart Corrêa[a], Heron Werner[a], Osvaldo J.M. Nascimento[b], and Luiz Celso Hygino da Cruz Júnior[a]

[a]*Department of Radiology, Clinic of Diagnostic Imaging/DASA, Rio de Janeiro, Rio de Janeiro, Brazil,* [b]*Department of Neurology, Federal Fluminense University, Niterói, Rio de Janeiro, Brazil*

Abbreviations

ADEM	acute disseminated encephalomyelitis
CSF	cerebrospinal fluid
CT	computed tomography
CMV	cytomegalovirus
CZS	congenital Zika syndrome
FLAIR	fluid-attenuated inversion recovery
GBS	Guillain-Barré syndrome
MRI	magnetic resonance imaging
rRT-PCR	real-time reverse transcription polymerase chain reaction
RNA	ribonucleic acid
SD	standard deviations
SWI	susceptibility-weighted imaging
TORCH	(toxoplasmosis, rubella, cytomegalovirus, herpes simplex, and other pathogens, including syphilis, parvovirus, human immunodeficiency virus, and varicella zoster)
US	ultrasound
ZIKV	Zika virus

Introduction

Various severe neurologic complications are associated with Zika virus (ZIKV) infection (Reid, Rimmer, & Thakur, 2018). Since the first majors reported outbreaks of ZIKV infection in French Polynesia in 2013 and 2015, neurologic complications of the infection were described (Hygino da Cruz, Nascimento, Lopes, & da Silva, 2018). Since then, some neurologic disabilities in infants of infected pregnant women were also observed. The first reported case of Guillain-Barré syndrome (GBS) associated with ZIKV occurred during the French Polynesia outbreak in November 2013 (Reid et al., 2018). Subsequently, scientific evidence has linked congenital infection by ZIKV with microcephaly (Ribeiro, Muniz, Gasparetto, Ventura, & Marchiori, 2017; Ribeiro, Werner, Lopes, et al., 2017).

Later, further evidence regarding the relationship between ZIKV infection and other neurological complication was also reported. In addition to congenital Zika syndrome (CZS) and GBS and its variants, other neurologic manifestations associated with ZIKV infections have been identified, including encephalitis/meningoencephalitis, myelitis, sensory neuropathy, acute disseminated encephalomyelitis (ADEM) and optic neuropathy, epilepsy, and childhood arterial ischemic stroke (Medina & Medina-Montoya, 2017).

Neuroimaging is useful to detect central nervous system damage secondary to ZIKV infection. Ultrasound (US) is frequently the first imaging examination performed in cases of congenital infections, because it is an inexpensive, easy-to-access, and safe method for screening pregnant women (Zare Mehrjardi, Carteaux, Poretti, et al., 2017; Zare

Zika Virus Impact, Diagnosis, Control, and Models. https://doi.org/10.1016/B978-0-12-820267-8.00021-2

Mehrjardi, Poretti, Huisman, et al., 2017). Postnatal computed tomography (CT) can detect calcifications with accuracy. However, magnetic resonance imaging (MRI) provides a better soft tissue contrast resolution than CT and US. Then, MRI is the imaging modality that better evaluates the neuronal lesions secondary to the infection, contributing to the assessment of lesions extension, in both adult and congenital infection (Carmo, Alves Simão, Amaral, et al., 2019).

In this chapter, we describe the pathophysiology and the neuroimaging findings of the main neurological complications related to ZIKV infection, with emphasis on MRI.

Congenital Zika syndrome

CZS results from vertical transmission of ZIKV from an infected woman to the fetus during pregnancy. The viral transmission occurs through the placenta (Sejvar, 2018). The neurological complications of CZS are caused by affection of the neural progenitor cells of the human fetal brain by the virus (Medina & Medina-Montoya, 2017).

In infected pregnant women, ZIKV crosses the placental barrier to infect the developing fetus. In the developing brain, ZIKV disrupts proliferation and migration of neural progenitor cells. The consequent cell death of neural progenitor cells is the development arrest of the fetal brain.

CZS is characterized by five main features: severe microcephaly; decreased brain tissue with a specific pattern of brain damage, including subcortical calcifications; damage to the back of the eye, including macular scarring and focal pigmentary retinal mottling; congenital contractures; or arthrogryposis and hypertonia restricting body movement soon after birth. Other abnormalities associated with CZS are brain atrophy and asymmetry, abnormally formed or absent brain structures, hydrocephalus, neuronal migration disorders, excessive and redundant scalp skin, hyperreflexia, irritability, tremors, seizures, brainstem dysfunction, dysphagia, focal pigmentary mottling and chorioretinal atrophy in the macula, optic nerve hypoplasia, colobomas, congenital glaucoma, microphthalmia, lens subluxation, cataracts, and intraocular calcifications (Reid et al., 2018).

Neuroimaging findings

Prenatal ultrasound

A baseline US examination should be performed upon all suspicion of ZIKV infection or during the evaluation of pregnant women in epidemic areas. US assessment should be performed every 4–6 weeks. If any abnormality is found, US evaluations should be performed every 2–3 weeks (Papageorghiou, Thilaganathan, Bilardo, et al., 2016).

The most frequent finding on prenatal US is microcephaly (head circumference 2 standard deviations (SD) below the mean for the gestational age or under the 2nd percentile for age and sex) (Zare Mehrjardi, Carteaux, et al., 2017; Zare Mehrjardi, Poretti, et al., 2017).

Microcephaly depends on the timing of prenatal ZIKV infection. Infections that occur early in pregnancy (first and early second trimesters) are related to microcephaly. Fetal microcephaly is most likely due to increased developmental neuronal apoptosis secondary to early ZIKV infection of progenitor cells. The risk of ZIKV infection associated fetal microcephaly is estimated to be 95:10,000 pregnant women infected in the first trimester of pregnancy (Carod-Artal, 2018).

The pathogenesis of microcephaly from ZIKV is believed to be a consequence of in utero destruction of neural progenitor cells and persistence of the inflammatory response-associated molecules. These signaling pathways can be activated by the ZIKV, resulting in the downregulation of the cell cycle, impaired cell-cycle progression, decreased proliferation, and neuronal apoptosis (Hussain, Ali, Latiwesh, & Hussain, 2018). Consequently, microcephaly is only the tip of the iceberg at prenatal imaging, because other fetal findings can be seen (Ribeiro, Werner, et al., 2017).

Fetal US is also able to detect subcortical brain calcifications; cerebral atrophy; posterior fossa alterations; agenesis/hypoplasia of the corpus callosum, with or without interhemispheric cyst; absent septum pellucidum cavum; unilateral of bilateral ventriculomegaly; craniofacial disproportion; and microphthalmia. Prenatal US with Doppler has also demonstrated placental insufficiency (Ribeiro, Werner, et al., 2017; Zare Mehrjardi, Carteaux, et al., 2017; Zare Mehrjardi, Keshavarz, Poretti, & Hazin, 2016; Zare Mehrjardi, Poretti, et al., 2017) (Fig. 1).

Prenatal magnetic resonance imaging

Fetal brain MRI better detects brain abnormalities and may be performed when fetal US shows brain abnormalities suggestive of ZIKV infection. Findings are similar to prenatal ultrasound. MRI better evaluates malformations of cortical development, such as polymicrogyria, opercular dysplasia, and lissencephaly-pachygyria, as well as white matter and cortical abnormalities such as dysmyelination and cortical laminar necrosis. Compared with fetal US, prenatal MRI is,

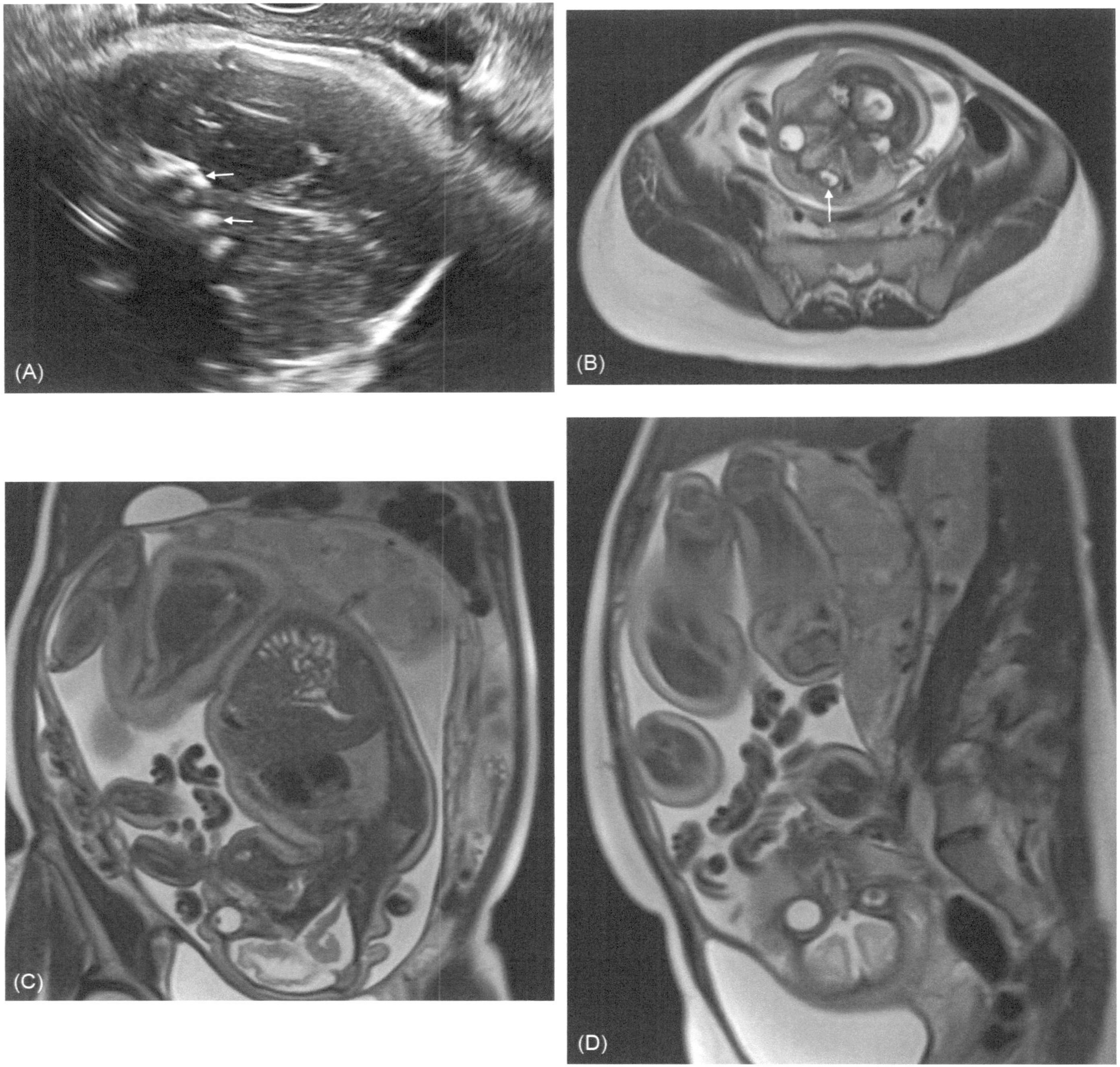

FIG. 1 Congenital Zika syndrome. (A) Obstetric US showing brain calcifications *(white arrows)*. Fetal MRI in the axial (B), sagittal (C), and coronal (D) planes demonstrating microphthalmia (*arrow* in B), microcephaly, a simplified pattern of brain gyri, and a redundant scalp skin in the occipital region.

however, less sensitive in detecting discrete/dispersed calcifications (Guillemette-Artur, Besnard, Eyrolle-Guignot, et al., 2016; Werner, Sodré, Hygino, et al., 2016).

In addition to microcephaly and malformations of cortical development, fetal MRI may also demonstrate agenesis/hypoplasia of the corpus callosum; absent cavum septum pellucidum, ventriculomegaly, and subependymal pseudocysts, particularly near the occipital horns of lateral ventricles; cerebellar and vermian hypoplasia; brainstem abnormalities; and enlarged extra-axial spaces. Another typical finding at fetal MRI is redundant scalp skin in the occipital region (Ribeiro, Werner, et al., 2017; Zare Mehrjardi, Poretti, et al., 2017) (Fig. 1).

Postnatal transfontanellar ultrasound

US may be used as a safe imaging method for the first-line investigation of suspected neonates, in which both brain parenchyma and the ventricular system can be evaluated, providing images in real time and may be performed at the bedside (Ribeiro, Werner, et al., 2017; Zare Mehrjardi, Poretti, et al., 2017). However, performing transfontanellar US may be a challenge in patients with congenital ZIKV infection who have microcephaly, as the anterior fontanelle is closed in up to 87% of these infants at the first examination (de Fatima Vasco Aragao, van der Linden, Brainer-Lima, et al., 2016). The extensive brain malformation presented by these patients contributes to an upper cranial bone collapse over the occipital area, causing the overlap of sutures and the closure of fontanels (Dain Gandelman Horovitz, da Silva Pone, Moura Pone, Dias Saad Salles, & Bastos Boechat, 2016).

Transfontanellar US may show parenchymal calcifications, typically located at or just below the corticomedullary junction; neuronal migrational abnormalities, with malformations of cortical development; parenchymal atrophy with secondary ventriculomegaly; dysgenesis of the corpus callosum, subependymal cysts; and cerebellar and brainstem hypoplasia (Ximenes, Pires, Werner, et al., 2019).

Postnatal computed tomography

Bain CT is sensitive for parenchymal calcification detection. It also shows associated skull deformities well, particularly using three-dimensional reconstructions (Zare Mehrjardi, Poretti, et al., 2017).

Parenchymal calcifications are the most common CT finding in congenital ZIKV infection and have been reported in almost all affected patients so far (Ribeiro, Werner, et al., 2017; Zare Mehrjardi, Carteaux, et al., 2017; Zare Mehrjardi, Poretti, et al., 2017). Parenchymal calcifications at different numbers and locations can be seen, characterized by foci of hyperattenuation on CT scans. Calcifications are usually seen in the junction between the cortex and the subcortical white matter, with a punctate or band-like shape, or even with a confluent arrangement (Fig. 2). However, deep and periventricular white matter and deep gray matter calcifications can also be seen (Peixoto Filho, de Freitas, Ciosaki, Oliveira, & Dos Santos Júnior, 2018). Moreover, in some cases, the brain atrophy is so great that it is difficult to determine if the calcifications are subcortical or periventricular (Fig. 3).

It is possible to observe microcephaly and an exuberant external occipital protuberance, caused by collapsed bone and constrained intracranial contents with associated redundant occipital scalp skin (Ribeiro, Muniz, et al., 2017).

Although MRI is more sensitive, CT can also depict cortical atrophy, enlarged supratentorial subarachnoid space, widely open Sylvian fissures, ventriculomegaly, and a simplified gyral pattern, suggesting malformation of cortical development. Other findings include hypoplasia of the cerebellum and brainstem (Hazin, Poretti, Turchi Martelli, et al., 2016; Soares de Oliveira-Szejnfeld, Levine, Melo, et al., 2016).

Postnatal magnetic resonance imaging

MRI is the best imaging method to identify brain lesions in infants born to women infected by the ZIKV. It allows a better analysis of the brain parenchyma and identifies satisfactorily congenital abnormalities. However, disadvantages of performing brain MRI in these patients are that the total exam time is long and the child must remain still during the entire scan. MRI pulse sequences less sensitive to motion artifacts can minimize the need for sedation in an already severe neurologically affected patient (Ribeiro, Werner, et al., 2017). In our center the MRI scan is performed with the infant sleeping, after feeding. Our protocol includes sagittal T1-weighted imaging, axial T2-weighted imaging, axial fluid-attenuated inversion recovery (FLAIR), axial susceptibility-weighted imaging (SWI), and axial diffusion-weighted imaging. We do not use intravenous gadolinium routinely in these patients.

MRI can detect skull deformities, characterized by a collapse of the frontal and parietal bones, rendering a flattened appearance, associated with scalp protrusion and redundant skin in the occipital region, secondary to a loss of brain parenchymal volume (Soares de Oliveira-Szejnfeld et al., 2016).

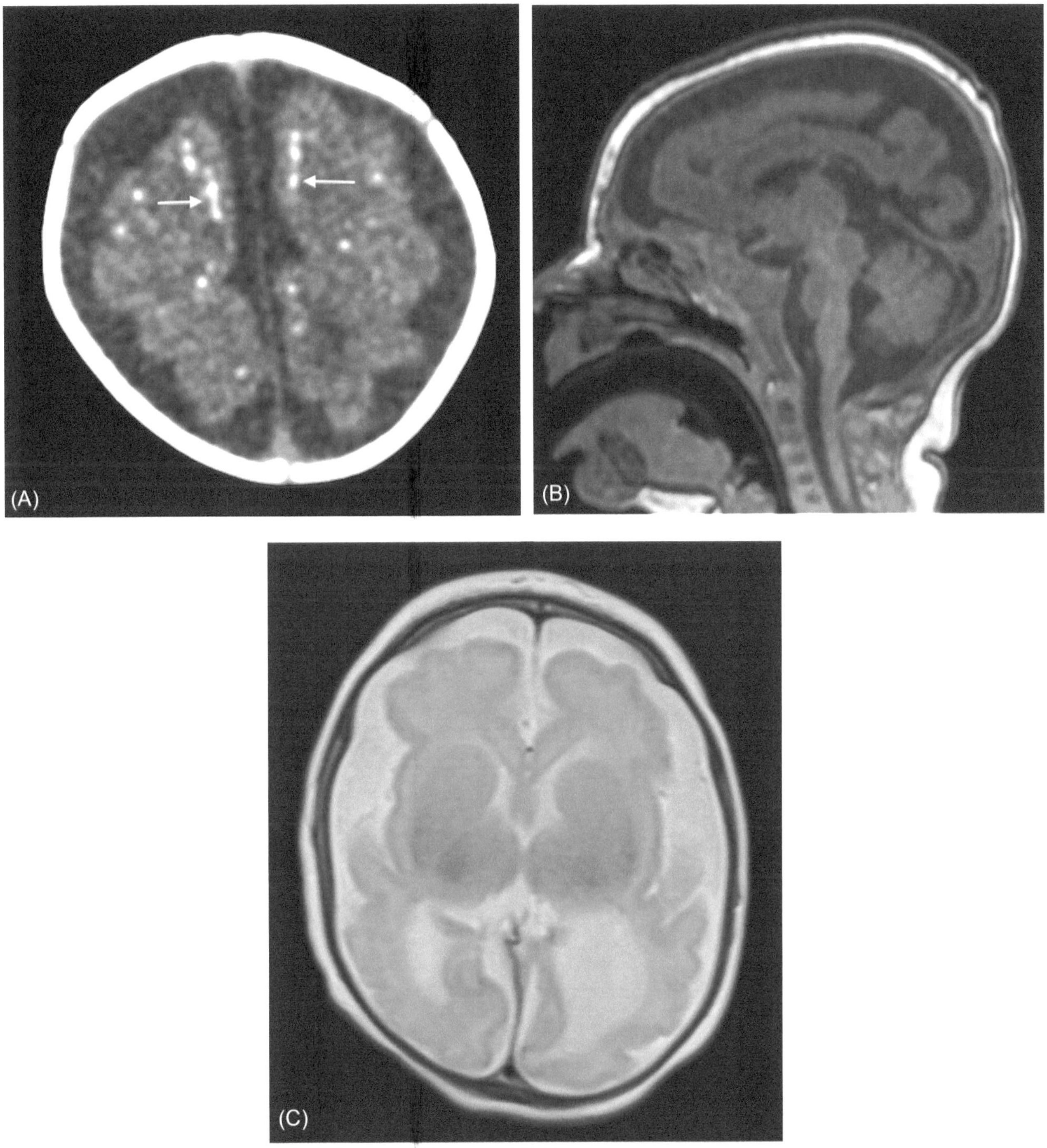

FIG. 2 Congenital Zika syndrome. Brain CT of a 2-month-old boy with congenital Zika syndrome, showing microcephaly and cerebral atrophy, with simplified cerebral gyrus and subcortical calcifications (*arrows* in A). (B) T1-weighted brain MRI, in the sagittal plane, showing craniofacial disproportion due to microcephaly. (C) Axial T2-weighted imaging showing a smooth cortical surface with a thickened cortex, due to pachygyria.

Although less sensitive than CT in detecting calcifications, MRI can also show the calcifications secondary to CZS. Usually, these calcifications are hypointense in SWI and hyperintense at T1-weighted imaging (Ribeiro, Werner, et al., 2017). It is important to remember that although the most common region to be affected by the calcification is the cortical-subcortical junction, some authors have showed that ZIKV-infected infants may also manifest calcifications in the brainstem and periventricular regions (Aragão, Holanda, Brainer-Lima, et al., 2017; Cavalheiro, Lopez, Serra, et al., 2016).

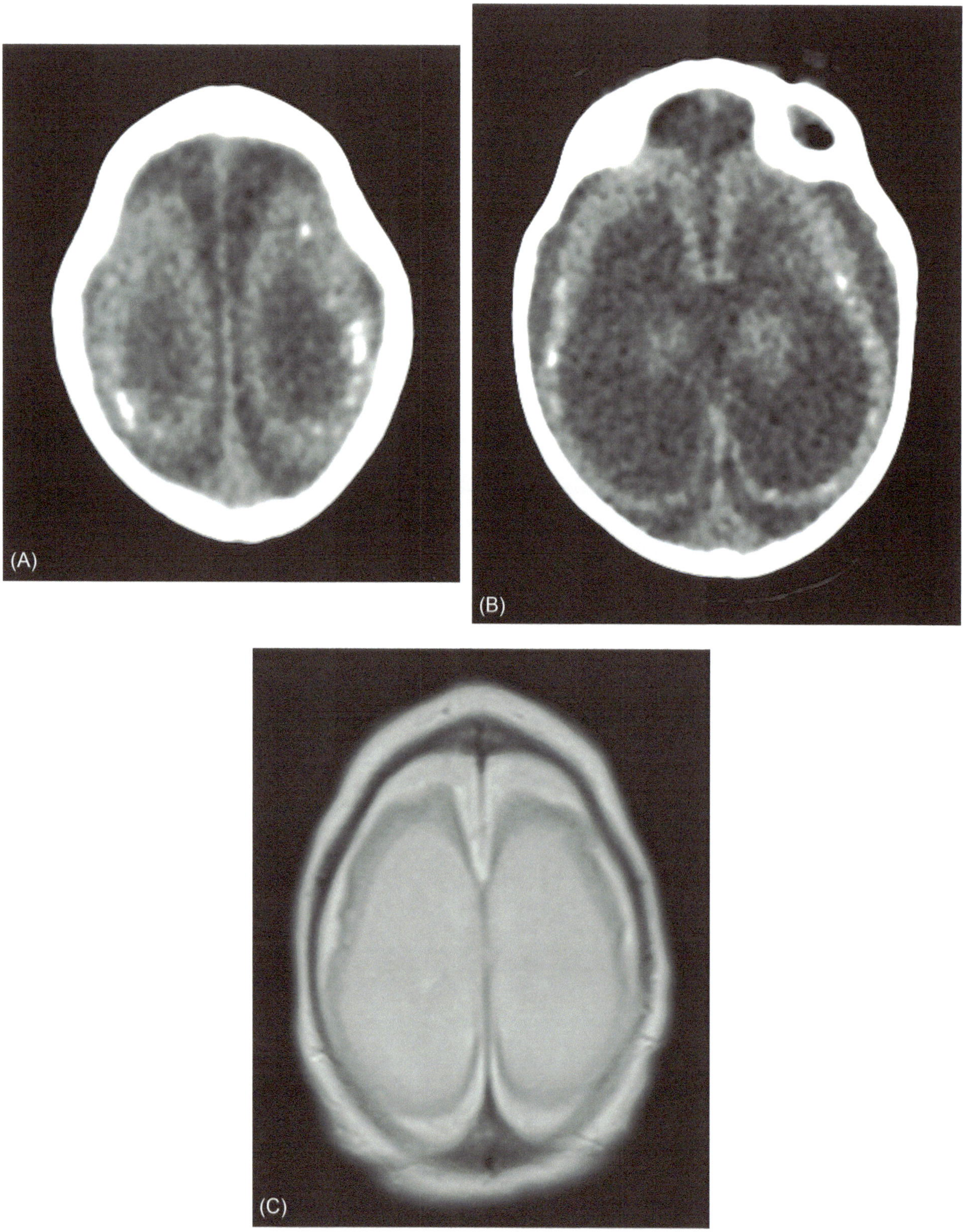

FIG. 3 Congenital Zika syndrome. (A and B) Brain CT of a 2-month-old boy with congenital Zika syndrome, showing microcephaly, cerebral atrophy, and brain calcifications. However, the marked brain atrophy makes it difficult to determine if the calcifications are subcortical or periventricular. (C) T2-weighted brain MRI, in the axial plane, showing a smooth cortical surface, due to lissencephaly and ventriculomegaly.

MRI is better in detecting malformations of cortical development, such as polymicrogyria and lissencephaly-pachygyria, which are common findings that probably vary according to the stage of cortical development in which the infection occurs. The involvement is usually diffuse, predominantly occurring in the frontal, insular, and parietal lobes, with varying degrees of severity. It is frequently accompanied by an increase in the extraaxial cerebrospinal fluid (CSF) space, ventriculomegaly, and loss of brain volume (Ribeiro, Muniz, et al., 2017) (Figs. 2 and 3).

Polymicrogyria is related to a late migrational failure, resulting in abnormal development of the deep layers of cerebral cortex. Imaging findings include innumerable disorganized gyri, with apparent irregular cortical thickening owing to packing of microgyri, due to neuronal disorganization. Lissencephaly results from either decreased cell production or increased apoptosis in the germinal zone of the cerebral cortex. It is characterized by a smooth cortical surface, with a thickened cortex (Abdel Razek, Kandell, Elsorogy, Elmongy, & Basett, 2009).

Delayed myelination is observed in some cases, associated with secondary thinning of the corpus callosum (Ribeiro, Werner, et al., 2017).

Congenital Zika virus infection without microcephaly

Although microcephaly is the most prominent feature of congenital Zika syndrome, a spectrum with less severe cases is starting to be recognized (Aragão et al., 2017). Brain abnormalities can also be seen in children with congenital ZIKV infection without congenital microcephaly. Neuroimaging findings in these cases include malformations of cortical development, such as pachygyria and polymicrogyria, calcifications at the cortex/subcortical white matter junction; hypoplasia of the corpus callosum, cerebellum, and brainstem; decreased brain volume; and ventriculomegaly (van der Linden, Pessoa, Dobyns, et al., 2016).

Statistically, patients with microcephaly present more frequently with reduced brain volume, calcifications outside the cortical-subcortical white matter junction, corpus callosum abnormalities, moderate to severe ventriculomegaly, enlarged extraaxial CSF space, and an enlarged cisterna magna, when compared with patients with CZS without microcephaly (Aragão et al., 2017; de Freitas Ribeiro, Muniz, Gasparetto, & Marchiori, 2018). Additionally, in a study by Aragão et al. (2017), polymicrogyria was not present in patients with microcephaly at birth, a finding that could be seen in some patients without microcephaly at birth.

These evidences suggest that there is a wide spectrum of findings associated with CZS and that we know so far only the most severe end of the spectrum. Future studies in less symptomatic children may reveal the whole spectrum of lesions and neuroimaging findings associated with CZS (Zare Mehrjardi, Poretti, et al., 2017).

Differential diagnosis

Infants with suspected congenital ZIKV infection should have a comprehensive evaluation, as the differential diagnosis of CZS includes both infectious and genetic etiologies (Moore, Staples, Dobyns, et al., 2017). Some congenital infections such as TORCH (toxoplasmosis, rubella, cytomegalovirus, herpes simplex, and other pathogens, including syphilis, parvovirus, human immunodeficiency virus, and varicella zoster) may disrupt fetal brain development and result in various brain injuries and congenital brain abnormalities. Congenital cytomegalovirus (CMV) infection has the most similar features to CZS among TORCH infections. The location of calcifications may help to differentiate congenital ZIKV from CMV infections, since the calcifications are mostly at the cortical-subcortical white matter junction in ZIKV infection and mostly periventricular in CMV infection. In addition, hepatic, renal, and hematologic abnormalities are common in congenital CMV infection. To date, no hematologic, hepatic, or renal laboratory abnormalities have been documented in infants with congenital ZIKV infection (Zare Mehrjardi, Poretti, et al., 2017).

Several genetic etiologies share some similarities with CZS including Aicardi-Goutières syndrome, cystic leukoencephalopathy without megacephaly, type 1 interferonopathies, RNASET2-related leukodystrophy, Baraitser-Reardon syndrome, and mutations in the JAM3, NDE1, and ANKLE2 genes (Moore et al., 2017; Ribeiro, Werner, et al., 2017).

In general, clinical history, neuroimaging features, laboratory studies, and/or genetic testing all together will differentiate CZS and its neuroimaging mimickers (Moore et al., 2017).

Neurological complications of postnatal acquired Zika virus infection

Usually, postnatal acquired ZIKV infection is asymptomatic or mildly symptomatic. However, in a small number of patients, postnatal ZIKV infection may cause neurological complications including GBS, myelitis, meningoencephalitis, and ADEM. Postnatal ZIKV-related neurological complications are most likely caused by neuronal damage and demyelination. Clinical history and neurophysiology studies, including electromyography and nerve conduction study, as well as serology and molecular testing of blood, urine, and/or CSF and real-time reverse transcription polymerase chain reaction (rRT-PCR) for the virus ribonucleic acid (RNA) play a key role in the diagnosis of postnatal ZIKV-related neurological complications (Zare Mehrjardi, Carteaux, et al., 2017).

Guillain-Barré syndrome

GBS is a complex acute polyradiculoneuropathy, embracing several diseases that share some similarities but differ in relation to their underlying pathogenesis and clinical manifestations (Araujo, Silva, & Araujo, 2016). Usually, GBS is preceded by a recent immune stimulation, often with sensory and cranial nerve involvement that proceeds to its peak of clinical symptoms in some weeks. GBS is clinically characterized by the presence of distal and bilateral flaccid weakness, loss of reflexes, and the presence of albuminocytologic dissociation in the CSF (Carod-Artal, 2018; Muñoz, Barreras, & Pardo, 2016).

GBS was the first reported neurological complication of ZIKV infection in adults. The link between ZIKV and GBS was confirmed by a case-control study in French Polynesia, in which an increased number of GBS cases associated with ZIKV infection were reported (Cao-Lormeau, Blake, Mons, et al., 2016). Several pathophysiological theories were proposed to explain the link between GBS and ZIKV infection:

(1) A humoral response in which ZIKV antigens resemble host nerve components, triggering humoral immune responses in which ZIKV-induced neutralizing antibodies react with the host peripheral nerve proteins, such as myelin, with subsequent immune-mediated damage of the nerve (molecular mimicry).
(2) ZIKV triggers a memory immune response in those subjects who have been exposed to other flavivirus (e.g., dengue) (secondary infection).
(3) Autoreactive T lymphocyte activation induced by ZIKV mediates peripheral nerve damage.
(4) Direct ZIKV infection of lower motor neurons, peripheral nerve, and/or Schwann cells inducing cytopathic changes and direct cytotoxicity (Carod-Artal, 2018; Muñoz et al., 2016).

Most patients who develop GBS secondary to ZIKV infection live or have recently been to endemic areas, are men, under 50 years old, and have had a recent symptomatic ZIKV infection. The mean duration between the onset of ZIKV-related symptoms and GBS development is about 7 days (Zare Mehrjardi, Carteaux, et al., 2017). Neurologic symptoms are present in almost half of the cases during or immediately after the development of the clinical viral syndrome itself.

The clinical presentations in most cases of ZIKV-related GBS are compatible with an acute motor axonal neuropathy, which is a pure motor neuropathy with mild loss of reflexes and usually shows rapid recovery. Nevertheless, GBS secondary to ZIKV seems to present with a similar clinical outcome and prognosis compared with more classic triggers for this polyneuropathy, with a fairly similar response to traditional therapies, such as intravenous immunoglobulin (Hygino da Cruz et al., 2018).

Brain and spine MRI are increasingly used in the diagnostic workup of GBS. Imaging is mainly performed to rule out other clinical conditions presenting with acute flaccid paralysis, such as transverse myelitis (another potential complication in ZIKV infection) (Zare Mehrjardi, Carteaux, et al., 2017). Non-gadolinium-enhanced imaging may appear completely normal. Then, administration of gadolinium-based contrast agents is essential. Brain and spine MRI findings in ZIKV-related GBS are similar to those in patients with GBS due to other etiologies:

(1) Post contrast enhancement of the conus medullaris and cauda equina nerve roots, with more prominent and common enhancement of the ventral roots compared with the dorsal ones. However, the involvement of the posterior nerve roots is commonly observed.
(2) Post contrast enhancement of cranial nerves, particularly the trigeminal and facial nerves.
(3) T2 hyperintensity and contrast enhancement of the lumbar spinal ganglia bilaterally. The findings are secondary to autoimmune inflammation and demyelination, along with blood-neuronal barrier breakdown (Fontes, Dos Santos, & Marchiori, 2016; Hygino da Cruz et al., 2018). So far, there is no imaging difference between ZIKV-related GBS and GBS secondary to other triggers (Fig. 4).

It is also important to remember that there are other causes of nerve root enhancement in spine MRI, such as leptomeningeal tumor dissemination, sarcoidosis, lymphoma, and infections, such as syphilis and Lyme disease, among others. Therefore MRI alone does not diagnose GBS, and the correlation with other complementary diagnostic methods is required.

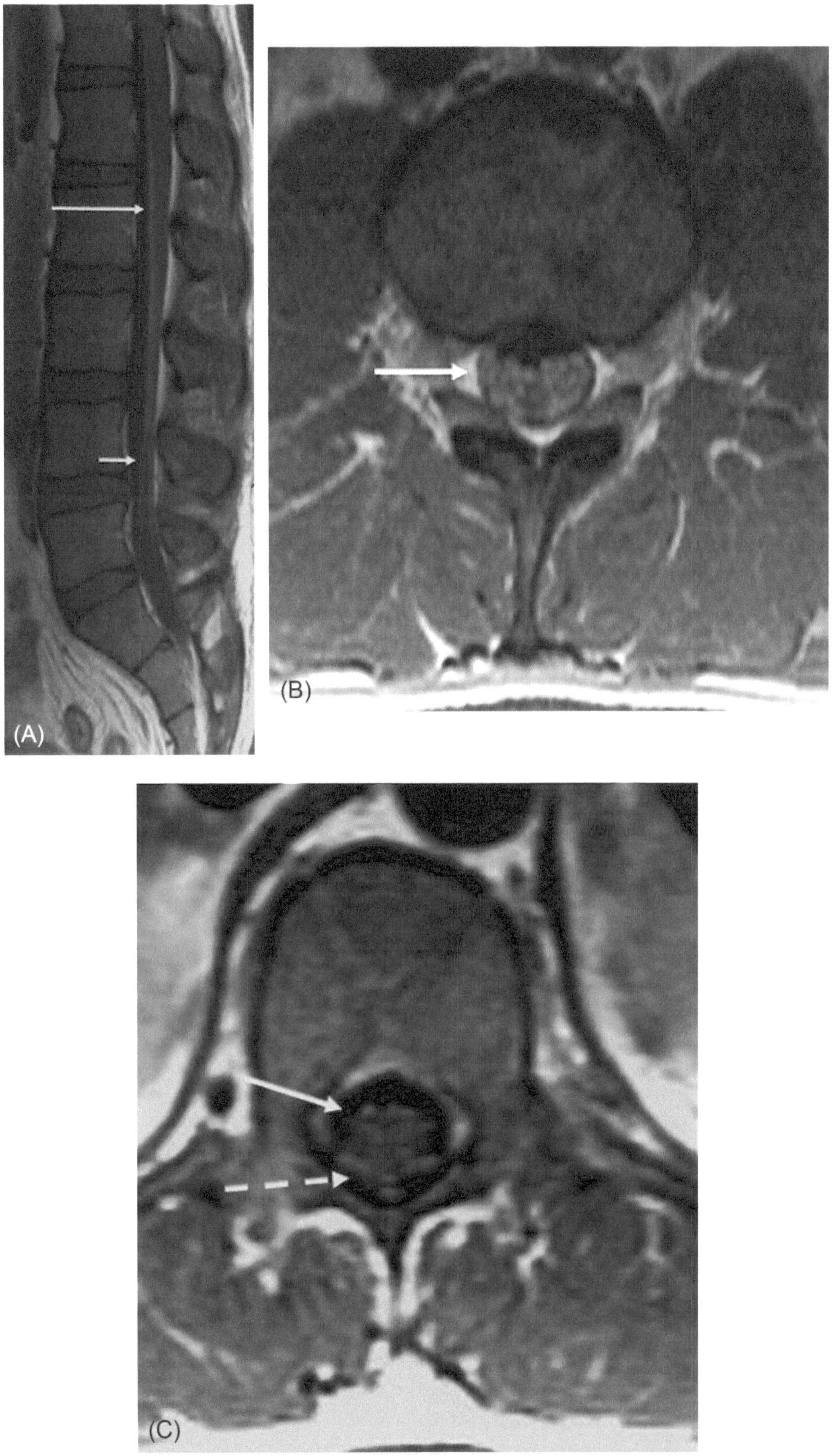

FIG. 4 GBS secondary to Zika virus infection. Lumbar spine MRI showing post contrast enhancement in the cauda equina nerve roots on sagittal (*arrows* in A) and axial (*arrow* in B and C) T1-weighted images. Note that both anterior (*solid arrow* in B) and posterior (*dashed arrow* in C) nerve roots enhance by gadolinium contrast. The other sequences were normal. *(From Hygino da Cruz, L. C., Jr., Nascimento, O. J. M., Lopes, F. P. P. L., da Silva, I. R. F. (2018). Neuroimaging findings of Zika virus-associated neurologic complications in adults.* AJNR. American Journal of Neuroradiology, 39, *1967–1974, with permission.)*

Encephalitis/meningoencephalitis

Encephalitis is defined as inflammation of the brain parenchyma associated with neurologic dysfunction (Venkatesan, Tunkel, Bloch, et al., 2013). Other flaviviruses such as Japanese encephalitis and West Nile viruses are known causes of encephalitis. Although there is no population-based or well-controlled studies evaluating the association between ZIKV infection and encephalitis, making this relation still presumptive, there are some case reports of encephalitis potentially related to ZIKV (Muñoz et al., 2016; Song, Yun, Woolley, & Lee, 2017). In these sporadic reported cases of encephalitis, ZIKV infection was confirmed by rRT-PCR in CSF, and other infectious causes for encephalitis were ruled out by PCR (da Silva et al., 2017).

Generally, ZIKV-associated encephalitis has been reported with an onset varying from 1 to 8 days since the beginning of the infection symptoms (Reid et al., 2018). Clinically the patients usually present with seizures, progressive somnolence, headache, and vomiting due to increased intracranial pressure, motor and sensory dysfunction, cranial nerve palsy, nuchal rigidity, focal deficits, and may evolve to deep coma and brain death (Hygino da Cruz et al., 2018; Zare Mehrjardi, Carteaux, et al., 2017).

The limited data from previously case reports suggest that encephalitis secondary to ZIKV infection may present with asymmetric subcortical FLAIR and T2-hyperintense lesions, which may show restricted diffusion, suggesting cytotoxic edema (Hygino da Cruz et al., 2018; Zare Mehrjardi, Carteaux, et al., 2017). ZIKV may also have a predilection for cortical gray matter and subcortical white matter, particularly for the cingulate and superior frontal gyrus, brainstem, and middle cerebellar peduncles (Hygino da Cruz et al., 2018; Zare Mehrjardi, Carteaux, et al., 2017). Other authors have reported a diffuse involvement of white matter and deep gray matter nuclei in a case of ZIKV-related encephalitis leading to death (Zare Mehrjardi, Carteaux, et al., 2017). Due to the few case reports relating ZIKV infection to encephalitis, there is no specific neuroimaging pattern of ZIKV-related encephalitis. In addition, it is important to remember that cases of viral encephalitis, irrespective of the cause, may show no abnormalities on brain MRI (Hygino da Cruz et al., 2018).

Also, there is evidence that different forms of ZIKV-related neurological complications may coexist in the same patient, with concomitant central and peripheral nervous system involvement (da Silva et al., 2017) (Fig. 5).

Myelitis

Myelitis is a focal inflammatory disorder of the spinal cord resulting in motor, sensory, and/or autonomic dysfunction, depending on the lesion location (Transverse Myelitis Consortium Working Group, 2002). The association between ZIKV infection and myelitis is also still presumptive, through case reports or small case series, and the causal relationship has not been confirmed yet (Muñoz et al., 2016; Zare Mehrjardi, Carteaux, et al., 2017).

Clinically, patients often present with fast-developing acute flaccid paralysis, with a distinctive presence of a sensory level. Spinal cord involvement in ZIKV is variable, and previous reports have described different forms of involvement. MRI usually shows the lesions as hyperintense signal in T2-weighted imaging and in short tau inversion recovery (STIR), with or without spine enlargement due to edema. Contrast enhancement may be seen in acute lesions (Zare Mehrjardi, Carteaux, et al., 2017).

Myelitis can be identified as diffuse lesions with varying sizes that may affect any portion of the spinal cord. Other form of spinal involvement has been described, such as longitudinal extensive transverse myelitis, involving three or more vertebral body segments in length and more than two-thirds of its surface area. Also, ZIKV-related myelitis may also specifically affect the anterior horns of the spinal cord, leading to an isolated motor neuron syndrome (Hygino da Cruz et al., 2018; Palacios, Clavijo-Prado, Ruiz, Arias Antun, & Julián Duran, 2019) (Fig. 6).

Acute disseminated encephalomyelitis

ADEM is an immune-mediated inflammatory demyelinating disorder targeting mostly the white matter of the brain and spine and less frequently the gray matter, particularly the basal ganglia. It is characterized by an acute or subacute encephalopathy with multiple neurologic deficits and is typically monophasic and self-limiting.

Usually, clinical symptoms begin within 3 weeks after a viral infection or vaccination. However, in some cases, it is not possible to prove such a relationship. The diagnosis is based on clinical and imaging findings, as well as on excluding other conditions that can mimic ADEM. Similar to observations of other viral infections, ZIKV may also be related to ADEM (Brito Ferreira, Antunes de Brito, Moreira, et al., 2017). Neuroimaging findings in ZIKV-related ADEM do not differ from those caused by other etiologies, and no specific imaging finding for ZIKV-related ADEM distinguishing it from other causes of ADEM has been reported yet (Zare Mehrjardi et al., 2016).

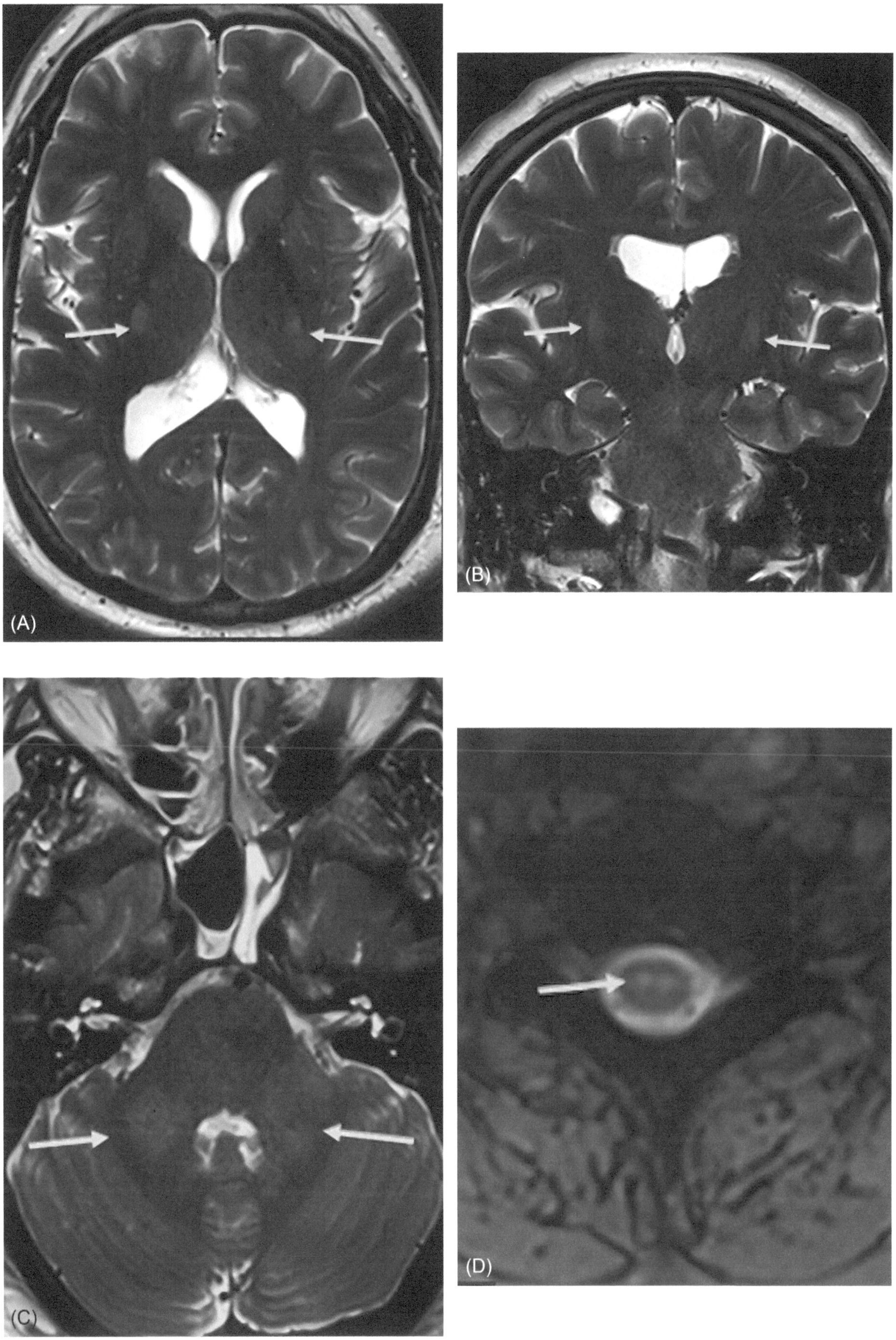

FIG. 5 *(Continued)*

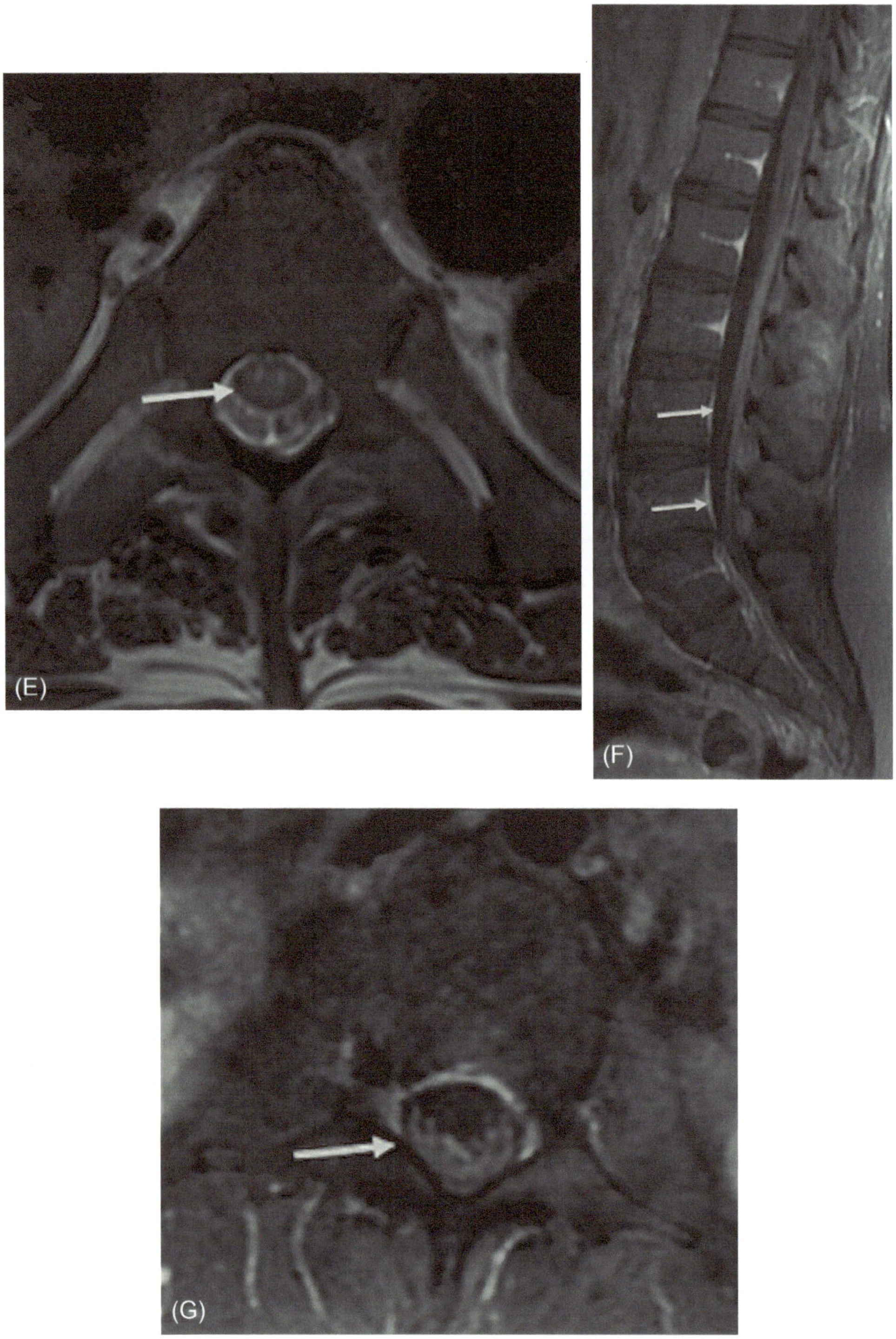

FIG. 5, cont'd Association of encephalitis, myelitis, and GBS, secondary to Zika virus infection in a 35-year-old man. Brain MRI showing hyperintense lesions, in T2-weighted imaging, in the corticospinal tracts (*arrows* in A and B), and in the middle cerebellar peduncles (*arrows* in C), due to encephalitis. Spine MRI showing hyperintense lesions in the anterior horns of the cervical (*arrow* in D) and thoracic spinal cord (*arrow* in E), in the axial gradient-echo T2-weighted (D) and T2-weighted (E) images. Post contrast enhancement is also seen in the cauda equina nerve roots on sagittal (*arrows* in F) and axial (*arrow* in G) T1-weighted spine images following gadolinium intravenous injection. *(From Hygino da Cruz, L. C., Jr., Nascimento, O. J. M., Lopes, F. P. P. L., da Silva, I. R. F. (2018). Neuroimaging findings of Zika virus-associated neurologic complications in adults.* AJNR. American Journal of Neuroradiology, 39, *1967–1974, with permission.)*

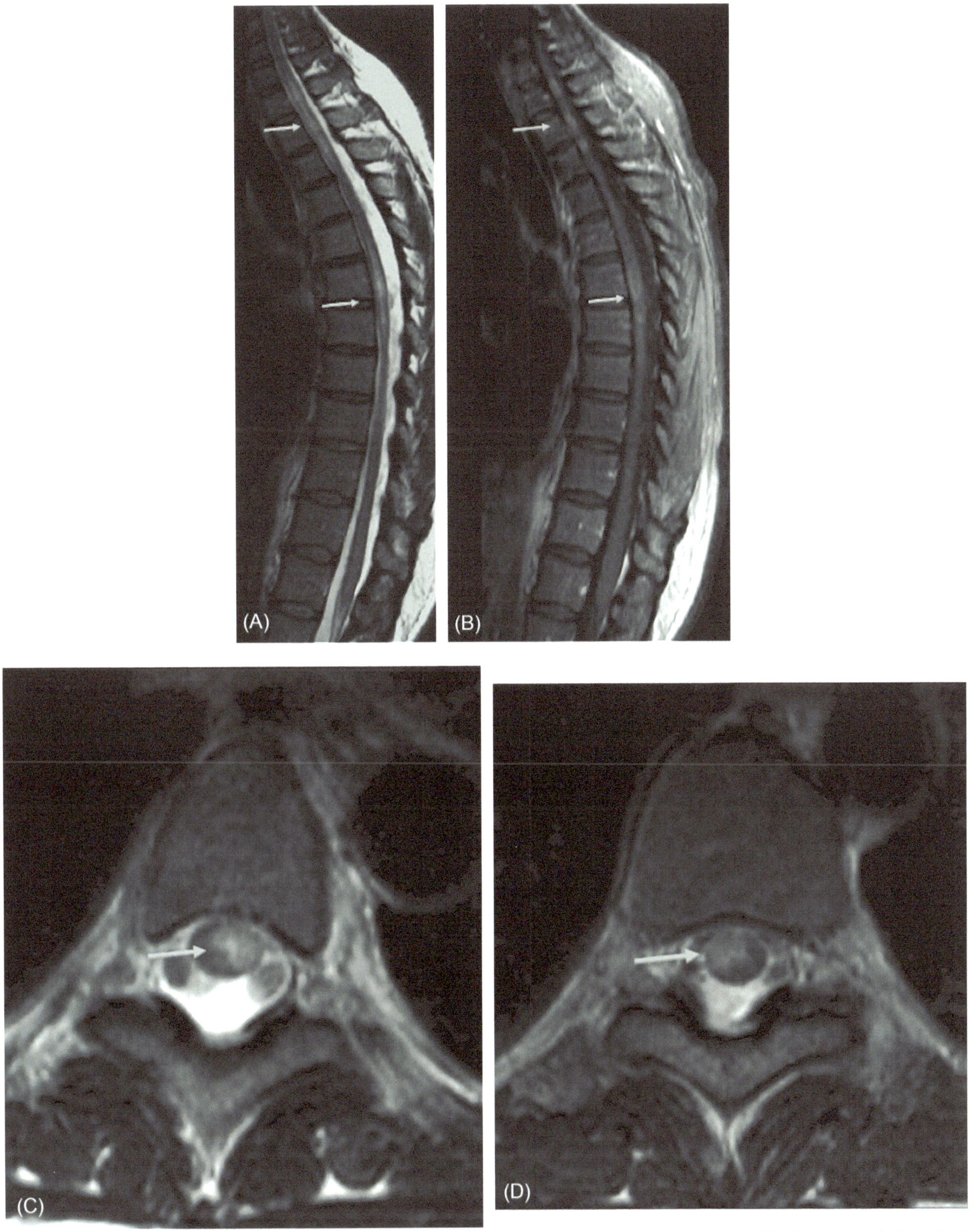

FIG. 6 Acute myelitis secondary to Zika virus infection in a 38-year-old woman. Spine MRI showing hyperintense, ill-defined lesions in the sagittal (*arrows* in A) and axial (*arrows* in C and D) T2-weighted images of the cervical and thoracic spinal cord, causing mild expansion of the cord. The lesions demonstrate contrast enhancement in the postgadolinium sagittal T1-weighted image (*arrows* in B). *(From Hygino da Cruz, L. C., Jr., Nascimento, O. J. M., Lopes, F. P. P. L., da Silva, I. R. F. (2018). Neuroimaging findings of Zika virus-associated neurologic complications in adults.* AJNR. American Journal of Neuroradiology, 39, *1967–1974, with permission.)*

The imaging findings described are like those observed in other related causes of ADEM, such as multifocal and asymmetric lesions, affecting the brain and the spinal cord. The clinical course of ADEM secondary to ZIKV infection is very similar to the typical ADEM cases, characterized as a self-limited and monophasic course, appearing weeks after the viral infection. MRI findings of ADEM typically include multiple hyperintense and irregular lesions in T2-weighted and FLAIR, with an asymmetrical distribution, involving the white matter and deep gray matter. Peripheral contrast enhancement is common, but complete-ring or punctate enhancement and lack of enhancement are also possible. In the acute phase, peripheral restricted diffusion is typically. Spinal MRI may show a longitudinal extensive transverse myelitis (Hygino da Cruz et al., 2018; Zare Mehrjardi et al., 2016; Zare Mehrjardi, Carteaux, et al., 2017).

Differential diagnosis

To date, there is no specific imaging finding to distinguish GBS, encephalitis, myelitis, and ADEM secondary to ZIKV infection to their other causes. Then, it is not recommended to consider ZIKV as the etiology of these neurologic complications just based on the neuroimaging findings. However, the combination of typical imaging finding of GBS, encephalitis, or myelitis, with the patient history, may be of help to suspect of ZIKV infection as the pathogenic agent (Hygino da Cruz et al., 2018; Zare Mehrjardi, Carteaux, et al., 2017).

Conclusion

ZIKV infection is related to a broad spectrum of congenital and acquired neurological complications. Imaging findings play a significant role in the diagnosis of ZIKV infection, and knowledge of its manifestations is of fundamental importance for an early and correct diagnosis.

Policy and procedures

A baseline obstetric US should be performed upon all suspicion of congenital ZIKV infection, and then every 4–6 weeks, in endemic areas. If any abnormality is found, ultrasound evaluations should be performed every 2–3 weeks. Fetal brain MRI better detects brain abnormalities and may be performed when fetal ultrasound shows brain abnormalities suggestive of ZIKV infection.

Postnatal transfontanellar US may be a challenge in patients with CZS, as the anterior fontanels may be closed. Postnatal brain CT is sensitive for parenchymal calcifications, whereas MRI is more sensitive for other brain alterations.

Although postnatal acquired ZIKV infection is usually asymptomatic or mildly symptomatic, a small number of patients develop neurological complications. Postnatal ZIKV-related neurological complications are most likely caused by neuronal damage and demyelination.

The main neurological complications are GBS, encephalitis/meningoencephalitis, myelitis, and ADEM.

In suspicious cases of GBS, imaging is performed to rule out other clinical conditions presenting with acute flaccid paralysis, and the intravenous administration of a gadolinium-based contrast agent is essential.

There is no specific imaging pattern of ZIKV-related encephalitis or myelitis. In addition, it is important to remember that cases of viral encephalitis and myelitis, irrespective of the cause, may show no abnormalities on MRI.

Mini-dictionary of terms

Congenital Zika syndrome. A group of symptoms secondary to Zika virus infection during fetal life, generally composed by microcephaly, skull collapse and occipital protrusion, decreased brain volume, and brain calcifications.

Microcephaly. Abnormally small head size. Head circumference 2 standard deviations below the mean for gestational age or under the 2nd percentile for age and sex.

Guillain-Barré syndrome. An acute inflammatory immune-mediated syndrome, typically characterized by polyradiculoneuropathy, with progressive ascendant paresthesia, pain, and weakness.

Encephalitis. An acute inflammation of the brain.

Myelitis. An acute inflammation of the spinal cord.

Key facts of imaging of neurological complications of Zika virus infection

- A baseline obstetric ultrasound should be performed in all suspicion cases of congenial Zika virus infection. If any abnormality is found, ultrasound evaluations should be performed every 2–3 weeks.
- The most striking finding of congenital Zika syndrome (CZS) on prenatal magnetic resonance imaging and ultrasound is microcephaly.
- Brain calcifications are very common in (CZS) and have been reported in almost all patients.
- Some patients with CZS do not present with microcephaly.
- There is no specific imaging pattern of Zika virus-related Guillain-Barré syndrome, encephalitis, myelitis, or acute disseminated encephalomyelitis.

Summary points

- Severe neurologic complications are associated with Zika virus infection, both when acquired before birth and after birth.
- Congenital Zika syndrome (CZS) is characterized by microcephaly, decreased brain volume, parenchymal calcifications, malformations of cortical development, damage to the back of the eye, congenital contractures, and hypertonia, which can be detected prenatally and postnatally.
- There are differential diagnosis of CZS, which can course with microcephaly and/or brain calcifications, such as toxoplasmosis, cytomegalovirus, and rubella.
- The main neurological complications of postnatal acquired Zika virus infection are Guillain-Barré syndrome (GBS), encephalitis, myelitis, and acute disseminated encephalomyelitis (ADEM).
- Although there is no specific imaging finding to distinguish GBS, encephalitis, myelitis, and ADEM secondary to Zika virus from other causes, the combination of typical imaging finding and clinical history may lead to the correct pathogenic agent.

References

Abdel Razek, A. A., Kandell, A. Y., Elsorogy, L. G., Elmongy, A., & Basett, A. A. (2009). Disorders of cortical formation: MR imaging features. *AJNR. American Journal of Neuroradiology*, *30*, 4–11.

Aragão, M. F. V. V., Holanda, A. C., Brainer-Lima, A. M., et al. (2017). Nonmicrocephalic infants with congenital Zika syndrome suspected only after neuroimaging evaluation compared with those with microcephaly at birth and postnatally: How large is the Zika virus "iceberg"? *AJNR. American Journal of Neuroradiology*, *38*, 1427–1434.

Araujo, A. Q., Silva, M. T., & Araujo, A. P. (2016). Zika virus-associated neurological disorders: A review. *Brain*, *139*, 2122–2130.

Brito Ferreira, M. L., Antunes de Brito, C. A., Moreira, A. J. P., et al. (2017). Guillain-Barre syndrome, acute disseminated encephalomyelitis and encephalitis associated with Zika virus infection in Brazil: Detection of viral RNA and isolation of virus during late infection. *The American Journal of Tropical Medicine and Hygiene*, *97*, 1405–1409.

Cao-Lormeau, V. M., Blake, A., Mons, S., et al. (2016). Guillain-Barré syndrome outbreak caused by Zika virus infection in French Polynesia: A case control study. *Lancet*, *387*, 1531–1539.

Carmo, R. L. D., Alves Simão, A. K., Amaral, L. L. F. D., et al. (2019). Neuroimaging of emergent and reemergent infections. *Radiographics*, *39*, 1649–1671.

Carod-Artal, F. J. (2018). Neurological complications of Zika virus infection. *Expert Review of Anti-Infective Therapy*, *16*, 399–410.

Cavalheiro, S., Lopez, A., Serra, S., et al. (2016). Microcephaly and Zika virus: Neonatal neuroradiological aspects. *Child's Nervous System*, *32*, 1057–1060.

da Silva, I. R. F., Frontera, J. A., Bispo de Filippis, A. M., Nascimento, O. J. M., & RIO-GBS-ZIKV Research Group. (2017). Neurologic complications associated with the Zika virus in Brazilian adults. *JAMA Neurology*, *74*, 1190–1198.

Dain Gandelman Horovitz, D., da Silva Pone, M. V., Moura Pone, S., Dias Saad Salles, T. R., & Bastos Boechat, M. C. (2016). Cranial bone collapse in microcephalic infants prenatally exposed to Zika virus infection. *Neurology*, *87*, 118–119.

de Fatima Vasco Aragao, M., van der Linden, V., Brainer-Lima, A. M., et al. (2016). Clinical features and neuroimaging (CT and MRI) findings in presumed Zika virus related congenital infection and microcephaly: Retrospective case series study. *BMJ*, *353*, i1901.

de Freitas Ribeiro, B. N., Muniz, B. C., Gasparetto, E. L., & Marchiori, E. (2018). Congenital involvement of the central nervous system by the Zika virus in a child without microcephaly—Spectrum of congenital syndrome by the Zika virus. *Journal of Neuroradiology*, *45*, 152–153.

Fontes, C. A., Dos Santos, A. A., & Marchiori, E. (2016). Magnetic resonance imaging findings in Guillain-Barré syndrome caused by Zika virus infection. *Neuroradiology*, *58*, 837–838.

Guillemette-Artur, P., Besnard, M., Eyrolle-Guignot, D., et al. (2016). Prenatal brain MRI of fetuses with Zika virus infection. *Pediatric Radiology*, *46*, 1032–1039.

Hazin, A. N., Poretti, A., Turchi Martelli, C. M., et al. (2016). Computed tomographic findings in microcephaly associated with Zika virus. *The New England Journal of Medicine*, *374*, 2193–2195.

Hussain, A., Ali, F., Latiwesh, O. B., & Hussain, S. (2018). A comprehensive review of the manifestations and pathogenesis of Zika virus in neonates and adults. *Cureus*, *10*, e3290.

Hygino da Cruz, L. C., Jr., Nascimento, O. J. M., Lopes, F. P. P. L., & da Silva, I. R. F. (2018). Neuroimaging findings of Zika virus-associated neurologic complications in adults. *AJNR. American Journal of Neuroradiology*, *39*, 1967–1974.

Medina, M. T., & Medina-Montoya, M. (2017). New spectrum of the neurologic consequences of Zika. *Journal of the Neurological Sciences*, *383*, 214–215.

Moore, C. A., Staples, J. E., Dobyns, W. B., et al. (2017). Congenital Zika syndrome: Characterizing the pattern of anomalies for pediatric healthcare providers. *JAMA Pediatrics*, *171*, 288–295.

Muñoz, L. S., Barreras, P., & Pardo, C. A. (2016). Zika virus-associated neurological disease in the adult: Guillain-Barré syndrome, encephalitis, and myelitis. *Seminars in Reproductive Medicine*, *34*, 273–279.

Palacios, E., Clavijo-Prado, C., Ruiz, A., Arias Antun, A., & Julián Duran, E. (2019). Longitudinal extensive transverse myelitis and Zika virus: A diagnostic challenge in a hospital in Colombia. *Neurología*, *34*, 204–206.

Papageorghiou, A. T., Thilaganathan, B., Bilardo, C. M., et al. (2016). ISUOG interim guidance on ultrasound for Zika virus infection in pregnancy: Information for healthcare professionals. *Ultrasound in Obstetrics & Gynecology*, *47*, 530–532.

Peixoto Filho, A. A. A., de Freitas, S. B., Ciosaki, M. M., Oliveira, L. N. E., & Dos Santos Júnior, O. T. (2018). Computed tomography and magnetic resonance imaging findings in infants with microcephaly potentially related to congenital Zika virus infection. *Radiologia Brasileira*, *51*, 119–122.

Reid, S., Rimmer, K., & Thakur, K. (2018). Zika virus and neurologic disease. *Neurologic Clinics*, *36*, 767–787.

Ribeiro, B. G., Werner, H., Lopes, F. P. P. L., et al. (2017). Central nervous system effects of intrauterine Zika virus infection: A pictorial review. *Radiographics*, *37*, 1840–1850.

Ribeiro, B. N. F., Muniz, B. C., Gasparetto, E. L., Ventura, N., & Marchiori, E. (2017). Congenital Zika syndrome and neuroimaging findings: What do we know so far? *Radiologia Brasileira*, *50*, 314–322.

Sejvar, J. J. (2018). Zika virus and other emerging arboviral central nervous system infections. *Continuum (Minneap Minn)*, *24*, 1512–1534.

Soares de Oliveira-Szejnfeld, P., Levine, D., Melo, A. S., et al. (2016). Congenital brain abnormalities and Zika virus: What the radiologist can expect to see prenatally and postnatally. *Radiology*, *281*, 203–218.

Song, B. H., Yun, S. I., Woolley, M., & Lee, Y. M. (2017). Zika virus: History, epidemiology, transmission, and clinical presentation. *Journal of Neuroimmunology*, *308*, 50–64.

Transverse Myelitis Consortium Working Group. (2002). Proposed diagnostic criteria and nosology of acute transverse myelitis. *Neurology*, *59*, 499–505.

van der Linden, V., Pessoa, A., Dobyns, W., et al. (2016). Description of 13 infants born during October 2015–January 2016 with congenital Zika virus infection without microcephaly at birth—Brazil. *MMWR. Morbidity and Mortality Weekly Report*, *65*, 1343–1348.

Venkatesan, A., Tunkel, A. R., Bloch, K. C., et al. (2013). Case definitions, diagnostic algorithms, and priorities in encephalitis: Consensus statement of the international encephalitis consortium. *Clinical Infectious Diseases*, *57*, 1114–1128.

Werner, H., Sodré, D., Hygino, C., et al. (2016). First-trimester intrauterine Zika virus infection and brain pathology: Prenatal and postnatal neuroimaging findings. *Prenatal Diagnosis*, *36*, 785–789.

Ximenes, A. S. F. C., Pires, P., Werner, H., et al. (2019). Neuroimaging findings using transfontanellar ultrasound in newborns with microcephaly: A possible association with congenital Zika virus infection. *The Journal of Maternal-Fetal & Neonatal Medicine*, *32*, 493–501.

Zare Mehrjardi, M., Carteaux, G., Poretti, A., et al. (2017). Neuroimaging findings of postnatally acquired Zika virus infection: A pictorial essay. *Japanese Journal of Radiology*, *35*, 341–349.

Zare Mehrjardi, M., Keshavarz, E., Poretti, A., & Hazin, A. N. (2016). Neuroimaging findings of Zika virus infection: A review article. *Japanese Journal of Radiology*, *34*, 765–770.

Zare Mehrjardi, M., Poretti, A., Huisman, T. A., et al. (2017). Neuroimaging findings of congenital Zika virus infection: A pictorial essay. *Japanese Journal of Radiology*, *35*, 89–94.

Chapter 22

Magnetic modulation biosensing: How it works and how it can be used to detect the Zika virus

Shira Roth and Amos Danielli

Faculty of Engineering, The Institute of Nanotechnology and Advanced Materials, Bar-Ilan University, Ramat Gan, Israel

Abbreviations

DENV	dengue virus
ELISA	enzyme-linked immunosorbent assay
EUA	emergency use authorization
IgG	immunoglobulin G
IgM	immunoglobulin M
LoD	limit of detection
MMB	magnetic modulation biosensing
NS1	nonstructural 1 protein
PSO	post symptom onset
qRT-PCR	quantitative reverse transcription polymerase chain reaction
ROC	receiver operating characteristic
SERS	surface-enhanced Raman spectroscopy
WNV	West Nile virus
ZIKV	Zika virus

Introduction

Infection by flaviviruses can lead to inflammation of the brain, vascular leakage and hypotension, acute liver failure, and death (Leyssen, De Clercq, & Neyts, 2000; Wilder-Smith et al., 2016). The Zika virus (ZIKV) can cause severe neurological phenotypes, including Guillain-Barré syndrome, meningoencephalitis (Carteaux et al., 2016; Miner et al., 2016; Oehler et al., 2014), and fetal abnormalities, such as microcephaly, spontaneous abortion, and intrauterine growth restriction due to placental insufficiency (Brasil et al., 2016; Miner et al., 2016; Petersen, Jamieson, Powers, & Honein, 2016). Diagnosing ZIKV accurately and timely, particularly in pregnant women, can provide compelling personal and social benefits (Petersen et al., 2016).

Similar to other flaviviruses, ZIKV is a spherical, enveloped virus with a positive-strand, nonsegmented RNA genome that produces 10 mature viral proteins, including three structural proteins (envelope (E) protein, capsid (C) protein, and protein prM) and seven nonstructural proteins [e.g., nonstructural 1 (NS1) protein] (Hollidge, Gonzalez-Scarano, & Soldan, 2010). ZIKV diagnostic tests include either direct viral detection tests or serological tests. Direct viral detection tests detect the viral components (e.g., the viral nucleic acids or proteins) in the body fluids (Lustig et al., 2017) by virus isolation (Peeling et al., 2010), antigenemia assays (Bedin et al., 2017; Sanchez-Purra et al., 2017), or reverse transcription-PCR (RT-PCR) (Moureau et al., 2007; Rastogi, Sharma, & Singh, 2016). These methods are considered to be specific and to indicate acute-phase infections (Petersen et al., 2016; Wilder-Smith et al., 2016). However, because ZIKV usually disappears from the blood a few days after the onset of symptoms (Venturi et al., 2016), these tests are applicable only for a short period of time. In addition, they are either time consuming, laborious, or not suitable for point-of-care requirements.

Zika Virus Impact, Diagnosis, Control, and Models. https://doi.org/10.1016/B978-0-12-820267-8.00022-4

Serological tests, which are among the most common techniques used in virology laboratories, detect an immune response agent, such as the antiviral antibodies immunoglobulin M (IgM) and immunoglobulin G (IgG), by enzyme-linked immunosorbent assay (ELISA) tests (L'Huillier et al., 2017; Peeling et al., 2010) or by plaque-reduction neutralization tests (L'Huillier et al., 2017). Because these tests detect the antibodies produced by the body in response to the ZIKV infection, they can be used at any time, even long after the virus has been greatly diminished in the bloodstream.

Due to several inherent weaknesses, serological tests to diagnose ZIKV are less reliable than direct detection tests. First, antibody concentrations vary between individuals and for some people may be very low, below the detectable level of standard ELISA tests. Low concentrations of IgM antibodies may result from previous exposure to other flaviviruses (Peeling et al., 2010) or the presence of secondary infections (Hunsperger et al., 2009). Second, there is extensive cross-reactivity among flaviviruses' antibodies that leads to false positive results (Venturi et al., 2016). Antibodies to other flaviviruses can originate from natural exposure in some geographic locations. For example, Israel is endemic to West Nile virus (WNV) but not to ZIKV, and therefore Israeli travelers who were previously exposed to WNV may be falsely identified as ZIKV positive. Other sources of flavivirus antibodies are vaccines for dengue, yellow fever, tick-borne encephalitis, and Japanese encephalitis (Huzly, Hanselmann, Schmidt-Chanasit, & Panning, 2016). Overall, serological tests lack specificity (Huzly et al., 2016) and suffer from reduced clinical sensitivity (Peeling et al., 2010), high cross-reactivity (Cleton et al., 2015), and lengthy protocols. Hence, new technologies and diagnostic serological assays must be explored (Singh et al., 2017).

Here, we will review state-of-the-art serological and antigenemia assays that are either commercially available or under development. We will focus on a novel technology, termed magnetic modulation biosensing (MMB), which has high sensitivity and specificity, and minimal cross-reactivity with WNV and dengue virus (DENV). We will present the principles of the MMB technology as well as clinical data.

State-of-the-art ZIKV serological and antigenemia assays

On February 26, 2016, the Secretary of Health and Human Services (HHS) declared that "circumstances exist justifying the authorization of the emergency use of *in vitro* diagnostics for detection of Zika virus and/or diagnosis of Zika virus infection" (U.S. Food and Drug Administration. Zika virus emergency use authorization, 2016). Since then the FDA has issued an Emergency Use Authorization (EUA) for several molecular- and serological-based assays for ZIKV. As of November 2019, there are 18 active EUAs for tests to diagnose Zika virus infection. Four of them detect ZIKV antibodies, and 14 of them detect ZIKV nucleic acids. Most of the nucleic acid tests have an analytical sensitivity of 500–5000 units/mL or better. However, even when fully or partially automated, these tests cannot achieve high throughput, and their turnaround time per run is long (2–7 h).

ELISA

The first FDA authorization for marketing a serological assay that detects the ZIKV IgM antibodies targeting the ZIKV envelope glycoproteins was granted in May 2019 for "ZIKV Detect 2.0 IgM Capture ELISA" (InBios, Washington, USA). Previously, tests for detecting ZIKV IgM antibodies—including the ZIKV Detect 2.0 IgM Capture ELISA—had been authorized for emergency use only. This test can be used in patients with clinical signs and symptoms consistent with ZIKV infection and/or who meet the "Centers for Disease Control and Prevention" (CDC) epidemiological criteria, such as geographic region of residence.

Two additional commercially available ELISA tests for detecting both IgM and IgG ZIKV antibodies are EUROIMMUN ZIKV IgM and IgG ELISAs (PerkinElmer, Luebeck, Germany) and Dia.Pro ZIKV IgM and IgG ELISAs (Diagnostic Bioprobes s.r.l, Milan, Italy). In both assays the detected antigen is the NS1 protein. These assays were tested on patients from endemic areas for flaviviruses (Matheus et al., 2019). The sensitivity and specificity of the EUROIMMUN ZIKV IgM test were 49% and 99%, respectively, and for the Dia.Pro IgM test, they were 69% and 96%, respectively. In both the EUROIMMUN and Dia.Pro IgM tests, 3% of the samples gave inconclusive results. The sensitivity and specificity of the EUROIMMUN ZIKV IgG test were 71% and 70%, respectively; for the Dia.Pro IgG test, they were 79% and 62%, respectively. In the EUROIMMUN and Dia.Pro IgG tests, inconclusive results were obtained in 8% and 4% of the samples, respectively. These results fit other studies that reported sensitivities of 37%–68% for IgM and 79% for IgG using the EUROIMMUN ZIKV assays (Lustig et al., 2017; Safronetz et al., 2017). The relatively low sensitivity and specificity highlight the complexity of serologic diagnosis of ZIKV in regions endemic for flaviviruses.

Immune-magnetic assay

In an immune-magnetic antigenemia assay (Blusense Diagnostics, Copenhagen, Denmark), blood flows into a microfluidic device that separates the plasma and transfers it into a chamber containing dried magnetic nanoparticles, which are coupled to anti-ZIKV antibodies. A strong magnetic field aligns the magnetic particles in chains. In a positive ZIKV sample, the target antigens glue the nanoparticles to each other, and therefore when the magnetic field is removed, chain-like clusters of nanoparticles remain. In a negative sample, where there is no antigen, no clusters are formed. To quantify the amount of target molecules, an oscillating magnetic field is applied to the sample and causes the nanoclusters to rotate. When a polarized light shines on the sample, it produces an oscillating signal, which is proportional to the size of the nanoclusters (i.e., to the amount of target molecules) in the sample (Blusense Diagnostics, 2019).

Lateral flow assays

Lateral flow-based assays provide results in a relatively short period of time. For example, the DPP Zika IgM/IgG assay (Chembio Diagnostic Systems, New York, USA) detects both ZIKV IgM and IgG antibodies in less than 15 min and has been authorized by the FDA under an EUA for use by authorized laboratories. To eliminate subjective interpretation the assay is read by a specifically designed optical reader. According to the manufacturer's data, the assay has a specificity of 94%–98%, no cross-reactivity with IgM and 0%–46% for IgG, and good agreement (91%–95%) with conventional laboratory enzyme immunoassays.

Biocan Diagnostic's Tell Me Fast ZIKV IgG/IgM assay (Biocan Diagnostics Inc. Coquitlam, Canada) is a rapid finger-prick assay to detect the IgM and IgG antibodies in minutes. According to the manufacturer's data, it has sensitivity of 92% and specificity of 100%. However, the test readout relies on visual assessment and a qualitative conclusion, which limit the objectivity and information of the assay.

Microfluidic paper-based analytical devices

Further reduction in processing time and assay complexity is provided by microfluidic paper-based analytical devices (μPADs). For example, Bedin et al. developed a microfluidic paper-based antigenemia assay targeting the ZIKV NS1 antigen (Bedin et al., 2017). The test area is functionalized by anti-ZIKV NS1 antibodies, and thus if the ZIKV NS1 is present in the sample, a blue line appears in the test area after 4–6 min. The readout and analysis are performed using a smartphone. The technique is applicable for both whole blood and plasma spiked with recombinant ZIKV NS1 protein, and the limit of detection (LoD) is 10,000 ng/L. The device is easy to produce, but the authors reported a high background noise from nonspecific binding. In addition, the clinical performance of the device was not evaluated (Bedin et al., 2017).

Time-resolved Förster resonance energy transfer

Another serological ZIKV assay targeting the anti-NS1 IgM/IgG antibodies is based on time-resolved Förster resonance energy transfer (FRET) (Kareinen et al., 2019). In this assay, NS1 recombinant protein labeled with europium (Eu) is mixed in a serum sample with a fluorescently labeled (Alexa Fluor 647) Protein L. Protein L binds to any antibody, and therefore if the sample contains ZIKV anti-NS1 IgG or IgM antibodies, both the Alexa Fluor 647-labeled Protein L and the Eu-labeled NS1 protein will bind to them, bringing the Eu (i.e., the donor) and the Alexa Fluor 647 (i.e., the acceptor) labels close to each other. When exposed to UV light, the proximity between the two labels ($\ll$10 nm) allows the donor and acceptor to produce a FRET signal. Moreover, chelated lanthanides, such as the Eu, exhibit a delay between their absorption and emission. Hence, time-resolved detection of the Alexa Fluor 647 signal further improves the signal-to-noise ratio of the measurement.

Three FRET-based assays have been developed: (1) LFRET, measuring all anti-ZIKV-NS1 antibodies; (2) acute-LFRET, measuring IgM and IgA antibodies; and (3) immunity-LFRET, measuring IgG antibodies. The sensitivity, specificity, and cross-reactivity of these assay were evaluated in Finnish and German travelers. The sensitivities of these assays were 92%, 100%, and 83%, respectively. Their specificities were 88%, 95%, and 100%, respectively. Cross-reactivity with DENV, yellow fever, and tick-borne encephalitis of the LFRET ranged between 0% and 13%. The LFRET results correlated ($r = 85$) with the results of the EUROIMMUN ZIKV IgM and IgG kit.

Surface-enhanced Raman spectroscopy (SERS)-based sandwich immunoassays

Surface-enhanced Raman spectroscopy occurs when a metal surface, such as gold, is irradiated at the plasmonic resonance of the metal. The resulted strong local electromagnetic field enhances the Raman scattering of molecules absorbed near the surface (Sanchez-Purra et al., 2017). Utilizing this phenomenon, Sanchez et al. developed a sandwich-based antigenemia assay targeting the ZIKV NS1 protein. Capture ZIKV anti-NS1 antibodies were conjugated to a nitrocellulose strip, and detection ZIKV anti-NS1 antibodies was conjugated to gold nanostars. Immunoassays were performed using the nitrocellulose strip as a dipstick. The gold nanostars mix with the serum samples, and in the present of ZIKV NS1, a sandwich with the nitrocellulose strip is created at the test line. The test line is read using surface-enhanced Raman spectroscopy. For ZIKV NS1 detection the LoD was determined to be 720 ng/L, which is ~15-fold better than a conventional colorimetric readout for the same pair of antibodies (Sanchez-Purra et al., 2017).

Magnetic modulation biosensing (MMB)-based assay

Magnetic modulation biosensing (MMB) is a novel technology that can rapidly detect very low concentrations of biomarkers (Danielli, Porat, Ehrlich, & Arie, 2010). For example, the MMB system can detect pM concentrations of specific nucleic acid fragments in human serum (Margulis, Ashri, Cohen, & Danielli, 2019) or as little as 0.08 ng/L of the human protein interleukin-8 (IL-8) in plasma (Verbarg, Hadass, Olivo, & Danielli, 2017). The MMB system was also demonstrated to identify interactions between proteins in the sub-pM range (Roth et al., 2020).

In an MMB-based ZIKV serological assay (Fig. 1A), magnetic beads are coated with ZIKV NS1 protein that captures ZIKV IgM or IgG antibodies in serum samples. The detection is done using a fluorescently labeled antibody. To increase the sensitivity of fluorescence detection, two electromagnets, one on each side of the glass sample cell, apply an alternating magnetic field gradient to the sample, attracting the beads from the entire solution volume and concentrating them into a small detection area (Fig. 1B and C). To separate the fluorescent signal from the background noise of unbound fluorescent molecules, the oscillating magnetic field gradient moves the bead aggregate from side to side, in and out of a laser beam, creating a flashing signal. The amplitude of the oscillating signal is proportional to the concentration of target molecules. Isolation of the signal from the background eliminates the need for separation and washing steps in the bioassays and therefore shortens the detection time and enables simple operation in non-ab settings.

ZIKV IgM and IgG MMB-based assay vs EUROIMMUN ELISA (analytical sensitivity and dynamic range)

To determine the analytical sensitivity, LoD, and dynamic range of the human IgM and IgG anti-Zika NS1 assays, we measured increasing concentrations of recombinant ZIKV antibodies (Fig. 2A). The same concentrations of recombinant ZIKV antibodies were measured using the EUROIMMUN ZIKV ELISA (Fig. 2B).

The limits of detection of the IgM and IgG serological assays using the MMB system were calculated as three standard deviations from the blank measurement and found to be 99 and 102 ng/L, with dynamic ranges of 3-logs and 2-logs, respectively (Fig. 2A).

In the EUROIMMUN ELISA, only values above 1.1 are regarded as positive (Steinhagen et al., 2016), and therefore it can positively detect IgM and IgG concentrations of 6×10^5 and 5×10^4 ng/L, respectively. In comparison, according to the ROC cutoffs of the MMB assays (Michelson, Lustig, Avivi, Schwartz, & Danielli, 2019), the MMB system can detect IgM and IgG concentrations of 7×10^3 ng/L (an ~85-fold improvement over EUROIMMUN ELISA) and 9×10^3 (a ~fivefold improvement over EUROIMMUN ELISA), respectively.

ZIKV IgM and IgG MMB-based assay vs EUROIMMUN ELISA (clinical sensitivity, specificity, and cross-reactivity)

To determine the clinical sensitivity and specificity of the MMB-based ZIKV assays, we blindly tested serum samples from healthy and ZIKV-positive patients (Michelson et al., 2019). The samples were also tested using the EUROIMMUN ELISA. To determine the cross reactivity of the MMB-based ZIKV assays, we tested them using samples from DENV- and WNV-positive patients.

The sensitivities of the EUROIMMUN ZIKV IgM and IgG assays were 38% and 74%, respectively, whereas the sensitivities of the MMB assays were 88% and 97% (Fig. 3). The specificity of the MMB assays was 100% for both IgM and IgG. The cross-reactivities to WNV and DENV of the IgM and IgG assays were 0% and 4%, respectively (Fig. 3).

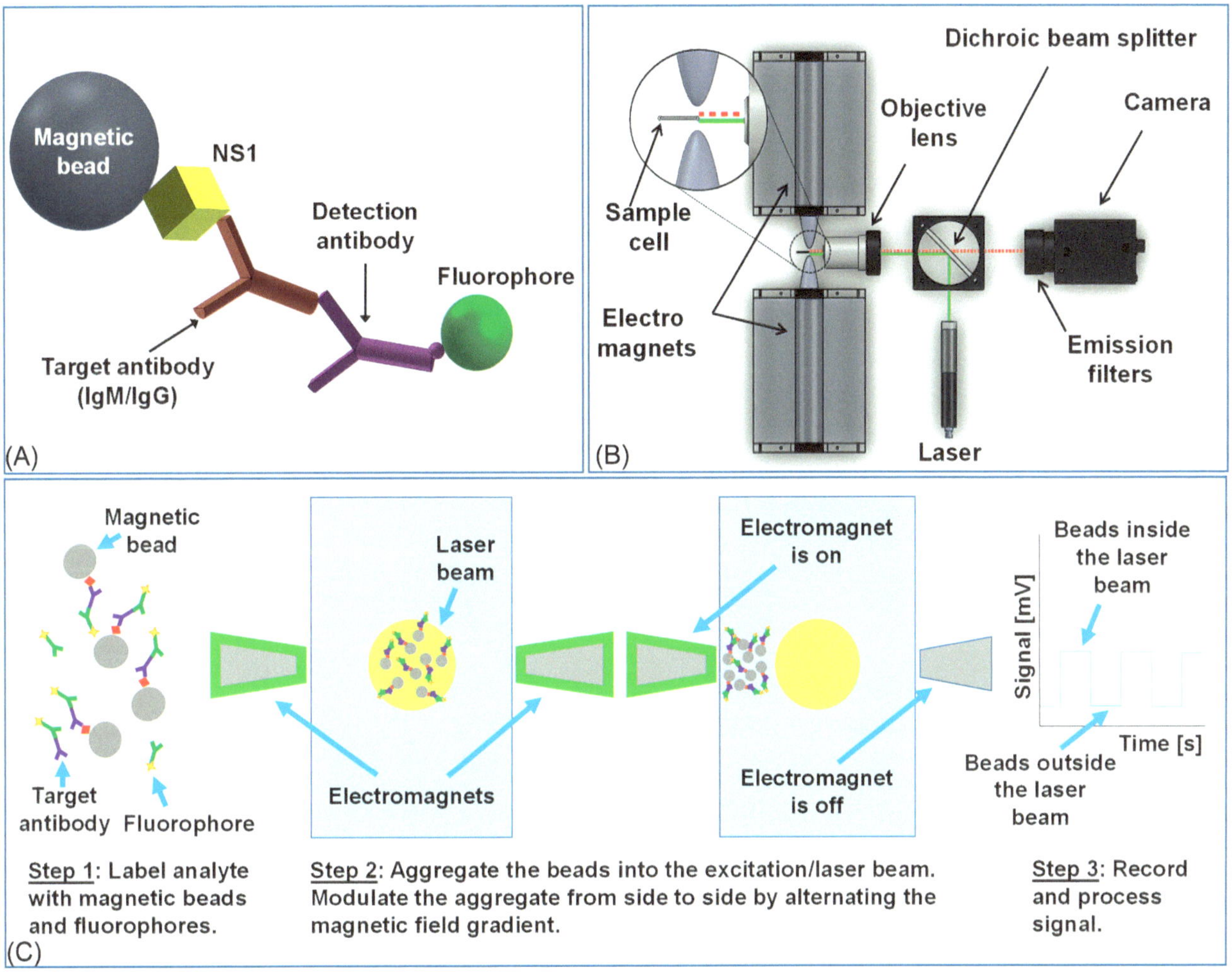

FIG. 1 **Magnetic modulation biosensing system.** (A) A schematic representation of an MMB-based Zika serologic assay. The assay consists of a magnetic bead coated with Zika NS1, a target IgM or IgG antibody, and a fluorescently-labeled detection antibody. (B) An illustration of a magnetic modulation biosensing system. A 532-nm laser beam is reshaped using two planoconvex lenses (not shown in the figure), reflected by a dichroic mirror, and focused by an objective lens onto the sample of magnetic beads aggregate inside a borosilicate cuvette. The laser excites the magnetic beads in the sample, and the emitted fluorescence is collected by the same objective lens, filtered by emission filters, and detected by a CMOS camera. (C) Schematic of the MMB system's assay in three steps: 1. Mix and label analyte with fluorescent probes and magnetic beads. 2. Place sample in the oscillating magnetic field gradient. The oscillating magnetic field gradient aggregates the magnetic beads (still attached to the analyte and fluorophores) in front of the optical detector and modulates them from side to side. 3. Record the time-varying signal and analyze the data. *(From Michelson, Y., Lustig, Y., Avivi, S., Schwartz, E., & Danielli, A. (2019). Highly sensitive and specific Zika virus serological assays using a magnetic modulation biosensing system.* The Journal of Infectious Diseases, 219*(7), 1035–1043. doi: https://doi.org/10.1093/infdis/jiy606 with permission.)*

The MMB ZIKV assays and the EUROIMMUN ELISA assays were also compared for their time-dependent detection limits. The positive samples were placed into five groups according to their days post symptom onset (PSO): 1–7, 8–15, 16–26, 27–59, and ≥60 days. The MMB IgM assay detected IgM antibodies from day 5 to 180 days PSO. In comparison the EUROIMMUN IgM ELISA assay detected IgM antibodies only from day 7 to 29 (Fig. 4A). The MMB IgG assay detected IgG antibodies as early as day 7 PSO, whereas the EUROIMMUN IgG ELISA assay detected IgG antibodies only from day 16 (Fig. 4B). The MMB IgM assay identified 100% of the positive samples for the first three groups (1–7, 8–15, and 16–26), 77% for days 27–59, and 30% for samples ≥60 days PSO. In comparison the EUROIMMUN IgM ELISA assay identified 33%, 100%, 60%, 9%, and 0% of the same groups (Fig. 4A). The MMB IgG assay identified 33% of the positive samples for days 1–7, 80% of the positive samples for days 8–15, and 100% of the positive samples for day 16 onward. In comparison the EUROIMMUN IgG ELISA assay identified 0%, 0%, 80%, 92%, and 0% of the same groups, respectively (Fig. 4B).

Compared with the EUROIMMUN ELISA, the MMB assays exhibited a significantly lower incidence of false positive results. Eleven serum samples that were negative for qRT-PCR and neutralization tests, but positive in the EUROIMMUN

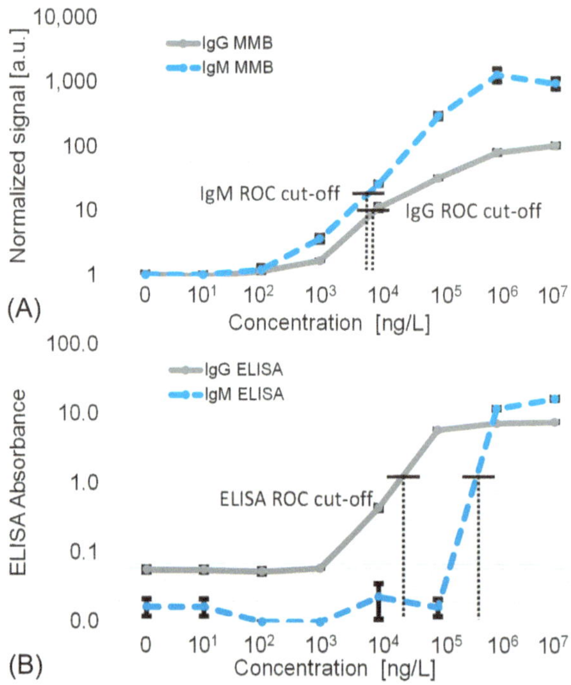

FIG. 2 **Dose responses of recombinant human IgM and human IgG anti-Zika NS1.** (A) MMB-based assays and (B) EUROIMMUN ELISAs. Error bars represent the standard deviation of blank measurements ($n = 6$) and all other samples ($n = 3$). The ROC cutoffs for MMB-based Zika IgG and IgM are 10.85 and 19.6, respectively. The ROC cutoff for EUROIMMUN ELISA Zika IgG and Zika IgM is 1.1. *(From Michelson, Y., Lustig, Y., Avivi, S., Schwartz, E., & Danielli, A. (2019). Highly sensitive and specific Zika virus serological assays using a magnetic modulation biosensing system.* The Journal of Infectious Diseases, *219(7), 1035–1043. doi: https://doi.org/10.1093/infdis/jiy606 with permission.)*

IgM or IgG assays, were tested by the MMB assays. The MMB-based IgM and IgG assays identified 10 out of the 11 samples as negative, and only one sample was falsely identified as positive (Table 1). It should be noted that both the EUROIMMUN ELISA and the MMB-based assays utilize the same ZIKV NS1 capture antigen, but they have different detection antibodies. Hence the higher rate of false positive incidences in the EUROIMMUN ELISA could be attributed to a difference in the specificity of the detection antibodies. Other causes could be the type of capture surface, its functionality, and blocking treatment used in each assay (Michelson et al., 2019).

Future research

Overall, MMB-based serological assays give many fewer false positive results than the EUROIMMUN ELISA. It should be noted that the MMB-based assays were evaluated using serum samples from Israeli travelers returning from Zika endemic areas. Future research should evaluate the sensitivity of the MMB-based assays on Zika-positive samples collected from patients living in Zika endemic areas. In addition, the MMB-based assays were evaluated on a relatively small number of ZIKV-positive and healthy patients. A large-scale clinical study would provide an improved estimate of the assay's sensitivity and specificity, as well as its false positive and false negative rates.

Policy and procedures

MMB system

The MMB system (Fig. 1B) uses a 532-nm laser diode module (ThorLabs, CPS532), working at 0.25 mW, with a 3.5-mm-diameter beam. The beam diameter is focused to a 150 μm spot size by two planoconvex lenses (ThorLabs, 200 mm and 50 mm focal lengths) and an objective lens (Newport, M-10X, 0.25NA). The excitation beam is diverted into an objective lens (M-10X, 0.25NA Newport) using a dichroic mirror (Semrock, BrightLine Di02-R532). The beam exiting the objective lens is focused on a rectangular, borosilicate sample cell, which contains the fluorescently tagged beads and has inner dimensions of 8 mm × 0.4 mm × 70 mm (Vitrocom, W2540). The emitted fluorescence is collected using the same objective

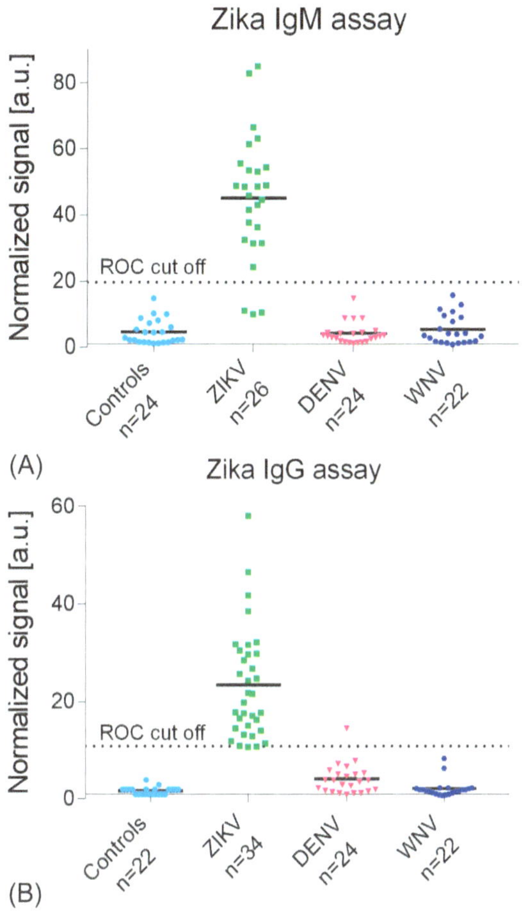

FIG. 3 **Sensitivity, specificity, and cross-reactivity of the MMB Zika IgG and IgM serological assays.** (A) Zika IgM and (B) Zika IgG assay results. All of the Zika-positive samples were taken from qRT-PCR/neutralization-positive patients. The IgM and IgG samples were obtained on days 1–60 and 7–180 PSO, respectively. DENV- and WNV-positive samples were taken from ELISA positive patients. *(From Michelson, Y., Lustig, Y., Avivi, S., Schwartz, E., & Danielli, A. (2019). Highly sensitive and specific Zika virus serological assays using a magnetic modulation biosensing system.* The Journal of Infectious Diseases, 219*(7), 1035–1043. doi: https://doi.org/10.1093/infdis/jiy606 with permission.)*

lens, passes through the dichroic mirror and two emission filters (Semrock, FF03-575/25), and is detected by a digital camera (FLIR, GS3-U3-23S6M-C).

Two electromagnets, one on each side of the glass sample cell, apply an alternating magnetic field gradient to the sample, causing the magnetic beads to group into a small aggregate and move together in a periodic lateral motion inside the sample cell. The modulation frequency of the electromagnets is set to 1 Hz. As the bead aggregate passes in front of the optical laser beam, the fluorescence emitted from the assay complex creates a flashing signal that is easily distinguishable from the constant background of the sample matrix or from unbound fluorescent molecules. The laser beam remains fixed while the beads are moving, thus causing the beads to form an "on" signal when they are in the laser beam and an "off" signal when they have moved out of the laser beam.

A sequence of 600 images is acquired over a period of 12 s. For each image the mean gray value from the laser beam area is extracted and presented as a function of time. The peak-to-peak differences over time are calculated and averaged. The MMB system is enclosed in a lightproof box, which allows for the detection of low analyte concentrations and greatly reduces fluctuations in measurement (Michelson et al., 2019).

Mini-dictionary of terms

Antigenemia test. The detection of a viral antigen (e.g., an envelope or a nonstructural 1 protein) in the blood.

Serological test. The detection of an immune response agent (e.g., immunoglobulin M or immunoglobulin G antibodies) in serum.

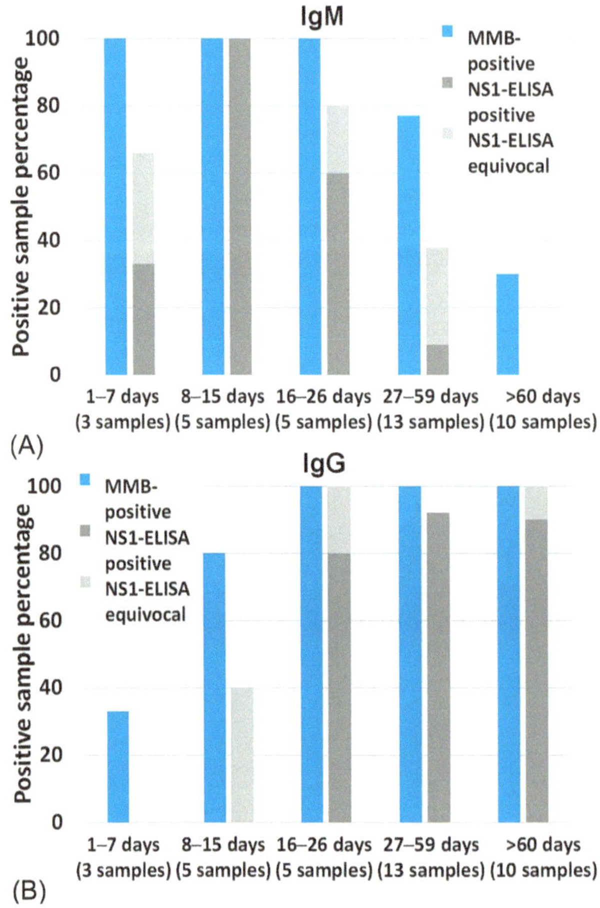

FIG. 4 **Comparison between the MMB Zika immunoassay and the EUROIMMUN ELISA.** Percentage of Zika-positive (A) IgM and (B) IgG samples detected by the MMB and EUROIMMUN ELISA assays. The tested groups included a total of 36 samples from qRT-PCR/neutralization-positive tests of Israeli patients, returning from endemic areas. The samples were obtained 1–7, 8–15, 16–26, 27–59, and ≥60 days after the onset of symptoms. *(From Michelson, Y., Lustig, Y., Avivi, S., Schwartz, E., & Danielli, A. (2019). Highly sensitive and specific Zika virus serological assays using a magnetic modulation biosensing system.* The Journal of Infectious Diseases, 219*(7), 1035–1043. doi: https://doi.org/10.1093/infdis/jiy606 with permission.)*

Diagnostic sensitivity. The percentage of positive patients that were identified as positive. $\frac{\text{True positive}}{\text{True positive} + \text{False negative}}$

Specificity. The percentage of patients that are not sick with the particular pathogen and are identified as negative by an assay. $\frac{\text{True negative}}{\text{True negative} + \text{False positive}}$

Cross-reactivity. The percentage of patients that are tested for a specific flavivirus and are wrongly identified as positive for another flavivirus.

Limit of detection. The lowest amount of analyte that can reliably be detected and differentiated from a blank sample with no analyte.

Key facts of ZIKV serological assays

- Serological assays detect an immune response agent, such as the antiviral antibodies [immunoglobulin M (IgM) and immunoglobulin (IgG)], in serum.
- Due to previous exposure to other flaviviruses or the presence of secondary infections, Zika virus (ZIKV) antibody concentrations vary between individuals. For some individuals, they may be very low, below the detectable level of standard serological tests.
- ZIKV serological assays suffer from high cross-reactivity to other flaviviruses' antibodies, leading to false positive results.

TABLE 1 Characteristics of 10 Israeli patients with Zika-positive NS1-based ELISA results, but negative qRT PCR/ neutralization tests (total of 11 samples, $n=11$).

							IgM		IgG	
Patient#	Country of possible exposure	Gender/ age (year)	RT-PCR result	Neutralization result	Sample number	Time from onset (days)	ELISA	MMB	ELISA	MMB
21	Colombia	M/31	Neg	Neg	1st	5	Pos	Neg	Neg	Neg
22	Mexico	F/28	Neg	Neg	1st	~20	Pos	Neg	Neg	Neg
					2nd	~37	Pos	Neg	Neg	Neg
23	Thailand	F/33	Neg	Neg	1st	~90	Pos	Pos	Neg	Neg
24	Thailand	F/33	Neg	Neg	1st	~45	Pos	Neg	Neg	Neg
25	Philippines	F/28	Neg	Neg	1st	~50	Neg	Neg	Pos	Neg
26	Thailand	F/34	Neg	Neg	1st	8	Equ	Neg	Neg	Neg
27	Central America	M/25	Neg	Neg	1st	N.A.	Pos	Neg	Neg	Neg
28	Thailand	F/25	Neg	Neg	1st	~50	Pos	Neg	Neg	Neg
29	Vietnam/ Thailand/ Singapore	M/34	Neg	Neg	1st	~150	Equ	Neg	Pos	Neg
30	Thailand	M/25	Neg	Neg	1st	15	Pos	Neg	Neg	Neg

Abbreviations: *Equ*, equivocal; *F*, female; *M*, male; *N.A.*, not applicable; *Neg*, negative; *Pos*, positive. All of the samples were negative for dengue IgM and IgG. The first sample of patient no. 22 was positive to WNV IgG. The sample of patient no. 27 was equivocal to WNV IgG.
From Michelson, Y., Lustig, Y., Avivi, S., Schwartz, E., & Danielli, A. (2019). Highly sensitive and specific Zika virus serological assays using a magnetic modulation biosensing system. *The Journal of Infectious Diseases, 219*(7), 1035–1043. doi: https://doi.org/10.1093/infdis/jiy606 with permission.

- Exposure to flaviviruses (e.g., naturally or from vaccines) increases the prevalence of antibodies to these viruses in the blood. This prevalence leads to extensive cross-reactivity between ZIKV and other flaviviruses' antibodies, which may generate a high rate of false positive results.
- The first FDA authorization for marketing a serological assay that detects ZIKV IgM antibodies targeting the ZIKV envelope glycoproteins was granted in May 2019 for "ZIKV Detect 2.0 IgM Capture ELISA" (InBios, Washington, USA).
- As of November 2019, there are four active emergency use authorizations for serological tests detecting ZIKV antibodies.

Summary points

- Current Zika virus (ZIKV) serological tests suffer from reduced sensitivity, high cross-reactivity, low specificity, and lengthy protocols.
- Here, we reviewed a novel ZIKV serological assay—named magnetic modulation biosensing (MMB)—that utilizes the highly-specific nonstructural 1 (NS1) antigen and a highly sensitive detection technology and significantly improves clinical performance.
- In MMB, by combining magnetic and fluorescent labeling of target ZIKV immunoglobulin G (IgG) and immunoglobulin M (IgM) antibodies and then applying an alternating external magnetic field gradient, the sensitivity of fluorescence detection is enhanced.
- MMB-based IgG and IgM assays were blindly tested using 60 reverse transcription polymerase chain reaction Zika-positive samples and healthy patients' serum samples, as well as 44 serum samples from enzyme-linked immunosorbent assay (ELISA) West Nile- and dengue-positive patients. The Zika-positive samples were collected from Israeli travelers returning from Zika endemic areas.

- The MMB-based IgG and IgM assays provided a 5–85-fold improvement over EUROIMMUN ELISA assays in detecting recombinant ZIKV anti-non-structural 1 protein IgG and IgM antibodies.
- The MMB-based ZIKV assays had 88%–97% sensitivity, much higher than the EUROIMMUN ELISA assays (38%–74%). Additionally, their specificity was 100%, and cross-reactivity with West Nile and dengue viruses was minimal (0%–4%).
- The MMB-based IgM and IgG assays significantly broadened the period of time in which the antibodies could be detected. The MMB IgM assay detected IgM antibodies from 5 to 180 days post symptom onset, whereas the EUROIMMUN IgM ELISA assay detected IgM antibodies only from day 7 to 29.
- Compared with the EUROIMMUN ELISA, the MMB-based assays gave significantly fewer false positive results.
- The improved sensitivity and specificity of the MMB-based assay, together with its simplicity, makes it a superior alternative method for ZIKV diagnosis.

References

Bedin, F., Boulet, L., Voilin, E., Theillet, G., Rubens, A., & Rozand, C. (2017). Paper-based point-of-care testing for cost-effective diagnosis of acute flavivirus infections. *Journal of Medical Virology*, *89*(9), 1520–1527. https://doi.org/10.1002/jmv.24806.

Blusense Diagnostics. (2019). From https://blusense-diagnostics.com/technology/.

Brasil, P., Pereira, J. P., Jr., Moreira, M. E., Ribeiro Nogueira, R. M., Damasceno, L., Wakimoto, M., … Nielsen-Saines, K. (2016). Zika virus infection in pregnant women in Rio de Janeiro. *The New England Journal of Medicine*, *375*(24), 2321–2334. https://doi.org/10.1056/NEJMoa1602412.

Carteaux, G., Maquart, M., Bedet, A., Contou, D., Brugieres, P., Fourati, S., … Mekontso Dessap, A. (2016). Zika Virus associated with meningoencephalitis. *The New England Journal of Medicine*, *374*(16), 1595–1596. https://doi.org/10.1056/NEJMc1602964.

Cleton, N. B., Godeke, G. J., Reimerink, J., Beersma, M. F., van Doorn, H. R., Franco, L., … Reusken, C. (2015). Spot the difference-development of a syndrome based protein microarray for specific serological detection of multiple flavivirus infections in travelers. *PLoS Neglected Tropical Diseases*, *9*(3). https://doi.org/10.1371/journal.pntd.0003580, e0003580.

Danielli, A., Porat, N., Ehrlich, M., & Arie, A. (2010). Magnetic modulation biosensing for rapid and homogeneous detection of biological targets at low concentrations. *Current Pharmaceutical Biotechnology*, *11*(1), 128–137.

Hollidge, B. S., Gonzalez-Scarano, F., & Soldan, S. S. (2010). Arboviral encephalitides: Transmission, emergence, and pathogenesis. *Journal of Neuroimmune Pharmacology*, *5*(3), 428–442. https://doi.org/10.1007/s11481-010-9234-7.

Hunsperger, E. A., Yoksan, S., Buchy, P., Nguyen, V. C., Sekaran, S. D., Enria, D. A., … Peeling, R. W. (2009). Evaluation of commercially available anti-dengue virus immunoglobulin M tests. *Emerging Infectious Diseases*, *15*(3), 436–439. https://doi.org/10.3201/eid1503.080923.

Huzly, D., Hanselmann, I., Schmidt-Chanasit, J., & Panning, M. (2016). High specificity of a novel Zika virus ELISA in European patients after exposure to different flaviviruses. *Eurosurveillance*, *21*(16).

Kareinen, L., Hepojoki, S., Huhtamo, E., Korhonen, E. M., Schmidt-Chanasit, J., Hedman, K., … Vapalahti, O. (2019). Immunoassay for serodiagnosis of Zika virus infection based on time-resolved Forster resonance energy transfer. *PLoS One*, *14*(7). https://doi.org/10.1371/journal.pone.0219474, e0219474.

Leyssen, P., De Clercq, E., & Neyts, J. (2000). Perspectives for the treatment of infections with Flaviviridae. *Clinical Microbiology Reviews*, *13*(1), 67–82.

L'Huillier, A. G., Hamid-Allie, A., Kristjanson, E., Papageorgiou, L., Hung, S., Wong, C. F., … Gubbay, J. B. (2017). Evaluation of Euroimmun anti-Zika virus ELISAs (IgM & IgG) for Zika virus serologic testing. *Journal of Clinical Microbiology*. https://doi.org/10.1128/jcm.00442-17.

Lustig, Y., Zelena, H., Venturi, G., Van Esbroeck, M., Rothe, C., Perret, C., … Schwartz, E. (2017). Sensitivity and kinetics of an NS1-based Zika virus enzyme-linked immunosorbent assay in Zika virus-infected travelers from Israel, the Czech Republic, Italy, Belgium, Germany, and Chile. *Journal of Clinical Microbiology*, *55*(6), 1894–1901. https://doi.org/10.1128/JCM.00346-17.

Margulis, M., Ashri, S., Cohen, M., & Danielli, A. (2019). Detecting nucleic acid fragments in serum using a magnetically modulated sandwich assay. *Journal of Biophotonics*, *12*(11). https://doi.org/10.1002/jbio.201900104, e201900104.

Matheus, S., Talla, C., Labeau, B., de Laval, F., Briolant, S., Berthelot, L., … Rousset, D. (2019). Performance of 2 commercial serologic tests for diagnosing Zika Virus infection. *Emerging Infectious Diseases*, *25*(6), 1153–1160. https://doi.org/10.3201/eid2506.180361.

Michelson, Y., Lustig, Y., Avivi, S., Schwartz, E., & Danielli, A. (2019). Highly sensitive and specific Zika virus serological assays using a magnetic modulation biosensing system. *The Journal of Infectious Diseases*, *219*(7), 1035–1043. https://doi.org/10.1093/infdis/jiy606.

Miner, J. J., Cao, B., Govero, J., Smith, A. M., Fernandez, E., Cabrera, O. H., … Diamond, M. S. (2016). Zika virus infection during pregnancy in mice causes placental damage and fetal demise. *Cell*, *165*(5), 1081–1091. https://doi.org/10.1016/j.cell.2016.05.008. S0092-8674(16)30556-6 [pii].

Moureau, G., Temmam, S., Gonzalez, J. P., Charrel, R. N., Grard, G., & de Lamballerie, X. (2007). A real-time RT-PCR method for the universal detection and identification of flaviviruses. *Vector Borne and Zoonotic Diseases*, *7*(4), 467–477. https://doi.org/10.1089/vbz.2007.0206.

Oehler, E., Watrin, L., Larre, P., Leparc-Goffart, I., Lastere, S., Valour, F., … Ghawche, F. (2014). Zika virus infection complicated by Guillain-Barre syndrome—Case report, French Polynesia, December 2013. *Euro Surveillance*, *19*(9), 20720.

Peeling, R. W., Artsob, H., Pelegrino, J. L., Buchy, P., Cardoso, M. J., Devi, S., … Yoksan, S. (2010). Evaluation of diagnostic tests: Dengue. *Nature Reviews Microbiology*, S30–S37. https://doi.org/10.1038/nrmicro2459.

Petersen, L. R., Jamieson, D. J., Powers, A. M., & Honein, M. A. (2016). Zika virus. *New England Journal of Medicine*, *374*(16), 1552–1563. https://doi.org/10.1056/NEJMra1602113.

Rastogi, M., Sharma, N., & Singh, S. K. (2016). Flavivirus NS1: A multifaceted enigmatic viral protein. *Virology Journal*, *13*(1), 131.

Roth, S., Zander, I., Michelson, Y., Ben-David, Y., Banin, E., & Danielli, A. (2020). Identification of protein-protein interactions using a magnetic modulation biosensing system. *Sensors and Actuators B: Chemical*, *303*, 127228. https://doi.org/10.1016/j.snb.2019.127228.

Safronetz, D., Sloan, A., Stein, D. R., Mendoza, E., Barairo, N., Ranadheera, C., … Drebot, M. (2017). Evaluation of 5 commercially available Zika virus immunoassays. *Emerging Infectious Diseases*, *23*(9), 1577–1580. https://doi.org/10.3201/eid2309.162043.

Sanchez-Purra, M., Carre-Camps, M., de Puig, H., Bosch, I., Gehrke, L., & Hamad-Schifferli, K. (2017). Surface-enhanced Raman spectroscopy-based sandwich immunoassays for multiplexed detection of Zika and dengue viral biomarkers. *ACS Infectious Diseases*, *3*(10), 767–776. https://doi.org/10.1021/acsinfecdis.7b00110.

Singh, R. K., Dhama, K., Karthik, K., Tiwari, R., Khandia, R., Munjal, A., … Bueno-Mari, R. (2017). Advances in diagnosis, surveillance, and monitoring of Zika virus: An update. *Frontiers in Microbiology*, *8*, 2677. https://doi.org/10.3389/fmicb.2017.02677.

Steinhagen, K., Probst, C., Radzimski, C., Schmidt-Chanasit, J., Emmerich, P., van Esbroeck, M., … Schlumberger, W. (2016). Serodiagnosis of Zika virus (ZIKV) infections by a novel NS1-based ELISA devoid of cross-reactivity with dengue virus antibodies: A multicohort study of assay performance, 2015 to 2016. *Euro Surveillance*, *21*(50). https://doi.org/10.2807/1560-7917.ES.2016.21.50.30426. 30426 [pii].

U.S. Food and Drug Administration. (2016). *Zika virus emergency use authorization*.

Venturi, G., Zammarchi, L., Fortuna, C., Remoli, M. E., Benedetti, E., Fiorentini, C., … Bartoloni, A. (2016). Authors' reply: Diagnostic challenges to be considered regarding Zika virus in the context of the presence of the vector *Aedes albopictus* in Europe. *Euro Surveillance*, *21*(10). https://doi.org/10.2807/1560-7917.ES.2016.21.10.30163 30163 [pii].

Verbarg, J., Hadass, O., Olivo, P. D., & Danielli, A. (2017). High sensitivity detection of a protein biomarker interleukin-8 utilizing a magnetic modulation biosensing system. *Sensors and Actuators B: Chemical*, *241*, 614–618. https://doi.org/10.1016/j.snb.2016.10.089.

Wilder-Smith, A., Gubler, D. J., Weaver, S. C., Monath, T. P., Heymann, D. L., & Scott, T. W. (2016). Epidemic arboviral diseases: Priorities for research and public health. *The Lancet Infectious Diseases*. https://doi.org/10.1016/s1473-3099(16)30518-7.

Chapter 23

RNA extraction techniques of different body fluids for Zika virus: Blood, genitourinary specimens, saliva, and other relevant fluids

Shannon E. Ronca, Bonnie E. Gulas-Wroblewski, Rebecca B. Kairis, and Kristy O. Murray
Department of Pediatrics, Section of Tropical Medicine, Baylor College of Medicine and Texas Children's Hospital, Houston, TX, United States

Abbreviations

DNA deoxyribonucleic acid
DPO days post symptom onset
PBS phosphate-buffered saline
PCR polymerase chain reaction
RNA ribonucleic acid
WB whole blood
ZIKV Zika virus

Introduction

A wide variety of body fluids have tested positive for Zika virus (ZIKV) ribonucleic acid (RNA), including blood products, genitourinary and gastrointestinal specimens, saliva, breast milk, and cerebrospinal and conjunctival fluids. Although ZIKV was first discovered in human patients in 1952, research on the molecular evaluation of viral shedding in body fluids started during the 2007 outbreak in Yap State, Federated State of Micronesia, with serum being the first body fluid tested for ZIKV by polymerase chain reaction (PCR). The most recent identification of a ZIKV-positive body fluid was anorectal mucus in 2018, underscoring the importance of continued surveillance of all possible body fluids for viral RNA (Table 1). Efficient and timely detection of ZIKV is critical for diagnosing and managing infections and for defining the international burden of disease.

RNA extractions and PCR are critical techniques for viral detection. Although extractions of nucleic acids can be completed with standard reagents, scientific supply companies produce kits for optimizing the process. ZIKV studies discussed herein used extraction kits from QIAGEN (Valencia, CA, USA), bioMérieux (Marcy l'Etoile, France), Bioneer (Alameda, CA, USA), Stratec (Birkenfeld, Germany), Roche Diagnostics (Risch-Rotkreuz, Switzerland), Promega (Madison, WI, USA), Zymo Research (Irvine, CA, USA), and TransGen Biotech (Beijing, China).

Blood products

Blood products, such as serum, plasma, and whole blood (WB), are the most commonly tested body fluids for diagnostics and epidemiological research. Since blood contains both replicating virus and circulating antibodies, testing its components facilitates evaluation of active infections and immunological responses. The first blood transfusion-related case of ZIKV infection was reported in 2013 in French Polynesia (Musso et al., 2014). Following this case, a subset of serum samples from blood donors were tested, and 2.8% were found positive for ZIKV, demonstrating the need for accurate surveillance to

Zika Virus Impact, Diagnosis, Control, and Models. https://doi.org/10.1016/B978-0-12-820267-8.00023-6

TABLE 1 Summary of Zika virus detection using PCR by body fluid.

Body fluid	Representative peer-reviewed study with PCR-positive results	First PCR positive
Serum	Atkinson et al. (2017), Davidson, Slavinski, Komoto, Rakeman, and Weiss (2016), Musso, Roche, Robin, et al. (2015), Murray et al. (2017), Paz-Bailey et al. (2018), Gourinat, O'Connor, Calvez, Goarant, and Dupont-Rouzeyrol (2015), Lanciotti et al. (2008), Heang et al. (2012), Cao-Lormeau et al. (2014), Kutsuna et al. (2014), Besnard, Lastere, Teissier, Cao-Lormeau, and Musso (2014), Musso et al. (2014), Fonseca et al. (2014), Pyke et al. (2014), Waehre, Maagard, Tappe, Cadar, and Schmidt-Chanasit (2014), Campos, Bandeira, and Sardi (2015), Zanluca et al. (2015), Buathong et al. (2015), Pacheco et al. (2016), Tognarelli et al. (2016), Barjas-Castro et al. (2016), Bingham et al. (2016), Campos Rde et al., 2016, Swaminathan, Schlaberg, Lewis, Hanson, and Couturier (2016), Freour et al. (2016), Dupont-Rouzeyrol, Biron, O'Connor, Huguon, and Descloux (2016), Waggoner et al. (2016), Suy et al. (2016), Faria et al. (2017), Joguet et al. (2017), de Araujo et al. (2018), Woon et al. (2019), Sun et al. (2016)	Lanciotti et al.
Urine	Atkinson et al. (2017), Davidson et al. (2016), Murray et al. (2017), Paz-Bailey et al. (2018), Gourinat et al. (2015), Kutsuna et al. (2014), Besnard et al. (2014), Fonseca et al. (2014), D'Ortenzio et al. (2016), Mansuy, Dutertre, et al. (2016), Mansuy, Suberbielle, et al. (2016), Bingham et al. (2016), Campos Rde et al. (2016), Swaminathan et al. (2016), Freour et al. (2016), Arsuaga et al. (2016), Roze et al. (2016), Mead et al. (2018), Joguet et al. (2017), Woon et al. (2019), Botto-Menezes et al. (2019), Barzon et al. (2018), Reyes et al. (2019)	Kutsuna et al.
Breast milk	Besnard et al. (2014), Dupont-Rouzeyrol et al. (2016)	Besnard et al.
Saliva	Murray et al. (2017), Paz-Bailey et al. (2018), Musso, Roche, Nhan, et al. (2015), Besnard et al. (2014), D'Ortenzio et al. (2016), Bingham et al. (2016), Botto-Menezes et al. (2019), Barzon et al. (2018), Reyes et al. (2019)	Besnard et al.
Semen	Atkinson et al. (2017), Musso, Roche, Robin, et al. (2015), Paz-Bailey et al. (2018), Atkinson et al. (2016), D'Ortenzio et al. (2016), Mansuy, Dutertre, et al. (2016), Deckard et al. (2016), Mansuy, Suberbielle, et al. (2016), Freour et al. (2016), Arsuaga et al. (2016), Mead et al. (2018), Joguet et al. (2017), Russell et al. (2017), Barzon et al. (2018)	Musso et al.
Whole blood	Musso, Roche, Robin, et al. (2015), Murray et al. (2017), Prisant et al. (2016), Melo et al. (2016), Arsuaga et al. (2016), Faria et al. (2017), Mansuy et al. (2017), Joguet et al. (2017), Barzon et al. (2018), Reyes et al. (2019)	Musso et al.
Vaginal secretions	Murray et al. (2017), Paz-Bailey et al. (2018), Prisant et al. (2016), Botto-Menezes et al. (2019), Reyes et al. (2019)	Prisant et al.
Plasma	Atkinson et al. (2017), Mansuy, Dutertre, et al. (2016), Mansuy, Suberbielle, et al. (2016), Mansuy et al. (2017), Botto-Menezes et al. (2019), Barzon et al. (2018)	Mansuy et al.
Amniotic fluid	Calvet et al. (2016), Driggers et al. (2016), Melo et al. (2016), Sarno et al. (2016), Jouannic, Friszer, Leparc-Goffart, Garel, & Eyrolle-Guignot (2016), Suy et al. (2016)	Sarno et al.
Cerebrospinal fluid	Sarno et al. (2016), de Araujo et al. (2018)	Sarno et al.
Conjunctival fluid/tears	Sun et al. (2016)	Sun et al.
Anorectal mucus	Botto-Menezes et al. (2019), Li et al. (2018)	Li et al.

First published report of PCR-positive sample for various body fluids and summary of representative peer-reviewed studies with PCR-positive results. *PCR*, polymerase chain reaction.

TABLE 2 Viral shedding in blood products.

Blood fraction	Range (DPO)	Median (DPO)	Mean (DPO)	References
Plasma	1–37	30	19	Mansuy, Suberbielle, et al. (2016), Mansuy et al. (2017), Botto-Menezes et al. (2019), Barzon et al. (2018)
Serum	0–107	21	53	Murray et al. (2017), Paz-Bailey et al. (2018), Lanciotti et al. (2008), Waehre et al. (2014), Driggers et al. (2016), Pacheco et al. (2016), Bingham et al. (2016), Campos Rde et al. (2016), Suy et al. (2016), Faria et al. (2017), Joguet et al. (2017), Tozetto-Mendoza et al. (2019)
Whole blood	3–120	100	61	Murray et al. (2017), Mansuy et al. (2017), Joguet et al. (2017)

Range, median, and mean values for the viral shedding of Zika virus in various blood products. *DPO*, days post symptom onset.

prevent viral transmission and to safeguard donor-supplied blood products, especially with the high rate of asymptomatic infections.

Serum

Isolation and characterization of ZIKV was first based on a serum sample collected during the 2007 epidemic in Yap State (Lanciotti et al., 2008). Serum has continued to be the most frequently used blood product for diagnostic purposes, predominately because the fluid is valuable for both molecular and serological testing (Table 2). Successful RNA extractions from serum can be achieved using various kits. Serum can be stored at 4–8°C for no more than 72 h or at −20°C or lower for longer. Notably, serum should be screened for other arboviruses as coinfection with chikungunya or dengue viruses can lower ZIKV viral loads (Waggoner et al., 2016).

Plasma

Plasma has been used for detection of ZIKV in patients, with viral RNA detection reported from 1 to 37 days post symptom onset (DPO; Table 2). As with serum, RNA extraction is successful using manufacturer's protocols after short-term storage in 4–8°C or long-term storage at −80°C. The short window of viral detection and low levels of viremia make plasma a less ideal candidate for molecular diagnostic testing than other blood products (Table 2).

Whole blood

WB is arguably ideal for diagnostics as RNA is detectable in WB up to 120 DPO (Table 2), remaining positive when plasma and serum samples consistently test negative (Murray et al., 2017). Although manufacturer's protocols have been applied to WB extractions, modified procedures recommended to improve the RNA yield are described elsewhere (Gorchakov, Berry, et al., 2019; Gorchakov, Gulas-Wroblewski, et al., 2019). Extractions can be performed on fresh blood collections in EDTA-preserved tubes or on frozen aliquots. Heparin blood collection tubes should not be used since the lithium additive interferes with PCR.

Genitourinary and gastrointestinal specimens

ZIKV detection is not limited to analyses of blood fractions; genitourinary and gastrointestinal specimens are critical, especially when evaluating the potential for sexual transmission. ZIKV detection in these fluids ranges widely, providing a variety of options for diagnostics at differing stages of infection (Table 3).

Urine

Viral RNA extraction from urine is plagued by challenges associated with its complexity of specimen matrices (Deng, Wang, Ward, Beckham, & McKenna, 2005; El Bali, Diman, Bernard, Roosens, & De Keersmaecker, 2014; El-Khoury, Pierson, Kaoma, Bernardin, & Berchem, 2016; Gibney et al., 2011; Tan, Sahoo, Milligan, Taylor, & Pinsky, 2017) and difficulties related to RNA degradation in banked samples (Gorchakov, Berry, et al., 2019; Gorchakov,

TABLE 3 Viral shedding in genitourinary and gastrointestinal specimens.

Fluid type	Range (DPO)	Median (DPO)	Mean (DPO)	References
Amniotic fluid	56–126	77	91	Calvet et al. (2016), Driggers et al. (2016), Sarno et al. (2016), Suy et al. (2016)
Anorectal mucus	2–14	4	8	Botto-Menezes et al. (2019)
Breast milk	4–20	6	12	Besnard et al. (2014), Dupont-Rouzeyrol et al. (2016), Tozetto-Mendoza et al. (2019)
Semen	7–370	96	188	Atkinson et al. (2017), Musso, Roche, Robin, et al. (2015), Paz-Bailey et al. (2018), Atkinson et al. (2016), D'Ortenzio et al. (2016), Mansuy, Suberbielle, et al. (2016), Arsuaga et al. (2016), Mead et al. (2018), Joguet et al. (2017), Russell et al. (2017), Barzon et al. (2018)
Urine	1–37	20	19	Davidson et al. (2016), Murray et al. (2017), Gourinat et al. (2015), Besnard et al. (2014), D'Ortenzio et al. (2016), Mansuy, Suberbielle, et al. (2016), Bingham et al. (2016), (Campos Rde et al. (2016), Swaminathan et al. (2016), Arsuaga et al. (2016), Roze et al. (2016), Joguet et al. (2017), Botto-Menezes et al. (2019), Barzon et al. (2018), Tozetto-Mendoza et al. (2019)
Vaginal secretions	2–180	12	91	Murray et al. (2017), Paz-Bailey et al. (2018), Prisant et al. (2016), Botto-Menezes et al. (2019), Tozetto-Mendoza et al. (2019), Reyes et al. (2019)

Range, median, and mean values for the viral shedding of Zika virus in various genitourinary and gastrointestinal specimens. *DPO*, days post symptom onset.

Gulas-Wroblewski, et al., 2019; Murray et al., 2017; Tan et al., 2017). Despite these limitations, ZIKV RNA has been detected in urine in multiple studies (Table 1). Viral detection is most successful during the acute stage of infection, with some studies identifying viral RNA years after infection (Table 3). Urine samples are frequently positive even when serum (Bingham et al., 2016; Davidson, Slavinski, Komoto, Rakeman, & Weiss, 2016; Freour et al., 2016; Gourinat, O'Connor, Calvez, Goarant, & Dupont-Rouzeyrol, 2015; Kutsuna et al., 2014; Paz-Bailey et al., 2018; Swaminathan, Schlaberg,

TABLE 4 Example extraction method for urine.

Step	Method
1	To 1 mL of frozen sample, add 70 μL of Zymo Urine Conditioning Buffer and thaw by loading into a mini tube rotator.
2	Once fully thawed, spin at 3000 × *g* for 15 min. This will form a pellet.
3	Discard the supernatant and resuspend pellet in the 200 μL QIAGEN AL + 200 μL QIAGEN AVE + 25 μL protease mixture.
4	Incubate at 56°C for 15 min.
5	Add 250 μL ethanol. Vortex. Let incubate at room temperature for 5 min.
6	Load sample onto the QIAGEN MinElute filter column. Spin at maximum speed. Repeat until all sample used. Discard flow-through.
7	Wash with 500 μL QIAGEN AW1. Discard flow-through.
8	Wash with 500 μL QIAGEN AW2. Discard flow-through.
9	Wash with 500 μL ethanol. Discard flow-through.
10	Dry-spin at maximum speed for 2 min to dry filter.
11	Incubate at 56°C for 3 min to ensure filter is dry and ethanol is evaporated.
12	Load eluate onto the filter column. Incubate at room temperature for 5 min. Spin at maximum speed to elute.

Protocol for the extraction of Zika virus RNA from urine.

Lewis, Hanson, & Couturier, 2016), WB (Musso, Roche, Robin, et al., 2015), plasma (D'Ortenzio et al., 2016; Roze et al., 2016), and saliva specimens test negative (Bingham et al., 2016; D'Ortenzio et al., 2016). Optimization of urine extractions has been conducted, including those recovering RNA from samples in long-term frozen storage (Table 4). Inclusion of urine testing, particularly in the earlier stages of infection, is highly recommended to increase the sensitivity of molecular diagnostic testing for ZIKV. Urine collection is easy and noninvasive, facilitating testing of young children, the elderly, and those providing self-collected specimens.

Semen

Semen has a complex cellular and biochemical composition that impedes recovery of complete and PCR-inhibitor-free genetic material (Christopher-Hennings, Dammen, Nelson, Rowland, & Oberst, 2006; Fieni et al., 2013; Georgiadis et al., 2015; Gorchakov, Berry, et al., 2019). ZIKV RNA has been recovered in semen from patients who display negative PCR results for plasma (D'Ortenzio et al., 2016; Mansuy, Suberbielle, et al., 2016), serum (Atkinson et al., 2016; Freour et al., 2016; Musso, Roche, Robin, et al., 2015), WB (Musso, Roche, Robin, et al., 2015), urine (Atkinson et al., 2016; Mansuy, Suberbielle, et al., 2016), and saliva (D'Ortenzio et al., 2016), most often during mid to late stages of disease. Patients are more likely to have a positive test on semen than they are via urine or serum over time (Joguet et al., 2017). When various samples are collected on the same day, the viral loads of semen are substantially higher than those for serum (Atkinson et al., 2017; Joguet et al., 2017), plasma (Atkinson et al., 2017; Mansuy, Dutertre, et al., 2016), WB (Joguet et al., 2017), and urine (Atkinson et al., 2017; Joguet et al., 2017; Mansuy, Dutertre, et al., 2016). Even semen samples from ZIKV-infected patients with vasectomies and azoospermic conditions have been found positive for ZIKV RNA (Arsuaga, Bujalance, Diaz-Menendez, Vazquez, & Arribas, 2016; Freour et al., 2016), supporting a distal location of ZIKV replication as well as the probability for viral RNA shedding in preejaculate fluids (Arsuaga et al., 2016; D'Ortenzio et al., 2016). Older age, less frequent ejaculations, and symptomatic presentations of infection have all been correlated with prolonged shedding of ZIKV RNA (Mansuy, Suberbielle, et al., 2016; Mead et al., 2018).

High DNA content of semen and its viscosity necessitate optimizations for successful viral detection (Gorchakov, Berry, et al., 2019) (Table 5). Examples of successful adjustments to standard protocols are pretreatment of thawed semen with dithiothreitol, dilution of samples in equal amounts of lysis buffer, and reduction of input cell volumes (Arsuaga et al., 2016; Atkinson et al., 2017; Deckard et al., 2016). The recovery of 29% ZIKV RNA from semen samples processed using a higher volume of RLT Plus buffer (QIAGEN) and RNeasy Mini Spin Columns (QIAGEN) was reported to be acceptably equivalent to an average recovery of 31% ZIKV RNA from semen samples extracted via magnetic bead-based automated sample RNA extraction (Gorchakov, Berry, et al., 2019; Paz-Bailey et al., 2018). The use of dithiothreitol on and the

TABLE 5 Example extraction method for semen.

Step	Method
1	To 200 μL sample, add 1.6 mL of QIAGEN RLT buffer/beta-mercaptoethanol (βME) mixture (instead of QIAGEN Buffer AL/cRNA) and 50 μL protease. Vortex.
2	Incubate at 56°C for 30 min.
3	Add 1400 μL ethanol. Vortex. Let incubate at room temperature for 5 min.
4	Load sample onto the QIAGEN Mini Spin filter column. Spin at maximum speed. Repeat until all sample used. Discard flow-through.
5	Wash with 500 μL QIAGEN AW1. Discard flow-through.
6	Wash with 500 μL QIAGEN AW2. Discard flow-through.
7	Wash with 500 μL ethanol. Discard flow-through.
8	Dry-spin at maximum speed for 2 min to dry filter.
9	Incubate at 56°C for 3 min to ensure filter is dry and ethanol is evaporated.
10	Load eluate onto the filter column. Incubate at 56°C for 5 min. Spin at maximum speed to elute. Reload the eluate onto the column and repeat spin for final product.

Protocol for the extraction of Zika virus RNA from semen.

TABLE 6 Example extraction method for vaginal secretion swabs, saliva, and breast milk.

Step	Method
1	To 200 μL sample, add 200 μL QIAGEN AL Buffer and 50 μL protease. Vortex.
2	Incubate at 56°C for 15 min.
3	Add 250 μL ethanol. Vortex. Let incubate at room temperature for 5 min.
4	Load sample onto the QIAGEN MinElute filter column. Spin at 3000x *g*. Repeat until all sample used. Discard flow-through.
5	Wash with 500 μL QIAGEN AW1. Discard flow-through.
6	Wash with 500 μL QIAGEN AW2. Discard flow-through.
7	Wash with 500 μL ethanol. Discard flow-through.
8	Dry-spin at max speed for 2 min to dry filter.
9	Incubate at 56°C for 3 min to ensure filter is dry and ethanol is evaporated.
10	Load eluate onto the filter column. Incubate at 56°C for 5 min. Spin at maximum speed to elute.

Protocol for the extraction of Zika virus RNA from vaginal secretion swabs, saliva, and breast milk samples.

presence of certain immunosuppressive drugs in semen specimens may reduce recovery of ZIKV RNA and/or inhibit PCR reactions (Atkinson et al., 2017; Deckard et al., 2016).

Vaginal secretions

ZIKV RNA shedding occurs for a prolonged period in vaginal fluids in comparison with other body fluids (Table 3). For instance, a patient's cervical mucus specimen tested positive 11 DPO, while samples of WB and urine tested negative (Prisant et al., 2016). Similar to semen, several challenges exist in the utilization of vaginal secretions for viral RNA extraction. Blood contamination is possible during menstruation, and the high cellular content may hinder collection of sufficient cellular nucleic acids from the column filter during elution. To address this, extended incubation at a temperature above that recommended by manufacturer's guides are recommended to maximize release of this cellular content (Gorchakov, Berry, et al., 2019) (Table 6).

Vaginal fluid specimens have been collected with the use of genital, endocervical, and cervical mucosal swabs (Botto-Menezes et al., 2019; Gorchakov, Berry, et al., 2019; Murray et al., 2017; Reyes et al., 2019; Tozetto-Mendoza et al., 2019). Following collection, swabs were processed in a variety of means prior to storage. When the BBL CultureSwab Collection and Transport System (Becton Dickinson) was used, specimens were placed within the provided swab transport media for storage (in one case up to 4 months at 4°C) until RNA extraction (Murray et al., 2017; Reyes et al., 2019). Other samples were diluted in 1 mL of sterile Hank's Balanced Salt Solution or submerged in DNAse-free water and slowly homogenized, ultimately for prolonged storage at −80°C (Botto-Menezes et al., 2019; Tozetto-Mendoza et al., 2019). An extended incubation phase prior to extraction was described in two studies: in one, vaginal excretions were incubated in 250 μL of AL/carrier RNA lysis buffer for 10 min at room temperature followed by the addition of 200 μL of phosphate-buffered saline (PBS) (Murray et al., 2017). In another, swabs were incubated in 140 μL of PBS at room temperature for 10 min (Reyes et al., 2019).

Amniotic fluid

Positive amniotic fluid samples have been recovered at 8–18 weeks of pregnancy following the onset of ZIKV symptoms (Calvet et al., 2016; Driggers et al., 2016; Sarno et al., 2016; Suy et al., 2016). Amniotic fluid testing provides an opportunity for molecular diagnosis well after the acute phase of ZIKV infection, and amniotic fluids remain positive when urine (Calvet et al., 2016; Suy et al., 2016), serum (Calvet et al., 2016), and vaginal secretions test negative on PCR (Suy et al., 2016). The QIAamp MinElute Virus Spin Kit (QIAGEN) has been the only kit reported for use in RNA extraction from amniotic fluids (Calvet et al., 2016; Melo et al., 2016). Modifications include preextraction of ZIKV RNA by filtering a 0.5-mL sample through 0.4-μm filters, centrifugation at 21,130 × *g* and 15,000 rpm for 90 min at 4°C, and then incubation of the resulting pellet with deoxyribonuclease and ribonuclease A at 37°C for 90 min. The authors also suggest omitting carrier RNA from the manufacturer's extraction protocol (Calvet et al., 2016).

Anorectal mucus

Although sampling of ZIKV RNA from anorectal mucus may be hampered by patient unease with the procedure, recovery of viral RNA from this secretion provides an additional means through which to perform molecular diagnoses during the early phases of infection (Table 3) and support the potential for ZIKV transmission via anal intercourse. Rectal swabs were diluted in 1 mL of sterile Hank's Balanced Salt Solution (Thermo Fisher Scientific) and stored at −80°C until RNA extraction was executed with the Maxwell 16 Viral Total Nucleic Acid Purification Kit developed by Promega Corporation (Botto-Menezes et al., 2019). Anal lavage was facilitated by the flushing of 150-μL PBS per mouse, while human and mouse feces were suspended in DMEM. Both mixtures were processed equivalently using manufacturer's instructions and the EasyPure Viral DNA/RNA Kit (TransGen Biotech) to extract ZIKV RNA (Li et al., 2018).

Additional fluids

Saliva samples

ZIKV RNA was first detected in saliva samples that were collected from a mother and her infant during a French Polynesian outbreak of the flavivirus (Besnard, Lastere, Teissier, Cao-Lormeau, & Musso, 2014). Subsequently, saliva-extracted ZIKV RNA has been recovered by multiple investigations (Table 1), and the transmission of ZIKV via saliva has been proposed for a variety of otherwise-inexplicable ZIKV cases, such as that of an Australian traveler who likely became infected with the virus through a monkey bite in Bali, Indonesia (Leung, Baird, Druce, & Anstey, 2015). Viral shedding of ZIKV in saliva is restricted to the acute phase of the disease (2–21 DPO) (Barzon et al., 2018; Bingham et al., 2016; Botto-Menezes et al., 2019; D'Ortenzio et al., 2016; Murray et al., 2017; Musso, Roche, Nhan, et al., 2015), and saliva may be positive when plasma (D'Ortenzio et al., 2016) and serum (Bingham et al., 2016; Leung et al., 2015; Musso, Roche, Nhan, et al., 2015) are negative. Molecular investigations of ZIKV shedding in saliva versus blood products across 182 patients suggest that sampling saliva on day three of symptom onset optimizes diagnosis of ZIKV (Musso, Roche, Nhan, et al., 2015). Although performing PCR-based testing on saliva samples fails to extend the temporal frame for diagnosis beyond the acute phase of infection, saliva-recovered RNA can significantly increase the rate of molecular detection of ZIKV during early stages of the disease (Musso, Roche, Nhan, et al., 2015). Saliva specimens are easily and painlessly collected from patients.

Saliva samples with detectable RNA are typically collected via oral swabs (Dudley et al., 2016; Murray et al., 2017; Musso, Roche, Nhan, et al., 2015). Preextraction steps have been recommended, including vortexing the swab in 2 mL of lysis buffer prior to following manufacturer's instructions for kit usage (Musso, Roche, Nhan, et al., 2015), storage in either RNAlater (Thermo Fisher) or viral transport media followed by vortexing and centrifugation (Gorchakov, Berry, et al., 2019; Dudley et al., 2016) or incubation at room temperature in 250 μL of Buffer AL/carrier RNA lysis buffer (QIAGEN) for 10 min, followed by the addition of 200 μL of PBS pre-RNA extraction (Gorchakov, Berry, et al., 2019; Murray et al., 2017). The recommended extraction method follows a modified kit protocol with a more stringent elution step to optimize a more reliable means for the recovery of significant ZIKV RNA from samples with variable DNA/RNA content (Gorchakov, Berry, et al., 2019) (Table 6).

Breast milk

Researchers first described ZIKV shedding in breast milk of infected mothers during the 2013–2014 outbreak in French Polynesia (Besnard et al., 2014) (Table 1). Viral loads peak in the early stages of disease when even serum specimens test negative (4–20 DPO) (Besnard et al., 2014). For this reason, breast milk should be incorporated into ZIKV RNA sampling procedures whenever possible to ensure the most sensitive screening for infection as possible.

When testing breast milk, samples should be collected before breastfeeding or after thorough cleaning of the nipple area to avoid possible contamination from the neonate's saliva (Dupont-Rouzeyrol, Biron, O'Connor, Huguon, & Descloux, 2016). Reminiscent of vaginal secretions, breast milk has a high cellular content, mandating extended incubation at high temperatures during the elution step to facilitate the release of viral RNA from column filters (Gorchakov, Berry, et al., 2019) (Table 6). Previous investigations have employed both the EasyMag automated system (bioMérieux) and the

QIAamp MinElute Virus Spin Kit (QIAGEN) for RNA extraction (Gorchakov, Berry, et al., 2019; Tozetto-Mendoza et al., 2019).

Policy and procedures

Accurate and timely diagnoses of ZIKV throughout the acute and convalescent stages of infection can be enhanced by the strategic testing of multiple body fluids. Given the restricted timeframe of viral detection in serum, molecular testing of other specimen types would optimize diagnostic efficacy. Prolonged detection described in WB and genitourinary specimens provide a wider diagnostic window. Inclusion of urine samples, especially in the earlier stages of infection, is highly recommended to substantially increase the sensitivity of molecular diagnostic testing for ZIKV in patients. Although performing PCR-based testing on saliva samples fails to extend the temporal frame for diagnosis beyond the acute phase of infection, saliva-recovered RNA can significantly increase the rate of molecular detection of ZIKV during early stages of the disease (Musso, Roche, Nhan, et al., 2015). The discovery of elevated viral loads of ZIKV in all constituents of semen verifies the effectiveness of semen sampling for molecular diagnostics even among men with vasectomies and other semen-component-inhibiting conditions. Addition of vaginal secretion testing to molecular diagnostic protocols for ZIKV will increase the probability of detection, most notably in the convalescent phase of infection. RNA extraction from amniotic fluids offers an additional option for molecular-based ZIKV diagnosis of both mother and fetus and should be incorporated into surveillance programs whenever possible. Importantly the methods discussed in this chapter have not been approved by regulatory agencies in various countries for standard or emergency use and are typically for research purposes only. It is critical to incorporate these methods for regular use in clinical applications in conjunction with currently approved procedures to expand not only diagnostic effectiveness but also our understanding of ZIKV transmission dynamics.

Mini-dictionary of terms

Elution. Release of a material from a matrix, such as adding water to a column for ribonucleic acid extraction.
Extraction. Procedure to isolate deoxyribonucleic acid and/or ribonucleic acid from a specific sample.
PCR. Polymerase chain reaction, a procedure used to amplify and detect genetic material in a sample.
Shedding. The release of viral particles.
Viremia. The presence of virus in blood products.

Key facts of viral diagnostics

- Zika virus has been detected in most body fluids throughout the course of infection.
- Viral ribonucleic acid (RNA) is only present in the serum for a short period of time, making the window of detection small.
- Viral RNA can be detected in the whole blood (WB) and genitourinary specimens for up to 3 months, making these ideal for detection.
- Optimized methods must be applied for appropriate detection in different body fluids.
- There are no approved diagnostics that use WB or other body fluids; all approved diagnostics depend on serum.

Summary points

- A variety of methods are available to extract Zika virus ribonucleic acid (RNA) from bodily fluids.
- Although manufacturers' protocols may be effective, modified extraction techniques may lead to higher yields of RNA.
- Zika virus RNA has been detected in most body fluids.
- The length of time Zika virus RNA can be detected varies by sample type.
- Zika virus RNA is detectable for the longest time in whole blood (WB) and genitourinary specimens.

References

Arsuaga, M., Bujalance, S. G., Diaz-Menendez, M., Vazquez, A., & Arribas, J. R. (2016). Probable sexual transmission of Zika virus from a vasectomised man. *The Lancet Infectious Diseases*, *16*(10), 1107. https://doi.org/10.1016/S1473-3099(16)30320-6.

Atkinson, B., Hearn, P., Afrough, B., Lumley, S., Carter, D., Aarons, E. J., … Hewson, R. (2016). Detection of Zika virus in semen. *Emerging Infectious Diseases*, *22*(5), 940. https://doi.org/10.3201/eid2205.160107.

Atkinson, B., Thorburn, F., Petridou, C., Bailey, D., Hewson, R., Simpson, A. J., … Aarons, E. J. (2017). Presence and persistence of Zika virus RNA in semen, United Kingdom, 2016. *Emerging Infectious Diseases*, *23*(4), 611–615. https://doi.org/10.3201/eid2304.161692.

Barjas-Castro, M. L., Angerami, R. N., Cunha, M. S., Suzuki, A., Nogueira, J. S., Rocco, I. M., … Addas-Carvalho, M. (2016). Probable transfusion-transmitted Zika virus in Brazil. *Transfusion*, *56*(7), 1684–1688. https://doi.org/10.1111/trf.13681.

Barzon, L., Percivalle, E., Pacenti, M., Rovida, F., Zavattoni, M., Del Bravo, P., … Baldanti, F. (2018). Virus and antibody dynamics in travelers with acute Zika virus infection. *Clinical Infectious Diseases*, *66*(8), 1173–1180. https://doi.org/10.1093/cid/cix967.

Besnard, M., Lastere, S., Teissier, A., Cao-Lormeau, V., & Musso, D. (2014). Evidence of perinatal transmission of Zika virus, French Polynesia, December 2013 and February 2014. *Euro Surveillance*, *19*(13). Retrieved from https://www.ncbi.nlm.nih.gov/pubmed/24721538.

Bingham, A. M., Cone, M., Mock, V., Heberlein-Larson, L., Stanek, D., Blackmore, C., & Likos, A. (2016). Comparison of test results for Zika virus RNA in urine, serum, and saliva specimens from persons with travel-associated Zika virus disease—Florida, 2016. *MMWR. Morbidity and Mortality Weekly Report*, *65*(18), 475–478. https://doi.org/10.15585/mmwr.mm6518e2.

Botto-Menezes, C. H. A., Neto, A. M., Calvet, G. A., Kara, E. O., Lacerda, M. V. G., Castilho, M. D. C., … Team, Z. S. (2019). Zika virus in rectal swab samples. *Emerging Infectious Diseases*, *25*(5), 951–954. https://doi.org/10.3201/eid2505.180904.

Buathong, R., Hermann, L., Thaisomboonsuk, B., Rutvisuttinunt, W., Klungthong, C., Chinnawirotpisan, P., … Plipat, T. (2015). Detection of Zika virus infection in Thailand, 2012-2014. *The American Journal of Tropical Medicine and Hygiene*, *93*(2), 380–383. https://doi.org/10.4269/ajtmh.15-0022.

Calvet, G., Aguiar, R. S., Melo, A. S. O., Sampaio, S. A., de Filippis, I., Fabri, A., … de Filippis, A. M. B. (2016). Detection and sequencing of Zika virus from amniotic fluid of fetuses with microcephaly in Brazil: A case study. *The Lancet Infectious Diseases*, *16*(6), 653–660. https://doi.org/10.1016/S1473-3099(16)00095-5.

Campos, G. S., Bandeira, A. C., & Sardi, S. I. (2015). Zika virus outbreak, Bahia, Brazil. *Emerging Infectious Diseases*, *21*(10), 1885–1886. https://doi.org/10.3201/eid2110.150847.

Campos Rde, M., Cirne-Santos, C., Meira, G. L., Santos, L. L., de Meneses, M. D., Friedrich, J., … Ferreira, D. F. (2016). Prolonged detection of Zika virus RNA in urine samples during the ongoing Zika virus epidemic in Brazil. *Journal of Clinical Virology*, *77*, 69–70. https://doi.org/10.1016/j.jcv.2016.02.009.

Cao-Lormeau, V. M., Roche, C., Teissier, A., Robin, E., Berry, A. L., Mallet, H. P., … Musso, D. (2014). Zika virus, French Polynesia, South Pacific, 2013. *Emerging Infectious Diseases*, *20*(6), 1085–1086. https://doi.org/10.3201/eid2006.140138.

Christopher-Hennings, J., Dammen, M., Nelson, E., Rowland, R., & Oberst, R. (2006). Comparison of RNA extraction methods for the detection of porcine reproductive and respiratory syndrome virus from boar semen. *Journal of Virological Methods*, *136*(1–2), 248–253. https://doi.org/10.1016/j.jviromet.2006.03.012.

Davidson, A., Slavinski, S., Komoto, K., Rakeman, J., & Weiss, D. (2016). Suspected female-to-male sexual transmission of Zika virus—New York City, 2016. *MMWR. Morbidity and Mortality Weekly Report*, *65*(28), 716–717. https://doi.org/10.15585/mmwr.mm6528e2.

de Araujo, T. V. B., Ximenes, R. A. A., Miranda-Filho, D. B., Souza, W. V., Montarroyos, U. R., de Melo, A. P. L., … State Health Department of Pernambuco. (2018). Association between microcephaly, Zika virus infection, and other risk factors in Brazil: Final report of a case-control study. *The Lancet Infectious Diseases*, *18*(3), 328–336. https://doi.org/10.1016/S1473-3099(17)30727-2.

Deckard, D. T., Chung, W. M., Brooks, J. T., Smith, J. C., Woldai, S., Hennessey, M., … Mead, P. (2016). Male-to-male sexual transmission of Zika virus—Texas, January 2016. *MMWR. Morbidity and Mortality Weekly Report*, *65*(14), 372–374. https://doi.org/10.15585/mmwr.mm6514a3.

Deng, M. Y., Wang, H., Ward, G. B., Beckham, T. R., & McKenna, T. S. (2005). Comparison of six RNA extraction methods for the detection of classical swine fever virus by real-time and conventional reverse transcription-PCR. *Journal of Veterinary Diagnostic Investigation*, *17*(6), 574–578. https://doi.org/10.1177/104063870501700609.

D'Ortenzio, E., Matheron, S., Yazdanpanah, Y., de Lamballerie, X., Hubert, B., Piorkowski, G., … Leparc-Goffart, I. (2016). Evidence of sexual transmission of Zika virus. *The New England Journal of Medicine*, *374*(22), 2195–2198. https://doi.org/10.1056/NEJMc1604449.

Driggers, R. W., Ho, C. Y., Korhonen, E. M., Kuivanen, S., Jaaskelainen, A. J., Smura, T., … Vapalahti, O. (2016). Zika virus infection with prolonged maternal viremia and fetal brain abnormalities. *The New England Journal of Medicine*, *374*(22), 2142–2151. https://doi.org/10.1056/NEJMoa1601824.

Dudley, D. M., Aliota, M. T., Mohr, E. L., Weiler, A. M., Lehrer-Brey, G., Weisgrau, K. L., … O'Connor, D. H. (2016). A rhesus macaque model of Asian-lineage Zika virus infection. *Nature Communications*, *7*. https://doi.org/10.1038/ncomms12204, 12204.

Dupont-Rouzeyrol, M., Biron, A., O'Connor, O., Huguon, E., & Descloux, E. (2016). Infectious Zika viral particles in breastmilk. *Lancet*, *387*(10023), 1051. https://doi.org/10.1016/S0140-6736(16)00624-3.

El Bali, L., Diman, A., Bernard, A., Roosens, N. H., & De Keersmaecker, S. C. (2014). Comparative study of seven commercial kits for human DNA extraction from urine samples suitable for DNA biomarker-based public health studies. *Journal of Biomolecular Techniques*, *25*(4), 96–110. https://doi.org/10.7171/jbt.14-2504-002.

El-Khoury, V., Pierson, S., Kaoma, T., Bernardin, F., & Berchem, G. (2016). Assessing cellular and circulating miRNA recovery: The impact of the RNA isolation method and the quantity of input material. *Scientific Reports*, *6*. https://doi.org/10.1038/srep19529, 19529.

Faria, N. R., Quick, J., Claro, I. M., Theze, J., de Jesus, J. G., Giovanetti, M., … Pybus, O. G. (2017). Establishment and cryptic transmission of Zika virus in Brazil and the Americas. *Nature*, *546*(7658), 406–410. https://doi.org/10.1038/nature22401.

Fieni, F., Stone, M., Ma, Z. M., Dutra, J., Fritts, L., & Miller, C. J. (2013). Viral RNA levels and env variants in semen and tissues of mature male rhesus macaques infected with SIV by penile inoculation. *PLoS One*, *8*(10), e76367. https://doi.org/10.1371/journal.pone.0076367.

Fonseca, K., Meatherall, B., Zarra, D., Drebot, M., MacDonald, J., Pabbaraju, K., ... Tellier, R. (2014). First case of Zika virus infection in a returning Canadian traveler. *The American Journal of Tropical Medicine and Hygiene*, *91*(5), 1035–1038. https://doi.org/10.4269/ajtmh.14-0151.

Freour, T., Mirallie, S., Hubert, B., Splingart, C., Barriere, P., Maquart, M., & Leparc-Goffart, I. (2016). Sexual transmission of Zika virus in an entirely asymptomatic couple returning from a Zika epidemic area, France, April 2016. *Euro Surveillance*, *21*(23). https://doi.org/10.2807/1560-7917.ES.2016.21.23.30254.

Georgiadis, A. P., Kishore, A., Zorrilla, M., Jaffe, T. M., Sanfilippo, J. S., Volk, E., ... Yatsenko, A. N. (2015). High quality RNA in semen and sperm: Isolation, analysis and potential application in clinical testing. *The Journal of Urology*, *193*(1), 352–359. https://doi.org/10.1016/j.juro.2014.07.107.

Gibney, K. B., Lanciotti, R. S., Sejvar, J. J., Nugent, C. T., Linnen, J. M., Delorey, M. J., ... Fischer, M. (2011). West nile virus RNA not detected in urine of 40 people tested 6 years after acute West Nile virus disease. *The Journal of Infectious Diseases*, *203*(3), 344–347. https://doi.org/10.1093/infdis/jiq057.

Gorchakov, R., Berry, R. M., Patel, S. M., El Sahly, H. M., Ronca, S. E., & Murray, K. O. (2019). Optimizing PCR detection of Zika virus from various body fluids. *The American Journal of Tropical Medicine and Hygiene*, *100*(2), 427–433. https://doi.org/10.4269/ajtmh.18-0755.

Gorchakov, R., Gulas-Wroblewski, B. E., Ronca, S. E., Ruff, J. C., Nolan, M. S., Berry, R., ... Murray, K. O. (2019). Optimizing PCR detection of West Nile virus from body fluid specimens to delineate natural history in an infected human cohort. *International Journal of Molecular Sciences*, *20*(8). https://doi.org/10.3390/ijms20081934.

Gourinat, A. C., O'Connor, O., Calvez, E., Goarant, C., & Dupont-Rouzeyrol, M. (2015). Detection of Zika virus in urine. *Emerging Infectious Diseases*, *21*(1), 84–86. https://doi.org/10.3201/eid2101.140894.

Heang, V., Yasuda, C. Y., Sovann, L., Haddow, A. D., Travassos da Rosa, A. P., Tesh, R. B., & Kasper, M. R. (2012). Zika virus infection, Cambodia, 2010. *Emerging Infectious Diseases*, *18*(2), 349–351. https://doi.org/10.3201/eid1802.111224.

Joguet, G., Mansuy, J. M., Matusali, G., Hamdi, S., Walschaerts, M., Pavili, L., ... Bujan, L. (2017). Effect of acute Zika virus infection on sperm and virus clearance in body fluids: A prospective observational study. *The Lancet Infectious Diseases*, *17*(11), 1200–1208. https://doi.org/10.1016/S1473-3099(17)30444-9.

Jouannic, J. M., Friszer, S., Leparc-Goffart, I., Garel, C., & Eyrolle-Guignot, D. (2016). Zika virus infection in French Polynesia. *Lancet*, *387*(10023), 1051–1052. https://doi.org/10.1016/S0140-6736(16)00625-5.

Kutsuna, S., Kato, Y., Takasaki, T., Moi, M., Kotaki, A., Uemura, H., ... Ohmagari, N. (2014). Two cases of Zika fever imported from French Polynesia to Japan, December 2013 to January 2014 [corrected]. *Euro Surveillance*, *19*(4). https://doi.org/10.2807/1560-7917.es2014.19.4.20683.

Lanciotti, R. S., Kosoy, O. L., Laven, J. J., Velez, J. O., Lambert, A. J., Johnson, A. J., ... Duffy, M. R. (2008). Genetic and serologic properties of Zika virus associated with an epidemic, Yap State, Micronesia, 2007. *Emerging Infectious Diseases*, *14*(8), 1232–1239. https://doi.org/10.3201/eid1408.080287.

Leung, G. H., Baird, R. W., Druce, J., & Anstey, N. M. (2015). Zika virus infection in Australia following a monkey bite in Indonesia. *The Southeast Asian Journal of Tropical Medicine and Public Health*, *46*(3), 460–464. Retrieved from https://www.ncbi.nlm.nih.gov/pubmed/26521519.

Li, C., Deng, Y. Q., Zu, S., Quanquin, N., Shang, J., Tian, M., ... Qin, C. F. (2018). Zika virus shedding in the stool and infection through the anorectal mucosa in mice. ***Emerging Microbes & Infections***, *7*(1), 169. https://doi.org/10.1038/s41426-018-0170-6.

Mansuy, J. M., Dutertre, M., Mengelle, C., Fourcade, C., Marchou, B., Delobel, P., ... Martin-Blondel, G. (2016). Zika virus: High infectious viral load in semen, a new sexually transmitted pathogen? *The Lancet Infectious Diseases*, *16*(4), 405. https://doi.org/10.1016/S1473-3099(16)00138-9.

Mansuy, J. M., Mengelle, C., Pasquier, C., Chapuy-Regaud, S., Delobel, P., Martin-Blondel, G., & Izopet, J. (2017). Zika virus infection and prolonged viremia in whole-blood specimens. *Emerging Infectious Diseases*, *23*(5), 863–865. https://doi.org/10.3201/eid2305.161631.

Mansuy, J. M., Suberbielle, E., Chapuy-Regaud, S., Mengelle, C., Bujan, L., Marchou, B., ... Martin-Blondel, G. (2016). Zika virus in semen and spermatozoa. *The Lancet Infectious Diseases*, *16*(10), 1106–1107. https://doi.org/10.1016/S1473-3099(16)30336-X.

Mead, P. S., Duggal, N. K., Hook, S. A., Delorey, M., Fischer, M., Olzenak McGuire, D., ... Hinckley, A. F. (2018). Zika virus shedding in semen of symptomatic infected men. *The New England Journal of Medicine*, *378*(15), 1377–1385. https://doi.org/10.1056/NEJMoa1711038.

Melo, A. S., Aguiar, R. S., Amorim, M. M., Arruda, M. B., Melo, F. O., Ribeiro, S. T., ... Tanuri, A. (2016). Congenital Zika virus infection: Beyond neonatal microcephaly. *JAMA Neurology*, *73*(12), 1407–1416. https://doi.org/10.1001/jamaneurol.2016.3720.

Murray, K. O., Gorchakov, R., Carlson, A. R., Berry, R., Lai, L., Natrajan, M., ... Mulligan, M. J. (2017). Prolonged detection of Zika virus in vaginal secretions and whole blood. *Emerging Infectious Diseases*, *23*(1), 99–101. https://doi.org/10.3201/eid2301.161394.

Musso, D., Nhan, T., Robin, E., Roche, C., Bierlaire, D., Zisou, K., ... Broult, J. (2014). Potential for Zika virus transmission through blood transfusion demonstrated during an outbreak in French Polynesia, November 2013 to February 2014. *Euro Surveillance*, *19*(14). https://doi.org/10.2807/1560-7917.es2014.19.14.20761.

Musso, D., Roche, C., Nhan, T. X., Robin, E., Teissier, A., & Cao-Lormeau, V. M. (2015). Detection of Zika virus in saliva. *Journal of Clinical Virology*, *68*, 53–55. https://doi.org/10.1016/j.jcv.2015.04.021.

Musso, D., Roche, C., Robin, E., Nhan, T., Teissier, A., & Cao-Lormeau, V. M. (2015). Potential sexual transmission of Zika virus. *Emerging Infectious Diseases*, *21*(2), 359–361. https://doi.org/10.3201/eid2102.141363.

Pacheco, O., Beltran, M., Nelson, C. A., Valencia, D., Tolosa, N., Farr, S. L., ... Ospina Martinez, M. L. (2016). Zika virus disease in Colombia—Preliminary report. *The New England Journal of Medicine*. https://doi.org/10.1056/NEJMoa1604037.

Paz-Bailey, G., Rosenberg, E. S., Doyle, K., Munoz-Jordan, J., Santiago, G. A., Klein, L., ... Sharp, T. M. (2018). Persistence of Zika virus in body fluids—Final report. *The New England Journal of Medicine*, *379*(13), 1234–1243. https://doi.org/10.1056/NEJMoa1613108.

Prisant, N., Bujan, L., Benichou, H., Hayot, P. H., Pavili, L., Lurel, S., ... Joguet, G. (2016). Zika virus in the female genital tract. *The Lancet Infectious Diseases*, *16*(9), 1000–1001. https://doi.org/10.1016/S1473-3099(16)30193-1.

Pyke, A. T., Daly, M. T., Cameron, J. N., Moore, P. R., Taylor, C. T., Hewitson, G. R., ... Gair, R. (2014). Imported zika virus infection from the cook islands into australia, 2014. *PLoS Currents*, *6*. https://doi.org/10.1371/currents.outbreaks.4635a54dbffba2156fb2fd76dc49f65e.

Reyes, Y., Bowman, N. M., Becker-Dreps, S., Centeno, E., Collins, M. H., Liou, G. A., & Bucardo, F. (2019). Prolonged shedding of Zika virus RNA in vaginal secretions, Nicaragua. *Emerging Infectious Diseases*, *25*(4), 808–810. https://doi.org/10.3201/eid2504.180977.

Roze, B., Najioullah, F., Ferge, J. L., Apetse, K., Brouste, Y., Cesaire, R., ... GBS Zika Working Group. (2016). Zika virus detection in urine from patients with Guillain-Barre syndrome on Martinique, January 2016. *Euro Surveillance*, *21*(9), 30154. https://doi.org/10.2807/1560-7917.ES.2016.21.9.30154.

Russell, K., Hills, S. L., Oster, A. M., Porse, C. C., Danyluk, G., Cone, M., ... Brooks, J. T. (2017). Male-to-female sexual transmission of Zika virus-United States, January-April 2016. *Clinical Infectious Diseases*, *64*(2), 211–213. https://doi.org/10.1093/cid/ciw692.

Sarno, M., Sacramento, G. A., Khouri, R., do Rosario, M. S., Costa, F., Archanjo, G., ... de Almeida, A. R. (2016). Zika virus infection and stillbirths: A case of hydrops fetalis, hydranencephaly and fetal demise. *PLoS Neglected Tropical Diseases*, *10*(2). https://doi.org/10.1371/journal.pntd.0004517, e0004517.

Sun, J., Wu, D., Zhong, H., Guan, D., Zhang, H., Tan, Q., & Ke, C. (2016). Presence of Zika virus in conjunctival fluid. *JAMA Ophthalmology*, *134*(11), 1330–1332. https://doi.org/10.1001/jamaophthalmol.2016.3417.

Suy, A., Sulleiro, E., Rodo, C., Vazquez, E., Bocanegra, C., Molina, I., ... Carreras, E. (2016). Prolonged Zika virus viremia during pregnancy. *The New England Journal of Medicine*, *375*(26), 2611–2613. https://doi.org/10.1056/NEJMc1607580.

Swaminathan, S., Schlaberg, R., Lewis, J., Hanson, K. E., & Couturier, M. R. (2016). Fatal Zika virus infection with secondary nonsexual transmission. *The New England Journal of Medicine*, *375*(19), 1907–1909. https://doi.org/10.1056/NEJMc1610613.

Tan, S. K., Sahoo, M. K., Milligan, S. B., Taylor, N., & Pinsky, B. A. (2017). Stability of Zika virus in urine: Specimen processing considerations and implications for the detection of RNA targets in urine. *Journal of Virological Methods*, *248*, 66–70. https://doi.org/10.1016/j.jviromet.2017.04.018.

Tognarelli, J., Ulloa, S., Villagra, E., Lagos, J., Aguayo, C., Fasce, R., ... Fernandez, J. (2016). A report on the outbreak of Zika virus on Easter Island, South Pacific, 2014. *Archives of Virology*, *161*(3), 665–668. https://doi.org/10.1007/s00705-015-2695-5.

Tozetto-Mendoza, T. R., Avelino-Silva, V. I., Fonseca, S., Claro, I. M., Paula, A. V., Levin, A. S., ... Mayaud, P. (2019). Zika virus infection among symptomatic patients from two healthcare centers in Sao Paulo State, Brazil: Prevalence, clinical characteristics, viral detection in body fluids and serodynamics. *Revista do Instituto de Medicina Tropical de São Paulo*, *61*. https://doi.org/10.1590/S1678-9946201961019, e19.

Waehre, T., Maagard, A., Tappe, D., Cadar, D., & Schmidt-Chanasit, J. (2014). Zika virus infection after travel to Tahiti, December 2013. *Emerging Infectious Diseases*, *20*(8), 1412–1414. https://doi.org/10.3201/eid2008.140302.

Waggoner, J. J., Gresh, L., Vargas, M. J., Ballesteros, G., Tellez, Y., Soda, K. J., ... Pinsky, B. A. (2016). Viremia and clinical presentation in Nicaraguan patients infected with Zika virus, chikungunya virus, and dengue virus. *Clinical Infectious Diseases*, *63*(12), 1584–1590. https://doi.org/10.1093/cid/ciw589.

Woon, Y. L., Lim, M. F., Tg Abd Rashid, T. R., Thayan, R., Chidambaram, S. K., Syed Abdul Rahim, S. S., ... Sivasampu, S. (2019). Zika virus infection in Malaysia: An epidemiological, clinical and virological analysis. *BMC Infectious Diseases*, *19*(1), 152. https://doi.org/10.1186/s12879-019-3786-9.

Zanluca, C., Melo, V. C., Mosimann, A. L., Santos, G. I., Santos, C. N., & Luz, K. (2015). First report of autochthonous transmission of Zika virus in Brazil. *Memórias do Instituto Oswaldo Cruz*, *110*(4), 569–572. https://doi.org/10.1590/0074-02760150192.

Chapter 24

Saliva and urine analysis of Zika virus using loop-mediated isothermal amplification (LAMP)

Talita Castro

School of Dentistry, University of Santo Amaro, Sao Paulo, SP, Brazil

Abbreviations

CDC	Centers for Disease Control and Prevention
DNA	deoxyribonucleic acid
GBS	Guillain-Barré syndrome
LAMP	loop-mediated isothermal amplification
PCR	polymerase chain reaction
qPCR	quantitative polymerase chain reaction
RNA	ribonucleic acid
RT-LAMP	reverse transcriptase loop-mediated isothermal amplification
WHO	World Health Organization
ZIKV	Zika virus

Introduction

Zika virus (ZIKV) is a single-stranded RNA virus, member of the *Flaviviridae* family (Kuno, Chang, Tsuchiya, Karabatsos, & Cropp, 1998). It is transmitted among humans usually by *Aedes* mosquito species, but sexual transmission (Musso, Roche, Nhan, et al., 2015; Musso, Roche, Robin, et al., 2015) and perinatal (Besnard, Lastere, Teissier, Cao-Lormeau, & Musso, 2014) and blood transfusion (Musso, Stramer, & Busch, 2016) have also been reported. During the acute phase, symptoms appear in 20% of infected individuals and include fever, cutaneous rash, headache, conjunctivitis, myalgia, and arthralgia (Barzon et al., 2016).

After the American ZIKV outbreak occurred in 2016, a sudden increase in microcephaly and Guillain-Barré syndrome (GBS) related to ZIKV infection prompted the World Health Organization (WHO) to declare a public health emergency of international concern (Mlakar et al., 2016; Oehler et al., 2014). In Brazil the high number of microcephaly and deaths of newborns of mothers infected during pregnancy made WHO to include ZIKV as a teratogenic agent (Rasmussen, Jamieson, Honein, & Petersen, 2016).

According to CDC (Centers for Disease Control and Prevention) recommendation, the diagnosis of ZIKV infection in symptomatic patients should be done through nucleic acid detection by PCR (polymerase chain reaction) in paired samples of serum and urine. ZIKV RNA disappears from blood in about 1 week after infection, but the virus remains sequestered up to 29 days in saliva (Barzon et al., 2016) and urine (Gourinat, O'Connor, Calvez, Goarant, & Dupont-Rouzeyrol, 2015).

Since the first report of ZIKV in saliva (Musso, Roche, Nhan, et al., 2015; Musso, Roche, Robin, et al., 2015), studies have suggested that saliva could be an important alternative sample to detect ZIKV (Bonaldo et al., 2016; Castro et al., 2018; Sabalza et al., 2018). Saliva collection is easy, noninvasive, and is safe to handle and requires less processing than blood, greatly simplifying the assay. Urine collection is also easy and noninvasive, and studies reported that ZIKV RNA can be frequently detected in urine during the acute phase (Paz-Bailey et al., 2017).

Zika Virus Impact, Diagnosis, Control, and Models. https://doi.org/10.1016/B978-0-12-820267-8.00024-8

Loop-mediated isothermal amplification (LAMP) has high sensitivity and specificity and can detect RNA and DNA without thermal cycling. The assay uses specific primers and generate real-time amplification that can be detected in less than 30 min. These characteristics make LAMP assays ideal to use in portable devices, which can be configured to diagnose diseases at point-of-care facilities (Calvert, Biggerstaff, Tanner, Lauterbach, & Lanciotti, 2017; Notomi, Mori, Tomita, & Kanda, 2015). Therefore the detection of ZIKV RNA by LAMP provides the basis of a test that can readily facilitate the diagnosis of new cases.

Zika virus in saliva

The first detection of ZIKV in saliva occurred during the French Polynesia outbreak in 2013. It was observed that individuals presenting specific symptoms of ZIKV infection had a negative blood test through qPCR (quantitative polymerase chain reaction) (Musso, Roche, Nhan, et al., 2015; Musso, Roche, Robin, et al., 2015). The researchers decided to collect a sample of saliva from a 1-year-old child with specific ZIKV symptoms, and the sample was positive. Since then, studies have been suggested that saliva could be an important alternative sample to detect ZIKV infection (Bonaldo et al., 2016; Castro et al., 2018; Sabalza et al., 2018).

ZIKV can be detected at high concentration in saliva in the acute phase (i.e., 5–7 days after infection) during manifestation of symptoms (Gourinat et al., 2015; Musso, Roche, Nhan, et al., 2015; Musso, Roche, Robin, et al., 2015). A study using qPCR assay stated a persistence of ZIKV of up to 29 days in saliva and urine and of up to 10 days in blood and reported that the viral load was higher in saliva than in urine and blood (Barzon et al., 2016).

Saliva sample collection is easy and noninvasive and requires less processing than blood. In a few minutes, using a simple sterile collection tube, like Salivette (Fig. 1), the patient independently provides a sample material that could be readily stored and frozen without the need of a professional assistance or any additional sample handling.

The transmission of Zika virus infection through saliva is greatly low; however, in vitro studies reported the presence of viable virus in saliva samples, suggesting that it could be a transmission route among humans in cases of oral wounds or severe periodontal diseases (Bonaldo et al., 2016; Musso, Roche, Nhan, et al., 2015; Musso, Roche, Robin, et al., 2015).

Recent studies related that Zika virus saliva is a feasible alternative to diagnose Zika virus and that the RNA could be more often detected in saliva than in urine samples (Castro et al., 2018; Sabalza et al., 2018).

Zika virus in urine

Urine testing is noninvasive and is an appealing sample for diagnostic in which blood collection can be difficult. This may include young children, neonates, elderly, or patients who present with small, dehydrated, or elusive veins.

Urine sample collection can also be done at field locations where trained medical personal or facilities may be lacking and allow for self-collection within communities for surveillance or epidemiology studies during an outbreak (Lamb et al., 2016).

ZIKV is detectable for a longer time postinfection in urine than serum. The CDC recommends that urine samples should be collected in patients suspected of ZIKV infection and that molecular testing of urine should be performed in conjunction with serum testing (Lamb et al., 2016; Paz-Bailey et al., 2017).

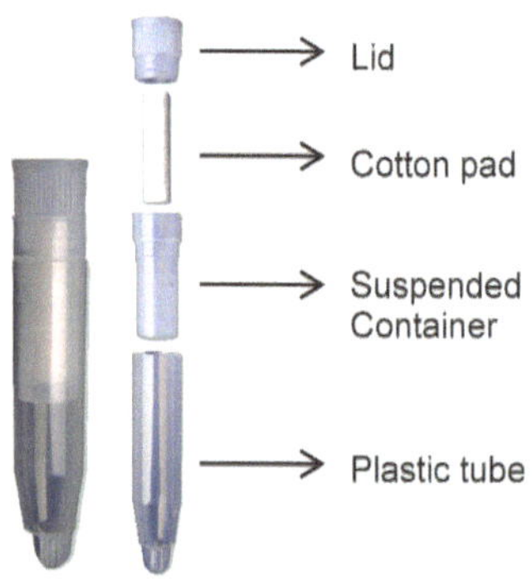

FIG. 1 **Salivette used to collect saliva sample**. Salivette is a tube that contains a sterile cotton pad that is placed under the tongue of the subject for a few minutes and then returned to the suspended container of tube. After centrifuged the saliva of the cotton pad goes to the bottom of the plastic tube. The suspended container and the cotton pad are discarded, and the sample of saliva inside the plastic tube can be stored.

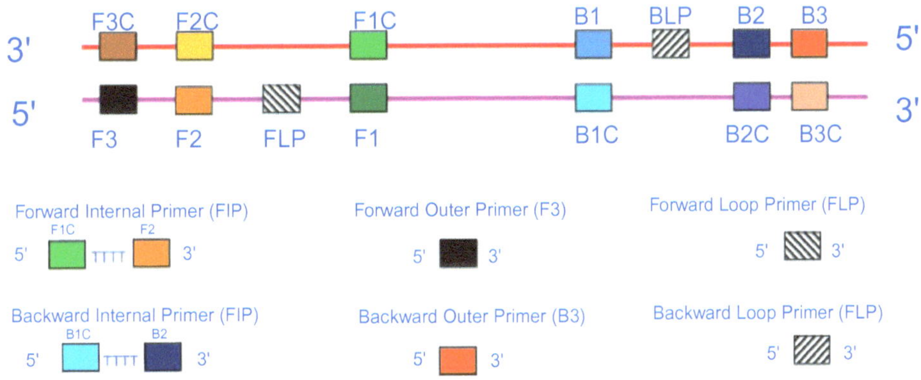

FIG. 2 **LAMP primers for the sequences of the target gene (Parida, Sannarangaiah, Dash, Rao, & Morita,** 2008). Loop-mediated isothermal amplification (LAMP) employs a set of six specially designed primers spanning eight distinct sequences of a target gene. Two outers primers (F3 and B3), two inner primers (FIP and BIP), and two loop primers (Loop F and Loop B) (Parida et al., 2008).

Loop-mediated isothermal amplification (LAMP)

Loop-mediated isothermal amplification, abbreviated LAMP, is a high sensitivity and specificity method for RNA and DNA amplification without thermal cycling. The assay was firstly described in 2000 by Notomi et al., and through the use of primers designed to target the capsid gene, LAMP generates real-time amplification that can be detected in less than 30 min (Hayes, 2009).

The fact that LAMP does not require thermal cycling makes the assay simple, fast, and cost effective compared with PCR assay (Zanoli & Spoto, 2013). These characteristics make LAMP ideal to use in portable devices, which can be configured to diagnose diseases at point-of-care facilities (Calvert et al., 2017; Notomi et al., 2000).

Previously published studies demonstrated the efficacy of RT-LAMP compared with qPCR (quantitative polymerase chain reaction) for Zika virus detection (Cadioli et al., 2015; Chen et al., 2011; Fallahi, Seyyed Tabaei, Pournia, Zebardast, & Kazemi, 2014; Wang, Seo, Lee, & Choi, 2014). The LAMP assay has shown to be useful to detect the ZIKV RNA in saliva and urine samples (Castro et al., 2018; Song et al., 2016). LAMP has better sensitivity and specificity compared with qPCR, offering a higher limit of detection and eliminating the risk of cross-reactivity with other flaviviruses (Castro et al., 2018; Chen et al., 2011).

Under isothermal conditions, LAMP employs a set of six specially designed primers spanning eight distinct sequences of a target gene. Two outers primers (F3 and B3), two inner primers (FIP and BIP), and two loop primers (Loop F and Loop B) (Fig. 2) (Parida et al., 2008).

The reverse transcriptase loop-mediated isothermal amplification (RT-LAMP) is initiated by adding internal primers (FIP or BIP) that anneal to regions (F2c or B2c) within the target RNA. The outer primer (F3 or B3) then hybridizes to its priming site (F3c or B3c) on the target RNA and initiates the formation of self-hybridizing loop structures by strand invasion of the DNA sequences already extended from the inner primers (FIP and BIP), resulting in a dumbbell structure. Moreover, RT-LAMP process can be accelerated by loop primers (LF and LB) (Yaren et al., 2017).

The LAMP assay is a real-time amplification that provides important information of the reaction: time to positivity (Tp), which is the time required for the assay to start amplifying the target sequence, and annealing temperature (At), which is a specific and unique temperature of the primers binding with the target sequence.

LAMP has rapidly gained popularity for point-of-care (POC) testing due to relative simplicity of building an isothermal device. One example of a device used for LAMP reaction is Genie III (OptiGene, UK, 2014), a compact and portable device that supports any isothermal DNA/RNA amplification method. It is a small, lightweight (1.75 kg), and self-contained device that reduces the contamination risk of LAMP reactions (Fig. 3). Genie III accepts a single 8-microtube strip that is measured by a dual-channel fluorescence detection system.

Zika virus detection through LAMP

Recent studies (Castro et al., 2018; Sabalza et al., 2018) have demonstrated the efficacy of Zika virus detection through LAMP reaction. The detection of ZIKV can be done through the use of primers targeting a unique sequence in the ZIKV, most of the times for conserved regions of the capsid gene. A set of three primer pairs is usually employed including two

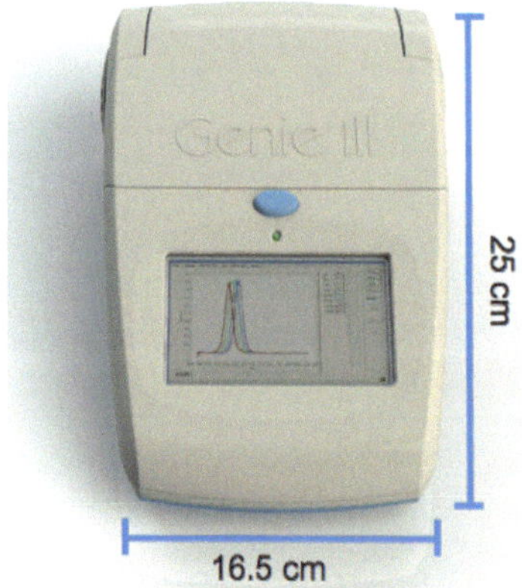

FIG. 3 **Genie III (OptiGene, UK, 2014), device used for LAMP assays.** Genie III (OptiGene, UK, 2014) is a compact and portable device (16.5 × 25 cm) that supports any isothermal DNA/RNA amplification method. It is a small, lightweight (1.75 kg), and self-contained device.

outer primers (forward primer F3 and backward primer B3), two inner primers (forward inner primer FIP and backward inner primer BIP), and two loop primers (forward loop primer Loop F and backward loop primer Loop B).

By using specific primers, LAMP analysis demonstrated specificity to ZIKV and had no cross-reaction with dengue virus, a common *flavivirus* usually presenting diagnostic challenges due to cross-reactivity with ZIKV (Castro et al., 2018). The fact that LAMP does not require thermal cycling makes the assay simpler, faster, and more cost effective compared with PCR assay (Zanoli & Spoto, 2013).

Castro et al. (2018) compared the detection rate of ZIKV in saliva and urine samples of subjects confirmed positive for ZIKV and exhibiting signs and symptoms of ZIKV infection (acute phase). The analysis of 131 samples (68 of saliva and 63 of urine) was done through RT-LAMP. The results demonstrated that ZIKV can be detected more frequently (1.7 times more) in saliva than in urine (Fig. 4).

The WHO demonstrated that 48 countries were affected with confirmed cases of ZIKV infection and estimated that over U$100 million would be necessary to implement a prevention and treatment planning for the medical complications plan caused by ZIKV. The development of simple and cost-effective diagnostic techniques such as LAMP associated with saliva would be helpful to use in epidemics.

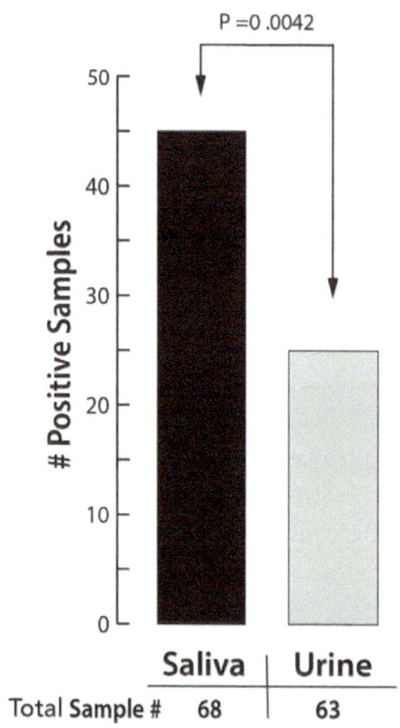

FIG. 4 **Number of positive samples of saliva and urine and the *P*-value (Castro et al., 2018).** The study of Castro et al. (2018) compared samples of saliva and urine of subjects positive for Zika virus infection. The study observed 45 (66.2%) saliva samples and 24 (39.7%) urine samples positive for Zika virus, resulting in a statistically significant difference between them ($P = 0.0042$).

Conclusions

The mechanism of excretion of ZIKV through body fluids is still not totally understood. Recent findings diagnosing ZIKV through saliva contribute to the knowledge of how is the virus behavior in the human organism.

ZIKV may not always be excreted through saliva. However, during the acute phase of Zika infection, the virus can be more often detected in saliva than in urine. Saliva is a useful alternative to rapidly diagnose Zika virus infection by loop-mediated isothermal amplification (LAMP).

Policy and procedures

Policy and procedures: Positive samples

Saliva and urine samples used in the study of Castro et al. (2018) were collected in 2016 in Brazil during the Zika virus epidemic. All subjects were confirmed positive for Zika virus infection through blood analysis by PCR. Saliva and urine samples were stored and freeze at −80°C and then analyzed.

Policy and procedures: Positive controls

Positive and negative controls were used in all LAMP reactions in the study of Castro et al. (2018). The positive control was a sample of saliva artificially infected with Zika virus RNA with a known RNA quantification.

Policy and procedures: Saliva collection

Saliva was collected using a cotton pad (Salivette) placed under the tongue and left for a few minutes. The samples were collected during the acute phase of the disease. The professional used all personal protective equipment (PPE) goggles, gloves, surgical face mask, cap, and long-sleeved disposable gown.

Policy and procedures: Zika virus analysis through LAMP

In the study of Castro et al. (2018), the RNA purification and reverse transcriptase loop-mediated isothermal amplification (RT-LAMP) were performed in a Biosafety Level 2+ facility with appropriate safety procedures, as described in the CDC/NIH Biosafety (BMBL 5th Edition). RNA isolation was performed with a commercial kit (ZR Viral RNA kit ZYMO Research). Primers targeting a unique sequence in the ZIKV capsid were designed by using the Primer Explorer V4 software (Eiken Chemical Co., Tokyo, Japan) for conserved regions of the capsid gene. The RT-LAMP reaction mixture contained Master Mix (OptiGene, UK), reverse transcriptase (WarmStart RT, New England Biolabs Inc.), primers, and purified RNA. The amplification step (30 min at 65°C) was followed by an annealing curve analysis step (98–60°C, ramping at 0.05°C/s; Genie III, OptiGene, UK).

Mini-dictionary of terms

Target sequence. Specific region of the RNA virus that amplifies when it binds to the primers.

Annealing temperature. Specific and unique temperature of the primers binding with the target sequence.

Isothermal device. Device that amplifies the target sequence with a specific temperature without the need of cycles of different temperatures.

Primers. A short, single-stranded RNA sequence used in the amplification. A pair of primers is used to hybridize with the RNA sample and define the region that will be amplified.

Loop primers. Specific primers for the loop area, a dumbbell structure formed in a LAMP reaction during amplification.

Thermal cycling. Cycles of repeated heating and cooling of the reaction for RNA/DNA melting and enzymatic replication. PCR method requires a high number of thermal cycle steps to create thousands of strands of RNA/DNA sequencing for analysis.

Key facts

Key facts of saliva

- Saliva sample collection is easy and noninvasive.
- ZIKV can be detected at high concentration in saliva during the acute phase.

- ZIKV may not always be excreted through saliva.
- The transmission of Zika virus infection through saliva is greatly low.
- Salivary diagnostics is an emerging field using technology and molecular diagnostics to aid in the diagnosis of infections and diseases.

Key facts of LAMP

- LAMP is an abbreviation of loop-mediated isothermal amplification.
- LAMP is a high sensitivity and specificity method for RNA and DNA amplification.
- LAMP does not require thermal cycling makes the assay simpler, faster, and more cost effective compared with PCR assay.
- LAMP assay has shown to be useful to detect the ZIKV RNA in saliva and urine samples.
- The characteristics of LAMP make the assays ideal to use in portable devices, which can be configured to diagnose diseases at point-of-care facilities.

Summary points

- Saliva is a feasible alternative to rapidly diagnose Zika virus infection by loop-mediated isothermal amplification (LAMP) in acute phase individuals.
- During the acute phase of Zika infection, the virus can be more often detected in saliva than in urine.
- Saliva requires less processing than blood, which greatly simplifies the assay process.
- LAMP is a molecular test with high sensibility and specificity for rapid detection of DNA or RNA of pathogens, including ZIKV.
- The recent findings are contributing to the knowledge of the ZIKV behavior in the human organism improving the diagnostic methods.

References

Barzon, L., Pacenti, M., Berto, A., Sinigaglia, A., Franchin, E., Levezzo, E., … Palù, G. (2016). Isolation of infectious Zika virus from saliva and prolonged viral RNA shedding in a traveller returning from the Dominican Republic to Italy, January 2016. *Euro Surveillance*, *2016*(21), 10.

Besnard, M., Lastere, S., Teissier, A., Cao-Lormeau, V., & Musso, D. (2014). Evidence of perinatal transmission of Zika virus, French Polynesia, December 2013 and February 2014. *Euro Surveillance*, *19*(13).

Bonaldo, M. C., Ribeiro, I. P., Lima, N. S., Dos Santos, A. A., Menezes, L. S., da Cruz, S. O., et al. (2016). Isolation of infective Zika virus from urine and saliva of patients in Brazil. *PLoS Neglected Tropical Diseases*, *10*(6), e0004816.

Cadioli, F. A., Fidelis Junior, O. L., Sampaio, P. H., dos Santos, G. N., Andre, M. R., Castilho, K. J., et al. (2015). Detection of *Trypanosoma vivax* using PCR and LAMP during aparasitemic periods. *Veterinary Parasitology*, *214*(1–2), 174–177.

Calvert, A. E., Biggerstaff, B. J., Tanner, N. A., Lauterbach, M., & Lanciotti, R. S. (2017). Rapid colorimetric detection of Zika virus from serum and urine specimens by reverse transcription loop-mediated isothermal amplification (RT-LAMP). *PLoS One*, *12*(9), e0185340.

Castro, T., Sabalza, M., Barber, C., Abrams, W., Da Costa, A. C., De Pádua Milagres, F. A., … Gallottini, M. (2018). Rapid diagnosis of Zika virus through saliva and urine by Loop-mediated isothermal amplification (LAMP). *Journal of Oral Microbiology*, *10*(1). https://doi.org/10.1080/20002297.2018.1510712, 1510712.

Chen, Z., Liao, Y., Ke, X., Zhou, J., Chen, Y., Gao, L., et al. (2011). Comparison of reverse transcription loop-mediated isothermal amplification, conventional PCR and real-time PCR assays for Japanese encephalitis virus. *Molecular Biology Reports*, *38*(6), 4063–4070.

Fallahi, S., Seyyed Tabaei, S. J., Pournia, Y., Zebardast, N., & Kazemi, B. (2014). Comparison of loop-mediated isothermal amplification (LAMP) and nested-PCR assay targeting the RE and B1 gene for detection of *Toxoplasma gondii* in blood samples of children with leukaemia. *Diagnostic Microbiology and Infectious Disease*, *79*(3), 347–354.

Gourinat, A. C., O'Connor, O., Calvez, E., Goarant, C., & Dupont-Rouzeyrol, M. (2015). Detection of Zika virus in urine. *Emerging Infectious Diseases*, *21*(1), 84–86.

Hayes, E. B. (2009). Zika virus outside Africa. *Emerging Infectious Diseases*, *15*(9), 1347–1350.

Kuno, G., Chang, G. J., Tsuchiya, K. R., Karabatsos, N., & Cropp, C. B. (1998). Phylogeny of the genus flavivirus. *Journal of Virology*, *72*(1), 73–83.

Lamb, L. E., Bartolone, S. N., Kutluay, S. B., Robledo, D., Porras, A., Plata, M., et al. (2016). Advantage of urine based molecular diagnosis of Zika virus. *International Urology and Nephrology*, *48*(12), 1961–1966.

Mlakar, J., Korva, M., Tul, N., Popovic, M., Poljsak-Prijatelj, M., Mraz, J., et al. (2016). Zika virus associated with microcephaly. *The New England Journal of Medicine*, *374*(10), 951–958.

Musso, D., Roche, C., Nhan, T. X., Robin, E., Teissier, A., & Cao-Lormeau, V. M. (2015). Detection of Zika virus in saliva. *Journal of Clinical Virology*, *68*, 53–55.

Musso, D., Roche, C., Robin, E., Nhan, T., Teissier, A., & Cao-Lormeau, V. M. (2015). Potential sexual transmission of Zika virus. *Emerging Infectious Diseases*, *21*(2), 359–361.

Musso, D., Stramer, S. L., & Busch, M. P. (2016). Zika virus: A new challenge for blood transfusion. *Lancet*, *387*(10032), 1993–1994.

Notomi, T., Mori, Y., Tomita, N., & Kanda, H. (2015). Loop-mediated isothermal amplification (LAMP): Principle, features, and future prospects. *Journal of Microbiology (Seoul, Korea)*, *53*(1), 1–5.

Notomi, T., Okayama, H., Masubuchi, H., Yonekawa, T., Watanabe, K., Amino, N., et al. (2000). Loop-mediated isothermal amplification of DNA. *Nucleic Acids Research*, *28*(12), E63.

Oehler, E., Watrin, L., Larre, P., Leparc-Goffart, I., Lastere, S., Valour, F., et al. (2014). Zika virus infection complicated by Guillain-Barre syndrome—Case report, French Polynesia, December 2013. *Euro Surveillance*, *19*(9).

Parida, M., Sannarangaiah, S., Dash, P. K., Rao, P. V., & Morita, K. (2008). Loop mediated isothermal amplification (LAMP): A new generation of innovative gene amplification technique; perspectives in clinical diagnosis of infectious diseases. *Reviews in Medical Virology*, *18*(6), 407–421.

Paz-Bailey, G., Rosenberg, E. S., Doyle, K., Munoz-Jordan, J., Santiago, G. A., Klein, L., et al. (2017). Persistence of Zika virus in body fluids—Preliminary report. *The New England Journal of Medicine*.

Rasmussen, S. A., Jamieson, D. J., Honein, M. A., & Petersen, L. R. (2016). Zika virus and birth defects—Reviewing the evidence for causality. *The New England Journal of Medicine*, *374*(20), 1981–1987.

Sabalza, M., Yasmin, R., Barber, C. A., Castro, T., Malamud, D., Kim, B. J., et al. (2018). Detection of Zika virus using reverse-transcription LAMP coupled with reverse dot blot analysis in saliva. *PLoS One*, *13*(2), e0192398.

Song, J., Mauk, M. G., Hackett, B. A., Cherry, S., Bau, H. H., & Liu, C. (2016). Instrument-free point-of-care molecular detection of Zika virus. *Analytical Chemistry*, *88*(14), 7289–7294.

Wang, X., Seo, D. J., Lee, M. H., & Choi, C. (2014). Comparison of conventional PCR, multiplex PCR, and loop-mediated isothermal amplification assays for rapid detection of Arcobacter species. *Journal of Clinical Microbiology*, *52*(2), 557–563.

Yaren, O., Alto, B. W., Gangodkar, P. V., Ranade, S. R., Patil, K. N., Bradley, K. M., et al. (2017). Point of sampling detection of Zika virus within a multiplexed kit capable of detecting dengue and chikungunya. *BMC Infectious Diseases*, *17*(1), 293.

Zanoli, L. M., & Spoto, G. (2013). Isothermal amplification methods for the detection of nucleic acids in microfluidic devices. *Biosensors (Basel)*, *3*(1), 18–43.

Chapter 25

Graphene-based biosensors for the detection of Zika virus

Ameya Chaudhari[a] and Prajakta Dandekar[b]

[a]*Department of Biomedical Engineering, Duke University, Durham, NC, United States,* [b]*Department of Pharmaceutical Sciences and Technology, Institute of Chemical Technology, Mumbai, India*

Abbreviations

csDNA	complementary strand DNA
CVD	chemical vapor deposition
DMTMM	4-(4,6-dimethoxy-1,3,5-triazin-2-yl)-4-methylmorpholinium chloride
EDC	1-ethyl-3-(3-dimethylaminopropyl)carbodiimide hydrochloride
EIS	electrochemical impedance spectroscopy
FAM	carboxy fluorescein
FET	field-effect transistor
GO	graphene oxide
HPA	hybridization protection assay
JEV NS1	Japanese encephalitis NS1
LoD	limit of detection
MAC-ELISA	IgM antibody capture enzyme-linked immunosorbent assay
MMLV	Moloney murine leukemia virus
NHS	*N*-hydroxysuccinimide
PFU	plaque-forming units
PRNT	plaque reduction neutralization test
RT-qPCR	real-time polymerase chain reaction
ssDNA	single stranded
SIPs	surface-imprinted polymers
SPE	screen-printed gold electrode
TMA	transcription-mediated amplification
ZIKV	Zika virus infection

Introduction

Biosensors are devices that generate biochemical or optical responses in response to the presence of biological analytes (Thvenot, Toth, Durst, & Wilson, 1999). The biological components of the biosensors, which specifically interact with the target analyte, may be enzymes, cells, tissues, nucleic acids, and proteins. Till date, numerous biosensors have been fabricated using diverse materials such as metals, polymers, ceramics, composites, and carbon materials based on their compatibility with biomolecules, the desired working conditions, and the required sensitivity or selectivity. Since the development of the first glucose biosensor the by Clark and Lyons, based on the enzymatic reaction of glucose oxidase with the blood glucose (Clark & Lyons, 1962), scientists have developed an arena of biosensors to detect analytes ranging from small molecules to complex macromolecules present in human fluids (Ivnitski, Abdel-Hamid, Atanasov, & Wilkins, 1999; Yu, Blankert, Viré, & Kauffmann, 2005). Viral diseases usually display generic symptoms, which delay the quick and accurate diagnosis of these conditions. Biosensors have been found to result in simple, rapid, and precise detection of such challenging medical conditions, including the Zika virus infection (ZIKV).

The ZIKV is a mosquito-borne disease (Basu & Tumban, 2016) that exhibited its first occurrence in humans in the year 1952 in Uganda and the United Republic of Tanzania. Since then, outbreaks have been recorded in Africa, the Americas,

Zika Virus Impact, Diagnosis, Control, and Models. https://doi.org/10.1016/B978-0-12-820267-8.00025-X

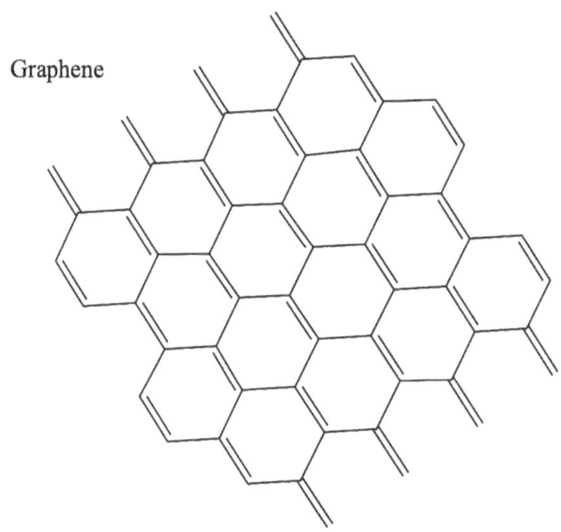

FIG. 1 Structure of graphene monolayer. It is a network of sp^2-hybridized carbon atoms.

Asia, and the Pacific (Zika Virus, n.d.). It was observed that patients infected with ZIKV did not exhibit any symptoms until days or weeks (Zika Virus, n.d.), which underlined the need for a rapid, efficient, and cost-effective diagnosis method for detecting this grave condition. Here, we have reviewed the recent advancements in graphene-based biosensors that have been explored to detect this pathologically lethal virus.

Graphene, as represented in Fig. 1, is a planer single atom layer, made of sp^2-hybridized carbon. In 2004 K. S. Novoselov and A. K. Geim developed the first ever graphene monolayer that resembled a chicken wire (Novoselov et al., 2004). Since its discovery in 2004, graphene has received tremendous attention from scientists due to its unique properties. This material is extremely strong, despite being thin and lightweight. It is approximately 1 million times thinner than paper and yet stronger than a steel sheet. It is impermeable to gases and liquids. Furthermore, it is a good conductor of heat and electricity and possesses a high surface area, which renders it as a suitable material for designing biosensors.

Some of the typical properties of this material that have been largely exploited to design various biosensors include (1) its robust nature and its ability to operate as a field-effect transistor (FET) in liquids; (2) its high surface area; (3) its ability to be easily functionalized with biological molecules; (4) its high electrical conductivity while maintaining low noise levels; (5) its electronic conductivity that that can be easily manipulated to develop biosensors for specific applications; (6) its properties such as band gap, carrier type, and conductivity; and (7) its high mechanical strength, enabling fabrication of miniature biosensors that display a high spatial resolution (Jalilian, Jauregui, Lee, Chen, & Irudayaraj, 2012).

Currently, graphene is mainly synthesized using chemical vapor deposition, epitaxial growth method, thermal exfoliation, and chemical reduction with hydrazine, methylhydrazine, sodium borohydride, and hydroquinone. A detailed overview about the various synthesis methods of graphene has already been presented by Choi, Lahiri, Seelaboyina, and Kang (2010).

Graphene-based biosensors

Biosensors consist of two elements, namely, a receptor and a transducer. The receptor binds to the target molecule, both of which may be organic or inorganic in nature. Further the transducer converts this chemical or biological interaction into a quantifiable signal (Peña-Bahamonde, Nguyen, Fanourakis, & Rodrigues, 2018). In graphene-based biosensors, graphene acts as the transducer. Biological moieties, such as single stranded (ssDNA) and antibodies, can be immobilized onto graphene sheets or its derivatives like graphene oxide and reduced graphene oxide. Immobilization of biomolecules on graphene is usually accomplished using 1-ethyl-3-(3-dimethylaminopropyl)carbodiimide hydrochloride (EDC)/*N*-hydroxysuccinimide (NHS)/4-(4,6-dimethoxy-1,3,5-triazin-2-yl)-4-methylmorpholinium chloride (DMTMM) chemistry, which has been depicted in Fig. 2. This is the most common approach of bioconjugation on graphene that results due to the formation of a covalent bond between the carboxylic acid groups of graphene and amine groups of antibody (Hermanson, 2013). Enzymes are usually immobilized by physisorption technique (Peña-Bahamonde et al., 2018). Physisorption or physical adsorption involves adsorption of adsorbates on substrates due to weak intermolecular forces, such as dispersion and dipolar or van der Waals interactions (Stine, 2017).

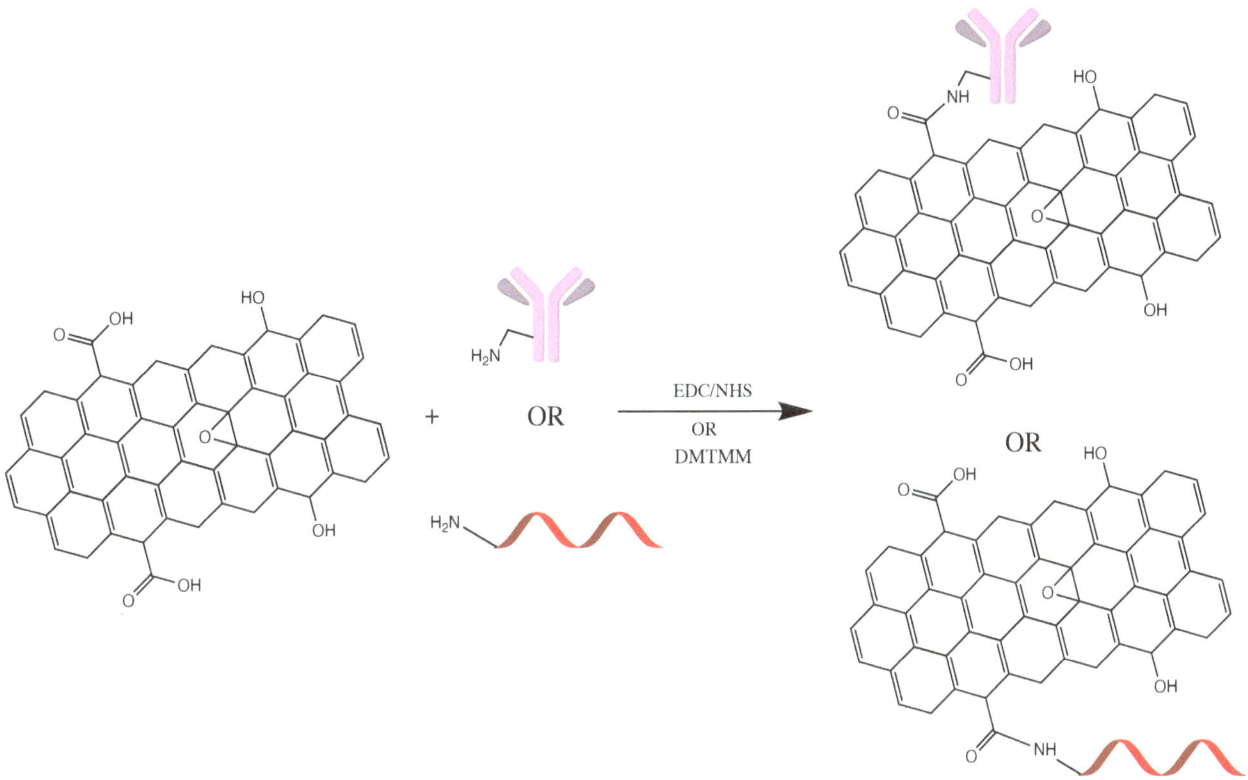

FIG. 2 Illustration of the EDC/NHS and DMTMM chemistry for functionalization of graphene layer with antibodies or single-strand DNA (ssDNA). Graphene layer has function moieties such as carboxylic acid that can be reacted with amine groups present on the antibodies/ssDNA in the presence of a carboxylic acid activator.

Various parameters are considered while fabricating graphene-based biosensors, which include orientation between graphene, graphene oxide, and reduced graphene oxide layers and the bioreceptors, the number of bioreceptors and their layers, and the oxidation states of graphene, all of which affect the performance of the biosensors (Peña-Bahamonde et al., 2018). The number of functional groups also affects the selectivity of the graphene-based biosensors, and therefore any nonspecific binding sites on the graphene surface are blocked by using blockers, such as bovine albumin serum and casein. Thus any false positive signal due to the nonspecific binding of an analyte is completely eliminated.

Graphene-based biosensors can be broadly divided into two classes, namely, the electrochemical and the fluorescent sensors (Peña-Bahamonde et al., 2018). In case of the latter, the target molecules are labeled with a fluorescence tag, the intensity of which is used to measure and quantify the amount of target molecules' presence. For example, for detecting different DNA sequences, multicolor single-strand DNA probes were synthesized, and the analyte solution was added to it in the presence of a graphene oxide (GO) layer. These probes were made by tagging carboxyfluorescein (FAM) to ssDNA. In the absence of target complementary strand DNA (csDNA), the GO layer quenched the fluorescence. However, in the presence of the target csDNA, fluorescence was emitted to detect the target analyte (He et al., 2010).

In the case of the electrochemical sensors, field-effect transistors are used to sense the biomolecules. The biosensors are fabricated by transferring graphene onto silicon/silicon dioxide, wherein the latter acts a carrier for the graphene layer. Electrical contacts are attached to the graphene layer by electron beam or optical lithography. This enables the sensor to respond to the presence of the target analyte by eliciting an electrical output via fluctuations in the voltage or the current.

Diagnostics methods for detecting Zika virus

Traditional diagnostic methods for detecting Zika virus

Since its outbreak in May 2015, scientists have tried to develop diagnostic techniques for sensitive, accurate, and early detection of ZIKV. This virus is member of the flavivirus genus and shares common genes and protein structures with dengue virus, West Nile virus, yellow fever virus, etc. (Barba-Spaeth et al., 2016; Bollati et al., 2010). Currently the most common methods used for laboratory-based detection of this virus include real-time polymerase chain reaction (RT-qPCR),

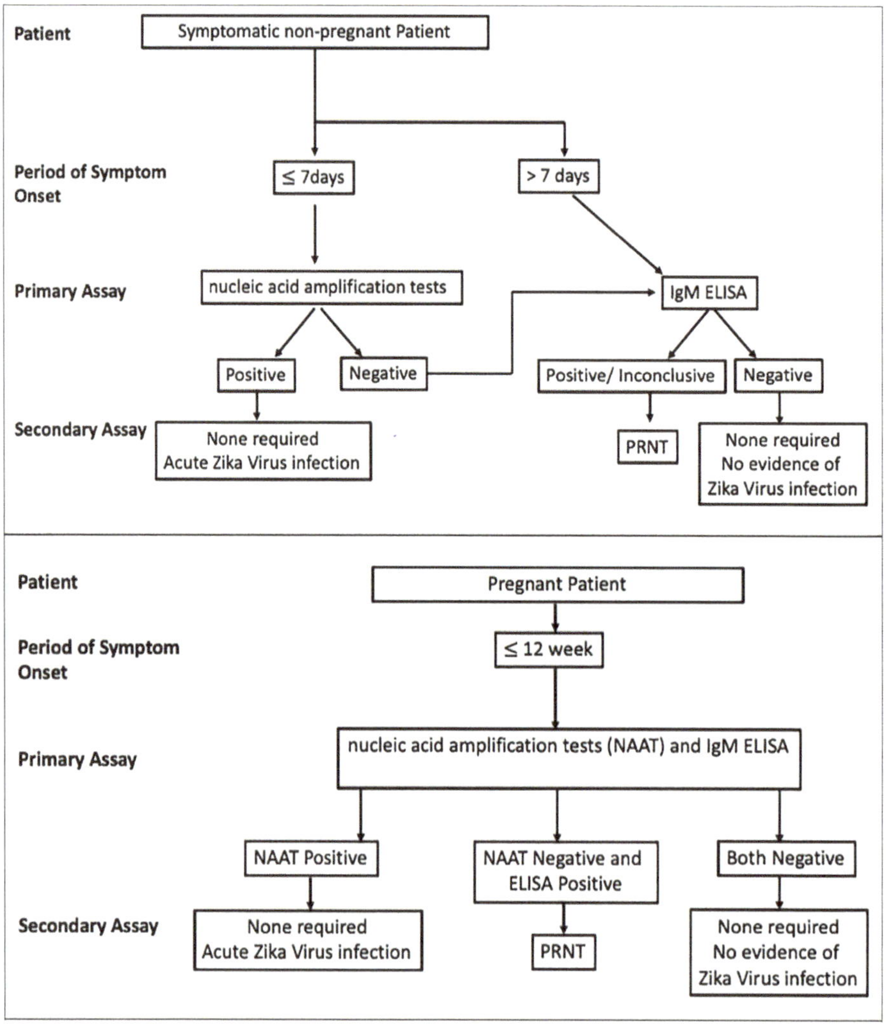

FIG. 3 Current testing guidance by Centers for Disease Control and Prevention for the diagnosis of Zika virus. This flow demonstrates the diagnostic steps that are taken for the detecting Zika virus for symptomatic nonpregnant patients (*upper chart*) and pregnant patient (*lower chart*). *IgM ELISA*, IgM antibody capture enzyme-linked immunosorbent assay; *PRNT*, plaque reduction neutralization test.

IgM antibody capture enzyme-linked immunosorbent assay (MAC-ELISA), or plaque reduction neutralization test (PRNT) (Testing for Zika Virus Infections | Zika Virus | CDC, n.d.; Nicolini, McCracken, & Yoon, 2017). The current protocol being followed for the detection of ZIKV has been represented in Fig. 3.

Real-time polymerase chain reaction

The RT-qPCR is a single-step method in which the sample is placed in a thermal cycler, along with reverse transcriptase enzyme, such as SuperScript IV Reverse Transcriptase by Thermo Fisher, ZIKV target primers, and other PCR reagents required for the amplification of the genomic RNA of ZIKV, with a fluorescent reporter such as FAM probes. The absolute or relative fluorescence is then used to quantify the amplified target (CDC, 2020). Several samples, such as serum, whole blood, cerebrospinal fluid, urine, and amniotic fluid, can be analyzed using this versatile technique. This method is generally used for early diagnosis of the disease when the patient is experiencing symptoms for less than 2 weeks. Due to the temporal nature of ZIKV in the event of a negative test result, ELISA is performed to confirm the outcomes of RT-qPCR technique (Testing for Zika Virus Infections | Zika Virus | CDC, n.d.).

IgM antibody capture enzyme-linked immunosorbent assay (MAC-ELISA)

IgM ELISA is a very common technique that is based on the detection of anti-ZIKV IgM antibodies in the patient's samples. In this method, patient's blood sample is added to a well plate that is coated with antibodies to capture the IgM antibodies.

Then a ZIKV-specific antigen, which can bind to the IgM antibody, is added. Only in the event of anti-ZIKV antibodies being present in the patient's blood will they bind to the antigen. Further an antibody linked to an enzyme, such horseradish peroxidase, is added to specifically bind to the previously added ZIKV antigen. Finally a chromogenic substrate is used for the quantification of the IgM antibodies in the patients' blood to yield an optical readout (Musso & Gubler, 2016). This detection technique is usually adopted when the patient has suffered from the symptoms of the Zika virus infection for more than 2 weeks. The disadvantage of this method is that it yields false positive results, and hence in case of a positive result, the same is conformed using PRNT (Testing for Zika Virus Infections | Zika Virus | CDC, n.d.).

Plaque reduction neutralization test

PRNT is used as a confirmatory test to validate the positive samples of the IgM ELISA test. In this method the patients' serum samples are serially diluted. These dilutions are then added to viral suspension separately, after which these viral solutions are applied to confluent host cells, such as Vero cells. Further the cells are incubated, and the resulting plaque-forming units (PFU) are counted (Schmidt, Dennis, & Lennette, 1976). If the patients' serum has antibodies against ZIKV, a reduced PFU count is obtained and vice versa. This method has higher sensitivity than ELISA but is time and labor intensive and hence is adopted as a secondary test. Table 1 states the sensitivities of various assays used for detection of the Zika virus.

However, these methods are cumbersome, time consuming, and expensive. Thus there is an urgent need to develop simple, rapid, and accurate methods for detecting this virus.

Recent diagnostic methods for detecting Zika virus

Recently a company known as Hologic, Inc., Massachusetts, United States, has developed a proprietary device, known as Aptima ZIKV assay (Aptima Zika Virus Assay | Hologic, n.d.). The limit of detection (LoD) of this technique has been reported as 5.9 copies/mL, at 95% detection probability, in a plasma sample, while it is 13.4 copies/mL, at 95% detection probability, in an in vitro transcript sample. This device is a completely automated assay system based on proprietary transcription-mediated amplification (TMA). It involves three main steps: (1) sample preparation, (2) amplification of RNA by TMA, and (3) detection of the amplified RNA using hybridization protection assay (HPA). The samples are prepared by isolating RNA from the sample by using a target capture. This is conducted by using a detergent to denature the viral envelope to release the genomic RNA. Further, capture oligonucleotides are used to hybridize with the RNA samples. These oligonucleotides are homologous to a highly conserved region in the ZIKV genome. The hybridized RNA is then captured onto magnetic beads and separated. RNA amplification is conducted with TMA, utilizing two enzymes, namely, the MMLV (Moloney murine leukemia virus) reverse transcriptase and T7 RNA polymerase. Here the MMLV reverse transcriptase generates DNA copies, while the T7 RNA polymerase amplifies them. Finally the amplified DNA is detected using HPA, in which single-stranded nucleic acid probes with chemiluminescent labels are added, these probes being complementary to the amplified DNA. Finally the unhydrized probes are inactivated using a proprietary selection reagent, allowing for the detection of the ZIKV. Despite the claim of ease of use, the method requires expensive laboratory equipment and considerable laboratory space.

Furthermore, extensive work has been done in the area of antibody-based assays. However, the problem associated with most of these detection techniques is that the ZIKV is similar to other flaviviruses, which has led to instances of cross-reactivity. However, work by Dai et al. has enabled characterization of surface proteins of ZIKV at the angstrom levels (Dai et al., 2016). This research indicated the possible mode of binding of antibody, specifically to the ZIKV E protein. A cryoelectron microscopy image of ZIKV revealed its structural similarity with other flaviviruses

TABLE 1 Sensitivities of the tests for the detection of ZIKV.

Assay	Sensitivity ($TCID_{50}$)
PCR	5–100
ELISA	10^5
PRNT	Not reported

These values are of tissue culture infection dose at 50% (TCID50) endpoint. *PCR*, real-time polymerase chain reaction; *ELISA*, enzyme-linked immunosorbent assay; *PRNT*, plaque reduction neutralization test.

(Kostyuchenko et al., 2016; Sirohi et al., 2016). The E protein of ZIKV was displayed on the envelope of the virus, along with membrane protein (Kuhn et al., 2002; Zhang et al., 2013). Upon endocytosis, acidic endosomal environment altered the conformation of the protein E, which formed a trimer from a dimer, which facilitated its membrane fusion (Modis, Ogata, Clements, & Harrison, 2004). The protein E was also responsible for the virus entry into the cells. This protein was employed as a target for the development of monoclonal antibodies (Dejnirattisai et al., 2015; Oliphant et al., 2005; Rouvinski et al., 2015; Smith et al., 2013). These investigations revealed the crystal structures of the dimeric free ZIKV-E ectodomain protein and its complex with a murine antibody 2A10G6. Thus this antibody has been proposed as a possible component for biosensors to detect the presence of ZIKV in samples. Further, Badawi et al. and Zhao et al. have reported that this protein is found in about 50 different ZIKV strains and that mouse antibodies, such as murine antibody 2A10G6, were able to bind specifically to this protein (Badawi et al., 2016; Zhao et al., 2016). These studies presented the possibility of protein E being used as a target for biosensing of ZIKV using these mAbs.

Graphene-based biosensors for the detection of ZIKV

Graphene-based biosensors have been developed for detecting pathogens such dengue virus and Ebola virus, among others (Chaudhari, Jagdale, Goswami, & Kerawalla, 2018). Various polymers such as polyurethane and poly(vinyl chloride) have been used to detect these viruses by functionalizing them with antibodies, aptamers, etc. However, this technology has created materials with a high electrical resistance. Scientists have overcome this problem through the use of nanocarbon materials, such as carbon nanotubes, graphene, or derivatives of graphene (Ma, Shen, Zeng, Wang, & Wang, 2017; Navakul et al., 2017; Sukjee, Tancharoen, Yenchitsomanus, Gleeson, & Sangma, 2017). The major advantages exhibited by the graphene-based biosensors include their simplicity, speed, economy, and lower detection limits due to their high sensitivity, which leads to faster and cheaper diagnosis of infections (Chaudhari et al., 2018).

Field effect biosensing with monoclonal antibodies

In February 2018, Afsahi et al. published a study that demonstrated immobilization of a monoclonal antibody on a graphene matrix for real-time detection of ZIKV (Afsahi et al., 2018). Here the biosensor was fabricated by photolithography and plasma-enhanced chemical vapor deposition (CVD). This method was employed to deposit thin films from gaseous to solid state. The graphene films were grown on copper, and then, bubbling transferring was used to transfer the film onto silicon wafers (Gao, Hao, Liang, Zheng, & Hu, 2012; Kybert et al., 2014). This graphene biosensor was read using Agile R100 system, wherein C-response read, that is, the capacitance of biosensor to the liquid, was used for detecting Zika NS1, the protein that allows for virus propagation. The C-response was finalized as the readout since anti-Zika NS1 exhibited a greater impact on the capacitance than the source-drain resistance. A monoclonal antibody, anti-Zika NS1 mouse mAb 6B1, was immobilized on the biosensor using EDC/NHS ester surface chemistry (Afsahi et al., 2017), which in turn conjugated with the native Zika viral antigens. The formation of antibody-antigen complex thus constituted the principle of functioning of this biosensor. Here, polyethylene glycol was used to block all the nonspecific interactions with the graphene surface (Gao et al., 2016; Liu et al., 2013). The design of this biosensor has been represented in Fig. 4.

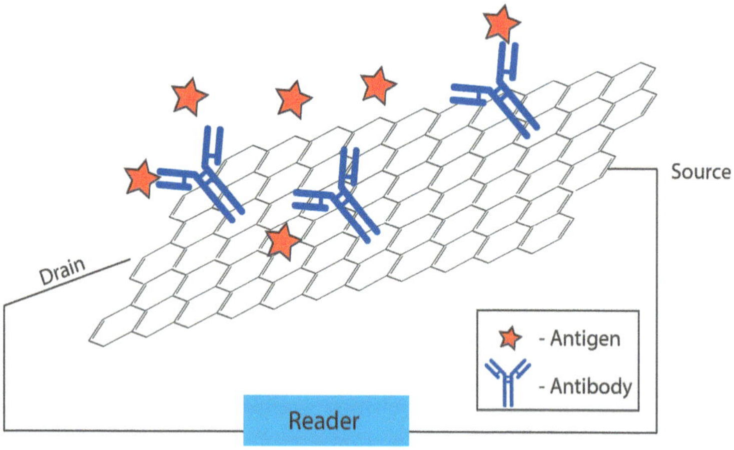

FIG. 4 Design of biosensor with antibody for the detection of ZIKV antigen in the analyte sample. In this method the antibodies are conjugated on the graphene surface that can bind to the antigen and output an electric response to detect the presence of the virus.

This biosensor enabled a detection limit of 0.45 nM of ZIKV NS1, which was a threefold improvement over previous analytical methods (Afsahi et al., 2017). It was also found to be selective toward Zika NS1, when C-response of ZIKV NS1 was compared with that of Japanese encephalitis NS1(JEV NS1), which is another type of flavivirus. Thus the developed biosensor demonstrated a high selectivity toward ZIKV NS1.

Surface imprint–based electrochemical biosensor

Further, Tancharoen et al. designed a biosensor that worked on the principle of surface-imprinted polymers (SIPs) on a gold electrode to form a screen-printed gold electrode (SPE) biosensor, which has been depicted in Fig. 5 (Tancharoen et al., 2019). SIPs are known to form biorecognition cavities that are specific and sensitive and have long-term stability. In this study the researchers used a polymer mix that was prepolymerized at its gel point. The polymer mix consisted of four monomers, namely, acrylamide, methacrylic acid, methyl methacrylate, and *N*-vinylpyrrolidone. Upon attaining the gel point of this polymer combination, graphene oxide was added. Thereafter the mixture was spin coated on a gold electrode. Finally the ZIKV sample (Zika virus RNA) was added, exposed to UV light, and then removed. The SIP was washed to obtain Zika virus sensing cavities in the SIP.

Further, electrochemical impedance spectroscopy (EIS) was used to verify the recognition of Zika virus. In this method the biosensor recorded the resistance and capacitance of the graphene with the analyte sample. Further an impedance spectrum was created by varying the frequency of the sinusoidal AC excitation signal, which was used to detect the binding of the antibody (Honeychurch, 2012). This biosensor was able to distinguish between different concentrations of ZIKV RNA. There was a decrease in impedance with an increase in concentration of the ZIKV RNA. This was attributed to the increase in conductivity when ZIKV was absorbed on the SIP surface. The limit of detection of this biosensor was 2×10^{-4} and 2×10^{-2} PFU/mL in the presence of dengue virus and media, respectively. This biosensor was used to test the presence of Zika virus in serum. Although a noise was observed at high serum concentrations, the biosensor still demonstrated a detection limit of 2×10^{-3} and 5×10^{-2} PFU/mL in presence of 1% and 10% serum, respectively.

These two studies demonstrate the possibility of developing sensitive and specific point-of-care graphene-based diagnostic devices for the detection of Zika virus. These newer diagnostics may permit the virus detection in in early stages of infection, allowing timely initiation of medical interventions. However, investigations in this area have been limited due to certain challenges that have been listed in the next section of this manuscript.

Current challenges and conclusion

Graphene, also referred to as "wonder material" (Chakraborty & Hashmi, 2018), exhibits several desirable properties that makes it a suitable material for developing newer biosensors. Although graphene has outstanding properties, the technologies for synthesis and processing of graphene are not suitable for large-scale manufacture, due to the high cost of graphene production (Peña-Bahamonde et al., 2018). Also, few of the top-down approaches used for graphene production have resulted in high polydispersibility and hence lack of reproducibility. Moreover, graphene biosensors still exhibit nonspecific binding of unwanted molecules, in turn resulting in poor detection capabilities and false positive detections (Contreras-Naranjo & Aguilar, 2019; Szunerits & Boukherroub, 2018). Till date, no correlation that has been established between the biological behavior and physical properties of graphene, such as its oxygen content, number of layers, and the

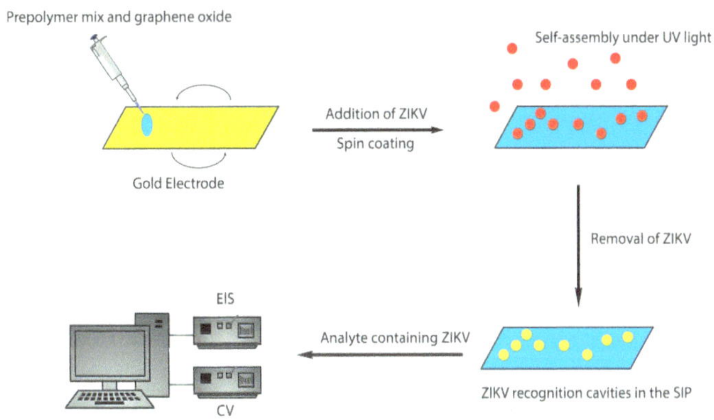

FIG. 5 Systematic development of SIP-graphene oxide biosensor for ZIKV sensing using electrochemical impedance spectroscopy (EIS) and cyclic voltammetry (CV). This method involves spin-coating prepolymer mix and graphene oxide (GO) on a gold electrode and then allowing template ZIKV to self-assemble with the polymer under UV light. On complete polymerization, ZIKV is removed, leaving recognition cavities behind that will allow for the detection ZIKV in samples.

lateral size. Standardized method for characterization is required to identify the critical parameters governing the properties of graphene (Hull & Bowman, 2014). Thus in-depth knowledge of the physics and surface chemistry of graphene and its interactions with biomolecules is warranted to make significant advances in this field. Overall, graphene and its derivative have exhibited the potential of being the material of choice for fabricating biosensors, provided the aforementioned challenges are addressed and the material may be clinically approved for ZIKV diagnosis.

Policy and procedure: Testing for Zika virus infections

According to CDC guidelines for laboratories, there are three tests that are recommended. Nucleic acid amplification test, or NAAT, is done to detect viral genomic material. Due to highly temporal nature of the Zika virus RNA in serum and urine, there could be false negatives and therefore Zika virus immunoglobulin (Ig) M antibody testing in certain case as seen in Fig. 3. The IgM antibody persists from first week after onset of symptoms to 12 weeks post symptom onset or exposure. However, due to cross-reactivity with other flaviviruses, it is difficult to get conclusive results. Hence, with IgM testing, there are higher chances of false positives than NAAT. PRNT should be used to resolve false positive IgM antibody results due to cross-reactivity. This method measures virus-specific neutralizing antibody titers. Confirmatory PRNT assay should be performed if IgM assay has a positive result as seen in Fig. 3. However, PRNT may not be able to differentiate between flaviviruses antibodies in cases of secondary flavivirus infections and in areas with high prevalence of dengue and Zika virus neutralizing antibodies. Therefore informed decision needs to be made about the coherence of the positive results from PRNT and IgM assay based on the prevalence of dengue and Zika virus neutralizing antibodies.

Source: Sharp, T. M. et al. Dengue and Zika virus diagnostic testing for patients with a clinically compatible illness and risk for infection with both viruses. MMWR Recomm. Reports 68, 1–10 (2019).

Mini-dictionary of terms

Graphene. It is an allotropic form of carbon, which is a single layer of carbon atoms, forming a hexagonal "honeycomb-like" lattice.

Real-time polymerase chain reaction (RT-qPCR). When RNA is the analyte to be measured, this technique is used in which RNA is first reverse-transcribed to complementary DNA (cDNA). cDNA is then used as a template to perform polymerase chain reaction (PCR), and then the amplification can be measured by dye or probe-based system, measuring the fluorescence.

Enzyme-linked immunosorbent assay (ELISA). It is a solid-phase enzyme immunoassay that can detect analyte in a liquid sample using antibodies against the analyte. The qualities of the bound antibody can be measured for the detection of the analyte.

Electrochemical impedance spectroscopy (EIS). This is a detection method in which the impedance of a system is measured dependent on the frequency of the AC potential.

Surface-imprinted polymers (SIP). These are polymeric materials that self-assemble with template molecules (analyte to be measured). Once the template molecules leave, the cavities left behind act as recognition sites for the analyte.

Key facts of graphene biosensors

- Graphene's favorable chemical, physical, and electrical properties make it a suitable candidate for the designing biosensors.
- Graphene surface harbors a large surface area and is compatible with a large number of biomolecules.
- Graphene can be functionalized with proteins, DNA, and RNA easily for biosensing, either by forming covalent bonds or physisorption.
- For covalent conjugation, usually carbodiimide/DMTMM chemistry is used.
- Broadly, there are two types of graphene biosensors, electrochemical and fluorescent sensors. Electrochemical sensors work by eliciting an electrical output in response to sensing the analyte. Fluorescent sensors work by labeling targeting molecules with a fluorescence tag so that the analyte can be detected based on the fluorescence readout.

Summary points

- Graphene is a good conductor of heat and electricity and possesses a high surface area, making it an excellent material for biosensing.
- Graphene can be easily functionalized with biomolecules for detecting the analyte.

- Traditional diagnostic methods involve using real-time polymerase chain reaction (RT-qPCR), IgM antibody capture enzyme-linked immunosorbent assay (MAC-ELISA), or plaque reduction neutralization test (PRNT). However, these methods are slow, expensive, and labor intensive.
- Graphene-based biosensors solve these problem. There are two studies that have reported the development of the graphene-based biosensor for the detection of Zika virus.
- The first study discusses a strategy in which the monoclonal antibody is used against Zika virus for the detection of the virus.
- The second study used a surface imprinted polymer strategy in which the template of the analyze to be detected in added to a prepolymerized solution of monomers. On polymerization and removal of the template, cavities are left behind that can detect the analyte.

References

Afsahi, S., Lerner, M. B., Goldstein, J. M., Lee, J., Tang, X., Bagarozzi, D. A., … Goldsmith, B. R. (2018). Novel graphene-based biosensor for early detection of Zika virus infection. *Biosensors and Bioelectronics*, *100*, 85–88. https://doi.org/10.1016/j.bios.2017.08.051.

Afsahi, S. J., Locascio, L. E., Pan, D., Gao, Y., Walker, A. E., Barron, F. E., … Lerner, M. B. (2017). Towards novel graphene-enabled diagnostic assays with improved signal-to-noise ratio. *MRS Advances*, *2*(60), 3733–3739. https://doi.org/10.1557/adv.2017.431.

Aptima Zika Virus Assay | Hologic. (n.d.). Retrieved 30 November 2019, from https://www.hologic.com/package-inserts/diagnostic-products/aptima-zika-virus-assay.

Badawi, M. M., Osman, M. M., Alla, A. A. F., Ahmedani, A. M., Abdalla, M. H., Gasemelseed, M. M., … Salih, M. A. (2016). Highly conserved epitopes of ZIKA envelope glycoprotein may act as a novel peptide vaccine with high coverage: Immunoinformatics approach. *American Journal of Biomedical Research*, *4*(3), 46–60. https://doi.org/10.12691/AJBR-4-3-1.

Barba-Spaeth, G., Dejnirattisai, W., Rouvinski, A., Vaney, M. C., Medits, I., Sharma, A., … Rey, F. A. (2016). Structural basis of potent Zika-dengue virus antibody cross-neutralization. *Nature*, *536*(7614), 48–53. https://doi.org/10.1038/nature18938.

Basu, R., & Tumban, E. (2016). Zika virus on a spreading spree: What we now know that was unknown in the 1950's. *Virology Journal*, *13*(1), 1–9. https://doi.org/10.1186/s12985-016-0623-2.

Bollati, M., Alvarez, K., Assenberg, R., Baronti, C., Canard, B., Cook, S., … Bolognesi, M. (2010). Structure and functionality in flavivirus NS-proteins: Perspectives for drug design. *Antiviral Research*, *87*(2), 125–148. https://doi.org/10.1016/j.antiviral.2009.11.009.

CDC. (2020). *Trioplex real-time RT-PCR assay for use under an emergency use authorization only instructions for use*. https://www.cdc.gov/zika/pdfs/trioplex-real-time-rt-pcr-assay-instructions-for-use.pdf. (Accessed 12 April 2020).

Chakraborty, M., & Hashmi, M. S. J. (2018). Wonder material graphene: Properties, synthesis and practical applications. *Advances in Materials and Processing Technologies*, *4*(4), 573–602. https://doi.org/10.1080/2374068X.2018.1484998.

Chaudhari, A., Jagdale, P., Goswami, P., & Kerawalla, M. A. K. (2018). Recent advancements in graphene biosensors for the detection of pathogens. *Indian Drugs*, *55*, 7–17.

Choi, W., Lahiri, I., Seelaboyina, R., & Kang, Y. S. (2010). Synthesis of graphene and its applications: A review. *Critical Reviews in Solid State and Materials Sciences*, *35*(1), 52–71. https://doi.org/10.1080/10408430903505036.

Clark, L. C., & Lyons, C. (1962). Electrode systems for continuous monitoring in cardiovascular surgery. *Annals of the New York Academy of Sciences*, *102*(1), 29–45. https://doi.org/10.1111/j.1749-6632.1962.tb13623.x.

Contreras-Naranjo, J. E., & Aguilar, O. (2019). Suppressing non-specific binding of proteins onto electrode surfaces in the development of electrochemical immunosensors. *Biosensors*, *9*(1). https://doi.org/10.3390/bios9010015.

Dai, L., Song, J., Lu, X., Deng, Y. Q., Musyoki, A. M., Cheng, H., … Gao, G. F. (2016). Structures of the Zika virus envelope protein and its complex with a flavivirus broadly protective antibody. *Cell Host & Microbe*, *19*(5), 696–704. https://doi.org/10.1016/j.chom.2016.04.013.

Dejnirattisai, W., Wongwiwat, W., Supasa, S., Zhang, X., Dai, X., Rouvinski, A., … Screaton, G.R. (2015). A new class of highly potent, broadly neutralizing antibodies isolated from viremic patients infected with dengue virus. *Nature Immunology*, *16*, 170–177. https://doi.org/10.1038/ni.3058.

Gao, N., Gao, T., Yang, X., Dai, X., Zhou, W., Zhang, A., & Lieber, C. M. (2016). Specific detection of biomolecules in physiological solutions using graphene transistor biosensors. *Proceedings of the National Academy of Sciences of the United States of America*, *113*(51), 14633–14638. https://doi.org/10.1073/pnas.1625010114.

Gao, C., Hao, Y., Liang, F., Zheng, K., & Hu, H. (2012). Structure and phase transition of Nd2Fe14.5Cr2.5 compound. *Zhongguo Xitu Xuebao/Journal of the Chinese Rare Earth Society*, *30*(6), 699–702.

He, S., Song, B., Li, D., Zhu, C., Qi, W., Wen, Y., … Fan, C. (2010). A graphene nanoprobe for rapid, sensitive, and multicolor fluorescent DNA analysis. *Advanced Functional Materials*, *20*(3), 453–459. Retrieved 27 February 2020, from https://onlinelibrary.wiley.com/doi/abs/10.1002/adfm.200901639.

Hermanson, G. T. (2013). *Bioconjugate techniques* (3rd ed.). Elsevier Inc. https://doi.org/10.1016/C2009-0-64240-9.

Honeychurch, K. C. (2012). Printed thick-film biosensors. In *Printed films: Materials science and applications in sensors, electronics and photonics* (pp. 366–409). Elsevier Ltd. https://doi.org/10.1533/9780857096210.2.366.

Hull, M., & Bowman, D. M. (2014). *Nanotechnology enviromental health and safety : Risks, regulation, and management*. William Andrew.

Ivnitski, D., Abdel-Hamid, I., Atanasov, P., & Wilkins, E. (1999). Biosensors for detection of pathogenic bacteria. *Biosensors and Bioelectronics*, *14*(7), 599–624. https://doi.org/10.1016/S0956-5663(99)00039-1.

Jalilian, R., Jauregui, L. A., Lee, K., Chen, Y. P., & Irudayaraj, J. M. K. (2012). Graphene for biosensing applications. In *Biomedical nanosensors* (pp. 153–186). Pan Stanford Publishing Pte. Ltd. https://doi.org/10.4032/9789814303040.

Kostyuchenko, V., Lim, E., Zhang, S., Fibriansah, G., Ng, T.-S., Ooi, J. S. G., … Lok, S.-M. (2016). Structure of the thermally stable Zika virus. *Nature*, *533*, 425–428. Retrieved 21 February 2020, from https://www.nature.com/articles/nature17994.

Kuhn, R., Zhang, W., Rossmann, M., Pletnev, S. V., Corver, J., Lenches, E., … Strauss, J. H. (2002). Structure of dengue virus: Implications for flavivirus organization, maturation, and fusion. *Cell*, *108*(5), 717–725. Retrieved 21 February 2020, from https://www.sciencedirect.com/science/article/pii/S0092867402006608.

Kybert, N. J., Han, G. H., Lerner, M. B., Dattoli, E. N., Esfandiar, A., & Charlie Johnson, A. T. (2014). Scalable arrays of chemical vapor sensors based on DNA-decorated graphene. *Nano Research*, *7*(1), 95–103. https://doi.org/10.1007/s12274-013-0376-9.

Liu, B., Huang, P.-J. J., Zhang, X., Wang, F., Pautler, R., Ip, A. C., & Liu, J. (2013). Parts-per-million of polyethylene glycol as a non-interfering blocking agent for homogeneous biosensor development. *Analytical Chemistry*, *85*(21), 10045–10050. https://doi.org/10.1021/ac4024654.

Ma, Y., Shen, X. L., Zeng, Q., Wang, H. S., & Wang, L. S. (2017). A multi-walled carbon nanotubes based molecularly imprinted polymers electrochemical sensor for the sensitive determination of HIV-p24. *Talanta*, *164*, 121–127. https://doi.org/10.1016/j.talanta.2016.11.043.

Modis, Y., Ogata, S., Clements, D., & Harrison, S. C. (2004). Structure of the dengue virus envelope protein after membrane fusion. *Nature*, *427*, 313–319. Retrieved 21 February 2020, from https://www.nature.com/articles/nature02165.

Musso, D., & Gubler, D. J. (2016). Zika virus. *Clinical Microbiology Reviews*, *29*(3), 487–524. https://doi.org/10.1128/CMR.00072-15.

Navakul, K., Warakulwit, C., Yenchitsomanus, P.t., Panya, A., Lieberzeit, P. A., & Sangma, C. (2017). A novel method for dengue virus detection and antibody screening using a graphene-polymer based electrochemical biosensor. *Nanomedicine: Nanotechnology, Biology, and Medicine*, *13*(2), 549–557. https://doi.org/10.1016/j.nano.2016.08.009.

Nicolini, A. M., McCracken, K. E., & Yoon, J.-Y. (2017). Future developments in biosensors for field-ready Zika virus diagnostics. *Journal of Biological Engineering*, *11*(1), 7. https://doi.org/10.1186/s13036-016-0046-z.

Novoselov, K. S., Geim, A. K., Morozov, S. V., Jiang, D., Zhang, Y., Dubonos, S. V., … Firsov, A. A. (2004). Electric field in atomically thin carbon films. *Science*, *306*(5696), 666–669. https://doi.org/10.1126/science.1102896.

Oliphant, T., Engle, M., Nybakken, G., Doane, C., Johnson, S., Huang, L., … Diamond, M. S. (2005). Development of a humanized monoclonal antibody with therapeutic potential against West Nile virus. *Nature Medicine*, *11*, 522–530. Retrieved 21 February 2020, from https://www.nature.com/articles/nm1240.

Peña-Bahamonde, J., Nguyen, H. N., Fanourakis, S. K., & Rodrigues, D. F. (2018). Recent advances in graphene-based biosensor technology with applications in life sciences. *Journal of Nanobiotechnology*, *16*(1). https://doi.org/10.1186/s12951-018-0400-z.

Rouvinski, A., Guardado-Calvo, P., Barba-Spaeth, G., Duquerroy, S., Vaney, M.-C., Kikuti, C. M., … Shepard, W. E. (2015). Recognition determinants of broadly neutralizing human antibodies against dengue viruses. *Nature*, *520*, 109–113. Retrieved 21 February 2020, from https://www.nature.com/articles/nature14130.

Schmidt, N. J., Dennis, J., & Lennette, E. H. (1976). Plaque reduction neutralization test for human cytomegalovirus based upon enhanced uptake of neutral red by virus-infected cells downloaded from. *Journal of Clinical Microbiology*, *4*(1). http://jcm.asm.org/.

Sirohi, D., Chen, Z., Sun, L., Klose, T., Pierson, T. C., Rossmann, M. G., & Kuhn, R. J. (2016). The 3.8 Å resolution cryo-EM structure of Zika virus. *Science*, *352*(6284), 467–470. https://doi.org/10.1126/science.aaf5316.

Smith, S. A., Ruklanthi De Alwis, A., Kose, N., Harris, E., Ibarra, K. D., Kahle, K. M., … Lipkin, E. W. I. (2013). The potent and broadly neutralizing human dengue virus-specific monoclonal antibody 1C19 reveals a unique cross-reactive epitope on the bc loop of domain II of the envelope protein downloaded from. *mBio*. https://doi.org/10.1128/mBio.00873-13.

Stine, K. J. (2017). Enzyme immobilization on nanoporous gold: A review. *Biochemistry Insights*, *10*. https://doi.org/10.1177/1178626417748607, 117862641774860.

Sukjee, W., Tancharoen, C., Yenchitsomanus, P. T., Gleeson, M. P., & Sangma, C. (2017). Small-molecule dengue virus co-imprinting and its application as an electrochemical sensor. *ChemistryOpen*, *6*(3), 340–344. https://doi.org/10.1002/open.201700037.

Szunerits, S., & Boukherroub, R. (2018). Graphene-based biosensors. *Interface Focus*, *8*(3). https://doi.org/10.1098/rsfs.2016.0132.

Tancharoen, C., Sukjee, W., Thepparit, C., Jaimipuk, T., Auewarakul, P., Thitithanyanont, A., & Sangma, C. (2019). Electrochemical biosensor based on surface imprinting for Zika virus detection in serum. *ACS Sensors*, *4*(1), 69–75. https://doi.org/10.1021/acssensors.8b00885.

Testing for Zika Virus Infections | Zika Virus | CDC. (n.d.). Retrieved November 30, 2019, from https://www.cdc.gov/zika/laboratories/types-of-tests.html.

Thvenot, D. R., Toth, K., Durst, R. A., & Wilson, G. S. (1999). Electrochemical biosensors: Recommended definitions and classification (technical report). *Pure and Applied Chemistry*, *71*(12), 2333–2348. https://doi.org/10.1351/pac199971122333.

Yu, D., Blankert, B., Viré, J. C., & Kauffmann, J. M. (2005). Biosensors in drug discovery and drug analysis. *Analytical Letters*, *38*(11), 1687–1701. https://doi.org/10.1080/00032710500205659.

Zhang, X., Ge, P., Yu, X., Brannan, J., Bi, G., Zhang, Q., … Zhou, Z. H. (2013). Cryo-EM structure of the mature dengue virus at 3.5-Å resolution. *Nature Structural & Molecular Biology*, *20*, 105–110. Retrieved 21 February 2020, from https://www.nature.com/nsmb/journal/v20/n1/abs/nsmb.2463.html.

Zhao, H., Fernandez, E., Dowd, K. A., Speer, S. D., Platt, D. J., Gorman, M. J., … Fremont, D. H. (2016). Structural basis of Zika virus-specific antibody protection. *Cell*, *166*(4), 1016–1027. https://doi.org/10.1016/j.cell.2016.07.020.

Zika virus (n.d.). Retrieved February 22, 2020, from https://www.who.int/news-room/fact-sheets/detail/zika-virus.

Chapter 26

The ZIKV *Detect* IgM Capture ELISA

Alison Jane Basile and Holly R. Hughes

Diagnostic and Reference Laboratory, Arboviral Diseases Branch, Division of Vector-Borne Diseases, Centers for Disease Control and Prevention, Fort Collins, CO, United States

Abbreviations

CCA	cross-reactivity control antigen
CHIKV	chikungunya virus
CV	coefficient of variation
DENV	dengue virus
E	envelope protein
ELISA	enzyme-linked immunosorbent assay
EUA	emergency use authorization
FDA	Food and Drug Administration
HRP	horseradish peroxidase
IFU	instructions for use
IgG	immunoglobulin G
IgM	immunoglobulin M
ISR	immune status ratio
IVD	in vitro diagnostic
NAAT	nucleic acid amplification testing
NCA	normal control antigen
NS1	nonstructural protein 1
OD	optical density
PrM	premembrane
RNA	ribonucleic acid
RT-PCR	reverse transcriptase-polymerase chain reaction
TMB	3,3′,5,5′-tetramethylbenzidine
WNV	West Nile virus
YFV	yellow fever virus
ZIKV	Zika virus

Introduction

Clinical diagnosis of Zika virus (ZIKV) is best done early in the course of the disease using nucleic acid amplification testing (NAAT) methods such as real-time reverse transcription-polymerase chain reaction (RT-PCR) (Lanciotti et al., 2008). This is because positive NAAT results are confirmatory and no corroborating evidence is required for diagnosis (Sharp et al., 2019). The restrictive kinetics of ZIKV viral load in patients often reduces the practicality of NAAT methods (Musso et al., 2017). The reality is that many patients fail to present for testing within the acute phase of the disease where RNA may be detected. In addition, dates after onset of symptoms may be inaccurate, and a negative result by RT-PCR in a symptomatic patient does not rule out ZIKV infection. Therefore serology is the means by which presence of a ZIKV infection can be detected under these circumstances.

Flaviviruses are enveloped viruses and are infamously cross-reactive in serological tests (Lindsey et al., 2018). This stems from the fact that the envelope protein (E) contains the majority of the immunoreactive epitopes, which are highly conserved among viruses within the genus (Barba Spaeth et al., 2016). Cross-reactivity can be mitigated by the use of other

Zika Virus Impact, Diagnosis, Control, and Models. https://doi.org/10.1016/B978-0-12-820267-8.00026-1

proteins in serological tests, but this often results in a loss of sensitivity (Basile et al., 2018). Therefore tests such as ELISA are likely to be highly sensitive if they utilize either whole viral antigen or the E protein but may suffer from a lack of specificity due to the presence of highly cross-reactive epitopes in the antigen.

Immunoglobulin M (IgM) is the first major class of antibodies to be elicited by a Zika viral infection, typically developing during the first 4 days after onset of symptoms (Griffin et al., 2019). The presence of IgM does not generally last a lifetime as does immunoglobulin G (IgG). A recent ZIKV infection is suggested when IgM to ZIKV is detected in the serum, and IgM testing forms the mainstay of ZIKV serological diagnosis although evidence exists that ZIKV IgM may persist for greater than 1 year in some cases (CDC, 2017; Griffin et al., 2019). A positive result is ideally confirmed by the complex but more specific plaque reduction neutralization test (PRNT) (Beaty, Calisher, & Shope, 1995). To provide an informative result, cross-PRNT's should be performed on ZIKV plus at least one other flavivirus, most commonly dengue virus (DENV), to confirm the presence of antibodies to only one of the viruses. The usefulness of such a confirmatory test is mostly limited to primary ZIKV infections (where the patient has not been previously exposed to other flaviviruses) (Calvert et al., 2018). In geographic regions where other flaviviruses regularly circulate, secondary flavivirus infections are likely to occur. The utility of PRNT is greatly diminished in this situation due to anamnestic responses where immune systems particularly of adults are primed to form neutralizing cross-reactive IgG antibodies when exposed to viruses that share the same epitopes. The infecting virus therefore cannot be easily identified in secondary flaviviral infections (Lindsey et al., 2018).

The ZIKV *Detect* 2.0 IgM Capture ELISA and its predecessor ZIKV *Detect* IgM Capture ELISA are E-based and are intended for the presumptive detection of Zika virus IgM in human serum. They were designed to take advantage of the sensitivity of E-based tests but mitigate some of the cross-reactivity problems that are inherent in such assays.

Composition and assay protocol of the ZIKV *Detect* 2.0 IgM Capture ELISA

The ZIKV *Detect* 2.0 IgM Capture ELISA is a qualitative assay adapted from a traditional IgM capture ELISA format (Basile et al., 2019). Here, IgM is captured from a serum sample by an antihuman IgM antibody attached to the plate, and a Zika viral antigen is added to the test, which reacts with any Zika viral-specific IgM that has been captured from the sample. The antigen in turn is reacted with an antiflavivirus mouse monoclonal antibody 6B6C-1 (secondary antibody), which is detected with horseradish peroxidase (HRP)–conjugated antimouse IgG (the original kit used 6B6C-1 HRP). The kit consists of 96-microtiter wells precoated with antihuman IgM, supplied as 8-well strips, and up to 28 samples can be assayed per 96-well test with a run time of 4–5 h. All reagents and controls necessary to perform the test are provided in the kit. Positive Zika IgM and negative validation controls are used in duplicate with each assay, and test samples are typically assayed in single replicates. Controls and test serum samples are diluted 1:100 in sample diluent, and 50 μL is added to each of three wells. After a covered incubation for 1 h at 37°C, wells are washed using diluted wash buffer (a 10 × concentrate is provided). Fifty microliters of three antigen preparations—Zika antigen, cross-reactivity control antigen (CCA), and normal control antigen (NCA)—are placed in one of each of the three wells per sample. The Zika antigen used in the kit is a recombinant prM/E viral-like particle (Chao et al., 2019). The CCA is a combination of DENV and West Nile virus (WNV) antigens (Chang et al., 2003; Davis et al., 2001). The NCA is a mock antigen preparation produced in normal COS-1 cells. This is included to control for nonspecific background reactions in the serum samples. The test is covered and incubated for 1 h at 37°C and then washed. Fifty microliters of ready-to-use antiflavivirus mouse monoclonal antibody (Tsai et al., 1987) is added to each well and incubated for 30 min at 37°C, followed by washing. The reactions are detected using 50 μL/well of a freshly prepared 1:100 dilution of antimouse IgG-HRP in conjugate diluent. After a 30-min incubation at 37°C, wells are washed, and 75 μL of TMB substrate is added. Wells are incubated in the dark for 20 min at room temperature, after which 50 μL of stop solution is added to complete the reaction. The unblanked optical densities are read at 450 nm.

Results classification

The immune status ratio (ISR) is the starting point for classifying ZIKV *Detect* 2.0 IgM Capture ELISA results. It is calculated for the controls and the test samples by dividing the Zika antigen OD value by that of the CCA OD value. Six quality control criteria must be met prior to results analysis, and these are listed in the instructions for use (IFU) (InBios, 2019). These include optical density limits for the negative control when reacted on the antigens, a minimum value for the ISR of the positive control, and maximum ISR and CCA OD/NCA OD (CCA/NCA) values for the negative control. In addition, a Zika antigen OD test threshold must be exceeded for a sample to be classified as positive, where the threshold is equal to 0.13 plus the average OD value of the negative control.

Any test results that exceed the Zika antigen OD test threshold can be examined for Zika positivity. Samples with ISRs ≥1.9 are considered "presumptive Zika positive." If a sample has an ISR between 1.5 and 1.9, it is retested, and if the ISR is ≥1.7 after repetition, the sample is considered presumptive Zika positive. If the repeat test yields an ISR of 1.5 to <1.7, the CCA/NCA is evaluated. If this is ≥5, the sample is classified as "presumptive other flavivirus positive" (i.e., non-Zika), and if it is <5, it is classified as "negative." All samples that do not exceed the test threshold for Zika antigen OD are examined for other flavivirus positivity.

Performance of the InBios ZIKV *Detect* 2.0 IgM Capture ELISA compared with the original version

A study published in 2018 (Basile et al., 2018) was undertaken in three geographically separate laboratories to evaluate the performance of the original version of the assay, ZIKV *Detect* IgM Capture ELISA, which had several major differences in format from its successor. The test was evaluated alongside two other commercial Zika IgM ELISAs, one of which used an associated IgG test in the results analysis. Both additional assays employed nonstructural protein 1 (NS1) as the antigen and thus did not suffer E-associated cross-reactivity but lacked sensitivity. A subsequent study (Basile et al., 2019) focused solely on the updated InBios assay. The ZIKV *Detect* 2.0 IgM Capture ELISA was evaluated in two of the three laboratories involved in the 2018 study using the same sample sets. Data from each of the studies in the two laboratories were compared, where test results were measured against the Centers for Disease Control and Prevention (CDC) M-antibody capture ELISA (Lanciotti et al., 2008). The largest sample set consisted of 289 blinded Zika IgM-positive and IgM-negative sera from the continental United States and thus was composed chiefly of primary infections obtained during travel. Smaller sets were employed to look specifically at the use of the tests at different dates after onset of symptoms and to evaluate the usefulness of the assays when used with primary Zika versus secondary flavivirus infections or primary/secondary DENV infections.

Several improvements in performance were identified for the InBios ZIKV *Detect* 2.0 IgM Capture ELISA compared with its predecessor. Sensitivity, specificity, accuracy, and concordance of results across the two laboratories were evaluated for the updated kit using the continental US sample set (Table 1). Sensitivity and specificity were measured at 92.5% and 97.4% respectively. The sensitivity represented a small and statistically insignificant increase over the original kit version (90.4%), but specificity improved significantly, by almost 18%. The original version of the kit included a results

TABLE 1 Performance of the ZIKV *Detect* 2.0 IgM Capture ELISA compared with the ZIKV *Detect* IgM Capture ELISA across two laboratories.

		ZIKV *Detect* 2.0 IgM Capture ELISA	ZIKV *Detect* IgM Capture ELISA
Measure	*N*	% (95% CI)	
Sensitivity	114	92.5 (88.4, 95.3)	90.4 (85.8, 93.5)
Specificity[a]	78	97.4 (93.6, 99.0)	79.5 (72.5, 85.1)
Specificity[b]	205	94.9 (92.3, 96.6)	56.6 (51.7, 61.3)
Accuracy[c]		89.5 (86.8, 91.7)	63.9 (60.1, 67.6)
Controls		CV (95% CI)	
Positive		15.7 (12.5, 21.4)	34.3 (26.0, 51.1)
Negative		15.6 (12.4, 21.3)	22.9 (17.5, 33.1)
		% agreement (Fleiss' Kappa statistic: 95% CI)	
Agreement		97.4 (0.96: 0.93, 0.99)	79.5 (0.59: 0.51, 0.67)

The ZIKV *Detect* 2.0 IgM Capture ELISA was compared with the ZIKV *Detect* IgM Capture ELISA at two independent reference laboratories. Each evaluation used the same set of clinical validation samples. ZIKV *Detect* 2.0 IgM Capture ELISA showed improved performance and agreement between the labs.

[a]*Specificity denominator consists of true negatives only.*

[b]*Specificity denominator consists of samples containing IgM antibodies to the following viruses: 10 yellow fever, 10 West Nile, 10 chikungunya, 97 dengue, and 78 negative samples.*

[c]*Accuracy based on proportion correctly identified by samples containing IgM to the following viruses according to these categories: Zika= "presumptive Zika positive," dengue and West Nile= "presumptive other flavivirus," and chikungunya and NEG= "negative" (except for one chikungunya sample that was classified as "presumptive other flavivirus" due to apparent dual infection).*

Adapted from Basile, A., Ao, J., Horiuchi, K., Semenova, V., Steward Clark, E., et al. (2019). Performance of InBios ZIKV Detect™ 2.0 IgM Capture ELISA in two reference laboratories compared to the original ZIKV Detect™ IgM Capture ELISA. *Journal of Virological Methods, 271*, 113671.

classification category of "possible Zika positive," into which samples with low Zika IgM levels tended to fall, along with samples containing IgM to both DENV and ZIKV. The inclusion of an OD threshold that samples had to exceed to be classed as "presumptive Zika positive" was responsible for the improvement in statistical specificity (based on true negatives). The accuracy of the kit increased to 89.5%, up from 63.9% for the former version. The agreement of results across the two laboratories also improved significantly from 79.5% (Fleiss' kappa statistic (K) = 0.59) to 97.4% (K = 0.96), and the few false negative results observed were all seen within the first two weeks of onset of symptoms. Sample sets where reference results allowed the differentiation of probable primary ZIKV or DENV infections and ZIKV or DENV infections secondary to previous flavivirus exposure were tested using the ZIKV *Detect* IgM Capture ELISA and the ZIKV *Detect* 2.0 IgM Capture ELISA (Table 2). Overall the sensitivity of primary ZIKV specimens was 91.5% using the updated kit compared with 85.2% with the original version. Sensitivity of detection of secondary infections dropped to 94.2% from 98.8% between the updated and original kits. The specificity of version 2.0 of the kit was 96.7% and 82.6% when primary and secondary DEN infections were tested, respectively. The specificity was far inferior with the original kit (35.0% for primary and 9.3% for secondary DEN infections). This vast improvement was due to the elimination of the "possible Zika positive" classification category and the conversion of samples with minimal Zika IgM levels but significant CCA reactivity to being classified as "presumptive other flavivirus positive," thus effectively leveraging the inclusion of CCA in the assay.

Other studies using InBios ZIKV *Detect* IgM Capture ELISA

Additional studies using the InBios ZIKV *Detect* IgM Capture ELISA have been reported in the literature. A 2019 publication from a Canadian arbovirus reference laboratory (Mendoza et al., 2019) describes the establishment of a high-throughput testing algorithm using this assay to allow all samples to be screened, including those collected after 3 months of onset. In this algorithm the InBios IgM assay was used in combination with an NS1-based Zika IgG assay from EUROIMMUN. Both sensitivity and specificity of this combined approach were high and exhibited excellent concordance with PRNT, illustrating the potential to eliminate the need for PRNT, which is a time-consuming and technically demanding assay. The results in Basile et al. (2019) corroborated findings from earlier the same year (Granger & Theel, 2019), where it was concluded that the replacement of InBios ZIKV *Detect* IgM Capture ELISA with InBios ZIKV *Detect* 2.0 IgM Capture ELISA would lead to fewer samples needing PRNT and thus minimize anxiety in patients ultimately shown to be free of ZIKV infection.

TABLE 2 Sensitivity (A) and specificity (B) of InBios ZIKV *Detect* 2.0 IgM Capture ELISA compared to the earlier version of the kit using samples from probable primary and secondary infections.

(A) Sensitivity using confirmed Zika[a] positive samples		**InBios ZIKV *Detect* 2.0 IgM Capture ELISA**	**InBios ZIKV *Detect* IgM Capture ELISA**
Probable status	***N*[b]**	**Mean sensitivity (95% CI)**	**Mean sensitivity (95% CI)**
Primary	71	91.5 (85.8, 95.1)	85.2 (78.4, 90.1)
Secondary	43	94.2 (87.1, 97.5)	98.8 (93.7, 99.9)
(B) Specificity using confirmed dengue[a] positive samples		**InBios ZIKV *Detect* 2.0 IgM Capture ELISA**	**InBios ZIKV *Detect* IgM Capture ELISA**
Probable status	***N*[c]**	**Mean specificity (95% CI)**	**Mean specificity (95% CI)**
Primary	30	96.7 (88.6, 99.1)	35.0 (24.2, 47.6)
Secondary	43	82.6 (73.2, 89.1)	9.3 (4.8, 17.3)

Primary versus secondary infection status was determined by plaque reduction neutralization test results using Zika and dengue viruses on single convalescent samples where a lack of titer to one of the viruses indicated a primary infection. Viral RNA was detected in associated acute samples for all but six of the secondary dengue infections to identify the infecting virus. Mean sensitivity and specificity values were derived from data generated at two laboratories. The InBios ZIKV *Detect* 2.0 showed a marked improvement in specificity when measured using dengue infections compared with the original version of the kit.
[a]*Virus responsible for current infection.*
[b]*Samples classified as "presumptive Zika positive" by InBios ZIKV* Detect *2.0 IgM Capture ELISA and samples classified as "presumptive Zika positive" and "possible Zika positive" by InBios ZIKV* Detect *IgM Capture ELISA were classified as positive for purposes of calculating sensitivity.*
[c]*Samples classified as "presumptive other flavivirus positive" and "negative" were classified as negative for purposes of calculating specificity.*
Adapted from Basile, A., Ao, J., Horiuchi, K., Semenova, V., Steward Clark, E., et al. (2019). Performance of InBios ZIKV Detect™ 2.0 IgM Capture ELISA in two reference laboratories compared to the original ZIKV Detect™ IgM Capture ELISA. *Journal of Virological Methods, 271*, 113671.

Applicability of the InBios ZIKV *Detect* 2.0 IgM Capture ELISA to clinical diagnostic testing

The InBios ZIKV *Detect* 2.0 IgM Capture ELISA is intended to be used as a screening assay for Zika virus infections. Positive IgM results should typically be confirmed using a confirmatory method such as plaque reduction neutralization assay (PRNT). Diagnostic testing algorithms for ZIKV vary around the world according to the prevalence of Zika and other flaviviruses. In the United States, as of November 2019, IgM testing is no longer recommended for clinically compatible symptomatic pregnant women with risk of exposure, for symptomatic nonpregnant persons having risk of infection, or for asymptomatic pregnant or nonpregnant women with the risk of exposure (CDC, 2019). The InBios ZIKV *Detect* 2.0 IgM Capture ELISA exhibited a specificity of 97.4%, which is considered excellent for this type of test. However, as with any test with less than 100% specificity, the positive predictive value decreases when the disease prevalence decreases (Altman & Bland, 1994). This results in costly and needless workups beyond the initial screening assay; hence the current recommendation against Zika IgM testing especially in a nonoutbreak situation.

Adult residents of South and Central America and the Caribbean are likely to have been exposed to DENV at some time in their lives. While the cross-reactivity of the InBios ZIKV *Detect* 2.0 IgM Capture ELISA is far less than that of the original version of the kit, more than 10% of positive results are likely to be false. This would indicate the continued need to confirm results using another serological method such as plaque reduction neutralization test (PRNT). The inability of PRNT to confirm the presence of ZIKV-specific neutralizing antibody for secondary flavivirus infections, or in some cases being unable to detect neutralizing antibodies at all (Lustig et al., 2018), further indicates that the ZIKV *Detect* 2.0 IgM Capture ELISA is of limited use in these geographic regions. In addition, IgM levels to ZIKV are sometimes diminished in a secondary flavivirus exposure compared with primary due to "original antigenic sin" (Lustig et al., 2018; Vatti et al., 2017). The InBios ZIKV *Detect* 2.0 IgM Capture ELISA is, however, appropriate for screening of babies born to mothers who may have been exposed during their pregnancy, provided that the samples are taken within 2 days of birth (WHO, 2017).

Regulatory status and labeling

The InBios ZIKV *Detect* IgM Capture ELISA was developed in 2016 and was granted emergency use authorization (EUA) status. This is a regulatory requirement for any in vitro diagnostic (IVD) marketed in the United States that is used in clinical diagnosis during a declared public health emergency, where an equivalent 510(k) cleared product is unavailable. In May of 2018, the US Food and Drug Administration (FDA) concurred with the modifications to the instructions for use and updated the EUA for the ZIKV *Detect* IgM 2.0 Capture ELISA. This version of the kit was subsequently granted FDA 510(k) clearance on May 23, 2019, and was the first Zika IgM assay to receive marketing authorization in the United States.

According to the instructions for use (IFU) (InBios, 2019), the InBios ZIKV *Detect* IgM 2.0 Capture ELISA is intended for the qualitative detection of Zika virus IgM antibodies in human serum. The assay is only intended for use in patients with clinical symptoms consistent with Zika virus infection and/or patients that meet regionally appropriate Zika virus epidemiological criteria. The IFU states that the performance of the assay has not been established for sample matrices other than serum or for use with serum of children under the age of 5. Some limitations listed include (1) the possibility of false positive results from cross-reactivity with DENV, West Nile virus, yellow fever virus, chikungunya virus, babesia, Lyme disease, and malaria infections; (2) the occurrence of false negative results in patients with high levels of human antimurine antibody; (3) possible misclassification of samples with high OD450 values for both Zika Ag and CCA as "presumptive other flavivirus positive" rather than "presumptive Zika positive"; and (4) the possibility of Zika infection in patients with negative results on samples collected outside the response window of 7–84 days post symptom onset. The IFU therefore cautions that the results of InBios ZIKV *Detect* IgM 2.0 Capture ELISA are not intended to be used as the sole basis for diagnosis of Zika virus infection. Further, they advise that all results should be interpreted by a trained professional and in conjunction with clinical symptoms, patient history, epidemiological factors, and other laboratory tests such as confirmatory PRNT.

Performance of the InBios ZIKV *Detect* IgM Capture ELISA in proficiency testing

The US Centers for Disease Control and Prevention, Division of Vector-Borne Diseases, provides annual arboviral serological panels for US public health laboratory system and military laboratory proficiency testing. The panels for Zika virus IgM proficiency testing consist of Zika virus IgM confirmed samples (Zika virus PRNT >10 and dengue virus PRNT < 10) and true IgM negative human serum samples (PRNT <10 to all arboviruses of interest). Annually the panels contain at least one high IgM positive, at least one midrange positive, and at least one low positive sample as determined by the reference CDC Zika MAC ELISA (Lanciotti et al., 2008), as well as at least two true negatives. Participating public health laboratories receive the panels, process, and test following each individual laboratory's standard operating procedures. Results are reported back to the CDC in Fort Collins, CO, United States, for comparison to reference results.

InBios ZIKV *Detect* IgM 2.0 Capture ELISA displayed an improved performance in proficiency testing in the US public health system in 2019 when compared with the performance of its predecessor in 2018. Proficiency testing exists to evaluate the abilities of the participants to perform assays correctly. Nevertheless, the fact that the low positive panel sample tested negative in 23/26 of the laboratories using the original kit in 2018 suggested the assay was not performing well. Laboratories using the InBios ZIKV *Detect* IgM 2.0 Capture ELISA in 2019 correctly identified all samples in the evaluation with 100% sensitivity, specificity, and accuracy (Table 3). In 2018 the InBios ZIKV *Detect* IgM Capture ELISA resulted in 69.2% sensitivity and 81.5% accuracy with a negative predictive value of 68.4%. The improvement in performance of the InBios ZIKV *Detect* IgM 2.0 Capture ELISA in proficiency testing was also seen in a reduction in the coefficient of variation (CV) across testing laboratories (Table 4). The CV of raw optical density readings ranged from 154% to 215% with the InBios ZIKV *Detect* IgM Capture ELISA, while CV for InBios ZIKV *Detect* IgM 2.0 Capture ELISA was noticeably reduced and ranged from 19.8% to 29.8%. The improved CV of optical density values reflect the improved agreement and consistency as previously described by Basile and colleagues (Basile et al., 2019).

TABLE 3 Performance of the InBios ZIKV *Detect* 2.0 IgM Capture ELISA compared with the InBios ZIKV *Detect* IgM Capture ELISA in US public health labs participating in Zika IgM proficiency testing.

	Zika *Detect* 2.0 IgM capture				Zika *Detect* IgM capture			
Measure	*N*	Positive	Negative	% (95% CI)	*N*	Positive	Negative	% (95% CI)
Sensitivity	54	54	0	100 (93.3, 100)	78	54	24	69.2 (58.3, 78.4)
Specificity	36	0	36	100 (90.3, 100)	52	0	52	100 (93.1, 100)
PPV[a]				100 (93.3, 100)				100 (93.4, 100)
NPV[b]				100 (90.3, 100)				68.4 (57.3, 77.7)
Accuracy				100				81.5

Unpublished sensitivity and specificity of Zika IgM proficiency testing in US public health labs. Annual proficiency testing using the InBios ZIKV *Detect* 2.0 IgM Capture ELISA shows improved sensitivity and specificity compared with the predecessor test.
[a]*Positive predictive value.*
[b]*Negative predictive value.*

TABLE 4 Variation of the InBios ZIKV *Detect* 2.0 IgM Capture ELISA optical density values across US public health laboratories.

		Coefficient of variation (optical density)				
	N	Zika (high positive) (%)	Zika (mid positive) (%)	Zika (low positive) (%)	True negative (%)	True negative (%)
ZIKV *Detect* 2.0 IgM	18	21.00	19.80	20.10	29.80	29.60
Zika *Detect* IgM	26	168	154	171	215	205

Unpublished coefficient of variation (CV) across US public health labs participating in the Zika IgM proficiency project. Annual proficiency testing using the InBios ZIKV *Detect* 2.0 IgM Capture ELISA shows improved CV across participating labs compared with the predecessor test. The CV is given for each proficiency sample. Each sample is defined by the qualitative measure of positive signal as determined by the reference assay (e.g., high, mid, and low positive). *N*, the number of participating labs reporting the use of each respective kit.

Policy and procedures

Recommendations for Zika diagnostic testing and diagnostic algorithms vary throughout the world dependent on current transmission and cocirculation of multiple flaviviruses. Local health administrations provide testing guidance. The InBios ZIKV *Detect* IgM 2.0 Capture ELISA is labeled for use with serum samples collected between 4 days post onset of symptoms and 84 days post onset of symptoms. Guidance for sample collection and shipment in the United States is available through the Centers for Disease Control and Prevention (Collecting and Submitting Body Fluid Specimens for Zika virus testing; https://www.cdc.gov/zika/laboratories/test-specimens-bodyfluids.html). Serum samples collected in the United States should be kept cold (2–8°C) or frozen (−20°C). According to the InBios ZIKV *Detect* IgM 2.0 Capture ELISA instructions for use, the serum samples should be equilibrated to room temperature (20–25°C) prior to testing and should not be heat inactivated. Additionally, avoiding repeated freeze-thaw cycles of the serum samples is recommended. Zika virus positive cases are reported according to local guidance. In the United States, Zika virus is a nationally notifiable condition. Healthcare providers are encouraged to report suspected Zika virus disease to the local, territorial, or state health department (Sharp et al., 2019). State and territorial health departments report Zika virus cases meeting the Council of State and Territorial Epidemiological case definitions to the Centers for Disease Control and Prevention (Sharp et al., 2019).

Mini-dictionary of terms

Negative predictive value. The probability that samples with a negative test result will truly be negative (e.g., truly be absent of Zika IgM in the sample).

Original antigenic sin. The serological response to the secondary flavivirus infection often results in a robust anamnestic antibody response to the primary infecting virus potentiating the development of a cross-reactive, nonspecific immune response to the current infecting virus.

Plaque reduction neutralization test (PRNT). Testing the ability of antibodies in the sample to inhibit live virus from infecting cells, thus reducing the number of virus-induced cell death plaques compared with the control.

Positive predictive value. The probability that samples with a positive test result will truly be positive (e.g., truly have Zika IgM present in the sample).

Primary flavivirus infection. An individual who was previously serologically naïve to flaviviruses prior to the current flavivirus infection.

Secondary flavivirus infection. A subsequent infection with additional flaviviruses, frequently resulting in generation of IgM and IgG reactive with multiple flaviviruses.

Sensitivity. Diagnostic sensitivity is the ability of a test to identify true positive samples, thus minimizing the presence of false negative results.

Specificity. Diagnostic specificity is the ability of a test to identify true negative samples, thus minimizing the presence of false positive results.

Key facts of InBios ZIKV *Detect* IgM 2.0 Capture ELISA

- The InBios ZIKV *Detect* IgM 2.0 Capture ELISA tests for the presence of ZIKV IgM in a clinical sample. Presence of IgM is identified by capturing serum IgM and binding of captured ZIKV-reactive IgM to ZIKV antigen. Bound viral antigen reacts with a flaviviral group-reactive monoclonal antibody, which is then reacted with an enzyme-conjugated mouse IgG antibody. The reporter enzyme elicits a colorimetric response when treated with a substrate. The absorbance is measured and used to calculate positivity.
- The InBios ZIKV *Detect* IgM 2.0 Capture ELISA uses an immune status ratio (ISR) to determine positivity. The ISR is calculated by dividing the raw optical density 450 (OD450) value of the sample tested on the Zika antigen by the raw OD450 value of the sample tested on the cross-reactivity control antigen (CCA).
- Dividing the OD450 value of the Zika antigen by the CCA antigen considers cross-reactive antibodies to dengue and West Nile viruses and "normalizes" the ISR to be more specific to Zika antibodies.
- The InBios ZIKV *Detect* IgM 2.0 Capture ELISA is improved compared with its predecessor by the inclusion of the "Zika Ag OD450 threshold," which is the Zika Ag OD450 of the negative control serum plus a constant value (0.130) as determined by true negative samples. This value determines a positivity threshold for samples to be considered "presumptive Zika positive."

- The InBios ZIKV *Detect* IgM 2.0 Capture ELISA is improved compared with its predecessor by the removal of the "possible Zika positive" classification, whereby classification of samples with low Zika IgM levels but significant CCA reactivity are now classified as "presumptive other flavivirus positive," thus efficiently utilizing the CCA and improving the specificity of the assay.

Summary points

- The InBios ZIKV *Detect* 2.0 IgM Capture ELISA is suitable for use in serological diagnosis of Zika infection.
- The performance of version 2.0 of the kit is superior to the original version.
- ZIKV *Detect* 2.0 IgM Capture ELISA has high sensitivity and specificity.
- Results are classified as presumptive Zika IgM positive, other flavivirus positive, or negative.
- The InBios ZIKV *Detect* 2.0 IgM Capture ELISA is approved for use as an in vitro diagnostic assay in the United States by the Food and Drug Administration.
- The InBios ZIKV *Detect* 2.0 IgM Capture ELISA performed well in the Centers for Disease Control and Prevention 2019 Zika proficiency project.

References

Altman, D. G., & Bland, J. M. (1994). Diagnostic tests 2: Predictive values. *British Medical Journal, 309*(6947), 102.

Barba Spaeth, G., Dejnirattisai, W., Rouvinski, A., Vaney, M.-C., Medits, I., Sharma, A., et al. (2016). Structural basis of potent Zika-dengue virus antibody cross-neutralization. *Nature, 536*(7614), 48–53.

Basile, A., Ao, J., Horiuchi, K., Semenova, V., Steward Clark, E., & Schiffer, J. (2019). Performance of InBios ZIKV Detect™ 2.0 IgM Capture ELISA in two reference laboratories compared to the original ZIKV Detect™ IgM Capture ELISA. *Journal of Virological Methods, 271*, 113671.

Basile, A., Goodman, C., Horiuchi, K., Sloan, A., Johnson, B., Kosoy, O., et al. (2018). Multi-laboratory comparison of three commercially available Zika IgM enzyme-linked immunosorbent assays. *Journal of Virological Methods, 260*, 26–33.

Beaty, B. J., Calisher, C. H., & Shope, R. E. (1995). Arboviruses. In E. H. Lennette, & D. A. Lennette (Eds.), *Diagnostic procedures for viral, rickettsial and chlamydial infections* (7th ed.). Washington, DC: American Public Health Association.

Calvert, A., Boroughs, K., Laven, J., Stovall, J., Luy, B., Kosoy, O., et al. (2018). Incorporation of IgG depletion in a neutralization assay facilitates differential diagnosis of Zika and dengue in secondary flavivirus infection cases. *Journal of Clinical Microbiology, 56*(6), 1–13.

CDC. (2017). *Prolonged IgM antibody response in people infected with Zika virus: Implications for interpreting serologic testing results for pregnant women*. CDC Health Advisory. Available from: https://emergency.cdc.gov/han/han00402.asp (Accessed).

CDC. (2019). *NEW Zika and dengue testing guidance (updated November 2019)*. Available from: https://www.cdc.gov/zika/hc-providers/testing-guidance.html (Accessed).

Chang, G.-J., Hunt, A., Holmes, D., Springfield, T., Chiueh, T.-S., Roehrig, J., et al. (2003). Enhancing biosynthesis and secretion of premembrane and envelope proteins by the chimeric plasmid of dengue virus type 2 and Japanese encephalitis virus. *Virology (New York, N.Y. Print), 306*(1), 170–180.

Chao, D.-Y., Whitney, M., Davis, B., Medina, F., Munoz, J., & Chang, G.-J. (2019). Comprehensive evaluation of differential serodiagnosis between Zika and dengue viral infections. *Journal of Clinical Microbiology, 57*(3), 1–14.

Davis, B. S., Chang, G. J., Cropp, B., Roehrig, J. T., Martin, D. A., Mitchell, C. J., et al. (2001). West Nile virus recombinant DNA vaccine protects mouse and horse from virus challenge and expresses in vitro a noninfectious recombinant antigen that can be used in enzyme-linked immunosorbent assays. *Journal of Virology, 75*(9), 4040–4047.

Granger, D., & Theel, E. (2019). Evaluation of a rapid immunochromatographic assay and two enzyme-linked immunosorbent assays for detection of IgM-class antibodies to Zika virus. *Journal of Clinical Microbiology, 57*(3), 1–9.

Griffin, I., Martin, S. W., Fischer, M., Chambers, T. V., Kosoy, O., Falise, A., et al. (2019). Zika virus IgM detection and neutralizing antibody profiles 12–19 months after illness onset. *Emerging Infectious Diseases, 25*(2), 299–303.

InBios. (2019). *ZIKV Detect™ 2.0 IgM Capture ELISA instructions for use*. Available from: https://inbios.com/zikv-detecttm-igm-capture-elisa-kit-usa/ (Accessed).

Lanciotti, R., Kosoy, O., Laven, J., Velez, J., Lambert, A., Johnson, A., et al. (2008). Genetic and serologic properties of Zika virus associated with an epidemic, Yap State, Micronesia, 2007. *Emerging Infectious Diseases, 14*(8), 1232–1239.

Lindsey, N., Staples, J. E., Powell, K., Rabe, I., Fischer, M., Powers, A., et al. (2018). Ability to serologically confirm recent Zika virus infection in areas with varying past incidence of dengue virus infection in the United States and U.S. Territories in 2016. *Journal of Clinical Microbiology, 56*(1).

Lustig, Y., Cotar, A., Ceianu, C., Castilletti, C., Zelena, H., Burdino, E., et al. (2018). Lack of Zika virus antibody response in confirmed patients in non-endemic countries. *Journal of Clinical Virology, 99–100*, 31–34.

Mendoza, E., Makowski, K., Barairo, N., Holloway, K., Dimitrova, K., Sloan, A., et al. (2019). Establishment of a comprehensive and high throughput serological algorithm for Zika virus diagnostic testing. *Diagnostic Microbiology and Infectious Disease, 94*(2), 140–146.

Musso, D., Rouault, E., Teissier, A., Lanteri, M. C., Zisou, K., Broult, J., et al. (2017). Molecular detection of Zika virus in blood and RNA load determination during the French Polynesian outbreak. *Journal of Medical Virology, 89*(9), 1505–1510.

Sharp, T., Fischer, M., Muñoz Jordán, J., Paz Bailey, G., Staples, J. E., Gregory, C., et al. (2019). Dengue and Zika virus diagnostic testing for patients with a clinically compatible illness and risk for infection with both viruses. *Morbidity and Mortality Weekly Report: Recommendations and Reports*, *68*(1), 1–10.

Tsai, T. F., Bolin, R. A., Montoya, M., Bailey, R. E., Francy, D. B., Jozan, M., et al. (1987). Detection of St. Louis encephalitis virus antigen in mosquitoes by capture enzyme immunoassay. *Journal of Clinical Microbiology*, *25*(2), 370–376.

Vatti, A., Monsalve, D., Pacheco, Y., Chang, C., Anaya, J.-M., & Gershwin, M. E. (2017). Original antigenic sin: A comprehensive review. *Journal of Autoimmunity*, *83*, 12–21.

WHO. (2017). *Standardized Protocol: 11 Prospective longitudinal cohort study of 12 women and newborns exposed to Zika 13 virus during the course of pregnancy*. Available from: http://origin.who.int/reproductivehealth/zika/Pregnant-women-cohort-full-protocolV1.13-2.pdf (Accessed).

Chapter 27

Quantum dot-based fluoroassays for Zika

Jéssika F.F. Ribeiro[a], Maria I.A. Pereira[a], Paulo E. Cabral Filho[a], Giovannia A.L. Pereira[b], Beate S. Santos[c], Goreti Pereira[b], and Adriana Fontes[a]

[a]*Biophysics and Radiobiology Department, Federal University of Pernambuco (UFPE), Biomedical Nanotechnology Research Group (NanoBio), Recife, Pernambuco, Brazil,* [b]*Department of Fundamental Chemistry, Federal University of Pernambuco (UFPE), Biomedical Nanotechnology Research Group (NanoBio), Recife, Pernambuco, Brazil,* [c]*Pharmaceutical Sciences Department, Federal University of Pernambuco (UFPE), Biomedical Nanotechnology Research Group (NanoBio), Recife, Pernambuco, Brazil*

Abbreviations

AgNPs	silver nanoparticles
AI	avian influenza virus
AuNPs	gold nanoparticles
Aβ1–42	β-amyloid peptide
BSA	bovine serum albumin
C	capsid
CB	conduction band
CdS	cadmium sulfide
CdSe	cadmium selenide
CdSeTeS	cadmium selenide telluride sulfide
CdTe	cadmium tellurite
DENV	dengue virus
E protein	envelope protein
EDC	1-ethyl-3-(3-dimethylaminopropyl)carbodiimide
E_g	band gap energy
ELISA	enzyme-linked immunosorbent assays
FLISA	fluorescence-linked immunosorbent assays
IgG	immunoglobulin G
IgM	immunoglobulin M
InP	indium phosphide
LFIA	lateral flow immunoassay
LSPR	localized surface plasmon resonance
MAC-ELISA	IgM antibody-capture enzyme-linked immunosorbent assay
MB	molecular beacon
MSA	mercaptosuccinic acid
NC	nitrocellulose
NHS	*N*-hydroxysuccinimide
NoV-LP	norovirus-like particles
NPs	nanoparticles
NS	nonstructural protein
PCR	polymerase chain reaction
prM	premembrane
PRNT	plaque reduction neutralization test
QDs	quantum dots
RNA	ribonucleic acid
Sulfo-NHS	*N*-hydroxysulfosuccinimide sodium salt
Tris	tris(hydroxymethyl)aminomethane
VB	valence band

Zika Virus Impact, Diagnosis, Control, and Models. https://doi.org/10.1016/B978-0-12-820267-8.00027-3

ZIKV	Zika virus
ZnS	zinc sulfide
ZnSeS	zinc selenide sulfide

Introduction

Zika is an infectious disease declared as an international public health emergency. The infection by Zika virus (ZIKV) has been associated with microcephaly in newborns and Guillain-Barré syndrome in adults (De Fatima Vasco Aragao et al., 2016; Miller et al., 2017). Zika clinical symptoms are usually similar to those of other viral diseases such as dengue and chikungunya (Shan et al., 2016). Some methods developed for Zika acute and postinfection diagnostics so far are polymerase chain reaction (PCR), and IgM antibody-capture enzyme-linked immunosorbent assay (MAC-ELISA) followed by plaque reduction neutralization test (PRNT) (Fischer et al., 2018; Nicolini, McCracken, & Yoon, 2017; Shukla, Hong, Chung, & Kim, 2016). Electrochemical biosensing has also been explored (Faria & Mazon, 2019; Kaushik et al., 2018). Nevertheless, Zika detection is challenging and still requires the development of less expensive and less laborious methods, with higher sensitivity (Fischer et al., 2018; Nicolini et al., 2017).

In the last two decades, nanotechnology has emerged as a new field with the potential to modernize procedures from electronics to biomedical sciences (Brkić, 2018). Nanoparticles (NPs) have particularly attracted the interest of researchers, mainly due to their unique size-dependent properties. Nanotechnology has been contributing to new assemblies of different materials, varying from metallic to fluorescent NPs. These nanostructures present interesting properties for improving methodologies to detect ZIKV and its antigens and antibodies (immunoglobulin G (IgG) and immunoglobulin M (IgM)) (Adegoke et al., 2017; Camacho, Sobral-Filho, Aoki, Constantino, & Brolo, 2018; Draz, Lakshminaraasimulu, et al., 2018; Draz, Venkataramani, et al., 2018; Ribeiro et al., 2019). Among fluorescent NPs, quantum dots (QDs) have been used for biomedical applications since 1998. Together with their high brightness and resistance to photodegradation, QDs are versatile probes to be conjugated with biomolecules, surfaces, and even other NPs, expanding their possibilities for bioanalysis (Pereira et al., 2019). QDs have been extensively studied, and their applications, in biological sciences, range from the labeling of fixed/living cells and tissues, in vivo monitoring, all the way to fluoro(immuno)assays, rendering innovative (bio)detection methodologies (Brkić, 2018; Fontes et al., 2012; Sukhanova et al., 2002).

Therefore this chapter starts with a brief introduction to the properties, advantages, and applications of QDs as fluorescent materials for biosensing. Then, studies on innovative QD-based assays for detecting biomolecules associated with ZIKV infection are reported in more detail.

Basic principles of QDs and their potential to bioanalysis

Brief introduction to QDs

QDs are fluorescent semiconductor nanocrystals, with typical sizes from 2 to 10nm, which possess unique photophysical properties. Usually, QDs present a core/shell structure. The core is responsible for the emission, whereas the shell determines the intensity and quality of the fluorescence profile (Brkić, 2018). QDs can have several compositions, based mainly on binary combinations of atoms from II–VI, III–V, or IV–VI groups of the periodic table (Pleskova, Mikheeva, & Gornostaeva, 2018). Cadmium telluride (CdTe), cadmium selenide (CdSe), and indium phosphide (InP) are among the most popular QDs (Skorupska & Grabowska-Jadach, 2019).

The QD fluorescence results from the return of excited electrons from the conduction band (CB) to the valence band (VB). As a consequence of their nanometric size, QDs present band gap energy (E_g) inversely proportional to their size (Fig. 1A). Thus the QD emission depends not only on the core composition but also on its size. Smaller QDs present emission at lower wavelengths, due to a higher E_g, while larger QDs present emission at higher wavelengths, due to the lower E_g (Brkić, 2018). Furthermore, QDs possess (i) high quantum yields; (ii) broad absorption bands, allowing exciting the fluorescence of different QDs with a single light source; (iii) narrow emission bands with size-tunable fluorescence; and (iv) active surface allowing the QD conjugation with (bio)molecules and even other nanomaterials (Fig. 1B) (Chandan, Schiffman, & Balakrishna, 2018; Li et al., 2017).

Besides the core/shell structure, the QD surface possesses an organic layer composed of a stabilizing/functionalizing agent that confers chemical stability, the surface charge, and functional groups for conjugation (Fontes et al., 2012).

The conjugation between QDs and biomolecules, or even nanoparticles (NPs), depends on the functional groups associated with the involved species. This process is not simple, since it is necessary to maintain, for example, the fluorescence properties of the QDs and the biological function of the molecules. The most common strategies used to achieve this goal

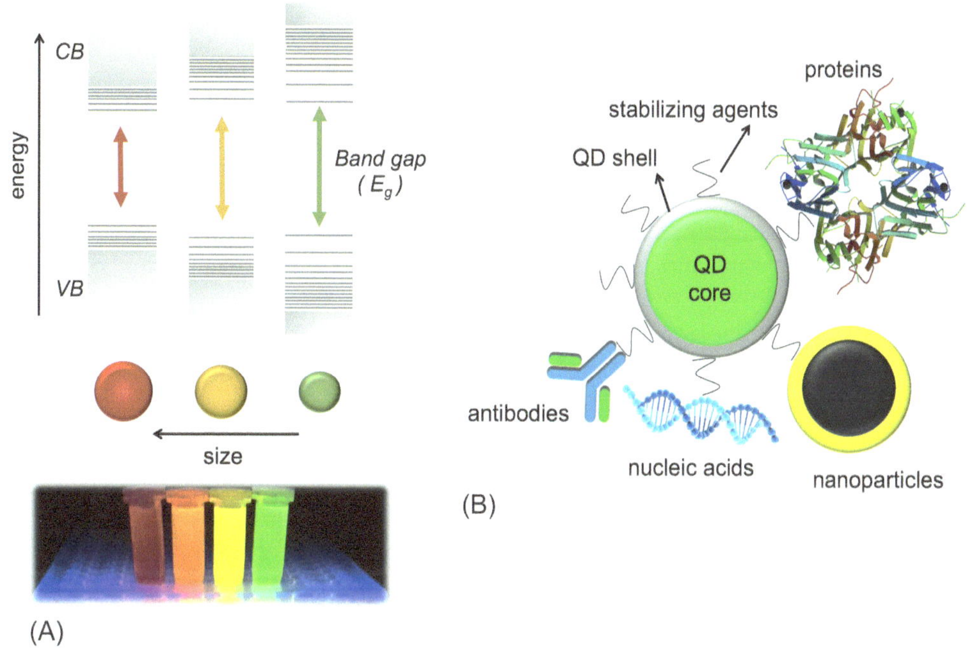

FIG. 1 Schematic representations of quantum dot (QD) features. (A) The relation between the size, the band gap (E_g), and the emission of QDs. *CB*, conduction band; *VB*, valence band. (B) The QD core/shell structure. The stabilizing agents allow for the QD conjugation with molecules and other nanoparticles. *(Unpublished images from Biomedical Nanotechnology Research Group (NanoBio/UFPE).)*

are (i) adsorption, involving electrostatic and hydrophobic interactions; (ii) covalent bond using carbodiimide agents; and (iii) (strept)avidin-biotin affinity (Pereira et al., 2019).

QDs have been applied to improve the comprehension of biological phenomena and diagnostics of diseases and even for therapeutic approaches. They have been involved in imaging (Fontes et al., 2012) and flow cytometry analyses (Cabral Filho et al., 2015), photodynamic therapy (Liu et al., 2018), electro- and photoelectrochemical sensors, and (fluoro)biosensing (Chandan et al., 2018; Mohamadi, Moghaddasi, Farahbakhsh, & Kazemi, 2017; Ribeiro et al., 2019; Yao et al., 2017).

QDs and optical biosensing

To better understand and diagnose diseases, the development of complementarily, novel, reliable, more sensitive, and less laborious assays is still required (Giljohann & Mirkin, 2009). Conventional enzyme-linked immunosorbent assays (ELISA) present simplicity and adaptability, but lower sensitivity. Furthermore, enzyme-based labels can be unstable due to possible processes of denaturation and degradation (Zhang et al., 2017). Fluorescence-linked immunosorbent assays (FLISA) usually employ organic dyes, which present low photostability. Thus it is essential to search for new labels to replace organic dyes and improve the long-term stability of FLISA. QDs have been employed as probes in fluorescence-based methods, overcoming disadvantages and complementing features of conventional ELISA and FLISA (Tully, Hearty, Leonard, & O'Kennedy, 2006). Due to the QD intense fluorescent signal and its high photostability, which ensures higher sensitivity, better detection limits are being obtained when compared with the conventional ELISA (Chen et al., 2010; Zhang et al., 2017). Furthermore, fluoro(immuno)assays using QDs are less susceptible to time and temperature than those based on enzymes (Yao et al., 2017).

An attractive feature of these nanocrystals to bioanalysis is the versatility for conjugation with oligonucleotides, amino acids, proteins (like antibodies), or even nanostructures, such as gold NPs (AuNPs) (Pereira et al., 2019). While the conjugation with biomolecules provides specificity for the target under investigation and flexibility to study various diseases, the association with other NPs allows the combination of different signals/properties in a single nanoplatform, which can further improve the bioevaluation (Adegoke & Park, 2017; Ribeiro et al., 2019; Tang et al., 2018).

QDs have already been applied, for example, to detect foodborne pathogens, such as *Staphylococcus aureus* and *Escherichia coli* (Mohamadi et al., 2017; Xue, Pan, Xie, Wang, & Zhang, 2009). Furthermore, Tang et al. (2018) also developed a sandwich immunoassay based on QDs, which was applied to detect the Aβ1–42 (a β-amyloid peptide).

This β-amyloid peptide has been associated with Alzheimer's disease. Streptavidin-coated QDs conjugated to biotinylated antibodies were employed to detect the concentration of those peptides through fluorescence intensity in the cerebrospinal fluid sample. This method presented high sensitivity and selectivity (Tang et al., 2018).

Although the avian influenza virus (AI) is occasionally transmitted to humans, AI can infect them, causing a mild to severe disease, especially in the case of highly contagious H5N1 virus infection, which has a mortality rate of approximately 60%. Yeo et al. (2018) developed a smartphone-based rapid dual fluorescence lateral flow system for quantitative, simultaneous, and specific detection of influenza A and its H5 subtype (Yeo et al., 2018). The authors used CdSe/CdS/ZnS carboxyl-coated QDs with emission at 580nm (for influenza A) or 650nm (for H5 subtype) bioconjugated simultaneously with latex beads and monoclonal antibodies. The multiplex systems were applied onto strips, and the detection was achieved through the evaluation of the fluorescence intensity.

For dengue virus (DENV) the diagnosis using cell culture still is the "gold standard" (Peeling et al., 2010; Shu & Huang, 2004). Nevertheless, this method presents low sensitivity and long detection time. Then, Adegoke and Park (2017), for example, developed a novel biosensor that presented high sensitivity and specificity for discriminating each serotype of DENV (1–4). The developed biosensor was based on multicolor glutathione-functionalized CdSe/ZnSeS core/alloyed shell QDs conjugated with AuNPs and molecular beacons (MBs). The authors explored the increase of the QD fluorescence intensity trigged by the localized surface plasmon resonance (LSPR) effect of AuNPs after hybridization. The increase was proportional to the concentration of the target virus, allowing the detection (Adegoke & Park, 2017).

Therefore QDs have been presenting a crescent potential to bioassays, to detect biomolecules related to pathogenic species or other diseases. Thus why not use QDs for Zika antigen and antibody detection?

Zika QD-based detection methods

Like all flaviviruses, ZIKV is a single-stranded genome RNA virus that is made up of three structural proteins: the capsid (C), the premembrane (prM), and the envelope (E) protein, plus seven nonstructural proteins (NS1, NS2A, NS2B, NS3, NS4A, NS4B, and NS5) (Sironi, Forni, Clerici, & Cagliani, 2016). Among these, the E protein is responsible for the virus binding to the target cells and it is present in the patient circulation in the viremia phase, which is usually short (from a few days to 1 week). Additionally, E protein is considered as the principal antigenic target for the human antibody response (Cavalcanti et al., 2012; Kaushik et al., 2018). The NS1 protein is also an important biomarker since its presence in the bloodstream is high in the first days of infection (Bosch et al., 2017; Song, Qi, Haywood, Shi, & Gao, 2016). Thus E protein and NS1 have been explored to develop assays to ZIKV detection (Afsahi et al., 2018; Ahmadivand et al., 2017; Kaushik et al., 2018; Lee & Zeng, 2017). An alternative is to analyze the immune response, detecting IgM (present especially in the acute phase) and IgG (associate mainly to the postinfection phase) antibodies. The detection of IgG antibodies can be important, especially for pregnant women and to help in the comprehension/treatment of this disease. The quantification of antibodies can also be important to monitor the immune response for vaccine trials (Ali, Chambers, Ali, & Tamfu, 2017; Sapparapu et al., 2016; Wong et al., 2017).

In this context, new fluorescent QD-based techniques, for the detection of RNA virus, antigens, or antibodies associated with Zika, have been proposed to bring specificity and higher sensitivity to diagnose this disease at different stages (Adegoke et al., 2017; Ribeiro et al., 2019; Rong et al., 2019; Takemura, Adegoke, Suzuki, & Park, 2019).

Zika antigen detection

Rong et al. (2019) developed a fluorescent lateral flow immunoassay (LFIA) platform (Fig. 2A), employing CdSe/ZnS QDs encapsulated in polymer microspheres, to detect the NS1 protein (Rong et al., 2019). The platform assembly was based on the formation of sandwich immune complexes on nitrocellulose (NC) membrane strips in the presence of NS1. QD-polymer microspheres were conjugated with mouse anti-ZIKV NS1 antibodies through covalent binding, mediated by 1-ethyl-3-(3-dimethylaminopropyl)carbodiimide (EDC) and *N*-hydroxysulfosuccinimide sodium salt (sulfo-NHS) coupling agents. Two antibodies were used in strips: (i) a mouse anti-ZIKV NS1 monoclonal antibody, as the test line for the sandwich detection of NS1, and (ii) a goat anti-mouse IgG antibody used as the control line (Fig. 2B). In the presence of the antigen, the bioconjugate (microsphere QD/antibody) was capable of recognizing and capturing the analyte target, forming sandwich immunocomplexes in about 20min. Fluorescence images of strips were recorded using a smartphone under UV light excitation, and fluorescence intensities were determined. The antibodies were effectively conjugated to the surface of QD microspheres, and their fluorescence was maintained. The platform showed sensitivity and was capable of detecting NS1 in the range of 0.045 and 0.15 $ng\,mL^{-1}$ in buffer and serum samples enriched with ZIKV NS1 antigens (Fig. 2C). The platform showed specificity since it did not detect DENV NS1.

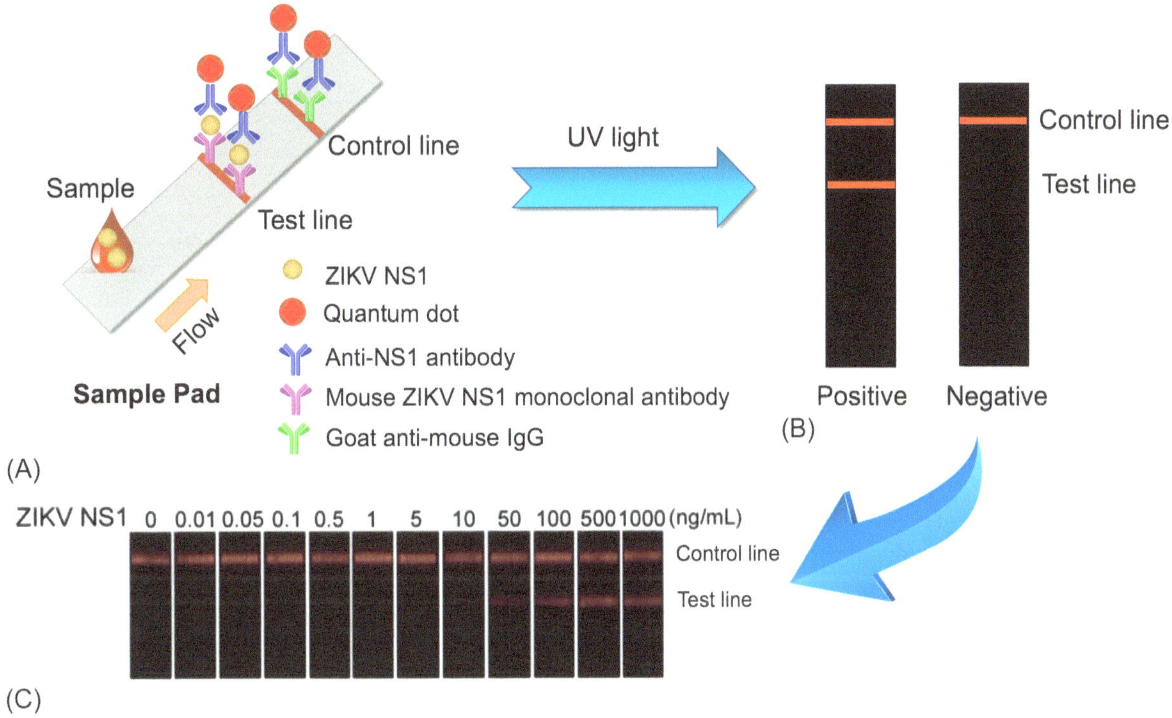

FIG. 2 Quantum dot-based fluorescence lateral flow immunoassay (LFIA). Representative schemes of the LFIA platform (A) and nitrocellulose membrane strips (B). In (C) an example of results for ZIKV NS1 detection under different concentrations is presented. *UV*, ultraviolet; *ZIKV*, Zika virus; *NS1*, nonstructural protein 1; *IgG*, immunoglobulin G. *(Adapted from Rong, Z., Wang, Q., Sun, N., Jia, X., Wang, K., Xiao, R., & Wang, S. (2019). Smartphone-based fluorescent lateral flow immunoassay platform for highly sensitive point-of-care detection of Zika virus nonstructural protein 1.* Analytica Chimica Acta, 1055, *140–147. doi:10.1016/j.aca.2018.12.043, with permission from Elsevier.)*

Takemura et al. (2019) also reported the development of a biosensor for the detection of ZIKV NS1 protein based on AuNPs and QDs (Takemura et al., 2019). Anti-ZIKV NS1 antibodies were covalently conjugated, via EDC and *N*-hydroxysuccinimide (NHS), to carboxylic groups of CdSeTeS QDs and AuNPs. The conjugation of anti-NS1 antibodies to QDs and AuNPs was confirmed using ELISA. Fluorescence immunoassay for ZIKV NS1 protein was performed in a 96-well plate. When antigens were present, both QDs and AuNPs conjugated to anti-NS1 antibodies bound to antigens, and the LSPR signal from AuNPs amplifies the fluorescence intensity of the QDs, proportional to the antigen concentration. The biosensor showed a detection limit of 1.28 fg mL^{-1} for NS1 (or 8.2 RNA copies·mL^{-1}), which shows that this sensor presents a great sensitivity. The platform specificity for ZIKV was also investigated using bovine serum albumin (BSA), norovirus-like particles (NoV-LP), and influenza A virus (H7N7) as negative controls, and those analytes were unable to modify the fluorescence intensity of QDs.

ZIKV RNA detection

Other authors have also explored the LSPR effect to increase the platform sensitivity for ZIKV diagnosis. Adegoke et al. (2017) prepared fluorescent nanohybrid biosensors, associating AuNPs, silver NPs (AgNPs), core/shell Au/Ag NPs, or alloyed AgAuNPs with glutathione-capped CdSeS QDs, using EDC and NHS (Adegoke et al., 2017). The nanohybrids were then conjugated covalently, using the same coupling agents, to MBs used for ZIKV RNA detection (Fig. 3A). The MB consisted of an oligonucleotide DNA sequence, complementary to the viral RNA, with a dabcyl (organic dye) attached to its end. When the ZIKV RNA was not present, the fluorescence of the probe NP-QD was quenched by dabcyl, and the MB kept a hairpin stem-loop structure. On the other way around, when hybridization occurred, a RNA/DNA heteroduplex was formed, and the dabcyl no longer quenched the QD fluorescence (Fig. 3B). Then the QD emission intensity was enhanced by the LSPR effect, being also proportional to the target concentration (Fig. 3C). The assay was carried out in 96-well plates. Higher sensitivity was obtained for the nanosystems assembled with alloyed AgAuNPs, which showed a detection limit of 1.7 copies·mL^{-1}. The probe specificity was investigated using influenza virus H3N2 RNA and norovirus GII RNA. In the presence of these RNA viruses, the fluorescence signal remained quenched, indicating the specificity of the

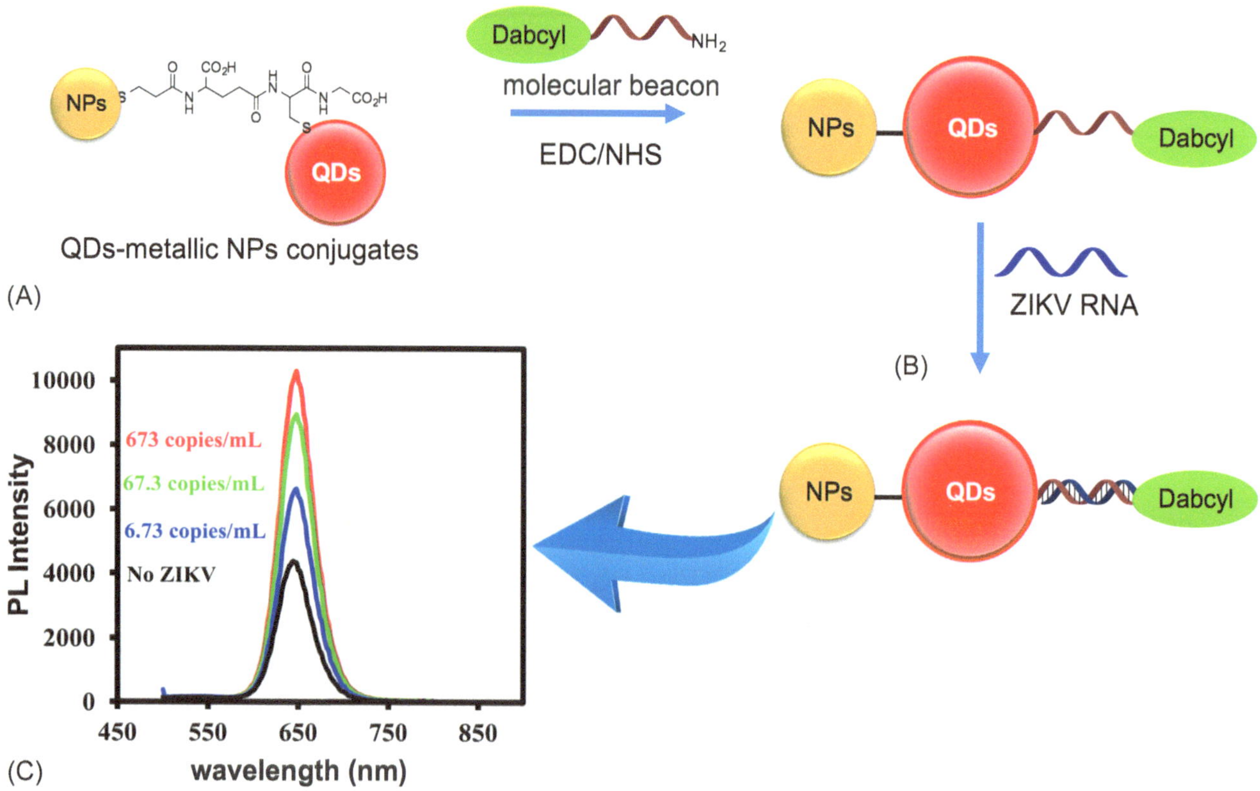

FIG. 3 Biosensing with quantum dots (QDs) for ZIKV RNA. Schematic representations of (A) the conjugation between QDs, metallic nanoparticles, and molecular beacons (MBs) and (B) hybridization with the Zika virus ribonucleic acid (ZIKV RNA). In (C) an example of results for ZIKV RNA detection under different concentrations is presented. EDC [(1-ethyl-3-(3-dimethylaminopropyl) carbodiimide] and NHS (*N*-hydroxysuccinimide) are coupling agents used for covalent conjugation. *PL*, photoluminescence. *(Adapted from Adegoke, O., Morita, M., Kato, T., Ito, M., Suzuki, T., & Park, E. Y. (2017). Localized surface plasmon resonance-mediated fluorescence signals in plasmonic nanoparticle-quantum dot hybrids for ultrasensitive Zika virus RNA detection via hairpin hybridization assays.* Biosensors and Bioelectronics, 94*(March), 513–522. doi:10.1016/j.bios.2017.03.046, with permission from Elsevier.)*

biosensor. The proposed system, therefore, provided a rapid (3 min to hybridization), sensitive, and specific detection platform for the ZIKV diagnosis. The authors reported that the major limitation of this platform is the cost-effectiveness.

ZIKV antibody detection

Another complementary strategy that can be employed for Zika detection is the evaluation of antibodies responsive to the virus infection. For this, two classes of immunoglobulins can be detected, IgG and IgM, which are the antibodies most involved in the infectious process. IgM is a circulating pentameric molecule presented in the acute phase of the infection. However, IgG is a monomeric molecule that identifies postinfections, since it remains in circulation for longer periods. In this context, Ribeiro et al. (2019) developed an indirect immunofluorescence assay for the detection of human anti-ZIKV E protein IgG antibodies using QDs (Ribeiro et al., 2019). For this, mercaptosuccinic acid (MSA) stabilized CdTe QDs were covalently conjugated to secondary anti-human IgG antibodies [QD-(anti-IgG)] using EDC and sulfo-NHS as coupling agents. A complete optimization, searching for the best conditions of the assay, was performed in black polystyrene 96-well plates using IgG antibodies, varying the well blocking and the conjugate quantity. The specificity of bioconjugates was also evaluated and proved using human anti-A antibodies (belonging to the IgM class). Then the wells were sensitized with ZIKV E protein, and the QD-(anti-IgG) bioconjugates were used to detect the anti-ZIKV E protein IgG antibodies (Fig. 4A).

The QD-(anti-IgG) conjugates presented promising results (Fig. 4B), detecting an amount of at least 100 times lower than the reported IgG concentration expected for E protein immune response (Dai et al., 2016). Moreover, QD-(anti-IgG) conjugates maintained their optical properties and functionality for at least 4 months. Thus the QD-based fluoroimmuno-platform was considered promising to detect Zika postinfection IgG antibodies and/or monitoring immune response in vaccine trials (Ribeiro et al., 2019).

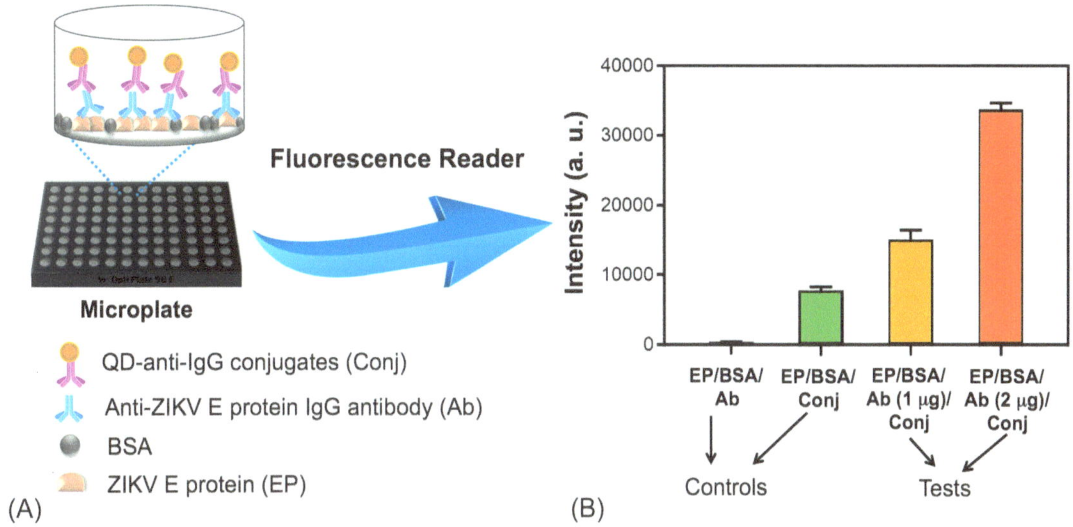

FIG. 4 Quantum dot–based indirect fluoroimmunoassay. (A) Representative scheme of the platform. (B) Results for anti-ZIKV E protein IgG antibodies detection under different concentrations (in *yellow and orange bars*). *ZIKV*, Zika virus; *IgG*, immunoglobulin G; *BSA*, bovine serum albumin; *anti-IgG*, anti-immunoglobulin G secondary antibodies; *QD*, quantum dot; *a.u.*, arbitrary units. *(Adapted from Ribeiro, J. F. F., Pereira, M. I. A., Assis, L. G., Cabral Filho, P. E., Santos, B. S., Pereira, G. A. L., ... Fontes, A. (2019). Quantum dots-based fluoroimmunoassay for anti-Zika virus IgG antibodies detection.* Journal of Photochemistry and Photobiology B: Biology, 194, *135–139. doi:10.1016/j.jphotobiol.2019.03.019, with permission from Elsevier.)*

Policy and procedures

One of the most important features, for the development of a sensitive and specific QD-based bioassay, is related to the association of these fluorescent NPs with the biomolecules of interest. During the conjugation process, it is necessary to standardize protocols to obtain stable conjugates without precipitation or agglomeration, maintaining also the optical properties of QDs and the functional activity of biomolecules. To keep the biomolecules active, it is recommended to store them at the appropriate temperature, and the stock solution should be fractionated to prevent the freezing and thawing process that can denature the biomolecules. A first requirement for the preparation of conjugates can be the pH adjustment of the QD suspension, depending on the coupling agents used, which is also associated with the functional groups of the components involved. For example, using EDC and NHS as mediators, the reaction is more efficient in the 5.5–6.0 pH range. The pH control is also important to avoid the precipitation, agglomeration, and loss of fluorescence of the QD suspension. Moreover the pH can also influence the performance of biomolecules. After conjugation, it is important to keep the conjugates at 4°C, mainly for the biomolecule preservation. In the detection stage, it is fundamental to guarantee that the interaction between the target and the probe is specific. Thus tris(hydroxymethyl)aminomethane (TRIS) can be used to inactivate the remained reactive functional groups from QDs or biomolecules. Moreover, BSA, for example, should be used to block microplate wells or NC membranes, avoiding unspecific interactions. It is very important to prevent false-positive and false-negative results. It is also noteworthy that the high concentration of blockers may also lead to a decrease or loss of the fluorescence signal.

Mini-dictionary of terms

Bioconjugation. Covalent (using coupling agents) or noncovalent (by adsorption) strategy to bind biomolecules, such as proteins (lectins or antibodies), carbohydrates, or nucleic acid sequences to quantum dots.

Fluorescence. Fast process of light emission from a compound or a nanoparticle as a result of singlet-singlet electronic relaxation after light absorption.

Fluorescent lateral flow immunoassay. Fluorescence assay based on sample flow through a membrane strip of polymeric material, containing reagents that are activated when immune complex formation occurs.

Fluoro(immuno)assays. Assays involving antigens and antibodies labeled with organic dyes or fluorescent nanoparticles, used to produce a light signal when recognizing the target (complementary) molecules. When the reaction is between an antigen and antibody, the assay can be divided in direct and indirect, depending on if it is used fluorescent-labeled primary or secondary antibodies. This assay can be performed in 96-well plates, for instance.

Functionalization. Process that confers functional groups, such as amine, carboxylic, or thiol groups on the nanoparticle surface, to make them capable of binding to (bio)molecules or even with other nanostructures. The functionalizing agent can also be the stabilizing agent of the nanoparticle colloidal suspension, conferring also charge to the nanostructure surface.

Localized surface plasmon resonance (LSPR). Optical phenomenon characterized by a collective oscillation of the conduction band electrons (localized plasmon) on the metallic nanoparticle surfaces under resonance with specific light incident wavelengths. LSPR is related with the nanostructure composition, size, and shape and can result in the enhancement of electromagnetic signals such as fluorescence, which depends also on the distance between the nanosystem and the fluorophore, for example.

Quantum dots. Fluorescent semiconductor nanocrystals with a few nanometers in size that have unique physical-chemical properties, such as high photostability, tunable fluorescence emission with their size, and an active surface for conjugation.

Key facts of quantum dot-based assays

- Quantum dots are nanocrystals with unique physical-chemical properties, such as narrow fluorescence spectra, high resistance to photobleaching, and active surface for conjugations, which make them perfectly suited for biological applications.
- Nowadays, applications of quantum dots involve not only living cells, bioassays, and therapy but also in vivo studies in small animals.
- Quantum dots have been presenting a crescent potential for biomolecule detections, related to pathogenic species or other diseases.
- Quantum dots have been employed as probes in fluorescence-based methods, overcoming disadvantages and complementing features of enzyme-linked immunosorbent assay (ELISA) and other detection methods.
- Novel quantum dot-based fluorescent approaches were developed, bringing high specificity and greater sensitivity to detect Zika-associated biomolecules (ribonucleic acid, antigens, and antibodies) at different stages of this disease.

Summary points

- This chapter presented a brief introduction to the properties and potentialities of quantum dots for biosensing.
- This review describes studies involving quantum dot–based assays for the detection of biomolecules associated with Zika.
- As far as our knowledge goes, there are so far four studies reported on the literature related to Zika quantum dot–based fluoroassays.
- A fluorescence lateral flow immunoassay based on quantum dots was used to detect the nonstructural NS1 protein.
- Quantum dots combined with metallic nanoparticles were employed to develop a biosensor for Zika virus (ZIKV) nonstructural NS1 protein.
- A fluorescence nanohybrid biosensor composed by metallic nanoparticles, quantum dots, and molecular beacons was applied for the ZIKV ribonucleic acid (RNA) detection.
- Anti-immunoglobulin G (anti-IgG) secondary antibodies conjugated with quantum dots were used to develop an indirect fluoroimmunoassay for anti-ZIKV envelope E protein IgG antibodies.
- The intense fluorescent signal and the high photostability of the quantum dots provide great sensitivity and detection limits for Zika-associated biomolecule detections.

References

Adegoke, O., Morita, M., Kato, T., Ito, M., Suzuki, T., & Park, E. Y. (2017). Localized surface plasmon resonance-mediated fluorescence signals in plasmonic nanoparticle-quantum dot hybrids for ultrasensitive Zika virus RNA detection via hairpin hybridization assays. *Biosensors and Bioelectronics*, *94*(March), 513–522. https://doi.org/10.1016/j.bios.2017.03.046.

Adegoke, O., & Park, E. Y. (2017). Bright luminescent optically engineered core/alloyed shell quantum dots: An ultrasensitive signal transducer for dengue virus RNA via localized surface plasmon resonance-induced hairpin hybridization. *Journal of Materials Chemistry B*, *5*(16), 3047–3058. https://doi.org/10.1039/c7tb00388a.

Afsahi, S., Lerner, M. B., Goldstein, J. M., Lee, J., Tang, X., Bagarozzi, D. A., … Goldsmith, B. R. (2018). Novel graphene-based biosensor for early detection of Zika virus infection. *Biosensors and Bioelectronics*, *100*, 85–88. https://doi.org/10.1016/j.bios.2017.08.051.

Ahmadivand, A., Gerislioglu, B., Manickam, P., Kaushik, A., Bhansali, S., Nair, M., & Pala, N. (2017). Rapid detection of infectious envelope proteins by magnetoplasmonic toroidal metasensors. *ACS Sensors*, *2*(9), 1359–1368. https://doi.org/10.1021/acssensors.7b00478.

Ali, I., Chambers, J. P., Ali, M., & Tamfu, R. C. (2017). Quantitation of IgG and IgM response to recombinant Zika virus proteins. *Journal of Vaccines and Vaccination*, *08*(05). https://doi.org/10.4172/2157-7560.1000368.

Bosch, I., de Puig, H., Hiley, M., Carré-Camps, M., Perdomo-Celis, F., Narváez, C. F., … Gehrke, L. (2017). Rapid antigen tests for dengue virus serotypes and Zika virus in patient serum. *Science Translational Medicine*, *9*(409), eaan1589. https://doi.org/10.1126/scitranslmed.aan1589.

Brkić, S. (2018). Applicability of quantum dots in biomedical science. In *Ionizing radiation effects and applications* (1st ed., pp. 21–39). InTech. https://doi.org/10.5772/intechopen.71428.

Cabral Filho, P. E., Pereira, M. I. A., Fernandes, H. P., de Thomaz, A. A., Cesar, C. L., Santos, B. S., … Fontes, A. (2015). Blood group antigen studies using CdTe quantum dots and flow cytometry. *International Journal of Nanomedicine*, *10*(1), 4393–4404. https://doi.org/10.2147/IJN.S84551.

Camacho, S. A., Sobral-Filho, R. G., Aoki, P. H. B., Constantino, C. J. L., & Brolo, A. G. (2018). Zika immunoassay based on surface-enhanced Raman scattering nanoprobes. *ACS Sensors*, *3*(3), 587–594. https://doi.org/10.1021/acssensors.7b00639.

Cavalcanti, I. T., Silva, B. V. M., Peres, N. G., Moura, P., Sotomayor, M. D. P. T., Guedes, M. I. F., & Dutra, R. F. (2012). A disposable chitosan-modified carbon fiber electrode for dengue virus envelope protein detection. *Talanta*, *91*, 41–46. https://doi.org/10.1016/j.talanta.2012.01.002.

Chandan, H. R., Schiffman, J. D., & Balakrishna, R. G. (2018). Quantum dots as fluorescent probes: Synthesis, surface chemistry, energy transfer mechanisms, and applications. *Sensors and Actuators B: Chemical*. https://doi.org/10.1016/j.snb.2017.11.189.

Chen, L., Sheng, Z., Zhang, A., Guo, X., Li, J., Han, H., & Jin, M. (2010). Quantum-dots-based fluoroimmunoassay for the rapid and sensitive detection of avian influenza virus subtype H5N1. *Luminescence: The Journal of Biological and Chemical Luminescence*, *25*(6), 419–423. https://doi.org/10.1002/bio.1167.

Dai, L., Song, J., Lu, X., Deng, Y.-Q., Musyoki, A. M., Cheng, H., … Gao, G. F. (2016). Structures of the Zika virus envelope protein and its complex with a flavivirus broadly protective antibody. *Cell Host & Microbe*, *19*(5), 696–704. https://doi.org/10.1016/j.chom.2016.04.013.

De Fatima Vasco Aragao, M., Van Der Linden, V., Brainer-Lima, A. M., Coeli, R. R., Rocha, M. A., Da Silva, P. S., … Valenca, M. M. (2016). Clinical features and neuroimaging (CT and MRI) findings in presumed Zika virus related congenital infection and microcephaly: Retrospective case series study. *BMJ*, 353. https://doi.org/10.1136/bmj.i1901.

Draz, M. S., Lakshminaraasimulu, N. K., Krishnakumar, S., Battalapalli, D., Vasan, A., Kanakasabapathy, M. K., … Shafiee, H. (2018). Motion-based immunological detection of Zika virus using Pt-nanomotors and a cellphone. *ACS Nano*, *12*(6), 5709–5718. https://doi.org/10.1021/acsnano.8b01515.

Draz, M. S., Venkataramani, M., Lakshminarayanan, H., Saygili, E., Moazeni, M., Vasan, A., … Shafiee, H. (2018). Nanoparticle-enhanced electrical detection of Zika virus on paper microchips. *Nanoscale*, *10*(25), 11841–11849. https://doi.org/10.1039/c8nr01646a.

Faria, A. M., & Mazon, T. (2019). Early diagnosis of Zika infection using a ZnO nanostructures-based rapid electrochemical biosensor. *Talanta*, *203*, 153–160. https://doi.org/10.1016/j.talanta.2019.04.080.

Fischer, C., Pedroso, C., Mendrone, A., de Filippis, A. M. B., Vallinoto, A. C. R., Morais Ribeiro, B., … Drexler, J. F. (2018). External quality assessment for Zika virus molecular diagnostic testing, Brazil. *Emerging Infectious Diseases*, *24*(5), 888–892. https://doi.org/10.3201/eid2405.171747.

Fontes, A., de Lira, R. B., Lopes Seabra, M. A. B., da Silva, T. G., de Castro Neto, A. G., & Saegesser, B. (2012). Quantum dots in biomedical research. In R. Hudak, M. Penhaker, & J. Majernik (Eds.), *Biomedical engineering—Technical applications in medicine* (pp. 269–290). IntechOpen.

Giljohann, D. A., & Mirkin, C. A. (2009). Drivers of biodiagnostic development. *Nature*. https://doi.org/10.1038/nature08605.

Kaushik, A., Yndart, A., Kumar, S., Jayant, R. D., Vashist, A., Brown, A. N., … Nair, M. (2018). A sensitive electrochemical immunosensor for label-free detection of Zika-virus protein. *Scientific Reports*, *8*(1), 9700. https://doi.org/10.1038/s41598-018-28035-3.

Lee, K. H., & Zeng, H. (2017). Aptamer-based ELISA assay for highly specific and sensitive detection of Zika NS1 protein. *Analytical Chemistry*, *89*(23), 12743–12748. https://doi.org/10.1021/acs.analchem.7b02862.

Li, X., Yan, Z., Xiao, J., Liu, G., Li, Y., & Xiu, Y. (2017). Cytotoxicity of CdSe quantum dots and corresponding comparison with FITC in cell imaging efficiency. *International Journal of Clinical and Experimental Medicine*, *10*(1), 753–759. Retrieved from www.ijcem.com/.

Liu, Y., Xu, Y., Geng, X., Huo, Y., Chen, D., Sun, K., … Tao, K. (2018). Synergistic targeting and efficient photodynamic therapy based on graphene oxide quantum dot-upconversion nanocrystal hybrid nanoparticles. *Small*, *14*(19). https://doi.org/10.1002/smll.201800293, 1800293.

Miller, E., Becker, Z., Shalev, D., Lee, C. T., Cioroiu, C., & Thakur, K. (2017). Probable Zika virus-associated Guillain-Barré syndrome: Challenges with clinico-laboratory diagnosis. *Journal of the Neurological Sciences*, *375*, 367–370. https://doi.org/10.1016/j.jns.2017.02.029.

Mohamadi, E., Moghaddasi, M., Farahbakhsh, A., & Kazemi, A. (2017). A quantum-dot-based fluoroassay for detection of food-borne pathogens. *Journal of Photochemistry and Photobiology B: Biology*, *174*, 291–297. https://doi.org/10.1016/j.jphotobiol.2017.08.005.

Nicolini, A. M., McCracken, K. E., & Yoon, J. Y. (2017). Future developments in biosensors for field-ready Zika virus diagnostics. *Journal of Biological Engineering*. https://doi.org/10.1186/s13036-016-0046-z.

Peeling, R. W., Artsob, H., Pelegrino, J. L., Buchy, P., Cardosa, M. J., Devi, S., … Yoksan, S. (2010). Evaluation of diagnostic tests: Dengue. *Nature Reviews Microbiology*, *12*, S30–S37. https://doi.org/10.1038/nrmicro2459.

Pereira, G., Monteiro, C. A. P., Albuquerque, G. M., Pereira, M. I. A., Cabrera, M. P., Cabral Filho, P. E., … Santos, B. S. (2019). (Bio)conjugation strategies applied to fluorescent semiconductor quantum dots. *Journal of the Brazilian Chemical Society*, 1–25. https://doi.org/10.21577/0103-5053.20190163.

Pleskova, S., Mikheeva, E., & Gornostaeva, E. (2018). Using of quantum dots in biology and medicine. *Advances in Experimental Medicine and Biology*, *1048*, 323–334. https://doi.org/10.1007/978-3-319-72041-8_19.

Ribeiro, J. F. F., Pereira, M. I. A., Assis, L. G., Cabral Filho, P. E., Santos, B. S., Pereira, G. A. L., … Fontes, A. (2019). Quantum dots-based fluoroimmunoassay for anti-Zika virus IgG antibodies detection. *Journal of Photochemistry and Photobiology B: Biology*, *194*, 135–139. https://doi.org/10.1016/j.jphotobiol.2019.03.019.

Rong, Z., Wang, Q., Sun, N., Jia, X., Wang, K., Xiao, R., & Wang, S. (2019). Smartphone-based fluorescent lateral flow immunoassay platform for highly sensitive point-of-care detection of Zika virus nonstructural protein 1. *Analytica Chimica Acta*, *1055*, 140–147. https://doi.org/10.1016/j.aca.2018.12.043.

Sapparapu, G., Fernandez, E., Kose, N., Cao, B., Fox, J. M., Bombardi, R. G., … Crowe, J. E. (2016). Neutralizing human antibodies prevent Zika virus replication and fetal disease in mice. *Nature*, *540*(7633), 443–447. https://doi.org/10.1038/nature20564.

Shan, C., Xie, X., Barrett, A. D. T., Garcia-Blanco, M. A., Tesh, R. B., Vasconcelos, P. F.d. C., … Shi, P. Y. (2016). Zika virus: Diagnosis, therapeutics, and vaccine. *ACS Infectious Diseases*, *2*(3), 170–172. https://doi.org/10.1021/acsinfecdis.6b00030.

Shu, P., & Huang, J. (2004). Current advances in dengue. *Clinical and Diagnostic Laboratory Immunology*, *11*(4), 642–650. https://doi.org/10.1128/CDLI.11.4.642.

Shukla, S., Hong, S. Y., Chung, S. H., & Kim, M. (2016). Rapid detection strategies for the global threat of Zika virus: Current state, new hypotheses, and limitations. *Frontiers in Microbiology*. https://doi.org/10.3389/fmicb.2016.01685.

Sironi, M., Forni, D., Clerici, M., & Cagliani, R. (2016). Nonstructural proteins are preferential positive selection targets in Zika virus and related flaviviruses. *PLoS Neglected Tropical Diseases*, *10*(9). https://doi.org/10.1371/journal.pntd.0004978, e0004978.

Skorupska, S., & Grabowska-Jadach, I. (2019). Cytotoxicity studies of quantum dots with the electroporation method. *Bioelectrochemistry*, *126*, 86–91. https://doi.org/10.1016/j.bioelechem.2018.11.011.

Song, H., Qi, J., Haywood, J., Shi, Y., & Gao, G. F. (2016). Zika virus NS1 structure reveals diversity of electrostatic surfaces among flaviviruses. *Nature Structural & Molecular Biology*, *23*(5), 456–458. https://doi.org/10.1038/nsmb.3213.

Sukhanova, A., Venteo, L., Devy, J., Artemyev, M., Oleinikov, V., Pluot, M., & Nabiev, I. (2002). Highly stable fluorescent nanocrystals as a novel class of labels for immunohistochemical analysis of paraffin-embedded tissue sections. *Laboratory Investigation*, *82*(9), 1259–1261. https://doi.org/10.1097/01.LAB.0000027837.13582.E8.

Takemura, K., Adegoke, O., Suzuki, T., & Park, E. Y. (2019). A localized surface plasmon resonance-amplified immunofluorescence biosensor for ultra-sensitive and rapid detection of nonstructural protein 1 of Zika virus. *PLoS One*, *14*(1). https://doi.org/10.1371/journal.pone.0211517.

Tang, M., Pi, J., Long, Y., Huang, N., Cheng, Y., & Zheng, H. (2018). Quantum dots-based sandwich immunoassay for sensitive detection of Alzheimer's disease-related Aβ 1–42. *Spectrochimica Acta Part A: Molecular and Biomolecular Spectroscopy*, *201*, 82–87. https://doi.org/10.1016/j.saa.2018.04.060.

Tully, E., Hearty, S., Leonard, P., & O'Kennedy, R. (2006). The development of rapid fluorescence-based immunoassays, using quantum dot-labelled antibodies for the detection of Listeria monocytogenes cell surface proteins. *International Journal of Biological Macromolecules*, *39*(1–3), 127–134. https://doi.org/10.1016/j.ijbiomac.2006.02.023.

Wong, S. J., Furuya, A., Zou, J., Xie, X., Dupuis, A. P., Kramer, L. D., & Shi, P. Y. (2017). A multiplex microsphere immunoassay for Zika virus diagnosis. *eBioMedicine*, *16*, 136–140. https://doi.org/10.1016/j.ebiom.2017.01.008.

Xue, X., Pan, J., Xie, H., Wang, J., & Zhang, S. (2009). Fluorescence detection of total count of *Escherichia coli* and *Staphylococcus aureus* on water-soluble CdSe quantum dots coupled with bacteria. *Talanta*, *77*(5), 1808–1813. https://doi.org/10.1016/j.talanta.2008.10.025.

Yao, J., Xing, G., Han, J., Sun, Y., Wang, F., Deng, R., … Zhang, G. (2017). Novel fluoroimmunoassays for detecting ochratoxin A using CdTe quantum dots. *Journal of Biophotonics*, *10*(5), 657–663. https://doi.org/10.1002/jbio.201600005.

Yeo, S. J., Kang, H., Dao, T. D., Cuc, B. T., Nguyen, A. T. V., Tien, T. T. T., … Park, H. (2018). Development of a smartphone-based rapid dual fluorescent diagnostic system for the simultaneous detection of influenza A and H5 subtype in avian influenza A-infected patients. *Theranostics*, *8*(22), 6132–6148. https://doi.org/10.7150/thno.28027.

Zhang, H., Xu, T., Gao, L., Liu, X., Liu, J., & Yu, B. (2017). A novel fluoroimmunoassay for detecting ruscogenin with monoclonal antibodies conjugated with CdSe/ZnS quantum dots. *Molecules*, *22*(8). https://doi.org/10.3390/molecules22081250.

Chapter 28

Serological detection of specific IgA antibodies against Zika virus nonstructural protein 1 contributes to diagnosis of acute Zika virus infections

Katja Steinhagen, Viola Borchardt-Lohölter, Konstanze Stiba, Julia Maria Klemens, Erik Lattwein, Sandra Saschenbrecker, and Wolfgang Schlumberger

Institute for Experimental Immunology, EUROIMMUN Medizinische Labordiagnostika AG, Lübeck, Germany

Abbreviations

DENV	dengue virus
dpso	days post symptom onset
ELISA	enzyme-linked immunosorbent assay
IgA	immunoglobulin A
IgG	immunoglobulin G
IgM	immunoglobulin M
IIFT	indirect immunofluorescence test
NAAT	nucleic acid amplification test
NS1	nonstructural protein 1
PCR	polymerase chain reaction
PRNT	plaque reduction neutralization test
RT-PCR	reverse transcription polymerase chain reaction
ZIKV	Zika virus

Introduction

Diagnosis of ZIKV infections by means of laboratory methods

The clinical manifestations of Zika virus (ZIKV) disease are often indistinguishable from those caused by coendemic flaviviruses, especially dengue virus (DENV). To correctly distinguish between the infectious agents, specific laboratory tests allowing adequate clinical management are required, especially given the differences in the disease course and outcome (e.g., ZIKV: teratological sequelae in the fetus, DENV: hemorrhage).

The selection of the appropriate diagnostic method depends on the time point of specimen collection and the clinical scenario (Landry & St George, 2017). There is no single laboratory test capable of accurately diagnosing ZIKV infections throughout the whole course of the disease. Therefore a sequential combination of direct pathogen detection and indirect determination of specific antibodies is preferable.

Viremia is usually limited to the first week of illness, reducing the diagnostic time window for direct viral detection using reverse transcription polymerase chain reaction (RT-PCR) to a couple of days post symptom onset (dpso). During viremia, ZIKV can be isolated using cell culture. The viral genome may be detectable for a prolonged time window in saliva, urine, semen, or amniotic fluid (Barzon et al., 2018; Bingham et al., 2016).

In the clinical reality, patients with a relatively mild disease course present for diagnosis only when the brief period of ZIKV shedding has passed. In this frequent case, direct pathogen detection methods yield negative results and are thus of

Zika Virus Impact, Diagnosis, Control, and Models. https://doi.org/10.1016/B978-0-12-820267-8.00028-5

limited practical utility. Importantly a negative RT-PCR result does not exclude an acute ZIKV infection, since it is possible that the patient has misreported the onset of symptoms and viremia has already declined.

After viremia, ZIKV diagnostics rely on serological testing for specific IgM, IgG, and IgA antibodies synthesized during the immune system response. Specific antibodies against ZIKV can be detected from few dpso to beyond convalescence (Rabe et al., 2016). Hence, serological methods provide a larger diagnostic window than direct methods and are therefore of major practical importance. Serological examination is most relevant for serum samples obtained after the first week following symptom onset and for samples that were tested negative by nucleic acid amplification test (NAAT) (Sharp et al., 2019). Serological tests are also suitable for disease monitoring.

In 2016 EUROIMMUN presented a new nonstructural protein 1 (NS1)-based anti-ZIKV IgM ELISA designed as a sensitive and highly specific tool for the serodiagnosis of ZIKV infections. At the same time the assay virtually eliminates cross-reactions with antibodies against DENV and other flaviviruses (Huzly, Hanselmann, Schmidt-Chanasit, & Panning, 2016; Steinhagen, Probst, et al., 2016). This assay was the first commercial serological Zika test to receive CE (Europe) and ANVISA (Brazil) marks. Patent applications have been filed (BR1120180141224, CN108650889A, DE112017000924.3, EP17707171.9, SG11201805496P, US16/077.757).

Kinetic profiles of specific anti-ZIKV antibodies of classes IgM and IgG

During the course of an infection, the titers of the immunoglobulin isotypes fluctuate, and their respective levels can be used to differentiate between acute and convalescent infection phases. Most of the commercially available serological diagnostic assays are based on either detection of specific IgM or IgG antibodies, while detection of IgA antibodies is not yet common.

IgM is a typical carrier of the early immune response. IgM antibodies are usually produced within the first few dpso, and their levels increase sharply. IgG is a typical marker of the late immune response. Compared with IgM levels, IgG titers are usually higher and persist longer (Lustig et al., 2017; Zhao et al., 2019). In ZIKV infections, IgG antibodies are usually present at the same time or shortly after IgM antibodies and provide the majority of antibody-based immunity against invading viruses.

Recent data suggest that ZIKV infection might induce immunological protection against reinfection (Siedner, Ryan, & Bogoch, 2018; Zambrana et al., 2018).

Regarding interpretation of serological test results for nonpregnant patients, the detection of anti-ZIKV IgM, seroconversion, or a twofold increase in the IgG titer in paired serum samples is indicative of an acute infection. For specimens collected from 7 days to 12 weeks after onset of symptoms, a negative anti-ZIKV IgM result rules out recent infection according to current diagnostic guidelines (Sharp et al., 2019). In the case of inconclusive anti-ZIKV IgM antibody results, PRNTs should, for example, be performed (Sharp et al., 2019). A negative IgM antibody test in the absence of NAAT might reflect sample collection before the development of detectable antibodies and thus does not rule out infection (Sharp et al., 2019).

But then, most residents of ZIKV-endemic areas already exhibit antibodies against DENV or other flaviviruses when their immune system is confronted with ZIKV as a secondary flavivirus infection or even as a coinfection. This can be problematic as preexisting DENV immunity can enhance the symptoms of ZIKV infection and vice versa (Bardina et al., 2017; Kawiecki & Christofferson, 2016). Many patients might not be aware of having had a previous flavivirus infection, since the proportion of asymptomatic cases is high (Landry & St George, 2017). If several flaviviruses are cocirculating in a certain region, interpretation of the serological responses can be complicated due to alterations of the aforementioned reaction kinetics resulting from past immune responses.

During flavivirus-primed secondary ZIKV infections, specific IgM levels might be decreased, occur with a time delay, or even be undetectable (Dittmar, Cleary, & Castro, 1979; Lanciotti et al., 2008) (Fig. 1). Thus testing solely for anti-ZIKV IgM antibodies bears a high risk of misdiagnosis. The synthesis of specific IgG, however, is immediately stimulated, as is a boost of reactive antibodies against the first infecting flavivirus (Andrade & Harris, 2018). The IgG titer quickly rises to levels even higher than during primary infection (Halstead, Rojanasuphot, & Sangkawibha, 1983; Rabe et al., 2016). Shortly after the high IgG titer levels off and is indistinguishable from IgG titers in primary infections (Barzon et al., 2018; Panning, 2017; Rabe et al., 2016).

The interpretation of the ZIKV immune response can further be complicated due to cross-reactivity resulting from structural homology of antigenic targets among genetically closely related flaviviruses (Rabe et al., 2016). Serologically, cross-reactive and cross-neutralizing antibodies bear the risk of false-positive test results and misdiagnosis, but clinically they may confer cross-protection (Gordon et al., 2019; Priyamvada et al., 2016; Swanstrom et al., 2016). For example, prior DENV immunity might cross-protect against symptomatic ZIKV infection (Gordon et al., 2019).

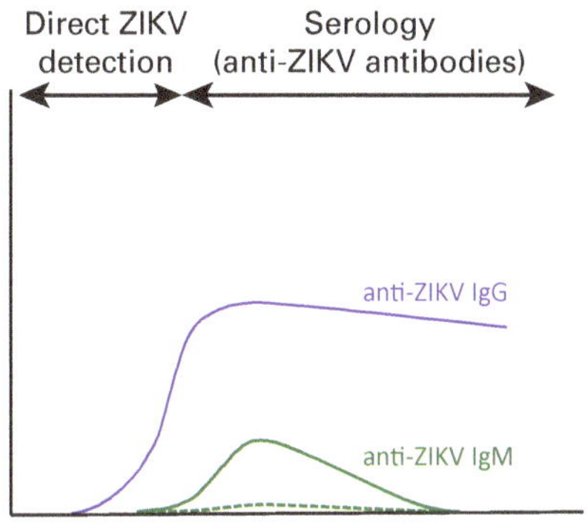

FIG. 1 Time points of appearance of the major laboratory diagnostic markers in ZIKV infection. Exemplary titers of anti-ZIKV antibodies of classes IgM and IgG are displayed in *green* and *purple*, respectively. Patients with flavivirus-primed secondary ZIKV infection typically develop high anti-ZIKV IgG titers early after the onset of symptoms, while the anti-ZIKV IgM response is often low or undetectable *(dashed green line)*.

While there is generally no or only low-grade cross-reactivity in a primary flavivirus infection, high-grade cross-reactivity is typical for flavivirus-primed secondary infections, which is usually stronger for IgG than for IgM antibodies.

Moreover, serological test results can be false positive or uninterpretable due to vaccination against other flaviviruses, for example, yellow fever virus, Japanese encephalitis virus, DENV, or Kyasanur forest disease virus.

The challenge: Diagnosing ZIKV infection based on the serological evidence obtained with a single serum sample

Patients with a relatively mild disease course usually seek medical advice when the symptoms have persisted for several days. In this frequently occurring case, indirect detection methods are the only diagnostic tools with practical applicability. Despite the fact that diagnostic guidelines recommend iterative testing to confirm ZIKV infection, in the clinical reality, there is often only one serum sample available per patient. If the immunoglobulin titers are measured only once, it is impossible to obtain information about their respective time courses, which could be helpful for the diagnosis. In the case of suspected acute ZIKV infection, the diagnostic guidelines further advise to test initially for anti-ZIKV IgM (Sharp et al., 2019). Thus the physician faces the challenge to diagnose the patient based on the symptoms—which are very similar across arboviral diseases—and the anti-ZIKV IgM titer as the only laboratory test result. Another aspect is that a serological result negative for anti-ZIKV IgM does not exclude an infection. It must be taken into account that patients who had previous contact with a flavivirus and have now come into contact again with a representative of this virus genus (secondary flavivirus infection) may only form a small or even undetectable amount of specific IgM antibodies. In this case, specific IgM antibodies are not detected by serological methods despite the fact that there is an acute infection. The share of serum samples from patients with an acute symptomatic ZIKV infection that can be negative for anti-ZIKV IgM is shown in the left panel of Fig. 2. Here, half of the serum samples taken in the acute phase of ZIKV infection were negative for anti-ZIKV IgM. A similar pattern was observed by Barzon and colleagues in serum samples from ZIKV patients with previous DENV infection, who did not develop detectable anti-ZIKV IgM after infection but had already high-level anti-ZIKV IgG at the time of diagnosis (Barzon et al., 2018).

In the same vein, Sa-Ngasang et al. reported detection of anti-DENV IgM in only approximately 50% of serum samples from patients with secondary DENV infection using an IgM-capture ELISA (Sa-Ngasang et al., 2006). Chanama et al. detected anti-DENV IgM in a large panel consisting of serum samples from patients with secondary DENV infection using an IgM-capture ELISA and reported a sensitivity of 28% (Chanama et al., 2004).

The absence of an IgM antibody response in patients who contracted ZIKV infection after having had a previous infection with DENV can be ascribed to the high similarities between ZIKV and DENV structural proteins. The prior exposure to an antigen (DENV) induces either no or an ineffective response to a related antigen (ZIKV), which is called "original antigenic sin" phenomenon.

To overcome this diagnostic challenge of ZIKV diagnostics in the serological time window, it would be beneficial for accurate and efficient diagnostics to obtain more information from one serum sample only. Hence, exploiting the diagnostic potential of immunoglobulins of other classes, for example, IgA, is of great interest and practical relevance.

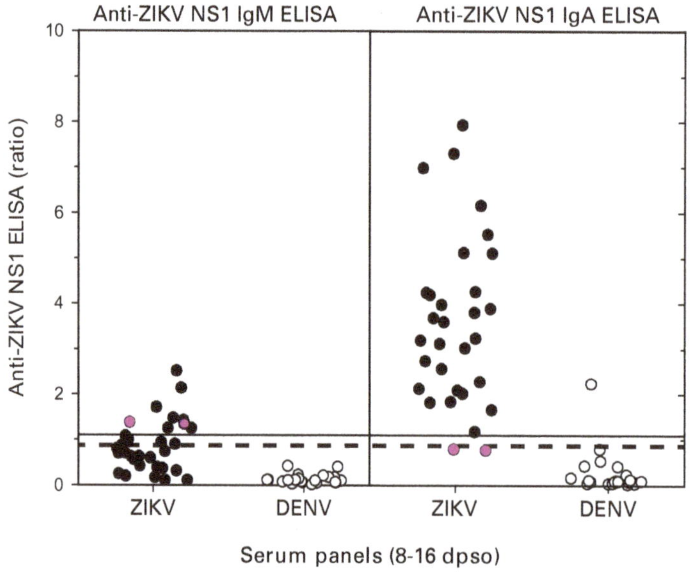

FIG. 2 Diagnostic performances of the anti-ZIKV NS1 IgM and IgA ELISAs in serum samples from ZIKV-infected patients compared with that from DENV-infected patients. Anti-ZIKV NS1 IgA and IgM antibodies were measured using ELISA in serum samples obtained 8–16 dpso from 31 Dominican patients with confirmed ZIKV infection and from 22 Vietnamese patients with confirmed DENV infection. *The continuous and the dashed horizontal lines* represent the cutoff ratios for positive and borderline results, respectively. Of all serum samples from patients with confirmed ZIKV infection, 16 were IgM negative/IgA positive, and 13 were IgM positive/IgA positive. Serum samples from two patients with ZIKV infection (samples marked in *pink*) were positive for anti-ZIKV NS1 IgM antibodies (left panel) but negative for anti-ZIKV NS1 IgA antibodies (right panel). One of the 22 samples from patients with confirmed DENV infection was positive for anti-ZIKV NS1 IgA, illustrating the high specificities of the two anti-ZIKV NS1 ELISAs.

Kinetics of specific IgA antibodies in acute ZIKV infection

To date, only few studies have investigated the diagnostic relevance of IgA in ZIKV infections. In 2016 serological evidence from two Colombian ZIKV patients provided early indication that determination of anti-ZIKV IgA may be helpful in both diagnosis of acute ZIKV infection and its discrimination from past infection when IgM is not detectable (Steinhagen, Schmidt-Chanasit, et al., 2016). In this investigation, serum samples were taken from two Colombians at days −16, 6, 24, and 66 (patient A) and at days 3, 15, 38, 52, 66, 76, and 95 (patient B) with respect to onset of first symptoms. ZIKV infection was confirmed for patient A and B by RT-PCR at days 6 and 3, respectively. Measurement of anti-ZIKV IgM was negative in all serum samples. Anti-ZIKV IgG titers started rising from the first day of blood withdrawal and reached their peaks at days 24 and 15, before they decreased, clearly remaining above the cutoff. Anti-ZIKV IgA titers also revealed first a constant increase and then a constant decrease but only peaking above the cutoff at days 24 and 15.

Zhang and colleagues reported high levels of ZIKV NS1 IgG and IgA antibodies detected in serum samples taken only days after illness onset from acutely infected ZIKV patients from DENV-endemic Colombia (Zhang et al., 2017). Given that robust IgG and IgA responses can occur early on during secondary infection, it was assumed that the early presence of ZIKV IgG and IgA could be due to prior exposure of the patients to DENV. Moreover the authors observed that IgG and IgA antibodies against ZIKV NS1 antigen were specific to ZIKV infection, whereas anti-ZIKV IgM showed a high degree of cross-reactivity. The authors further suggested that the presence of anti-ZIKV NS1 IgA might support diagnosis of a recent ZIKV infection.

Zhao et al. analyzed longitudinal serum samples from two Chinese patients with primary ZIKV infection (Zhao et al., 2019). NS1-specific IgA antibodies were detected within the first week of disease onset reaching peak levels at 10–12 dpso and declined rapidly to undetectable levels within 1 month.

Amaro et al. reported detection of anti-ZIKV NS1 IgA from 3 to 27 dpso in acutely infected patients returning from ZIKV-endemic areas and from 7 to 19 dpso in patients with serological suspicion of ZIKV infection. Anti-ZIKV IgA thereby contributed to the diagnosis of both early acute ($\leq$5 dpso) and acute (6–20 dpso) stages of ZIKV infection (Amaro et al., 2019). In general, anti-ZIKV IgA seemed to appear promptly and persist only for a short time.

It has also been reported that the diagnostic sensitivity was increased in patients with ZIKV infection when using an anti-ZIKV NS1 IgA ELISA (53%) instead of an NS1-based IgM ELISA (33%) (Bozza et al., 2019). All patients positive for anti-ZIKV IgM also showed anti-ZIKV IgA. The anti-ZIKV IgA titer increased during the acute phase of infection and decreased during convalescence.

Can anti-ZIKV IgA antibodies contribute to the detection of acute ZIKV infection in an endemic setting?

To scrutinize the diagnostic relevance of specific IgA in a flavivirus-endemic setting, we measured anti-ZIKV NS1 IgA, IgG, and IgM titers in paired serum samples taken at different time points from 31 patients with RT-PCR-confirmed ZIKV infection (Warnecke et al., 2019). In our study, anti-ZIKV IgA, IgG, IgM, and IgAM ELISA kits based on microplates coated with recombinant ZIKV NS1 antigen were applied according to the manufacturer's instructions (Steinhagen, Probst, et al., 2016).

The first aim was to measure the number of serum samples positive for IgG, IgM, and, especially, IgA in acute ZIKV infection.

In the early acute phase (2–5 dpso), all of the 31 initial serum samples were negative for anti-ZIKV NS1 IgA and IgM, whereas IgG was detected in 21 serum samples (Warnecke et al., 2019).

In the acute phase (8–16 dpso), 29, 15, and 31 of the 31 follow-up serum samples were positive for IgA, IgM, and IgG, respectively. Among all ZIKV serum samples taken in the acute phase, 16 were IgA positive/IgM negative, two were IgA negative/IgM positive, and 13 were IgA positive/IgM positive by NS1-based ZIKV ELISA (Fig. 2). Interestingly the distribution of test results revealed that IgA and IgM could be mutually complementing markers of acute ZIKV infection. The most surprising aspect of these results was the low number of samples positive for IgM but negative for IgA (see Fig. 2, samples marked in pink). Hence a diagnostic algorithm based on anti-ZIKV NS1 IgA alone would bear the risk of missing samples in the acute phase of a ZIKV infection.

To assess the kinetic profile of specific IgA on a longitudinal time scale, serum samples from patients with confirmed ZIKV infection residing in Colombia (3 patients) or the Dominican Republic (5 patients) were tested for anti-ZIKV NS1 IgA, IgM, and IgG. The time course analysis revealed that an increase in specific IgA titers in the acute phase of an infection was followed by a relatively fast decline to low positive or negative levels in the majority of cases (Warnecke et al., 2019). Fig. 3 shows selected kinetic profiles of specific IgA, IgM, and IgG, reflecting typical and untypical cases.

The next aim was to determine the diagnostic sensitivities of the different ELISAs.

Among the follow-up samples (8–16 dpso), the ELISAs showed sensitivities of 100%, 48.4%, and 93.5% for the detection of IgG, IgM, and IgA, respectively (Table 1). Combined anti-ZIKV NS1 IgAM testing resulted in a sensitivity of 100%, increasing the detection rate of acute ZIKV infection in the serological time window by 51.6% compared with the investigation of IgM only (Table 1).

To allow for a more detailed analysis of diagnostic sensitivities, the patients were distinguished as having a primary or secondary infection (Warnecke et al., 2019).

Pursuant to the criteria of virus-specific IgG being present in the early serum sample, a positive RT-PCR and IgM becoming positive at a later stage or, occasionally, remaining undetectable (Cordeiro et al., 2007), 21 patients were classified as having suspected (flavivirus-primed) secondary ZIKV infection. In these patients, seroconversion for anti-ZIKV IgA was demonstrated in all (100%) and for anti-ZIKV IgM in eight (38.1%) cases in the acute phase. In the 21 samples, combined anti-ZIKV NS1 IgAM testing increased the detection rate of acute ZIKV infection by 61.9%–100% compared with the detection of IgM only (38.1%).

In serum samples from 10 patients with suspected primary ZIKV infection (based on positive ZIKV RT-PCR in the absence of ZIKV-specific antibodies in the viremic samples), combined anti-ZIKV NS1 IgAM testing increased the detection rate of acute ZIKV infection by 30%–100% compared with the detection of IgM only (70%) in the acute phase.

Then turning to the diagnostic specificity of the test systems, we focused on potential cross-reactivity with anti-DENV antibodies, the most prevalent flavivirus antibodies in ZIKV-endemic areas (Warnecke et al., 2019).

Among a control group consisting of 139 serum samples from 95 DENV-infected Vietnamese patients, no anti-ZIKV NS1 reactivity was detected in 132 (IgA) and 135 (IgM) samples, corresponding to specificities of the ELISA of 95.0% (IgA) and 97.1% (IgM). There was no overlap between anti-ZIKV IgA- and IgM-positive individuals. In the subset of samples obtained in the acute phase (8–16 dpso), the specificity amounted to 95.5% (IgA) and 100% (IgM). Anti-ZIKV NS1 IgA and IgM antibody levels in the DENV group were significantly lower than in ZIKV patients (Fig. 2).

Added value of specific IgA antibodies to the serodiagnosis of acute ZIKV infection

Recent studies provided strong evidence that the determination of anti-ZIKV NS1 IgA can contribute to the diagnosis of acute ZIKV infection at points in time when viremia has decreased to an extent that RT-PCR can no longer be used to detect the virus. Taken together, our results revealed a high prevalence of specific IgA among patients with acute ZIKV infection (Warnecke et al., 2019). Anti-ZIKV IgA antibodies appear rapidly in the acute phase of the infection, followed by a

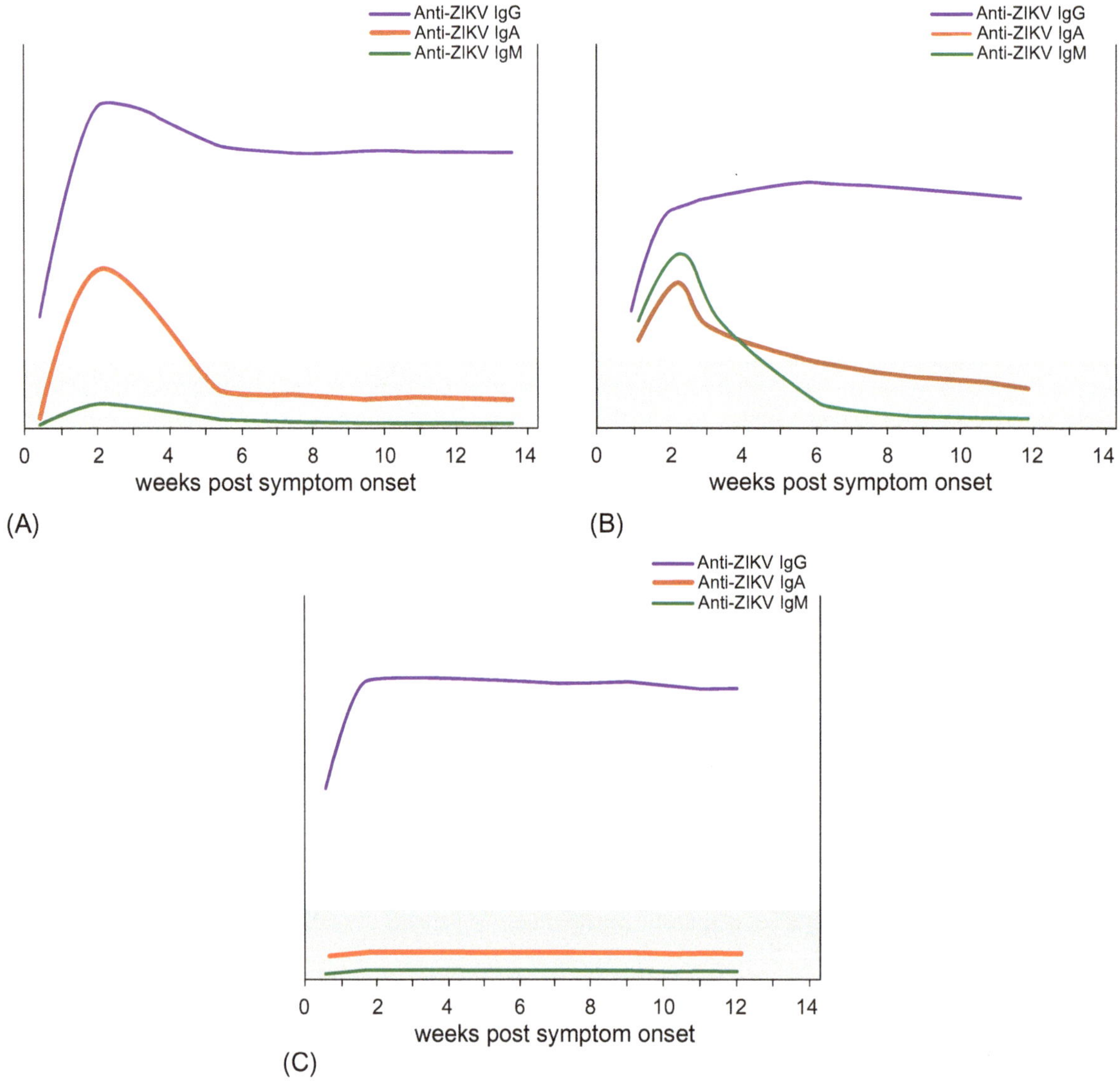

FIG. 3 Examples for longitudinal kinetics of anti-ZIKV NS1 IgA, IgM, and IgG. Consecutive serum samples from three selected patients with acute ZIKV infection were tested for specific IgG, IgM, and IgA. Their time courses are displayed schematically above (A–C). The *gray area* represents negative ELISA results. Cases A and B represent kinetic profiles and respective antibody levels that we frequently observed in clinical reality. Case C, however, represents a rather untypical kinetic profile showing negative IgA and IgM but positive IgG throughout the infection.

TABLE 1 Sensitivity of anti-ZIKV NS1 antibody detection using ELISA in acutely infected ZIKV patients.

Patients with acute ZIKV infection		Early acute phase (2–5 dpso)		Acute phase (8–16 dpso)	
(*n* = 31)		*n* positive	Sensitivity (%)	*n* positive	Sensitivity (%)
Anti-ZIKV NS1 ELISA	IgG	21	67.7	31	100.0
	IgM	0	0	15	48.4
	IgA	0	0	29	93.5
	IgAM	0	0	31	100.0

Combined anti-ZIKV NS1 IgAM testing resulted in a sensitivity of 100% in the serological time window, increasing the detection rate of acute ZIKV infection by 51.6% compared with the single detection of specific IgM.

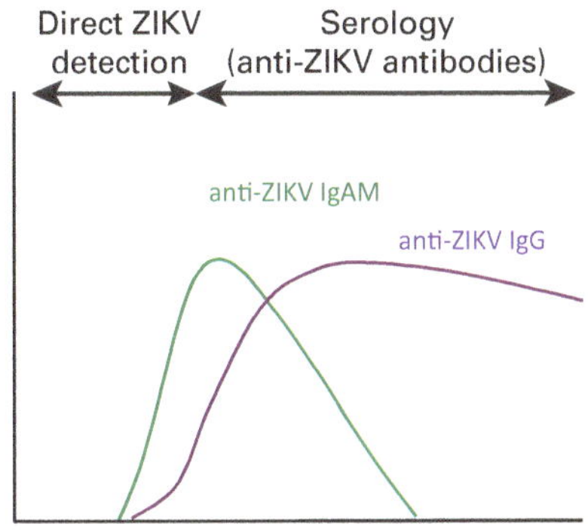

FIG. 4 Time course of appearance of anti-ZIKV NS1 IgAM and IgG. Anti-ZIKV NS1 IgA and IgM represent mutually complementing serological markers of acute ZIKV infection. At time points after the virus can no longer be detected by PCR, their combined detection (anti-ZIKV NS1 IgAM ELISA) can improve the accuracy in diagnosing acute ZIKV infection in flavivirus-endemic regions in the context of both primary and secondary infection, especially when IgM is undetectable.

relatively fast decline. Anti-ZIKV IgA antibodies may be detectable before or in parallel with specific IgM (Amaro et al., 2019; Zhao et al., 2019). The advantage of the detection of specific IgA antibodies is its potential to compensate for the frequent absence of specific IgM (Fig. 4). Especially in flavivirus-endemic regions, where more and more patients have a history of flavivirus infection and where IgG and IgM testing might be insufficient to differentiate acute from past infections, combined testing for anti-ZIKV IgAM proved its value for diagnosing acute ZIKV infection. Nevertheless, during interpretation of anti-ZIKV IgA negativity in the presence of anti-ZIKV IgM, it should be taken into account that immunoglobulin class switching may not have occurred yet, in addition to the possibility of selective IgA deficiency. It is of note that the remarkable strength of the combined testing approach is most prominent when only a single serum sample is available. In that case, additional determination of specific IgA supports accurate diagnosis, especially when a follow-up sample is not (yet) available. Including anti-ZIKV IgA in the diagnostic algorithm will hopefully reduce the rate of misinterpreted ZIKV infections in the future.

Policy and procedures

Principle of the anti-Zika virus ELISA (IgAM)

The anti-Zika virus ELISA (IgAM) from EUROIMMUN Medizinische Labordiagnostika AG (Lübeck, Germany) provides antibody detection. The assay is suitable for diagnosing acute infections and for disease monitoring. The test kit contains microplate strips each with 8 break-off reagent wells coated with recombinant ZIKV NS1 protein, which has proven to be a highly specific target structure for the immune response. Despite great homologies within the flavivirus genus, cross-reactions can be virtually excluded when the NS1 protein is used. This enables delimitation of ZIKV infections from infections with other viruses, such as DENV. In the first reaction step, diluted serum samples are incubated in the wells. Specific IgA and IgM (also IgG) antibodies will bind to the antigens in the case of a positive reaction. To detect the bound antibodies, a second incubation is carried out using an enzyme-labeled antihuman IgA/IgM (enzyme conjugate) catalyzing a color reaction. The extinction of the calibrator defines the upper limit of the reference range of noninfected persons (cutoff). Results are evaluated semiquantitatively by calculating a ratio of the extinction of the control or patient sample over the extinction of the calibrator. Readings above or equal the ratio of 1.1 are considered as positive, and those below 0.8 as negative. Results $\geq$0.8 and $<$1.1 are interpreted as borderline. A negative serological result does not exclude an infection. Particularly in the early phase of an infection, antibodies may not yet be present or are only present in such small quantities that they are not detectable. In the case of a borderline result, a secure evaluation is not possible. If there is a clinical suspicion and a negative test result, clarification by means of other diagnostic methods and/or the serological investigation of a follow-up sample is recommended. A positive result indicates that there has been contact with the pathogen. In the determination of pathogen-specific IgM antibodies, polyclonal stimulation of the immune system or antibody persistence may affect the diagnostic relevance of positive findings. An IgG titer increase (exceeding factor 2) and/or seroconversion in a follow-up sample taken after at least 7–10 days can indicate an acute infection. To investigate titer changes, the initial sample and the follow-up sample should be incubated on the ELISA microplate within the same test run. For diagnosis the clinical picture of the patient always needs to be taken into account along with the serological findings.

Mini-dictionary of terms

Primary infection. Infection without prior contact to the virus.
Secondary infection. Infection with prior contact to the same virus or another virus of the same genus.
Acute infection. Early state of an infection (includes viremia).
Convalescent infection. Late state of an infection when symptoms usually have disappeared and seroconversion has finished.
Seroconversion. A specific antibody develops and becomes detectable in blood serum or plasma.
Past infection. Prior contact with a (flavi)virus has led to an immune reaction.
Cross-reactivity. Analyte-dependent interference in immunoassays referring to the interaction between compounds and the antibody in the sample due to structural similarities despite chemical differences. Cross-reactivity leads to unspecific positive reactions in serological tests.
Coinfection. Infection with more than one (flavi)virus at the same time.
Selective immunoglobulin deficiency. Levels of a specific immunoglobulin class are permanently decreased, while immunoglobulins of other classes show normal reaction kinetics.

Key facts

- Acute infection with ZIKV can be diagnosed by virus isolation, direct detection of viral nucleic acid (e.g., by PCR), or detection of the viral antigen (during viremia).
- Another option is indirect detection, that is, determination of the host-antibody response by means of serological tests such as ELISA, IIFT, or PRNT (following viremia).
- The most suitable method for the in vitro detection of anti-ZIKV antibodies depends on the disease stage.
- The immune response differs in primary and flavivirus-primed secondary ZIKV infections.
- Drawbacks of antibody testing are as follows: (1) Results can be negative in the viremic phase of infection due to the kinetics of the immune response. (2) A positive anti-ZIKV IgG result in the first serum sample can indicate either a convalescent or past infection.
- The structural similarity of flaviviral antigens complicates the interpretation of serological results due to cross-reactivity.

Summary points

- Anti-ZIKV IgA, IgM, and IgG show different kinetic profiles for primary respectively flavivirus-primed secondary ZIKV infections.
- Detection of specific IgA, IgM, and IgG antibodies supports the diagnosis of ZIKV infections and allows differentiation between acute and convalescent stages of the infection as well as between primary and secondary infections.
- This text focuses on serological detection of anti-ZIKV antibodies of immunoglobulin class A.
- In flavivirus-primed secondary ZIKV infections, the proportion of cases with negative anti-ZIKV IgM but positive anti-ZIKV IgA is higher than in primary ZIKV infections.
- Combined detection of IgAM yields highest sensitivity in the acute phase of ZIKV infections.
- Serological detection of anti-ZIKV IgA antibodies supports accurate diagnosis of acute ZIKV infections in both primary and flavivirus-primed secondary infections, especially when IgM is undetectable.

References

Amaro, F., Sánchez-Seco, M. P., Vázquez, A., Alves, M. J., Zé-Zé, L., Luz, M. T., … De Ory, F. (2019). The application and interpretation of IgG avidity and IgA ELISA tests to characterize Zika virus infections. *Viruses*, *11*(2). https://doi.org/10.3390/v11020179.

Andrade, D. V., & Harris, E. (2018). Recent advances in understanding the adaptive immune response to Zika virus and the effect of previous flavivirus exposure. *Virus Research*, *254*(June 2017), 27–33. https://doi.org/10.1016/j.virusres.2017.06.019.

Bardina, S. V., Bunduc, P., Tripathi, S., Duehr, J., Frere, J. J., Brown, J. A., … Lim, J. K. (2017). Enhancement of Zika virus pathogenesis by preexisting antiflavivirus immunity. *Science*, *356*(6334), 175–180. https://doi.org/10.1126/science.aal4365.

Barzon, L., Percivalle, E., Pacenti, M., Rovida, F., Zavattoni, M., Del Bravo, P., … Baldanti, F. (2018). Virus and antibody dynamics in travelers with acute Zika virus infection. *Clinical Infectious Diseases*, *66*(8), 1173–1180. https://doi.org/10.1093/cid/cix967.

Bingham, A. M., Cone, M., Mock, V., Heberlein-Larson, L., Stanek, D., Blackmore, C., & Likos, A. (2016). Comparison of test results for Zika virus RNA in urine, serum, and saliva specimens from persons with travel-associated Zika virus disease – Florida, 2016. *MMWR Morbidity and Mortality Weekly Report*, *65*(18), 475–478. https://doi.org/10.15585/mmwr.mm6518e2.

Bozza, F. A., Moreira-Soto, A., Rockstroh, A., Fischer, C., Nascimento, A. D., Calheiros, A. S., … Drexler, J. F. (2019). Differential shedding and antibody kinetics of Zika and chikungunya viruses, Brazil. *Emerging Infectious Diseases*, *25*(2), 311–315. https://doi.org/10.3201/eid2502.180166.

Chanama, S., Anantapreecha, S., A-Nuegoonpipat, A., Sa-Gnasang, A., Kurane, I., & Sawanpanyalert, P. (2004). Analysis of specific IgM responses in secondary dengue virus infections: Levels and positive rates in comparison with primary infections. *Journal of Clinical Virology*, *31*(3), 185–189. https://doi.org/10.1016/j.jcv.2004.03.005.

Cordeiro, M. T., Silva, A. M., Brito, C. A. A., Nascimento, E. J. M., Magalhães, M. C. F., Guimarães, G. F., … Marques, E. T. A. (2007). Characterization of a dengue patient cohort in Recife, Brazil. *American Journal of Tropical Medicine and Hygiene*, *77*(6), 1128–1134. https://doi.org/10.4269/ajtmh.2007.77.1128.

Dittmar, D., Cleary, T. J., & Castro, A. (1979). Immunoglobulin G- and M-specific enzyme-linked immunosorbent assay for detection of dengue antibodies. *Journal of Clinical Microbiology*, *9*(4), 498–502.

Gordon, A., Gresh, L., Ojeda, S., Katzelnick, L. C., Sanchez, N., Mercado, J. C., … Harris, E. (2019). Prior dengue virus infection and risk of Zika: A pediatric cohort in Nicaragua. *PLoS Medicine*, *16*(1), 1–16. https://doi.org/10.1371/journal.pmed.1002726.

Halstead, S., Rojanasuphot, S., & Sangkawibha, N. (1983). Original antigenic sin in dengue. *American Journal of Tropical Medicine and Hygiene*, *32*(1), 154–156.

Huzly, D., Hanselmann, I., Schmidt-Chanasit, J., & Panning, M. (2016). High specificity of a novel Zika virus ELISA in European patients after exposure to different flaviviruses. *Eurosurveillance*, *21*(16), 1–4. https://doi.org/10.2807/1560-7917.ES.2016.21.16.30203.

Kawiecki, A. B., & Christofferson, R. C. (2016). Zika virus-induced antibody response enhances dengue virus serotype 2 replication in vitro. *Journal of Infectious Diseases*, *214*(9), 1357–1360. https://doi.org/10.1093/infdis/jiw377.

Lanciotti, R. S., Kosoy, O. L., Laven, J. J., Velez, J. O., Lambert, A. J., Johnson, A. J., … Duffy, M. R. (2008). Genetic and serologic properties of Zika virus associated with an epidemic, Yap State, Micronesia, 2007. *Emerging Infectious Diseases*, *14*(8), 1232–1239. https://doi.org/10.3201/eid1408.080287.

Landry, M. L., & St George, K. (2017). Laboratory diagnosis of Zika virus infection. *Archives of Pathology and Laboratory Medicine*, *141*(1), 60–67. https://doi.org/10.5858/arpa.2016-0406-SA.

Lustig, Y., Zelena, H., Venturi, G., Van Esbroeck, M., Rothe, C., Perret, C., … Schwartz, E. (2017). Sensitivity and kinetics of an NS1-based Zika virus enzyme-linked immunosorbent assay in Zika virus-infected travelers from Israel, the Czech Republic, Italy, Belgium, Germany, and Chile. *Journal of Clinical Microbiology*, *55*(6), 1894–1901.

Panning, M. (2017). Zika virus serology: More diagnostic targets, more reliable answers? *eBioMedicine*, *16*, 12–13. https://doi.org/10.1016/j.ebiom.2017.01.045.

Priyamvada, L., Quicke, K. M., Hudson, W. H., Onlamoon, N., Sewatanon, J., Edupuganti, S., … Wrammert, J. (2016). Human antibody responses after dengue virus infection are highly cross-reactive to Zika virus. *Proceedings of the National Academy of Sciences of the United States of America*, *113*(28), 7852–7857. https://doi.org/10.1073/pnas.1607931113.

Rabe, I. B., Staples, J. E., Villanueva, J., Hummel, K. B., Johnson, J. A., Rose, L., … Powers, A. M. (2016). Interim guidance for interpretation of Zika virus antibody test results. *Morbidity and Mortality Weekly Report*, *65*(21), 543–546. https://doi.org/10.15585/mmwr.mm6521e1.

Sa-Ngasang, A., Anantapreecha, S., A-Nuegoonpipat, A., Chanama, S., Wibulwattanakij, S., Pattanakul, K., … Kurane, I. (2006). Specific IgM and IgG responses in primary and secondary dengue virus infections determined by enzyme-linked immunosorbent assay. *Epidemiology and Infection*, *134*(4), 820–825. https://doi.org/10.1017/S0950268805005753.

Sharp, T., Fischer, M., Muñoz-Jordán, J., Paz-Bailey, G., Staples, J., Gregory, C., & Waterman, S. (2019). Dengue and Zika virus diagnostic testing for patients with a clinically compatible illness and risk for infection with both viruses. *Morbidity and Mortality Weekly Report: Recommendations and Reports*, *68*(1), 1–10. Retrieved from https://www.cdc.gov/mmwr/volumes/68/rr/rr6801a1.htm?s_cid=rr6801a1_w.

Siedner, M. J., Ryan, E. T., & Bogoch, I. I. (2018). Gone or forgotten? The rise and fall of Zika virus. *The Lancet Public Health*, *3*(3), e109–e110. https://doi.org/10.1016/S2468-2667(18)30029-X.

Steinhagen, K., Probst, C., Radzimski, C., Schmidt-Chanasit, J., Emmerich, P., Van Esbroeck, M., … Schlumberger, W. (2016). Serodiagnosis of Zika virus (ZIKV) infections by a novel NS1-based ELISA devoid of cross-reactivity with dengue virus antibodies: A multicohort study of assay performance, 2015 to 2016. *Eurosurveillance*, *21*(50), 1–16. https://doi.org/10.2807/1560-7917.ES.2016.21.50.30426.

Steinhagen, K., Schmidt-Chanasit, J., Emmerich, P., Lattwein, E., Fraune, J., Schlumberger, W., & Warnecke, J. M. (2016). Anti-Zika virus IgA as putative indicator of acute infection in anti-Zika virus IgM-negative patients. *International Journal of Infectious Diseases*, *53S*(2016), 93–94. https://doi.org/10.1016/j.ijid.2016.11.235.

Swanstrom, J. A., Plante, J. A., Plante, K. S., Young, E. F., Mcgowan, E., Gallichotte, E. N., … Heise, M. T. (2016). Dengue virus envelope dimer epitope monoclonal antibodies isolated from Dengue patients are protective against Zika Virus. *American Society for Microbiology*, *7*(4), 1–8. https://doi.org/10.1128/mBio.01123-16.Editor.

Warnecke, J., Lattwein, E., Saschenbrecker, S., Stöcker, W., Schlumberger, S., & Steinhagen, K. (2019). Added value of IgA antibodies against Zika virus non-structural protein 1 in the diagnosis of acute Zika virus infections. *Journal of Virological Methods*, *267*, 8–15.

Zambrana, J. V., Carrillo, F. B., Burger-Calderon, R., Collado, D., Sanchez, N., Ojeda, S., … Harris, E. (2018). Seroprevalence, risk factor, and spatial analyses of Zika virus infection after the 2016 epidemic in Managua, Nicaragua. *Proceedings of the National Academy of Sciences of the United States of America*, *115*(37), 9294–9299. https://doi.org/10.1073/pnas.1804672115.

Zhang, B., Pinsky, B. A., Ananta, J. S., Zhao, S., Arulkumar, S., Wan, H., … Dai, H. (2017). Diagnosis of Zika virus infection on a nanotechnology platform. *Nature Medicine*, *23*(5), 548–550. https://doi.org/10.1038/nm.4302.

Zhao, L. Z., Hong, W. X., Wang, J., Yu, L., Hu, F. Y., Qiu, S., … Zhang, F. C. (2019). Kinetics of antigen-specific IgM/IgG/IgA antibody responses during Zika virus natural infection in two patients. *Journal of Medical Virology*, *91*(5), 872–876. https://doi.org/10.1002/jmv.25366.

Chapter 29

Serological algorithms: How they can be used for differentiating ZIKV from DENV infection

Day-Yu Chao[a] and Gwong-Jen J. Chang[b]

[a]*Graduate Institute of Microbiology and Public Health, College of Veterinary Medicine, National Chung Hsing University, Taichung, Taiwan,* [b]*Division of Vector-Borne Diseases, Centers for Disease Control and Prevention, US Department of Health and Human Services, Fort Collins, CO, United States*

Abbreviations

Abs	antibodies
CR	complex-reactive
DENV	dengue virus
E	envelope
FL	fusion loop
GR	group-reactive
JEV	Japanese encephalitis virus
HI	hemagglutination inhibition
MAC-ELISA	IgM-capture ELISA
PRNT	plaque reduction neutralization test
prM	pre-membrane
NS1	nonstructural protein 1
RT-PCR	reverse-transcriptase polymerase chain reaction
TS	type-specific
TBEV	tick-borne encephalitis virus
VLP	virus-like particles
YFV	yellow fever virus
ZIKV	Zika virus

Introduction

Zika virus (ZIKV) and dengue virus (DENV), members of the Flaviviridae family, are associated with the resurgence of mosquito-transmitted diseases worldwide (Musso & Gubler, 2016). The viruses belonging to this family are enveloped, single-stranded, and positive-sense RNA genomes of approximately 11 kb in length that encode polyprotein precursors of about 3400 amino acid residues (Chambers, Hahn, Galler, & Rice, 1990). Co- and posttranslational processing by cellular and viral proteases generate three structural proteins [capsid (C), premembrane (prM), and envelope (E)] and seven nonstructural proteins (NS1, NS2A, NS2B, NS3, NS4A, NS4B, and NS5) (Chambers et al., 1990; Lindenbach & Rice, 2003). While DENV continues to impose the greatest economic and public health burden in tropical and subtropical countries, the recent emergence of ZIKV, which continuously circulates in Central and South America since 2013 (Faria et al., 2016), has resulted in terrifying outbreaks with severe health outcomes, including Guillain-Barré syndrome in adults as well as microcephaly, congenital neurologic malformations, and fetal demise (Baud, Gubler, Schaub, Lanteri, & Musso, 2017; Rasmussen, Jamieson, Honein, & Petersen, 2016). Considering the similar clinical symptoms between DENV and ZIKV infections, overlapping geographical distribution, and transmission cycles between humans and *Aedes aegypti* mosquitoes,

Zika Virus Impact, Diagnosis, Control, and Models. https://doi.org/10.1016/B978-0-12-820267-8.00029-7

these viruses are expected to elicit cross-reactive immunity among infected individuals (Calvet, Santos, & Sequeira, 2016; Saron et al., 2018). A confirmatory diagnosis can be obtained by virus isolation or viral RNA detection in serum and other body fluids. Currently, at least 14 different molecular and five serological assays are available for ZIKV diagnosis (Theel & Hata, 2018). Despite the availability of commercial assays, the timing of specimen collection and the high proportion of mild or asymptomatic ZIKV infections affect the diagnostic accuracy of these assays. These factors are due to the low viral titers, short duration of viremia during ZIKV infection, and the cross-reactivity of ZIKV antibodies with antibodies elicited from other flavivirus infections. In particular the development of commercial ZIKV serologic tests for use in clinical laboratories and at the point of care requires careful evaluation of test specificity, particularly in populations with a high prevalence of DENV and other flaviviral infections, including ZIKV. Therefore a ZIKV-specific serological assay with the proper algorithm is essential for the accurate diagnosis of patients who tested negative by virus isolation or viral RNA detection (Singh et al., 2018; Waggoner & Pinsky, 2016).

Antibody response after flavivirus infection

Mosquito-borne flaviviruses can be serologically classified into several serocomplexes, which include but are not limited to some of the medically important viruses such as the Japanese encephalitis virus (JEV), DENV, tick-borne encephalitis virus (TBEV), yellow fever virus (YFV), and the recently emerged ZIKV (Heinz & Stiasny, 2017). During natural infection the majority of elicited antibodies (Abs) recognize the structural proteins, premembrane (prM) and envelope (E), and the nonstructural protein 1 (NS1) (Dejnirattisai et al., 2010; Heinz & Stiasny, 2017; Sapparapu et al., 2016; Stettler et al., 2016). On the other hand, a recent study suggested that nonstructural proteins NS2b and NS3 could be the targets of anti-ZIKV IgG in patients' sera (Ravichandran et al., 2019; Song et al., 2018). Anti-E antibodies that recognize all members of the flavivirus group, members from different serocomplexes, or members within a serocomplex are classified as group-reactive (GR), complex-reactive (CR), or type-specific (TS) Abs, respectively (Calisher et al., 1989; Priyamvada et al., 2016; Robbiani et al., 2017).

The identification of flavivirus-specific neutralizing antibodies in human serum is the confirmation of flaviviral infections in humans (Dick, 1952; Smithburn, 1954; Waggoner & Pinsky, 2016). However, serological overlap or cross-reactivity within different flaviviruses is common for both plaque reduction neutralization tests (PRNT) and IgM-capture ELISA (MAC-ELISA). Serologic surveys conducted in Nigeria noted that sera capable of neutralizing ZIKV in mouse protection experiments were also strongly associated with the neutralization of DENV and YFV, suggesting the occurrence of serologic overlap or cross-reactivity (Macnamara, Horn, & Porterfield, 1959). Serologic evaluation of the Yap outbreak specimens at the Arbovirus Diagnostic Laboratory, United States-Centers for Disease Control and Prevention (US-CDC), revealed that acute-phase samples, collected within the first 9 days of illness from patients with a previous flavivirus infection, demonstrated an extensive cross-reactivity with other flaviviruses, particularly DENV and Japanese encephalitis virus (JEV), in both MAC-ELISA and PRNT (Lanciotti et al., 2008). The extensive antibody cross-reactivity among flaviviruses significantly reduced the value of performing PRNT analysis, mainly when a traditional paired serum algorithm with viral neutralization titering was used (Lee et al., 2018). An IgG depletion step before performing IgM-specific PRNT was proposed, which significantly enhances the diagnostic accuracy (Calvert et al., 2018). However, performing PRNT assay is labor intensive and requires to function under a Biosafety Level 2 environment. As ZIKV spreads throughout the areas of the world, this presents a diagnostic dilemma for patients living in regions in which both ZIKV and DENV or other flaviviruses are endemic, whose infections require significantly different medical management strategies (Oduyebo et al., 2017).

Serological methods used to detect anti-ZIKV antibodies

Traditionally, neutralization (Nt) assay or hemagglutination inhibition (HI) assay are used to measure the anti-E antibodies against flavivirus infections (Clarke & Casals, 1958; Roehrig, Hombach, & Barrett, 2008); on the contrary, complement-fixation assay is used to determine the anti-NS1 antibodies (Brandt, Cardiff, & Russell, 1970; Brandt, Chiewslip, Harris, & Russell, 1970). Although there were published articles using nanoparticles or biosensors to detect ZIKV antibodies (Cabral-Miranda, Cardoso, Ferreira, Sales, & Bachmann, 2018; Camacho, Sobral-Filho, Aoki, Constantino, & Brolo, 2018; Zhang et al., 2017), the most common type of IgM/IgG detection is enzyme-linked immunosorbent assays (ELISA). Depending on the target and the detection methods, methods detecting anti-E antibodies (E-based) or anti-NS1 antibodies (NS1-based) can be further classified into a direct or an antibody-capture sandwich method.

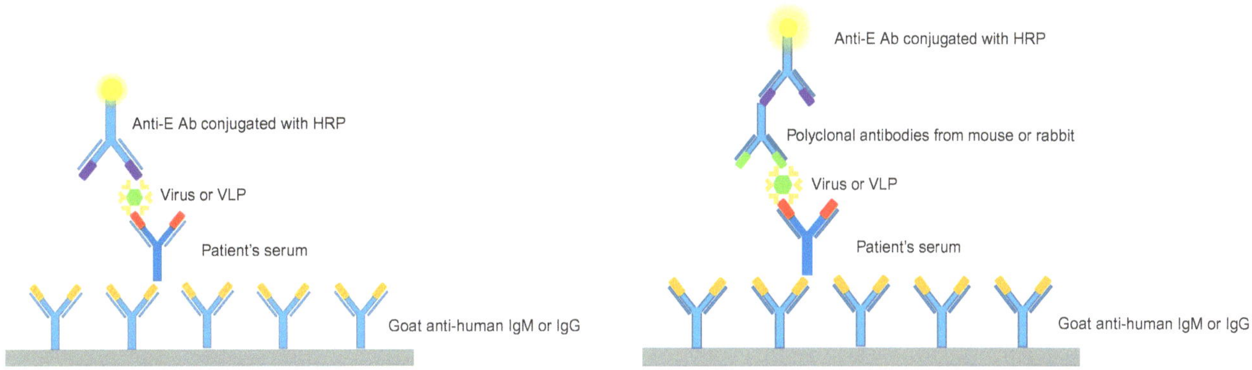

FIG. 1 E-specific MAC/GAC-ELISA.

1. E-based ELISA
   (a) MAC/GAC ELISA (Fig. 1A)

The MAC-ELISA for serological diagnosis of flavivirus infection was first published in 1979 for TBEV (Heinz, Roggendorf, Hofmann, Kunz, & Deinhardt, 1981). Since then, it has become the standard method to detect recent infections replacing the traditional HI method. In MAC-ELISA the total IgM in patients' sera is captured by anti-μ chain specific antibodies (specific to human IgM) coated onto a microplate. Either suckling mouse brain or cell culture-derived viral antigens, or virus-like particles (VLP) of different flaviviruses, from one to four serotypes of DENV (DEN-1, DEN-2, DEN-3, and DEN-4), or ZIKV, are bound to the captured antiflavivirus-specific IgM antibodies and are detected by monoclonal or polyclonal flavivirus-specific antibodies directly or indirectly conjugated with an enzyme that will transform a noncolored substrate into colored products. The optical density is measured by a spectrophotometer. IgM isotype-specific antibody-capture ELISA eliminates the IgG competition for the same antigenic sites and thus enhances the sensitivity of the assay. Several commercially available anti-prM/E ELISA kits for both IgM and IgG detection are commonly used for DENV clinical serodiagnosis, and the results from comprehensive evaluations suggest a good correlation of these assays with the gold standard Nt test (Blacksell et al., 2012; Namekar et al., 2013; Welch, Chang, & Litwin, 2014).

However, due to the nature of cross-reactivity of anti-E antibodies induced after flaviviral infections, several DENV IgM assays demonstrated false-positive results in ZIKV infection including Focus Diagnostics DENV IgM Capture (Fonseca et al., 2014) and SD Bioline Dengue Duo NS1 Ag/Ab Combo (Tappe et al., 2014). The antibody cross-reactivities can be improved by using mutated, cross-reactivity reduced recombinant E protein or VLPs (Roberson, Crill, & Chang, 2007; Rockstroh et al., 2017, 2019). The predominant target for cross-reactive antibodies is the fusion loop (FL) domain in the flavivirus envelope (E) protein, a short amino acid sequence that is almost identical in many pathogenic flaviviruses (Crill & Chang, 2004; Crill, Hughes, Delorey, & Chang, 2009). E106/107 mutations within this domain reduce the binding of GR cross-reactive antibodies (Roberson et al., 2007).

Although MAC-ELISA has been established useful in early diagnosis of viral infections, IgM antibodies are often produced only in low amounts and offer shorter half-lives in secondary flavivirus infections or after vaccination. Hence the detection of IgG antibodies offers an alternative advantage in diagnostics. GAC-ELISA works complementarily with MAC-ELISA in detecting past flaviviral infections for surveillance purposes or serological diagnosis of recent infections; collection of paired sera within the correct time frame during the acute and convalescent phase after the infection is necessary. Similarly, human Fc-specific antibodies are coated onto the microtiter plate to capture human IgG antibodies in the serum followed by the addition of the same antigens as performed in MAC-ELISA. Results of the GAC-ELISA correlate well with the HI test. This method also permits the identification of a case as either primary or secondary dengue infection. The disadvantage of using GAC-ELISA lies on the waiting time for the collection of convalescent serum after the patient has been discharged; this renders little use for early clinical diagnosis or patient management and provides only a retrospective outcome.

(b) IgM/IgG indirect ELISA (Fig. 2)

A similar sandwich ELISA can be used to detect ZIKV-specific IgM or IgG by directly coating E antigens to the microtiter plate. Diluted human sera are bound to antigen-coated wells and are detected using enzyme-conjugated antihuman secondary antibody, specifically to IgM or IgG. Once added a chromogenic substrate acts as a visualizing agent, allowing

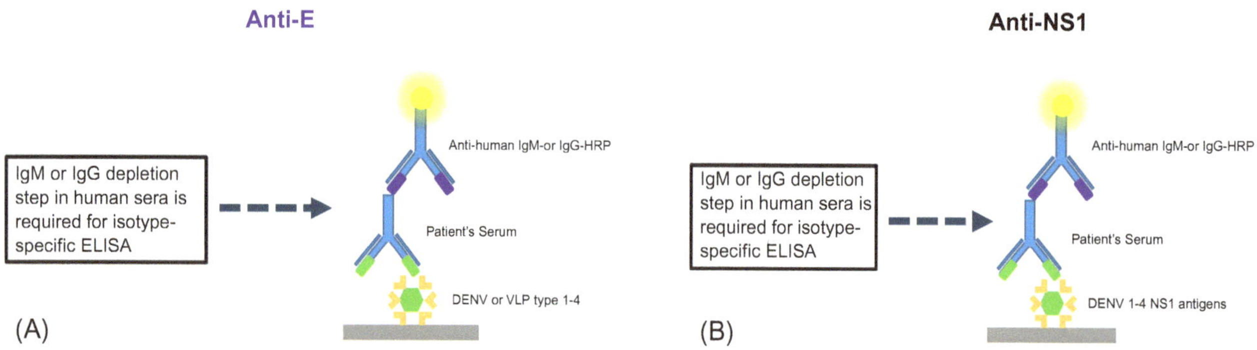

FIG. 2 IgM/IgG indirect ELISA.

the generation of colored products, which are measurable by a spectrophotometer. The commonly used E antigens are recombinant E antigens, including full-length prM/E-expressed virus-like particles, and full-length ectodomain E or domain I/II or III E antigens (Premkumar et al., 2018). The pitfall of this type of assay is the high background signal due to the need for using a higher concentration of serum (i.e., 1:20–1:100). Another disadvantage is the competition between IgM and IgG for binding to the E antigens, which requires the depletion of either IgM or IgG for isotype-specific detection (Tyson, Tsai, Tsai, Brites, et al., 2019).

(c) IgM/IgG inhibition ELISA

An ELISA inhibition method (EIM) is also used to detect dengue IgG antibodies for serological diagnosis and surveillance of DENV- or ZIKV-infected cases (Balmaseda et al., 2018). This system is based on the competition for the antigen binding sites between anti-ZIKV IgG antibodies in the serum sample and the enzyme-conjugated anti-ZIKV human IgG monoclonal antibody. The pitfall of this type of assay is the requirement for a serial dilution of human sera. Besides the presence of such epitope-specific IgG is not guaranteed for all ZIKV-infected patients, which might lead to the low detectability of this assay.

2. NS1-based ELISA

Although GR or CR anti-NS1 antibodies are found in other flaviviral infections, recent studies suggested that the majority of the anti-NS1 antibodies from primary ZIKV infections are type specific and can be used as serological markers to differentiate ZIKV from DENV infections (Balmaseda et al., 2017; Huzly, Hanselmann, Schmidt-Chanasit, & Panning, 2016; Probst et al., 2016; Stettler et al., 2016). The NS1-based ELISA has also been suggested as an important biomarker to differentiate antibodies induced by vaccination or natural infection (Chao, Galula, Shen, Davis, & Chang, 2015; Nascimento et al., 2018; Sridhar et al., 2018).

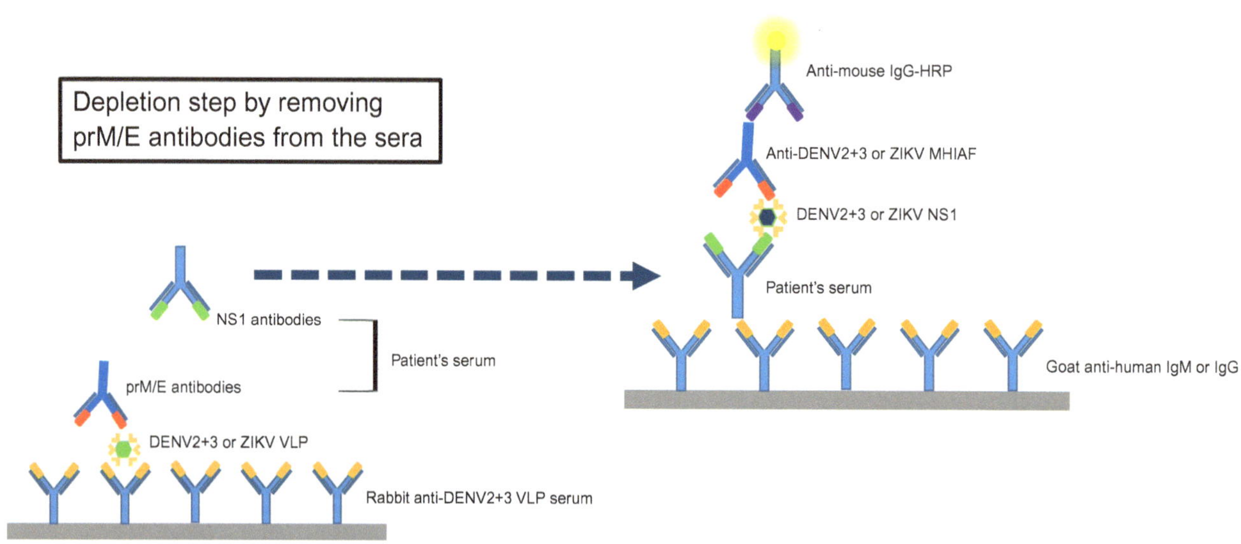

FIG. 3 NS1-specific MAC/GAC_ELISA.

(a) MAC/GAC ELISA (Fig. 3)

The main purpose of developing NS1-based MAC/GAC-ELISA is to replace E antigens by NS1 antigens with a similar approach of E-based MAC/GAC-ELISA. However, low sensitivities of NS1 antibody detection were observed in previous studies due to the relative abundance of anti-prM/E antibodies in the serum of infected individuals. Thus a procedure to deplete anti-prM/E antibodies using prM/E-containing virus-like particles (VLPs) was necessary to improve the sensitivity of NS1-based MAC/GAC-ELISA. The sensitivities and specificities between the NS1- and VLP-MAC/GAC-ELISAs were comparable after the depletion step (Chao et al., 2015, 2019; Galula, Chang, Chuang, & Chao, 2016).

(b) IgM/IgG indirect detection (Fig. 2)

Anti-NS1 IgM/IgG indirect ELISA has been proposed and developed for the detection of anti-NS1 antibodies as serological markers of natural infection among Japanese encephalitis virus (JEV) vaccinated populations (Konishi & Kitai, 2009; Konishi, Shoda, Ajiro, & Kondo, 2004; Konishi & Suzuki, 2002). However, this type of NS1-ELISA required that the microtiter plates be sensitized with purified NS1 or let NS1-specific murine monoclonal antibodies capture the NS1 antigen before adding the test human serum samples. As a result the sensitivity of NS1-specific Ig detection, particularly IgG, was largely reduced probably due to the competition between two antibody isotypes, that is, IgM and IgG, for the same antigenic sites.

(c) BOB (blockade of binding) ELISA

Flavivirus NS1 antigens are first coated onto the plates, and then human sera (1:10 dilution) are added onto individual wells. After 1 h of incubation, an equal volume of anti-NS1 monoclonal antibodies, which are either biotinylated and detected by enzyme-conjugated streptavidin or directly conjugated with an enzyme, is added to the individual wells. After the addition of a chromogenic substrate, colored products result in a detectable signal, which is then measured by a spectrophotometer. BOB values were calculated as follows: $\{1 - [(OD_{sample} - OD_{minimum\ value}/delta\ value)]\} \times 100$, where OD represents optical density. A BOB value of >50 was considered positive, and a BOB value of <50 was considered negative. Although the procedure is similar to E-based EIM, it has been reported that NS1-based BOB ELISA is more sensitive in detecting ZIKV-specific NS1 antibodies among primary or secondary ZIKV-infected patients (Balmaseda et al., 2018).

Cross-reactivity of anti-E and NS1 antibody between ZIKV and DENV

In the differential diagnosis of ZIKV infections, evaluation for other vector-borne viruses, especially DENV, should be included because of their clinical similarity, overlapping geographic locations, and similar transmission cycles. Table 1 summarizes the publications performing differential diagnosis of ZIKV infections while simultaneously using E or NS1 antigens from both ZIKV and DENV. Briefly the cross-reactivity of ZIKV E-based ELISA with DENV-infected sera ranged from 20.8% to 63.6%. E106/107 mutated VLPs reduced the cross-reactivity of DENV sera recognizing ZIKV mutant VLP from 63.6% to 38.6%. Similarly, 95% of ZIKV-infected sera recognized DENV-2/3 VLP antigens, and the cross-reactivity remained the same even when the mutant DENV-2/3 VLP antigens were used. The results of highly cross-reactive anti-prM/E antibodies were consistent with focus reduction neutralization test results (Chao et al., 2019). The low cross-reactivity in the study of Rockstroh et al. (2019) could be due to the low sensitivity resulting from the type of E-protein and indirect ELISA used. Of note, preabsorption by heterologous antigens can reduce the cross-reactivity and increase the detection specificity. Unfortunately, only IgG ELISA results were shown in the publication, which again implies the low sensitivity of such assay to detect IgM (Rockstroh et al., 2019).

Although anti-NS1 antibodies from primary ZIKV infections are dominantly type specific and the ZIKV NS1-based assay was suggested to be capable of differentiating ZIKV from DENV infections, cross-reactivity of anti-NS1 antibodies between ZIKV and DENV infected sera remains. Using MAC-ELISA, 73% of DENV sera were cross-reactive with ZIKV NS1 antigen, and 57% of ZIKV sera were cross-reactive with DENV NS1 antigen (Chao et al., 2019). Meanwhile a 10% cross-reactivity was observed by using ZIKV NS1 BOB ELISA (Balmaseda et al., 2018). Notably the low cross-reactivity between DENV and ZIKV in Tsai or Tyson's study might be due to the low sensitivity of indirect ELISA, which performs poorly in detecting IgM from patients' sera (Tsai et al., 2017; Tyson, Tsai, Tsai, Brites, et al., 2019). Similarly, high cross-reactivity of anti-NS1 IgG was found among ZIKV patients' sera with a prior history of DENV infection in Tsai or Tyson's study (Tsai et al., 2017; Tyson, Tsai, Tsai, Brites, et al., 2019) and in our unpublished results.

In summary the cross-reactivity of IgM or IgG after ZIKV and DENV infections is commonly observed with a higher proportion of DENV sera cross-reacting with ZIKV antigen than ZIKV sera with DENV antigen after secondary infections. Early identification of ZIKV or DENV infection is critical for patients' management and epidemiologic monitoring of outbreaks. Because of the early appearance of IgM in the patient's serum, it is a preferred diagnostic antibody marker, especially when viral nucleic acids and NS1 antigens are no longer detectable.

TABLE 1 Studies detecting anti-ZIKV antibodies using various serological assays.

Type	References	Serological assay	Serum dilution	Type of antigens	Cross-reactivity			
					Primary ZIKV	**Secondary ZIKV**	**Primary DENV**	**Secondary DENV**
Anti-E	Balmaseda et al. (2018)	NIAID-BEI MAC-ELISA CDNR MAC-ELISA Inhibition ELISA	1:400 1:20 1: 10 serial dilution	ZIKV-VLP from CDC Mouse brain-derived ZIKV Mouse brain-derived ZIKV	62/62 (100%) 60/62 (96.8%) 55.7%	49/65 (75.4%) 61/65 (93.8%) 80%	**24/97 (24.7%)** **25/97 (25.8%)** **20.8%**	
	Rockstroh et al. (2019)	IgM indirect ELISA IgG indirect ELISA IgM indirect ELISA IgG indirect ELISA	1:100	Quadruple mutant ZIKV E-protein +DENV competition Quadruple mutant D1–4 E-protein +ZIKV competition	14/16 (87.5%) 18/18 (100%) **0/54 (0%)** 1/18 (5.6%)	21/21 (100%) 18/21 (85.7%)	**5/54 (9.3%)** 0/23 (0%) 49/54 (90.7%) 23/23 (100%)	2/55 (3.6%) 55/55 (100%)
	Chao et al. (2019)	MAC-ELISA	1:2000(E-specific Ig depletion)	ZIKV VLP DENV-2/3 VLP	21/21 (100%) **20/21 (95.2%)**		**28/44 (63.6%)** 42/44 (95.5%)	
	Chao et al. (2019)	MAC-ELISA	1:2000(E-specific Ig depletion)	ZIKV fl-mutant VLP DENV-2fl-mutant VLP	21/21 (100%) **20/21 (95.2%)**		**17/44 (38.6%)** 38/44 (86.4%)	
Anti-NS1	Balmaseda et al. (2018)	NS1 BOB ELISA	1:10	ZIKV rNS1	58/62 (93.5%)	60/65 (92.3%)	**10/97 (10.3%)**	
	Tsai et al. (2017)	IgM indirect ELISA	1:400	ZIKV rNS1	90%	55%	**0%**	**4%**
		IgG indirect ELISA		DENV-1 rNS1	**0%**	**85%**	81.3%	95.8%
	Tyson, Tsai, Tsai, Brites, et al. (2019)	IgG indirect ELISA	1:400 (results from <3 months)	ZIKV rNS1 DENV-1-4 rNS1	50% **85.7%**	97.1% **85.7%**	**0%** 81.4%	**47.8%** 95.5%
		IgM indirect ELISA	1:400 (IgG depletion)	ZIKV rNS1 DENV-1 rNS1	90% **0%**	55% **15%**	**0%** 55.6%	**4%** 16.7%
	Chao et al. (2019)	MAC-ELISA	1:2000(E-specific Ig depletion)	ZIKV rNS1 DENV-2 rNS1	100% **57.1%**		**72.7%** 90.9%	

Current algorithms proposed for differentiation

Traditionally, flaviviruses are classified into different serocomplexes, composed of members that are cross-neutralized by polyclonal sera (Calisher et al., 1989). Such seroclassification is correlated with the similarity in amino acid sequences of prM/E (Heinz & Stiasny, 2017). ZIKV clustered with the Spondweni virus and showed an intermediate position with viruses from JEV and DENV serocomplexes in the phylogenetic tree (based on the complete genome, E, or NS1 gene sequences). The overall picture of flavivirus serocomplexes indicates that cross-reactivity and cross-neutralizing ability of antibodies are usually lost when the amino acid sequence divergence of E is more than 50% (Heinz & Stiasny, 2017). Therefore it is suggested that ZIKV, together with the viruses from the Spondweni viral group, could form an independent flaviviral serocomplex. A similar concept can be applied to anti-NS1 antibodies from a different serocomplex of flaviviruses. The major difference of anti-E antibodies from other antibodies, elicited after flavivirus infections, is that their cross-reactive antibody responses are immune dominant. Thus the overall cross-reactivity of anti-prM/E antibodies induced after ZIKV infection was significantly higher than that of the anti-NS1 antibodies. The majority of cross-reactive anti-prM/E antibodies during flaviviral infections are GR (4G2- and 6B6C-1-like) antibodies recognizing FP-mutated VLP (FP-VLP), with the potential to enhance viral infection and induce a low-to-moderate neutralizing activity (Heinz & Stiasny, 2017). The use of FP-VLP for the detection of anti-E antibodies is intended to reduce the binding of cross-reactive GR antibodies while maintaining the binding of CR or TS antibodies (Crill et al., 2012; Hughes, Crill, & Chang, 2012). However, the remaining antibodies induced after infection by different serotypes of DENV can still bind to DENV-2/3-FP-VLP due to the presence of DENV complex-specific, cross-reactive antibodies.

Since CR cross-reactive antibodies recognizing FP-VLP antigens from the different serocomplexes will be fewer with weaker affinity, the algorithm to distinguish IgM antibodies induced by ZIKV from DENV infection was proposed by Chao et al. (2019) using the ratio test of ZIKV/DENV in FP-VLP-MAC-ELISA. This method successfully differentiated between ZIKV and DENV infection, with 90%–100% accuracy (Chao et al., 2019). The NS1 and E proteins share similar degrees of amino acid sequence divergence. Thus a similar IgM ratio test to detect VLP-antibody-depleted anti-NS1 antibodies by ZIKV/DENV NS1-MAC-ELISA could also be used to distinguish ZIKV or DENV infection between viruses belonging to different serocomplexes. The use of FP-VLP in MAC-ELISA has several advantages, including the avoidance of a predepletion step before detecting anti-NS1 antibodies and the reduced binding of cross-reactive fusion-peptide antibodies, to significantly enhance the specificity and accuracy of using a ZIKV/DENV ratio test in differentiating ZIKV and DENV infections.

A proposal was made to use the algorithm of ZIKV/DENV ratio in performing indirect ELISA to distinguish IgG antibodies induced by ZIKV from DENV infections (Tsai et al., 2017; Tyson, Tsai, Tsai, Brites, et al., 2019); however, this method performs poorly in detecting IgM due to its low abundance, thus resulting in low sensitivity. By far the only way to sensitively detect anti-NS1 IgM with ELISA is by using VLP-depleted NS1-MAC-ELISA. IgG indirect ELISA can be used to distinguish the past infection between different flaviviral serocomplexes but not within the same serocomplex, which lacks specificity (i.e., differentiating different serotypes of DENV infection). Therefore IgG indirect ELISA has the potential to reduce the need for the labor-intensive PRNT for seroprevalence study and the ability to possibly identify recent infections by testing a second sample collected at a later time point during infection, particularly for sera obtained at 8 days post disease onset. Notably the CDC currently recommends testing by ZIKV RT-PCR during the first 14 days post onset of symptoms or 14 days after departing a ZIKV-endemic region (Oduyebo et al., 2017), which has the potential to close the gap during the first 8 days when an immune response is mounting. Where RT-PCR results are negative, testing of appropriately time-paired sera is necessary to demonstrate seroconversion or differentiation of recent from past infection with prior exposure to ZIKV.

Similar cross-reactivity of anti-E or NS1 antibodies using microsphere immunoassay can also be found and can be differentiated by calculating the ratio of relative median fluorescence intensity (rMFI) of ZIKV NS1 to that of DENV1 NS1 and determining the cutoff of rMFI by receiver operating characteristics (ROC) analysis (Taylor et al., 2018; Tyson, Tsai, Tsai, Mässgård, et al., 2019). Antibody avidity is used in a few laboratories to discriminate primary from secondary dengue infections. Following viral infections, newly produced antibodies are less avid than antibodies produced from the previous infection. Such a test has been proposed for the differentiation between ZIKV and DENV infections but are neither in wide use nor available commercially (Furuya et al., 2019; Rönnberg et al., 2017).

We have summarized the algorithm based on testing IgM or IgG for the differentiation of ZIKV from DENV infections (Fig. 4). IgM is vital for the diagnosis of acute and recent infections. An ideal IgM test would be one that allows specific detection of ZIKV antibodies at high sensitivity to distinguish infection from DENV and other flaviviruses in asymptomatic, symptomatic, and postsymptomatic individuals, particularly in pregnant women and women of childbearing age. However, IgG is particularly helpful for pregnant women whose ZIKV IgM antibody response has waned 2–3 months after infection and may be erroneously dismissed as noninfected, potentially culminating in adverse fetal outcomes.

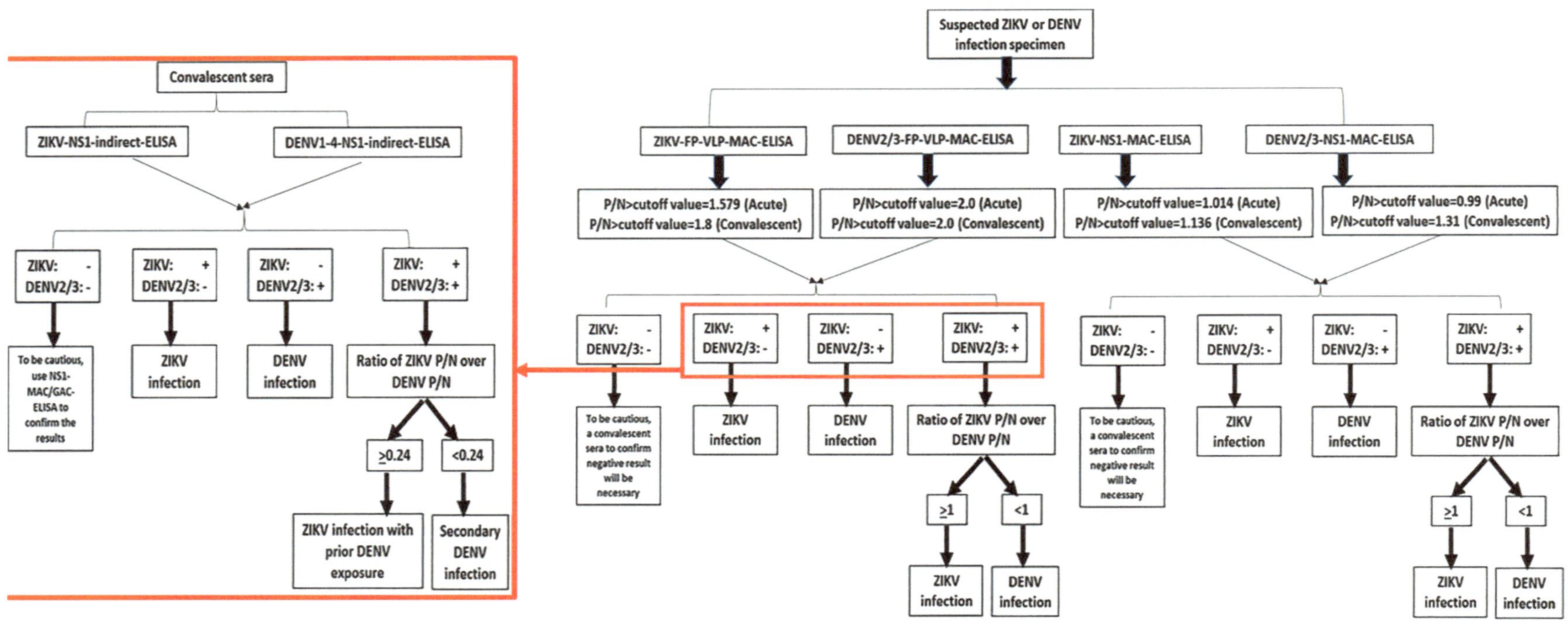

FIG. 4 Algorithm differentiating ZIKV from DENV infection.

Future perspectives and challenges

Molecular methods to detect arboviruses are ineffectual when samples have low viral loads, which occur during the short duration of viremia caused by ZIKV. Serology extends the window of identification but is often cross-reactive; that is, antibodies triggered by one flavivirus infection may react with other various flaviviruses. Serological results are sometimes confounded with the presence of preexisting antibodies attributed to previous flavivirus infection or vaccination. Several recent publications have described global efforts to resolve this issue to determine the status of ZIKV infection (Fritzell et al., 2018; Theel & Hata, 2018). Given the association of ZIKV with fetal microcephaly, improved accessibility to ZIKV-specific serology is essential to assess the risk of fetal transmission and to identify women requiring regular ultrasound or amniocentesis for confirmation of infection, in analogy to the use of serologic testing for congenital cytomegalovirus and *Toxoplasma gondii* infections. Currently the best algorithm to achieve 100% accuracy requires the combination of both anti-E and anti-NS1 IgM and IgG assay. The multiplex assay, such as the microsphere or nanotechnology platform, might be the best fit to perform the assay (Zhang et al., 2017). With the incremental cocirculation of viruses causing clinically indistinguishable acute febrile symptoms by the emergence and reemergence of antigenically similar viruses, the diagnostic tools should be developed according to the ASSURED (affordable, sensitive, specific, user-friendly, rapid, equipment-free, delivered) criteria to suit the needs of the developing world. With the rapid advances in digital technology, these criteria have been revised to include real-time connectivity and ease of specimen collection (REASSURED) (Land, Boeras, Chen, Ramsay, & Peeling, 2019).

To achieve the aforementioned criteria, it is also pivotal to establish the serum bank with the collection of patient sera with a confirmed history of prior exposure to different flaviviruses. The study of false-positive results with sera from malaria-infected persons has been reported for an NS1-based ZIKV assay (Van Esbroeck, Meersman, Michiels, Ariën, & Van den Bossche, 2016). Serum collected under different time points with confirmed RT-PCR and PRNT results will be essential for developing the best diagnostic tools applied in different time course following the infection.

Mini-dictionary of terms

Flavivirus. A genus of vector-borne viruses within the family Flaviviridae, containing mosquito-borne, tick-borne, and insect-only viruses.

ELISA. Enzyme-linked immunosorbent assay is the serological assay commonly used to detect either the antigen or the antibody with the advantages of being low cost, high throughput, and speediness.

IgM. Immunoglobulin M has higher cross-reactivity, usually short-lived and used as a biomarker for recent infection.

IgG. Immunoglobulin G has different antigen specificities, usually has long half-lives, and used as a biomarker for infection history.

Algorithm. A decision tree to determine the individual infection from a specific flavivirus by determining the timing of specimen collection, positivity of ELISA, and ratio of ZIKV/DENV.

Key facts

Window of detectability using different diagnostic tools

During an acute infection the viral nucleic acid can be detected in serum, urine, or other bodily fluids. At 4–5 days after infection, when the virus titer is low and the immune cells initiated viral clearance, the virus-specific antibodies from the immune reaction can already be detected in different body fluids.

Serocomplex

Traditionally, flaviviruses can be classified based on the cross-reactivity of antibodies toward the virus particles within the same group, termed as serocomplex. At least three serocomplexes have been proposed, including JEV, DENV, and TBEV. ZIKV and YFV are also treated as independent serocomplexes.

Reassured

The ASSURED (*a*ffordable, *s*ensitive, *s*pecific, *u*ser-friendly, *r*apid, *e*quipment-free, *d*elivered) criteria were first proposed in 2017 by Kosack and colleagues to describe the ideal point-of-care diagnostic test to meet the needs of the developing world (*Bulletin of the World Health Organization*, 2017; 95:639–645). In 2019 Land and colleagues added two more criteria which are *r*eal-time connectivity and *e*ase of specimen collection, which gave rise to REASSURED (Nat Microbiol., 2019 Jan; 4(1):46–54).

Cross-reactivity of flavivirus-specific antibody

Within a serocomplex, it is difficult to differentiate viral infections serologically. However, serological differentiation of viral infections between serocomplexes can be done based on the avidity of antibodies against different viruses using OD readings or P/N ratio calculation. Cross-reactive antibodies tend to have a lower avidity toward different serocomplexes.

Antibody kinetics

Different Ig isotypes display different kinetics after infection such that IgM and IgA appear at the earliest phase and are then followed by IgG. Different IgG isotypes also show different half-lives.

Summary points

1. In dengue-naive and ZIKV-infected sera, VLP- or NS1-based MAC-ELISA can differentiate a ZIKV from a DENV infection.
2. In dengue-exposed and currently ZIKV-infected sera, 73% of DENV sera were cross-reactive with ZIKV NS1 antigen, and 57% of ZIKV sera were cross-reactive with DENV NS1 antigen using MAC-ELISA.
3. Applying the ratio of ZIKV/DENV, VLP- and NS1-based MAC-ELISA perform equally well in terms of sensitivity and specificity.
4. NS1-based indirect ELISA performs poorly in detecting ZIKV-specific anti-NS1 IgM due to the relative abundance of anti-E IgM during acute flavivirus infections.
5. Using the ratio of ZIKV/DENV, NS1-based indirect IgG-ELISA performs equally well with GAC-ELISA.
6. This chapter introduces an algorithm in differentiating ZIKV from DENV infections. The detection of ZIKV IgM using MAC-ELISA is particularly helpful for pregnant women and is recommended for ZIKV-infected patients with asymptomatic, symptomatic, and postsymptomatic infections. However, IgG detection using GAC-ELISA or indirect ELISA is especially suitable for individuals whose antibody response has waned 2–3 months after the infection. In the latter case, individuals may be erroneously dismissed as noninfected, which can potentially culminate in adverse fetal outcomes.

References

Balmaseda, A., Stettler, K., Medialdea-Carrera, R., Collado, D., Jin, X., Zambrana, J., et al. (2017). Antibody-based assay discriminates Zika virus infection from other flaviviruses. *Proceedings of the National Academy of Sciences of the United States of America*, *114*(31), 8384–8389.

Balmaseda, A., Zambrana, J., Collado, D., García, N., Saborío, S., Elizondo, D., et al. (2018). Comparison of four serological methods and two reverse transcription-PCR assays for diagnosis and surveillance of Zika virus infection. *Journal of Clinical Microbiology*, *56*(3). pii:e01785-17.

Baud, D., Gubler, D., Schaub, B., Lanteri, M., & Musso, D. (2017). An update on Zika virus infection. *The Lancet*, *390*(10107), 2099–2109.

Blacksell, S., Lee, S., Chanthongthip, A., Taojaikong, T., Thongpaseuth, S., Hübscher, T., et al. (2012). Comparison of performance of serum and plasma in panbio dengue and Japanese encephalitis virus enzyme-linked immunosorbent assays. *The American Journal of Tropical Medicine and Hygiene*, *87*(3), 573–575.

Brandt, W., Cardiff, R., & Russell, P. (1970). Dengue virions and antigens in brain and serum of infected mice. *Journal of Virology*, *6*(4), 500–506.

Brandt, W., Chiewslip, D., Harris, D., & Russell, P. (1970). Partial purification and characterization of a dengue virus soluble complement-fixing antigen. *Journal of Immunology*, *105*(6), 1565–1568.

Cabral-Miranda, G., Cardoso, A., Ferreira, L., Sales, M., & Bachmann, M. (2018). Biosensor-based selective detection of Zika virus specific antibodies in infected individuals. *Biosensors & Bioelectronics*, *113*, 101–107.

Calisher, C., Karabatsos, N., Dalrymple, J., Shope, R., Porterfield, J., Westaway, E., et al. (1989). Antigenic relationships between flaviviruses as determined by cross-neutralization tests with polyclonal antisera. *The Journal of General Virology*, *70*(Pt 1), 37–43.

Calvert, A., Boroughs, K., Laven, J., Stovall, J., Luy, B., Kosoy, O., et al. (2018). Incorporation of IgG depletion in a neutralization assay facilitates differential diagnosis of Zika and Dengue in secondary flavivirus infection cases. *Journal of Clinical Microbiology*, *56*(6). pii:e00234-18.

Calvet, G. A., Santos, F. B. D., & Sequeira, P. C. (2016). Zika virus infection: Epidemiology, clinical manifestations and diagnosis. *Current Opinion in Infectious Diseases*, *29*(5), 459–466.

Camacho, S., Sobral-Filho, R., Aoki, P., Constantino, C., & Brolo, A. (2018). Zika immunoassay based on surface-enhanced Raman scattering nanoprobes. *ACS Sensors*, *3*(3), 587–594.

Chambers, T., Hahn, C., Galler, R., & Rice, C. (1990). Flavivirus genome organization, expression, and replication. *Annual Review of Microbiology*, *44*, 649–688.

Chao, D., Galula, J., Shen, W., Davis, B., & Chang, G. (2015). Nonstructural protein 1-specific immunoglobulin M and G antibody capture enzyme-linked immunosorbent assays in diagnosis of flaviviral infections in humans. *Journal of Clinical Microbiology*, *53*(2), 557–566.

Chao, D., Whitney, M., Davis, B., Medina, F., Munoz, J., & Chang, G. (2019). Comprehensive evaluation of differential serodiagnosis between Zika and Dengue viral infections. *Journal of Clinical Microbiology*, *57*(3). pii:e01506-18.

Clarke, D., & Casals, J. (1958). Techniques for hemagglutination and hemagglutination-inhibition with arthropod-borne viruses. *The American Journal of Tropical Medicine and Hygiene*, *7*(5), 561–573.

Crill, W., & Chang, G. (2004). Localization and characterization of flavivirus envelope glycoprotein cross-reactive epitopes. *Journal of Virology*, *78*(24), 13975–13986.

Crill, W., Hughes, H., Delorey, M., & Chang, G. (2009). Humoral immune responses of dengue fever patients using epitope-specific serotype-2 virus-like particle antigens. *PLoS One*, *4*(4), e4991.

Crill, W., Hughes, H., Trainor, N., Davis, B., Whitney, M., & Chang, G. (2012). Sculpting humoral immunity through dengue vaccination to enhance protective immunity. *Frontiers in Immunology*, *3*, 334.

Dejnirattisai, W., Jumnainsong, A., Onsirisakul, N., Fitton, P., Vasanawathana, S., Limpitikul, W., et al. (2010). Cross-reacting antibodies enhance dengue virus infection in humans. *Science*, *328*(5979), 745–748.

Dick, G. (1952). Zika virus. II. Pathogenicity and physical properties. *Transactions of the Royal Society of Tropical Medicine and Hygiene*, *46*(5), 521–534.

Faria, N., Azevedo, R., Kraemer, M., Souza, R., Cunha, M., Hill, S., et al. (2016). Zika virus in the Americas: Early epidemiological and genetic findings. *Science*, *352*(6283), 345–349.

Fonseca, K., Meatherall, B., Zarra, D., Drebot, M., MacDonald, J., Pabbaraju, K., et al. (2014). First case of Zika virus infection in a returning Canadian traveler. *The American Journal of Tropical Medicine and Hygiene*, *91*(5), 1035–1038.

Fritzell, C., Rousset, D., Adde, A., Kazanji, M., Van Kerkhove, M., & Flamand, C. (2018). Current challenges and implications for dengue, chikungunya and Zika seroprevalence studies worldwide: A scoping review. *PLoS Neglected Tropical Diseases*, *12*(7), e0006533.

Furuya, A., Hunt, D., George, K., An, D., Kramer, L., Shi, P., et al. (2019). Use of the immunoglobulin G avidity assay to differentiate between recent Zika and past dengue virus infections. *Clinical Science (London, England)*, *133*(7), 859–867.

Galula, J., Chang, G., Chuang, S., & Chao, D. (2016). Establishment of an algorithm using prM/E- and NS1-specific IgM antibody-capture enzyme-linked immunosorbent assays in diagnosis of Japanese encephalitis virus and West Nile virus infections in humans. *Journal of Clinical Microbiology*, *54*(2), 412–422.

Heinz, F., Roggendorf, M., Hofmann, H., Kunz, C., & Deinhardt, F. (1981). Comparison of two different enzyme immunoassays for detection of immunoglobulin M antibodies against tickborne encephalitis virus in serum and cerebrospinal fluid. *Journal of Clinical Microbiology*, *14*, 141–146.

Heinz, F. X., & Stiasny, K. (2017). The antigenic structure of Zika virus and its relation to other flaviviruses: Implications for infection and immunoprophylaxis. *Microbiology and Molecular Biology Reviews*, *81*(1). e00055-16.

Hughes, H., Crill, W., & Chang, G. (2012). Manipulation of immunodominant dengue virus E protein epitopes reduces potential antibody-dependent enhancement. *Virology Journal*, *9*, 115.

Huzly, D., Hanselmann, I., Schmidt-Chanasit, J., & Panning, M. (2016). High specificity of a novel Zika virus ELISA in European patients after exposure to different flaviviruses. *Euro Surveillance*, *21*(16). pii:30203.

Konishi, E., & Kitai, Y. (2009). Detection by ELISA of antibodies to Japanese encephalitis virus nonstructural 1 protein induced in subclinically infected humans. *Vaccine*, *27*(50), 7053–7058.

Konishi, E., Shoda, M., Ajiro, N., & Kondo, T. (2004). Development and evaluation of an enzyme-linked immunosorbent assay for quantifying antibodies to Japanese encephalitis virus nonstructural 1 protein to detect subclinical infections in vaccinated horses. *Journal of Clinical Microbiology*, *42*(11), 5087–5093.

Konishi, E., & Suzuki, T. (2002). Ratios of subclinical to clinical Japanese encephalitis (JE) virus infections in vaccinated populations: Evaluation of an inactivated JE vaccine by comparing the ratios with those in unvaccinated populations. *Vaccine*, *21*(1–2), 98–107.

Lanciotti, R., Kosoy, O., Laven, J., Velez, J., Lambert, A., Johnson, A., et al. (2008). Genetic and serologic properties of Zika virus associated with an epidemic, Yap state, Micronesia, 2007. *Emerging Infectious Diseases*, *14*(8), 1232–1239.

Land, K., Boeras, D., Chen, X., Ramsay, A., & Peeling, R. (2019). Reassured diagnostics to inform disease control strategies, strengthen health systems and improve patient outcomes. *Nature Microbiology*, *4*(1), 46–54.

Lee, W., Wong, S., Kulas, K., Dupuis, A. P., Payne, A., Kramer, L., et al. (2018). Development of Zika virus serological testing strategies in New York State. *Journal of Clinical Microbiology*, *56*(3). pii:e01591-17.

Lindenbach, B., & Rice, C. (2003). Molecular biology of flaviviruses. *Advances in Virus Research*, *59*, 23–61.

Macnamara, F., Horn, D., & Porterfield, J. (1959). Yellow fever and other arthropod-borne viruses; a consideration of two serological surveys made in South Western Nigeria. *Transactions of the Royal Society of Tropical Medicine and Hygiene*, *53*(2), 202–212.

Musso, D., & Gubler, D. J. (2016). Zika virus. *Clinical Microbiology Reviews*, *22*(6), 494–496.

Namekar, M., Ellis, E., O'Connell, M., Elm, J., Gurary, A., Park, S., et al. (2013). Evaluation of a new commercially available immunoglobulin M capture enzyme-linked immunosorbent assay for diagnosis of dengue virus infection. *Journal of Clinical Microbiology*, *51*(9), 3102–3106.

Nascimento, E., George, J., Velasco, M., Bonaparte, M., Zheng, L., DiazGranados, C., et al. (2018). Development of an anti-dengue NS1 IgG ELISA to evaluate exposure to dengue virus. *Journal of Virological Methods*, *257*, 48–57.

Oduyebo, T., Polen, K., Walke, H., Reagan-Steiner, S., Lathrop, E., Rabe, I., et al. (2017). Update: Interim guidance for health care providers caring for pregnant women with possible Zika virus exposure—United States (including U.S. territories), July 2017. *Morbidity and Mortality Weekly Report*, *66* (29), 781–793.

Premkumar, L., Collins, M., Graham, S., Liou, G., Lopez, C., Jadi, R., et al. (2018). Development of envelope protein antigens to serologically differentiate Zika virus infection from dengue virus infection. *Journal of Clinical Microbiology*, *56*(3), e01504–e01517.

Priyamvada, L., Quicke, K., Hudson, W., Onlamoon, N., Sewatanon, J., Edupuganti, S., et al. (2016). Human antibody responses after dengue virus infection are highly cross-reactive to Zika virus. *Proceedings of the National Academy of Sciences of the United States of America*, *113*(28), 7852–7857.

Rasmussen, S. A., Jamieson, D. J., Honein, M. A., & Petersen, L. R. (2016). Zika virus and birth defects—Reviewing the evidence for causality. *The New England Journal of Medicine*, *374*(20), 1981–1987.

Ravichandran, S., Hahn, M., Belaunzarán-Zamudio, P., Ramos-Castañeda, J., Nájera-Cancino, G., Caballero-Sosa, S., et al. (2019). Differential human antibody repertoires following Zika infection and the implications for serodiagnostics and disease outcome. *Nature Communications*, *10*(1), 1943.

Robbiani, D., Bozzacco, L., Keeffe, J., Khouri, R., Olsen, P., Gazumyan, A., et al. (2017). Recurrent potent human neutralizing antibodies to Zika virus in Brazil and Mexico. *Cell*, *169*(4), 597–609.

Roberson, J., Crill, W., & Chang, G. (2007). Differentiation of West Nile and St. Louis encephalitis virus infections by use of noninfectious virus-like particles with reduced cross-reactivity. *Journal of Clinical Microbiology*, *45*(10), 3167–3174.

Rockstroh, A., Barzon, L., Kumbukgolla, W., Su, H., Lizarazo, E., Vincenti-Gonzalez, M., et al. (2019). Dengue virus IgM serotyping by ELISA with recombinant mutant envelope proteins. *Emerging Infectious Diseases*, *25*(1), 1111–1115.

Rockstroh, A., Moges, B., Barzon, L., Sinigaglia, A., Palù, G., Kumbukgolla, W., et al. (2017). Specific detection of dengue and Zika virus antibodies using envelope proteins with mutations in the conserved fusion loop. *Emerging Microbes & Infections*, *6*(11), e99.

Roehrig, J., Hombach, J., & Barrett, A. (2008). Guidelines for plaque-reduction neutralization testing of human antibodies to dengue viruses. *Viral Immunology*, *21*(2), 123–132.

Rönnberg, B., Gustafsson, Å., Vapalahti, O., Emmerich, P., Lundkvist, Å., Schmidt-Chanasit, J., et al. (2017). Compensating for cross-reactions using avidity and computation in a suspension multiplex immunoassay for serotyping of Zika versus other flavivirus infections. *Medical Microbiology and Immunology*, *206*(5), 383–401.

Sapparapu, G., Fernandez, E., Kose, N., Bin, C., Fox, J., Bombardi, R., et al. (2016). Neutralizing human antibodies prevent Zika virus replication and fetal disease in mice. *Nature*, *540*(7633), 443–447.

Saron, W., Rathore, A., Ting, L., Ooi, E., Low, J., Abraham, S., et al. (2018). Flavivirus serocomplex cross-reactive immunity is protective by activating heterologous memory CD4 T cells. *Science Advances*, *4*(7). eaar4297.

Singh, R., Dhama, K., Karthik, K., Tiwari, R., Khandia, R., Munjal, A., et al. (2018). Advances in diagnosis, surveillance, and monitoring of Zika virus: An update. *Frontiers in Microbiology*, *8*, 2677.

Smithburn, K. (1954). Neutralizing antibodies against arthropod-borne viruses in the sera of long-time residents of Malaya and Borneo. *American Journal of Hygiene*, *59*(2), 157–163.

Song, G., Rho, H., Pan, J., Ramos, P., Yoon, K., Medina, F., et al. (2018). Multiplexed biomarker panels discriminate Zika and dengue virus infection in humans. *Molecular & Cellular Proteomics*, *17*(2), 349–356.

Sridhar, S., Luedtke, A., Langevin, E., Zhu, M., Bonaparte, M., Machabert, T., et al. (2018). Effect of dengue serostatus on dengue vaccine safety and efficacy. *The New England Journal of Medicine*, *379*(4), 327–340.

Steinhagen, K., Probst, C., Radzimski, C., Schmidt-Chanasit, J., Emmerich, P., van Esbroeck, M., et al. (2016). Serodiagnosis of Zika virus (ZIKV) infections by a novel NS1-based ELISA devoid of cross-reactivity with dengue virus antibodies: A multicohort study of assay performance, 2015 to 2016. *Euro Surveillance*, *21*(50). pii:30426.

Stettler, K., Beltramello, M., Espinosa, D., Graham, V., Cassotta, A., Bianchi, S., et al. (2016). Specificity, cross-reactivity, and function of antibodies elicited by Zika virus infection. *Science*, *353*(6301), 823–826.

Tappe, D., Rissland, J., Gabriel, M., Emmerich, P., Gunther, S., Held, G., et al. (2014). First case of laboratory-confirmed Zika virus infection imported into Europe, November 2013. *Euro Surveillance*, *19*(4). pii:20685.

Taylor, C., Mackay, I., McMahon, J., Wheatley, S., Moore, P., Finger, M., et al. (2018). Detection of specific ZIKV IgM in travelers using a multiplexed Flavivirus microsphere immunoassay. *Viruses*, *10*(5), E253.

Theel, E., & Hata, D. (2018). Diagnostic testing for Zika virus: A postoutbreak update. *Journal of Clinical Microbiology*, *56*(4). pii:e01972-17.

Tsai, W., Youn, H., Brites, C., Tsai, J., Tyson, J., Pedroso, C., et al. (2017). Distinguishing secondary dengue virus infection from Zika virus infection with previous dengue by a combination of 3 simple serological tests. *Clinical Infectious Diseases*, *65*(11), 1829–1836.

Tyson, J., Tsai, W., Tsai, J., Brites, C., Mässgård, L., Ha Youn, H., et al. (2019). Combination of nonstructural protein 1-based enzyme-linked immunosorbent assays can detect and distinguish various Dengue Virus and Zika Virus infections. *Journal of Clinical Microbiology*, *57*(2). pii: e01464-18.

Tyson, J., Tsai, W., Tsai, J., Mässgård, L., Stramer, S., Lehrer, A., et al. (2019). A high-throughput and multiplex microsphere immunoassay based on non-structural protein 1 can discriminate three flavivirus infections. *PLoS Neglected Tropical Diseases*, *13*(8), e0007649.

Van Esbroeck, M., Meersman, K., Michiels, J., Ariën, K., & Van den Bossche, D. (2016). Letter to the editor: Specificity of Zika virus ELISA: Interference with malaria. *Euro Surveillance*, *21*(21). https://doi.org/10.2807/1560-7917.ES.2016.21.21.30237.

Waggoner, J. J., & Pinsky, B. A. (2016). Zika virus: Diagnostics for an emerging pandemic threat. *Journal of Clinical Microbiology*, *54*(4), 860–867.

Welch, R., Chang, G., & Litwin, C. (2014). Comparison of a commercial dengue IgM capture ELISA with dengue antigen focus reduction microneutralization test and the centers for disease control dengue IgM capture-ELISA. *Journal of Virological Methods*, *195*, 247–249.

Zhang, B., Pinsky, B., Ananta, J., Zhao, S., Arulkumar, S., Wan, H., et al. (2017). Diagnosis of Zika virus infection on a nanotechnology platform. *Nature Medicine*, *23*(5), 548–550.

Section F

Control, vaccines, and treatments

Chapter 30

Aedes aegypti and the use of natural molecules for its control: Implications in the decrease of Zika disease

Stelia Mendez-Sanchez[a], Duverney Chaverra-Rodriguez[b], and Jonny Duque[c]

[a]*Group for Research in Biochemistry and Microbiology (Grupo de Investigación en Bioquímica y Microbiología-GIBIM). School of Chemistry, Universidad Industrial de Santander, Bucaramanga, Colombia,* [b]*Department of Cell and Developmental Biology, University of California, San Diego, La Jolla, CA, United States,* [c]*Center for Research on Tropical Diseases (Centro de Investigaciones en Enfermedades Tropicales-CINTROP), Faculty of Health, School of Medicine, Department of Basic Sciences, Universidad Industrial de Santander, Bucaramanga, Colombia*

Abbreviations

ChC	chemical compounds
ChS	chemical synthesis
EOs	essential oils
ExP	extract of plants
IPM	integrated pest management
IRAC	Insecticide Resistance Action Committee
ISy	industrial synthesis
MC	mosquito control
MCP	mosquito control programs
NaI	natural insecticides
NaM	natural molecules
NaP	natural pesticides
SyI	synthetic insecticides
SyM	synthetic molecules
SyP	synthetic pesticides
VCP	vector control programs
VEx	vegetal extracts
ZIKV	Zika virus

Introduction

Aedes aegypti and *Ae. albopictus* are mosquitoes that can be vectors of Zika virus (ZIKV) in urban and suburban neotropical regions in the world (Plourde & Bloch, 2016). Despite the countless scientific papers suggesting multiple strategies to control these mosquitoes, there is a gap between the theory and the application of such strategies in the field. Additionally a unique prevention method against viral diseases such as the one caused by ZIKV or other arbovirus has not been developed yet. Even in a hypothetical situation that a useful vaccine existed, the vector control programs (VCP) must be yet mandatory because the vaccination coverage does not always reach total efficacy and it is always possible that the viral etiological agent mutates generating resistance to the vaccine (Achee et al., 2015; Barrett, 2018; Roiz et al., 2018).

Despite the effects on human health caused by the arboviruses transmitted by *Ae. aegypti*, today we could say, sadly, that the efforts to reduce the mosquito populations have been null in all the tropical areas where the mosquito is distributed. As if this was not enough, global projections of the geographical distribution of this mosquito and the vector *Ae. albopictus* for the year 2050 are frightening. Both species will have broader dispersion areas and will spread arboviral diseases in places where humans are susceptible (Kraemer et al., 2019). According to these predictions the mosquito control programs (MCP) should have considerable changes to avoid the transmission of diseases caused by arboviruses such as ZIKV.

Zika Virus Impact, Diagnosis, Control, and Models. https://doi.org/10.1016/B978-0-12-820267-8.00030-3

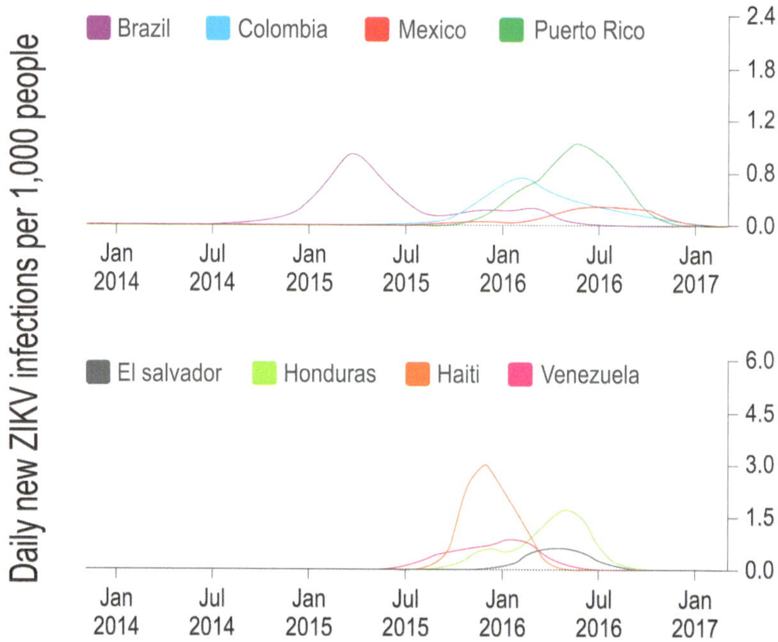

FIG. 1 Behavior of ZIKV infections (per 1000 people) in different countries of South America. From 2014 to 2017 the epidemiological outbreak of ZKV was present in the countries of the Caribbean, Central America, and South America. According to these data, Brazil was the first place where the ZKV transmission started. *(Data taken from Zhang, Q., Sun, K., Chinazzi, M., Piontti, A. P. Y., Dean, N. E., Rojas, Di. P., … Vespignani, A. (2017). Spread of Zika virus in the Americas.* Proceedings of the National Academy of Sciences of the United States of America, 114*(22), E4334–E4343. https://doi.org/10.1073/pnas.1620161114.)*

Some of the available current control measures that can be applied without restrictions to mosquito control are environmental sanitation (elimination of breeding sites), chemical control (use of classical insecticides), biological control (biological insecticides), social control and environmental education (involves the affected population), vector monitoring (studies of mosquito population growth), and others. From these, chemical insecticides (chemical control) are the most popular and include many options against the mosquitoes *Ae. aegypti* and *Ae. albopictus*. The main products are organophosphates (malathion and temephos), pyrethroids (lambda-cyhalothrin, deltamethrin, and cyfluthrin), and growth regulators (pyriproxyfen and fenoxycarb). Pyrethroids are used for the control of the adults, while growth regulators and temephos are used in immature stages. Interestingly, despite these alternatives, chemical control has not yet been able to stop the increase of cases of viral diseases, nor the dispersion of the vectors to new areas, but it has generated mosquito populations resistant to the molecules used for its control and has increased pollution in the environment (Bona et al., 2016; Moyes et al., 2017; Rivero, Vézilier, Will, Read, & Gandon, 2010).

Unfortunately, even the mentioned control methods have not been able to stop the beginning of a disease epidemics, as the one observed during 2014–17 for ZIKV. Fig. 1 shows the development of this epidemic in Brazil and how it rapidly spread in an event inevitable for neighboring countries. Expansion of the cases could be associated with the excellent adaptation of the vector to tropical areas, with the existing faults in mosquito control, and with the poor participation of people from the local communities in strategies to protect themselves (Duque, Navarro-Silva, & Trejos, 2009). The World Health Organization (WHO) highlights the necessity of developing novel control methods that may contribute to the reduction of the mosquito populations and disease burden. Proposed innovative methods include the use of genetically modified mosquitoes, producing new traps or baits, and finding new insecticide and repellent molecules that are safe and environmentally friendly (WHO, 2017).

The major goal of this chapter is to present basic information about the realistic uses of natural molecules as a source of new repellents and insecticides for mosquito control. We will present the definition of natural molecules (NaM) and their characteristics, commercial availability, examples of molecules with potential to be used in the control of *Ae. aegypti* or *Ae. albopictus*, search for new molecules, and how this could have an impact on the reduction of transmission for ZIKV.

What should be considered as a natural molecule with pesticide activity?

Natural molecules with pesticide activity (e.g., insecticide or repellent) could be any substance or compound synthesized by the metabolism of any organism but are mainly extracted from plants and microorganisms. This definition does not correspond with synthetic molecules manufactured by humans, even if these molecules are structural models copied from chemical compounds present in living organisms. Most importantly, to be considered NaM, they must be biosynthesized by one organism, either as a way to protect against pests or simply because it is a subproduct of their metabolism. An important feature is that these molecules are categorized as "green biopesticides" because they come from a biological source and their production processes are eco-friendly (Benelli & Pavela, 2018).

Good examples are NaM from plant or microbial source. Plants synthesize molecules called phytotoxins, which protect them from predatory or parasitizing species. On the other hand, in microorganisms such as bacteria, the production of toxins is not necessarily due to a defense system against competitors or predators, but basically, it is produced as a metabolic subproduct. However, in both cases, the biosynthesis of molecules can result in the production of powerful toxins. The technical and regulatory processes required to start producing and using those compounds in pest control is not simple. Before extracting the molecules from their source or starting a synthesis process, these molecules must have proven tests for chemical characterization, biological activity, mechanisms of action, and biosafety profiles in humans and other biological models. A critical aspect after acquiring the maximum possible information about the molecules is obtaining a method to synthesize them on a big scale (Clardy & Walsh, 2004; Duke et al., 2010; Khater et al., 2019).

Insecticidal and repellents are toxins, no matter its origin

One extremely important aspect in the control of insects with substances of natural origin is that these molecules cause mortality or repellence because the ingredients that make them are toxic. Remarkably insecticides or repellents, whether natural or produced by chemical synthesis, have similar effects that of a poison, toxungen, or venom. It means that those natural bioactive substances are composed of one or many toxic molecules causing dose-dependent physiological damage in any organism that makes contact with them (Nelsen et al., 2014). In this sense, once it is known that there are possible dangerous effects on nontarget organisms with the use of natural pesticides (NaP), the recommendation is to run laboratory studies with a minimum of experiments that would help to predict the toxicity levels of the substance in question for different organisms. In the same way, this observation should be applied to substances that have no insecticidal action like repellents, antifeedant, or other deterrent effects. Insects avoid having contact with any substance that their sensory system alerts them to be likely dangerous to them, basically because they recognize the substance as something similar to a poison (Navarro-Silva, Marques, & Duque, 2009).

In any case, if a product made by chemical synthesis (ChS) is compared against one produced with natural ingredients, not industrially synthesized, the main benefit observed is that NaM do not bioaccumulate in living organisms. Synthetic pesticides (SyP) such as malathion, methomyl, chlorpyrifos, pirimiphos-methyl, dimethoate, and cyfluthrin can accumulate in tissues such as adipose tissue or liver and could create metabolic problems as a consequence of an extensive exposure. However, it is important to clearly highlight that not all the molecules taken from biological sources have the absolute category of being totally safe and neither they are going to be useful for the control of insects. Some examples are arsenic and nicotine, which cause mortality in insects as insecticides, but do not comply with the safety standards to be considered as marketable insecticides. In exact terms, they are just poisons, as discussed previously (Mossa, Mohafrash, & Chandrasekaran, 2018; World Health Organization (WHO), 2009). Accordingly, we recommend that experts do the design and application of new insecticides.

Multieffect of natural molecules with potential use against *Ae. aegypti*

One distinctive feature of this kind of molecules that should stimulate their use in vector control programs is that they have a lot of utilities to protect against mosquitoes. For example, essential oils (EOs) and vegetal extracts (VEx) have properties that go beyond the classic insecticidal and repellent effect. In addition, a single plant can produce many molecules with different effects such as fumigants, antifeedants, inhibitors of oviposition, attractants, or antiviral agents that could be incorporated into a single formulation (Fig. 2).

To illustrate how plants offer substances with multiple bioactive effects, let us bring the case of two plants that should be considered as a source of molecules with potential use as insecticides or repellent effects of mosquitoes. *Citrus sinensis* (Rutaceae) and *Lippia origanoides* (Verbenaceae) produce EOs that have been proven to be larvicidal, pupicidal, and adulticidal and also have effects on oviposition deterrence in *Ae. aegypti*. It is clear that the multiple effects on the biology of this mosquito are due to the richness of chemical compounds (CHC) that are present in both of these two plants (Vera et al., 2014).

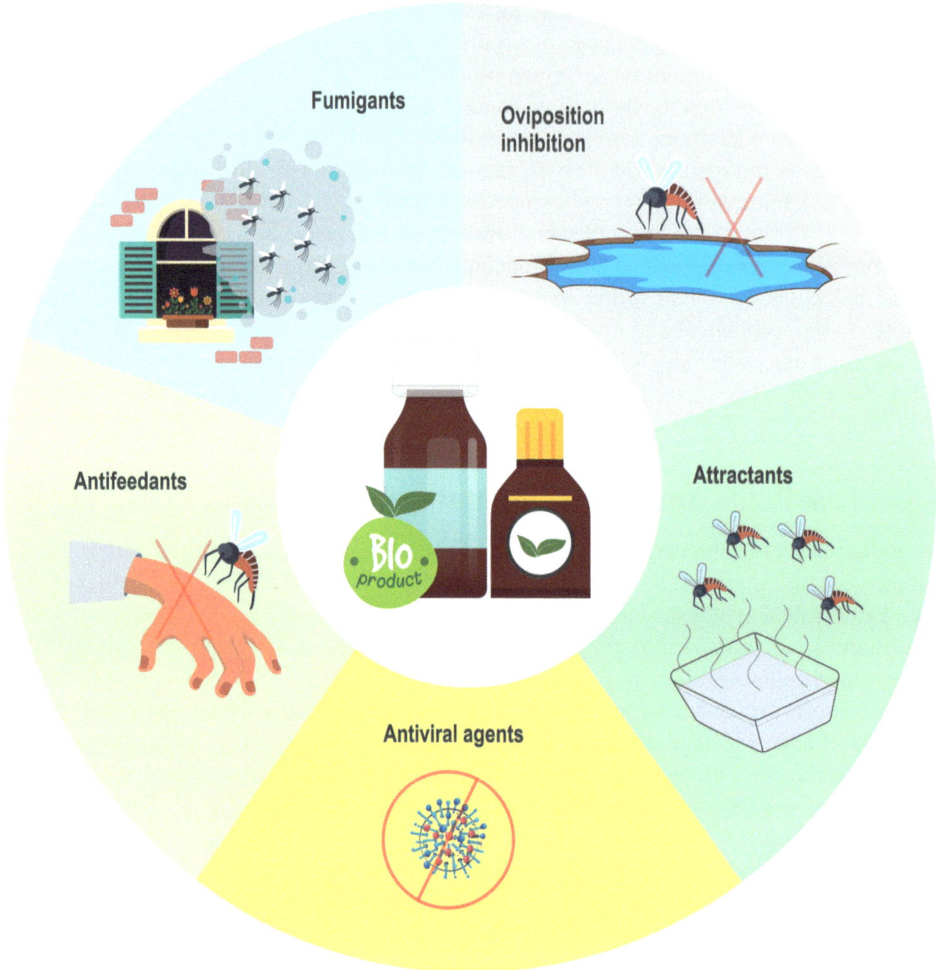

FIG. 2 Multieffect of natural molecules against *Ae. aegypti*. One of the most important aspects when using natural molecules in mosquito control is that they are multiuse. This means that a single source of natural molecules can offer diverse activities against mosquitoes. For example, some activities are deterrence (antifeedants, oviposition inhibition, or repellence), lure actions (attractants), and virucidal molecules (antiviral agents).

Market availability of pesticides made with NaM

Insecticides and commercial repellents made exclusively with EOs or VEx are uncommon. Some biopesticides registered as insecticides are *Bacillus thuringiensis* var. israelensis (Bti), *B. sphaericus* (Bs), *Saccharopolyspora spinosa* (spinosad), *Azadirachta indica (*neem*), and Cymbopogon* spp. (*cymbopogon*) (Koul, Walia, & Dhaliwal, 2008; Mazid, 2011; Pérez et al., 2007).

The active ingredients obtained from natural sources are not used in the industrial production of insecticides or repellents massively, probably because of the low concentration of NaM in their source. By way of speculation, in practical terms, it is much easier for any pharmaceutical industry to standardize the ChS of a single molecule and produce it in large quantities than extract it from its primary source. Another problematic aspect of the industrialization of pesticides with natural ingredients is that their active compounds are generally volatile. Because of this, their biocidal or repellent effect is quickly lost. Fortunately, essential oils and bioactive molecules are currently being stabilized in nanoparticles, increasing their lifespan (Benelli, 2016; Misni, Nor, & Ahmad, 2017; Tavares et al., 2018).

On the other hand, gathering data regarding safety issues and obtaining a chemical characterization are more complicated for substances that have multiple compounds. These requirements are easier and cheaper to achieve for one active ingredient than for many active compounds. It is supremely important to remember that national programs of vector-borne diseases in many countries are allowed to use only "safe substances." However, it is well known that the typical insecticides manufactured by artificial ChS are not so safe when compared with the ones obtained from natural sources as plants (Mossa et al., 2018; World Health Organization (WHO), 2009).

Challenges of producing marketable products from NaM

The selection of a molecule or substance with insecticidal or repellent effect depends on many criteria. Pesticide substances must be characterized according to the dosage or concentration applied in optimal conditions. It is essential to clarify that not all insecticides and repellents reported in the scientific literature will become a product to be used in real-life conditions. In fact, for one molecule to become an insecticide or repellent, it must pass a series of toxicological tests before being tested in field conditions. Once they have passed all tests and are on the track for commercialization, it is necessary to know the production costs. At this point, many molecules are dropped from the process because the costs required for production are too big, so it is not profitable to manufacture by pharmaceutical companies (Clardy & Walsh, 2004; Goodyer et al., 2010; Koul et al., 2008).

The development of insecticides worldwide, especially in countries located in tropical areas, is a prolonged process that has different stages (design new molecules or synthesis of them—bioassays—mechanism of action—formulation—toxicological tests—field tests). Only when it goes through all the rigor tests can it be commercialized, of course, if it is also economically viable (Insecticide Resistance Action Committee (IRAC)). As expected, and due to the large costs involved in the discovery of an insecticide, this task is almost exclusive to large companies such as BASF, Monsanto, Sumitomo, DuPont Agro, Syngenta, and FMC.

Search of new compounds with insecticidal or repellent activity

The use of NaM is not new; undoubtedly at the beginning of history, humans had to acquire this knowledge for its benefit in medicine and agriculture (Azwanida, 2015). Searching for new molecules is not easy, and it is similar to looking for a needle in a haystack. A common starting point for discovering a new bioactive compound is producing basic knowledge about where it is more likely to find a source of bioactive molecules. Fortunately, today there is much information in scientific publications about different organisms and their chemical characterization. That means it is possible to identify molecules with the potential to be used in mosquito control (MoC) and to obtain the bioactive compound to manufacture pesticides. The more important sources for these compounds include plants, bacteria, and fungi (Dias & Moraes, 2014; Kishore, Mishra, Tiwari, & Tripathi, 2011; Pavela, 2015; Pérez et al., 2007).

Traditionally, once the biological model to get molecules is selected, for example, plants, the next step is to evaluate the expected biocidal effect of VEx or EOs through trial-and-error experiments. Current advances in the science of NaM have produced methods that contribute to identifying compounds (with insecticide or repellent activity) against one specific pharmacological target. Also, these methods allow creating mixtures of natural molecules with SyM and creating substances with multiple biological targets against insects. Fig. 3 shows several steps for designing insecticides from NaM. We use this approach in our current research focused on the mitochondria as a pharmacological target for the design of insecticides. Without a doubt, this will be one of the trends in the industrial manufacturing of new products to eliminate

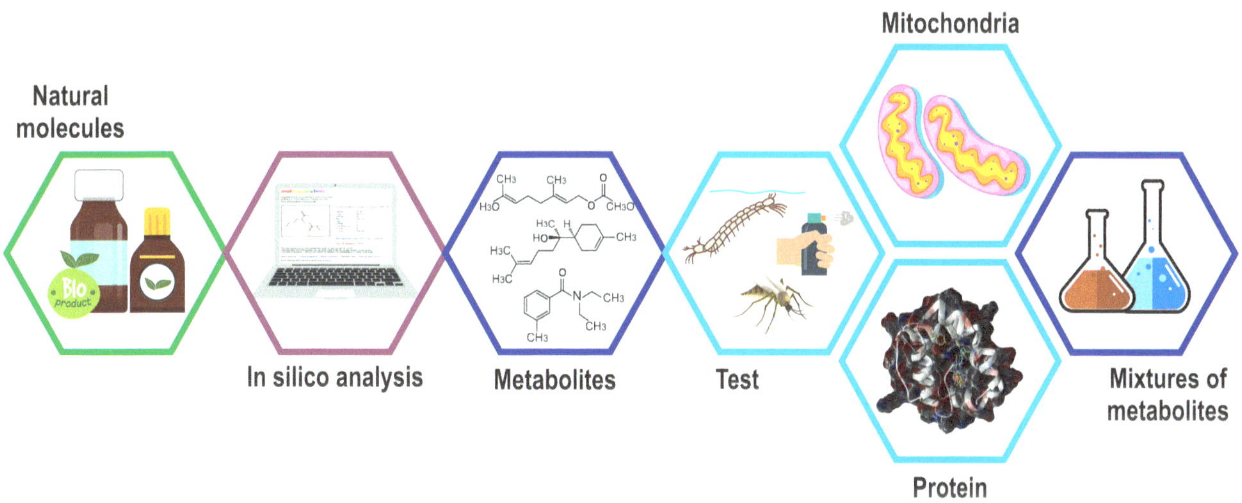

FIG. 3 Design of insecticides from natural molecules. The picture describes the process of how to design insecticides from NaM. From left to right, with the list of the components of oils and extracts, the researcher proceeds to perform in silico analysis to determine which are the potential metabolites that must be tested in vivo. With the metabolites that present insecticidal activity, it studied the mechanism of action on the specific target (in this case, the mitochondria), and then it established mixtures that can enhance the insecticidal effect.

mosquitoes and protect ourselves against their bites (Borrero-Landazabal, Duque, & Mendez-Sanchez, 2020; Portilla-Pulido, Castillo-Morales, Barón-Rodríguez, Duque, & Mendez-Sanchez, 2019).

The implications of using natural molecules on the reduction of the incidence of Zika disease

Systematic and experimental studies that explicitly evaluate the decreasing effect of a particular disease as Zika, using only NaM, do not exist, as it is also not precisely known how insecticides (pyrethroids and organophosphorus) impact on the reduction of vector-borne diseases. Today, it is well understood that decreasing the incidence of diseases transmitted by vectors depends on multiple factors that must be working synergistically. In this way, NaM are just one of the multiple available options to contribute to the control of the mosquito vectors, either as insecticide or as protection against bites (Bowman, Donegan, & McCall, 2016; Wilson et al., 2015).

Regrettably, around the world, official mosquito control programs do not use essential oils, vegetal extracts, or another source of molecules with biocidal action. Basically, this happens because there is still little knowledge describing the advantages offered by those molecules, which, in turn, creates distrust about their use on a large scale in control programs. Most people, even few scientists, believe that these products are classified exclusively as insecticides or repellents, ignoring other benefits that are useful for vector control. NaM could be articulated into different applications of the integrated pest management (IPM) programs, especially if we take advantage of the multiple protective options against mosquitoes that come from EO and EV, either as an insecticide, repellent, deterrent, or dissuasive effect.

To make a specific recommendation of plants or their essential oils to use against *Ae. aegypti* like insecticide or repellent, it is difficult because the chemical composition of plants can vary according to many factors, such as the locality and the environmental conditions to which the plant was exposed. Nevertheless, we would like to present a list of plants that have been shown promisourious (with compounds having lethal concentration 50 (CL50) equal or lower than 20 parts per million ($CL_{50} < 20$ ppm)) as a source of compounds for the development of new insecticides and repellents: *Callitris glaucophylla* ($CL_{50} = 0.7$ ppm), *Cinnamomum microphyllum* ($CL_{50} = 7$ ppm), *Auxemma glazioviana* ($CL_{50} = 3$ ppm), *Cinnamomum impressicostatum* ($CL_{50} = 11$ ppm), *Tagetes patula* ($CL_{50} = 13$ ppm), *Cymbopogon flexuosus* ($CL_{50} = 17.1$ ppm), *Hyptis martiusii* ($CL_{50} = 18.2$ ppm), *Lippia sidoides* ($CL_{50} = 20$ ppm), *Citrus sinensis* ($CL_{50} = 20.6$ ppm), *Allium monanthum* ($CL_{50} = 23$ ppm), and *Coriandrum sativum* ($CL_{50} = 23$ ppm) (Dias & Moraes, 2014; Diaz, 2016; Vera et al., 2014).

An approximation of how the use of manufactured products from natural molecules could impact on the reduction of the disease such as Zika can be estimated empirically by simulation of mathematical models (Bañuelos, Martinez, Mitchell, & Prieto-Langarica, 2019; Duque et al., 2009; Zhang et al., 2017). In these models, variables such as natural insecticides (NaI), synthetic insecticides (SyI), use of repellents, and community participation in the elimination of mosquito breeding sites were introduced into an *Ae. aegypti* control system to determine their effects on disease transmission. According to the simulations, integrated control with different strategies for the elimination of mosquitoes in both the larval and adult states must be used in parallel to reduce the cases of viral diseases like Zika or dengue (Fig. 4).

Although today there is much information about new natural molecules and also synthetic molecules designed from natural compounds, it is necessary to stimulate the search for new molecules and create new formulations to make new

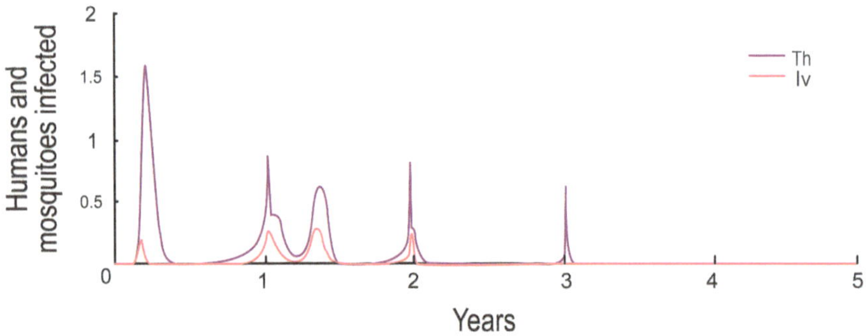

FIG. 4 A mathematical model simulation to estimate how the use of natural molecules in an integrated vector control program against *Ae. aegypti* impacts the reduction of diseases such as Zika. ***Th*** are humans infectious and ***Iv*** are vector infectious. Different scenarios to mosquito control on *Ae. aegypti*. The first year without any control strategy, the second year with control on both larvae and adults with low effectiveness on infectious vectors, the third year with a control with high efficiency on infectious vectors (ideal control system), fourth and fifth years of perfect control.

pesticides. These new insecticides or repellents should be designed to be applied by people who suffer from the presence of mosquitoes in their homes. In this way, they would be the first control agents. Undoubtedly, involving the community in control programs that use natural molecules will positively impact the reduction of mosquito-borne diseases.

Probably when an epidemic event of a disease like Zika arises, it cannot be controlled because people have a stronger dependence on traditional SyI supported on expectations of effectiveness. Regarding this, and having clear that mosquito control must be carried out from different combat fronts and that local communities can play an essential role in this process, it is necessary that wherever a person is living with *Ae. aegypti*, he or she has the opportunity to participate actively in its elimination using natural insecticides or natural repellents. A good starting point to achieve this is generating understandable information for academics and laypeople and creating the availability of these products in chain markets in the same way as products of synthetic origin.

Finally, natural substances must be incorporated into pest control systems anywhere in the world without fear, not only because their compounds are friendly to the environment but also because they are proven effective as insecticides and repellents (Dias & Moraes, 2014; Koul et al., 2008; Pavela, 2008). For millennia, humans used products derived from other organisms for their benefit in medicine and also in pest control. However, our modern society relies totally on synthetic products that have failed to control diseases transmitted by mosquitoes. Also the use of living organism to extract their molecules can help in mosquito control systems when traditional pesticides have lost their insecticide effect due to resistance. In this way, essential oils, plant extracts, or their compounds can be mixed with conventional insecticides and get beneficial effects on mosquito control (Norris, Gross, Bartholomay, & Coats, 2019).

Mini-dictionary of terms

Active ingredient. A molecule, molecules, or mixture of them that provides the biological action in a formulation.

Antifeedant. A molecule, molecules, or mixture of them that trigger in an organism the desire do not to feed.

Attractant. A molecule, molecules, or mixture of them that trigger in an organism the desire to go to the source that issues it.

Antiviral. A molecule, molecules, or a mixture of them that can kill the virus and also can suppress its replication capacity and reproduce.

Deterrent. A molecule, molecules, or a mixture of them that have a triggering effect in an organism that prevents or inhibits them from entering a place where is the deterrent.

Green biopesticides (bioinsecticides). A molecule, molecules, or a mixture of them with a biocidal activity that has been developed exclusively from natural ingredients.

Insecticidal activity. A molecule, molecules, or a mixture of them that trigger mortality in insects.

Fumigant. It is referred to as a pesticide that is intentionally dispersed or applied in the environment, producing fumes.

Mosquito control. Systematic activity to eliminate or control mosquitoes.

Natural molecules. Is a substance or compound synthesized by metabolic way from whatever kind of organism, principally plants, and microorganisms.

Natural pesticides. Is a product made of natural molecules with pesticidal properties.

Vector control. Systematic activity to eliminate or control any insect that transmits diseases to man or other animals.

Repellent activity. It is a behavioral response activated only in the presence of a chemical compound, in which an individual (any kind) avoids or moves away from the place where he feels his presence.

Synthetic molecules. It refers to a molecule that has been produced from other precursor elements by various chemical reactions.

Pesticide. It is a chemical that can negatively affect individuals (e.g., animals, fungi, and plants) that have grown populationally to a level harmful to man or other animals.

Poison. It is a molecule or different molecules that cause physiological damage and can be quantified in a dose-dependent way.

Phytotoxins. It is a poison from plants that can be a toxin or venom; this specific classification depends if it included a mechanism of delivery and injury.

Toxins. It is a poison that can be a toxungen or venom; this specific classification depends if it included a mechanism of delivery and injury.

Toxungen. It is a substance with poison properties that can be supplied away actively from one organism to another. In this case, without a mechanism of delivery and injury.

Venom. It is a substance with poison properties that can be supplied away passively or actively from one organism to another. In this case, there is a mechanism of delivery and injury.

Key facts of the use of natural molecules for mosquito control

- It is possible to use NaM for the design of fumigant, oviposition inhibitors, attractants, antifeedants, and antiviral agents.
- Essential oils and extracts contain several compounds, so it is important to know the biological activities and the possible effects on humans and other animals of each of these compounds.
- There are many methodologies to determine which natural compounds or an NaM or ExP has insecticidal activity and then design mixtures that can enhance the insecticidal effect.
- An effective control of the mosquito population using natural insecticides and natural repellents can only be carried out with active participation from the community.

Summary points

- Even in a hypothetical situation that a useful vaccine existed, the vector control programs (VCP) must be yet mandatory because the vaccination coverage does not always get a total efficacy and there is the possibility that the viral etiological agent mutates generating resistance to the vaccine.
- NaM with a pesticide activity (e.g., insecticide or repellent) could be any substance or compound synthesized by metabolic way from any living organism.
- To be used as an insecticide or repellent, NaM must have studies aiming to describe their chemical characterization, biological activity, mechanisms of action, and biosafety profiles in humans and other biological models.
- Although there is a lot of information about new natural molecules and also synthetic molecules designed from natural compounds, it is necessary to stimulate the search for new molecules and create new formulations to make new pesticides.
- Involving the community in control programs that use natural molecules will positively impact the reduction of mosquito-borne diseases.

Acknowledgments

This work was carried out thanks to financing provided by the Ministerio de Ciencia, Tecnología e Innovación, Ministerio de Educación Nacional, Ministerio de Industria, Comercio y Turismo e ICETEX; Grant Ecosistema Científico—Colombia Científica; Fondo Francisco José de Caldas, Contrato RC-FP44842-212-2018; and Programa Bio-Reto XXI-15:50. Also, thanks to Maria Fernanda Vidal for her support designing the images.

References

Achee, N. L., Gould, F., Perkins, T. A., Reiner, R. C., Morrison, A. C., Ritchie, S. A., ... Scott, T. W. (2015). A critical assessment of vector control for dengue prevention. *PLoS Neglected Tropical Diseases*, *9*(5). https://doi.org/10.1371/journal.pntd.0003655, e0003655.

Azwanida, N. (2015). A review on the extraction methods use in medicinal plants, principle, strength and limitation. *Medicinal & Aromatic Plants*, *4*(3), 1–6. https://doi.org/10.4172/2167-0412.1000196.

Bañuelos, S., Martinez, M. V., Mitchell, C., & Prieto-Langarica, A. (2019). Using mathematical modelling to investigate the effect of the sexual behaviour of asymptomatic individuals and vector control measures on Zika. *Letters in Biomathematics*, *6*(1), 1–19. https://doi.org/10.1080/23737867.2019.1624631.

Barrett, A. D. T. (2018). Current status of Zika vaccine development: Zika vaccines advance into clinical evaluation. *Npj Vaccines*, *3*(1), 1–4. https://doi.org/10.1038/s41541-018-0061-9.

Benelli, G. (2016). Plant-mediated biosynthesis of nanoparticles as an emerging tool against mosquitoes of medical and veterinary importance: A review. *Parasitology Research*, *115*(1), 23–34. https://doi.org/10.1007/s00436-015-4800-9.

Benelli, G., & Pavela, R. (2018). Beyond mosquitoes—Essential oil toxicity and repellency against bloodsucking insects. *Industrial Crops and Products*, *117*(February), 382–392. https://doi.org/10.1016/j.indcrop.2018.02.072.

Bona, A. C. D., Chitolina, R. F., Fermino, M. L., De Castro Poncio, L., Weiss, A., Lima, J. B. P., ... Maori, E. (2016). Larval application of sodium channel homologous dsRNA restores pyrethroid insecticide susceptibility in a resistant adult mosquito population. *Parasites and Vectors*, *9*(1), 1–14. https://doi.org/10.1186/s13071-016-1634-y.

Borrero-Landazabal, M. A., Duque, J. E., & Mendez-Sanchez, S. C. (2020). Model to design insecticides against Aedes aegypti using in silico and in vivo analysis of different pharmacological targets. *Comparative Biochemistry and Physiology – Part C*, *229*(October 2019), 108664. https://doi.org/10.1016/j.cbpc.2019.108664.

Bowman, L. R., Donegan, S., & McCall, P. J. (2016). Is dengue vector control deficient in effectiveness or evidence?: Systematic review and meta-analysis. *PLoS Neglected Tropical Diseases*, *10*(3), 1–24. https://doi.org/10.1371/journal.pntd.0004551.

Clardy, J., & Walsh, C. (2004). Lessons from natural molecules. *Nature*, *432*(7019), 829–837. https://doi.org/10.1038/nature03194.

Dias, C. N., & Moraes, D. F. C. (2014). Essential oils and their compounds as *Aedes aegypti* L. (Diptera: Culicidae) larvicides: Review. *Parasitology Research*, *113*(2), 565–592. https://doi.org/10.1007/s00436-013-3687-6.

Diaz, J. H. (2016). Chemical and plant-based insect repellents: Efficacy, safety, and toxicity. *Wilderness and Environmental Medicine*, *27*(1), 153–163. https://doi.org/10.1016/j.wem.2015.11.007.

Duke, S. O., Cantrell, C. L., Meepagala, K. M., Wedge, D. E., Tabanca, N., & Schrader, K. K. (2010). Natural toxins for use in pest management. *Toxins*, *2*(8), 1943–1962. https://doi.org/10.3390/toxins2081943.

Duque, J. E. L., Navarro-Silva, M., & Trejos, D. Y. (2009). Simulando manejo de *Aedes aegypti* (Diptera: Culicidae) y sus efectos en una epidemia de dengue. *Revista Colombiana de Entomología*, *35*(1), 66–72.

Goodyer, L. I., Croft, A. M., Frances, S. P., Hill, N., Moore, S. J., Onyango, S. P., & Debboun, M. (2010). Expert review of the evidence base for arthropod bite avoidance. *Journal of Travel Medicine*, *17*(3), 182–192. https://doi.org/10.1111/j.1708-8305.2010.00402.x.

Khater, H. F., Selim, A. M., Abouelella, G. A., Abouelella, N. A., Murugan, K., Vaz, N. P., & Govindarajan, M. (2019). Commercial mosquito repellents and their safety concerns. In *Malaria* (p. 4133). IntechOpen.

Kishore, N., Mishra, B., Tiwari, V., & Tripathi, V. (2011). A review on natural products with mosquitocidal potentials. In *Vol. 661. Opportunity, challenge and scope of natural products in medicinal chemistry* (pp. 335–365). Kerala, India: Research Signpost. Retrieved from http://www.trnres.com/ebook/uploads/tiwari/T_1302158830Tiwari-11.pdf.

Koul, O., Walia, S., & Dhaliwal, G. S. (2008). Essential oils as green pesticides: Potential and constraints. *Biopesticides International*, *4*(1), 63–84.

Kraemer, M. U. G., Reiner, R. C., Brady, O. J., Messina, J. P., Gilbert, M., Pigott, D. M., … Golding, N. (2019). Past and future spread of the arbovirus vectors *Aedes aegypti* and *Aedes albopictus*. *Nature Microbiology*, *4*(5), 854–863. https://doi.org/10.1038/s41564-019-0376-y.

Mazid, S. (2011). A review on the use of biopesticides in insect pest management. *International Journal of Science and Advanced Technology*, *1*(7), 169–178. https://doi.org/10.1007/s10886-005-4244-2.

Misni, N., Nor, Z. M., & Ahmad, R. (2017). Repellent effect of microencapsulated essential oil in lotion formulation against mosquito bites. *Journal of Vector Borne Diseases*, *54*(1), 44–53.

Mossa, A. T. H., Mohafrash, S. M. M., & Chandrasekaran, N. (2018). Safety of natural insecticides: Toxic effects on experimental animals. *BioMed Research International*, *2018*. https://doi.org/10.1155/2018/4308054.

Moyes, C. L., Vontas, J., Martins, A. J., Ng, L. C., Koou, S. Y., Dusfour, I., … Weetman, D. (2017). Contemporary status of insecticide resistance in the major *Aedes* vectors of arboviruses infecting humans. *PLoS Neglected Tropical Diseases*, *11*(7), 1–20. https://doi.org/10.1371/journal.pntd.0005625.

Navarro-Silva, M., Marques, F., & Duque, J. E. (2009). Review of semiochemicals that mediate the oviposition of mosquitoes: A possible sustainable tool for the control and monitoring of Culicidae. *Revista Brasileira de Entomologia*, *53*(1), 1–6. https://doi.org/10.1590/S0085-56262009000100002.

Nelsen, D. R., Nisani, Z., Cooper, A. M., Fox, G. A., Gren, E. C. K., Corbit, A. G., & Hayes, W. K. (2014). Poisons, toxungens, and venoms: Redefining and classifying toxic biological secretions and the organisms that employ them. *Biological Reviews*, *89*(2), 450–465. https://doi.org/10.1111/brv.12062.

Norris, E. J., Gross, A. D., Bartholomay, L. C., & Coats, J. R. (2019). Plant essential oils synergize various pyrethroid insecticides and antagonize malathion in *Aedes aegypti*. *Medical and Veterinary Entomology*, *33*(4), 453–466. https://doi.org/10.1111/mve.12380.

Pavela, R. (2008). Larvicidal effects of various Euro-Asiatic plants against *Culex quinquefasciatus* say larvae (Diptera: Culicidae). *Parasitology Research*, *102*(3), 555–559. https://doi.org/10.1007/s00436-007-0821-3.

Pavela, R. (2015). Essential oils for the development of eco-friendly mosquito larvicides: A review. *Industrial Crops and Products*, *76*, 174–187. https://doi.org/10.1016/j.indcrop.2015.06.050.

Pérez, C. M., Marina, C. F., Bond, J. G., Rojas, J. C., Valle, J., & Williams, T. (2007). Spinosad, a naturally derived insecticide, for control of *Aedes aegypti* (Diptera: Culicidae): Efficacy, persistence, and elicited oviposition response. *Journal of Medical Entomology*, *44*(4), 631–638. https://doi.org/10.1603/0022-2585(2007)44[631:sandif]2.0.co;2.

Plourde, A. R., & Bloch, E. M. (2016). A literature review of Zika virus. *Emerging Infectious Diseases*, *22*(7), 1185–1192. https://doi.org/10.3201/eid2207.151990.

Portilla-Pulido, J. S., Castillo-Morales, R. M., Barón-Rodríguez, M. A., Duque, J. E., & Mendez-Sanchez, S. C. (2019). Design of a repellent against *Aedes aegypti* (Diptera: Culicidae) using in silico simulations with AaegOBP1 protein. *Journal of Medical Entomology*, 1–14. https://doi.org/10.1093/jme/tjz171.

Rivero, A., Vézilier, J., Will, M., Read, A. F., & Gandon, S. (2010). Insecticide control of vector-borne diseases: When is insecticide resistance a problem? *PLoS Pathogens*, *6*(8), 1–9. https://doi.org/10.1371/journal.ppat.1001000.

Roiz, D., Wilson, A. L., Scott, T. W., Fonseca, D. M., Jourdain, F., Müller, P., … Corbel, V. (2018). Integrated *Aedes* management for the control of *Aedes*-borne diseases. *PLoS Neglected Tropical Diseases*, *12*(12), 1–21. https://doi.org/10.1371/journal.pntd.0006845.

Tavares, M., da Silva, M. R. M., de Oliveira de Siqueira, L. B., Rodrigues, R. A. S., Bodjolle-d'Almeira, L., dos Santos, E. P., & Ricci-Júnior, E. (2018). Trends in insect repellent formulations: A review. *International Journal of Pharmaceutics*, *539*(1–2), 190–209. https://doi.org/10.1016/j.ijpharm.2018.01.046.

Vera, S. S., Zambrano, D. F., Méndez-Sanchez, S. C., Rodríguez-Sanabria, F., Stashenko, E. E., & Duque, J. E. (2014). Essential oils with insecticidal activity against larvae of *Aedes aegypti* (Diptera: Culicidae). *Parasitology Research*, *113*(7), 2647–2654. https://doi.org/10.1007/s00436-014-3917-6.

Wilson, A. L., Boelaert, M., Kleinschmidt, I., Pinder, M., Scott, T. W., Tusting, L. S., & Lindsay, S. W. (2015). Evidence-based vector control? Improving the quality of vector control trials. *Trends in Parasitology*, *31*(8), 380–390. https://doi.org/10.1016/j.pt.2015.04.015.

World Health Organization (WHO). (2009). *The WHO recommended classification of pesticides by hazard and guidelines to classification.*

World Health Organization. (2017). *Guidelines for the registration of microbial, botanical and semiochemical pest control agents for plant protection and public health uses*. FAO. ISBN 978-92-5-130015-2.

Zhang, Q., Sun, K., Chinazzi, M., Piontti, A. P. Y., Dean, N. E., Rojas, D. P., … Vespignani, A. (2017). Spread of Zika virus in the Americas. *Proceedings of the National Academy of Sciences of the United States of America*, *114*(22), E4334–E4343. https://doi.org/10.1073/pnas.1620161114.

Chapter 31

Strategies of Zika virus control with larvicides and their toxic potential: A focus on pyriproxyfen

Patrícia e Silva Alves[a], Maria das Dores Alves de Oliveira[b], Teresinha De Jesus Aguiar Dos Santos Andrade[c], Nerilson Marques Lima[d], and Joaquim Soares da Costa Júnior[e]

[a]*Department of Chemistry, Chemistry Graduate Program, State University of Piauí, Teresina, PI, Brazil,* [b]*Chemistry Department, Federal University of Piauí, Teresina, PI, Brazil,* [c]*Nucleus of Applied Research to Sciences, Federal Institute of Maranhão, Timon, MA, Brazil,* [d]*Department of Chemistry—ICE, Federal University of Juiz de Fora, Juiz de Fora, Brazil,* [e]*Department of Chemistry, Federal Institute of Piauí, Teresina, PI, Brazil*

Abbreviations

CDC	Centers for Disease Control and Prevention
DENV	dengue virus
JEV	Japanese encephalitis virus
OMS	World Health Organization
PAHO	Pan American Health Organization
PPF	pyriproxyfen
UBV	ultra low volume
WNV	West Nile virus
YFV	yellow fever virus
ZIKV	Zika virus

Introduction

The Zika virus (ZIKV) is an arbovirus belonging to the family *Flaviviridae* transmitted by the bite of contaminated mosquitoes from species *Aedes aegypti*. Within the *Flavivirus* genus, ZIKV is transmitted by a species of mosquito phylogenetically similar to the Japanese encephalitis virus (JEV), West Nile virus (WNV), dengue virus (DENV), and yellow fever virus (YFV) (Weaver & Barrett, 2004).

The transmission of the virus (Fig. 1) occurs through arthropod vectors of the genus *Aedes* of multiple species (Al-Abdely & Zika, 2016). In addition, contamination can occur through vertical transmission in the uterus of infected mothers to the fetus (Brasil et al., 2016) or through other direct transmission routes such as sexually through predominantly male-to-female secretions (Mansuy et al., 2016; Musso, Roche, Robin, et al., 2015), breastfeeding (Dupont-Rouzeyrol, Biron, O'Connor, Huguon, & Descloux, 2016), blood transfusion (Aubry et al., 2015; Gallian et al., 2017), saliva (Musso, Roche, Nhan, et al., 2015) and urine (Musso & Gubler, 2016).

The ZIKV epidemic emerged in 2015 and came to generate public health chaos (Nunes et al., 2016). The most serious complication of Zika virus infection is the risk of microcephaly (Fig. 2) and other congenital diseases and brain abnormalities in fetuses of infected pregnant women (Agumadu & Ramphul, 2018). ZIKV can also cause neurological development abnormalities due to maternal infections (Kuno, Chang, & Tsuchiya, 1998).

Transmission of the virus in the mother's womb to the developing fetus has been proven after isolation of the virus from tissues of the placenta and fetus as well as amniotic fluid and cerebrospinal fluid from infected newborns (Mittal et al., 2017). The clinical diagnosis of the infection by the virus is complicated because its similarities with other arboviral fevers

Zika Virus Impact, Diagnosis, Control, and Models. https://doi.org/10.1016/B978-0-12-820267-8.00031-5

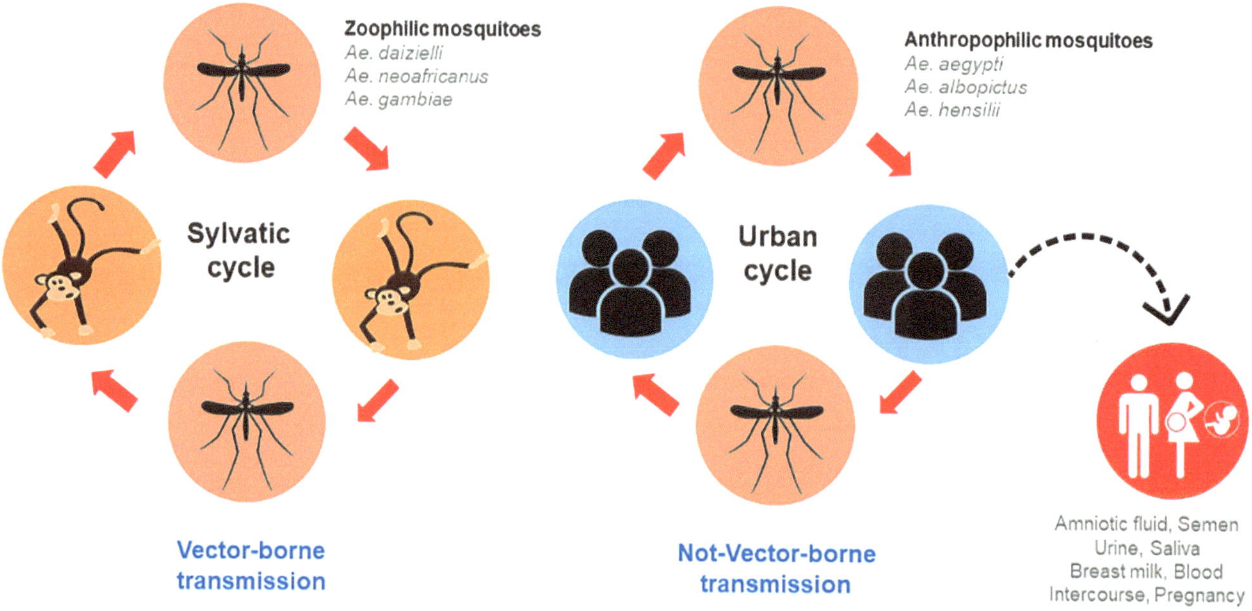

FIG. 1 Cycle of Zika virus vector and nonvector transmissions. It shows the Zika virus cycle, which includes a sylvatic cycle with viral circulation between nonhuman primates and zoophilic mosquitoes and an urban cycle with humans as reservoir hosts and amplification of anthropophilic mosquitoes as vectors. The nonvector transported mode involves direct human-to-human transmission through fluid contacts. Nonvector transmission may occur through blood transfusion and transplacental and sexual modes. Other means of virus transmission include breastfeeding, pregnancy, saliva, or urine. *(Adapted from Boyer, S., Calvez, E., Chouin-Carneiro, T., Diallo, D., & Failloux, A. B. (2018). An overview of mosquito vectors of Zika virus.* Microbes and Infection, 20, *646–660.)*

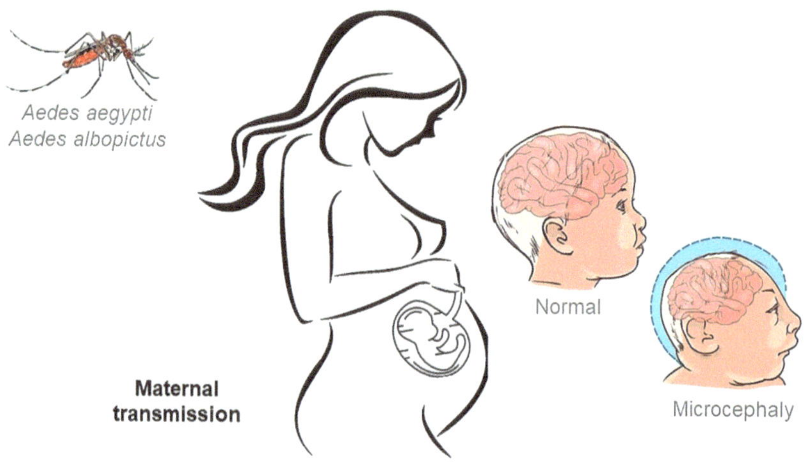

FIG. 2 Microcephaly associated with the Zika virus. When a pregnant woman is infected with Zika virus during pregnancy, maternal transmission of the virus to the developing fetus can result in microcephaly as a result of cell death in the fetus' brain. The primary transmission vectors include *Aedes aegypti* and *Aedes albopictus*. *(Adapted from Kuadkitkan, A., Wikan, N., Sornjai, W., & Smith, D. R. (2019). Zika virus and microcephaly in Southeast Asia: A cause for concern?* Journal of Infection and Public Health, 13, *11–15.)*

share insect vectors and the geographic area with dengue and chikungunya (Kelser, 2016). This is associated with a range of asymptomatic infections or self-limited febrile infections due to some fatal diseases such as hemorrhage, shock, meningitis, and encephalitis (Muller & Young, 2013).

One of the control strategies (Fig. 3) of the ZIKV vector is chemical control, which consists of the use of substances called insecticides or pesticides, which can act for both the larval phase of the vector and adult phase (Erlanger, Keiser, & Utzinger, 2008). The uses of these pesticides are focused on eliminating and preventing possible mosquito breeding sites such as water tanks, garbage dumps and tires, and places that can store standing water (Erlanger et al., 2008). However, with the synthesis of these pesticides, toxicity analyses are necessary for there to be confirmed protection of the environment and human health.

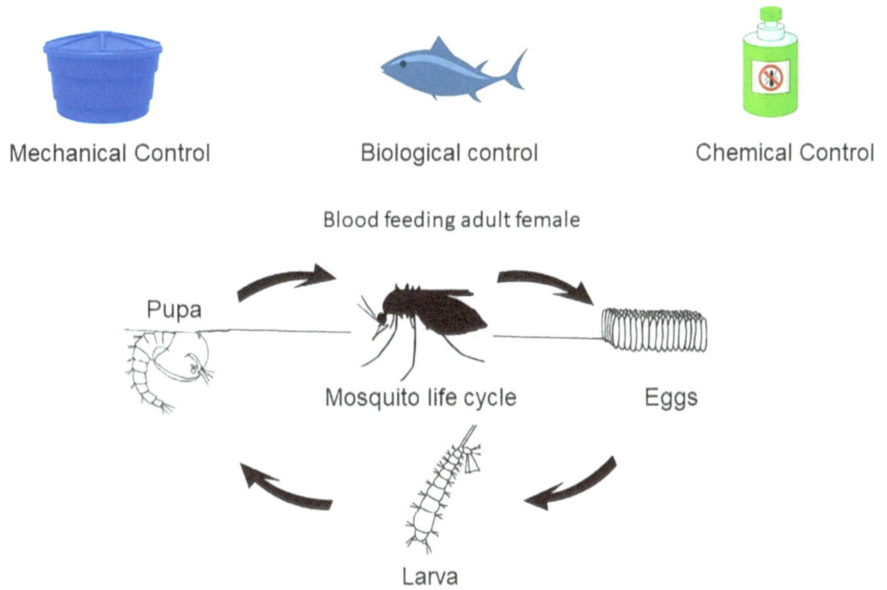

FIG. 3 Vector control strategies for mosquitoes. It shows the different mosquito control strategies, aiming at different stages of the mosquito's life cycle. Strategies include mechanical control, which is able to eliminate the vector and breeding sites or minimize the mosquito's contact with man; biological control, which uses predators or pathogens with effect to minimize the vector population; and, finally, chemical control, to use chemical compounds (insecticides or pesticides), which can act both for the larval phase of the vector and for the phase adult. *(Source: Personal data.)*

Pesticide pyriproxyfen

Among pesticides, pyriproxyfen (PPF), due to its low toxicity to mammals, stands out as promising in integrated pest management strategies of direct importance in the sanitary and health sector, including its interaction with *Musca domestica* (Shah, Shad, & Abbas, 2015). The dose recommended by WHO and the producer of the chemical PPF Sumitomo © (2012) is 0.01 ppm, which is sufficient to cause larval inhibition. At this dose, it is characterized as nontoxic, noncarcinogenic, or genotoxic. PPF has been used to control a variety of arthropods and disease vectors, and according to WHO, not only it is a growth inhibitory agent for *A. aegypti* (Dzieciolowska, Larroque, Kranjec, Drapeau, & Samarut, 2017), but also it has been used to control mosquitoes of *Aedes* sp., *Culex* sp., and *Anopheles* sp. (Bouhsira ct al., 2016).

In its larvicide (pesticide) evaluation performed by World Health Organization (2001), it was concluded that the laboratory and field data indicated that PPF will not adversely affect the vast majority of aquatic invertebrates and fish when applied at rates generally <50 ppb in mosquito control programs. In addition, according to Oo et al. (2018), another relevant fact of the PPF is that it requires a lower concentration than other larvicides, such as temephos and Israeli *Bacillus thuringiensis* (Oo et al., 2018), to achieve the same result.

PPF is used to control the vector *A. aegypti* L. through the spread of PPF particles attached to the mosquito (Fig. 4). This is a self-dissemination approach in the assembly of a dissemination station (ovitrampa) absorbed with PPF (Caputo et al., 2012), in which the female of the *A. aegypti* L. mosquito is stimulated to oviposit in this ovitrampa; in this way the mosquito must remain on the trap wall during this process and thus become impregnated with PPF particles. By carrying out several ovipositions in its gonotrophic period, the vector allows contamination of different breeding sites when it lands on them (Gaugler, Suman, & Wang, 2012). The mosquito then contaminates the water with PPF, exercising its larvicidal activity when the larvae reach the pupal stage (Abad-Franch, Zamora-Perea, Ferraz, Padilla-Torres, & Luz, 2015).

The use of this strategy is extremely positive for the integrated vector control because it offers low cost and easy maintenance. Another advantage of this type of strategy is the contamination of breeding sites that are not detectable by active searches by endemic control agents or community health agents. The PPF used in the trap does not pose a risk to the human population, and this contributes to the dissemination and acceptance of this approach.

The use of PPF cannot be uncontrolled; thus the concentration applied must be ideal so that it has an effect of larval control and does not offer possible conditions for the development of resistance by mosquitoes. When used in high concentrations, it can present significant toxicity (Ahemad & Khan, 2010; Alves et al., 2019) according to Boonstra and Post (2004), which indicates that pesticides can induce toxicity by increasing the production of reactive oxygen species through higher concentrations.

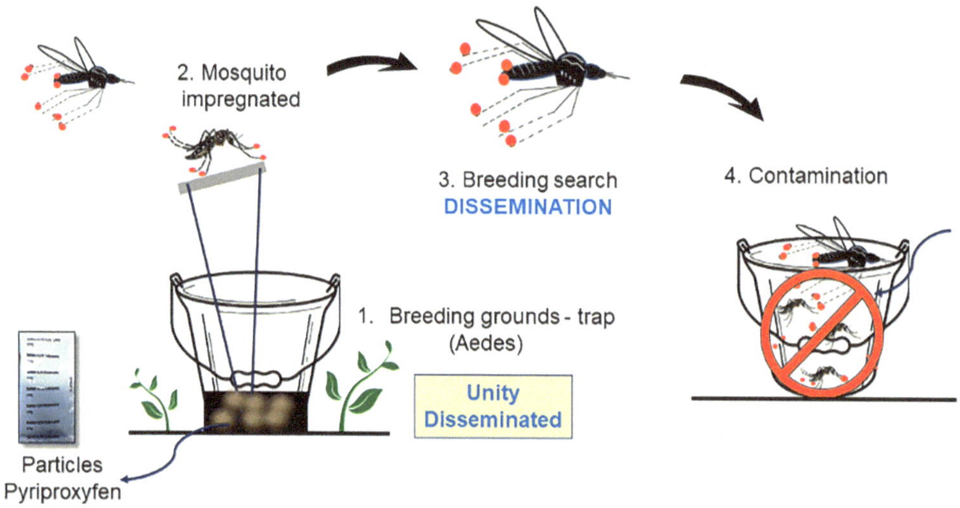

FIG. 4 Pyriproxyfen self-dissemination technique. Mechanism of action of pyriproxyfen on how to kill insects. It consists of the self-dissemination technique in the assembly of a dissemination station (ovitrampa) absorbed with pyriproxyfen in which the female of the *Aedes aegypti* L. mosquito is stimulated to oviposit in this ovitrampa; in this way the mosquito must remain on the trap wall during that process and so impregnate with pyriproxyfen particles. The mosquito then contaminates the water with larvicide, exercising its larvicidal activity when the larvae reach the pupal stage. *(Adapted from Alves, P. S., Oliveira, M. D. A., Almeida, P. M., Martins, F. A., Cavalcante, A. A. C. M., Andrade, T. J. A. S., … Júnior, J. S. C. (2019). Determination by chromatography and cytotoxotoxic and oxidative effects of pyriproxyfen and pyridalyl.* Chemosphere, *224, 398–406.)*

Thus the use of pesticides for mosquito control has several advantages where there is a potential for resistance to almost all known classes of pesticides currently used as adulticides; thus there is a need for greater use of larvicides such as pyriproxyfen whose compound has unique mode of action and no recorded field resistance. In addition, PPF has a low environmental impact, being suitable for the control of mosquito larvae (Invest & Lucas, 2008).

Policy and procedures

Aedes aegypti and *Aedes albopictus*

ZIKV is transmitted mainly by the infected *A. aegypti* and *A. albopictus* mosquitoes. However, there is also the possibility of sexual transmission and blood and neonatal transfusion, although the real role of these transmission routes in the spread of infection is unknown.

Originally from Africa, however, distributed in tropical and subtropical regions of the world (Fig. 5), *A. aegypti* is the main vector of transmission of the dengue virus (Brown et al., 2014). *A. albopictus*, which is one of the most rapidly spreading invasive species in the world, acts as a secondary vector (Kong et al., 2018). *A. albopictus* is considered a less efficient vector of arbovirus compared with *A. aegypti* for not being well adapted to urban conditions and domestic environments, and it is less anthropophilic than *A. aegypti* (Giovanni, 2012). Although both (Fig. 6) vectors cause microcephaly and Guillain-Barré syndrome (Jain, Biswas, & Daniel, 2018), *A. albopictus* has lesser vectorial competence for dengue, yellow fever, and Zika viruses (WHO, 2018). The transmission of urban yellow fever and Venezuelan equine encephalitis can be done through *A. albopictus* (Vega-Rúa, Zouache, Girod, Failloux, & Lourenço de Oliveira, 2014).

Vector control strategies

Mechanisms of action to combat the ZIKV emerge when countries begin to intervene with the use of discussions on preventing sexual pathways. Relevant principles including (1) vector control to limit the spread of Zika via mosquitoes, (2) sexual and reproductive health interventions, (3) knowledge and technology production, including diagnostic testing and a vaccine (4) preparing the health system to address needs of families affected by Zika. Each category is a crucial component of a comprehensive epidemic response (Rasanathan, MacCarthy, Diniz, Torreele, & Gruskin, 2017).

Thus vector control strategies emerge which traditionally focus on killing mosquitoes using a variety of insecticides. One of these strategies is environmental management, which consists of reducing or removing mosquito breeding sites, which has often been used together with chemical or microbiological ovicides, larvicides, and pupicides (Amer & Mehlhorn, 2006; Benelli, 2015) in areas where endemic diseases occur transmitted by mosquitoes.

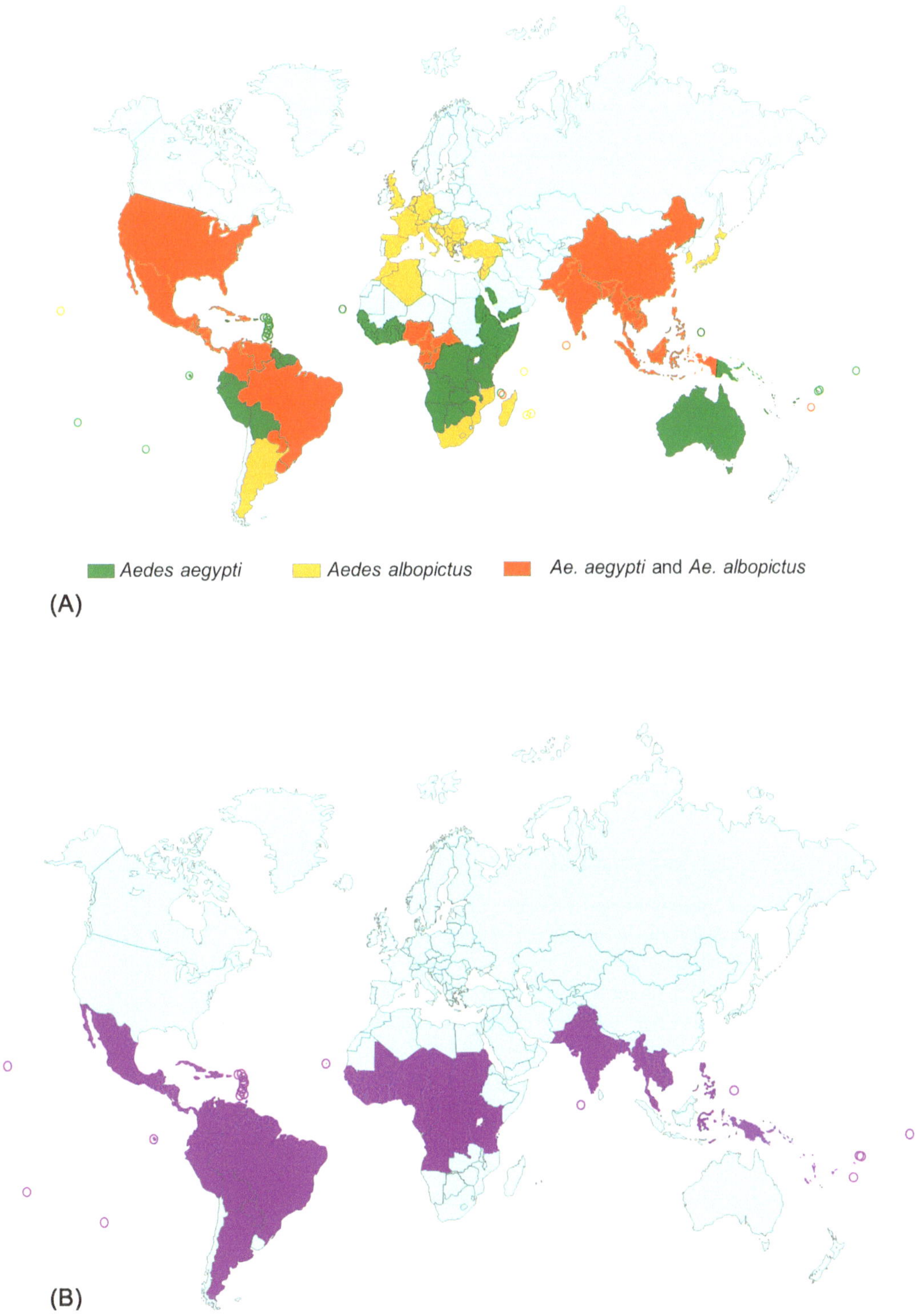

FIG. 5 Countries with symptomatic Zika virus case reports. Geographic distribution of *Aedes aegypti* and *Aedes albopictus*. (A) Represents the geographical distribution of *Aedes aegypti* and *Aedes albopictus*. (B) Shows the emergence of the *Zika virus*, first discovered in Uganda, then originated from Africa in Asia. It caused a great outbreak on Yap Island, spread to French Polynesia and other Pacific islands, then reached Latin America and affected more than 30 American countries. ZIKV then circulated on all 51 continents, with the exception of Europe. *(From Boyer, S., Calvez, E., Chouin-Carneiro, T., Diallo, D., & Failloux, A. B. (2018). An overview of mosquito vectors of Zika virus.* Microbes and Infection, 20, *646–660.)*

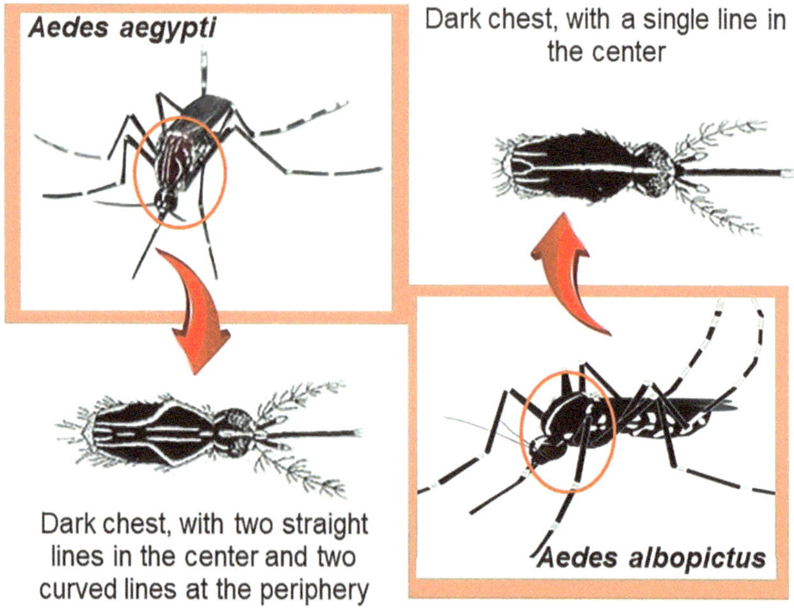

FIG. 6 Differences between *Aedes aegypti* and *Aedes albopictus* mosquitoes. It shows the morphological differences of *A. aegypti and A. albopictus*, in which *A. aegypti* presents a dark chest with two straight lines in the center and two curved lines in the periphery, whereas *A. albopictus* also has a dark chest, but only a single line in the center. *(Adapted from www.mdsaude.com.)*

Traditional control of *A. aegypti* focuses on visiting homes to manage or dispose of containers that house immature mosquitoes, in addition to the application of larvicides or adulticide spraying (WHO, 2009). These approaches have limited final impacts, as they require home visits, which results in low vector control coverage since many residents are absent or refuse to cooperate with surveillance agents (Chadee, 1988) or because enigmatic containers inside and outside the houses escape the inspection and control of the agents (Barrera, 2016). The low coverage of vector control means that the fraction of the treated *A. aegypti* population is not high enough to prevent or control the transmission site of the arbovirus or cause visible reductions in the adult mosquito population (Barrera, 2016; Lounibos, 2003).

Use of insecticides/pesticides to control ZIKV

The use of synthetic insecticides should be regulated since the development of resistance to insecticides is widespread (Liu, 2015; Naqqash, Gokce, Bakhsh, & Salim, 2016) and it generates a concern about damage to the environment and the effects on nontarget organisms. The use of insecticides to control mosquitoes, including organophosphates, carbamates, and pyrethroids, can also have negative effects on human health, depending on the dose tested and the exposure time. Insecticide-treated mosquito nets have played a very important role in reducing the prevalence of infection by *Plasmodium* in malaria-endemic sub-Saharan Africa, which has observed the incidence of clinical diseases falling 40% between 2000 and 2015 (Bhatt et al., 2015).

However, mosquito nets are only effective against mosquitoes that bite at nighttime, and concern about insecticide resistance is growing; particularly due to the most commonly used class of pyrethroids, it could reverse this trend and lead to an increased incidence of malaria and increased fatalities (Hemingway et al., 2016). As resistance to insecticides is now widespread in several species of mosquitoes (Hemingway & Ranson, 2000; Naqqash et al., 2016; Ranson & Lissenden, 2016), there is an increasing need for new, cheap, and reliable mosquito control strategies (Benelli, 2015; Jeffries & Walker, 2015).

It is important to note that the World Health Organization recommends alternative strategies in resistance management programs, such as mixtures, rotations, or mosaics, using pesticides that share different modes of action (Ahmad, Saleem, & Sayyed, 2009; WHO, 2003). The use of a mixture is favored when the effectiveness of the pesticide is high and the exposure is low (Levick, South, & Hastings, 2017).

Among the strategies for vector control are the UBV (ultra low volume) technique for the dissemination of PPF particles in areas of interest, which have shown positive results in terms of pesticide dissemination and mosquito mortality. Suman et al. (2014) used different sites for installing the stations, but analogous in socioeconomic characteristics. Abad-Franch et al. (2015) used the PPF in sentinel stations (100) scattered in Manaus (Amazonas), in which they presented contamination levels between 75.5% and 100% for 4 months, with a mortality rate of 100% in more than half of the sentinel stations.

Consequently, both authors (Abad-Franch et al., 2015; Suman et al., 2014) consider the technique of self-spreading PPF by mosquitoes to be a valid and advantageous strategy in controlling the vector of transmission through mosquitoes.

The technique using larvicides, to be effective against mosquitoes, must have low toxicity to mammals, low impact on the environment, broad spectrum of activity against all target species of mosquitoes, and a long duration of effect, thus reducing the frequency application (Invest & Lucas, 2008). It was found that these requirements are consistent with the pesticide proposed in the study. In addition, the method using PPF larvicide is effective and economically viable, thus not causing any deleterious effects on nontarget organisms.

Innovative strategies

On the impacts of mother-to-child transmission, these require innovative strategies, including expanded protective measures for sexual and reproductive health in most affected populations (Check, 2016; Meaney-Delman et al., 2016). Strategies for developing a vaccine are also imperative. Likewise a serological test is necessary to conduct a robust survey across the territory to assess the impact and epidemiology of ZIKV.

It is noteworthy that with the relationship between Zika infection and birth defects, the world's health authorities such as the Pan American Health Organization (PAHO), World Health Organization (WHO), and Centers for Disease Control and Prevention (CDC) issued guidelines for pregnant women, as well as those looking to become pregnant to consider delaying travel to areas affected by Zika, as well as those living in countries with widespread Zika transmission to avoid exposure to mosquito bites and/or having unprotected sex, to avoid risk situations prone to ZIKV (Check, 2016; Dyer, 2015).

However, investing in advanced research from ZIKV associated with *Aedes* is necessary to realign policies, legislation, and local and global rights in aligning international treaties and standards, as well as the unmet needs of communities affected by Zika. It is also necessary to have a better understanding of the natural history of ZIKV infection to establish effective control measures for this outbreak in Brazil and Latin America.

Mini-dictionary of terms

Japanese encephalitis virus. Acute infectious disease caused by a flavivirus capable of affecting the central nervous system. This virus is transmitted by mosquitoes and occurs mainly in Asia.

West Nile virus. It is a flavivirus transmitted through the bite of mosquitoes that circulates between different species, such as birds and horses, but can also infect humans.

Dengue virus. Flavivirus normally transmitted by female mosquitoes of the genus *Aedes*, with *Aedes aegypti* being the main vector.

Yellow fever virus. Virus transmitted by the bite of infected mosquitoes; these mosquitoes mainly of the species *Aedes aegypti*.

Mechanical control. It consists of adopting activities capable of eliminating the vector and breeding sites or minimizing the mosquito's contact with man.

Biological control. It is based on the use of predators or pathogens with effect to minimize the vector population.

Chemical control. It consists of the use of chemical compounds called insecticides or pesticides, which can act both for the larval phase of the vector and for the adult phase. It is a type of control recommended through rational and safe use for the environment and for the population, complementary to the actions of surveillance and environmental management.

Pesticides. Group of organic chemical compounds that have physicochemical properties used extensively in agriculture to control drugs and prevent weeds.

Pyriproxyfen. Is a pesticide classified as larvicide or insecticide, organic origin based on carbamate, and analog of juvenile hormone, used in the water supply in Brazil to combat the proliferation of mosquito larvae of *Aedes aegypti* L. and *A. albopictus* L.

Oviposition. Act of laying eggs by female invertebrate animals.

Ovitrampa. Economic trap method to monitor the population of *Aedes aegypti*.

Gonotrophic period. Cycle that has existed since the mosquito sucks the blood from the egg, until it feeds again.

Pupal stage. This phase goes from the formation of the cocoon until the hatching of the adult individual.

Sentinel. Technique created by researchers where there is a kind of monitored trap for the insect, causing it to carry the larvicide to where it breeds.

Birth defects. Disorders of genetic and/or environmental origin that occur in the embryonic-fetal period, that is, they are physical anomalies that occur before the baby is born.

Key facts of virus incidence and control strategies

- Zika outbreaks have occurred in areas of Africa, Southeast Asia, and Pacific islands and the Americas. Because the *Aedes* mosquito species that transmit the Zika virus are found worldwide, the outbreak is likely to spread to other countries.
- Serological and entomological investigations suggest that the Zika virus had an extensive geographic distribution in periods before 2007, with less than 20 cases in humans described, until then, with self-limiting clinical manifestations.
- The Zika virus epidemic began in 2012 in Brazil and later spread to other countries in South America, Central America, and the Caribbean.
- After the Zika virus outbreak, there was an abnormal increase in microcephaly among newborns, as well as an increase in the number of cases of Guillain-Barré syndrome.
- In 2013 and 2014 the Zika virus caused an outbreak in French Polynesia with about 28.000 people infected, including cases of Guillain-Barré syndrome.
- Brazil was the country in Latin America most affected by ZIKV, with approximately 1.500,000 cases between 2015 and 2016.
- Microcephaly and other fetal malformations potentially associated with Zika virus infection or suggestive of congenital infection have been reported in six countries or territories. Two cases, each linked to a stay in Brazil, were detected in Slovenia and the United States. An additional case, linked to a brief stay in Mexico, Guatemala, and Belize, was detected in a pregnant woman in the United States. A case in the Marshall Islands has also been reported recently and is pending confirmation.
- The global prevention and control strategy launched by the World Health Organization (WHO) as a Strategic Response Framework covers surveillance, response, and research activities. Key interventions are being carried out jointly by WHO and international, regional, and national partners in response to this public health emergency.

Summary points

- This chapter focuses on strategies to control the Zika virus with the use of larvicide, emphasizing the larvicide pyriproxyfen.
- One of the control strategies consists of the use of pesticides (insecticides), the use of these consists of the ultra low volume (UVB) technique for the dissemination of pyriproxyfen particles, and the focal treatment occurs with the application of the larvicidal product in the positive deposits for immature forms of mosquitoes that cannot be eliminated mechanically.
- The PPF used in the trap does not pose a risk to the human population, and this contributes to the dissemination and acceptance of this approach. However, the use of PPF cannot be uncontrolled, and the concentration applied must be ideal so that it has an effect of larval control and does not offer possible conditions for the development of resistance by mosquitoes.
- With the synthesis of these pesticides, toxicity analyses are necessary for confirmed protection of the environment and human health. Thus its use must be correctly guided by professionals in the area, respecting the current legislation and the health of the population.
- Despite advances in research in the pharmaceutical industry, the development of Zika vaccines is a complex task, where vector-borne diseases remain a global challenge to public health. Therefore several strategies must be strengthened, aiming at their elimination.

References

Abad-Franch, F., Zamora-Perea, E., Ferraz, G., Padilla-Torres, S. D., & Luz, S. L. B. (2015). Mosquito-disseminated pyriproxyfen yields high breeding-site coverage and boosts juvenile mosquito mortality at the neighborhood scale. *PLoS Neglected Tropical Diseases*, *9*, 1935–2735.

Agumadu, V. C., & Ramphul, K. (2018). Zika virus: A review of literature. *Cureus*, *10*, 1–5.

Ahemad, M., & Khan, M. S. (2010). Comparative toxicity of selected insecticides to pea plants and growth promotion in response to insecticide-tolerant and plant growth promoting *Rhizobium leguminosarum*. *Crop Protection*, *29*, 325–329.

Ahmad, M., Saleem, M. A., & Sayyed, A. H. (2009). Efficacy of insecticide mixtures against pyrethroid- and organophosphate-resistant populations of *Spodoptera litura* (Lepidoptera: Noctuidae). *Pest Management Science*, *65*, 266–274.

Al-Abdely, H., & Zika, M. (2016). An emerging teratogenic virus. *Saudi Medical Journal*, *3*, 831–833.

Alves, P. S., Oliveira, M. D. A., Almeida, P. M., Martins, F. A., Cavalcante, A. A. C. M., Andrade, T. J. A. S., ... Júnior, J. S. C. (2019). Determination by chromatography and cytotoxotoxic and oxidative effects of pyriproxyfen and pyridalyl. *Chemosphere*, *224*, 398–406.

Amer, A., & Mehlhorn, H. (2006). Larvicidal effects of various essential oils against Aedes, Anopheles, and Culex larvae (Diptera, Culicidae). *Parasitology Research*, *99*, 466–472.

Aubry, M., Finke, J., Teissier, A., Roche, C., Broult, J., Paulous, S., … Musso, D. (2015). Seroprevalence of arboviruses among blood donors in French Polynesia. *International Journal of Infectious Diseases*, *41*, 2011–2013.

Barrera, R. (2016). Considerations for disrupting dengue virus transmission; Ecology of *Aedes aegypti* and current (nongenetic) methods of control. In Z. N. Adelman (Ed.), *Genetic control of malaria and dengue* (pp. 103–124). Oxford, United Kingdom: Academic Press.

Benelli, G. (2015). Research in mosquito control: Current challenges for a brighter future. *Parasitology Research*, *114*, 2801–2805.

Bhatt, S., Weiss, D. J., Cameron, E., Bisanzio, D., Mappin, B., Dalrymple, U., … Eckhoff, P. A. (2015). The effect of malaria control on *Plasmodium falciparum* in Africa between 2000 and 2015. *Nature*, *526*, 207–211.

Boonstra, J., & Post, J. A. (2004). Molecular events associated with reactive oxygen species and cell cycle progression in mammalian cells. *Gene*, *337*, 1–13.

Bouhsira, E., Liénard, E., Lyazrhi, F., Jacquiet, P., Varloud, M., Deflandre, A., & Franc, M. (2016). Repellent and insecticidal efficacy of a combination of dinotefuran, pyriproxyfen and permethrin (Vectra® 3D) against *Culex pipiens* in dogs. *Parasite Epidemiology and Control*, *1*, 233–238.

Brasil, P., Sequeira, P. C., Freitas, A. D., Zogbi, H. E., Calvet, G. A., de Souza, R. V., … de Filippis, A. M. B. (2016). Guillain Barre syndrome associated with Zika virus infection. *Lancet*, 387–1482.

Brown, J. E., Evans, B. R., Zheng, W., Obas, V., Barreramartinez, L., & Egizi, A. (2014). Human impacts have shaped historical and recent evolution in *Aedes aegypti*, the dengue and yellow fever mosquito. *Evolution*, *68*, 514–525.

Caputo, B., Ienco, A., Cianci, D., Pombi, M., Petrarca, V., Baseggio, A., … Della Torre, A. (2012). The "auto-dissemination" approach: A novel concept to fight Aedes albopictus in urban areas. *PLoS Neglected Tropical Diseases*, *6*, 1935–2735.

Chadee, D. D. (1988). Effects of 'closed' houses on the *Aedes aegypti* eradication programme in Trinidad. *Medical and Veterinary Entomology*, *2*, 193–198.

Check, H. E. (2016). Proving Zika link to birth defects poses huge challenge. *Nature*, *530*, 142.

Dupont-Rouzeyrol, M., Biron, A., O'Connor, O., Huguon, E., & Descloux, E. (2016). Infectious Zika viral particles in breastmilk. *Lancet*, *387*, 1. https://doi.org/10.1016/S0140-6736(16)00624-3.

Dyer, O. (2015). Zika virus spreads across Americas as concerns mount over birth defects. *BMJ*, *351*, 1–2. https://doi.org/10.1136/bmj.h6983.

Dzieciolowska, S., Larroque, A. L., Kranjec, E. A., Drapeau, P., & Samarut, E. (2017). The larvicide pyriproxyfen blamed during the Zika virus outbreak does not cause microcephaly in zebrafish embryos. *Scientific Reports*, *7*, 01–11.

Erlanger, T. E., Keiser, J., & Utzinger, J. (2008). Effect of dengue vector control interventions on entomological parameters in developing countries: A systematic review and meta-analysis. *Medical and Veterinary Entomology*, *22*, 1365–2915.

Gallian, P., Cabie, A., Richard, P., Paturel, L., Charrel, R. N., Pastorino, B., … de Lamballerie, X. (2017). Zika virus in asymptomatic blood donors in Martinique. *Blood*, *129*, 263–266.

Gaugler, R., Suman, D., & Wang, Y. (2012). An autodissemination station for the transfer of an insect growth regulator to mosquito oviposition sites. *Medical and Veterinary Entomology*, *26*, 37–45.

Giovanni, R. (2012). *Aedes albopictus* and the reemergence of dengue. *BMC Public Health*, *12*, 72.

Hemingway, J., & Ranson, H. (2000). Insecticide resistance in insect vectors of human disease. *Annual Review of Entomology*, *45*, 371–391.

Hemingway, J., Ranson, H., Magill, A., Kolaczinski, J., Fornadel, C., Gimnig, J., … Hinzoumbe, C. K. (2016). Averting a malaria disaster: Will insecticide resistance derail malaria control? *Lancet*, *387*, 1785–1788.

Invest, J., & Lucas, J. (2008). Pyriproxyfen as a mosquito larvicide. In *Proceedings of the sixth international conference on urban pests, Veszprem, Hungary* (pp. 13–16).

Jain, N., Biswas, A., & Daniel, K. V. (2018). Review on Zika virus. *Pharmatutor*, *6*, 30–37.

Jeffries, C. L., & Walker, T. (2015). The potential use of Wolbachia-based mosquito biocontrol strategies for *Japanese encephalitis*. *PLoS Neglected Tropical Diseases*, *9*, e0003576.

Kelser, E. A. (2016). Meet dengue's cousin, Zika. *Microbes and Infection*, *18*, 163–166.

Kong, L., Wang, J., Li, Z., Lai, S., Liu, Q., & Wu, H. (2018). Modeling the heterogeneity of dengue transmission in a city. *International Journal of Environmental Research and Public Health*, *15*, 1128.

Kuno, G., Chang, G.-J., & Tsuchiya, K. R. (1998). Phylogeny of the genus Flavivirus. *Journal of Virology*, *72*, 73–83.

Levick, B., South, A., & Hastings, I. M. (2017). A two-locus model of the evolution of insecticide resistance to inform and optimise public health insecticide deployment strategies. *PLoS Computational Biology*, *13*, 1–35.

Liu, N. (2015). Insecticide resistance in mosquitoes: Impact, mechanisms, and research directions. *Annual Review of Entomology*, *60*, 537–559.

Lounibos, L. P. (2003). Genetic-control trials and the ecology of Aedes aegypti at the Kenya coast. In W. Takken, & T. W. Scott (Eds.), *Ecological aspects for application of genetically modified mosquitoes* (pp. 33–43). Dordrecht, The Netherlands: Kluwer Academic.

Mansuy, J. M., Dutertre, M., Mengelle, C., Fourcade, C., Marchou, B., Delobel, P., … Martin-Blondel, G. (2016). Zika virus: High infectious viral load in semen, a new sexually transmitted pathogen? *The Lancet Infectious Diseases*, *16*, 405.

Meaney-Delman, D., Hills, S. L., Williams, C., Galang, R. R., Iyengar, P., Hennenfent, A. K., … Jamieson, D. J. (2016). Zika virus infection among US pregnant travelers. *Morbidity and Mortality Weekly Report*, *65*, 211–214.

Mittal, R., Nguyen, D., Debs, L. H., Patel, A. P., Liu, G., Jhaveri, V. M., … Younis, R. T. (2017). Zika virus: An emerging global health threat. *Frontiers in Cellular and Infection Microbiology*, *7*, 486.

Muller, D. A., & Young, P. R. (2013). The flavivirus NS1 protein: Molecular and structural biology, immunology, role in pathogenesis and application as a diagnostic biomarker. *Antiviral Research*, *98*, 192–208.

Musso, D., & Gubler, D. J. (2016). Zika virus. *Clinical Microbiology Reviews*, *29*, 487–524.

Musso, D., Roche, C., Nhan, T. X., Robin, E., Teissier, A., & Cao-Lormeau, V. M. (2015). Detection of Zika virus in saliva. *Journal of Clinical Virology*, *68*, 53–55.

Musso, D., Roche, C., Robin, E., Nhan, T., Teissier, A., & Cao-Lormeau, V. M. (2015). Potential sexual transmission of Zika virus. *Emerging Infectious Diseases*, *21*, 359–361.

Naqqash, M. N., Gokce, A., Bakhsh, A., & Salim, M. (2016). Insecticide resistance and its molecular basis in urban insect pests. *Parasitology Research*, *115*, 1363–1373.

Nunes, M. L., Carlini, C. R., Marinowic, D., Kalil Neto, F., Fiori, H. H., Scotta, M. C., … Costa, J. C. (2016). Microcefalia e virus Zika: Um olhar clínico e epidemiológico do surto em vigência no Brasil. *The Journal of Pediatrics*, *92*, 230–240.

Oo, S. Z. M., Thaung, S., Maung, Y. N. M., Aye, K. M., Aung, Z. Z., Thu, H. M., … Minakawa, N. (2018). Effectiveness of a novel long-lasting pyriproxyfen larvicide (SumiLarv(R)2MR) against Aedes mosquitoes in schools in Yangon, Myanmar. *Parasite Vector*, *11*, 16.

Ranson, H., & Lissenden, N. (2016). Insecticide resistance in African Anopheles mosquitoes: A worsening situation that needs urgent action to maintain malaria control. *Trends in Parasitology*, *32*, 187–196.

Rasanathan, J. J., MacCarthy, S., Diniz, D., Torreele, E., & Gruskin, S. (2017). Engaging human rights in the response to the evolving Zika virus epidemic. *American Journal of Public Health*, *107*, 525–531.

Shah, R. M., Shad, S. A., & Abbas, N. (2015). Mechanism, stability and fitness cost of resistance to pyriproxyfen in the house fly, *Musca domestica* L. (Diptera: Muscidae). *Pesticide Biochemistry and Physiology*, *119*, 67–73.

Suman, D. S., Farajollahi, A., Healy, S., Williams, G. M., Wang, Y., Schoeler, G., & Gaugler, R. (2014). Point-source and area-wide field studies of pyriproxyfen autodissemination against urban container-inhabiting mosquitoes. *Acta Tropica*, *135*, 1873–6254.

Vega-Rúa, A., Zouache, K., Girod, R., Failloux, A. B., & Lourenço de Oliveira, R. (2014). High level of vector competence of *Aedes aegypti* and *Aedes albopictus* from ten American countries as a crucial factor in the spread of chikungunya virus. *Journal of Virology*, *88*, 6294–6306.

Weaver, S. C., & Barrett, A. D. (2004). Transmission cycles, host range, evolution and emergence of arboviral disease. *Nature Reviews. Microbiology*, *2*, 789–801.

World Health Organization (WHO). (2003). *Guidelines on the management of public health pesticides*. Document WHO/CDS/WHOPES/2003.7 Geneva, Switzerland: WHO.

World Health Organization (WHO). (2009). *Dengue guidelines for diagnosis, treatment, prevention and control* (p. 147). Geneva, Switzerland: World Health Organization.

World Health Organization (WHO). (2018). *Vacuna contra el dengue*. https://www.who.int/immunization/policy/position_papers/PP_dengue_2018_SP.pdf.

World Health Organization. (2001). Report of the fourth WHOPES working group meeting, WHO/HQ, Geneva, 4–5 December 2000. WHO/CDS/WHOPES/2001/2.

Chapter 32

Larvicides: Plant oils and Zika control

Taruna Kaura[a], Naveed Pervaiz[b], and Abhishek Mewara[a]
[a]*Department of Medical Parasitology, Postgraduate Institute of Medical Education and Research, Chandigarh, India,* [b]*Department of Zoology, Panjab University, Chandigarh, India*

Abbreviations

CDC	Centers for Disease Control and Prevention
DEET	*N,N*-diethyl-meta-toluamide
EO	essential oil
GRAS	Generally Recognized as Safe
PMD	*p*-menthane-3,8-diol
US EPA	United States Environmental Protection Agency
US FDA	United States Food and Drug Administration
WHO	World Health Organization
ZIKV	Zika virus

Introduction

Arthropod-borne viruses, also called as arboviruses, are a worldwide public health concern because of their association with several vector-borne neglected tropical diseases (Braack, de Almeida, Cornel, Swanepoel, & De Jager, 2018; Dias & Moraes, 2014). The arboviruses transmitted to human beings by the arthropods are divided into three genera—*Flavivirus*, *Togavirus*, and *Bunyavirus*. Flaviviruses transmit some important arboviruses, which include yellow fever virus, dengue virus, Japanese encephalitis virus, West Nile virus, and Zika virus (ZIKV) (Govindarajan, Hoti, & Benelli, 2016). The primary vector known for the transmission of some viruses such as dengue, yellow, chikungunya, and ZIKV is the mosquito species *Aedes aegypti* L. (Conti et al., 2013; Gould, Pettersson, Higgs, Charrel, & de Lamballerie, 2017). *Aedes aegypti* is a daytime feeder, and the infected female mosquitoes are involved in transmitting the viruses to human beings. Other arthropods responsible for transmission of these viruses include midges, ticks, flies, and fleas (Pohlit, Rezende, Baldin, Lopes, & de Andrade Neto, 2011).

ZIKV is one of the recent emerging pathogens that caused a great fear and worrying situation globally because of its quick spread and complicated clinical manifestations like microcephaly in infants and Guillain-Barré syndrome. ZIKV is a mosquito-borne positive-sense single-stranded and enveloped virus belonging to the genus *Flavivirus* and family *Flaviviridae* (Ferraris, Yssel, & Missé, 2019). It was first reported from the Zika forest of Uganda in 1947 (Dick, Kitchen, & Haddow, 1952) and was initially confined to Africa and Asia; however, it later spread worldwide and became a global threat (Chang, Ortiz, Ansari, & Gershwin, 2016; Singh et al., 2016; Vest, 2017; Zhang et al., 2017). Severe diseases like Guillain-Barré syndrome and meningoencephalitis in adults and congenital deformations have been linked to ZIKV infection, besides common clinical syndromes of flaviviruses such as headache, fever, and muscle pain (Cao-Lormeau et al., 2016; Parra et al., 2016). After the pandemic outbreak of Ebola virus, the global attention diverted toward ZIKV and its menacing disease profile. After staying harmless for almost six decades since 1947, the ZIKV outbreak forced the World Health Organization (WHO) to pronounce the situation as an emergency on February 1, 2016 (Chang et al., 2016; Gulland, 2016; Singh et al., 2016).

Unfortunately, neither effective vaccines nor specific drugs are available against these viruses; therefore direct targeting of mosquitoes is required to control their spread. To this end, tremendous efforts have been put to create powerful and efficacious repellents and larvicides to control the vector. However, innovation and improvement of effective and safe

Zika Virus Impact, Diagnosis, Control, and Models. https://doi.org/10.1016/B978-0-12-820267-8.00032-7

remedies against ZIKV still remains a challenge. Therefore, keeping these considerations in mind, this chapter aims to provide an update on the potential of essential oils (EO) and their derived fractions against control of ZIKV.

Requirement of plant-derived/essential oil-based insect repellents

From the advent of humanity, plants have been considered significant for use in diverse therapeutic and cosmetic applications. The basic chemical constituents of plants include EOs, which are volatile secondary metabolites fortified with terpenoids and related aromatic compounds (Maia & Moore, 2011). Being concoction of different bioactive compounds, plant EOs can be used as efficient bioinsecticides that are environment- and user-friendly with minimal propensity of development of resistance (Isman, 2000).

Over the course of history, more than 3000 EOs have been analyzed for their potential activity, and approximately 10% have been commercially marketed as repellents and insecticides (Sharifi-Rad et al., 2017). The chemical constituents in EOs are responsible for remarkable therapeutic, antioxidative, and insecticidal properties (Lupi, Hatz, & Schlagenhauf, 2013). The repellent effects of many EOs are exhibited by their common constituents such as α-pinene, citronellol, citronellal, and limonene (Soleimani-Ahmadi et al., 2017; Tong & Bloomquist, 2013). Monoterpenoids, sesquiterpenes, and alcohols present in EOs have also been shown to attribute to their repellent properties (Sathantriphop, Achee, Sanguanpong, & Chareonviriyaphap, 2015). Terpenoids and phenylpropanoids are the bioactive molecules that are extensively utilized in medicine, agronomy, biology, and pharmaceutical sciences (Benelli & Mehlhorn, 2016). These compounds possess some medicopharmaceutical properties like antitumor, antihelminthic, larvicidal, and insecticidal activity, as well as activity against arbovirus vectors (da Silva et al., 2017; Gonçalves et al., 2015; Voris et al., 2017). A terpene-rich derivative of EO from *Ayapana triplinervis* was found effective against ZIKV. A major phytocompound in *Ayapana triplinervis* EO is thymohydroquinone dimethyl ether, which also showed inhibition of ZIKV in human cells. Both EO and phytocompounds were found effective against the epidemic ZIKV strain decreasing the growth of the virus up to 3-log (Haddad et al., 2019).

Recently, it has been demonstrated that eucalyptol, a natural organic compound, and linalool, a naturally occurring terpene alcohol, activate the mosquito's receptors present in the antennal sensilla (Dickens & Bohbot, 2013; Taylor et al., 2012). The repellent screening on the basis of odor sensing may emerge as a novel approach toward development of compounds and repellents against arthropods. In several public health programs, larvicides and insecticides like the organophosphate temephos and pyrethroids have been found potent in wiping out mosquitoes and larvae, but their consistent use has rendered few mosquitoes resistant to them (Dias & Moraes, 2014; Oliveira et al., 2013). In addition, insecticides do not have selective action and therefore cause harm to other life forms, including humans (Manzoor, Samreen, & Parveen, 2013; Santos, Bandeira, Lemos, & Santiago, 2017). For example, from 1960 to 2000, temephos was monitored in 16 nations and was found to remain active in water for 48–72 h; however, in addition to *Aedes aegypti*, it was shown to interfere with the viability of other living organisms causing serious environmental impact (Navarro, da Silva, da Silva, Napoleao, & Paiva, 2013). Similarly, DEET (*N,N*-diethyl-meta-toluamide), a synthetic repellent, has been extensively utilized. It blocks the olfactory receptor of insects, responsible for detection of l-octen-3-ol, present in human sweat and breath (Swale, Sun, Tong, & Bloomquist, 2014). However, several human health and environment concerns have risen due to the use of DEET, especially in children (Khanikor, Parida, Yadav, & Bora, 2013).

In this scenario, it is important to recognize new larvicides with different modes of action to decrease the development of resistance due to selection pressure of pesticides and boost the public health programs (Dias & Moraes, 2014). Therefore plant EOs with better repellent properties, selectivity to the target, and low toxicities to the humans and environment are being encouraged as an alternative to the synthetic insecticides (Navarro et al., 2013; Nerio, Olivero-Verbel, & Stashenko, 2010).

Plant essential oils for mosquito repellency

Sometimes, natural products can cause adverse effects, and these products are not always safer than the synthetic compounds. In addition, for many products, the available information regarding safety and toxicity is limited and sometimes even contradictory to the notional beliefs about their safety (Dickens & Bohbot, 2013; Niroumand et al., 2016). Even though plant EOs are considered to be safe, however, few side effects like skin irritation are associated with some EOs. These side effects arise mainly because of their constituents and therefore limit their widespread usage (Benelli & Pavela, 2018; Govindarajan & Benelli, 2016). Some of the commercially available repellent EOs with their general properties, toxicity, and safety are described as follows:

Citronella

Citronella oil is an insect repellent derived from *Cymbopogon citratus* used topically with suitable precautions in children and other vulnerable populations under the US EPA (Environmental Protection Agency) guidance (Kongkaew, Sakunrag, Chaiyakunapruk, & Tawatsin, 2011; Sharma, Rao, Kumar, Mahant, & Khatkar, 2019). Citronellol or 3,7-dimethyloct-6-en-1-ol is the active component in citronella oil. The oil has been "Generally Recognized as Safe" (GRAS) by the United States Food and Drug Administration (US FDA). Even though citronella is less effective than DEET, but it can avert the insect biting approximately for a couple of hours. Contrary to the use in the United States, the use of citronella as an insect repellent is not popular in Canada and Europe, the reason being the lack of safety data for methyl eugenol, which one of its constituents (Tan & Nishida, 2012). Methyl eugenol is a carcinogenic substance and was categorized as a group 2B substance by International Agency for Research on Cancer in 2013. Its carcinogenicity was evidenced in in vivo animal experiments, but there are no data available for its effect in humans. In addition, US National Toxicology Program considered methyl eugenol as "reasonably anticipated to be a human carcinogen" because of its carcinogenic effects in other rodents (Tan & Nishida, 2012).

Clove oil

Syzygium aromaticum, *Eugenia caryophyllata*, and *E. aromaticum* are the main sources of clove oil with eugenol, eugenol acetate, and caryophyllene being the chief active constituents in it. It is of extensive use in food, cosmetic, and medicinal industry and is also widely used in insect repellents. US FDA approved clove oil as GRAS and thus can be used in human food directly in both natural and synthetic formulations. US FDA also approved its use as a painkiller and in dentistry for dental cements, in transdermal drug delivery systems, and in aromatherapy products as fragrance. Even though clove oil has been classified in the list of Minimum Risk Pesticides by the US Environmental Protection Agency (US EPA), some side effects like skin irritation, headache, and decreased blood coagulation are associated with its use. Further some reports have raised concerns about its carcinogenicity due to methyl eugenol, which is a trace constituent. As such the safety with the use of clove oil and its negative effects and carcinogenic potential need to be further established scientifically before determining its extensive utilization.

p-Menthane-3,8-diol

p-Menthane-3,8-diol (PMD) is an organic compound derived from the leaves of *Corymbia citriodora*. For utilization as commercial repellents, PMD is also chemically synthesized. It is a major ingredient of the repellent extracted from *Eucalyptus* with higher efficiency and provides better protection than DEET against ticks (Diaz, 2016). Due to its higher efficacy and protective properties, EPA has registered PMD as a biopesticide repellent. It is not a skin sensitizer but is classified as an eye irritant, and therefore PMD-containing products must carry a "warning" label. Even though small epidemiologic data are available about the effects of PMD, toxicological studies have however revealed low toxicity associated with its use. PMD as an active constituent is categorized as GRAS by the EPA and is used in flavoring foods and in medicines and component of many consumer products. The Centers for Disease Control and Prevention (CDC) has recommended the utilization of PMD against mosquito bites, but FDA has recommended that it should not be used in children under the age of 3 years (Lee, 2018).

Permethrin

It is the most commonly used insecticide acquired from the flowers of *Chrysanthemum cinerariifolium*. It is also chemically synthesized and used globally as a neurotoxic insecticide. As a neurotoxin, it causes repetitive nerve impulses due to depolarization of axonal sodium channels (Drago, Shah, & Shah, 2014). Severe type II poisoning that result in seizures is caused by higher doses of permethrin by affecting gamma-aminobutyric acid–gated chloride channels. Though permethrin causes toxic effects in poikilotherms, for example, insects, rather than homeotherms, exposure to its higher doses may cause toxicity in humans and animals. The direct contact with skin should be avoided and is therefore recommended for use as insect repellent on clothing only. Application of permethrin to beddings and bed nets delivers better safety against mosquitoes, ticks, etc. (Diaz, 2016). However, based on a murine reproduction test, EPA has classified permethrin as "likely human carcinogen."

Despite the uncharacterized constitution of many of them, the mode of action of various EOs has been postulated. Some show neurotoxicity and intoxicate the insects by obstructing the enzyme acetylcholinesterase (Houghton, Ren, & Howes,

TABLE 1 Characteristics of common plant-based repellents.

Repellent	Origin	Repellency	Safety/toxicity
Citronella	*Cymbopogon citratus*	It can deter mosquito biting from 10 min to 2 h	Potential eye and skin irritant Toxicity category by EPA: IV (nontoxic)
Clove oil	*Syzygium aromaticum* *Eugenia caryophyllata* *Eugenia aromaticum*	It provides about 1–1.5 h of protection against mosquitoes	Skin irritation, headache, and decreased blood clotting Toxicity category by EPA: minimum risk pesticides
p-Menthane-3,8-diol (PMD)	*Eucalyptus citriodora* *Corymbia citriodora*	PMD has better efficiency and protects from mosquito bites for about 2 h	Potential eye irritant with less skin sensitization Toxicity category by EPA: I (highly toxic)
Permethrin	*Chrysanthemum cinerariifolium*	Clothes treated with permethrin show efficiency against mosquitoes for 2–24 weeks	Likely skin irritant Toxicity category by EPA: possible human carcinogen
N,N-Diethyl-meta-toluamide (DEET)	Synthetic	Provides ~5 h of protection against mosquitoes	Neurotoxic Toxicity category by EPA: III (slightly toxic)

2006; Isman, 1999). Some studies have reported that some of the EOs show their effect by regulating octopamine, a peripheral transmitter in superior invertebrates (Enan, 2001; Woodring, McBride, & Fields, 1989). The studies carried out on *Drosophila melanogaster* determined that EOs with thymol terpene as a constituent work by modulating GABA receptors (Priestley, Williamson, Wafford, & Sattelle, 2003). The mode of action of various evaluated EOs and their constituents are associated with their physicochemical properties, but these properties need further validation. Due to lesser available data about evaluated compounds, these results should be considered as preliminaries. Table 1 provides an overview of the aforementioned repellents and their origin and associated safety and toxicity.

Methods of extraction of plant-based repellents

Essential oils can be obtained from different plant parts, namely, flowers, leaves, stem, bark, and roots, by using various solvents. Different techniques have been developed to extract the EOs as some methods are more suitable for certain plants than others depending on the plant's chemical composition (Hesham, Abdurahman, & Rosli, 2016). Table 2 lists the different methods of essential oil extraction.

Promising developments in plant-based repellents and efficacy of plant oils to control different vector-borne (mosquito) diseases

In recent years, various plant-driven products are being employed to check their effectiveness against mosquitoes. Different types of plants are employed for synthesis of nanomosquitocides for larvicidal purpose. *Aedes aegypti* larvae and nonmosquito predator *Artemia salina* was found to be extremely sensitive to the methanolic seed extract derived from *Myracrodruon urundeuva* Fr. Allemão, a Brazilian plant (Souza et al., 2011). Moreover, plant-based insecticides from *Persea americana*, *Callistemon rigidus*, and *Apium graveolens* and have also been shown to possess larvicidal activity against *Aedes aegypti* mosquito. Oil extracted from plant *Apium graveolens* have been reported to show 100% repellent efficacy against the mosquitoes for about 3h (Kumar, Mishra, Wahab, & Warikoo, 2014; Ramkumar, Karthi, Muthusamy, Natarajan, & Shivakumar, 2015). A medicinal plant called *Ayapana triplinervis* from Reunion Island has demonstrated its effectiveness as a plant-based mosquito control agent against ZIKV. The EOs derived from the plant with main active component thymohydroquinone dimethyl ether has shown effective for in vitro and in vivo prevention (Haddad et al., 2019). An EO extracted from *Syzygium lanceolatum* has been shown to exhibit larvicidal effects on *Aedes aegypti*, *Aedes albopictus*, *Anopheles stephensi*, *Anopheles subpictus*, *Culex quinquefasciatus*, and *Culex tritaeniorhynchus*. It can prove

TABLE 2 Extraction methods of essential oil extraction.

Name	Method
Hydrodistillation (HD)	The technique employed is isotropic distillation, which involves three main physicochemical processes: hydrodiffusion, hydrolysis, and decomposition by heat
Steam distillation	In this method, first the steam vaporizes volatile compounds of the plant material, followed by condensation and collection of the distillate
Solvent extraction	Also known as liquid-liquid extraction partitioning, where the compound is separated based on the solubility of its parts. First the plant material is treated with the solvents, and then a waxy aromatic compound called a "concrete" is produced. This concrete substance is then mixed with alcohol, and finally the oil particles are extracted
Soxhlet extraction	It is used when the compound has limited solubility in a solvent. The partially soluble components of a solid are transferred to the liquid phase using a Soxhlet extractor
Cold pressing	This method is also known as scarification method. In this method the whole plant part is mechanically ruptured and squeezed out of its juice and oil, followed by separation of the oil from the juice layer
Supercritical fluid extraction (SFE)	It is the process of separating one component (the extractant) from another (the matrix) using supercritical fluid state, usually from a solid matrix
Microwave-assisted hydrodistillation (MAHD)	In this technique the separation of the compound is based upon the direct impact of microwave-induced ionic conduction and dipole rotation on polar solvents
Ultrasound-assisted extraction (UAE)	In this method, ultrasonic waves are used to generate implosion of bubbles, which will induce macroturbulence, high-velocity interparticle collisions, and perturbation in the microporous particles of the sample, due to which the solute quickly diffuses from the solid phase to the solvent
Solvent-free microwave extraction (SFME)	This process consists of microwave-assisted dry distillation of a fresh plant matrix without adding water or any organic solvent
Microwave hydrodiffusion and gravity (MHG)	It is a combination of microwaves for hydrodiffusion of essential oils from the inside to the exterior of the biological material and earth gravity to collect and separate the material

an eco-friendly mosquito repellent because of its low toxic effects on water bugs and fishes (Benelli, Rajeswary, & Govindarajan, 2018).

Essential oil extracts from the plant *Casuarina equisetifolia* has been shown to possess toxicity levels against *Aedes aegypti* and *Anopheles gambiae*. Nineteen compounds with the chief constituent being fatty acids have been identified from this EO. Other compounds include *cis*-13-octadecenoic acid (17.83%), *n*-hexadecanoic acid (18.67%), undecane (10.45%), tridecane (11.84%), nonanal (8.62%), oxirane (2.43%), and hentriacontane (8.91%). The EO has effectively been shown to kill all the insects completely within 24 h (Adeosun, Bamidele, & Ayorinde, 2016). In another study, potent toxic larvicidal activity against *Aedes aegypti* and *Anopheles culicifacies* has been shown by the leaf and seed extracts of *Ricinus communis*, a widely distributed plant throughout the temperate and tropical regions of the world (Sogan et al., 2018). A study on the EOs extracted from *Origanum scabrum* showed potent activity against larvae of many mosquitoes, including *Anopheles stephensi*, *Aedes aegypti*, *Culex quinquefasciatus*, and *Culex tritaeniorhynchus*. Likewise the EO also showed toxicity against the ova of each species. It also inhibited the biting by *Anopheles stephensi*, *Culex tritaeniorhynchus*, *Culex quinquefasciatus*, and *Aedes aegypti* for approximately 210, 120, 150, and 180 h, respectively (Govindarajan, Kadaikunnan, Alharbi, & Benelli, 2016).

The larvicidal effects of other EOs from herbs, like *Heracleum sprengelianum*, have also been reported against *Aedes albopictus*, *Anopheles subpictus*, and *Culex tritaeniorhynchus* (Govindarajan & Benelli, 2016). The World Health Organization (WHO) evaluated the toxicity levels of different parts of the plant *Ipomoea cairica* (Railway Creeper) against larvae of *Aedes* utilizing standard larval susceptibility test method. The results showed a favorable efficacy of the extract of *Ipomoea cairica* leaves against *Aedes aegypti* and *Aedes albopictus* (AhbiRami, Zuharah, Thiagaletchumi, Subramaniam, & Sundarasekar, 2014). Similarly, toxic activity was shown by the leaf extracts obtained from *Glycosmis pentaphylla* against the larvae and adults of *Anopheles stephensi*, *Aedes aegypti*, and *Culex quinquefasciatus* mosquitoes. The results obtained were in accordance to the WHO test procedures recommended for bioassays of larval and adult efficacy of compounds (Ramkumar et al., 2016). Herbal extracts along with *Mesocyclops* sp. have been reported to synergistically increase the predation of *Aedes* sp. Similarly, from extraction of neem oil from the seeds of *Azadirachta indica*,

TABLE 3 Results of various studies on the use of eucalyptus oil.

Study findings	References
Kaura et al. studied the efficacy of eucalyptus oil for *Aedes aegypti* and *Aedes albopictus* and found that the LC50 (concentration at which 50% mortality is observed) for larvae and pupae was 93.3 and 144.5 ppm and LC90 (concentration at which 90% mortality is observed) was 707.9 and 741.3 ppm, respectively, while for neem oil, LC50 for larvae and pupae was 7852 and 19,054 ppm and LC90 was 10,092 and 19,952 ppm, respectively	Kaura et al. (2019)
Pugazhvendan et al. studied the larvicidal activity of eucalyptus oil. LC50 and LC90 against *Aedes* were found to be 68.18 and 248.37 ppm, respectively	Pugazhvendan and Elumali (2013)
Manimaran et al. recorded LC50 and LC90 to be 64 and 80 ppm for eucalyptus oil	Manimaran et al. (2012)

a by-product called neem cake is obtained that has also been used for biosynthesizing of AgNP nanoparticles. When *Carassius auratus* is treated with these nanoparticles, its predation efficiency has been shown to increase (Chandramohan et al., 2016). *Ichnocarpus frutescens* extract based on silver nanoparticle also showed larvicidal properties and is therefore considered to be a promising candidate for mosquito control (Govindarajan, Rajeswary, & Benelli, 2016).

In our laboratory, we evaluated commercially available oil preparations of *Eucalyptus* (*E. globulus*) and neem (*Azadirachta indica*) against larvae and pupae of *Aedes aegypti* and *Aedes albopictus*. We observed that eucalyptus oil was more effective against larvae and pupae at lower concentrations (Table 3) and was also found to have a residual efficacy up to 8 days (196 h); however, after 8 days, the efficacy declined sharply, thereby emphasizing the need to drain or replace the water at the earliest. Similar to our finding, other studies have also found eucalyptus oil to be efficacious against *Aedes* larvae; however, the effective concentration of the oil varies substantially for different oil preparations (Table 3).

In general the use of plant-based extracts is intriguing and promising to eradicate mosquitoes without any concern of environmental bioaccumulation. However, prior to their use, these must be profiled for any acute or chronic toxicity against the other unspecified or undesired insects or organisms.

Laboratory setup and equipment required for essential oil testing

The larvicidal activity of an oil preparation can be evaluated in a well-equipped insectary unit under ideal conditions. The insectary can be set up in a small but dedicated room with essential equipment (Table 4, Fig. 1). The identification of mosquito species needs training and experience:

- **Temperature:** A temperature of 27°C is required for growth of mosquitoes that can be obtained by heaters and air conditioners.
- **Humidity:** A relative humidity of 80% of required to maintain adult mosquitoes. A constant level of humidity is achieved by using a pair of humidifier and dehumidifier, which can maintain the desired temperature in combination.

TABLE 4 Basic laboratory equipment required for mosquito rearing.

- Mosquito trap
- Mouth aspirator/suction tube
- White plastic containers
- Larval rearing trays with tray covers
- Larval diet
- Adult mosquito cage
- Plastic disposable pipettes
- Plastic pitchers and bottles
- Dehumidifier
- Humidifier
- Stereomicroscope

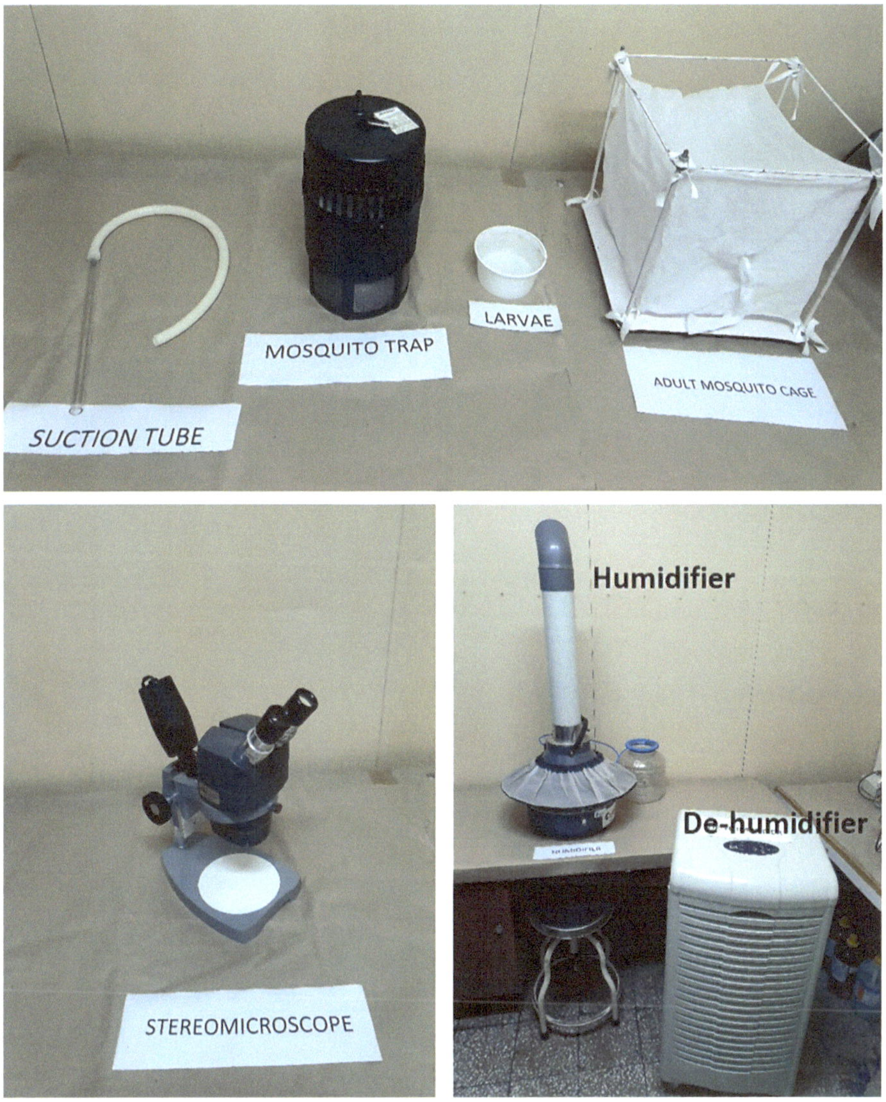

FIG. 1 Laboratory equipment required for mosquito rearing.

- **Lighting:** A 12-h day (light) and 12-h night (dark) schedule is followed by most insectaries.
- **Biosafety:** The escape of adult mosquitoes must be contained using adequate protection.

For evaluation of an essential oil, mosquito larvae can be collected from water collection in artificial containers, ditches, and other natural water bodies, followed by rearing of the larvae to adult stage. One-fifth of the collected larvae should be reared up to the adult stage (Fig. 2) so as to identify the freshly emerged adults up to species level by using standard morphological keys (Barraud, 1934). Table 5 shows the stepwise protocol for larvicidal testing.

Future research areas of using essential oils for mosquito control

The easy availability and less adverse environmental impact of EOs, which are complex mixtures of volatile organic compounds from plants, have developed an increased interest in using these as an alternative to chemical insecticides. Among the major limitations of using them as insect repellents is their volatile nature, that is, they tend to evaporate quickly and need to be reapplied frequently. Therefore there is a need to prepare better formulations for product delivery, besides the other research arenas in this field (Table 6). This is a potentially rich and interesting area for research scientists with direct benefit to human, animal, and environmental well-being.

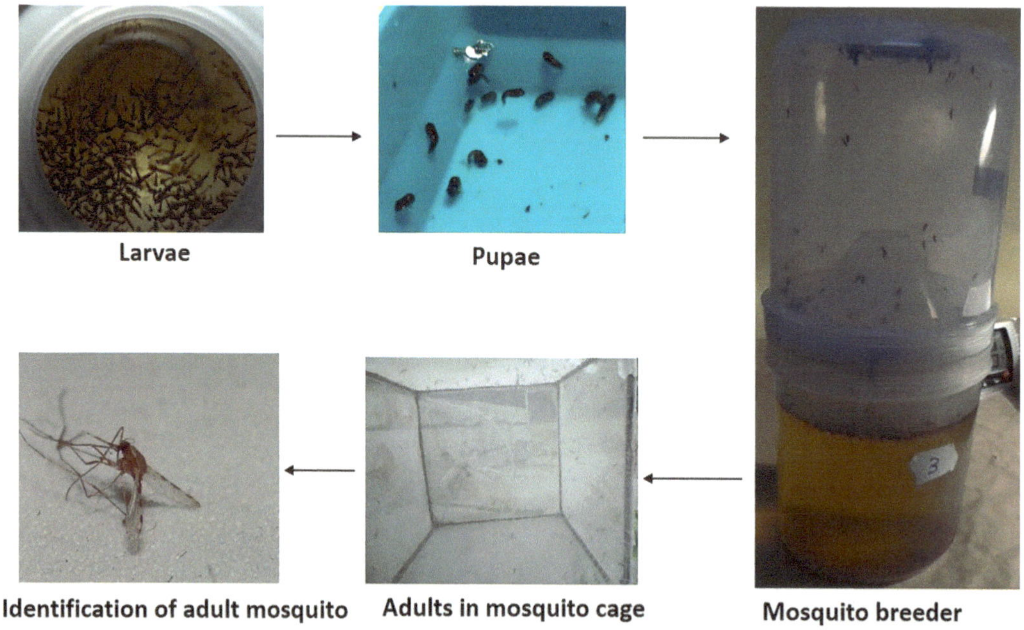

FIG. 2 Rearing of mosquito larva to adult stage.

TABLE 5 Protocol to study larvicidal activity of essential oils.

Mosquito larvae collected and reared
↓
Third or fourth instar larvae transferred to small disposable test cups or vessels each containing 100–200 mL of water by means of strainers or droppers
↓
Larval food added to each test cup for long exposures
↓
Small, unhealthy, or damaged larvae removed and replaced, if required
↓
Mosquito larvae exposed to a wide range of test concentrations and a control agent with known potency to find out the activity range of the larvicide under test
↓
After 24-h exposure, larval mortality recorded
↓
LC50 and LC90 values calculated

TABLE 6 Potential areas of future research on essential oils.

- Purification of the potent compounds in the essential oils may lead to the discovery of key effective molecules
- Cellular pathway responses to different doses of essential oils and mechanisms of action need to be unraveled so as to predict health risks
- Toxicological studies are required to evaluate the safety of essential oil repellents
- Improvements in water dispersibility of essential oils can improve insecticidal activity in natural water bodies
- Nanotechnology can be used to accelerate the efficacy and dispersion of essential oils
- Development of slow-release devices can make the essential oil preparations long acting
- Development of formulations that can be applied topically on human skin
- Development of patch formulations can render delivery of the repellents effective

Conclusion

Millions of humans are afflicted by the life-threatening effects of mosquito borne diseases. *Aedes aegypti* is responsible for the recent surge in the incidents of dengue, chikungunya, and yellow fever, and it has generated a lot of intrigue in finding effective and eco-friendly solutions to prevent the occurrence of mosquito bites and its related effects. The global attention freshly diverted to *Aedes* mosquito when it was identified that they are accountable for the transmission of ZIKV. The vital limitation in our armamentarium to combat these diseases is that there are no potential vaccines against these viruses nor any drugs to prevent their proliferation. Hence, it is better to prevent the problem at initial stages in terms of direct targeting the mosquito vectors. A wide variety of synthetic repellents have been used in the past, but they suffer on the health safety aspect. It is thus pertinent to turn toward nature and look for means and methods to develop bio-based repellents with long-lasting protection and efficacy. The accelerated advancement in the field of biotechnology can be exploited to cater to this need and develop natural repellents with improved properties. The plant-derived EOs are exciting substitutes to synthetic repellents. They can be scrutinized based on their toxicological profiles to discover novel repellents after duly monitoring the target exposure and cellular response along with the mode of action and last but not the least assessing the health risks associated with them.

Policy and procedures

This chapter focuses on providing information on the potential of essential oils (EO) and their derivatives for the control of ZIKV. The authors have reviewed and discussed the role of the derivatives of EO obtained from different plants against ZIKV. The general properties, toxicity, and safety of some of the commercially available EO were described in detail. Various methods used in extraction of EOs were tabulated. Previously published studies on the efficacy of EOs as mosquito repellents was reviewed and discussed along with the author's own work. The minimum equipment and conditions required for an entomology laboratory set up and the basic protocol to study larvicidal activity of EOs has been described. Finally, potential areas of future research required on EOs has been discussed.

Mini-dictionary of terms

Arbovirus. Any virus that is transmitted to vertebrates through any species of blood feeding arthropods like flies, mosquitoes, ticks, etc.

Terpenoids. A large and diverse class of naturally occurring organic chemicals derived from terpenes. Most of these compounds are multicyclic structures with oxygen-containing functional groups.

Essential oils. These volatile oils are natural metabolic secretions of plants considered to be true plant hormones and fluid components of the immune system of plants. Some essential oils give the fragrance of the aromatic plant and have medicinal properties, such as fever management, blood cleansing, and relieving pain.

Plant-based insecticides. Also called as botanical insecticides or natural insecticides. These are naturally occurring insect toxins derived from plants.

Nanomosquitocides. These are the nanoformulations or nanoparticles utilized for the killing of mosquitoes. A number of botanicals have been utilized for this purpose and are referred as green nanomosquitocides.

Key facts of plant essential oils (EOs)

- Plants continue to be one of the major raw materials for various compounds useful for human health.
- The abundance of Eos in nature and their uses in repelling mosquitoes, easy availability, low cost, acceptability for general household use, and rapid degradation in the environment can be a sustainable alternative to insecticides.
- Plant EOs are volatile metabolic secretions of plants and are stored in specific parts like leaves, fruits, flowers, or stem.
- Though EOs are widely present in nature, they are found in very low concentrations.
- EOs are capable of crossing the placenta to reach the fetus and may have effect on the development of its nervous system.
- EOs are also utilized as natural additives in food, cosmetics, and pharmaceuticals because of their antibacterial activities.
- EOS of many plants possess pharmacological and therapeutic properties like antiinflammatory, analgesic, antidiabetic, antidiarrheal, antioxidant, antipyretic, antiinflammatory, and insecticidal activities.

Summary points

- Arboviruses are a worldwide public health concern because of their association with several vector-borne diseases.
- ZIKV is an arbovirus belonging to genus *Flavivirus* and family *Flaviviridae*. It was first reported from the Zika forest of Uganda in 1947, later spread worldwide, and became a global threat.
- No effective vaccines or specific drugs are available against ZIKV; therefore direct targeting of mosquitoes is required to control the spread of this neglected tropical disease.
- Plant EOs, being concoction of different bioactive compounds, can be used as efficient bioinsecticides that are environment- and user-friendly.
- Even though plant EOs are considered to be safe, few side effects like skin irritation are associated with some EOs. The EOs with better repellent properties, selectivity to target, and low toxicities to humans and environment are required.
- Citronella, clove oil, PMD, and permethrin are some of the efficient EOs used currently as insect repellents.
- A number of plant extracts derived from different plants are utilized widely as such or even as nanoformulations as insect repellents or larvicides for mosquito control.

References

Adeosun, C., Bamidele, I., & Ayorinde, A. (2016). Essential oil from the stem bark of Casuarina equisetifolia exhibits mosquitocidal potential. *Journal of Phytopathology and Pest Management*, *3*, 53–62.

AhbiRami, R., Zuharah, W. F., Thiagaletchumi, M., Subramaniam, S., & Sundarasekar, J. (2014). Larvicidal efficacy of different plant parts of railway creeper, Ipomoea cairica extract against dengue vector mosquitoes, *Aedes albopictus* (Diptera: Culicidae) and *Aedes aegypti* (Diptera: Culicidae). *Journal of Insect Science*, *14*, 180.

Barraud, P. J. (1934). *Diptera*, family *Culicidae*, tribes *Megarhinni* and *Culicine*. In S. R. Christophers (Ed.), *The Fauna of British India including Ceylon and Burma* (pp. 221–260). London: Taylor and Francis.

Benelli, G., & Mehlhorn, H. (2016). Declining malaria, rising of dengue and Zika virus: Insights for mosquito vector control. *Parasitology Research*, *115*, 1747–1754.

Benelli, G., & Pavela, R. (2018). Beyond mosquitoes—Essential oil toxicity and repellency against bloodsucking insects. *Industrial Crops and Products*, *117*, 382–392.

Benelli, G., Rajeswary, M., & Govindarajan, M. (2018). Towards green oviposition deterrents? Effectiveness of Syzygium lanceolatum (Myrtaceae) essential oil against six mosquito vectors and impact on four aquatic biological control agents. *Environmental Science and Pollution Research*, *25*, 10218–10227.

Braack, L., de Almeida, A. P., Cornel, A. J., Swanepoel, R., & De Jager, C. (2018). Mosquito-borne arboviruses of African origin: Review of key viruses and vectors. *Parasites & Vectors*, *11*, 29.

Cao-Lormeau, V. M., Blake, A., Mons, S., Lastère, S., Roche, C., Vanhomwegen, J., et al. (2016). Guillain-Barré syndrome outbreak associated with Zika virus infection in French Polynesia: A case-control study. *The Lancet*, *387*, 1531–1539.

Chandramohan, B., Murugan, K., Panneerselvam, C., Madhiyazhagan, P., Chandirasekar, R., Dinesh, D., et al. (2016). Characterization and mosquitocidal potential of neem cake-synthesized silver nanoparticles: Genotoxicity and impact on predation efficiency of mosquito natural enemies. *Parasitology Research*, *115*, 1015–1025.

Chang, C., Ortiz, K., Ansari, A., & Gershwin, M. E. (2016). The Zika outbreak of the 21st century. *Journal of Autoimmunity*, *68*, 1–3.

Conti, B., Leonardi, M., Pistelli, L., Profeti, R., Ouerghemmi, I., & Benelli, G. (2013). Larvicidal and repellent activity of essential oils from wild and cultivated *Ruta chalepensis* L.(Rutaceae) against *Aedes albopictus* Skuse (Diptera: Culicidae), an arbovirus vector. *Parasitology Research*, *112*, 991–999.

da Silva, T. I., Alves, A. C., de Azevedo, F. R., Marco, C. A., dos Santos, H. R., & Alves, W. S. (2017). Efeito larvicida de óleos essenciais de plantas medicinais sobre larvas de *Aedes aegypti* L. (Diptera: Culicidae). *Revista Verde de Agroecologia e Desenvolvimento Sustentável*, *12*, 256–260.

Dias, C. N., & Moraes, D. F. (2014). Essential oils and their compounds as *Aedes aegypti* L. (Diptera: Culicidae) larvicides. *Parasitology Research*, *113*, 565–592.

Diaz, J. H. (2016). Chemical and plant-based insect repellents: Efficacy, safety, and toxicity. *Wilderness & Environmental Medicine*, *27*, 153–163.

Dick, G. W., Kitchen, S. F., & Haddow, A. J. (1952). Zika virus (I). Isolations and serological specificity. *Transactions of the Royal Society of Tropical Medicine and Hygiene*, *46*, 509–520.

Dickens, J. C., & Bohbot, J. D. (2013). Mini review: Mode of action of mosquito repellents. *Pesticide Biochemistry and Physiology*, *106*, 149–155.

Drago, B., Shah, N. S., & Shah, S. H. (2014). Acute permethrin neurotoxicity: Variable presentations, high index of suspicion. *Toxicology Reports*, *1*, 1026–1028.

Enan, E. (2001). Insecticidal activity of essential oils: Octopaminergic sites of action. *Comparative Biochemistry and Physiology Part C: Toxicology & Pharmacology*, *130*, 325–337.

Ferraris, P., Yssel, H., & Missé, D. (2019). Zika virus infection: An update. *Microbes and Infection*, *21*, 353–360.

Gonçalves, G. L., Ribeiro, L. D., Gimenes, L., Vieira, P. C., da Silva, M. F., Forim, M. R., et al. (2015). Lethal and sublethal toxicities of Annona sylvatica (Magnoliales: Annonaceae) extracts to Zabrotes subfasciatus (Coleoptera: Chrysomelidae: Bruchinae). *Florida Entomologist*, *98*, 921–928.

Gould, E., Pettersson, J., Higgs, S., Charrel, R., & de Lamballerie, X. (2017). Emerging arboviruses: Why today? *One Health*, *4*, 1–3.

Govindarajan, M., & Benelli, G. (2016). α-Humulene and β-elemene from Syzygium zeylanicum (Myrtaceae) essential oil: Highly effective and eco-friendly larvicides against *Anopheles subpictus*, *Aedes albopictus*, and *Culex tritaeniorhynchus* (Diptera: Culicidae). *Parasitology Research*, *115*, 2771–2778.

Govindarajan, M., Hoti, S. L., & Benelli, G. (2016). Facile fabrication of eco-friendly nano-mosquitocides: Biophysical characterization and effectiveness on neglected tropical mosquito vectors. *Enzyme and Microbial Technology*, *95*, 155–163.

Govindarajan, M., Kadaikunnan, S., Alharbi, N. S., & Benelli, G. (2016). Acute toxicity and repellent activity of the Origanum scabrum Boiss. & Heldr. (Lamiaceae) essential oil against four mosquito vectors of public health importance and its biosafety on non-target aquatic organisms. *Environmental Science and Pollution Research*, *23*, 23228–23238.

Govindarajan, M., Rajeswary, M., & Benelli, G. (2016). Chemical composition, toxicity and non-target effects of *Pinus kesiya* essential oil: An eco-friendly and novel larvicide against malaria, dengue and lymphatic filariasis mosquito vectors. *Ecotoxicology and Environmental Safety*, *129*, 85–90.

Gulland, A. (2016). Zika virus is a global public health emergency, declares WHO. *British Medical Journal*, *352*, i657.

Haddad, J. G., Picard, M., Bénard, S., Desvignes, C., Desprès, P., Diotel, N., et al. (2019). Essential oil and its main component thymohydroquinone dimethyl ether inhibit Zika virus at doses devoid of toxicity in zebrafish. *Molecules*, *24*, 3447.

Hesham, H. A. R., Abdurahman, H. N., & Rosli, M. Y. (2016). Techniques for extraction of essential oils from plants: A review. *Australian Journal of Basic and Applied Sciences*, *10*(16), 117–127.

Houghton, P. J., Ren, Y., & Howes, M. J. (2006). Acetylcholinesterase inhibitors from plants and fungi. *Natural Product Reports*, *23*, 181–199.

Isman, M. (1999). Pesticides based on plant essential oils. *Pesticide Outlook*, *10*, 68–72.

Isman, M. B. (2000). Plant essential oils for pest and disease management. *Crop Protection*, *19*, 603–608.

Kaura, T., Mewara, A., Zaman, K., Sharma, A., Agrawal, S. K., Thakur, V., et al. (2019). Utilizing larvicidal and pupicidal efficacy of *Eucalyptus* and neem oil against *Aedes* mosquito: An approach for mosquito control. *Tropical Parasitology*, *9*(1), 12–17.

Khanikor, B., Parida, P., Yadav, R. N., & Bora, D. (2013). Comparative mode of action of some terpene compounds against octopamine receptor and acetyl cholinesterase of mosquito and human system by the help of homology modeling and docking studies. *Journal of Applied Pharmaceutical Science*, *3*, 6.

Kongkaew, C., Sakunrag, I., Chaiyakunapruk, N., & Tawatsin, A. (2011). Effectiveness of citronella preparations in preventing mosquito bites: Systematic review of controlled laboratory experimental studies. *Tropical Medicine & International Health*, *16*, 802–810.

Kumar, S., Mishra, M., Wahab, N., & Warikoo, R. (2014). Larvicidal, repellent, and irritant potential of the seed-derived essential oil of Apium graveolens against dengue vector, *Aedes aegypti* L. (Diptera: Culicidae). *Frontiers in Public Health*, *2*, 147.

Lee, M. Y. (2018). Essential oils as repellents against arthropods. *BioMed Research International*, *1*, 1–9.

Lupi, E., Hatz, C., & Schlagenhauf, P. (2013). The efficacy of repellents against Aedes, Anopheles, Culex and Ixodes spp.—A literature review. *Travel Medicine and Infectious Disease*, *11*, 374–411.

Maia, M. F., & Moore, S. J. (2011). Plant-based insect repellents: A review of their efficacy, development and testing. *Malaria Journal*, *10*, S11.

Manimaran, A., Mary, J. J., Cruz, M., Muthu, C., Vincent, S., & Ignacimuthu, S. (2012). Larvicidal and knockdown effects of some essential oils against *Culex quinquefasciatus* (Say), *Aedes aegypti* (L.) and *Anopheles stephensi* (Liston). *Advance in Bioscience and Biotechnology*, *3*, 855–862.

Manzoor, F., Samreen, K. B., & Parveen, Z. (2013). Larvicidal activity of essential oils against *Aedes aegypti* and *Culex quinquefasciatus* larvae (Diptera: Culicidae). *Journal of Animal and Plant Sciences*, *23*, 420–424.

Navarro, D. M. D. A. F., da Silva, P. C. B., da Silva, M. F. R., Napoleao, T. H., & Paiva, P. M. G. (2013). Larvicidal activity of plant and algae extracts, essential oils and isolated chemical constituents against *Aedes aegypti*. *The Natural Products Journal*, *3*, 268–291.

Nerio, L. S., Olivero-Verbel, J., & Stashenko, E. (2010). Repellent activity of essential oils: A review. *Bioresource Technology*, *101*, 372–378.

Niroumand, M. C., Farzaei, M. H., Razkenari, E. K., Amin, G., Khanavi, M., Akbarzadeh, T., et al. (2016). An evidence-based review on medicinal plants used as insecticide and insect repellent in traditional Iranian medicine. *Iranian Red Crescent Medical Journal*, *18*, e22361.

Oliveira, G. L., Cardoso, S. K., Lara Junior, C. R., Vieira, T. M., Guimarães, E. F., Figueiredo, L. S., et al. (2013). Chemical study and larvicidal activity against *Aedes aegypti* of essential oil of Piper aduncum L. (Piperaceae). *Anais da Academia Brasileira de Ciências*, *85*, 1227–1234.

Parra, B., Lizarazo, J., Jiménez-Arango, J. A., Zea-Vera, A. F., González-Manrique, G., Vargas, J., et al. (2016). Guillain–Barré syndrome associated with Zika virus infection in Colombia. *New England Journal of Medicine*, *375*, 1513–1523.

Pohlit, A. M., Rezende, A. R., Baldin, E. L., Lopes, N. P., & de Andrade Neto, V. F. (2011). Plant extracts, isolated phytochemicals, and plant-derived agents which are lethal to arthropod vectors of human tropical diseases—A review. *Planta Medica*, *77*, 618–630.

Priestley, C. M., Williamson, E. M., Wafford, K. A., & Sattelle, D. B. (2003). Thymol, a constituent of thyme essential oil, is a positive allosteric modulator of human GABAA receptors and a homo-oligomeric GABA receptor from Drosophila melanogaster. *British Journal of Pharmacology*, *140*, 1363–1372.

Pugazhvendan, S. R., & Elumali, K. (2013). Larvicidal activity of selected plant essential oil against important vector mosquitoes: Dengue vector, *Aedes aegypti* (L.), malaria vector, *Anopheles stephensi* (Liston) and filarial vector, *Culex quinquefasciatus* (Say) (Diptera: Culicidae). *Middle East Journal of Science Research*, *18*, 91–95.

Ramkumar, G., Karthi, S., Muthusamy, R., Natarajan, D., & Shivakumar, M. S. (2015). Adulticidal and smoke toxicity of Cipadessa baccifera (Roth) plant extracts against *Anopheles stephensi*, *Aedes aegypti*, and *Culex quinquefasciatus*. *Parasitology Research*, *114*, 167–173.

Ramkumar, G., Karthi, S., Muthusamy, R., Suganya, P., Natarajan, D., Kweka, E. J., et al. (2016). Mosquitocidal effect of Glycosmis pentaphylla leaf extracts against three mosquito species (Diptera: Culicidae). *PLoS One*, *1*, e0158088.

Santos, H. S., Bandeira, P. N., Lemos, T. L., & Santiago, G. M. (2017). Chemical composition and larvicidal activity against *Aedes aegypti* L. (Diptera: Culicidae) of essential oils from leaves, stalks and roots of the Croton nepetaefolius Baill (Euphorbiaceae). *International Journal of Mosquito Research*, *4*, 19–22.

Sathantriphop, S., Achee, N. L., Sanguanpong, U., & Chareonviriyaphap, T. (2015). The effects of plant essential oils on escape response and mortality rate of *Aedes aegypti* and *Anopheles minimus*. *Journal of Vector Ecology*, *40*, 318–326.

Sharifi-Rad, J., Sureda, A., Tenore, G., Daglia, M., Sharifi-Rad, M., Valussi, M., et al. (2017). Biological activities of essential oils: From plant chemoecology to traditional healing systems. *Molecules*, *22*, 70.

Sharma, R., Rao, R., Kumar, S., Mahant, S., & Khatkar, S. (2019). Therapeutic potential of citronella essential oil: A review. *Current Drug Discovery Technologies*, *16*, 330–339.

Singh, R. K., Dhama, K., Malik, Y. S., Ramakrishnan, M. A., Karthik, K., Tiwari, R., et al. (2016). Zika virus—Emergence, evolution, pathology, diagnosis, and control: Current global scenario and future perspectives—A comprehensive review. *Veterinary Quarterly*, *36*, 150–175.

Sogan, N., Kapoor, N., Singh, H., Kala, S., Nayak, A., & Nagpal, B. N. (2018). Larvicidal activity of Ricinus communis extract against mosquitoes. *Journal of Vector Borne Diseases*, *55*, 282.

Soleimani-Ahmadi, M., Abtahi, S. M., Madani, A., Paksa, A., Abadi, Y. S., Gorouhi, M. A., et al. (2017). Phytochemical profile and mosquito larvicidal activity of the essential oil from aerial parts of Satureja bachtiarica Bunge against malaria and lymphatic filariasis vectors. *Journal of Essential Oil Bearing Plants*, *20*, 328–336.

Souza, T. M., Farias, D. F., Soares, B. M., Viana, M. P., Lima, G. P., Machado, L. K., et al. (2011). Toxicity of Brazilian plant seed extracts to two strains of *Aedes aegypti* (Diptera: Culicidae) and nontarget animals. *Journal of Medical Entomology*, *48*, 846–851.

Swale, D. R., Sun, B., Tong, F., & Bloomquist, J. R. (2014). Neurotoxicity and mode of action of N, N-diethyl-meta-toluamide (DEET). *PLoS One*, *9*, e103713.

Tan, K. H., & Nishida, R. (2012). Methyl eugenol: Its occurrence, distribution, and role in nature, especially in relation to insect behavior and pollination. *Journal of Insect Science*, *12*, 1–74.

Taylor, R. W., Romaine, I. M., Liu, C., Murthi, P., Jones, P. L., Waterson, A. G., et al. (2012). Structure–activity relationship of a broad-spectrum insect odorant receptor agonist. *ACS Chemical Biology*, *7*, 1647–1652.

Tong, F., & Bloomquist, J. R. (2013). Plant essential oils affect the toxicities of carbaryl and permethrin against *Aedes aegypti* (Diptera: Culicidae). *Journal of Medical Entomology*, *50*, 826–832.

Vest, K. G. (2017). Zika virus update: More on an emerging arboviral disease in the western hemisphere. *Disaster Medicine and Public Health Preparedness*, *11*, 163–167.

Voris, D. G., Afonso, C. H., Almeida Filho, C. A., Fernandes, C. O., Brito, D. Q., Moraes, C. S., et al. (2017). Estudos etnofarmacológicos de óleos essenciais com atividade larvicida contra o mosquito *Aedes aegypti*. *Semioses*, *11*, 86–94.

Woodring, J. P., McBride, L. A., & Fields, P. (1989). The role of octopamine in handling and exercise-induced hyperglycaemia and hyperlipaemia in Acheta domesticus. *Journal of Insect Physiology*, *35*, 613–617.

Zhang, Q., Sun, K., Chinazzi, M., Piontti, A. P. Y., Dean, N. E., Rojas, D. P., et al. (2017). Spread of Zika virus in the Americas. *Proceedings of the National Academy of Sciences*, *114*, E4334–E4343.

Chapter 33

Pyridobenzothiazolones as anti-flavivirus agents: Impact on Zika virus

Maria Sole Burali and Giuseppe Manfroni
Department of Pharmaceutical Sciences, University of Perugia, Perugia, Italy

Abbreviations

ADE	antibody-dependent enhancement
BBB	blood-brain barrier
DENV	dengue virus
JEV	Japanese encephalitis virus
MoA	mode of action
NNI	non-nucleoside inhibitors
PBTZ	pyridobenzothiazolone
PHEIC	Public Health Emergency of International Concern
PPI	protein-protein interaction
RC	replication complex
RdRp	RNA-dependent RNA polymerase
REMSA	RNA-electrophoretic mobility shift assay
SAR	structure-activity relationship
TBEV	tick-borne encephalitis virus
UTR	untranslated region
WNV	West Nile virus
YFV	yellow fever virus
ZIKV	Zika virus

A brief introduction

Zika virus (ZIKV) belongs to the *Flavivirus* genus together with more than 70 arthropod-borne RNA viruses including the most studied dengue (DENV), West Nile (WNV), yellow fever (YFV), Japanese encephalitis (JEV), and tick-borne encephalitis (TBEV). These viruses can be transmitted by infected arthropods and therefore can quickly spread when introduced into new regions, other than endemic ones, as exemplified by the Zika explosion in Brazil during 2016 Olympic Games (Ai, Zhang, & Zhang, 2016). To be noted the WHO declared ZIKV outbreak a Public Health Emergency of International Concern (PHEIC) since 2016, especially because of its rapid spread coupled with the sexual and vertical mother-fetus transmission (Paixão et al., 2016). The infection usually causes mild symptoms; however, more severe diseases can also occur such as microcephaly in children born from infected mothers or neurological complications in adults (Guillain-Barrè syndrome, meningoencephalitis, and myelitis) (Musso, Ko, & Baud, 2019).

After the Brazil outbreak, an intense research from both academia and pharmaceutical industries has been carried out with the hope to find suitable or prophylactic treatments before the explosion of another outbreak. There are currently no licensed drugs to treat or prevent ZIKV or other flavivirus infections, but several compounds inhibiting ZIKV replication were discovered over the last 4 years through drug discovery programs or repurposing approaches (Bernatchez et al., 2020). Regarding the repurposing strategy, only sofosbuvir (Bullard-Feibelman et al., 2017; Ferreira et al., 2017) and BCX-4430 (Warren et al., 2014) proved to be active against ZIKV. In particular, sofosbuvir is a nucleoside analog ProTide already on the market for the treatment of HCV infections, and BCX-4430 is an imino C-nucleoside with antiviral activity against Ebola virus, Marburg virus, and YFV that is in phase II clinical trial (see Fig. 1) (Julander et al., 2017). Unfortunately

Zika Virus Impact, Diagnosis, Control, and Models. https://doi.org/10.1016/B978-0-12-820267-8.00033-9

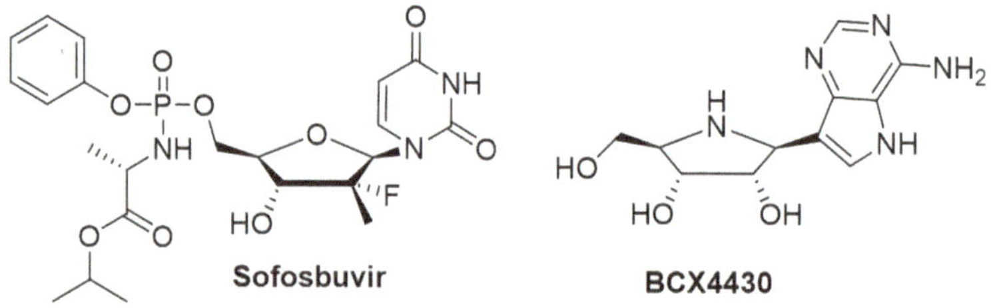

Sofosbuvir | **BCX4430**

$EC_{50(ZIKV)}$ =0.12-30.9 μM
CC_{50} = 360-512 μM

$EC_{50(ZIKV)}$ =3.8-11.7 μg/mL
CC_{50} > 89 μM

FIG. 1 Structure and anti-ZIKV activity of sofosbuvir and BCX-4430. The data show the variability of sofosbuvir and BCX-4430 antiviral activity and cytotoxicity in different cell lines.

the lack of extensive preclinical investigation of the aforementioned drugs is thus hampering the potential administration of these drugs for ZIKV treatment.

In the same way, all the developed ZIKV vaccines, which have the potential to protect more than 2 billion people living in endemic regions, are still in phase I/II of clinical trial, and this is due to the difficult recruitment of ZIKV-infected patients (low number of ZIKV cases in the world) and the potential antibody-dependent enhancement (ADE). Therefore both challenges have to be addressed for the development of an effective and safe ZIKV vaccine (Ngono & Shresta, 2019).

Regarding the antiviral strategy, one of the main approaches to develop a promising antiviral compound is to target a viral protein essential for virus survival such as the RNA-dependent RNA polymerase (RdRp) that is located at the C-terminal part of NS5. This domain represents a promising target (Lim et al., 2016; Lim, Noble, & Shi, 2015) because is highly conserved, is essential for flavivirus replication, and lacks a eukaryotic counterpart. These features can be exploited to identify broad-spectrum anti-flavivirus agents that should have an activity against multiple flavivirus. In the last years a great demand for anti-flavivirus compounds with a broad-spectrum activity is growing to prepare for the possible next epidemic, which could emerge from several neglected viruses (Boldescu, Behnam, Vasilakis, & Klein, 2017). Moreover, targeting NS5 RdRp has already been proved as a winning strategy if we consider the presence in the market of polymerase inhibitors for HCV and HIV (e.g., sofosbuvir and tenofovir disoproxil fumarate) (De Clercq & Li, 2016).

In this chapter, we report our medicinal chemistry efforts leading to the identification and the development of a novel class of anti-flavivirus agents having a 1*H*-pyrido[2,1-*b*][1,3]benzothiazol-1-one (PBTZ) core deriving from a repurposing strategy of anti-HCV NS5B non-nucleoside inhibitors (NNI) (Manfroni et al., 2012). During the hit explosion strategy, entailing modifications at different positions of the PBTZ core, several promising anti-flavivirus compounds active in the low μM range and coupled with low cytotoxicity were identified (Cannalire et al., 2019; Tarantino et al., 2016), and insights on the mechanism of action (MoA) have been obtained.

The discovery of PBTZ 1 (HeE1-2Tyr) as anti-flavivirus inhibitor

Through a scaffold hopping, we wanted to identify new HCV chemotypes having a replacer of benzimidazole core present in many known HCV NS5B NNIs (Ganta et al., 2019) that could achieve the inhibition of the viral polymerase (NS5B) by the binding to the allosteric site called thumb site I (Manfroni et al., 2012). The synthesized compounds showed an interesting NS5B inhibition (in particular compound **HeE1-2SO2**; see Fig. 2 for structure) but a poor anti-HCV activity in the cell-based assay probably due to permeability issue related to the presence of an acid moiety (Manfroni et al., 2012).

Some years after the PBTZ class was rediscovered through an in silico screening on DENV3 RdRp using the in-house library of NS5B HCV inhibitors. In fact, it is important to mention that DENV NS5 RdRp and HCV NS5B share a common structural architecture, a similar operating mechanism, as well as conserved motifs in the catalytic sites and palm subdomains (Yap et al., 2007). In particular, hit compounds **HeE1-2Tyr (1)** and **HeE1-2SO2**, not published yet at the time of the research, were selected as virtual hits worthy to be tested in biochemical settings. Structurally, they were characterized by the presence of the same C-8 cyclohexyloxy group and C-2 phenyl ring but differ for the presence of an amino acid moiety at the C-4 carboxylic group and for a sulfone group, respectively. Both derivatives **1** and **HeE1-2SO2** were validated as NS5 inhibitors in biochemical assays using DENV3 and WNV RdRps. In particular, derivative **1** displayed low-μM activity against the isolated target NS5 and most importantly was also effective against the replication of different flaviviruses in cells without significant toxicity and an encouraging selectivity index (Fig. 2) (Tarantino et al., 2016). To be noted,

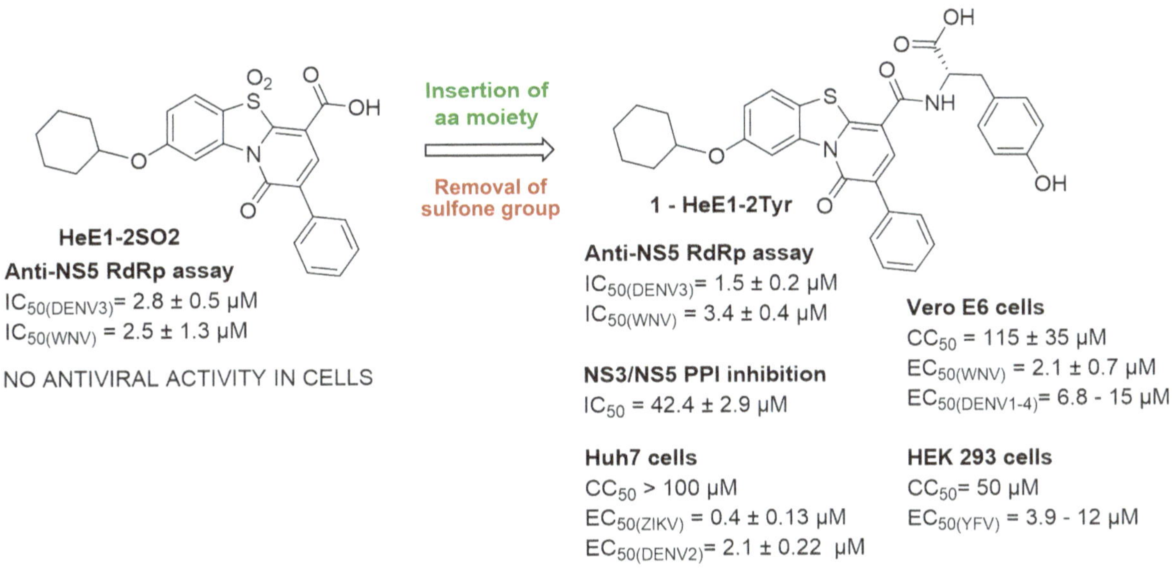

FIG. 2 Pyridobenzothiazolones hit development. Chemical modifications from **HeE1-2SO2** to hit compound **1**. The removal of the sulfone and the insertion of a Tyr allowed **1** to be active against different flaviviruses also in the cell-based assay. The data show the mean ± SD or the variable antiviral activity of **1** when tested in different cell lines.

compound **1** is structurally characterized by an aminoacyl moiety (Tyr) bound to C-4 position, and this moiety is probably indispensable for the antiviral activity; in fact, compound **HeE1-2SO2** lacking the C-4 amide side chain is not active at all in the cell context (Manfroni et al., 2012). However, at the time, the tyrosinyl derivative **1** was not tested against ZIKV since we did not have access to the assay and the virus was not available yet. Anyway the identification of a promising broad-spectrum anti-flavivirus compound prompted us to carry out several assays to in-depth investigate the MoA of compound **1**, such as kinetic, mutational, crystallographic, and docking studies that will be deeply described in the section "Unraveling the mechanism of action of PBTZ." Unfortunately, all the data were not conclusive to move through a structure-based drug design; therefore we decided to start from a ligand-based chemical optimization of PBTZ series with the aim to synthesize more potent compounds also useful for further MoA investigations.

From the hit identification to the hit explosion: Design of potent, selective, and broad-spectrum anti-flavivirus PBTZs

As a continuation of our effort in the identification of anti-flavivirus agents, a new PBTZ series based on hit **1** was designed and synthesized focusing on the C-2 phenyl moiety (R^2 in Fig. 3), C-4 amide side chain (R^4 in Fig. 3), and the C-7/C-8 cyclohexyloxy (R^8 in Fig. 3). In detail the C-2 position was explored by introducing electron-withdrawing or electron-donating substituents on different positions of the phenyl moiety to test whether the presence of different functionalities could influence the activity. The C-4 amide side chain was modified replacing the Tyr moiety with canonical (i.e., Trp, Ser, Phe, Ala, and Met) and non-canonical amino acids (i.e., β-Ala, (*m*-OH)Phe and (*p*-F)Phe). In addition, to explore the importance of the carboxylic function, its deletion (tyramine) and esterification (Tyr methyl ester) were also taken into account together with a replacement of Tyr with aromatic (*p*-OH)benzylamine, benzylamine, and aliphatic ethanolamine, all devoid of aminoacyl functionalities. The C-8 cyclohexyloxy moiety was replaced by smaller, bigger, and more polar alkyl/cycloalkyl substituents, desaturated to its aromatic counterpart, replaced with a fluorine and deleted. Finally a core size reduction due to a complete removal of C-4 substituent and the opening of the flat tricyclic PBTZ core led to an unsubstituted PBTZ and two more flexible pyridone derivatives, respectively (Fig. 3).

All the designed PBTZ compounds were then synthesized and evaluated for their anti-DENV1–4 activity and against a panel of other flaviviruses.

The results retrieved from the antiviral evaluation obtained merging the anti-DENV, anti-WNV, anti-YFV, and anti-ZIKV data helped us to build an appropriate structure-activity relationship (SAR) profile as showed in Fig. 3:

- At C-4 amide side chain, the presence of an amino acid moiety is considered a key chemical requirement. In particular the L-Tyr, (*m*-OH)Phe and β-Ala residues resulted the best substitutions, while the replacement of L-Tyr with its

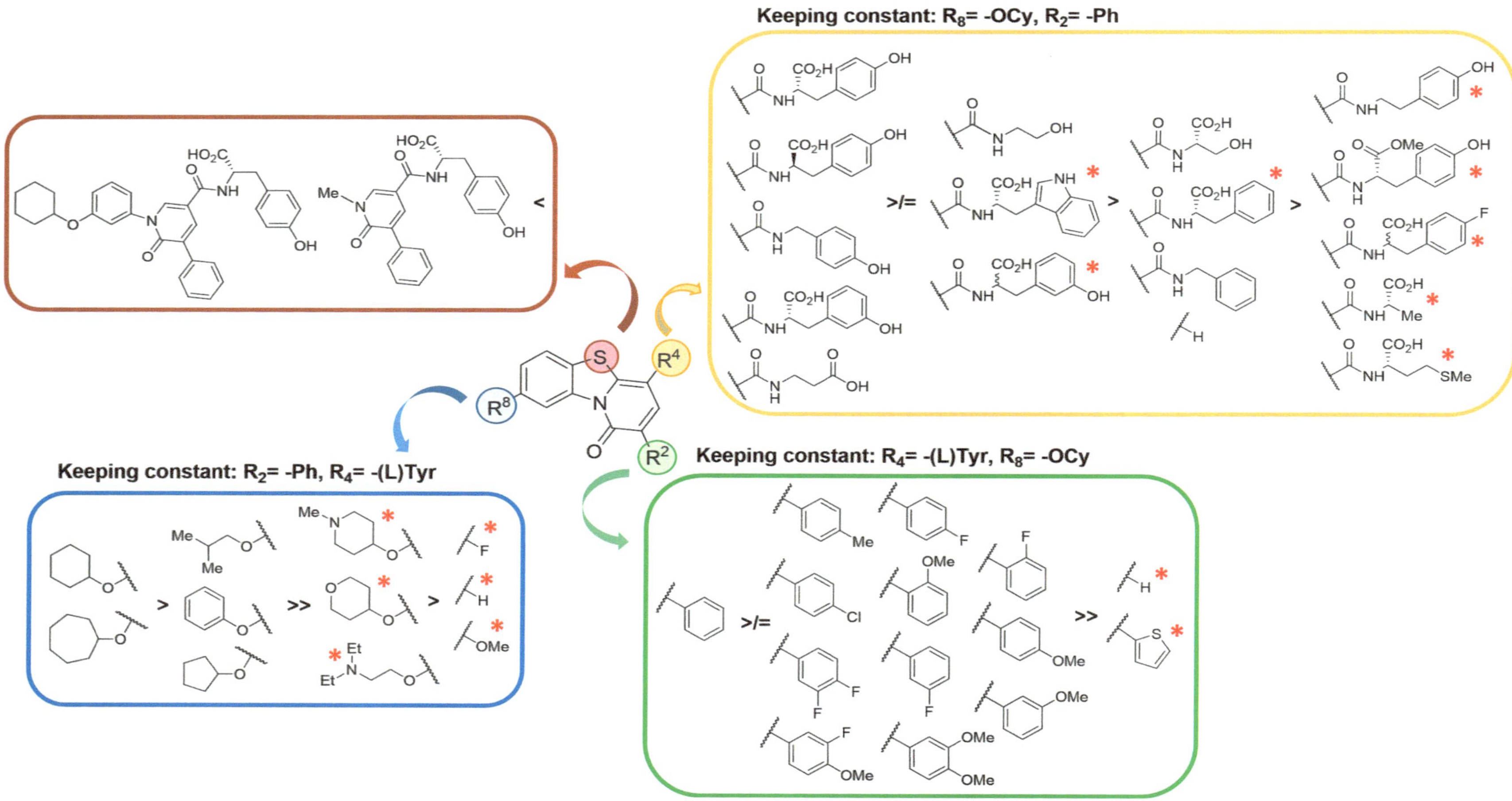

FIG. 3 Schematic SAR of PBTZ derivatives. The data derived from antiviral activity against WNV (Uganda), DENV2 (Martinique) replicating in simian Vero E6 cells and YFV (Bolivia) replicating in human HEK293 cells, and DENV2 (EDEN2 strain) and ZIKV (H/PF/2013 strain) replicating in Huh7 cells. The activity of compounds with asterisk (*) has not been published but included in this chapter to collect more detailed SAR.

(D) counterpart showed a comparable antiviral activity but resulted more toxic. On the other hand, the insertion of a more polar Ser or Phe and the complete removal of C-4 amide side chain resulted detrimental for antiviral activity. Moreover, derivatives bearing an ethanolamine or (*p*-OH)benzylamine could be considered as outsider compounds with respect to the canonical PBTZ series because they showed an excellent antiviral activity despite not having an amino acid moiety. Finally, other compounds previously synthetized and having the tyramine, Tyr methyl ester, Ala, Met, and (*p*-F)Phe are not considered in the SAR because of their high cytotoxicity, which results in unreliable antiviral activity.

- At the C-8 position, cyclohexyloxy is the best substituent with cycloheptyl and cyclopentyl substituents that could be considered suitable replacers since having a similar antiviral activity and selectivity indexes with respect to **1**. The presence of more polar substituents caused a drop in the antiviral activity, while the replacement with a phenyl ring or an isobutoxyl gave not selective antiviral activity. The presence of fluorine or methoxy or the removal of the substituent gave inactive compounds although devoid of any cytotoxicity (data not published).
- At the C-2 position, almost all modifications, entailing the introduction of electron-donating or electron-withdrawing substituents in the phenyl ring, gave compounds with biological activity in the same range of **1**. However, a slightly increase in compound **2** potency (Cannalire et al., 2019) against DENV serotypes was observed (compare the activity of compounds **1** and **2** in Figs. 1 and 4).

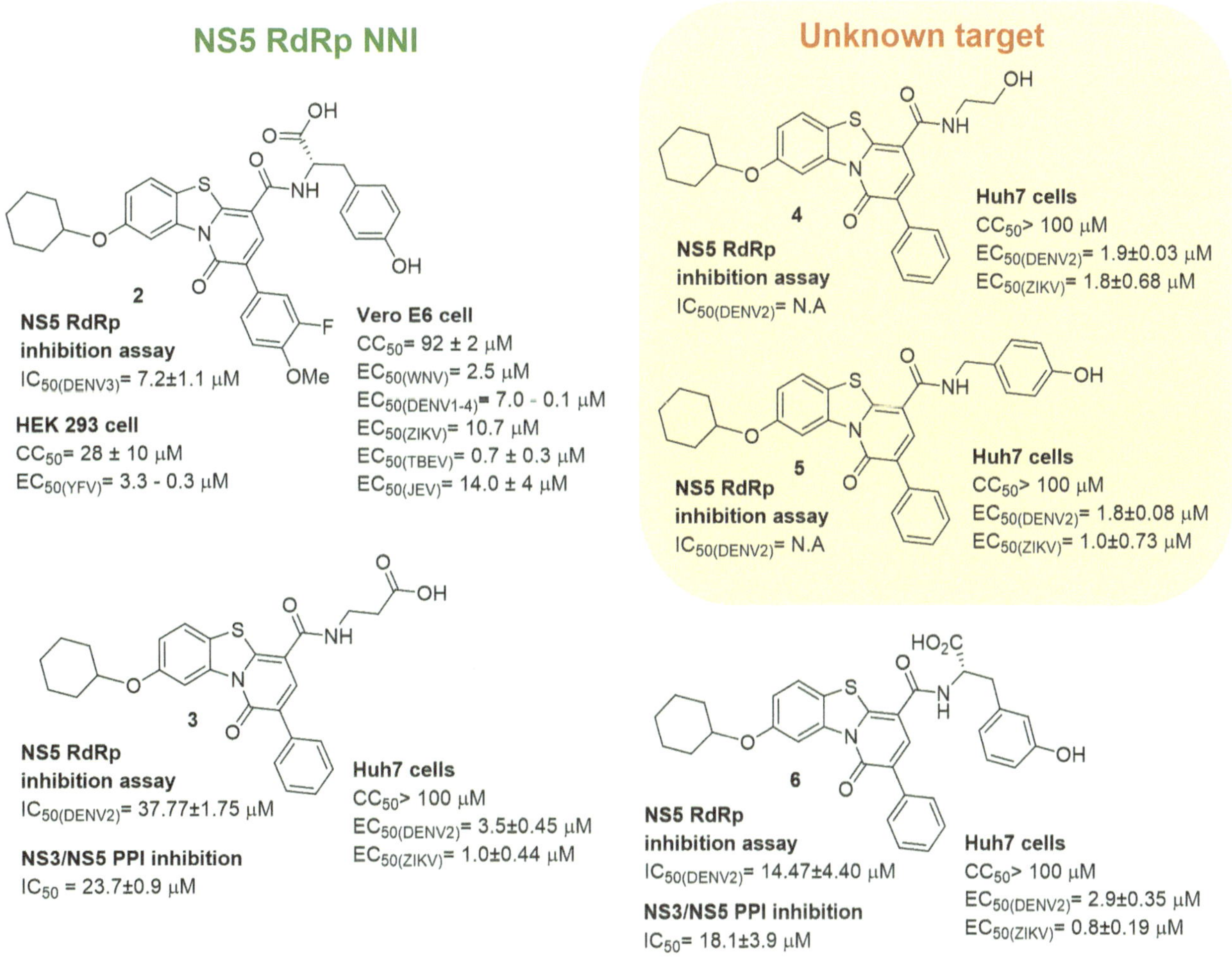

FIG. 4 Structure and activity of the most promising PBTZs. PBTZs **2**, **3**, and **6** are active against the known target NS5 RdRp, the NS5–NS3 PPI, and DENV2 and ZIKV replicating in cells. The NS5 RdRp IC_{50} of **2** is unpublished data since the compound was tested against the isolated target after the publication. On the other hand, PBTZs **5** and **6** act on an unknown target but are promising antiviral agents in cells. The data show the mean ± SD in all assays.

- Scaffold hopping approaches aimed at replacing the tricyclic core with pyridone derivatives resulted unsuccessful since compounds obtained from this strategy resulted completely inactive. Therefore the presence of a rigid and tricyclic PBTZ core resulted in an essential chemical requirement for the anti-flavivirus activity.

Several compounds previously described were also tested against DENV2/DENV3 NS5 RdRp in both de novo and elongation formats. The combination of biochemical and cell-based assays results highlighted a very intriguing behavior for the PBTZ class. Indeed, compounds bearing amino acids at the C-4 amide side chain (such as **2** and **3** (Cannalire et al., 2020)) resulted in promising anti-flavivirus compounds able to inhibit RdRp activity in the elongation phase, and their activity was comparable with hit **1**. On the contrary, non-aminoacidic derivatives such as ethanolamine (**4**) or (*p*-OH)benzylamine (**5**) (Cannalire et al., 2020) are two exhaustive examples of inactive compounds when tested on the isolate NS5 RdRp target but endowed with low-μM anti-flavivirus activity. Their target, cellular or viral, is thus still unknown and deserves further investigations (Fig. 4).

These results suggested a critical role for the carboxyl moiety of aminoacyl derivatives, emerging as a pharmacophore requirement for RdRp inhibition. Even if some PBTZs were able to inhibit the isolate enzyme NS5 RdRp, more in-depth studies were necessary to further understand the MoA for this compound class. All the virological efforts carried out to clarify the molecular target involved in the antiviral activity of PBTZs are described in the next section.

Unraveling the mechanism of action of PBTZs

After the identification of **1** (Tarantino et al., 2016) (Fig. 1), our team carried out several studies trying to elucidate a clear MoA. While crystallographic studies revealed compound **1** located between the finger domain and priming loop of NS5 RdRp, enzymatic assays and kinetic studies showed that a non-competitive inhibition with RNA and nucleotides happened during the elongation step. Since the priming loop is essential for the nucleotide recruitment and the finger domain harbors the RNA, the location of compound **1** in these regions appeared in contrast with respect to kinetic studies. To explain the findings, our team conducted a series of in silico studies to explore other possible allosteric binding sites within the NS5 RdRp. We hypothesized that during the catalytic cycle and assembly of the RdRp-RNA complex, PBTZ **1** might be hosted in a distinct site located in the thumb subdomain (Tarantino et al., 2016). Mutational studies consisting in replacing Cys753Glu and Arg749Ala located in the thumb subdomain of NS5 RdRp and considered key residues for the NS5/**1** interactions gave a functional enzyme less sensitive to the compound inhibitory activity. Not having in our hands straightforward results pushed us to expand our PBTZ series applying a ligand-based instead of structure-based hit explosion, and we synthesized more than 40 compounds with the aim to optimize the initial hit and to find useful leads to definitively clarify the MoA. Within the PBTZ series, two compounds (**2** and **3**) were selected for this purpose. Initially, **2** (Cannalire et al., 2019), endowed with a slightly better antiviral activity than reference compound **1**, was selected as chemical probe for virology studies. At first, our team decided to test **2** in an expanded virus panel including ZIKV, JEV, and TBEV, and we found that the compound possessed a very promising and broad anti-flavivirus activity. At this point, we wanted to ascertain if the antiviral activity was genus specific or other RNA viruses from a different family could also be inhibited. When compound **2** was tested against phlebovirus (Toscana virus) and alphavirus (chikungunya), it resulted completely inactive, thereby confirming the selectivity of PBTZs toward flaviviruses (Cannalire et al., 2019). To check the possible involvement of a host factor in the MoA, our team evaluated the anti-flavivirus activity of compound **2** in different cells (hamster BHK-21, mosquito C6/36, simian Vero E6, and human SW13) infected with YFV (Bolivia strain), used as easy-to-handle but a representative virus for the whole *Flavivirus* genus. Independently from the different cell lines, the activity was maintained with the same EC_{50}s observed in HEK and human Vero cells. Although this evidence did not exclude an involvement of a host target, it strongly supported the hypothesis of a specific viral factor inhibited by PBTZs. To identify the viral target, a resistant mutant selection was attempted exploiting compound **2,** but despite several attempts the selection failed. Then, another experiment was carried out with the aim to monitor the viral release in a single round of infection, and from this evaluation was important to note that PBTZs did not hamper the production of viral RNA and the release of viral particles, but rather the virus infectivity because, after the treatment with **2**, the recovered viruses were not able to perform a second round of infection. The experiment with benzonase, useful for assess the integrity of virions, showed that virions lacked of macroscopic defects on the packaging but were still endowed with a reduced infectivity, likely due to the production of a damaged viral phenotype (Cannalire et al., 2019). In the meantime, it was already known from the literature that two NS proteins, NS3 and NS5, were able to interact for the correct assembly of the replication complex (RC) in infected host cells. The RC is a multiprotein complex of viral and host proteins that assembles on endoplasmic reticulum–derived membranes and works in tight association for mainly viral RNA genome synthesis, methylation of the 5′-cap of viral genome, and polyprotein processing.

Although the exact interplays between RC components are not well understood yet, interactions of NS3 with NS5 that in turn interacts with the untranslated regions (UTRs) are essential for RNA replication and infectious virion production (Brooks et al., 2002; Duan et al., 2019;Lescar et al., 2011 ; Tay et al., 2015). Even if the crystal structure of NS3–NS5 complex is not available yet, an area of NS3 (Tay et al., 2015) and one of NS5, called cavity B (Lescar et al., 2011), were reported to be the main surfaces that take part in this interaction. Moreover, in cavity B of DENV2 NS5 RdRp, the conserved residues Leu327, Leu328, Lys330, Thr858, Trp859, Asn862, Ile863, and Ala866 were discovered as hotspots for NS3-NS5 protein-protein interaction (PPI) and essential for viral replication (Lescar et al., 2011; Tay et al., 2015). Noteworthy, Lys330 was identified as key amino acid because the mutation to Ala disrupted the interaction with NS3 and abolished RNA replication, although the in vitro RdRp activity was unaffected (Lescar et al., 2011).

Cavity B can be therefore considered as a suitable allosteric pocket for either NS5 RdRp inhibitors or NS3–NS5 PPI disruptors. Moreover, Vasudevan and coworkers, in 2015, demonstrated that NS3–NS5 interaction-defective mutants can impair infectious virus production, viral protein synthesis, and RNA replication, which is likely dependent on the presence of the key amino acids involved in NS3–NS5 interaction (Tay et al., 2015). This finding fits well with previous virological data that highlighted the ability of PBTZs to produce non-infective viral particles (Cannalire et al., 2019). Therefore in silico studies within cavity B carried out on selected PBTZs highlighted that all derivatives showed plausible and energetically favorable binding mode. Given that, the potential interactions with cavity B seemed consistent with the capability of the PBTZs to inhibit simultaneously RdRp and the NS3–NS5 PPI.

To assess if PBTZs were able to disrupt NS3–NS5 PPI, some compounds (e.g., **3** and **6**) were tested in a competitive NS3–NS5 ELISA interaction assay, and our previous hit **1** was also included (Cannalire et al., 2020). All the tested compounds blocked NS3–NS5 interaction with IC_{50} in the low μM range, thus resulting in the most potent compounds so far reported for this PPI inhibition and with **6** exerting the best activity (Fig. 4). To be noted, aminoacyl PBTZs **1**, **3**, and **6** that displayed a potent NS3–NS5 PPI inhibition were also able to inhibit RdRp activity, and this dual nature has been supported by in silico and biological results.

To acquire more insights in the MoA, PBTZs **1**, **3**, and **6** were chosen for resistant mutant selection assay, and only **3** was able to select a resistant clone bearing multiple mutations in the 3′UTR region of the genome. In our opinion, these multiple mutations could alter the RNA secondary structure that is important for interaction with viral NS proteins during RC formation, in particular NS5 RdRp (Gebhard, Filomatori, & Gamarnik, 2011).

Although the precise mechanism responsible for the broad-spectrum anti-flavivirus activity requires further in-depth studies, it is possible that the antiviral effect of PTBZ **3** is due to the disruption of NS5-RNA interactions since the 3′UTR is recognized by NS5 RdRp to start viral RNA replication (Filomatori, Iglesias, Villordo, Alvarez, & Gamarnik, 2011). With the aim to provide experimental data in support of this hypothesis, an electrophoresis (REMSA) between viral RNA and the viral protein NS5 was carried out to assess if PBTZs were able to bind NS5. Exploiting this new assay, it was clear that the addition of compound **3** to NS5/3′UTR RNA mixture caused the disruption of the protein-RNA interaction, leading to the appearance of unbound RNA (Cannalire et al., 2020).

At the end, with our studies, we can suppose that PBTZs act at the level of the RC by hampering multiple functions of NS5 polymerase. The effect of the compounds on the entire RC, during RNA replication, is not totally understood, and it is a very hard task to address; however, studies in this direction are ongoing.

Impact of PBTZs on ZIKV drug discovery and conclusions

In this book chapter, we reported all the synthetic and biological efforts made for the identification of PBTZ compounds as promising and virus-specific flavivirus replication inhibitors. PBTZs are also characterized from low to moderate cytotoxicity when tested in cell lines often with selectivity indexes above 10 (the threshold for a genuine antiviral compound that does not produce pleiotropic effects affecting virus replication). Even if PBTZ compound class is still in an early drug discovery phase, they can be considered as promising candidates for the treatment of ZIKV infections since their antiviral activity against this virus is in the low μM to high nM range. In particular, we identified compounds **3** and **6** (Fig. 4) whose potency against ZIKV was improved of about one log unit in comparison with compound **2** and thus occupying a special place between anti-ZIKV agents. Moreover, as observed with some compounds, PBTZ class does not act against all RNA viruses but seems directed to *Flavivirus* genus covering almost all the relevant flaviviruses. Pharmacological properties of PBTZs well fit with the features needed for an anti-ZIKV drug that should treat, simultaneously, multiple flavivirus infections, and this approach may represent an efficient and cost-effective strategy for dealing with co-infections with more than one flavivirus in endemic areas. At the end, this strategy could be able to control the explosion of new potential outbreaks. More in-depth, endemic ZIKV regions are usually also characterized by co-circulation of other flaviviruses such as DENV and YFV. Therefore the broad-spectrum anti-flavivirus PBTZ could be administered for (i) prophylaxis or early postexposure prophylaxis in the first phase of infection, (ii) accelerating viral clearance, and (iii)reducing disease severity (Bernatchez et al., 2020; Wilder-Smith et al., 2018). By applying this strategy, it would be possible to reduce ZIKV-associated diseases such as Guillain-Barré syndrome or other neurological complications at global level and to prevent the vertical ZIKV transmission that is responsible for microcephaly in newborns.

As observed in the section "Unraveling the mechanism of action of PBTZ," PBTZs (e.g., **1**, **3**, and **6**) were able to inhibit viral replication thanks to their peculiar behavior as dual inhibitors of NS5 RdRp and NS3–NS5 PPI. The interaction of PBTZs with NS5 disrupts the correct RC assembly, leading to the production of a defective viral genome. In our opinion the multiple effect can be considered an excellent result because this mixed behavior makes hard the production of escape mutants. Therefore, it will be tougher for the virus to create resistant strains during the compound administration, and this is completely in line with the difficulty to generate escape mutants we encountered during our studies. To the best of our knowledge, PBTZs are among the most active and selective anti-ZIKV compounds reported so far in literature.

However, there are other requirements that a potential ZIKV drug must have, for example, the compound have to cross the blood-brain barrier (BBB) to reduce the neurological complications of the disease and should be sufficiently safe to be administered in pregnant women or women trying to become pregnant. In fact, with the exception of adults with Guillain-Barré syndrome, pregnant women remain yet the major target of ZIKV infections, and due to the severity of the symptoms for both the mother and the fetus, it is necessary to find therapeutics that will help us to treat and eradicate this infection. Therefore the risk/benefit balance must be taken into account before treating pregnant women with anti-ZIKV compounds, and clinical trials should be guaranteed to assess if a certain antiviral compound is suitable to be administered to the mother for the treatment of ZIKV infections or if the same compound is going to harm the fetus.

Moreover the compound should not produce intense adverse reaction or interact with other drugs and most importantly should be cheap since the majority of endemic areas comprehend third world or developing countries. None of these data are available yet for PBTZ derivatives, but will be acquired in future studies focused on the improvement of the PK properties. All the information will help us decide if PBTZs could be promising anti-ZIKV compounds to be advanced to preclinical evaluation.

Policy and procedures

Pyridobenzothiazolones preparation

All the pyridobenzothiazolones and pyridones described in this book chapter were synthesized following a procedure developed in our laboratory and already reported in three publications (Cannalire et al., 2019; Cannalire et al., 2020; Tarantino et al., 2016).

In vitro assays for evaluating DENV and WNV polymerase inhibition

Two different assays have been used to assess the ability of PBTZs to inhibit NS5 RdRp. The first assay measured the RdRp activity following the synthesis of double-stranded RNA from a single-stranded RNA poly(C) template and using Pico-Green as a fluorescent nucleic acid stain as already reported (Tarantino et al., 2016). The second assay evaluated RdRp activity in both de novo and elongation phase with AttoPhos used as fluorescence indicator (Cannalire et al., 2020). The comparable results obtained from both assays confirmed PBTZs as real flavivirus polymerase inhibitors.

Competitive NS3–NS5 ELISA interaction assay

In the assay, NS5 RdRp in PBS was used to coat MaxiSorp Immunoplates, and the blocked wells were then incubated with $NS2B_{CF18}NS3$ full-length at fixed concentration and mixed with increasing concentrations of DENV3 NS3 helicase (aa 173–618) or inhibitors. After washing the bound NS2B/NS3 was detected by using antibodies (3F10 IgG). Colorimetric substrate TMB (3,3′,5,5′-tetramethylbenzidine) was added after washing and pretreatment with antihuman IgG HRP. Data were fitted to the sigmoidal dose-response equation (variable slope) by nonlinear regression using GraphPad Prism 5 and expressed as IC_{50}. This protocol has been already reported in Tay et al. (2015).

Cell assays for evaluating anti-flavivirus activity

The antiviral activity was determined by using RT-PCR or plaque reduction assay (Cannalire et al., 2019; Cannalire et al., 2020; Tarantino et al., 2016), while the compound cytotoxicity was evaluated employing CellTiter-Blue reagent to assess cell viability. The specific antiviral assay used for compounds reported in this book chapter are extensively described in the aforementioned publications.

Mini-dictionary of terms

Scaffold hopping. Modification of the central core structure of active compounds to give structurally different chemical entities that maintain some essential features of the template and exert the same biological activity.

Ligand-based drug design. Drug design strategy that relies on the knowledge of other compounds that bind/inhibit the biological target.

Structure-based drug design. Drug design approach that relies on the knowledge of the 3D structure of the biological target. The 3D structure can be obtained by different procedures such as NMR or X-ray studies.

Structure-activity relationship. The direct correlation between the compound chemical structure and its biological activity. The analysis of that correlation allows to understand to what degree a chemical group is responsible for the biological activity of that compound.

REMSA. Electrophoresis technique that studies protein-RNA interactions. This procedure can be used to assess if a protein is able to bind the RNA.

Key facts of Zika treatment

- Currently, there are no drugs against Zika virus infections, and thus there is a compelling need to find broad anti-flavivirus agents.
- Among structural and non-structural viral proteins, the polymerase emerges as one of the most promising targets for antiviral therapy.
- Many nucleoside and nonnucleoside inhibitors of Zika polymerase have been discovered, but none of them have passed clinical trials.
- Targeting protein-protein interaction is a new strategy for the development of potent anti-flavivirus compounds, and in particular the interaction between NS5 polymerase and NS3 helicase is the one that mainly draws the attention of research groups.
- Pyridobenzothiazolones are a class of well-characterized anti-flavivirus compounds, and some of them are active against Zika.

Summary points

- Pyridobenzothiazolone scaffold was discovered through a scaffold hopping approach with the aim to identify new anti-HCV chemotypes.
- Years later, some pyridobenzothiazolones were repurposed as flavivirus inhibitors thanks to an in silico screening on DENV3 polymerase using the in-house library of NS5B HCV inhibitors.
- Compound **1** inhibited the isolated enzyme NS5 polymerase, and it was endowed with a promising anti-flavivirus activity in cell lines.
- Starting from compound **1**, a new series of pyridobenzothiazolones was designed and synthesized identifying potent and selective anti-flavivirus agents.
- Conflicting results between crystallographic data and enzymatic and kinetic studies prompted us to perform more in-depth studies to understand the mechanism of action of this class.
- Viral particles treated with pyridobenzothiazolones were unable to perform a second round of infection, probably for the formation of a damaged viral phenotype.
- Some aminoacyl pyridobenzothiazolones exert their antiviral activity by the dual inhibition of NS5 RdRp and NS3–NS5 protein-protein interaction.
- Aminoacyl pyridobenzothiazolones are the most potent DENV NS3–NS5 protein-protein interaction inhibitors so far reported in the literature.
- Pyridobenzothiazolones are a good starting point to develop clinical candidates to fight Zika infections.

References

Ai, J., Zhang, Y., & Zhang, W. (2016). Zika virus outbreak: A perfect storm. *Emerging Microbes & Infections*, *5*, 1–3.

Bernatchez, J. A., Tran, L. T., Li, J., Luan, Y., Siqueira-Neto, J. L., & Li, R. (2020). Drugs for the treatment of Zika virus infection. *Journal of Medicinal Chemistry*, *63*, 470–489.

Boldescu, V., Behnam, M. A. M., Vasilakis, N., & Klein, C. D. (2017). Broad-spectrum agents for flaviviral infections: Dengue, Zika and beyond. *Nature Reviews Drug Discovery*, *16*, 565–586.

Brooks, A. J., Johansson, M., John, A. V., Xu, Y., Jans, D. A., & Vasudevan, S. G. (2002). The interdomain region of dengue NS5 protein that binds to the viral helicase NS3 contains independently functional importin β1 and importin α/β-recognized nuclear localization signals. *Journal of Biological Chemistry*, *277*, 36399–36407.

Bullard-Feibelman, K. M., Govero, J., Zhu, Z., Salazar, V., Veselinovic, M., Diamond, M. S., & Geiss, B. J. (2017). The FDA-approved drug sofosbuvir inhibits Zika virus infection. *Antiviral Research*, *137*, 134–140.

Cannalire, R., Tarantino, D., Piorkowski, G., Carletti, T., Massari, S., Felicetti, T., … Querat, G. (2019). Broad spectrum anti-flavivirus pyridobenzothiazolones leading to less infective virions. *Antiviral Research*, *167*, 6–12.

Cannalire, R., Wing, K., Chan, K., Burali, M. S., Gwee, C. P., Wang, S., … Manfroni, G. (2020). Pyridobenzothiazolones exert potent anti-dengue activity by hampering multiple functions of NS5 polymerase. *ACS Medicinal Chemistry Letters*. https://doi.org/10.1021/acsmedchemlett.9b00619.

De Clercq, E., & Li, G. (2016). Approved antiviral drugs over the past 50 years. *Clinical Microbiology Reviews*, *29*, 695–747.

Duan, Y., Zeng, M., Jiang, B., Zhang, W., Wang, M., Jia, R., … Cheng, A. (2019). Flavivirus RNA-dependent RNA polymerase interacts with genome UTRs and viral proteins to facilitate flavivirus RNA replication. *Viruses*, *11*, 929.

Ferreira, A. C., Zaverucha-Do-Valle, C., Reis, P. A., Barbosa-Lima, G., Vieira, Y. R., Mattos, M., … Souza, T. M. L. (2017). Sofosbuvir protects Zika virus-infected mice from mortality, preventing short- and long-term sequelae. *Scientific Reports*, *7*, 1–9.

Filomatori, C. V., Iglesias, N. G., Villordo, S. M., Alvarez, D. E., & Gamarnik, A. V. (2011). RNA sequences and structures required for the recruitment and activity of the dengue virus polymerase. *Journal of Biological Chemistry*, *286*, 6929–6939.

Ganta, N. M., Gedda, G., Rathnakar, B., Satyanarayana, M., Yamajala, B., Ahsan, M. J., … Balaraju, T. (2019). A review on HCV inhibitors: Significance of non-structural polyproteins. *European Journal of Medicinal Chemistry*, *164*, 576–601.

Gebhard, L. G., Filomatori, C. V., & Gamarnik, A. V. (2011). Functional RNA elements in the dengue virus genome. *Viruses*, *3*, 1739–1756.

Julander, J. G., Siddharthan, V., Evans, J., Taylor, R., Tolbert, K., Apuli, C., … Babu, Y. S. (2017). Efficacy of the broad-spectrum antiviral compound BCX4430 against Zika virus in cell culture and in a mouse model. *Antiviral Research*, *137*, 14–22.

Lescar, J., Zou, G., Yap, L. J., Shochat, S. G., Shi, P.-Y., Yau, Y. H., … Chen, Y. L. (2011). Functional analysis of two cavities in flavivirus NS5 polymerase. *Journal of Biological Chemistry*, *286*, 14362–14372.

Lim, S. P., Noble, C. G., Seh, C. C., Soh, T. S., El Sahili, A., Chan, G. K. Y., … Yokokawa, F. (2016). Potent allosteric dengue virus NS5 polymerase inhibitors: Mechanism of action and resistance profiling. *PLoS Pathogens*, *12*, 1–25.

Lim, S. P., Noble, C. G., & Shi, P. Y. (2015). The dengue virus NS5 protein as a target for drug discovery. *Antiviral Research*, *119*, 57–67.

Manfroni, G., Meschini, F., Barreca, M. L., Leyssen, P., Samuele, A., Iraci, N., … Cecchetti, V. (2012). Pyridobenzothiazole derivatives as new chemotype targeting the HCV NS5B polymerase. *Bioorganic and Medicinal Chemistry*, *20*, 866–876.

Musso, D., Ko, A. I., & Baud, D. (2019). Zika virus infection—After the pandemic. *New England Journal of Medicine*, *381*, 1444–1457.

Ngono, A. E., & Shresta, S. (2019). Cross-reactive T cell immunity to dengue and zika viruses: New insights into vaccine development. *Frontiers in Immunology*, *10*, 1–9.

Paixão, E. S., Barreto, F., Da Glória Teixeira, M., Da Conceição, N., Costa, M., & Rodrigues, L. C. (2016). History, epidemiology, and clinical manifestations of Zika: A systematic review. *American Journal of Public Health*, *106*, 606–612.

Tarantino, D., Cannalire, R., Mastrangelo, E., Croci, R., Querat, G., Barreca, M. L., … Milani, M. (2016). Targeting flavivirus RNA dependent RNA polymerase through a pyridobenzothiazole inhibitor. *Antiviral Research*, *134*, 226–235.

Tay, M. Y. F., Saw, W. G., Zhao, Y., Chan, K. W. K., Singh, D., Chong, Y., … Vasudevan, S. G. (2015). The C-terminal 50 amino acid residues of dengue NS3 protein are important for NS3-NS5 interaction and viral replication. *Journal of Biological Chemistry*, *290*, 2379–2394.

Warren, T. K., Wells, J., Panchal, R. G., Stuthman, K. S., Garza, N. L., Van Tongeren, S. A., … Bavari, S. (2014). Protection against filovirus diseases by a novel broad-spectrum nucleoside analogue BCX4430. *Nature*, *508*, 402–405.

Wilder-Smith, A., Vannice, K., Durbin, A., Hombach, J., Thomas, S. J., Thevarjan, I., & Simmons, C. P. (2018). Zika vaccines and therapeutics: Landscape analysis and challenges ahead. *BMC Medicine*, *16*, 84.

Yap, T. L., Xu, T., Chen, Y.-L., Malet, H., Egloff, M.-P., Canard, B., … Lescar, J. (2007). Crystal structure of the dengue virus RNA-dependent RNA polymerase catalytic domain at 1.85-angstrom resolution. *Journal of Virology*, *81*, 4753–4765.

Chapter 34

The development of human monoclonal antibodies against Zika virus

Cui Li and Zhiheng Xu

State Key Laboratory of Molecular Developmental Biology, Institute of Genetics and Developmental Biology, Chinese Academy of Sciences, Beijing, China

Abbreviations

ADE	antibody-dependent enhancement
DENV	dengue virus
DI	domain I
DII	domain II
DIII	domain III
E	envelope
IFN	interferon
mAbs	monoclonal antibodies
NS1	nonstructural protein 1
PBMCs	peripheral blood mononuclear cells
ZIKV	Zika virus

Introduction

Effective vaccines and therapy for Zika virus (ZIKV) are still not available. Since the devastating neurodevelopmental disorders caused by latest outbreak of ZIKV infection, scientists have developed many ZIKV antibodies from humans or mice within a short period of time (Barba-Spaeth et al., 2016; Robbiani et al., 2017; Sapparapu et al., 2016; Stettler et al., 2016; Wang et al., 2016; Wu et al., 2017; Yu et al., 2017). For immediate therapy the protective efficiency of convalescent serum derived from patients was verified in a mouse model (Wang, Hong, et al., 2017). Antibodies are the major component of protective immunity against viral infection in humans. Most of the antibodies studied so far are neutralizing monoclonal antibodies (mAbs) targeting the envelope (E) protein on the surface of the virus (Barba-Spaeth et al., 2016; Robbiani et al., 2017; Sapparapu et al., 2016; Stettler et al., 2016; Wang et al., 2016; Wu et al., 2017; Yu et al., 2017). Other viral proteins including nonstructural 1 (NS1) have also been reported as promising targets for antibodies (Bailey et al., 2018; Stettler et al., 2016).

Verification of the protective efficacy in animal models is an important index for evaluation of antibodies. Nowadays, available animal models for antibody tests include ZIKV-infected wild-type mice, immunodeficient mice, and even non-human primate macaques (Keeffe et al., 2018; Magnani et al., 2017). It is more convincing if the antibody can prevent the fetal or neonatal microcephaly defects, the core pathological feature caused by ZIKV. Available fetal microcephaly models include infection of immunocompromised pregnant mice lacking interferon (IFN) receptors or direct injection of the virus into the lateral ventricles of the fetal immunocompetent mouse brains (Li, Gao, et al., 2018; Li, Wang, et al., 2018; Sapparapu et al., 2016; Wang, Hong, et al., 2017).

Neutralization and epitope specificity are crucial for understanding the mechanism by which antibodies are effective. With the development of cryo-electron microscopy (cryo-EM), the structure of many ZIKV antibodies and their targeting epitopes have been analyzed in detail (Hasan et al., 2017; Robbiani et al., 2017; Wang et al., 2019; Wang et al., 2016). Evaluation of the protein structures is helpful for further development of antibody-based therapy.

Zika Virus Impact, Diagnosis, Control, and Models. https://doi.org/10.1016/B978-0-12-820267-8.00034-0

In addition, information about protein structure will be helpful for the design of immunogens, providing new vaccines for the prevention of virus infection.

Even though the study of mAbs against ZIKV has progressed rapidly in the past few years, antibody-dependent enhancement (ADE) might limit their use as therapeutic candidates. As a member of the flaviviruses, ZIKV is close to dengue virus (DENV), another threatening epidemic virus (Barrows et al., 2018). Neutralizing antibodies isolated from convalescent patients infected by DENV or ZIKV have been reported to show cross-neutralization (Barba-Spaeth et al., 2016; Wang et al., 2016). How to avoid or overcome the ADE effect is still a critical dilemma to be resolved during the development of mAbs against ZIKV.

Convalescent serum

During flavivirus infection, humoral immunity plays an important role in the host response to clear the virus. Human convalescent serum derived from recovered patients is a possible therapy for patients who are infected by an emergent virus infection. In the past few years, convalescent plasma had been used against emergent epidemic virus infections, such as severe acute respiratory syndrome, H5N1 influenza, Ebola virus disease, and Middle East respiratory syndrome (Marano et al., 2016; Winkler & Koepsell, 2015).

To explore immediately available therapeutics against ZIKV, Wang et al. designed experiments to test the protective effects of human convalescent serum (Wang, Hong, et al., 2017). They took advantage of an immunocompetent pregnant mouse model, in which ZIKV was injected directly into the lateral ventricles of fetal mouse brains (Li et al., 2016). Convalescent serum restricted virus replication and suppressed disruption of neural progenitor cell development and the loss of neurons due to cell death in infected fetal brains, thus preventing ZIKV-induced microcephaly. The effective ingredient of convalescent serum should be the neutralizing antibodies. Antibodies are Y-shaped proteins and produced by the human immune system to neutralize pathogens (Fig. 1). Convalescent serum contains mixed polyclonal antibodies, recognizing different epitopes and showing a comprehensive protective effect. This study indicated that antibodies could pass through both the placental barrier of pregnant mice and the blood-brain barrier of fetuses to neutralize ZIKV. This study was very promising because it revealed the potential use of human antibodies for prevention and treatment of ZIKV infection (Wang, Hong, et al., 2017). Due to the limited availability of convalescent sera, isolating and reproducing effective human neutralizing mAbs are essential for clinical treatment.

Human monoclonal antibodies (mAbs)

As mentioned earlier, a panel of human mAbs with potent neutralizing activities were isolated from convalescent patients and immunized animals (Barba-Spaeth et al., 2016; Stettler et al., 2016; Wang et al., 2016; Wang, Hong, et al., 2017;

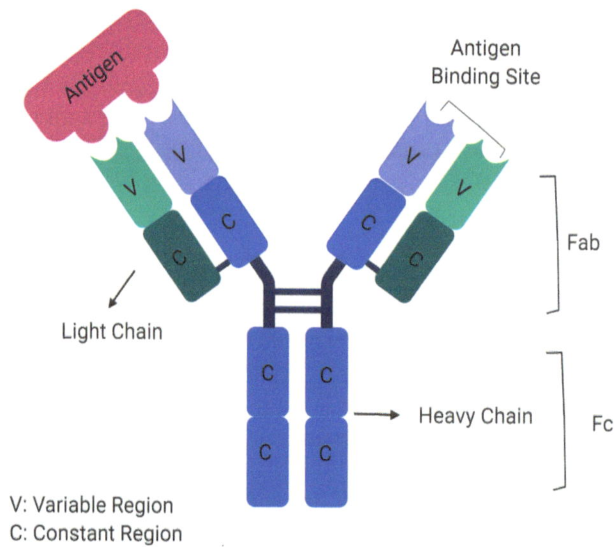

FIG. 1 Antibody structure and binding to antigen. Each antibody includes two large heavy chains and two small light chains. The light and heavy chains are composed of variable regions (V) and constant regions (C). The variable regions contain sites of antigen recognition. Abundant changes of variable regions enable antibodies to recognize specific antigens. *Fab*, fragment antigen binding. *Fc*, fragment crystallizable region.

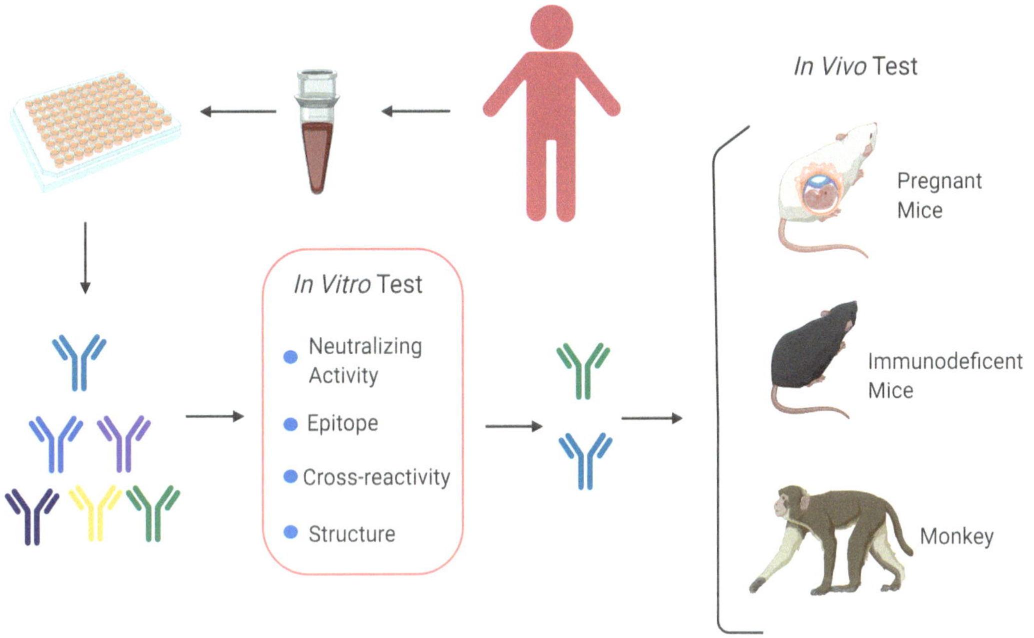

FIG. 2 Schematic diagram of human mAb screening. Human monoclonal antibodies were isolated from convalescent patients. In vitro and in vivo experiments were established to test the efficiency against ZIKV infection.

Yu et al., 2017; Zhao et al., 2016). Here, we focus on human mAbs that were isolated from convalescent patients. These antibodies were tested in vitro and in vivo against ZIKV infection (Fig. 2). Identification of the recognized epitopes of mAbs is crucial for better understanding of the protective antibody response and for future development of therapeutic agents and vaccines. According to the different binding epitopes, we classify the mAbs into two types: against either E or NS1 protein.

ZIKV E protein

Most of the potently neutralizing antibodies target the E surface glycoprotein of ZIKV. E protein plays an important role in virus entry and membrane fusion with the host cell. During ZIKV infection, E protein interacts with host cell receptors, although the exact receptor(s) for ZIKV is still debatable (Dai, Wang, Song, & Gao, 2018; Yang, Gong, & de Val, 2019). Following the close contact, ZIKV undergoes low pH-dependent endocytosis. Conformational rearrangement of E proteins happens in the endosomes. Next the viral membrane and host cell fuse, and the viral genome is released into the cytoplasm (Gerold, Bruening, Weigel, & Pietschmann, 2017; Harrison, 2008; Stiasny & Heinz, 2006). Given the important roles of the E proteins and their surface location on the virus, they are ideal targets for neutralizing mAbs. ZIKV E antibodies can neutralize the virus in multiple steps by blocking attachment to host cells, preventing rearrangement of E proteins, or affecting membrane fusion.

A compact particle of ZIKV contains 180 copies of E protein on the surface. The ectodomain of each E protomer is composed of three structurally distinct domains, DI–DIII. DI is a β-barrel structure, located between the other two domains; DII is a finger-like structure, containing a pH-sensitive fusion loop (FL) in the tip; and DIII contains a C-terminal domain of protein E and shows an IgC-like folding. DIII is assumed to contribute to viral attachment, while DII triggers fusion between the virus and target cell membrane. DI is located in the center to link DII and DIII, forming an E monomer. Each E monomer interacts with its adjacent one to form a dimer in an antiparallel way. The three E-dimers are arrayed in parallel to each other to form a structural unit known as a "raft" (Dai et al., 2016, 2018; Stiasny & Heinz, 2006). Several studies have shown that all three domains of ZIKV E protein can be targeted by neutralizing antibodies (Stettler et al., 2016; Wang et al., 2016; Yu et al., 2017).

Human mAbs targeting ZIKV E protein

Human mAbs were isolated from ZIKV convalescent patients by human hybridoma or sequencing the antigen-specific memory B cells. The mAbs are summarized in Table 1. Different antibodies vary with the recovery time of infected

TABLE 1 Summary of human mAbs against ZIKV.

	Name	Epitope	IC50	Test in vivo	Structure
Robbiani et al. (2017)	Z004	EDIII	0.7 ng/mL (Asian strain)	Prophylactic and therapeutic protection in IFNRA$^{-/-}$ mice	Yes
Sapparapu et al. (2016)	ZIKV117	EDII	5–25 ng/mL (African, Asian, and American strain)	Prophylactic and therapeutic protection in anti-Ifnar1 mAb pregnant and nonpregnant mice	Yes (Hasan et al., 2017)
Stettler et al. (2016)	ZKA64	EDIII	93 ng/mL (African strain)	Prophylactic and therapeutic protection in A129 mice (LALA mutation)	No
Wang, Bardelli, et al. (2017)	ZKA190 FIT-1	EDIII EDII/III	0.004–0.05 nM (African, Asian, and American strain)	Prophylactic and therapeutic protection in A129 mice (LALA mutation)	Yes
Wang et al. (2016)	Z3L1	EDI	0.17–0.24 mg/mL (Asian strain)	Therapeutic protection in type I interferon receptor knockout mice	Yes
	Z20	EDII	0.37–0.89 mg/mL		
	Z23	EDIII	0.37–0.56 mg/mL		
Yu et al. (2017)	ZK2B10	EDIII	0.04 μg/mL (Asian strain)	Prophylactic protection in adult AG6 mice; prophylactic and therapeutic protection in utero WT mice (Li, Gao, et al., 2018)	Yes (Wang et al., 2019)
	ZK7C3	EDI/II	0.11 μg/mL (Asian strain)		No
	ZK2C2	EDI/II/III	3.86 μg/mL (Asian strain)		
	ZK22F6	EDI/II	1.56 μg/mL (Asian strain)		
	ZK5-2	EDI/II	13.49 μg/mL (Asian strain)		
	ZK8-4	EDI/II	>40 μg/mL (Asian strain)		
Magnani et al. (2017)	MZAb1	EDIII	5.4 ng/mL (American strain)	A cocktail of these three mAbs prevent viremia in nonhuman primates	No
	SMZAb2	EDII	189 ng/mL (American strain)		
	SMZAb5	EDIII	3.8 ng/mL (American strain)		
Wu et al. (2017)	m301 m302	EDIII EDIII	~2 μM (Asian and American strain)	Prophylactic and therapeutic protection in AG6 mice	No
Bailey et al. (2018)	AA12	NS1	None detected	Prophylactic protection in Stat2$^{-/-}$ mice	No

EDI/II/III, envelope domain I/II/III; *IC50*, half maximal inhibitory concentration. AG6 mice: IFN-α/β and IFN-γ receptor knockout mice. Human mAb were summarized above, showing their epitope, IC50, in vivo test, and whether crystal structure was analyzed or not.

patients. Yu et al. recorded a longitudinal antibody response against ZIKV during natural infection in two convalescent individuals. They showed dynamic features of the antibodies followed a spatiotemporal hierarchy. mAbs isolated during early infection were largely DI/DII specific and weakly neutralizing. However, DIII-specific neutralizing activity increased over the course of infection, and the antibodies from later infection were more diverse in recognition and showed potent neutralizing activity. Moreover, mAbs against DI/II could strongly cross-neutralize DENV1 and DENV2, while mAbs from later infection showed no such cross-reactivity (Yu et al., 2017). Therefore, due to this cross-neutralize phenomenon, the use of mAbs targeting DIII is recommended for the development of a future antiviral agent. The dynamic features unveiled in this study will help us to understand better the pathogenesis of ZIKV infection and provide more information for the rational design of vaccines.

The efficacy of many human mAbs had been tested in animal models (Table 1), and importantly, some of the antibodies were verified in the microcephaly mouse models. The therapeutic efficacy of ZIKV-117 was evaluated in type I IFN-deficient pregnant mice. mAb treatment markedly reduced tissue pathology, placental and fetal infection, and mortality in mice (Sapparapu et al., 2016). Li et al. inspected six human mAbs in ZIKV-infected WT pregnant mice. The six antibodies conveyed distinct protection levels, which closely mirrored their respective in vitro neutralizing activities. ZK2B10 showed the most potent neutralization activity, completely protected infected fetal mouse brains before or right after infection, markedly reduced tissue pathology, and significantly delayed mortality even 2 days after infection (Li, Gao, et al., 2018). Therefore ZK2B10 could be an encouraging candidate for antibody-based therapy and instruct structure-based vaccine design against ZIKV infection in humans.

To better understand the mechanism of protection, the structural features of the mAbs and their binding epitopes were resolved by X-ray crystallography and cryo-EM (Barba-Spaeth et al., 2016; Hasan et al., 2017; Wang et al., 2016, 2019; Wang, Hong, et al., 2017; Zhao et al., 2016). For instance, the crystal structure of ZK2B10 revealed that the epitope is located on the lateral ridge of DIII of E protein. Cryo-EM structure with mature ZIKV showed that the antibody binds to DIIIs around the icosahedral two-, three-, and fivefold axes, a distinct feature compared with other reported DIII-specific antibodies (Wang et al., 2019). These studies will be useful for the development of antibody-based interventions and inform the rational design of ZIKV vaccine.

Human mAbs targeting NS1 protein

Other viral proteins including NS1 have emerged as promising targets for antibodies (Bailey et al., 2018). Antibodies execute their protective effects by virus neutralization or Fc-mediated effector functions, such as activating the classical complement pathway (Lu, Suscovich, Fortune, & Alter, 2018). NS1-specific mAbs are nonneutralizing and engage Fcγ receptors without inducing ADE of infection. As a nonstructural protein, NS1 is not part of the viral particle. mAbs that target Zika virus NS1 protein do not bind to the virion so they are unlikely to enhance the disease. Interestingly, 34.4% of virus-specific mAbs from ZIKV-infected patients targeted the NS1 protein (Stettler et al., 2016). It has been demonstrated that ZIKV NS1-based vaccines are protective against ZIKV in mice (Brault et al., 2017; Li, Yu, et al., 2018; Liu et al., 2018). These results show the possibility of using nonneutralizing antibodies targeting NS1 to prevent ZIKV infection in humans.

NS1 is an immunogenic glycoprotein and plays essential roles in viral RNA replication and immune evasion. NS1 protein is initially translated as a monomer and exists as a membrane-associated homodimer after translocation into the endoplasmic reticulum lumen (Dai et al., 2018; Shi & Gao, 2017). NS1 is secreted from infected cells into the extracellular space as a hexameric lipoprotein particle (sNS1) (Xu et al., 2016). NS1 also represents the major antigenic marker for viral infection and is used for diagnosis (Stettler et al., 2016). mAbs targeting NS1 isolated and characterized from infected patients have been tested in a mouse model. Protective efficacy of mAb AA12 was detected as it could reduce the lethality of Stat2$^{-/-}$ mice caused by ZIKV infection. A mutated antibody that was unable to activate known Fc effector, functions or complement, was not protective in vivo, demonstrating that the protection is Fc dependent. NS1-specific mAbs show protective efficacy in a nonneutralizing way and could provide information for novel vaccine development (Bailey et al., 2018).

Antibody-dependent enhancement of infection (ADE)

The use of mAbs as therapeutic candidates to prevent ZIKV is promising, but the existence of ADE might be an obstacle toward that aim. ADE refers to the increase in virus infection when virus-reactive antibodies are insufficient to neutralize virus or they are in subneutralizing concentrations. On this point, antibodies can bind efficiently to virus, but do not reach the neutralization threshold. Therefore these antibodies may, instead, help virus uptake by cells expressing Fc or

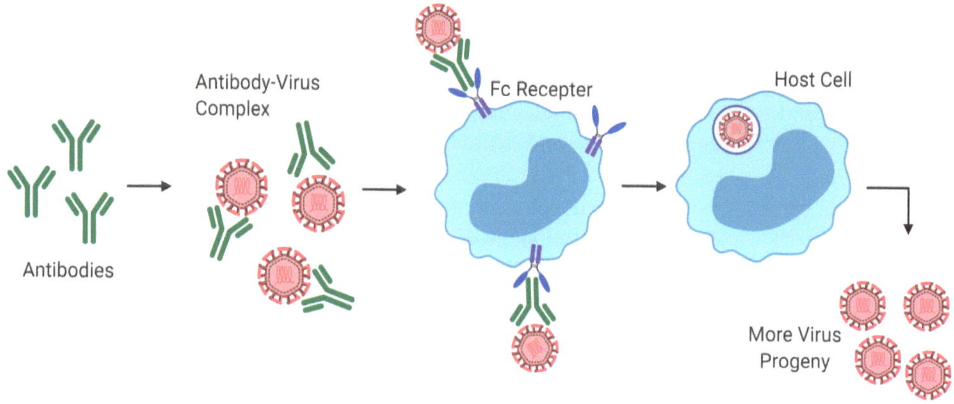

FIG. 3 ZIKV antibody-dependent enhancement (ADE). When antibodies are insufficient to neutralize virus completely or they are in subneutralizing concentrations, virus-antibody complex entry is facilitated through Fc-receptor-mediated endocytosis. Since incomplete neutralization, viruses initiate proliferation in host cells and subsequently produce more virus.

complement receptors and subsequently promote virus infectivity (Fig. 3) (Dai et al., 2018; Halstead, 2014; Katzelnick et al., 2017; Tirado & Yoon, 2003).

As a member of the *Flaviviridae* family of positive-strand RNA viruses, ZIKV is close to some other important human pathogens such as DENV, yellow fever virus, West Nile virus, Japanese encephalitis virus, and tick-borne encephalitis virus (Barrows et al., 2018). Among these flaviviruses, DENV is the closest one to ZIKV. Some studies have proved that neutralizing antibodies isolated from convalescent patients infected by DENV or ZIKV show cross-neutralizing ability (Barba-Spaeth et al., 2016; Wang et al., 2016). Acquired DENV or ZIKV antibodies could enhance infection with the other (Brown et al., 2019; Dejnirattisai et al., 2016; Fowler et al., 2018; Li, Zhao, et al., 2018; Zimmerman et al., 2018). Therefore finding the strategy to avoid ADE is an important question to be considered during development of antibody-based therapy and design of vaccines against ZIKV (Khandia et al., 2018). Scientists are trying to detect the specific neutralizing epitope or engineering the mAbs with Fc LALA mutations (i.e., the substitution of IgG-Fc residues at positions 234 and 235 from leucine (L) to alanine (A)) that abolish Fcγ receptor binding to eliminate the risk of ADE (Sapparapu et al., 2016; Stettler et al., 2016; Wang, Bardelli, et al., 2017). In addition, nonneutralizing epitopes, including NS1, are also a direction for the use of monoclonal antibody for therapy.

Summary

Since the outbreak of ZIKV infection, scientists have spent much effort trying to develop effective treatments, including human mAbs. For immediate therapy the protective effect of human convalescent serum was verified in pregnant mice. A series of mAbs were isolated from infected mice or humans. Their target epitopes, neutralizing activity in vitro, and protective effects in vivo were investigated. To study the potential protective mechanism of the mAbs, some of their structures and epitopes were analyzed in detail. Cross-neutralization with DENV drew much attention during the development of ZIKV antibodies in consideration of ADE. Therefore special epitopes or mutational methods have been used to avoid ADE. The progress made in human mAbs is not only important for the clinical application of mAbs against ZIKV but also provide important information for vaccine development.

Policy and procedures: Isolation of mAbs from infected humans

Peripheral blood mononuclear cells (PBMCs) were isolated with Ficoll-Histopaque by density gradient centrifugation. The cells were used immediately or stored in the liquid nitrogen until use. Ten million PBMCs were cultured in multiwell plates with culture medium for a few days. Supernatants of cultured B cells were screened in ELISA with different target proteins. The positive culture was subjected to cellular RNA extraction, reverse transcription, and sequencing of the heavy and light variable genes.

Mini-dictionary of terms

Microcephaly. Refers to a smaller head. The head circumference measurement value for microcephaly is usually <2 SDs below the average value of the same age and sex.

Convalescent serum. Sera from patients who have recovered from a particular infection and usually contain a large amount of specific antibodies.
Monoclonal antibodies. Antibodies that are made by unique parent immune cells and bind to the identical epitope.
Epitope. Refers to the part of an antigen that is recognized by the immune system.
Vaccine. Attenuated or killed microorganisms that are administered to produce or artificially increase immunity against a particular disease.

Key facts about neutralizing antibodies

- Neutralizing antibodies are produced naturally by immune response during infection.
- Neutralizing antibodies play an important role on defense against viral infection.
- Neutralizing antibodies not only bind the virus but also in a manner that blocks infection.
- Only a small subset of the antibodies that bind to the virus have neutralizing ability.
- Antibodies binding but insufficient to neutralize virus may increase virus infection.

Summary points

- Human convalescent serum is protective in mice.
- Zika virus envelope (E) proteins are ideal targets for neutralizing monoclonal antibodies.
- Nonstructural protein 1 (NS1) is a promising target for Zika virus antibodies.
- Verification of the epitope information is important for the development of antibody-based therapy and design of vaccines against Zika virus.
- Overcoming antibody-dependent enhancement (ADE) is important for the study of Zika virus monoclonal antibodies.

REFERENCES

Bailey, M. J., Duehr, J., Dulin, H., Broecker, F., Brown, J. A., Arumemi, F. O., … Tan, G. S. (2018). Human antibodies targeting Zika virus NS1 provide protection against disease in a mouse model. *Nature Communications*, *9*(1), 4560. https://doi.org/10.1038/s41467-018-07008-0.

Barba-Spaeth, G., Dejnirattisai, W., Rouvinski, A., Vaney, M. C., Medits, I., Sharma, A., … Rey, F. A. (2016). Structural basis of potent Zika-dengue virus antibody cross-neutralization. *Nature*, *536*(7614), 48–53. https://doi.org/10.1038/nature18938.

Barrows, N. J., Campos, R. K., Liao, K. C., Prasanth, K. R., Soto-Acosta, R., Yeh, S. C., … Garcia-Blanco, M. A. (2018). Biochemistry and molecular biology of flaviviruses. *Chemical Reviews*, *118*(8), 4448–4482. https://doi.org/10.1021/acs.chemrev.7b00719.

Brault, A. C., Domi, A., McDonald, E. M., Talmi-Frank, D., McCurley, N., Basu, R., … Guirakhoo, F. (2017). A Zika vaccine targeting NS1 protein protects immunocompetent adult mice in a lethal challenge model. *Scientific Reports*, *7*(1). https://doi.org/10.1038/s41598-017-15039-8, 14769.

Brown, J. A., Singh, G., Acklin, J. A., Lee, S., Duehr, J. E., Chokola, A. N., … Lim, J. K. (2019). Dengue virus immunity increases Zika virus-induced damage during pregnancy. *Immunity*, *50*(3), 751–762. e755 https://doi.org/10.1016/j.immuni.2019.01.005.

Dai, L., Song, J., Lu, X., Deng, Y. Q., Musyoki, A. M., Cheng, H., … Gao, G. F. (2016). Structures of the Zika virus envelope protein and its complex with a flavivirus broadly protective antibody. *Cell Host & Microbe*, *19*(5), 696–704. https://doi.org/10.1016/j.chom.2016.04.013.

Dai, L., Wang, Q., Song, H., & Gao, G. F. (2018). Zika virus envelope protein and antibody complexes. *Sub-Cellular Biochemistry*, *88*, 147–168. https://doi.org/10.1007/978-981-10-8456-0_7.

Dejnirattisai, W., Supasa, P., Wongwiwat, W., Rouvinski, A., Barba-Spaeth, G., Duangchinda, T., … Screaton, G. R. (2016). Dengue virus sero-cross-reactivity drives antibody-dependent enhancement of infection with zika virus. *Nature Immunology*, *17*(9), 1102–1108. https://doi.org/10.1038/ni.3515.

Fowler, A. M., Tang, W. W., Young, M. P., Mamidi, A., Viramontes, K. M., McCauley, M. D., … Shresta, S. (2018). Maternally acquired Zika antibodies enhance dengue disease severity in mice. *Cell Host & Microbe*, *24*(5), 743–750. e745 https://doi.org/10.1016/j.chom.2018.09.015.

Gerold, G., Bruening, J., Weigel, B., & Pietschmann, T. (2017). Protein interactions during the flavivirus and hepacivirus life cycle. *Molecular & Cellular Proteomics*, *16*(4 suppl 1), S75–S91. https://doi.org/10.1074/mcp.R116.065649.

Halstead, S. B. (2014). Dengue antibody-dependent enhancement: Knowns and unknowns. *Microbiology Spectrum*, *2*(6). https://doi.org/10.1128/microbiolspec.AID-0022-2014.

Harrison, S. C. (2008). The pH sensor for flavivirus membrane fusion. *Journal of Cell Biology*, *183*(2), 177–179. https://doi.org/10.1083/jcb.200809175.

Hasan, S. S., Miller, A., Sapparapu, G., Fernandez, E., Klose, T., Long, F., … Rossmann, M. G. (2017). A human antibody against Zika virus crosslinks the E protein to prevent infection. *Nature Communications*, *8*. https://doi.org/10.1038/ncomms14722, 14722.

Katzelnick, L. C., Gresh, L., Halloran, M. E., Mercado, J. C., Kuan, G., Gordon, A., … Harris, E. (2017). Antibody-dependent enhancement of severe dengue disease in humans. *Science*, *358*(6365), 929–932. https://doi.org/10.1126/science.aan6836.

Keeffe, J. R., Van Rompay, K. K. A., Olsen, P. C., Wang, Q., Gazumyan, A., Azzopardi, S. A., … Robbiani, D. F. (2018). A combination of two human monoclonal antibodies prevents Zika virus escape mutations in non-human primates. *Cell Reports*, *25*(6), 1385–1394. e1387 https://doi.org/10.1016/j.celrep.2018.10.031.

Khandia, R., Munjal, A., Dhama, K., Karthik, K., Tiwari, R., Malik, Y. S., … Chaicumpa, W. (2018). Modulation of dengue/Zika virus pathogenicity by antibody-dependent enhancement and strategies to protect against enhancement in Zika virus infection. *Frontiers in Immunology*, *9*, 597. https://doi.org/10.3389/fimmu.2018.00597.

Li, A., Yu, J., Lu, M., Ma, Y., Attia, Z., Shan, C., … Li, J. (2018). A Zika virus vaccine expressing premembrane-envelope-NS1 polyprotein. *Nature Communications*, *9*(1), 3067. https://doi.org/10.1038/s41467-018-05276-4.

Li, C., Gao, F., Yu, L., Wang, R., Jiang, Y., Shi, X., … Zhang, L. (2018). A single injection of human neutralizing antibody protects against Zika virus infection and microcephaly in developing mouse embryos. *Cell Reports*, *23*(5), 1424–1434. https://doi.org/10.1016/j.celrep.2018.04.005.

Li, C., Wang, Q., Jiang, Y., Ye, Q., Xu, D., Gao, F., … Xu, Z. (2018). Disruption of glial cell development by Zika virus contributes to severe microcephalic newborn mice. *Cell Discovery*, *4*, 43. https://doi.org/10.1038/s41421-018-0042-1.

Li, C., Xu, D., Ye, Q., Hong, S., Jiang, Y., Liu, X., … Xu, Z. (2016). Zika virus disrupts neural progenitor development and leads to microcephaly in mice. *Cell Stem Cell*, *19*(1), 120–126. https://doi.org/10.1016/j.stem.2016.04.017.

Li, M., Zhao, L., Zhang, C., Wang, X., Hong, W., Sun, J., … Jin, X. (2018). Dengue immune sera enhance Zika virus infection in human peripheral blood monocytes through Fc gamma receptors. *PLoS One*, *13*(7). https://doi.org/10.1371/journal.pone.0200478, e0200478.

Liu, X., Qu, L., Ye, X., Yi, C., Zheng, X., Hao, M., … Chen, L. (2018). Incorporation of NS1 and prM/M are important to confer effective protection of adenovirus-vectored Zika virus vaccine carrying E protein. *NPJ Vaccines*, *3*, 29. https://doi.org/10.1038/s41541-018-0072-6.

Lu, L. L., Suscovich, T. J., Fortune, S. M., & Alter, G. (2018). Beyond binding: Antibody effector functions in infectious diseases. *Nature Reviews Immunology*, *18*(1), 46–61. https://doi.org/10.1038/nri.2017.106.

Magnani, D. M., Rogers, T. F., Beutler, N., Ricciardi, M. J., Bailey, V. K., Gonzalez-Nieto, L., … Watkins, D. I. (2017). Neutralizing human monoclonal antibodies prevent Zika virus infection in macaques. *Science Translational Medicine*, *9*(410). https://doi.org/10.1126/scitranslmed.aan8184.

Marano, G., Vaglio, S., Pupella, S., Facco, G., Catalano, L., Liumbruno, G. M., & Grazzini, G. (2016). Convalescent plasma: New evidence for an old therapeutic tool? *Blood Transfusion*, *14*(2), 152–157. https://doi.org/10.2450/2015.0131-15.

Robbiani, D. F., Bozzacco, L., Keeffe, J. R., Khouri, R., Olsen, P. C., Gazumyan, A., … Nussenzweig, M. C. (2017). Recurrent potent human neutralizing antibodies to Zika virus in Brazil and Mexico. *Cell*, *169*(4), 597–609. e511 https://doi.org/10.1016/j.cell.2017.04.024.

Sapparapu, G., Fernandez, E., Kose, N., Bin, C., Fox, J. M., Bombardi, R. G., … Crowe, J. E. (2016). Neutralizing human antibodies prevent Zika virus replication and fetal disease in mice. *Nature*, *540*(7633), 443–447. https://doi.org/10.1038/nature20564.

Shi, Y., & Gao, G. F. (2017). Structural biology of the Zika virus. *Trends in Biochemical Sciences*, *42*(6), 443–456. https://doi.org/10.1016/j.tibs.2017.02.009.

Stettler, K., Beltramello, M., Espinosa, D. A., Graham, V., Cassotta, A., Bianchi, S., … Corti, D. (2016). Specificity, cross-reactivity, and function of antibodies elicited by Zika virus infection. *Science*, *353*(6301), 823–826. https://doi.org/10.1126/science.aaf8505.

Stiasny, K., & Heinz, F. X. (2006). Flavivirus membrane fusion. *Journal of General Virology*, *87*(Pt 10), 2755–2766. https://doi.org/10.1099/vir.0.82210-0.

Tirado, S. M., & Yoon, K. J. (2003). Antibody-dependent enhancement of virus infection and disease. *Viral Immunology*, *16*(1), 69–86. https://doi.org/10.1089/088282403763635465.

Wang, J., Bardelli, M., Espinosa, D. A., Pedotti, M., Ng, T. S., Bianchi, S., … Corti, D. (2017). A human bi-specific antibody against Zika virus with high therapeutic potential. *Cell*, *171*(1), 229–241. e215 https://doi.org/10.1016/j.cell.2017.09.002.

Wang, L., Wang, R., Wang, L., Ben, H., Yu, L., Gao, F., … Zhang, L. (2019). Structural basis for neutralization and protection by a Zika virus-specific human antibody. *Cell Reports*, *26*(12), 3360–3368. e3365 https://doi.org/10.1016/j.celrep.2019.02.062.

Wang, Q., Yang, H., Liu, X., Dai, L., Ma, T., Qi, J., … Gao, G. F. (2016). Molecular determinants of human neutralizing antibodies isolated from a patient infected with Zika virus. *Science Translational Medicine*, *8*(369), 369ra179. https://doi.org/10.1126/scitranslmed.aai8336.

Wang, S., Hong, S., Deng, Y. Q., Ye, Q., Zhao, L. Z., Zhang, F. C., … Xu, Z. (2017). Transfer of convalescent serum to pregnant mice prevents Zika virus infection and microcephaly in offspring. *Cell Research*, *27*(1), 158–160. https://doi.org/10.1038/cr.2016.144.

Winkler, A. M., & Koepsell, S. A. (2015). The use of convalescent plasma to treat emerging infectious diseases: Focus on Ebola virus disease. *Current Opinion in Hematology*, *22*(6), 521–526. https://doi.org/10.1097/MOH.0000000000000191.

Wu, Y., Li, S., Du, L., Wang, C., Zou, P., Hong, B., … Ying, T. (2017). Neutralization of Zika virus by germline-like human monoclonal antibodies targeting cryptic epitopes on envelope domain III. *Emerging Microbes & Infections*, *6*(10), e89. https://doi.org/10.1038/emi.2017.79.

Xu, X., Song, H., Qi, J., Liu, Y., Wang, H., Su, C., … Gao, G. F. (2016). Contribution of intertwined loop to membrane association revealed by Zika virus full-length NS1 structure. *EMBO Journal*, *35*(20), 2170–2178. https://doi.org/10.15252/embj.201695290.

Yang, C., Gong, R., & de Val, N. (2019). Development of neutralizing antibodies against Zika virus based on its envelope protein structure. *Virologica Sinica*, *34*(2), 168–174. https://doi.org/10.1007/s12250-019-00093-5.

Yu, L., Wang, R., Gao, F., Li, M., Liu, J., Wang, J., … Zhang, L. (2017). Delineating antibody recognition against Zika virus during natural infection. *JCI Insight*, *2*(12). https://doi.org/10.1172/jci.insight.93042.

Zhao, H., Fernandez, E., Dowd, K. A., Speer, S. D., Platt, D. J., Gorman, M. J., … Fremont, D. H. (2016). Structural basis of Zika virus-specific antibody protection. *Cell*, *166*(4), 1016–1027. https://doi.org/10.1016/j.cell.2016.07.020.

Zimmerman, M. G., Quicke, K. M., O'Neal, J. T., Arora, N., Machiah, D., Priyamvada, L., … Suthar, M. S. (2018). Cross-reactive dengue virus antibodies augment Zika virus infection of human placental macrophages. *Cell Host & Microbe*, *24*(5), 731–742. e736 https://doi.org/10.1016/j.chom.2018.10.008.

Chapter 35

The Zika virus NS1 protein as a vaccine target

Mark J. Bailey[a] and Gene S. Tan[b,c]

[a]*Department of Microbiology, Graduate School of Biomedical Sciences, Icahn School of Medicine at Mount Sinai, New York, NY, United States,* [b]*J. Craig Venter Institute, La Jolla, CA, United States,* [c]*Division of Infectious Diseases, Department of Medicine, University of California San Diego, La Jolla, CA, United States*

Abbreviations

ADE antibody-dependent enhancement
DENV dengue virus
ER endoplasmic reticulum
Fc fragment crystallizable
FcγR fragment crystallizable gamma receptor
JEV Japanese encephalitis virus
kDa kilodalton
KO knockout
NS1 nonstructural 1 protein
NS4A nonstructural 4A protein
NS4B nonstructural 4B protein
Stat2 signal transducer and activator of transcription 2
TBEV tick-borne encephalitis virus
WNV West Nile virus
YFV yellow fever virus
ZIKV Zika virus

Introduction

The nonstructural 1 (NS1) protein of Zika virus (ZIKV) and other flaviviruses encodes a 352-amino acid protein with a molecular weight of 46–55 kDa—existing in several isoforms and oligomeric states (Rastogi, Sharma, & Singh, 2016). A majority of our knowledge concerning ZIKV NS1 is extrapolated from previous work on NS1 proteins from other flaviviruses such as dengue virus (DENV), yellow fever virus (YFV), Japanese encephalitis virus (JEV), and West Nile virus (WNV). The NS1 protein has several functions related to viral replication cycle and pathogenesis that ultimately makes it an important viral factor to study. Intracellular dimeric NS1 is known to regulate viral transcription through its interactions with other viral nonstructural proteins NS4A and NS4B (Hilgenfeld, 2016). NS1 is also found as a dimeric complex, expressed extracellularly on the surface of infected cells through glycosylphosphatidylinositol linkers. Lastly, NS1 is found as a soluble hexameric complex found abundantly in the plasma of infected individuals. Unlike other nonstructural proteins of other viruses, such as influenza virus, the NS1 of flaviviruses is also found outside infected cells. The last two oligomeric states, namely, as a dimeric molecule expressed on the surface of cells or as a soluble hexameric complex found in plasma, can elicit antibody responses, important considerations when developing vaccines for flaviviruses. Past studies have demonstrated the ability of NS1-specific antibody to prevent gross disease against flavivirus infections. It should be noted that this class of antibody does not have neutralizing activity in vitro and does not provide sterilizing immunity. Rather, this subset of *nonneutralizing* antibodies provides protection through Fc-dependent cell-mediated activities and relies on the effector functions of innate immune cells in the clearance of infected cells. The type of innate immune cells, which mediate

Zika Virus Impact, Diagnosis, Control, and Models. https://doi.org/10.1016/B978-0-12-820267-8.00035-2

antiviral activity in humans, remains to be fully elucidated, but may depend on the physiological compartment where infection takes place. Nonetheless, prior studies highlight the potential contribution of other viral antigens such as non-structural proteins in an effective ZIKV vaccine.

Current available flavivirus vaccines

There are already several approved vaccines for flaviviruses, which may provide some insights into the development of a successful ZIKV vaccine. Vaccines for YFV, DENV, JEV, and tick-borne encephalitis virus (TBEV) are currently available and can be generally be divided into live-attenuated and inactivated formulations. The yellow fever virus vaccine is typically considered the ideal vaccine that provides robust protection coupled with long-lasting immunity. Based on an attenuated strain (17D) of the virus, a single subcutaneous or intramuscular administration (WHO, 2015) of the vaccine is reported to result in 80%–99% seroconversion (Belmusto-Worn et al., 2005; Gotuzzo, Yactayo, & Córdova, 2013). However, a recent study report that waning immunity in the years immediately after the vaccination of 9-month-old infants suggests that perhaps booster immunizations are necessary for optimal protection against reinfection with YFV (Domingo et al., 2019). This finding may upend the previous recommendation that a single administration of vaccine is enough to confer lifelong immunity and reestablish the recommendation of reimmunization every 10 years (Estofolete & Nogueira, 2018). The success of the YFV vaccine strain led to the development of chimeric live-attenuated vaccines based on the 17D vaccine strain for JEV and DENV. These live-attenuated vaccines are based on the YFV 17D genome with the premembrane and envelope proteins substituted with that of the target virus (e.g., JEV or DENV). For example, ChimeriVax-JE is a live-attenuated vaccine for JEV but is currently only approved in Australia (Halstead & Thomas, 2011). The first approved vaccine for all serotypes for DENV, marketed as Dengvaxia, is a quadrivalent vaccine based on the YFV 17D attenuated strain. Similar to ChimeriVax-JE, Dengvaxia expresses the premembrane and envelope proteins of one of the four DENV serotypes in the genetic background of YFV 17D. The idea of using a proven and successful vaccine such as the YFV 17D is important in establishing a starting point in the development of vaccines for other flavivirus pathogens. It is thus not surprising that a similar strategy is being utilized for WNV (in clinical trials) (Kaiser & Barrett, 2019) and ZIKV (preclinical studies) (Giel-Moloney et al., 2018). Inactivation of a virus is another approach and has resulted in many successful vaccines (Burrell, Howard, & Murphy, 2016). However, inactivated flavivirus vaccines are hypothetically less immunogenic requiring more than one administration (Kiran et al., 2015), and protection may be shorter than live-attenuated vaccines. There are currently inactivated vaccines for TBEV and JEV.

Correlates of protection

Induction of neutralizing antibodies against the flavivirus envelope protein is generally accepted as a correlate of protection and can provide sterilizing immunity against YFV infection (Plotkin, 2010). Common among all current available vaccines for flaviviruses is their ability to elicit neutralizing antibodies against the envelope protein. This is true for both live-attenuated and inactivated flavivirus vaccine formulations. The presence of this class of antibodies is the ideal antiviral response and theoretically sufficient to protect against infection and disease. However, due to shared antigenic regions in the envelope protein of all flaviviruses, cross-reactive antibodies can inadvertently pose a couple of problems (Zhang et al., 2017). For one, the phenomenon of antibody-dependent enhancement (ADE) occurs when poorly or non-neutralizing antibodies bind to the surface of a viral particle to enhance internalization of the virus-antibody complex into Fcγ receptor bearing myeloid cells. Through this antibody-mediated process, opsonized viral particles are not fully neutralized and are instead internalized into myeloid cells that support viral replication. More viral progenies are produced and as a result enhance disease. For example, ADE is implicated in dengue hemorrhagic fever and dengue shock syndrome, two severe forms of dengue disease (Smatti, Al Thani, & Yassine, 2018). Second, a recent report show that preexisting YFV antibodies can negatively affect the antibody response induced by TBEV vaccination and its neutralizing activity (Bradt et al., 2019). While this observation requires further investigation, it highlights that traditional correlates of protection against flavivirus infection may not provide a straightforward measurement of immunity.

The inherent complexities associated with cross-reactive antibodies targeting the flavivirus envelope protein can be a concern. Most available flavivirus vaccines mentioned earlier rely on the induction of highly neutralizing envelope-specific antibodies except for the YFV vaccine. Since the YFV vaccine is based on a full-length attenuated YFV strain, replication of an attenuated vaccine will induce immunity not only against the premembrane and envelope proteins but also to non-structural proteins, including the NS1. In contrast, inactivated virus vaccines or chimeric vaccines will typically generate a response against structural proteins, such as the premembrane and envelope proteins. In the latter case, chimeric vaccines, which have the YFV 17D core, only expresses the heterologous premembrane and envelope proteins of the target viral pathogen. The rest of the immune response will be directed toward the nonstructural proteins of YFV. The hypothetical advantage of the YFV vaccine is the ability to elicit responses to both the structural and nonstructural proteins of YFV.

Protection mediated by NS1-specific antibodies

The NS1 of ZIKV, like that of other flaviviruses, is a highly immunogenic protein making it an excellent candidate antigen for a vaccine. Unlike the nonstructural proteins of other viruses, the NS1 of flaviviruses are both expressed on the cell surface and secreted extracellularly from infected cells. The NS1 can elicit both B-cell and T-cell-mediated response (Reynolds et al., 2018). Since flavivirus NS1 proteins are only produced during infection and not typically incorporated into mature infectious virions, NS1-specific antibodies are nonneutralizing. Notably, NS1 is one of the chief flavivirus protein (the others being the premembrane and envelope proteins) targets of the humoral antibody response. During the mid-1980s, Schlesinger and colleagues first observed the ability of YFV NS1-specific antibodies to provide protection against encephalitis. They demonstrated protection by passively transferring monoclonal antibodies against NS1 (or gp48 as NS1 was previously named) in Swiss CD1 mice prophylactically followed by a lethal challenge with a lethal dose of 17D YFV (nonattenuated). Concomitantly, they also prime-boost immunized another group of mice followed by a lethal challenge of 17D YF. Of note, while mice administered with monoclonal antibodies did not provide complete protection, immunized mice were fully protected (Schlesinger, Brandriss, & Walsh, 1985). The results highlighted two important observations. Firstly, while NS1-specific antibodies are nonneutralizing in vitro, they do have a functional ability to protect against disease in vivo. Secondly, the finding that NS1-immunized mice provided full protection against encephalitis whereas monoclonal antibody-treated mice did not suggest a polyclonal response against a multitude of NS1 epitopes is required for optimal protection. A similar study by the same group demonstrated that mice immunized with the DENV NS1 can also provide protection against disease (Falgout, Bray, Schlesinger, & Lai, 1990). Specific to DENV, others have found that NS1 vaccination prevented NS1-mediated endothelial permeability and vascular leakage (Beatty et al., 2015). Others have found that antibodies targeting WNV NS1 also have cytolytic activity involved in protection (Krishna, Rangappa, & Satchidanandam, 2009). Our laboratory isolated several human monoclonal antibodies from an individual acutely infected with ZIKV several years ago. We found that similar to what others have found, ZIKV NS1-specific antibodies can provide protection against disease in vivo in the absence of in vitro neutralizing activity (Bailey et al., 2018). In our experiments, we passively transferred NS1-specific monoclonal antibodies into immunodeficient Stat2 knock out (KO) mice followed by a lethal challenge with ZIKV. Mice administered with NS1-specific monoclonal antibodies had lower viral burden and exhibited less clinical signs of disease. However, we only observed partial protection from death. We speculate that increasing the dose of antibody in our passive transfer studies may increase protection from succumbing to disease. In a subsequent study, we furthered investigated if a polyclonal antibody response to NS1 can also protect mice from ZIKV disease. In these experiments, we first immunized immunocompetent C57BL/6 mice with a vaccination schema with a DNA plasmid encoding NS1 followed by a NS1 protein boost. Mice were sacrificed and sera were collected and pooled. The pooled sera were passively transferred into Stat2 KO mice and challenged with a lethal dose of ZIKV. Again, we saw partial protection from death, which corroborated with our earlier experiments with NS1 monoclonal antibodies (Bailey, Broecker, Duehr, et al., 2019). What we and others have observed reinforces the idea that NS1 can provide another level of protection that should be considered in designing candidate ZIKV vaccines. However, questions remain as to why full protection against death was not achieved after passive transfer of NS1 antibodies in monoclonal or polyclonal form. In our own experiments, we speculate that the dosage of administered monoclonal antibodies may play a role in conferring optimal protection in addition to only targeting one epitope. Using a cocktail of monoclonal antibodies targeting multiple NS1 epitopes can potentially increase the effectiveness antibody-based protection. However, one must note that these speculations only addresses the issue of using monoclonal antibodies as prophylactics or therapeutics. In the context of vaccination, the significance of our experiments involving the passive transfer of polyclonal hyperimmune serum solely focused on the role of NS1-specific antibodies in protection. We hypothesize that T-cell responses against NS1, which we did not address, are also important in providing another dimension of immunity to ZIKV that should be further investigated.

The mechanism of NS1 antibody-based protection

The exact mechanism(s) of how NS1-specific antibodies provide protection remains to be fully understood. However, we and others have demonstrated that protection is mediated through Fc-dependent mechanisms. Effector functions activated by the interaction between Fc receptors bearing innate immune cells such as natural killer cells, neutrophils, or macrophages and antibodies can direct the killing or clearance of virus-infected cells. In one of the first studies, Schlesinger,

Foltzer, and Chapman (1993) noted that anti-NS1 antibodies that bind to cell surfaced-expressed NS1 "raises the possibility of involvement of an Fc piece-directed immune function." They discovered that the Fc region is a determinant of protection by demonstrating that a murine IgG2a variant (isotype with high affinity for activating FcγRs) (Stewart, Hammond, Oberst, & Wilkinson, 2014) of an NS1 antibody protected mice against YFV-induced encephalitis, while an IgG1 variant (isotype with affinity for an inhibitory FcγR) did not. Notably, they did not see the involvement of complement (C3) activity in protection. Subsequent studies on antibodies targeting the NS1 of WNV (Chung et al., 2006; Chung, Thompson, Fremont, & Diamond, 2007) also provided further evidences that FcγR engagement and activation of innate immune cell effector functions are important factors in protection against disease. These latter experiments utilized transgenic mice with FcγRs knocked out to demonstrate the involvement of Fc-mediated effector functions in protection. In our own studies (Bailey et al., 2018), we generated genetic variants of the human IgG1 that lack the ability to engage FcγRs by mutating residues 234, 235, and 329 in the Fc region (Schlothauer et al., 2016; Vafa et al., 2014). The mutated IgG1 or LALAPG variant of NS1-specific monoclonal antibodies failed to confer protection against lethal ZIKV challenge compared with its parental wild-type variant. The results from these sets of experiments demonstrated that antibodies recognizing the ZIKV NS1 are protective and mediate their antiviral activities through interactions with FcγRs on myeloid cells.

Fc-mediated effector functions of ZIKV glycoprotein-specific antibodies

The contribution of Fc-mediated effector functions in antiviral immunity is an important component in controlling virus replication and limiting severe disease (Nimmerjahn & Ravetch, 2010a, 2010b; Taylor et al., 2015). Work in the past decade has increasingly highlighted the importance and role of antibody-dependent cell-mediated mechanisms in antiviral activity against a number of viral pathogens such as influenza, human immunodeficiency, Ebola, and Lassa viruses (Chen et al., 2018; DiLillo, Palese, Wilson, & Ravetch, 2016; DiLillo, Tan, Palese, & Ravetch, 2014; He et al., 2017; Henry Dunand et al., 2016; Howell et al., 2017; Karasavvas et al., 2012; Tan et al., 2016; Tolbert, Sherburn, Van, & Pazgier, 2019). However, it is slightly more complicated for flaviviruses, and Fc-FcγR interactions are a double-edged sword in the context of developing an immune response to this group of viral pathogens. The major concern of premembrane- and envelope-specific antibodies is their potential to enhance internalization of viral particles via Fc-FcγR interactions instead of neutralizing them (Gan, Ting, & Chan, 2017). While ADE is mostly associated with DENV-specific antibodies, others have recently showed that sera and monoclonal antibodies from DENV-infected patients enhanced ZIKV replication in vitro (Dejnirattisai et al., 2016; Stettler et al., 2016). The converse was also shown as murine ZIKV-elicited antibodies enhanced DENV replication in vitro (Kawiecki & Christofferson, 2016). Bardina et al. (2017) showed that both WNV- and DENV-convalescent plasma increased the severity of ZIKV infection and rates of replication in mice. On the other hand, there is experimental evidence to show that nonneutralizing NS1-specific antibodies protect against disease via Fc-dependent mechanisms. This raises the question: can antibodies targeting the premembrane and envelope proteins also behave in the same manner as NS1-specific antibodies? Our own studies reveal that ZIKV premembrane and envelope-specific antibodies are poor at activating Fc-mediated effector functions. Utilizing a surrogate in vitro assay with a genetically modified Jurkat cell line expressing the human FcγRIIIa, we show that human monoclonal antibodies targeting ZIKV structural proteins are unable to activate the luciferase reporter. By contrast, we demonstrated that these same monoclonal antibodies can enhance ZIKV uptake into human leukemia K562 cells, suggesting that their Fc regions are not impaired in interacting with FcγRs (Bailey, Broecker, Freyn, et al., 2019). Perhaps these observations are not too surprising and can be easily explained by how flaviviruses replicate. Nascent flaviviruses bud internally through the endoplasmic reticulum membrane (ER), which are studded with viral glycoproteins (i.e., premembrane and envelope proteins). Viral particles are then transported to the surface of the infected cell via the secretory pathway (Apte-Sengupta, Sirohi, & Kuhn, 2014; Murray, Jones, & Rice, 2008) (Fig. 1). Antibodies targeting the premembrane and envelope proteins therefore do not have access to bind nascently translated ZIKV glycoproteins located in the lumen of the ER. In contrast, viruses such as influenza virus (Shaw & Palese, 2013) and Ebola virus (Noda et al., 2006) bud through the plasma membrane, where their respective nascent glycoproteins reside before being incorporated into mature virions. In the latter scenario, viral glycoproteins are presented on the surface of infected cells and allow for antibody recognition (Fig. 2). Our current understanding suggests that while NS1-specific antibodies do not neutralize virus, their ability to activate Fc-mediated effector functions provides them with an invaluable role in immunity. This type of Fc-mediated immunity is not demonstrated for premembrane- and envelope-specific antibodies. These observations have broad implications for vaccine design.

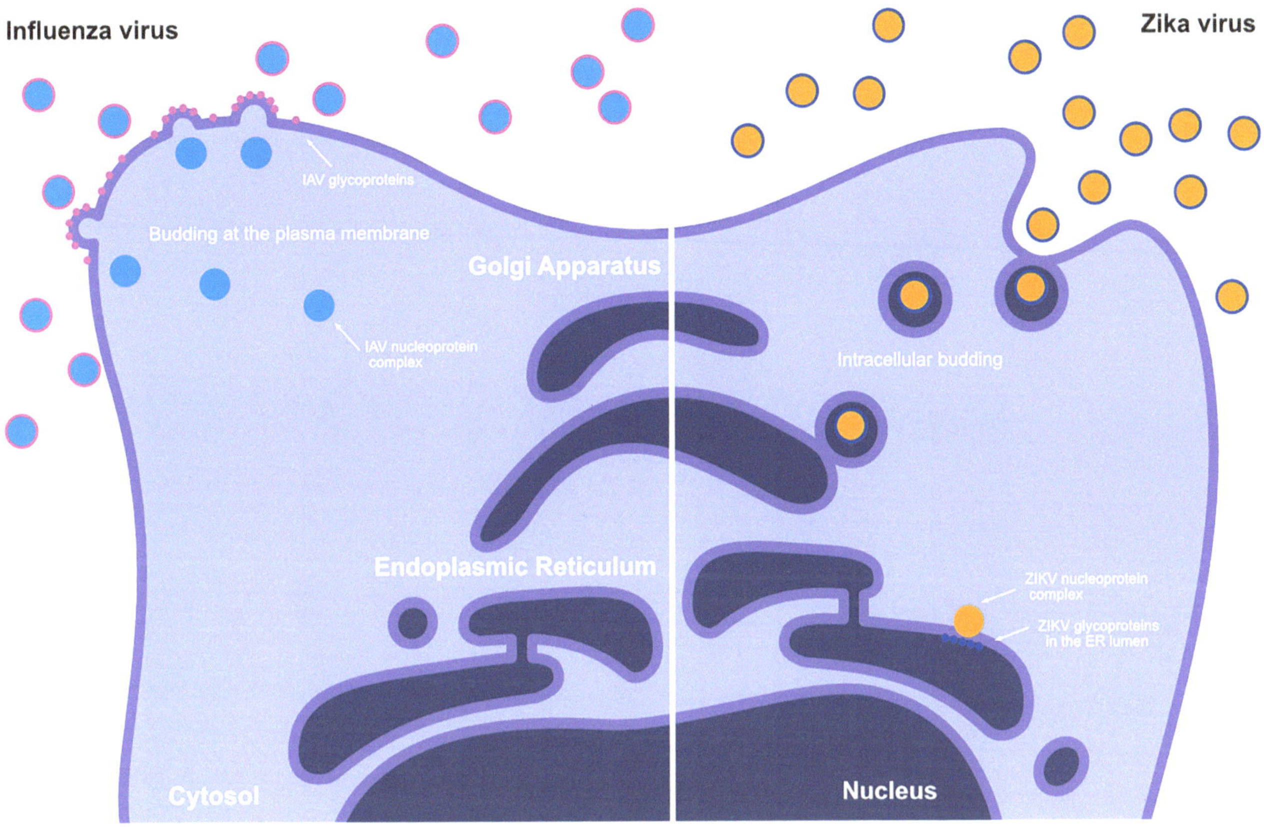

FIG. 1 A comparison in viral egress between influenza A virus (IAV) and Zika virus (ZIKV). Assembly and subsequent budding of IAV particles occur near the surface of the host cell, where the hemagglutinin and neuraminidase glycoproteins reside before incorporation into virions. Assembly and budding of ZIKV particles initially occur intracellularly. The ZIKV nucleoprotein complex putatively buds into the lumen of the endoplasmic reticulum and acquires premembrane and envelope proteins. Mature or immature ZIKV particles are then transported to the surface of the host cells in vesicles.

Candidate ZIKV vaccines

The majority of candidate ZIKV vaccines contain only the premembrane and envelope proteins from ZIKV (Table 1). Novel vaccine platforms include DNA plasmid vaccines expressing full-length M and E proteins (Abbink et al., 2016; Dowd et al., 2016; Larocca et al., 2016). Results from phase I trials show that the DNA vaccination was both tolerated and immunogenic (Gaudinski et al., 2018; Tebas et al., 2017). An inactivated ZIKV vaccine was found to be safe and immunogenic with 92% of volunteers having measurable microneutralization titers of 1:10 (Modjarrad et al., 2018). Live-attenuated viral vectors were also developed, including a replication-deficient construct with a 10-nucleotide deletion in the 3′ untranslated region of the ZIKV genome (Shan, Muruato, Jagger, et al., 2017; Shan, Muruato, Nunes, et al., 2017). Another group attenuated their virus by deleting a 9-amino acids in the capsid protein of ZIKV (Xie et al., 2018). Both live-attenuated approaches were protective in mouse models. Pardi et al. (2017) generated an mRNA vaccine for ZIKV and demonstrated its effectiveness in mice. This vaccine construct encoded the premembrane and envelope of ZIKV and protected mice and nonhuman primates against viremia. Another iteration of an mRNA vaccine construct was designed to avoid induction of ADE. Richner et al. (2017) mutated residues located in the fusion-loop epitope located in the domain II region of the ZIKV envelope glycoprotein. In vivo and in vitro experiments suggested that mutating this region diminished the induction of ADE in mice.

Lastly a few groups incorporated the ZIKV NS1 protein in their vaccine preparations, which also showed promising results. For example, Brault et al. (2017) demonstrated that a modified vaccinia Ankara poxvirus vector expressing NS1 can confer protection of vaccinated mice following intracranial viral challenge. Two groups have recently incorporated NS1 in addition to ZIKV glycoproteins in their vaccine design (Li et al., 2018; Liu et al., 2018). Data from these experiments show incorporation of NS1 contributed to protection against ZIKV challenge in mice beyond that provided by premembrane and envelope constructs alone. Notably, Liu et al. (2018) demonstrated that a recombinant vesicular stomatitis virus expressing NS1 alone was able to reduce viral titers compared with control mice.

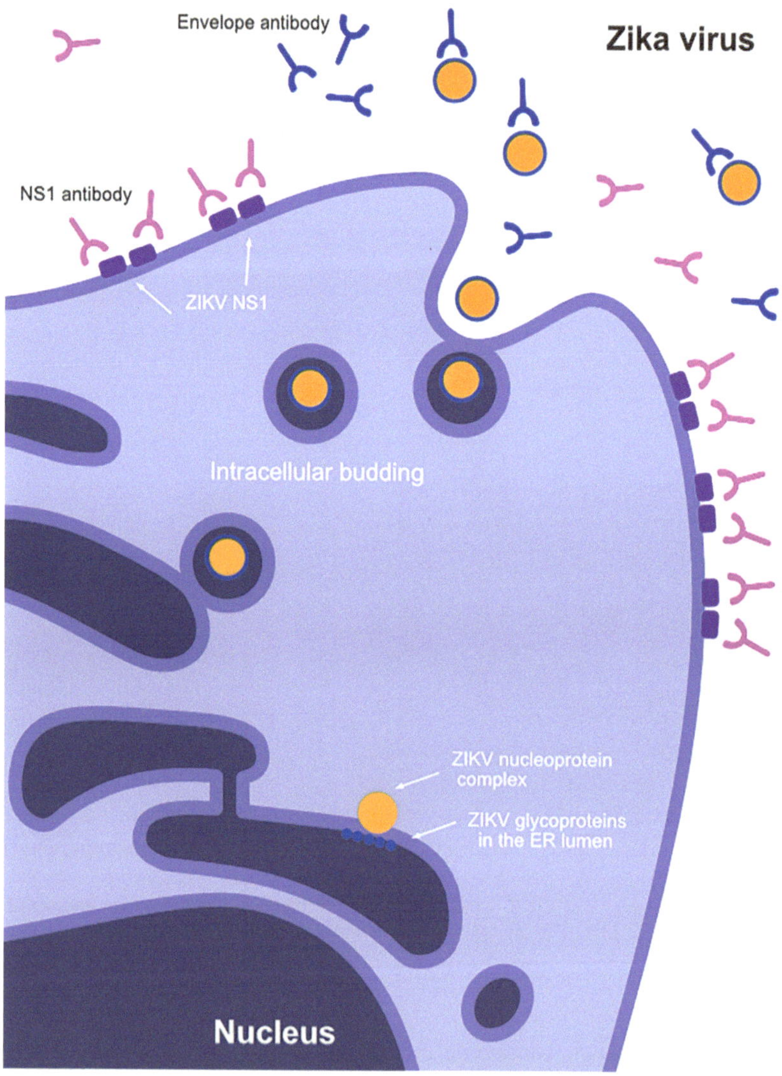

FIG. 2 Binding of NS1- and envelope-specific antibodies to their cognate antigens. NS1 proteins *(purple rectangles)* expressed on the surface of infected cells are readily recognized and bound by NS1-specific antibodies (*pink* antibodies). On the other hand, premembrane- or envelope-specific antibodies (*blue* antibodies) can only bind to the surface of ZIKV virions, since nascent premembrane and envelope glycoproteins are expressed on the luminal side of the endoplasmic reticulum.

Summary

The ability to induce highly neutralizing antibodies and confer sterilizing immunity is the ideal goal in vaccine design. However, due to the conserved epitopes present on flavivirus glycoproteins and the potential phenomenon of ADE, induction of antibodies against ZIKV glycoproteins can be a concern. Notably, Fc-FcγR interactions mediated by premembrane- and envelope-specific antibodies do not correlate to Fc-mediated effector functions. In the preceding text, we discussed the contributions of NS1-specific antibodies in protection and rationales for incorporating a second viral antigen in ZIKV design. We hypothesize that inclusion of NS1 can provide an additional line of defense that directs the killing or clearance of virus infected cells.

Policy and procedures

Not applicable.

Mini-dictionary of terms

Antibody-dependent enhancement of disease. An exacerbation of viral disease mediated by antibodies that bind to free-floating viral particles and internalized into cells that support viral replication instead of being neutralized.

TABLE 1 ZIKV vaccine candidates in clinical trials from https://clinicaltrials.gov.

Vaccine platform	Name	Sponsor	Antigen	Phase I identifiers	Phase II identifiers
DNA vaccine	VRC5283	NIAID	prM-E	NCT02996461	NCT03110770
	VRC5288	NIAID	prM-E	NCT02840487	
	GLS-5700	GeneOne Life Science	prM-E	NCT02809443 NCT02887482	
mRNA	mRNA-1325	Moderna Therapeutics	prM-E	NCT03014089	
Inactivated virion	ZPIV	NIAID	Virion	NCT02963909 NCT02952833 NCT02937233	
	PIZV	Takeda	Virion	NCT03343626	
	VLA1601	Valneva Austria	Virion	NCT03425149	
Live attenuated	rZIKV/D4Δ 30–713	NIAID	Virion	NCT03611946	
Viral vectored	MV-ZIKV	Themis Bioscience	Virion	NCT02996890	

Correlate of protection. A metric by which an immune response elicited by vaccination is associated with protection from infection or disease.

Fc-FcγR interactions. An intermolecular interaction that describes the engagement of the Fc region of an antibody to a receptor expressed on the surface of a myeloid cell such as a natural killer cell, macrophage, or neutrophil. Engagement may result in internalization of the opsonized macromolecule or activation of cellular effector function.

Live-attenuated virus vaccine. A weakened variant of a virus that hypothetical should not cause disease in its host, but can elicit an immune response similar to its wild-type version.

mRNA vaccine. A nucleic acid-based vaccine that encodes an antigen of interest. Administration with this class of vaccine allows for immediate translation of the antigen in the cytoplasm of the target cell.

Nascent. Newly generated or expressed.

Key facts of antibodies that target the NS1

- Antibodies that bind to NS1 do not neutralize virus.
- NS1-specific antibodies can bind to dimeric NS1 molecules expressed on the surface of ZIKV-infected cells.
- Passive transfer of polyclonal or monoclonal NS1 antibodies is known to confer partial protection against ZIKV disease.
- NS1 antibody-mediated protection requires Fc-FcγR interaction.

Key facts of flavivirus envelope glycoproteins

- Envelope-specific antibodies elicited during infection or vaccination can be highly neutralizing.
- Flavivirus envelope glycoproteins share conserved epitopes.
- The conserved epitopes present in the envelope glycoproteins can elicit cross-reactive antibodies.
- Nascent flavivirus glycoproteins are expressed on the membrane of the endoplasmic reticulum and not the plasma membrane.
- Flavivirus virions bud internally and not through the plasma membrane.

Summary points

- ZIKV NS1 is not incorporated into mature infectious virions.
- Antibodies that target ZIKV NS1 have no neutralizing activities in vitro.

- ZIKV NS1-specific antibodies can protect from lethal disease in the mouse model.
- ZIKV NS1-specific antibodies require FcγR engagement for optimal protection.
- Antibodies targeting flavivirus premembrane and envelope proteins can be highly neutralizing but are also implicated in enhancing internalization of viral particles.

REFERENCES

Abbink, P., Larocca, R. A., De La Barrera, R. A., Bricault, C. A., Moseley, E. T., Boyd, M., et al. (2016). Protective efficacy of multiple vaccine platforms against Zika virus challenge in rhesus monkeys. *Science*, *353*(6304), 1129–1132. https://doi.org/10.1126/science.aah6157.

Apte-Sengupta, S., Sirohi, D., & Kuhn, R. J. (2014). Coupling of replication and assembly in flaviviruses. *Current Opinion in Virology*, *9*, 134–142. https://doi.org/10.1016/j.coviro.2014.09.020.

Bailey, M. J., Broecker, F., Duehr, J., Arumemi, F., Krammer, F., Palese, P., et al. (2019). Antibodies elicited by an NS1-based vaccine protect mice against Zika virus. *MBio*, *10*(2), 294. https://doi.org/10.1128/mBio.02861-18.

Bailey, M. J., Broecker, F., Freyn, A., Choi, A., Brown, J. A., Fedorova, N., et al. (2019). Human monoclonal antibodies potently neutralize Zika virus and select for escape mutations on the lateral ridge of the envelope protein. *Journal of Virology*. https://doi.org/10.1128/JVI.00405-19. e00405-19.

Bailey, M. J., Duehr, J., Dulin, H., Broecker, F., Brown, J. A., Arumemi, F. O., et al. (2018). Human antibodies targeting Zika virus NS1 provide protection against disease in a mouse model. *Nature Communications*, *9*(1), 4560. https://doi.org/10.1038/s41467-018-07008-0.

Bardina, S. V., Bunduc, P., Tripathi, S., Duehr, J., Frere, J. J., Brown, J. A., et al. (2017). Enhancement of Zika virus pathogenesis by preexisting anti-flavivirus immunity. *Science*, *356*(6334), 175–180. https://doi.org/10.1126/science.aal4365.

Beatty, P. R., Puerta-Guardo, H., Killingbeck, S. S., Glasner, D. R., Hopkins, K., & Harris, E. (2015). Dengue virus NS1 triggers endothelial permeability and vascular leak that is prevented by NS1 vaccination. *Science Translational Medicine*, *7*(304), 304ra141. https://doi.org/10.1126/scitranslmed.aaa3787.

Belmusto-Worn, V. E., Sanchez, J. L., McCarthy, K., Nichols, R., Bautista, C. T., Magill, A. J., et al. (2005). Randomized, double-blind, phase III, pivotal field trial of the comparative immunogenicity, safety, and tolerability of two yellow fever 17D vaccines (Arilvax and YF-VAX) in healthy infants and children in Peru. *The American Journal of Tropical Medicine and Hygiene*, *72*(2), 189–197.

Bradt, V., Malafa, S., von Braun, A., Jarmer, J., Tsouchnikas, G., Medits, I., et al. (2019). Pre-existing yellow fever immunity impairs and modulates the antibody response to tick-borne encephalitis vaccination. *NPJ Vaccines*, *4*(1), 38. https://doi.org/10.1038/s41541-019-0133-5.

Brault, A. C., Domi, A., McDonald, E. M., Talmi-Frank, D., McCurley, N., Basu, R., et al. (2017). A Zika vaccine targeting NS1 protein protects immunocompetent adult mice in a lethal challenge model. *Scientific Reports*, *7*(1), 14769. https://doi.org/10.1038/s41598-017-15039-8.

Burrell, C. J., Howard, C. R., & Murphy, F. A. (2016). *Fenner and White's medical virology. Vol. 5*. London, United Kingdom: Academic Press.

Chen, X., Lin, M., Qian, S., Zhang, Z., Fu, Y., Xu, J., et al. (2018). The early antibody-dependent cell-mediated cytotoxicity response is associated with lower viral set point in individuals with primary HIV infection. *Frontiers in Immunology*, *9*, 509. https://doi.org/10.3389/fimmu.2018.02322.

Chung, K. M., Nybakken, G. E., Thompson, B. S., Engle, M. J., Marri, A., Fremont, D. H., et al. (2006). Antibodies against West Nile virus nonstructural protein NS1 prevent lethal infection through Fc gamma receptor-dependent and -independent mechanisms. *Journal of Virology*, *80*(3), 1340–1351. https://doi.org/10.1128/JVI.80.3.1340-1351.2006.

Chung, K. M., Thompson, B. S., Fremont, D. H., & Diamond, M. S. (2007). Antibody recognition of cell surface-associated NS1 triggers Fc-gamma receptor-mediated phagocytosis and clearance of West Nile virus-infected cells. *Journal of Virology*, *81*(17), 9551–9555. https://doi.org/10.1128/JVI.00879-07.

Dejnirattisai, W., Supasa, P., Wongwiwat, W., Rouvinski, A., Barba-Spaeth, G., Duangchinda, T., et al. (2016). Dengue virus sero-cross-reactivity drives antibody-dependent enhancement of infection with Zika virus. *Nature Structural and Molecular Biology*, *17*(9), 1102–1108. https://doi.org/10.1038/ni.3515.

DiLillo, D. J., Palese, P., Wilson, P. C., & Ravetch, J. V. (2016). Broadly neutralizing anti-influenza antibodies require Fc receptor engagement for in vivo protection. *The Journal of Clinical Investigation*, *126*(2), 605–610. https://doi.org/10.1172/JCI84428.

DiLillo, D. J., Tan, G. S., Palese, P., & Ravetch, J. V. (2014). Broadly neutralizing hemagglutinin stalk-specific antibodies require FcγR interactions for protection against influenza virus in vivo. *Nature Medicine*, *20*(2), 143–151. https://doi.org/10.1038/nm.3443.

Domingo, C., Fraissinet, J., Ansah, P. O., Kelly, C., Bhat, N., Sow, S. O., et al. (2019). Long-term immunity against yellow fever in children vaccinated during infancy: A longitudinal cohort study. *The Lancet Infectious Diseases*, *19*(12), 1363–1370. https://doi.org/10.1016/S1473-3099(19)30323-8.

Dowd, K. A., Ko, S.-Y., Morabito, K. M., Yang, E. S., Pelc, R. S., DeMaso, C. R., et al. (2016). Rapid development of a DNA vaccine for Zika virus. *Science*, *354*(6309), 237–240. https://doi.org/10.1126/science.aai9137.

Estofolete, C. F., & Nogueira, M. L. (2018). Is a dose of 17D vaccine in the current context of yellow fever enough? *Brazilian Journal of Microbiology*, *49*(3), 683–684. https://doi.org/10.1016/j.bjm.2018.02.003.

Falgout, B., Bray, M., Schlesinger, J. J., & Lai, C. J. (1990). Immunization of mice with recombinant vaccinia virus expressing authentic dengue virus nonstructural protein NS1 protects against lethal dengue virus encephalitis. *Journal of Virology*, *64*(9), 4356–4363.

Gan, E. S., Ting, D. H. R., & Chan, K. R. (2017). The mechanistic role of antibodies to dengue virus in protection and disease pathogenesis. *Expert Review of Anti-Infective Therapy*, *15*(2), 111–119. https://doi.org/10.1080/14787210.2017.1254550.

Gaudinski, M. R., Houser, K. V., Morabito, K. M., Hu, Z., Yamshchikov, G., Rothwell, R. S., et al. (2018). Safety, tolerability, and immunogenicity of two Zika virus DNA vaccine candidates in healthy adults: Randomised, open-label, phase 1 clinical trials. *Lancet*, *391*(10120), 552–562. https://doi.org/10.1016/S0140-6736(17)33105-7.

Giel-Moloney, M., Goncalvez, A. P., Catalan, J., Lecouturier, V., Girerd-Chambaz, Y., Diaz, F., et al. (2018). Chimeric yellow fever 17D-Zika virus (ChimeriVax-Zika) as a live-attenuated Zika virus vaccine. *Scientific Reports*, *8*(1), 13206–13211. https://doi.org/10.1038/s41598-018-31375-9.

Gotuzzo, E., Yactayo, S., & Córdova, E. (2013). Efficacy and duration of immunity after yellow fever vaccination: Systematic review on the need for a booster every 10 years. *The American Journal of Tropical Medicine and Hygiene*, *89*(3), 434–444. https://doi.org/10.4269/ajtmh.13-0264.

Halstead, S. B., & Thomas, S. J. (2011). New Japanese encephalitis vaccines: Alternatives to production in mouse brain. *Expert Review of Vaccines*, *10*(3), 355–364. https://doi.org/10.1586/erv.11.7.

He, W., Chen, C.-J., Mullarkey, C. E., Hamilton, J. R., Wong, C. K., Leon, P. E., et al. (2017). Alveolar macrophages are critical for broadly-reactive antibody-mediated protection against influenza A virus in mice. *Nature Communications*, *8*, 846. https://doi.org/10.1038/s41467-017-00928-3.

Henry Dunand, C. J., Leon, P. E., Huang, M., Choi, A., Chromikova, V., Ho, I. Y., et al. (2016). Both neutralizing and non-neutralizing human H7N9 influenza vaccine-induced monoclonal antibodies confer protection. *Cell Host & Microbe*, *19*(6), 800–813. https://doi.org/10.1016/j.chom.2016.05.014.

Hilgenfeld, R. (2016). Zika virus NS1, a pathogenicity factor with many faces. *The EMBO Journal*, *35*(24), 2631–2633. https://doi.org/10.15252/embj.201695871.

Howell, K. A., Brannan, J. M., Bryan, C., McNeal, A., Davidson, E., Turner, H. L., et al. (2017). Cooperativity enables non-neutralizing antibodies to neutralize Ebolavirus. *Cell Reports*, *19*(2), 413–424. https://doi.org/10.1016/j.celrep.2017.03.049.

Kaiser, J. A., & Barrett, A. D. T. (2019). Twenty years of progress toward West Nile virus vaccine development. *Viruses*, *11*(9), 823. https://doi.org/10.3390/v11090823.

Karasavvas, N., Billings, E., Rao, M., Williams, C., Zolla-Pazner, S., Bailer, R. T., et al. (2012). The Thai phase III HIV type 1 vaccine trial (RV144) regimen induces antibodies that target conserved regions within the V2 loop of gp120. *AIDS Research and Human Retroviruses*, *28*(11), 1444–1457. https://doi.org/10.1089/aid.2012.0103.

Kawiecki, A. B., & Christofferson, R. C. (2016). Zika virus-induced antibody response enhances dengue virus serotype 2 replication in vitro. *The Journal of Infectious Diseases*, *214*(9), 1357–1360. https://doi.org/10.1093/infdis/jiw377.

Kiran, S. K., Pasi, A., Kumar, S., Kasabi, G. S., Gujjarappa, P., Shrivastava, A., et al. (2015). Kyasanur Forest disease outbreak and vaccination strategy, Shimoga District, India, 2013-2014. *Emerging Infectious Diseases*, *21*(1), 146–149. https://doi.org/10.3201/eid2101.141227.

Krishna, V. D., Rangappa, M., & Satchidanandam, V. (2009). Virus-specific cytolytic antibodies to nonstructural protein 1 of Japanese encephalitis virus effect reduction of virus output from infected cells. *Journal of Virology*, *83*(10), 4766–4777. https://doi.org/10.1128/JVI.01850-08.

Larocca, R. A., Abbink, P., Peron, J. P. S., Zanotto, P. M.d. A., Iampietro, M. J., Badamchi-Zadeh, A., et al. (2016). Vaccine protection against Zika virus from Brazil. *Nature*, *536*(7617), 474–478. https://doi.org/10.1038/nature18952.

Li, A., Yu, J., Lu, M., Ma, Y., Attia, Z., Shan, C., et al. (2018). A Zika virus vaccine expressing premembrane-envelope-NS1 polyprotein. *Nature Communications*, *9*(1), 3067. https://doi.org/10.1038/s41467-018-05276-4.

Liu, X., Qu, L., Ye, X., Yi, C., Zheng, X., Hao, M., et al. (2018). Incorporation of NS1 and prM/M are important to confer effective protection of adenovirus-vectored Zika virus vaccine carrying E protein. *NPJ Vaccines*, *3*(1), 29. https://doi.org/10.1038/s41541-018-0072-6.

Modjarrad, K., Lin, L., George, S. L., Stephenson, K. E., Eckels, K. H., La, B., et al. (2018). Preliminary aggregate safety and immunogenicity results from three trials of a purified inactivated Zika virus vaccine candidate: Phase 1, randomised, double-blind, placebo-controlled clinical trials. *Lancet*, *391* (10120), 563–571. https://doi.org/10.1016/S0140-6736(17)33106-9.

Murray, C. L., Jones, C. T., & Rice, C. M. (2008). Architects of assembly: Roles of Flaviviridae non-structural proteins in virion morphogenesis. *Nature Reviews Microbiology*, *6*(9), 699–708. https://doi.org/10.1038/nrmicro1928.

Nimmerjahn, F., & Ravetch, J. V. (2010a). Antibody-mediated modulation of immune responses. *Immunological Reviews*, *236*(1), 265–275. https://doi.org/10.1111/j.1600-065X.2010.00910.x.

Nimmerjahn, F., & Ravetch, J. V. (2010b). FcγRs in health and disease. In *Vol. 350. Negative co-receptors and ligands* (pp. 105–125). Berlin, Heidelberg: Springer. https://doi.org/10.1007/82_2010_86.

Noda, T., Ebihara, H., Muramoto, Y., Fujii, K., Takada, A., Sagara, H., et al. (2006). Assembly and budding of Ebolavirus. *PLoS Pathogens*, *2*(9), e99. https://doi.org/10.1371/journal.ppat.0020099.

Pardi, N., Hogan, M. J., Pelc, R. S., Muramatsu, H., Andersen, H., DeMaso, C. R., et al. (2017). Zika virus protection by a single low-dose nucleoside-modified mRNA vaccination. *Nature*, *543*(7644), 248–251. https://doi.org/10.1038/nature21428.

Plotkin, S. A. (2010). Correlates of protection induced by vaccination. *Clinical and Vaccine Immunology*, *17*(7), 1055–1065. https://doi.org/10.1128/CVI.00131-10.

Rastogi, M., Sharma, N., & Singh, S. K. (2016). Flavivirus NS1: A multifaceted enigmatic viral protein. *Virology Journal*, 1–10. https://doi.org/10.1186/s12985-016-0590-7.

Reynolds, C. J., Suleyman, O. M., Ortega-Prieto, A. M., Skelton, J. K., Bonnesoeur, P., Blohm, A., et al. (2018). T cell immunity to Zika virus targets immunodominant epitopes that show cross-reactivity with other Flaviviruses. *Scientific Reports*, *8*(1), 672. https://doi.org/10.1038/s41598-017-18781-1.

Richner, J. M., Himansu, S., Dowd, K. A., Butler, S. L., Salazar, V., Fox, J. M., et al. (2017). Modified mRNA vaccines protect against Zika virus infection. *Cell*, *168*(6), 1114–1125. e10 https://doi.org/10.1016/j.cell.2017.02.017.

Schlesinger, J. J., Brandriss, M. W., & Walsh, E. E. (1985). Protection against 17D yellow fever encephalitis in mice by passive transfer of monoclonal antibodies to the nonstructural glycoprotein gp48 and by active immunization with gp48. *The Journal of Immunology*, *135*(4), 2805–2809.

Schlesinger, J. J., Foltzer, M., & Chapman, S. (1993). The Fc portion of antibody to yellow fever virus NS1 is a determinant of protection against YF encephalitis in mice. *Virology*, *192*(1), 132–141. https://doi.org/10.1006/viro.1993.1015.

Schlothauer, T., Herter, S., Koller, C. F., Grau-Richards, S., Steinhart, V., Spick, C., et al. (2016). Novel human IgG1 and IgG4 Fc-engineered antibodies with completely abolished immune effector functions. *Protein Engineering, Design & Selection, 29*(10), 457–466. https://doi.org/10.1093/protein/gzw040.

Shan, C., Muruato, A. E., Jagger, B. W., Richner, J., Nunes, B. T. D., Medeiros, D. B. A., et al. (2017). A single-dose live-attenuated vaccine prevents Zika virus pregnancy transmission and testis damage. *Nature Communications, 8*(1), 676–679. https://doi.org/10.1038/s41467-017-00737-8.

Shan, C., Muruato, A. E., Nunes, B. T. D., Luo, H., Xie, X., Medeiros, D. B. A., et al. (2017). A live-attenuated Zika virus vaccine candidate induces sterilizing immunity in mouse models. *Nature Medicine, 23*(6), 763–767. https://doi.org/10.1038/nm.4322.

Shaw, M. L., & Palese, P. (2013). Orthomyxoviridae: The viruses and their replication. In D. M. Knipe, & P. Howley (Eds.), *Fields virology* (6th ed., pp. 1691–1740). Philadelphia: Lippincott Williams and Wilkins.

Smatti, M. K., Al Thani, A. A., & Yassine, H. M. (2018). Viral-induced enhanced disease illness. *Frontiers in Microbiology, 9*, 2991. https://doi.org/10.3389/fmicb.2018.02991.

Stettler, K., Beltramello, M., Espinosa, D. A., Graham, V., Cassotta, A., Bianchi, S., et al. (2016). Specificity, cross-reactivity, and function of antibodies elicited by Zika virus infection. *Science, 353*(6301), 823–826. https://doi.org/10.1126/science.aaf8505.

Stewart, R., Hammond, S. A., Oberst, M., & Wilkinson, R. W. (2014). The role of Fc gamma receptors in the activity of immunomodulatory antibodies for cancer. *Journal for Immunotherapy of Cancer, 2*(1), 29. https://doi.org/10.1186/s40425-014-0029-x.

Tan, G. S., Leon, P. E., Albrecht, R. A., Margine, I., Hirsh, A., Bahl, J., et al. (2016). Broadly-reactive neutralizing and non-neutralizing antibodies directed against the H7 influenza virus hemagglutinin reveal divergent mechanisms of protection. *PLoS Pathogens, 12*(4), e1005578. https://doi.org/10.1371/journal.ppat.1005578.

Taylor, A., Foo, S.-S., Bruzzone, R., Dinh, L. V., King, N. J. C., & Mahalingam, S. (2015). Fc receptors in antibody-dependent enhancement of viral infections. *Immunological Reviews, 268*(1), 340–364. https://doi.org/10.1111/imr.12367.

Tebas, P., Roberts, C. C., Muthumani, K., Reuschel, E. L., Kudchodkar, S. B., Zaidi, F. I., et al. (2017). Safety and immunogenicity of an anti-Zika virus DNA vaccine—Preliminary report. *New England Journal of Medicine*. https://doi.org/10.1056/NEJMoa1708120. NEJMoa1708120.

Tolbert, W. D., Sherburn, R. T., Van, V., & Pazgier, M. (2019). Structural basis for epitopes in the gp120 cluster A region that invokes potent effector cell activity. *Viruses, 11*(1), 69. https://doi.org/10.3390/v11010069.

Vafa, O., Gilliland, G. L., Brezski, R. J., Strake, B., Wilkinson, T., Lacy, E. R., et al. (2014). An engineered Fc variant of an IgG eliminates all immune effector functions via structural perturbations. *Methods, 65*(1), 114–126. https://doi.org/10.1016/j.ymeth.2013.06.035.

WHO. (2015). Vaccines and vaccination against yellow fever: WHO position paper, June 2013—Recommendations. *Vaccine, 33*(1), 76–77. https://doi.org/10.1016/j.vaccine.2014.05.040.

Xie, X., Kum, D. B., Xia, H., Luo, H., Shan, C., Zou, J., et al. (2018). A single-dose live-attenuated Zika virus vaccine with controlled infection rounds that protects against vertical transmission. *Cell Host & Microbe, 24*(4), 487–499. e5 https://doi.org/10.1016/j.chom.2018.09.008.

Zhang, X., Jia, R., Shen, H., Wang, M., Yin, Z., & Cheng, A. (2017). Structures and functions of the envelope glycoprotein in flavivirus infections. *Viruses, 9*(11), 338. https://doi.org/10.3390/v9110338.

Chapter 36

Zika vaccines must prevent sexual transmission

Omar Bagasra[a] and Ewen McLean[b]
[a]*Department of Biology, South Carolina Center for Biotechnology, School of Natural Sciences and Mathematics, Claflin University, Orangeburg, SC, United States,* [b]*Aqua Cognoscenti, West Columbia, SC, United States*

Abbreviations

Abs	antibodies
ADCC	Ab-dependent cellular cytotoxicity
Fab	fragment antigen-binding
GBS	Guillain-Barré syndrome
GC	germ cell
IFN-I	type I interferon
JEV	Japanese encephalitis virus
MAC	IgM-capture
NK	natural killer
PHEIC	Public Health Emergency of International Concern
RNAi	RNA interference
WHO	World Health Organization
YFV	yellow fever virus

Introduction

In response to a cluster of Zika virus (ZIKV)-associated neurological disorders and neonatal deformities, centered around Brazil and Puerto Rico, the World Health Organization declared a *Public Health Emergency of International Concern* (PHEIC), on February 1, 2016 (WHO, 2005). Since then, uninterrupted ZIKV infections have been reported in 71 nations with a further 15 countries recording interrupted transmission (WHO, 2019). Another 62 states have populations of competent vectors but without records of previous ZIKV transmission; over 75% of the world's nations are thus vulnerable to ZIKV. Steady declines in the prevalence of ZIKV, in the hardest hit areas, throughout 2016, ultimately led the WHO to downgrade its PHEIC declaration, and in November 2016, ZIKV was consigned to a program of sustained activities but remained on the 2018 WHO *List of Blueprint Priority Diseases*.

ZIKV can be transmitted by at least nine species of *Aedes* mosquito (Boyer, Calvez, Chouin-Carneiroc, et al., 2018), with the Asian tiger *A. albopictus* and yellow fever *A. aegypti* mosquitoes being considered the most significant vectors for humans (Wican & Smith, 2016). The latter two species have extensive distribution, are endophilic in nature, and presently experience range extension due to climate change (Kumar, Jovel, Lopez-Orozco, et al., 2018; Rochlin, Ninivaggi, Hutchinson, & Farajollahi, 2013). Until comparatively recently, methods of ZIKV transmission were believed only to be sylvatic (nonhuman primate to mosquito), via human-mosquito cycle in urban environments, and from mother to fetus (Grischott, Puhan, Hatza, et al., 2016). This possibility was challenged in 2011 (Grischott et al., 2016), when a ZIKV-positive individual, returning from an endemic to nonendemic area, seemingly infected his sexual partner. This opening nonvector-borne ZIKV infection, in a nonendemic area, where *Aedes* mosquitoes nonetheless bred (i.e., the United States; Leta, Beyene, DeClercq, et al., 2018), provided serious notice of what might lie ahead. Indeed, subsequent to the Colorado infection (Foy, Kobylinski, Foy, et al., 2001), autochthonous sexual transmission of ZIKV in nonendemic areas was reported in 13 countries (Mead, Duggal, Hook, et al., 2018; Moreira, Peixoto, Siqueira, et al., 2017). From January 1, 2015 to December 31, 2018, there have been 52 confirmed cases of M-F, F-M, and M-M sexual transmission in the United States alone (CDC, 2019) through vaginal, rectal, and oral routes (Mead, Duggal, et al., 2018; Moreira et al., 2017).

Zika Virus Impact, Diagnosis, Control, and Models. https://doi.org/10.1016/B978-0-12-820267-8.00036-4

In general, ZIKV appears to infect host cells as other flaviviruses do. In this regard the multifarious roles of RNAs during viral infection and replication are meticulously considered by Göertz, Abbo, Fros, et al. (2018). Briefly the envelope protein of the virus couples with a variety of cell surface receptors and attachment factors (e.g., TIM-1, AXL, and Tyro3: Lee, Bos, Li, et al., 2018; Hamel, Dejarnac, Wichit, et al., 2015), and passage into the cell is facilitated by clathrin-mediated endocytosis. Viral particles, fused with endosomal membrane, release genomic RNA into compartments of rough endoplasmic reticulum wherein immature particles are assembled and transported through the trans-Golgi network, where fully infectious viruses are constructed. These are subsequently released into extracellular spaces to commence a new cycle of cell infection and replication (Ploudre & Bloch, 2016). During the acute phase of the illness, when ZIKV IgM may be imperceptible (Ploudre & Bloch, 2016), infection is confirmed by RT-PCR and its affiliates (Jiang, Loeb, Manzanas, et al., 2018; Rong, Wang, Sun, Wang, et al., 2019; Tancharoen, Sukjee, Thepparit, et al., 2019). After the acute phase has lapsed, 7–14 days post infection, recognition of ZIKV IgM antibodies by IgM capture (MAC)-ELISA, with confirmation by seroneutralization assay, is suggested (Petersen, Jamieson, Powers, et al., 2016). The need for a confirmatory step in MAC-ELISA is due to possible false positives caused by sample contamination and/or presence of other flaviviruses (Jääskeläinen, Korhonen, Huhtamo, et al., 2019; Roberts & Maslow, 2018).

Although ZIKV RNA has been detected in a wide variety of body fluids including, saliva, tears, sweat, blood, cerebrospinal fluid, breast milk, urine, seminal fluid, amniotic fluid, and other vaginal secretions (Dias, da Silva Sobreira, Martins, et al., 2019; Joguet, Mansuy, Matusali, et al., 2017), viral RNA is demonstrable, by RT-PCR, in semen for much longer than other body fluids (Medina, Torres, Acevedo, et al., 2019; Paz-Bailey, Rosenberg, Doyle, et al., 2018). Mead, Hills, and Brooks (2018) determined that symptomatic ZIKV infection was generally associated with persistent presence of ZIKV RNA in semen, but with viral shedding declining rapidly 3 months post onset of illness. Nonetheless, in some men, and especially in less sexually active, generally older individuals, shedding endured for over half a year, although no infective cultured ZIKV was obtained 30 days post initial onset of illness. Observations of others (Arsuaga, Bujalance, Díaz-Menéndez, et al., 2016; Nicastri, Castilletti, Liuzzi, et al., 2016; Sánchez-Montalvá, Pou, et al., 2018) extend ZIKV infectivity past 65 days, however. In mammalian cells, early viral infections may be suppressed by the innate type I interferon (IFN-I) response. Accordingly, viruses have evolved mechanisms to disrupt or evade host cell antiviral or RNA interference (RNAi) pathways. In general, transcriptome profiling indicates that Zika infection of Sertoli cells upregulates numerous genes involved in innate immunity, including IFN regulatory factors and those implicated in testicular immune privilege (Strange, Green, Siemann, et al., 2018). Zika and other flaviviruses suppress RNAi pathways with subgenomic flaviviral (sf) RNA, which have been demonstrated to quell the viral replication sensors RIG-I and MDA-5 and to upregulate the cytokine FGF2, leading to augmented viral pathogenicity and persistence (Göertz et al., 2018; Kumar et al., 2018). Down-regulation of a germ cell (GC) IFN-stimulated gene, *lfi44l*, likewise augments fabrication of ZIKV (Robinson, Chong, Ashbrook, et al., 2018), with mock infection of GC having no impact on cell proliferation or survival. These observations, together with those of Kumar et al. (2018), who observed that Sertoli cells, but not Leydig cells, were highly susceptible to ZIKV infection, lead to the conclusion that ZIKV can cross the blood-testis barrier and establish a persistent infection within seminiferous tubules—an immune-privileged site of the testis, where postmeiotic GC are shielded from systemic immune attack (Siemann, Strange, Maharaj, et al., 2017; Strange et al., 2018).

Although our current understanding of ZIKV infection and replication appears complete, specific aspects of the cycle remain imprecise. Some reported variations in the outcome to ZIKV infection likely spring from confusion due to the use of animal models. For example, immunocompetent mice do not express clinical symptoms of ZIKV infection due to a species-specific IFN-I response (NS5 binding to STAT2) that restricts ZIKV replication. To evade this complication, murine models that carry deficiencies in the IFN-pathway, or have had their IFN response artificially suppressed, have been employed (Kurscheidt, Mesquita, Damke, et al., 2019; Ma, Li, Ma, et al., 2016; Nazerai, Christensen, & Thomsen, 2019). However, these models have provided pivotal information on ZIKV transmission and infection, for example, murine models revealed ZIKV damaged seminiferous tubules and germ, Sertoli, and Leydig cells, thereby reducing fertility (Govero, Esakky, Scheaffer, et al., 2016; Griffin, Muthumani, Warner, et al., 2017; Robinson et al., 2018); results with immunosuppressed animals may differ significantly from those derived from healthy human adults. For example, Matusali, Houzet, Satie, et al. (2018) reported a consistently weaker infection of human seminiferous tubule cells when compared against the mouse model. The authors suggested that this difference likely reflected distinctions in susceptibility of mouse versus human Sertoli and germ cells and, perhaps, in the innate immune response. The use and limitations of murine and nonhuman primate models for ZIKV studies, especially with respect to the male genital tract, receive comprehensive consideration by Kurscheidt et al. (2019).

While mathematical models indicate that sexual spread of ZIKV would contribute only moderately to disease occurrences (Yakob, Kucharski, Hue, et al., 2016), sexual transmission could, nonetheless, frustrate attempts to control and prevent ZIKV outbreaks (Althaus & Low, 2016). Estimates of the contribution of sexual transmission of ZIKV range from

4.8% (Maxian, Neufeld, Talis, et al., 2017) to much higher levels, perhaps even order of magnitude, especially in "at-risk" populations (Allard, Althouse, Hebert-Dufresne, et al., 2017; Arsuaga, Bujalance, Díaz-Menéndez, et al., 2017; Augusto, Bewick, & Fagan, 2017). Also, it has been suggested that sexual ZIKV transmission is more likely to cause fetal infections than vector-borne spread (Duggal, McDonald, Ritter, et al., 2018; Russell, Hills, Oster, et al., 2017), further highlighting the need to develop effective vaccines.

Current vaccine effort to quell ZIKV

Unlike many other seasonally infectious diseases, such as influenza, Japanese encephalitis virus (JEV), and yellow fever virus (YFV), for which vaccines and antivirals exist, similar treatment options are unavailable in the drug cabinet for ZIKV. Although mortality due to ZIKV is low, the risk of Zika-associated neonatal deformities and development of Guillain-Barré syndrome (GBS) place high urgency on the development of vaccine(s) to protect populations in endemic regions of the globe. Indeed the original ZIKV PHEIC called for a global effort to develop ZIKV vaccines, and over 45 candidates have been evaluated since, with some entering phase I/II clinical trials (Wilder-Smith, Vannice, & Durbin, 2018). Candidate vaccines have included live attenuated strains, purified inactivated virus, mRNA vaccines, synthetic DNA vaccines, adenovirus vector-based subunit vaccines, and combinatorial types (Ghaffar, Ng, & Renia, 2018; Makhluf & Shresta, 2018). Several of these provide strong protection of mice and nonhuman primate models against ZIKV challenge (Vannice, Cassetti, Eisinger, et al., 2019). While lack of available phase III testing landscapes to check clinical efficacy of vaccines remains problematic, collated observations from animal and phase I/II trials indicate that prevention of ZIKV by vaccination is achievable. Use of a controlled human infection model combined with immune marker studies to assess protection conferred may, in the absence of suitable phase III sites, assist future licensure of vaccines (Das Neves Almeida, Racine, Magalhães, et al., 2018; França, Hess, Dufour, et al., 2016).

Vaccinologists, with experience in developing vaccines against Japanese encephalitis virus (JEV) and YFV, are well aware of the difficulties and challenges encountered in conceiving a ZIKV vaccine. Generally a live ZIKV vaccine would be contraindicated for pregnant women because it may infect the fetus and such vaccines may not meet safety requirements. Therefore a desirable ZIKV vaccine must elicit a strong anti-ZIKV immunity that would prevent neonatal deformities and GBS in vaccinated individuals (Das Neves Almeida et al., 2018). In addition, since ZIKV is mostly endemic in developing nations, vaccines must be cost effective and, preferably, express extended shelf life under rigorous conditions.

One of the most important factors that has been overlooked is the persistence of ZIKV in the semen and male reproductive system that allows sexual transmission. One critical unanswered question is why, in the presence of anti-ZIKV antibodies (Abs), infected men are still able to transmit the virus sexually? Examples of sexually transmitted flaviviruses are known for humans and animals, and their vertical transmission, from mother to fetus, like ZIKV, has also been recorded, resulting in augmented persistence and potential passage to new reservoirs (Medina et al., 2019; Paz-Bailey et al., 2018). It is well established that Abs does not effectively pass privileged sites such as the blood-brain barrier, vitreous humor, and Sertoli-blood barrier (França et al., 2016). However, we are aware (vide supra) that ZIKV crosses the blood-testis barrier to establish persistent infection within seminiferous tubules, both in human and in animal models (Atkinson, Thorburn, Petridou, et al., 2017; Bagasra, Addanki, Goodwin, et al., 2017). Contemporary evidence also indicates that ZIKV in semen is suppressed after 6 months (55). Accordingly, we believe that a certain subclass of IgG may play a role in penetrating the Sertoli-blood barrier (Mruk & Cheng, 2015). If this assumption is correct, it demands special attention during development and subsequent approval of Zika vaccines (Quanquin, Wang, & Cheng, 2017). We believe that a Zika vaccine could be designed to consider this specific goal. In the following section, we briefly describe the potential role of IgG subclasses and how these may subsequently impact the development of an effective ZIKV inoculation.

In humans and other mammals, IgG is present as a number of different subclasses that differ in their Fc regions, size, and other features at the molecular level (Schroeder & Cavacini, 2010), implying that each subclass expresses different functions. These IgG subclasses were numbered in reference to the rank order (IgG1 > IgG2 > IgG3 > IgG4) of the serum levels of these antibodies in the blood of normal, healthy individuals (Schroeder & Cavacini, 2010). The IgG subclasses exhibit different functional activities. Activation of the complement cascade is an important means of clearance of opsonized pathogens. IgG4 exhibits very low affinity to C1q, which is the first component of the complement pathway and binds to the CH2 domain of IgG; hence, IgG4 fails to fix complement. There are also defined differences in the affinity to the three classes of Fc receptors (FcγR) (I, II, and III). IgG1 and IgG3 bind to all three FcγR classes. IgG4 binds only FcγRII and III, albeit less strongly than the binding of IgG1. And IgG2 binds only to FcγRII.

Human IgG1 has strong opsonization properties and enhances phagocytosis by macrophages and neutrophils that possess Fc receptors and hence speed up the clearance of invading microorganisms (Bournazos & Ravetch, 2017). On the other hand, IgG3 has an ability to enhance complement fixation. Natural killer (NK) cells express Fc receptors for

IgG3, which sanction destruction of virally infected cells via an Ab-dependent cellular cytotoxicity (ADCC) mechanism (Bournazos & Ravetch, 2017; Schroeder & Cavacini, 2010). Noteworthy is that IgG1 and IgG3 antibodies are generally induced in response to protein antigens, whereas IgG2 and IgG4 are associated with polysaccharide antigens. The reaction to given antigens can also result in a skewed IgG subclass response, and this is frequently a source of investigation as it correlates with protection and in the design of vaccines. Therefore it would be of interest to evaluate if men who harbor ZIKV for a long time lack an IgG4 response to the virus (Bagasra et al., 2017). Specific subclasses may be associated with individual disease processes. For example, in pemphigus vulgaris, a mucocutaneous blistering disease, IgG4 antibodies to desmoglein 3 are pathogenic (Schroeder & Cavacini, 2010), whereas first-degree relatives with IgG1 autoantibodies to the same protein show no evidence of the disease.

IgG4 is unique in both structure and function (Schroeder & Cavacini, 2010) and is the least abundant immunoglobulin in human serum, accounting for less than 5% of total IgG in healthy individuals. While the concentrations of the other three subclasses of IgG can vary more than 100-fold, IgG4 levels are stable in healthy people (Koneczny, 2018). Even though the constant domains of IgG4 heavy chains are remarkably similar to those of other IgG subclasses (sharing more than 95% homology), <5% amino acid disparity within the second constant domain produces weak or negligible binding of IgG4 to both C1q and Fcγ receptors (Huijbers, Plomp, van der Maarel, et al., 2018). Thus IgG4 does not appear to play any role in activating the classical complement pathway and has been traditionally considered to play only a limited role in immune activation. IgG4 is the smallest of the IgG subclasses and has the potential to cross membranes of blood-Sertoli barrier; its ability to cross the placenta is well established (Aalberse, Stapel, Schuurman, et al., 2009). Not all immunoglobulins can penetrate the blood-Sertoli barrier, and there is an opportunity to uncover the mystery underlying the prolonged shedding of ZIKV.

Of note, in the serum of individuals immunized for several years with phospholipase (PLA) from honeybee venom, grass pollen allergen, or house dust mite allergen, more than 50% of the total antigen-binding capacity is due to IgG4. In the case of apiarists, during prolonged immunization, a shift in the IgG4/IgG1 ratio is observed, ultimately resulting in an IgG4-dominated response (Bozzalla Cassione & Stone, 2017). IgG4 also expresses a strong antiinflammatory effect, which may be of high significance when considering the prevention of GBS (Stone, Zen, & Deshpande, 2012). We believe that anti-ZIKV IgG4 titers should be evaluated to determine whether patients inflicted with GBS exhibit depressed titers of IgG4.

A unique characteristic of IgG4 is its half-antibody exchange reaction, also referred to as fragment antigen-binding (Fab) arm exchange (Stone et al., 2012). IgG4 readily forms disulfide bonds within the heavy chains in its hinge region because, in contrast to the other IgG subclasses, the disulfide bonds between the heavy chains of the IgG4 molecule are unstable. Therefore ~50% of IgG4 molecules consist of heavy chains linked weakly by noncovalent forces. The remainder of the IgG4 molecules presumably retain intact disulfide bonds between the heavy chains in vivo, but the actual percentages of intrachain isomers may vary with environmental conditions. In an IgG4 molecule without disulfide bonds between the heavy chains, dissociations of the noncovalent bonds permit the chains to separate and recombine randomly, such that asymmetric antibodies with two different antigen-combining sites are formed. The resulting bispecific (functionally monovalent) IgG4 molecules are unable to cross-link antigens, thereby losing the ability to form immune complexes (Stone et al., 2012).

The persistence of ZIKV in semen is a matter of great concern. Paz-Bailey et al. (2018) described the dynamics of ZIKV in body fluids in a cohort of participants who lived in Puerto Rico, an area where the virus is endemic, and in whom ZIKV infection was confirmed by RT-PCR. ZIKV RNA was detectable from 100 to 139 days post infection in semen but for much shorter periods in serum, saliva, and urine. Others have reported the presence of ZIKV RNA for a prolonged duration in semen, with the longest reported duration of detectable ZIKV RNA being up to 370 days after onset (Counotte, Kim, Wang, Bernstein, et al., 2018). ZIKV can replicate in the Sertoli but not in Leydig cell cultures in vitro (Mlera & Bloom, 2019). We suggest that one explanation for the dichotomous observations in ZIKV presence may be that men who are shedding Zika for a long period of time may have selective IgG subclass immunodeficiency or have delayed IgG4 (or other IgG subtype) response to the neutralizing IgG against ZIKV that is able to penetrate the blood-Sertoli barrier (Bagasra et al., 2017).

Chronic sinusitis in patients with selective IgG4 subclass deficiency can be controlled with enriched immunoglobulins (Aalberse et al., 2009). We suggest that the individuals who exhibit prolonged duration of ZIKV RNA in their semen should be investigated for IgG subclass levels in their serum, especially IgG4. Also, not all immunoglobulins can penetrate the blood-Sertoli barrier, and there is an opportunity to uncover the mystery behind the prolonged ZIKV shedding. A vaccine that is deficient in inducing only limited humoral immunity that is unable to penetrate the Sertoli-blood barrier may not be able to prevent sexual transmission of ZIKV effectively (Bagasra et al., 2017). Future studies with vaccines should take account also of patient ethnicity and age in efforts to explore potential differences in IgG4 deficiencies in ZIKV-infected individuals. The effects that coexisting sexually transmitted and other flaviviral diseases have on the duration of viral persistence and immune responses to vaccination, including antibody-dependent enhancement of infection (Martín-Acebes, Saiz, & Jiménez de Oya, 2018), are needed. More data are required to understand viral persistence and infectivity in the female reproductive tract.

Summary points

- ZIKV can be transmitted sexually.
- The virus can transmit infection even after it is undetectable from saliva, blood, and other bodily fluids and antiviral antibodies appear to have cleared the virus.
- Virus shedding from semen of infected men can be detected for several months even if it has been cleared from other bodily fluids.
- It is proposed that none of the normal immunoglobulins (IgM, IgG1, IgG2, IgG3, IgA1, and IgA2) can cross the Sertoli-blood barrier (SBB) due to their large size.
- Therefore individuals will most likely continue viral shedding until the smallest immunoglobulin, that is, IgG4, develops and penetrates the SBB to neutralize the virus.
- Accordingly, anti-ZIKV vaccines must stimulate the IgG4 immune response, along with other immune subclass responses, to prevent sexual transmission in ZIKV-vaccinated individuals.
- The ZIKV vaccine must evoke IgG4 in order to be effective against sexual transmission of the virus.

References

Aalberse, R. C., Stapel, S. O., Schuurman, J., et al. (2009). Immunoglobulin G4: An odd antibody. *Clinical and Experimental Allergy*, *39*, 469–477.

Allard, A., Althouse, B. M., Hebert-Dufresne, L., et al. (2017). The risk of sustained sexual transmission of Zika is underestimated. *PLoS Pathogens*, *13*(9), e1006633. https://doi.org/10.1371/journal.ppat.1006633.

Althaus, C. L., & Low, N. (2016). How relevant is sexual transmission of Zika virus? *PLoS Medicine*, *13*, e1002157. https://doi.org/10.1371/journal.pmed.1002157.

Arsuaga, M., Bujalance, S. G., Díaz-Menéndez, M., et al. (2016). Probable sexual transmission of Zika virus from a vasectomised man. *The Lancet Infectious Diseases*, *16*, 1107.

Arsuaga, M., Bujalance, S. G., Díaz-Menéndez, M., et al. (2017). Probable sexual transmission of Zika virus from a vasectomised man. *The Lancet Infectious Diseases*, *16*, 1107.

Atkinson, B., Thorburn, F., Petridou, C., et al. (2017). Presence and persistence of Zika virus RNA in semen, United Kingdom, 2016. *Emerging Infectious Diseases*, (4), 611–615. https://doi.org/10.3201/eid2304.161692.

Augusto, F. B., Bewick, S., & Fagan, W. F. (2017). Mathematical model for Zika virus dynamics with sexual transmission route. *Ecological Complexity*, *29*, 61–81.

Bagasra, O., Addanki, K. C., Goodwin, G. R., et al. (2017). Cellular targets and receptor of sexual transmission of Zika virus. *Applied Immunohistochemistry and Molecular Morphology*, *25*(10), 679–686.

Bournazos, S., & Ravetch, J. V. (2017). Diversification of IgG effector functions. *International Immunology*, *29*(7), 303–310. https://doi.org/10.1093/intimm/dxx025.

Boyer, S., Calvez, E., Chouin-Carneiroc, T., et al. (2018). An overview of mosquito vectors of Zika virus. *Microbes and Infection*, *20*, 646–660.

Bozzalla Cassione, E., & Stone, J. H. (2017). IgG4-related disease. *Current Opinion in Rheumatology*, *29*(3), 223–227. https://doi.org/10.1097/BOR.0000000000000383.

Centers for Disease Control and Prevention. (2019). *Zika virus*. https://www.cdc.gov/zika/reporting/case-counts.html. (Accessed 21 January 2019).

Counotte, M. J., Kim, C. R., Wang, J., Bernstein, K., et al. (2018). Sexual transmission of Zika virus and other flaviviruses: A living systematic review. *PLoS Medicine*, *15*(7), e1002611. https://doi.org/10.1371/journal.pmed.1002611.

Das Neves Almeida, R., Racine, T., Magalhães, K. G., et al. (2018). Zika virus vaccines: Challenges and perspectives. *Vaccine*, *6*, 62. https://doi.org/10.3390/vaccines6030062.

Dias, I. K. R., da Silva Sobreira, C. L., Martins, R. M. G., et al. (2019). Zika virus: A review of the main aspects of this type of arbovirosis. *Revista da Sociedade Brasileira de Medicina Tropical*, *51*, 261–269.

Duggal, N. K., McDonald, E. M., Ritter, J. M., et al. (2018). Sexual transmission of Zika virus enhances *in utero* transmission in a mouse model. *Scientific Reports*, *8*, 4510.

Foy, B. D., Kobylinski, K. C., Foy, J. L. C., et al. (2001). Probable non-vector-borne transmission of Zika virus, Colorado, USA. *Emerging Infectious Diseases*, *17*, 880–882.

França, L. R., Hess, R. A., Dufour, J. M., et al. (2016). The Sertoli cell: One hundred fifty years of beauty and plasticity. *Andrology*, *4*(2), 189–212.

Ghaffar, K. A., Ng, L. F. P., & Renia, L. (2018). Fast tracks and roadblocks for Zika vaccines. *Vaccine*, *6*(4), 77. https://doi.org/10.3390/vaccines6040077.

Göertz, G. P., Abbo, S. R., Fros, J. J., et al. (2018). Functional RNA during Zika virus infection. *Virus Research*, *254*, 41–53.

Govero, J., Esakky, P., Scheaffer, S. M., et al. (2016). Zika virus infection damages the testes in mice. *Nature*, *540*, 438–442.

Griffin, B. D., Muthumani, K., Warner, B. M., et al. (2017). DNA vaccination protects mice against Zika virus-induced damage to the testes. *Nature Communications*, *8*, 1–8.

Grischott, F., Puhan, M., Hatza, C., et al. (2016). Non-vector-borne transmission of Zika virus: A systematic review. *Travel Medicine and Infectious Disease*, *14*, 313–330.

Hamel, R., Dejarnac, O., Wichit, S., et al. (2015). Biology of Zika virus infection in human skin cells. *Journal of Virology*, *89*, 8880–8896.

Huijbers, M. G., Plomp, J. J., van der Maarel, S. M., et al. (2018). IgG4-mediated autoimmune diseases: A niche of antibody-mediated disorders. *Annals of the New York Academy of Sciences*, *1413*(1), 92–103. https://doi.org/10.1111/nyas.13561.

Jääskeläinen, A. J., Korhonen, E. M., Huhtamo, E., et al. (2019). Validation of serological and molecular methods for diagnosis of Zika virus infections. *Journal of Virological Methods*, *263*, 68–74.

Jiang, X., Loeb, J. C., Manzanas, C., et al. (2018). Valve-enabled sample preparation and RNA amplification in a coffee mug for Zika virus detection. *Angewandte Chemie International Edition*, *57*, 17211–17214.

Joguet, G., Mansuy, J.-M., Matusali, G., et al. (2017). Effect of acute Zika virus infection on sperm and virus clearance in body fluids: A prospective observational study. *The Lancet*, *17*, 1200–1208.

Koneczny, I. (2018). A new classification system for IgG4 autoantibodies. *Frontiers in Immunology*, *9*, 97. https://doi.org/10.3389/fimmu.2018.00097.

Kumar, A., Jovel, J., Lopez-Orozco, J., et al. (2018). Human sertoli cells support high levels of Zika virus replication and persistence. *Scientific Reports*, *8*, 5477.

Kurscheidt, F. A., Mesquita, C. S. S., Damke, G. M. Z. F., et al. (2019). Persistence and clinical relevance of Zika virus in the male genital tract. *Nature Reviews Urology*. https://doi.org/10.1038/s41585-019-0149-7.

Lee, I., Bos, S., Li, G., et al. (2018). Probing molecular insights into Zika virus–host interactions. *Viruses*, *10*, 233. https://doi.org/10.3390/v10050233.

Leta, S., Beyene, T. J., DeClercq, E. M., et al. (2018). Global risk mapping for major diseases transmitted by *Aedes aegypti* and *Aedes albopictus*. *International Journal of Infectious Diseases*, *67*, 25–35.

Ma, M., Li, S., Ma, S., et al. (2016). Zika virus causes testis damage and leads to male infertility in mice. *Cell*, *167*, 1511–1524.

Makhluf, H., & Shresta, S. (2018). Development of Zika virus vaccines. *Vaccine*, *6*, 7. https://doi.org/10.3390/vaccines6010007.

Martín-Acebes, M. A., Saiz, J.-C., & Jiménez de Oya, N. (2018). Antibody-dependent enhancement and Zika: Real threat or phantom menace? *Frontiers in Cellular and Infection Microbiology*, *8*, 44. https://doi.org/10.3389/fcimb.2018.00044.

Matusali, G., Houzet, L., Satie, A. P., et al. (2018). Zika virus infects human testicular tissue and germ cells. *The Journal of Clinical Investigation*, *128*(10), 4697–4710. https://doi.org/10.1172/JCI121735.

Maxian, O., Neufeld, A., Talis, E. J., et al. (2017). Zika virus dynamics: When does sexual transmission matter? *Epidemics*, *21*, 48–55.

Mead, P. S., Duggal, N. K., Hook, S. A., et al. (2018). Zika virus shedding in semen of symptomatic infected men. *The New England Journal of Medicine*, *378*, 1377–1385.

Mead, P. S., Hills, S. L., & Brooks, J. T. (2018). Zika virus as a sexually transmitted pathogen. *Current Opinion in Infectious Diseases*, *21*, 39–44.

Medina, F. A., Torres, G., Acevedo, J., et al. (2019). Duration of the presence of infectious Zika virus in semen and serum. *The Journal of Infectious Diseases*, *219*, 31–40.

Mlera, L., & Bloom, M. E. (2019). Differential Zika virus infection of testicular cell lines. *Viruses*, *11*(1). https://doi.org/10.3390/v11010042. pii: E42.

Moreira, J., Peixoto, T. M., Siqueira, A. M., et al. (2017). Sexually acquired Zika virus: A systematic review. *Clinical Microbiology and Infection*, *23*, 296–305.

Mruk, D. D., & Cheng, C. Y. (2015). The mammalian blood-testis barrier: Its biology and regulation. *Endocrine Reviews*, *36*(5), 564–591.

Nazerai, L., Christensen, J. P., & Thomsen, A. R. (2019). A 'furry-tale' of Zika virus infection: What have we learned from animal models? *Viruses*, *11*, 29. https://doi.org/10.3390/v11010029.

Nicastri, E., Castilletti, C., Liuzzi, G., et al. (2016). Persistent detection of Zika virus RNA in semen for six months after symptom onset in a traveller returning from Haiti to Italy, February 2016. *Euro Surveillance*, *21*, 21.

Paz-Bailey, G., Rosenberg, E. S., Doyle, K., et al. (2018). Persistence of Zika virus in body fluids—Final report. *The New England Journal of Medicine*, *379*, 1234–1243.

Petersen, L. R., Jamieson, D. J., Powers, A. M., et al. (2016). Zika virus. *The New England Journal of Medicine*, *374*, 1552–1563.

Ploudre, A. R., & Bloch, E. M. (2016). A literature review of Zika virus. *Emerging Infectious Diseases*, *22*, 1185–1192.

Quanquin, N., Wang, L., & Cheng, G. (2017). Potential for treatment and a Zika virus vaccine. *Current Opinion in Pediatrics*, *29*(1), 114–121. 31.

Roberts, C. C., & Maslow, J. N. (2018). Assay challenges for emerging infectious diseases: The Zika experience. *Vaccine*, *6*, 70. https://doi.org/10.3390/vaccines6040070.

Robinson, C. L., Chong, A. C. N., Ashbrook, A. W., et al. (2018). Male germ cells support long-term propagation of Zika virus. *Nature Communications*, *9*, 2090.

Rochlin, I., Ninivaggi, D. V., Hutchinson, M. L., & Farajollahi, A. (2013). Climate change and range expansion of the Asian tiger mosquito (*Aedes albopictus*) in northeastern USA: Implications for public health practitioners. *PLoS One*, *8*, e60874.

Rong, Z., Wang, Q., Sun, N., Wang, K., et al. (2019). Smartphone-based fluorescent lateral flow immunoassay platform for highly sensitive point-of-care detection of Zika virus nonstructural protein 1. *Analytica Chimica Acta*, *1055*, 140–147. https://doi.org/10.1016/j.aca.2018.12.043.

Russell, K., Hills, S. L., Oster, A. M., et al. (2017). Male-to-female sexual transmission of Zika virus—United States, January-April 2016. *Clinical Infectious Diseases*, *64*, 211–213.

Sánchez-Montalvá, A., Pou, D., et al. (2018). Zika virus dynamics in body fluids and risk of sexual transmission in a non-endemic area. *Tropical Medicine & International Health*, *23*, 92–100.

Schroeder, H. W., & Cavacini, L. (2010). Structure and function of immunoglobulins. *The Journal of Allergy and Clinical Immunology*, *125*(202), S41–S52.

Siemann, D. N., Strange, D. P., Maharaj, P. N., et al. (2017). Zika virus infects human Sertoli cells and modulates the integrity of the *in vitro* blood-testis barrier model. *Journal of Virology*, *91*. e00623-17.

Stone, J. H., Zen, Y., & Deshpande, V. (2012). IgG4-related disease. *The New England Journal of Medicine*, *366*(6), 539–551. https://doi.org/10.1056/NEJMra1104650.

Strange, D. P., Green, R., Siemann, D. N., et al. (2018). Immunoprofiles of human Sertoli cells infected with Zika virus reveals unique insights into host-pathogen crosstalk. *Scientific Reports*, *8*(1), 8702. https://doi.org/10.1038/s41598-018-27027-7.

Tancharoen, C., Sukjee, W., Thepparit, C., et al. (2019). Electrochemical biosensor based on surface imprinting for Zika virus detection in serum. *ACS Sensors*, *4*, 69–75.

Vannice, C. S., Cassetti, M. C., Eisinger, R. W., et al. (2019). Demonstrating vaccine effectiveness during a waning epidemic: A WHO/NIH meeting report on approaches to development and licensure of Zika vaccine candidates. *Vaccine*, *37*, 863–868.

WHO. (2005). *WHO statement on the first meeting of the International Health Regulations Emergency Committee on Zika virus and observed increase in neurological disorders and neonatal malformations*. https://www.who.int/news-room/detail/01-02-2016.

WHO. (2019). *Zika virus (ZIKV) classification table data as of 15 February 2018*. https://apps.who.int/iris/bitstream/handle/10665/260419/zika-classification-15Feb18-eng.pdf;jsessionid=7B1B3EAF783138C13ECF1281A7912767?sequence=1.

Wican, N., & Smith, D. R. (2016). Zika virus: History of a newly emerging arbovirus. *The Lancet Infectious Diseases*, *16*, PE119–PE126.

Wilder-Smith, A., Vannice, K., & Durbin, A. (2018). Zika vaccines and therapeutics: Landscape analysis and challenges ahead. *BMC Medicine*, *16*(1), 84. https://doi.org/10.1186/s12916-018-1067-x.

Yakob, L., Kucharski, A., Hue, S., et al. (2016). Low risk of a sexually transmitted Zika virus outbreak. *The Lancet Infectious Diseases*, *16*, 1100–1102.

Chapter 37

Nucleoside analogue inhibitors for Zika virus infection

Jean A. Bernatchez[a], Michael Coste[b], Byron W. Purse[b], and Jair L. Siqueira-Neto[a]

[a]*Centre for Discovery and Innovation in Parasitic Diseases, Skaggs School of Pharmacy and Pharmaceutical Sciences, University of California, San Diego, La Jolla, CA, United States,* [b]*Department of Chemistry and Biochemistry and the Viral Information Institute, San Diego State University, San Diego, CA, United States*

Abbreviations

DAST	diethylaminosulfur trifluoride
DENV	dengue virus
HBV	hepatitis B virus
HCMV	human cytomegalovirus
HCV	hepatitis C virus
HIV	human immunodeficiency virus
HSV	herpes simplex virus
MT	methyltransferase
NI	nucleoside analogue inhibitor
NS5	nonstructural protein 5
RdRp	RNA-dependent RNA polymerase

Introduction

Viral RNA-dependent RNA polymerases (RdRps) are among the most attractive target enzymes for antiviral drug development because their catalytic function is absent from the metabolism of normal human cells. Inspired by successful past drug development campaigns against hepatitis C virus (HCV), a global effort is underway to develop Zika antivirals by targeting its polymerase.

Zika virus (ZIKV) genome replication and structure/function of ZIKV NS5

ZIKV, a member of the *Flaviviridae* family, possesses a single-stranded, positive-sense RNA genome. Replication of viral nucleic acid is performed by the virally encoded nonstructural protein 5 (NS5), which possesses both RdRp and methyltransferase (MT) activities. Several structures of ZIKV NS5 have been published recently (Godoy et al., 2017; Wang et al., 2017; Zhao et al., 2017). The MT domain, responsible for methylation of the viral mRNA cap and internal adenosine residues, is located within the N-terminal portion of the protein, while the RdRp domain, required for viral RNA synthesis, is found in the C-terminal portion of the protein. The overall structure of NS5, like other viral RdRps and most nucleic acid polymerases, mimics the shape of a right hand, with fingers, thumb, and palm subdomains (Fig. 1). Upon binding of viral genomic RNA in the cell, the enzyme can initiate de novo RNA synthesis, whereby a single nucleoside triphosphate is used to prime RNA synthesis by the enzyme. After incorporation of a few nucleotides, the enzyme switches to a more processive elongation phase of RNA synthesis. The ability of the MT domain to bind RNA appears to assist the RdRp domain in the synthesis of RNA during elongation (Zhao et al., 2017). Catalysis in the polymerase domain occurs via three aspartate residues (Asp535, Asp665, and Asp666) coordinating two divalent metal ions. RNA synthesis proceeds via attack of the 3′-OH group of the sugar moiety of the primer terminus on the alpha phosphate of the incoming nucleotide and release of inorganic pyrophosphate as a by-product of the reaction. The catalytic aspartates assist in the coordination of the primer

Zika Virus Impact, Diagnosis, Control, and Models. https://doi.org/10.1016/B978-0-12-820267-8.00037-6

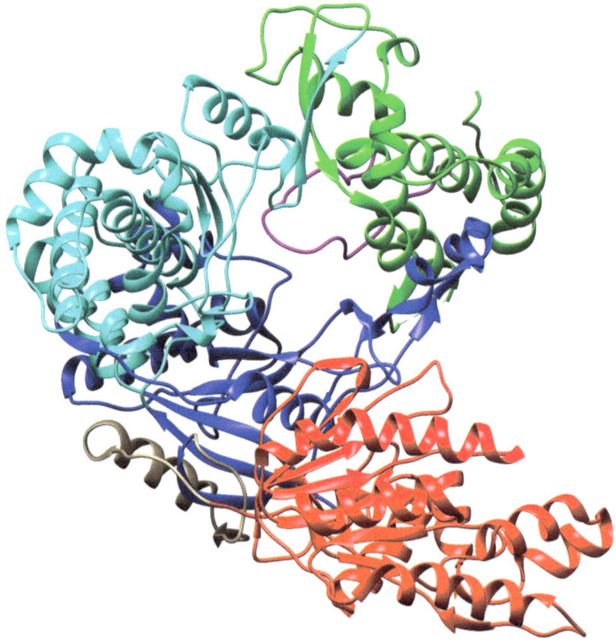

FIG. 1 Crystal structure of ZIKV NS5. The polymerase domain of ZIKV adopts a canonical structural arrangement of a right hand, with figures *(cyan)*, thumb *(green)*, priming loop *(magenta)*, and palm *(blue)*. Shown in *red* is the methyltransferase domain (MTase) of the protein. The structure presented is from data deposited in the Protein Data Bank (PDB; code 5TMH) from Wang et al. (2017). The figure was prepared using Chimera (UCSF).

terminus and the incoming nucleotide for phosphodiester bond formation to occur with the proper geometry and in the activation of the 3′-OH of the primer terminus for nucleophilic attack (Peersen, 2019).

Given the essentiality of ZIKV NS5 to the viral life cycle, this protein represents an excellent target for antiviral therapy.

Nucleoside analogues (NIs) as treatments for viral infections

RNA and DNA polymerases have been extensively studied as therapeutic targets for a host of viral infections, including the human immunodeficiency virus (HIV), herpes simplex virus (HSV), hepatitis B virus (HBV), HCV, human cytomegalovirus (HCMV), and others (Berdis, 2008; Deval, Symons, & Beigelman, 2014). One of the principal classes of inhibitors that have been designed to target these enzymes is nucleoside analogues. These compounds, after phosphorylation in cells, mimic the natural ribonucleotide or deoxyribonucleotide substrates of the viral polymerases, yet possess chemical modifications that preclude or impair further extension of the viral nucleic acid by the polymerase after the inhibitor has been incorporated into the nascent nucleic acid. One of the attractive features of compounds of this class is their relatively low host toxicity levels. This feature derives from the higher level of discrimination of the nucleotide binding pocket of host polymerases against unnatural nucleotides as compared with viral polymerases, in addition to the ability of host DNA polymerases to engage in proofreading, an exonuclease activity that often excises deoxynucleoside analogues. Viral polymerases have evolved more accommodating nucleotide binding pockets to increase the chance of nucleotide misincorporation events so as to vary the molecular composition of their antigens and evade the host immune response (Hemelaar, 2013; Lapa, Garbuglia, Capobianchi, & Del Porto, 2019). This property also makes them more susceptible to the incorporation of unnatural nucleotides, which can arrest nucleic acid synthesis.

Nucleoside analogue inhibitors of viral polymerases are generally grouped into three principal categories: obligate chain terminators, nonobligate chain terminators, and mutagens (Eyer, Nencka, de Clercq, Seley-Radtke, & Růžek, 2018). Obligate chain terminators lack the 3′-OH or a functional equivalent on their ribose moiety. Once these compounds are incorporated by NS5, they arrest nucleic acid synthesis; further elongation is chemically impossible. Nonobligate chain terminators, on the other hand, do possess the functional equivalent of a 3′-OH group. However, upon incorporation of the inhibitor into nucleic acid, synthesis is terminated due to steric conflict between the inhibitor and either the incoming nucleotide or the enzyme or by preventing the closing down of the fingers domain of the enzyme (Boehr, Arnold, Oh, Cameron, & Boehr, 2019). These features either preclude nucleotide binding to the polymerase active site or disrupt the chemical step of phosphodiester bond formation. Lastly, mutagens are readily incorporated by the polymerase. Once they are present in the templating nucleic acid strand, multiple possible base pairing interactions between the mutagen's nitrogenous base and an incoming nucleotide can scramble the genetic code in further rounds of replication. When this mutagenesis occurs at multiple positions in the viral genome, a hypermutagenesis and error catastrophe ultimately result in unviable viral progeny.

Nucleoside analogue prodrugs and ribonucleoside analogue phosphorylation

There are caveats that are of particular importance to the design of ribonucleoside analogues for the inhibition of RNA virus genome replication (Mehellou, Rattan, & Balzarini, 2018). First, the triphosphate active forms of nucleoside analogue inhibitors have poor cellular permeability due to their quadruple negative charge. Therefore, at least some cellular phosphorylation of the inhibitor must take place using host enzymes to bypass this bioavailability issue. Typically, the rate of metabolic installation of the first phosphate group is the most limited by structural modifications designed to inhibit RdRp. As such, this represents a rate-limiting step for conversion of these inhibitors to their triphosphorylated active form. 3′-Deoxyribonucleoside analogues are often especially poor substrates for cellular nucleoside kinases, which has limited the development of obligate ribonucleoside analogue chain terminators for RNA virus infections (Eldrup, Allerson, et al., 2004). Thus, nucleoside analogue development has generally focused on nonobligate chain terminators for these viruses.

Several strategies have been developed to install a single uncharged, masked phosphate on a chosen nucleoside analogue inhibitor. The resulting nucleotide prodrugs include a precursor to the α-phosphate that can be metabolically activated upon entry of the compound into the cell (Mehellou, Balzarini, & McGuigan, 2009; Slusarczyk, Serpi, & Pertusati, 2018). Given that the second and third phosphorylation events occur more easily, this approach dramatically improves the bioavailability of these inhibitors in the cell. One of the most successful of these prodrug strategies, termed ProTides, is used in the clinical drug sofosbuvir against HCV and has been investigated by several research groups for Zika inhibition.

Chemistry

Synthetic methods for antiviral nucleoside analogues offer many opportunities to modify ribose for chain termination, to select a desired nucleobase to influence drug metabolism, and to install masked phosphate groups that enable membrane permeation and subsequent prodrug activation. A few highlights most relevant to tested ZIKV inhibitors are discussed here, and, for further information, the reader is referred to recent reviews (Seley-Radtke & Yates, 2018; Yates & Seley-Radtke, 2019).

Ribose modifications

Nonobligate chain termination has been the most successful mechanism of antiviral nucleoside analogues that target viral RNA polymerization, and most compounds operating by this mechanism are 2′-modified. The most common starting materials for the synthesis of these compounds are protected 2′-oxoribose precursors, as shown in Fig. 2 (Benzaria et al., 2007; Harry-O'kuru, Smith, & Wolfe, 1997; McGuigan, Perrone, Madela, & Neyts, 2009). The stereogenic configuration at the 2′ position is set by the α or β facial selectivity of nucleophile addition to the ketone, which is determined primarily by steric effects. To prepare inhibitors with methyl or other alkyl groups on the β face, on the same side of ribose as the nucleobase, the 2′-oxoribose **1** can be used (Fig. 2A). The bottom face of its ketone is crowded by the OBz groups, directing organometallic nucleophiles to the top, β face. Protective group manipulation followed by the use of well-established silyl-Hilbert-Johnson conditions allows for the installation of any of the canonical nucleobases, or modified derivatives, as shown. This approach has been used to synthesize 2′-β-methyl, ethynyl, and other nucleotides that are among the most potent tested inhibitors of ZIKV RdRp (Bernatchez et al., 2018, 2019).

Compounds that include a fluorine substitution for the 2′-OH must be prepared using a different starting material because the available synthetic methods change the configuration of the 2′-*C* when fluorine is introduced (Fig. 2B). The 2′-oxoribose **2** has a more open α face, allowing an alkyl lithium nucleophile to add from below and resulting in an alcohol on the β face of ribose. A common alternative 2′-oxoribose that offers the same selectivity includes a protected nucleobase and a 5′,3′-disiloxane as protective group (Clark et al., 2005). The configuration of the 2′-*C* is inverted in a subsequent fluorination reaction with diethylaminosulfur trifluoride (DAST) (Clark, Christian Mason, Hobbs, Hollecker, & Schinazi, 2006). Generally the presence of a 2′-OH or 2′-F has been important for maintaining enzyme compatibility during nucleotide insertion, while the 2′-β substituent blocks further RNA elongation, causing chain termination (Boehr et al., 2019). 2′-α-F substituents have the added benefit of increasing the metabolic stability of these compounds by slowing deglycosylation (Liu, Sharon, & Chu, 2008).

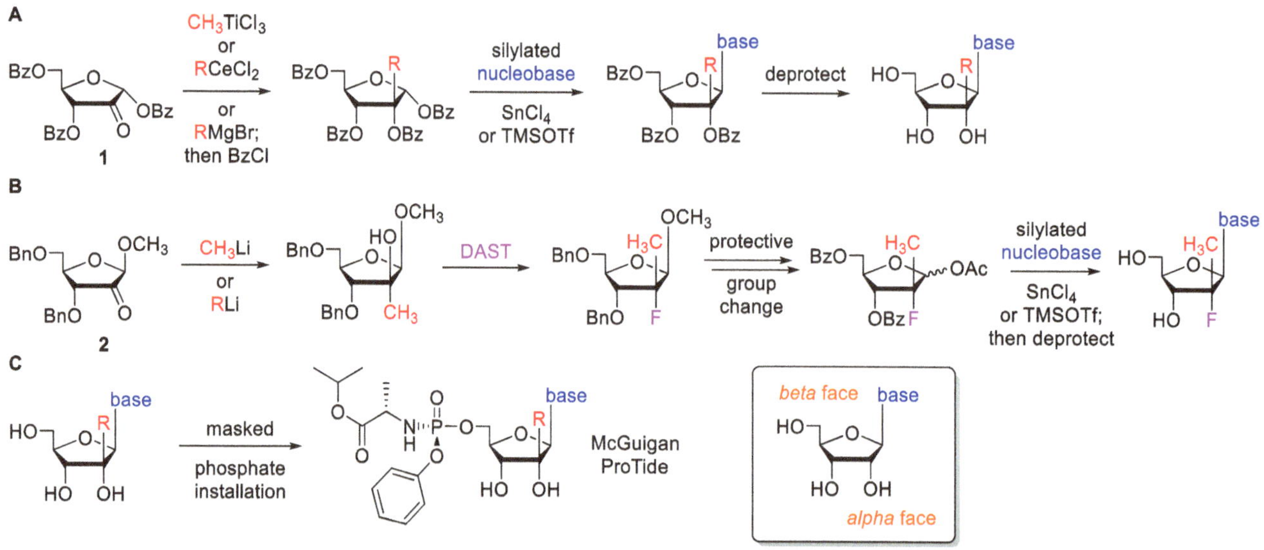

FIG. 2 General synthetic scheme for key nucleoside analogues. (A and B) Common synthetic methods for preparing 2′-modified nucleoside analogues as nonobligate terminators of Zika virus RdRp. (C) Installation of masked phosphate groups to give McGuigan ProTides. References are located in the section "Chemistry" of the text.

The versatility of these and similar 2′-oxoribose starting materials will enable many more structural modifications at the 2′ position in future drug development efforts.

Multiple chemical methods for replacing the 3′-OH are available to prepare obligate chain-terminating nucleoside analogues. These methods typically entail protection of the 2′- and 5′-OH groups, followed by activation of the 2′-OH groups (Furukawa, Kobayashi, Kanai, & Honjo, 1965; Lin et al., 1991; Pierra, Amador, Badaroux, Storer, & Gosselin, 2006). While the resulting compounds are sometimes very potent against polymerases when used as the corresponding triphosphates in biochemical assays, they have typically had poor activity in cell-based assays owing to poor phosphorylation or other metabolic effects (Shim, Larson, Lai, Naim, & Wu, 2003). Additional methods for modifying ribose at the 1′ and 4′ positions, plus methods for replacing the tetrahydrofuranyl oxygen, are established and have shown considerable promise against other RNA viruses, but few such compounds have yet been tested against ZIKV (Kirschberg et al., 2015; Smith et al., 2009).

Nucleobase modifications and selection

While the chain-terminating properties of nucleoside analogues are determined primarily by ribose modifications, the selection of nucleobase is important for optimizing bioavailability and metabolic properties, such as maximizing the activity of kinases and minimizing that of glycosylases. Because ribose modifications are typically performed in synthesis prior to the installation of the nucleobase (Fig. 2), all four canonical nucleobases plus nucleobase analogues can be tested for a given ribose derivative. For example, 7-deazapurines have been developed as adenine and guanine replacements (e.g., in 7DMA and NITD008); they allow for retention of potent activity against RdRps but are poor substrates for degradation by nucleobase deamination or purine nucleoside phosphorylase, which catalyzes deglycosylation (Eldrup, Prhavc, et al., 2004).

Phosphate modifications

Masked phosphate groups tested to date in nucleoside analogue ZIKV inhibitors are designs borrowed from drug development efforts against other viruses, especially HCV. The most tested of these, the McGuigan ProTide, can be installed onto a nucleoside analogue in a single stereoselective step using the commercially available reagent *N*-[(*S*)-(2,3,4,5,6-pentafluorophenoxy)phenoxyphosphinyl]-L-alanine 1-methylethyl ester (Ross, Reddy, Zhang, Rachakonda, & Sofia, 2011). Much opportunity exists to evaluate other successful phosphate masking strategies such as alkoxycarbonyloxymethyl phosphoesters (cf. tenofovir disoproxil and adefovir dipivoxil) in nucleoside analogue ZIKV inhibitors.

Enzymology

Several reports exploring the relative activities of nucleoside analogue triphosphates against recombinant ZIKV NS5 have been published. The structures of these compounds are shown in Fig. 3, and their activities against the enzyme are displayed in Table 1. Potisopon and colleagues have reported a bifurcation in nucleoside analogue preference for *Flaviviridae* polymerases: the HCV polymerase NS5B readily incorporates 2′-*C*-F-modified nucleoside analogue triphosphates (as found in the clinically approved drug sofosbuvir used to treat HCV), while the presence of this modification reduces incorporation of nucleotide analogues by ZIKV and dengue virus (DENV) polymerases (Potisopon, Ferron, Fattorini, Selisko, & Canard, 2017). However, the authors report that 2′-*C*-Me modified nucleotides are readily incorporated by both ZIKV and DENV polymerases, as is the case for the HCV RdRp. 2′-*C*-Me-CTP, 2′-*C*-Me-2′-F-CTP, and 2′-*C*-Me-2′-F-UTP were all found to act as nonobligate chain terminators for ZIKV and DENV RdRps. The activity of 2′-*C*-Me-2′-F-UTP (the active triphosphate metabolite of sofosbuvir) against ZIKV NS5 was confirmed in de novo RNA synthesis biochemical assays by Xu and coworkers (IC_{50} of 7.3 μM), although a large discrimination value of 661 relative to the natural substrate has been reported for this compound, suggesting room for improvement in terms of ZIKV NS5 selectivity (Lu et al., 2017; Xu et al., 2017). Indeed, Lu et al. have reported using biochemical assays that while 2′-*C*-Me-2′-F-UTP displays poor inhibition of ZIKV and DENV NS5 (IC_{50} of 90.76 and 55.13 μM, respectively), both 2′-*C*-Me-UTP (IC_{50} of 5.78 and 4.94 μM, respectively) and 2′-*C*-ethynyl-UTP (IC_{50} of 0.46 and 0.33 μM, respectively) are good inhibitors of polymerases from these two flaviviruses (Bernatchez et al., 2019; Lu et al., 2017). Low discrimination of ZIKV NS5 against the nucleoside analogue compared with the natural nucleotide substrate was observed for 2′-*C*-Me-ATP, 2′-*C*-ethynyl-7-deaza-ATP, and 2′-*C*-ethynyl-UTP (discrimination values of 3.7, 5.7, and 4.9, respectively), suggesting that these are efficiently incorporated by this enzyme and may represent interesting leads for antiviral development against ZIKV. Interestingly, conformationally locked carbocyclic adenosine triphosphate analogue inhibitors of ZIKV NS5 have been reported (Fig. 3) (Hercík et al., 2017). These novel structures may represent yet another unexplored scaffold for the development of new potent flavivirus inhibitors.

Virology

Activity of nucleoside analogues against ZIKV and related flaviviruses in cell model and animal models

Several nucleoside analogues have been reported to have antiviral activity against ZIKV in cell culture (structures shown in Fig. 4). One early report by Eyer et al. demonstrated that 2′-*C*-modified nucleoside analogues (2′-*C*-Me-adenosine, 7-deaza-2′-*C*-Me-adenosine (7DMA), 2′-*C*-Me-guanosine, 2′-*C*-Me-cytidine, and 2′-*C*-Me-uridine) had anti-ZIKV activity (MR766 viral strain) in Vero cells (Fig. 4) (Eyer et al., 2016). The activity of 2′-*C*-derivatized nucleoside analogues and their ProTides has been confirmed in other viral culture studies (Bernatchez et al., 2018, 2019). 7DMA has also been shown to protect mice from lethal ZIKV challenge in vivo (Zmurko et al., 2016). The HCV drug sofosbuvir has also been demonstrated by multiple groups to inhibit ZIKV in cell culture models and mouse models of infection (adult and vertical transfer) (Bullard-Feibelman et al., 2017; Ferreira et al., 2017; Mesci et al., 2018; Retallack et al., 2016; Sacramento et al., 2017). Recent work has suggested that ProTide anti-ZIKV activity, such as for sofosbuvir, is cell line dependent (Bernatchez et al., 2019; Mumtaz et al., 2017), making it clear that the use of biologically relevant cell lines will be essential for the future testing of ProTides against ZIKV using in vitro assays. The adenosine analogue 7-deaza-2′-*C*-ethynyladenosine (NITD008) was shown by multiple groups to suppress ZIKV replication (Deng et al., 2016; Shan et al., 2016; Xie et al., 2016). The compound was also able to protect against ZIKV in an A129 mouse model of infection (Deng et al., 2016). Ribavirin has also been shown to inhibit ZIKV replication in vitro and in a STAT-1-deficient mouse model of infection (Bernatchez et al., 2018; Kamiyama et al., 2017; Sacramento et al., 2017; Zmurko et al., 2016). The EC_{50} and CC_{50} values for these compounds are shown in Table 2. Overall, based on the presented results from these studies, sofosbuvir and related 2′-modified uridylate analogues may represent the current best candidates for future development as anti-ZIKV nucleoside analogues.

Conclusion

Given the rapid developments in the area of ZIKV nucleoside analogue research in the past few years, this class of inhibitors can be considered one of the most promising for ZIKV drug design. Although 7DMA and NITD008 have been discontinued as clinical drug candidates due to unacceptable toxicity (Munjal et al., 2017), a number of compounds described earlier have the potential to be moved quickly to clinical trials for ZIKV treatment. Sofosbuvir may be the compound that would be the easiest to move forward to phase I safety trials for pregnant women, given its safety track record as a clinically approved

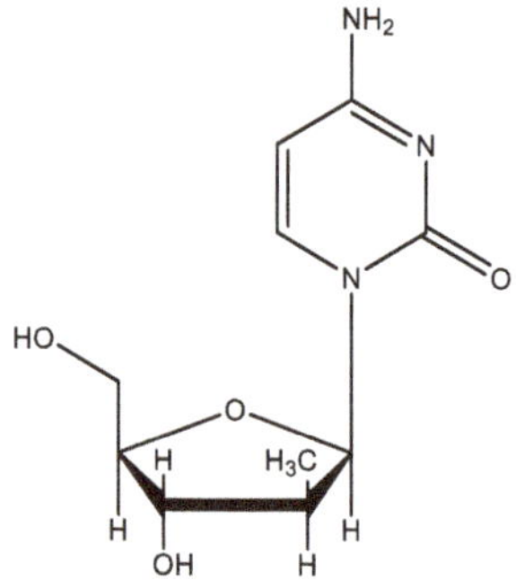

2'-C-Me-cytidine
Potisopon et al., 2017

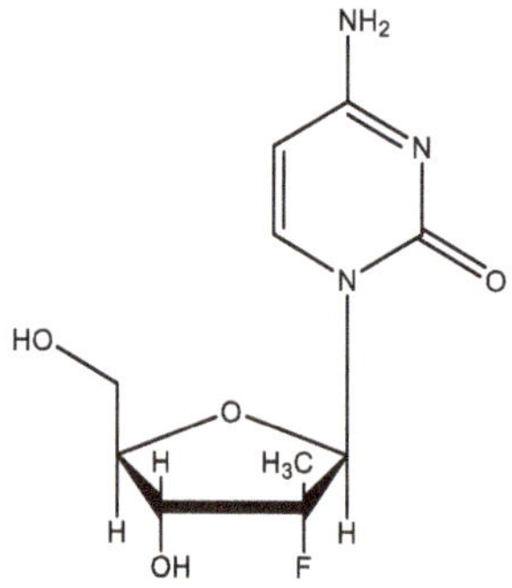

2'-C-Me-2'-F-cytidine
Potisopon et al., 2017

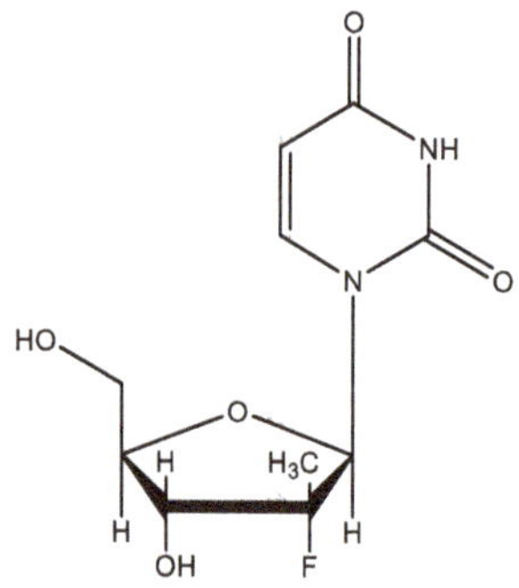

2'-C-Me-2'-F-uridine
Potisopon et al., 2017
Xu et al., 2017
Lu et al., 2017

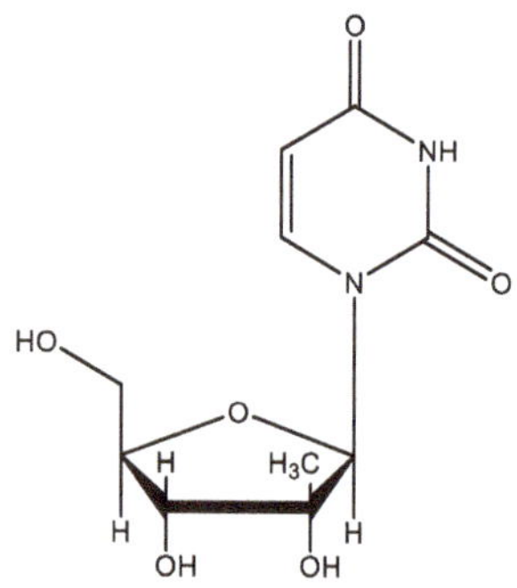

2'-C-Me-uridine
Lu et al., 2017
Bernatchez et al., 2018

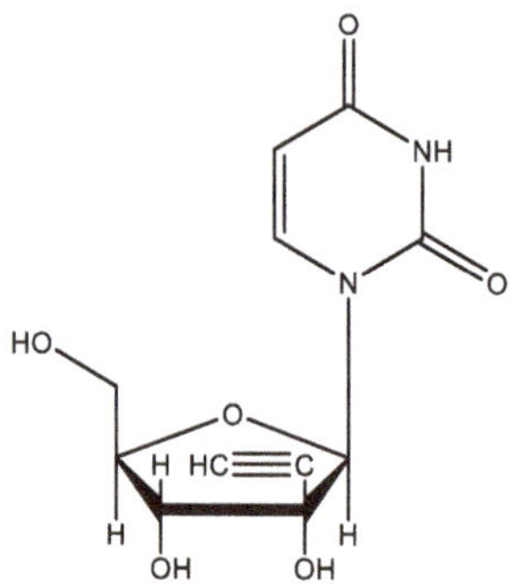

2'-C-ethynyluridine
Lu et al., 2017

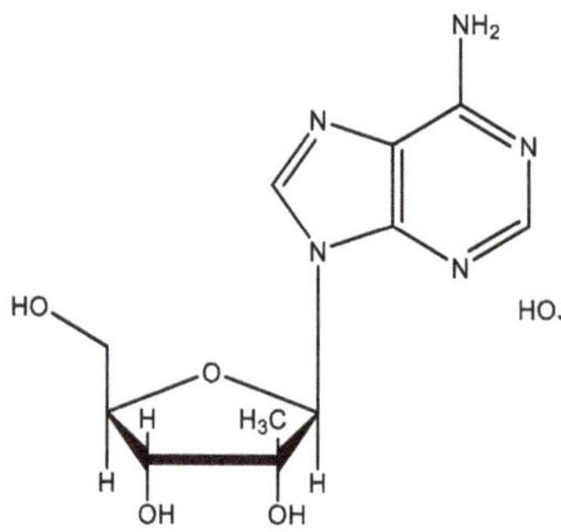

2'-C-Me-adenosine
Lu et al., 2017
Hercík et al., 2017

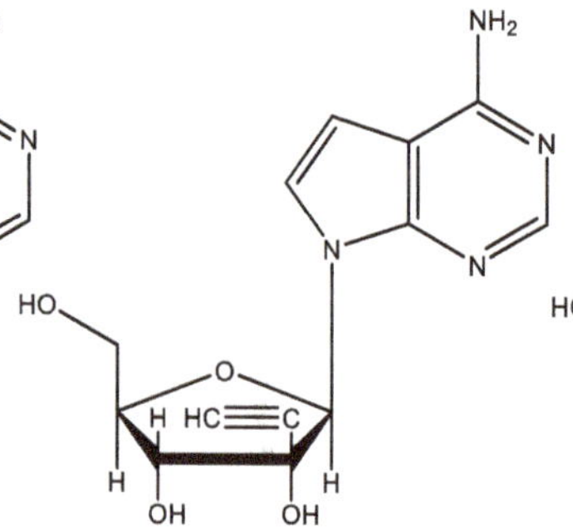

7-deaza-2'-C-ethynyladenosine
Lu et al., 2017

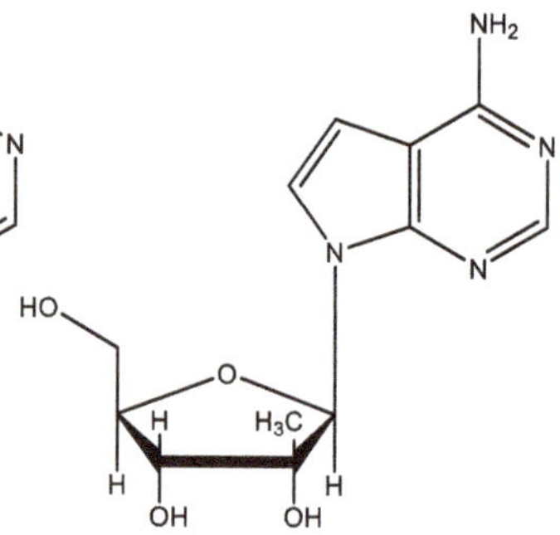

7-deaza-2'-C-Me-adenosine
Hercík et al., 2017

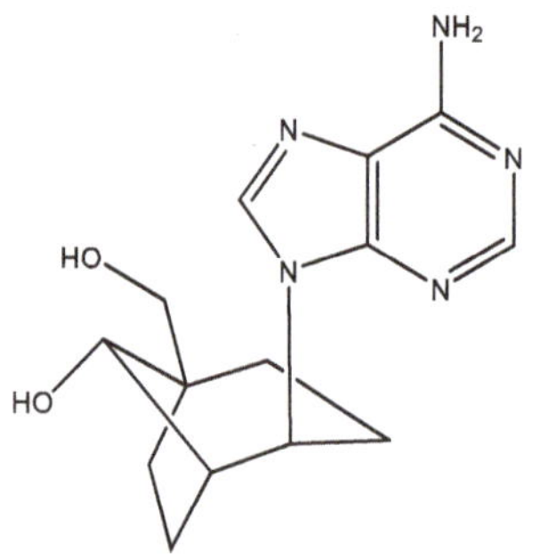

Compound **3**
Hercík et al., 2017

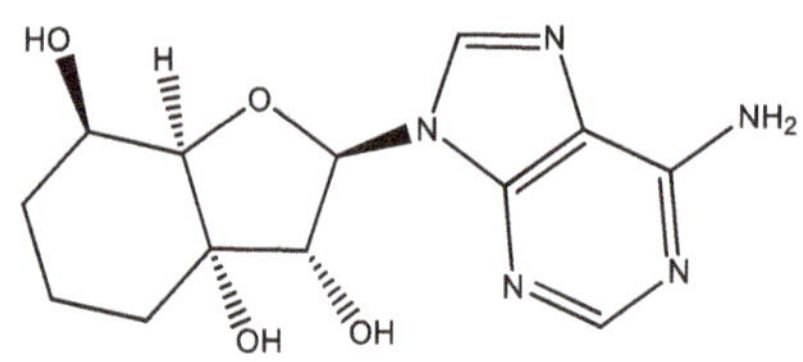

Compound **6**
Hercík et al., 2017

FIG. 3 Chemical structures of selected nucleoside analogues incorporated by recombinantly expressed ZIKV NS5. Shown are chemical structures of important nucleoside analogues that are incorporated by ZIKV NS5 RdRp in biochemical assays into nascent RNA (triphosphorylated active metabolites are used in these experiments). References associated with these compounds are shown in the figure beneath each structure.

TABLE 1 Activity values for selected nucleoside analogues active against recombinantly expressed ZIKV RdRp.

Compound	IC_{50} (μM)	References
2′-*C*-Me-2′-F-UTP	90.76 7.3	Lu et al. (2017) Xu et al. (2017)
2′-*C*-Me-UTP	5.78	Lu et al. (2017)
2′-*C*-Ethynyl-UTP	0.46	Lu et al. (2017)
2′-*C*-Me-ATP	5.6	Hercík et al. (2017)
7-Deaza-2′-*C*-Me-ATP	7.9	Hercík et al. (2017)
Compound 3	6.4	Hercík et al. (2017)
Compound 6	2.7	Hercík et al. (2017)

IC_{50} (inhibitory concentration 50) values are displayed for key active compounds against ZIKV NS5 in the literature. References containing these data are shown in the table for each compound.

drug for HCV. Other 2′-*C*-modified uridylate ProTides, which are also active in neural stem cell systems, as well as the reported carbocyclic nucleoside analogues with activity against ZIKV polymerase, should be tested in animal models of ZIKV infection to assess their appropriateness for future clinical development.

Nucleoside analogues have a history of success in the treatment of viral infections and are well positioned to be effective antivirals against ZIKV. Future studies refining the selectivity and pharmacokinetic properties of these compounds to avoid adverse effects in pregnancy and adequate biodistribution to the developing fetal brain will be needed to bring these compounds to term as ZIKV drugs.

Policy and procedures: Cell assays for evaluating anti-ZIKV compounds in vitro

Numerous assays have been developed to assess the activity of nucleoside analogues and other classes of inhibitors of ZIKV in cellular systems. Generally, these assays either directly measure viral replication via viral gene product generation or viable viral particle production or measure host cell survival throughout assessment of the virus-induced cytopathic effect. Examples of these assays include plaque reduction (Eyer et al., 2016), viral replicons (Xie et al., 2016), cell survival measurements (Pascoalino, Courtemanche, Cordeiro, Gil, & Freitas-Junior, 2016), qRT-PCR of ZIKV RNA (Lanko et al., 2017), luciferase-containing ZIKV (Shan et al., 2016), caspase activation (Xu et al., 2016), immunofluorescence for viral antigens (Zhou et al., 2017), and automated image analysis of cellular phenotypes (Bernatchez et al., 2018). These assays have led to the confirmation across assay types and research groups of the utility of nucleoside analogues as a potent class of anti-ZIKV agents and can be further expanded to other related viruses and cell systems for the screening of compounds against pathogens related to ZIKV.

Mini-dictionary of terms

ZIKV NS5. The RNA-dependent RNA polymerase of ZIKV that copies the viral genome. Essential for viral replication and a prime target for ZIKV drug development.

Nucleoside analogue. A synthetic mimic of a natural nucleoside that includes most of the canonical structure, but with modifications designed to alter biological activity or impart other artificial properties.

Nucleotide. Phosphorylated nucleoside.

ProTide. A prodrug form of a nucleotide analogue, including a peptide fragment that masks the phosphate's charge for membrane permeation and is removed metabolically upon entry into the cell.

Chain terminator. Nucleoside analogue that, when incorporated into the nascent nucleic acid by the viral polymerase, prevents incorporation of the next incoming nucleotide, thereby aborting nucleic acid synthesis by the virus.

Mutagenic nucleoside analogue. Nucleoside analogue that, when incorporated by the viral polymerase, is capable of promiscuous base pairing, resulting in the introduction of mutations into the viral nucleic acid. The accumulation of these mutations renders viral progeny unviable.

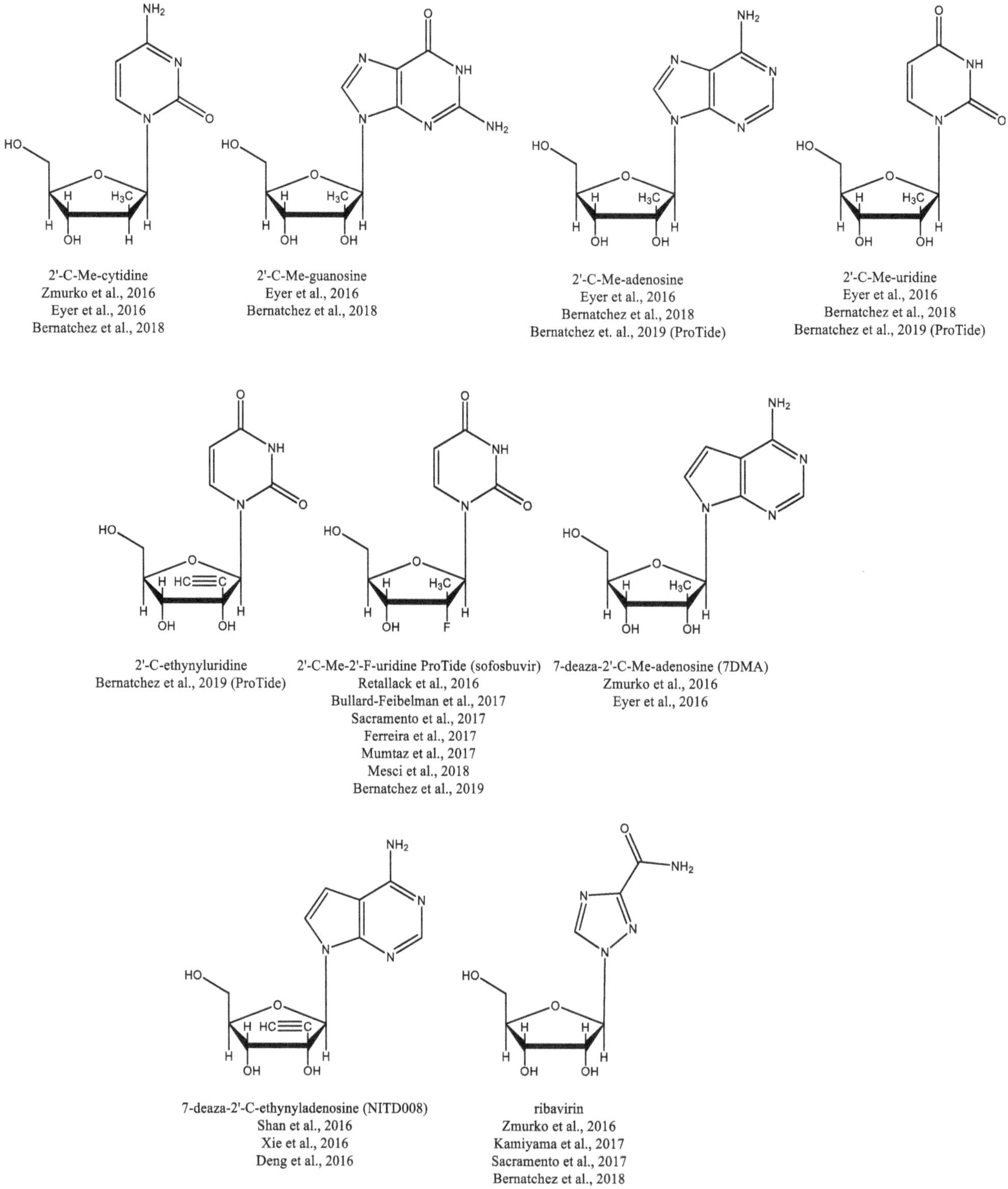

FIG. 4 Chemical structures of selected nucleoside analogues/ProTides active against ZIKV in either cellular or animal infection models. Shown are chemical structures of important nucleoside analogues or ProTides that are active against ZIKV cellular or mouse models of infection. References associated with these compounds are shown in the figure beneath each structure.

TABLE 2 Activity and toxicity values for selected nucleoside analogues/ProTides active against ZIKV in either cellular or animal infection models.

Compound	EC_{50} (μM)	CC_{50} (μM)	References
Sofosbuvir	12.4	NA	Retallack et al. (2016)
			Bullard-Feibelman et al. (2017)
	1–5	>200	Sacramento et al. (2017)
			Mumtaz et al. (2017)
	0.12–1.9 (U/mL)	360–421 (U/mL)	Mesci et al. (2018)
			Bernatchez et al. (2019)
	3.0–4.1	>100	
	13.6–30.9	NA	
	35–46	>50	
2′-*C*-Me-uridine	45.45	>100	Eyer et al. (2016)
	4.21	>100	Bernatchez et al. (2018)
2′-*C*-Me-guanosine	22.25	>100	Eyer et al. (2016)
	2.20	>100	Bernatchez et al. (2018)
2′-*C*-Me-adenosine	5.26	>100	Eyer et al. (2016)
	0.60	>100	Bernatchez et al. (2018)
2′-*C*-Me-cytidine	3.9–10	28	Zmurko et al. (2016)
	10.51	>100	Eyer et al. (2016)
	0.30	>100	Bernatchez et al. (2018)
7-Deaza-2′-*C*-Me-adenosine (7DMA)	1.3–20	>357	Zmurko et al. (2016)
	8.92	>100	Eyer et al. (2016)
7-Deaza-2′-*C*-ethynyladenosine (NITD008)	0.137–0.950	NA	Deng et al. (2016)
	0.8	>25	Xie et al. (2016)
	2.5–3.6	>100	Shan et al. (2016)
2′-*C*-Me-uridine aryloxyl phosphoramidate ProTide	1–2	47	Bernatchez et al. (2019)
2′-*C*-Ethynyl-uridine aryloxyl phosphoramidate ProTide	0.3–0.8	>50	Bernatchez et al. (2019)
2′-*C*-Me-adenosine aryloxyl phosphoramidate ProTide	10–18	42	Bernatchez et al. (2019)
Ribavirin	12–13	>409	Zmurko et al. (2016)
	2.9–6.9 (U/mL)	177–321 (U/mL)	Sacramento et al. (2017)
			Bernatchez et al. (2018)
	20.80	>100	

EC_{50} (effective concentration 50) and CC_{50} (cytotoxicity concentration 50) values are displayed for key active compounds against ZIKV NS5 in the literature. References containing these data are shown in the table for each compound.

Key facts of nucleoside analogues for ZIKV

- 2′-*C*-Methyl- and 2′-*C*-ethynyl-modified nucleoside analogues are among the most potent known RNA synthesis inhibitors by ZIKV NS5, and they use a nonobligate chain termination mechanism.
- Nucleoside analogues containing a 2′-*C*-fluoro group are less readily incorporated by ZIKV NS5 than has been reported for HCV NS5B.
- ProTides show cell line–dependent activity against ZIKV, highlighting the importance of using biologically relevant cell systems for the screening of these compounds.
- The choice of ProTide masking group is critical for compound activity against ZIKV in biologically relevant cell systems.
- Sofosbuvir, 7DMA, and NITD008 have been shown to be active against ZIKV in animal models.
- Sofosbuvir can prevent vertical transmission of Zika virus in a mouse model of disease.
- Additional nucleoside analogues and their ProTides need to be tested in animal models of infection to advance additional candidate molecules toward clinical trials.

Summary points

- Viral polymerases are ideal drug targets due to their essentiality for viral replication and their lower discrimination against unnatural nucleotides.
- Zika virus NS5, the virally encoded RNA-dependent RNA polymerase, has been successfully recombinantly expressed, crystallized, and used in biochemical activity assays.
- Nucleoside analogues with 2′-*C*-methyl, 2′-*C*-ethynyl, or carbocyclic sugar ring modifications have been shown to be active against Zika virus NS5 in biochemical assays.
- 2′-*C*-Methyl-modified nucleosides, 7-deaza-2′-*C*-methyl-adenosine, sofosbuvir, NITD008, and the phosphoramidate ProTides of 2′-*C*-methyluridine and 2′-*C*-ethynyluridine have been shown to have antiviral activity against Zika virus in cell culture.
- 7-Deaza-2′-*C*-methyladenosine, sofosbuvir, and NITD008 are capable of preventing mortality in mouse models of Zika virus infection.
- Several promising nucleoside analogues presented in biochemical and virological studies should be tested in animals to assess their appropriateness for further preclinical development.

References

Benzaria, S., Bardiot, D., Bouisset, T., Counor, C., Rabeson, C., Pierra, C., … La Colla, P. (2007). 2′-C-Methyl branched pyrimidine ribonucleoside analogues: Potent inhibitors of RNA virus replication. *Antiviral Chemistry and Chemotherapy*, *18*, 225–242. https://doi.org/10.1177/09563 2020701800406.

Berdis, A. J. (2008). DNA polymerases as therapeutic targets. *Biochemistry*, *47*, 8253–8260. https://doi.org/10.1021/bi801179f.

Bernatchez, J. A., Coste, M., Beck, S., Wells, G. A., Luna, L. A., Clark, A. E., … Siqueira-Neto, J. L. (2019). Activity of selected nucleoside analogue ProTides against Zika virus in human neural stem cells. *Viruses*, *11*. https://doi.org/10.3390/v11040365.

Bernatchez, J. A., Yang, Z., Coste, M., Li, J., Beck, S., Liu, Y., … Siqueira-Neto, J. L. (2018). Development and validation of a phenotypic high-content imaging assay for assessing the antiviral activity of small-molecule inhibitors targeting Zika virus. *Antimicrobial Agents and Chemotherapy*, *62*. https://doi.org/10.1128/AAC.00725-18.

Boehr, A. K., Arnold, J. J., Oh, H. S., Cameron, C. E., & Boehr, D. D. (2019). 2′-C-Methylated nucleotides terminate virus RNA synthesis by preventing active site closure of the viral RNA-dependent RNA polymerase. *Journal of Biological Chemistry*, *294*, 16897–16907. https://doi.org/10.1074/jbc. RA119.010214.

Bullard-Feibelman, K. M., Govero, J., Zhu, Z., Salazar, V., Veselinovic, M., Diamond, M. S., & Geiss, B. J. (2017). The FDA-approved drug sofosbuvir inhibits Zika virus infection. *Antiviral Research*, *137*, 134–140. https://doi.org/10.1016/j.antiviral.2016.11.023.

Clark, J. L., Christian Mason, J., Hobbs, A. J., Hollecker, L., & Schinazi, R. F. (2006). Synthesis of 2-deoxy-2-fluoro-2-*c*-methyl-d-ribofuranoses. *Journal of Carbohydrate Chemistry*, *25*, 461–470. https://doi.org/10.1080/07328300600859783.

Clark, J. L., Hollecker, L., Mason, J. C., Stuyver, L. J., Tharnish, P. M., Lostia, S., … Pankiewicz, K. W. (2005). Design, synthesis, and antiviral activity of 2′-deoxy-2′-fluoro-2′-C-methylcytidine, a potent inhibitor of hepatitis C virus replication. *Journal of Medicinal Chemistry*, *48*, 5504–5508. https://doi. org/10.1021/jm0502788.

Deng, Y.-Q., Zhang, N.-N., Li, C.-F., Tian, M., Hao, J.-N., Xie, X.-P., … Qin, C.-F. (2016). Adenosine analog NITD008 is a potent inhibitor of Zika virus. *Open Forum Infectious Diseases*, *3*. https://doi.org/10.1093/ofid/ofw175. ofw175.

Deval, J., Symons, J. A., & Beigelman, L. (2014). Inhibition of viral RNA polymerases by nucleoside and nucleotide analogs: Therapeutic applications against positive-strand RNA viruses beyond hepatitis C virus. *Current Opinion in Virology*, *9*, 1–7. https://doi.org/10.1016/j.coviro.2014.08.004.

Eldrup, A. B., Allerson, C. R., Bennett, C. F., Bera, S., Bhat, B., Bhat, N., … Xia, J. (2004). Structure-activity relationship of purine ribonucleosides for inhibition of hepatitis C virus RNA-dependent RNA polymerase. *Journal of Medicinal Chemistry*, *47*, 2283–2295. https://doi.org/10.1021/ jm030424e.

Eldrup, A. B., Prhavc, M., Brooks, J., Bhat, B., Prakash, T. P., Song, Q., … Olsen, D. B. (2004). Structure-activity relationship of heterobase-modified 2′-C-methyl ribonucleosides as inhibitors of hepatitis C virus RNA replication. *Journal of Medicinal Chemistry*, *47*, 5284–5297. https://doi.org/10.1021/ jm040068f.

Eyer, L., Nencka, R., de Clercq, E., Seley-Radtke, K., & Růžek, D. (2018). Nucleoside analogs as a rich source of antiviral agents active against arthropod-borne flaviviruses. *Antiviral Chemistry and Chemotherapy*, *26*. https://doi.org/10.1177/2040206618761299, 2040206618761299.

Eyer, L., Nencka, R., Huvarová, I., Palus, M., Joao Alves, M., Gould, E. A., … Růžek, D. (2016). Nucleoside inhibitors of Zika virus. *Journal of Infectious Diseases*, *214*, 707–711. https://doi.org/10.1093/infdis/jiw226.

Ferreira, A. C., Zaverucha-do-Valle, C., Reis, P. A., Barbosa-Lima, G., Vieira, Y. R., Mattos, M., … Souza, T. M. L. (2017). Sofosbuvir protects Zika virus-infected mice from mortality, preventing short- and long-term sequelae. *Scientific Reports*, *7*, 9409. https://doi.org/10.1038/s41598-017-09797-8.

Furukawa, Y., Kobayashi, K., Kanai, Y., & Honjo, M. (1965). Synthesis of 2′-O-methyluridine, 2′-O-methylcytidine and their relating compounds. *Chemical and Pharmaceutical Bulletin*, *13*, 1273–1278. https://doi.org/10.1248/cpb.13.1273.

Godoy, A. S., Lima, G. M. A., Oliveira, K. I. Z., Torres, N. U., Maluf, F. V., Guido, R. V. C., & Oliva, G. (2017). Crystal structure of Zika virus NS5 RNA-dependent RNA polymerase. *Nature Communications*, *8*, 14764. https://doi.org/10.1038/ncomms14764.

Harry-O'kuru, R. E., Smith, J. M., & Wolfe, M. S. (1997). A short, flexible route toward 2′-*C*-branched ribonucleosides. *Journal of Organic Chemistry*, *62*, 1754–1759. https://doi.org/10.1021/jo961893+.

Hemelaar, J. (2013). Implications of HIV diversity for the HIV-1 pandemic. *Journal of Infection*, *66*, 391–400. https://doi.org/10.1016/j.jinf.2012.10.026.

Hercík, K., Kozak, J., Šála, M., Dejmek, M., Hřebabecký, H., Zborníková, E., … Boura, E. (2017). Adenosine triphosphate analogs can efficiently inhibit the Zika virus RNA-dependent RNA polymerase. *Antiviral Research*, *137*, 131–133. https://doi.org/10.1016/j.antiviral.2016.11.020.

Kamiyama, N., Soma, R., Hidano, S., Watanabe, K., Umekita, H., Fukuda, C., … Kobayashi, T. (2017). Ribavirin inhibits Zika virus (ZIKV) replication in vitro and suppresses viremia in ZIKV-infected STAT1-deficient mice. *Antiviral Research*, *146*, 1–11. https://doi.org/10.1016/j.antiviral.2017.08.007.

Kirschberg, T. A., Mish, M., Squires, N. H., Zonte, S., Aktoudianakis, E., Metobo, S., … Kim, C. U. (2015). Synthesis of 1′-C-cyano pyrimidine nucleosides and characterization as HCV polymerase inhibitors. *Nucleosides, Nucleotides and Nucleic Acids*, *34*, 763–785. https://doi.org/10.1080/15257770.2015.1075550.

Lanko, K., Eggermont, K., Patel, A., Kaptein, S., Delang, L., Verfaillie, C. M., & Neyts, J. (2017). Replication of the Zika virus in different iPSC-derived neuronal cells and implications to assess efficacy of antivirals. *Antiviral Research*, *145*, 82–86. https://doi.org/10.1016/j.antiviral.2017.07.010.

Lapa, D., Garbuglia, A. R., Capobianchi, M. R., & Del Porto, P. (2019). Hepatitis C virus genetic variability, human immune response, and genome polymorphisms: Which is the interplay? *Cell*, *8*. https://doi.org/10.3390/cells8040305.

Lin, T. S., Yang, J. H., Liu, M. C., Shen, Z. Y., Cheng, Y. C., Prusoff, W. H., … Brankovan, V. (1991). Synthesis and anticancer activity of various 3′-deoxy pyrimidine nucleoside analogues and crystal structure of 1-(3-deoxy-beta-D-threo-pentofuranosyl)cytosine. *Journal of Medicinal Chemistry*, *34*, 693–701. https://doi.org/10.1021/jm00106a034.

Liu, P., Sharon, A., & Chu, C. K. (2008). Fluorinated nucleosides: Synthesis and biological implication. *Journal of Fluorine Chemistry*, *129*, 743–766. https://doi.org/10.1016/j.jfluchem.2008.06.007.

Lu, G., Bluemling, G. R., Collop, P., Hager, M., Kuiper, D., Gurale, B. P., … Kolykhalov, A. A. (2017). Analysis of ribonucleotide 5′-triphosphate analogs as potential inhibitors of Zika virus RNA-dependent RNA polymerase by using nonradioactive polymerase assays. *Antimicrobial Agents and Chemotherapy*, *61*. https://doi.org/10.1128/AAC.01967-16.

McGuigan, C., Perrone, P., Madela, K., & Neyts, J. (2009). The phosphoramidate ProTide approach greatly enhances the activity of beta-2′-C-methylguanosine against hepatitis C virus. *Bioorganic and Medicinal Chemistry Letters*, *19*, 4316–4320. https://doi.org/10.1016/j.bmcl.2009.05.122.

Mehellou, Y., Balzarini, J., & McGuigan, C. (2009). Aryloxy phosphoramidate triesters: A technology for delivering monophosphorylated nucleosides and sugars into cells. *ChemMedChem*, *4*, 1779–1791. https://doi.org/10.1002/cmdc.200900289.

Mehellou, Y., Rattan, H. S., & Balzarini, J. (2018). The protide prodrug technology: From the concept to the clinic. *Journal of Medicinal Chemistry*, *61*, 2211–2226. https://doi.org/10.1021/acs.jmedchem.7b00734.

Mesci, P., Macia, A., Moore, S. M., Shiryaev, S. A., Pinto, A., Huang, C.-T., … Muotri, A. R. (2018). Blocking Zika virus vertical transmission. *Scientific Reports*, *8*, 1218. https://doi.org/10.1038/s41598-018-19526-4.

Mumtaz, N., Jimmerson, L. C., Bushman, L. R., Kiser, J. J., Aron, G., Reusken, C. B. E. M., … van Kampen, J. J. A. (2017). Cell-line dependent antiviral activity of sofosbuvir against Zika virus. *Antiviral Research*, *146*, 161–163. https://doi.org/10.1016/j.antiviral.2017.09.004.

Munjal, A., Khandia, R., Dhama, K., Sachan, S., Karthik, K., Tiwari, R., … Joshi, S. K. (2017). Advances in developing therapies to combat Zika virus: Current knowledge and future perspectives. *Frontiers in Microbiology*, *8*, 1469. https://doi.org/10.3389/fmicb.2017.01469.

Pascoalino, B. S., Courtemanche, G., Cordeiro, M. T., Gil, L. H. V. G., & Freitas-Junior, L. (2016). Zika antiviral chemotherapy: Identification of drugs and promising starting points for drug discovery from an FDA-approved library. [version 1; peer review: 2 approved]. *F1000Res*, *5*, 2523. https://doi.org/10.12688/f1000research.9648.1.

Peersen, O. B. (2019). A comprehensive superposition of viral polymerase structures. *Viruses*, *11*. https://doi.org/10.3390/v11080745.

Pierra, C., Amador, A., Badaroux, E., Storer, R., & Gosselin, G. (2006). Synthesis of 2′-C-methylcytidine and 2′-C-methyluridine derivatives modified in the 3′-position as potential antiviral agents. *Collection of Czechoslovak Chemical Communications*, *71*, 991–1010. https://doi.org/10.1135/cccc20060991.

Potisopon, S., Ferron, F., Fattorini, V., Selisko, B., & Canard, B. (2017). Substrate selectivity of dengue and Zika virus NS5 polymerase towards 2′-modified nucleotide analogues. *Antiviral Research*, *140*, 25–36. https://doi.org/10.1016/j.antiviral.2016.12.021.

Retallack, H., Di Lullo, E., Arias, C., Knopp, K. A., Laurie, M. T., Sandoval-Espinosa, C., … DeRisi, J. L. (2016). Zika virus cell tropism in the developing human brain and inhibition by azithromycin. *Proceedings of the National Academy of Sciences of the United States of America*, *113*, 14408–14413. https://doi.org/10.1073/pnas.1618029113.

Ross, B. S., Reddy, P. G., Zhang, H.-R., Rachakonda, S., & Sofia, M. J. (2011). Synthesis of diastereomerically pure nucleotide phosphoramidates. *Journal of Organic Chemistry*, *76*, 8311–8319. https://doi.org/10.1021/jo201492m.

Sacramento, C. Q., de Melo, G. R., de Freitas, C. S., Rocha, N., Hoelz, L. V. B., Miranda, M., … Souza, T. M. L. (2017). The clinically approved antiviral drug sofosbuvir inhibits Zika virus replication. *Scientific Reports*, *7*, 40920. https://doi.org/10.1038/srep40920.

Seley-Radtke, K. L., & Yates, M. K. (2018). The evolution of nucleoside analogue antivirals: A review for chemists and non-chemists. Part 1: Early structural modifications to the nucleoside scaffold. *Antiviral Research*, *154*, 66–86. https://doi.org/10.1016/j.antiviral.2018.04.004.

Shan, C., Xie, X., Muruato, A. E., Rossi, S. L., Roundy, C. M., Azar, S. R., … Shi, P.-Y. (2016). An infectious cDNA clone of Zika virus to study viral virulence, mosquito transmission, and antiviral inhibitors. *Cell Host and Microbe*, *19*, 891–900. https://doi.org/10.1016/j.chom.2016.05.004.

Shim, J., Larson, G., Lai, V., Naim, S., & Wu, J. Z. (2003). Canonical 3′-deoxyribonucleotides as a chain terminator for HCV NS5B RNA-dependent RNA polymerase. *Antiviral Research*, *58*, 243–251. https://doi.org/10.1016/S0166-3542(03)00007-X.

Slusarczyk, M., Serpi, M., & Pertusati, F. (2018). Phosphoramidates and phosphonamidates (ProTides) with antiviral activity. *Antiviral Chemistry and Chemotherapy*, *26*. https://doi.org/10.1177/2040206618775243, 2040206618775243.

Smith, D. B., Kalayanov, G., Sund, C., Winqvist, A., Maltseva, T., Leveque, V. J.-P., … Smith, M. (2009). The design, synthesis, and antiviral activity of monofluoro and difluoro analogues of 4′-azidocytidine against hepatitis C virus replication: The discovery of 4′-azido-2′-deoxy-2′-fluorocytidine and 4′-azido-2′-dideoxy-2′,2′-difluorocytidine. *Journal of Medicinal Chemistry*, *52*, 2971–2978. https://doi.org/10.1021/jm801595c.

Wang, B., Tan, X.-F., Thurmond, S., Zhang, Z.-M., Lin, A., Hai, R., & Song, J. (2017). The structure of Zika virus NS5 reveals a conserved domain conformation. *Nature Communications*, *8*, 14763. https://doi.org/10.1038/ncomms14763.

Xie, X., Zou, J., Shan, C., Yang, Y., Kum, D. B., Dallmeier, K., … Shi, P.-Y. (2016). Zika virus replicons for drug discovery. *eBioMedicine*, *12*, 156–160. https://doi.org/10.1016/j.ebiom.2016.09.013.

Xu, H.-T., Hassounah, S. A., Colby-Germinario, S. P., Oliveira, M., Fogarty, C., Quan, Y., … Wainberg, M. A. (2017). Purification of Zika virus RNA-dependent RNA polymerase and its use to identify small-molecule Zika inhibitors. *Journal of Antimicrobial Chemotherapy*, *72*, 727–734. https://doi.org/10.1093/jac/dkw514.

Xu, M., Lee, E. M., Wen, Z., Cheng, Y., Huang, W.-K., Qian, X., … Tang, H. (2016). Identification of small-molecule inhibitors of Zika virus infection and induced neural cell death via a drug repurposing screen. *Nature Medicine*, *22*, 1101–1107. https://doi.org/10.1038/nm.4184.

Yates, M. K., & Seley-Radtke, K. L. (2019). The evolution of antiviral nucleoside analogues: A review for chemists and non-chemists. Part II: Complex modifications to the nucleoside scaffold. *Antiviral Research*, *162*, 5–21. https://doi.org/10.1016/j.antiviral.2018.11.016.

Zhao, B., Yi, G., Du, F., Chuang, Y.-C., Vaughan, R. C., Sankaran, B., … Li, P. (2017). Structure and function of the Zika virus full-length NS5 protein. *Nature Communications*, *8*. https://doi.org/10.1038/ncomms14762, 14762.

Zhou, T., Tan, L., Cederquist, G. Y., Fan, Y., Hartley, B. J., Mukherjee, S., … Chen, S. (2017). High-content screening in hPSC-neural progenitors identifies drug candidates that inhibit Zika virus infection in fetal-like organoids and adult brain. *Cell Stem Cell*, *21*, 274–283. e5 https://doi.org/10.1016/j.stem.2017.06.017.

Zmurko, J., Marques, R. E., Schols, D., Verbeken, E., Kaptein, S. J. F., & Neyts, J. (2016). The viral polymerase inhibitor 7-deaza-2′-C-methyladenosine is a potent inhibitor of in vitro Zika virus replication and delays disease progression in a robust mouse infection model. *PLoS Neglected Tropical Diseases*, *10*. https://doi.org/10.1371/journal.pntd.0004695, e0004695.

Chapter 38

Medicinal plants as promising source of natural antiviral substances against Zika virus

Juliano G. Haddad, Gilles Gadea, Philippe Desprès, and Chaker El Kalamouni

University of Reunion Island, UMR PIMIT, INSERM U1187, CNRS UMR 9192, IRD UMR 249, Technological platform CYROI, Sainte Clotilde, France

Abbreviations

BHK-21	baby hamster kidneycell lines
DENV	dengue virus
DR	human muscle rhabdomyosarcoma cell lines
EDIII	domain III of ZIKV E protein
EGCG	epigallocatechin gallate
FRET	fluorescence resonance energy transfer
HEK293T	human embryo kidney cell lines
HeLa	human cervical carcinoma cell lines
Huh7	human hepatoma cell lines
JEG-3	human placenta carcinoma cell lines
JEV	Japanese encephalitis virus
LLC-MK2	rhesus monkey kidney epithelial cell lines
NS	nonstructural protein
Q3G	isoquercitrin
RdRp	RNA-dependent RNA polymerase
SH-SY5Y	human neuroblastoma cell lines
THQ	thymohydroquinone dimethyl ether
Vero	kidney epithelial cell lines
WNV	West Nile virus
YFV	yellow fever virus
ZIKV	Zika virus

Introduction

Mosquito-borne Zika virus (ZIKV) belonging to *Flavivirus* genus (*Flaviviridae* family) became a major medical concern worldwide in the last decade (Chan, Choi, Yip, Cheng, & Yuen, 2016; Mittal et al., 2017; Yun & Lee, 2017). ZIKV is closely related to other medically-important flaviviruses such as dengue virus, yellow fever virus, West Nile virus and Japanese encephalitis virus (Laureti, Narayanan, Rodriguez-Andres, Fazakerley, & Kedzierski, 2018). Emergence of ZIKV has been associated with severe neurological complications in humans including congenital Zika syndrome and Guillain-Barré syndrome (Cao-Lormeau et al., 2016; Parra et al., 2016). Nonconventional ZIKV transmission routes such as sexual contact, blood transfusion, and intrauterine transmission have been documented during the recent epidemics in South America (Chen, Liang, et al., 2017; Duggal et al., 2017). An unexpected observation is that ZIKV can be detected in human biological fluids for prolonged periods after the acute phase of the disease (Chen, Yang, et al., 2017; Paz-Bailey et al., 2018). To date, effective preventive measures against Zika disease are still lacking.

Like other flaviviruses, ZIKV is an enveloped, single-stranded RNA virus (Perera-Lecoin, Meertens, Carnec, & Amara, 2013). The binding of virus particles to cell surface receptor(s) initiates their internalization inside the target cell followed

Zika Virus Impact, Diagnosis, Control, and Models. https://doi.org/10.1016/B978-0-12-820267-8.00038-8

by the delivering of genomic RNA into the cytosol (Chambers, Hahn, Galler, & Rice, 1990). Genomic RNA of about 11,000 nucleotides of length encodes a large polyprotein that is co- and posttranslationally processed into structural proteins capsid (C), premembrane (prM), and envelope (E) and nonstructural proteins NS1, NS2a, NS2b, NS3, NS4A, NS4B, and NS5 (Laureti et al., 2018). Both replication of viral RNA and assembly of virus particles occur in the close vicinity of the endoplasmic reticulum compartment. The proteins C, M, and E associated with genomic RNA compose the infectious virus particles (Chambers et al., 1990). The nonstructural proteins NS1 to NS5 not only contribute to the replication of viral RNA but also participate to the innate immune evasion mediated by ZIKV. There is mounting consensus to claim that efficient anti-Zika virus strategies based on antiviral compounds will have to block the main stages of infectious replication cycle of ZIKV without sorely inhibiting cell host functions (Wang & Shi, 2015).

Traditional herbal medicines and their purified phytochemicals can guide the development of new potential antivirals (Mani et al., 2020). The secondary metabolites extracted from plants can also give rise to the synthesis of analogues possessing better activity and/or better bioavailability. Thus natural products remain a significant source of potential drugs (Andersen, Ntie-Kang, & Tietjen, 2018; Lee, 2010). Albeit a huge amount of research has been conducted in the last 5 years, which extensively exceeded that of any other emerging virus, there is still a lack of effective anti-ZIKV compounds (Kang, Keller, & Luo, 2017). The knowledge about ZIKV molecular mechanisms of pathogenesis related to the neurological disorders including the elucidation of viral entry pathways and the elucidation of the crystal structure of some ZIKV protein (C, NS1, NS2b-NS3, NS3, and NS5) has certainly contributed to identify antiviral compounds (Byler, Ogungbe, & Setzer, 2016; Lee & Shin, 2019). Here, we present an overview of the latest advances made in the identification of antiviral from plant origin and the molecular tools that were used to assess anti-ZIKV compounds. This review will highlight the main finding on naturally derived antiviral against ZIKV; it was not intended to completely cover all the publications on anti-ZIKV compounds.

Molecular tools to assess anti-ZIKV compounds

Strategies have been developed to rapidly assess antiviral compound efficacy on Zika virus infection. This paragraph is not intended to give an exhaustive list of studies relating Zika virus screening tools but rather to highlight strategies that are useful for the discovery of anti-Zika virus effective drugs.

Full genome reporter

Identification of ZIKV-infected cells generally relies on antibody staining techniques that are resource and time consuming. Achievement of antiviral activities of drugs requires techniques that are rapid, money saving, and easy to automate. To this end, two different approaches based on reporter systems have been described in the literature, the reporter system being carried either by the virus or by the infected cells. The first Zika virus reporter full genome was generated soon after the beginning of the ZIKV outbreak (Gadea et al., 2016). This reporter virus was generated using the highly convenient ISA (Infectious Subgenomic Amplicon) method, which allows rapid recovery of recombinant viruses. Briefly the viral genomes are directly assembled into producing cells after introduction of overlapping synthetic fragments that cover the full genome RNA sequence. The recovered Zika virus GFP reporter full genome helped to rapidly identify antiviral extracts or drug compounds to tackle growing emergency (Clain et al., 2018; Gaudry et al., 2018). Other strategies including full-length cDNA clones have been used to generate reporter full genome viruses (Münster et al., 2018). Unfortunately, full genomes bearing a reporter gene generally lead to viruses that are attenuated and unstable, losing their reporter gene after a few passages. To tackle this issue, other studies have focused on the infected cells, where the cell itself becomes a reporter of the viral infection. To this end, studies have used the NS2B-NS3 protease activity that occurs during the viral infection to engineer fluorescent cell–based reporter system (Arias-Arias, MacPherson, Hill, Hardy, & Mora-Rodríguez, 2020; McFadden et al., 2018). These strategies are based on constitutive expression of a reporter system that becomes fluorescent or changes its localization upon infection and activation of the NS2B-NS3 protease. The intensity of the fluorescence can then be used to achieve the infection onset. Although this strategy allows the use of unmodified viruses or clinical isolates, expression of the reporter system into the host cell could lead to the selection of clonal cells that do not resemble original cell properties. Depending on the purpose of the screen, these different technologies could selectively be used to get the best of their performance and avoid their weaknesses.

Minimal replication units

Replicons are self-replicative subgenomic viral systems that allow antiviral screening without using infectious virus. As such, replicon systems focus on the analyzing of replication events in the absence of virus entry or virion assembly. Indeed, replicon systems are not fully reporting the infectious cycle. As for full genome reporters, several strategies have been used to generate replicon systems to screen anti-Zika virus compounds. Studies by (Xie et al., 2016) first described the development of a luciferase reporter gene–based stable replicon cell line. This first study paved the way to fully optimized studies that allow HTS of antiviral compounds (Li et al., 2018). Furthermore, by complementation with structural proteins, replicons can also lead to the production of single-round infectious particles for investigating the mechanisms of viral entry into the host cell in a safer way than infectious viruses (Lu et al., 2019; Mutso et al., 2017).

Isolated enzymes or catalytic domains

Zika virus genome comprises several enzymes whose activities could be targeted by antiviral compounds. To assess compound inhibitory effects on enzymatic activities, in vitro enzymatic assays have been designed based on the production of recombinant Zika virus enzymes and on the development of readout systems. The two main enzymatic activities that are classically targeted are the NS2B-NS3 protease and the RNA-dependent RNA polymerase (RdRp). The NS2B-NS3 protease activity is usually measured in vitro using fluorescence-based substrates. Substrates are double-labeled to allow a fluorescence resonance energy transfer (FRET) so that cleavage of the substrate by NS2B-NS3 protease enhances the fluorescence. This strategy was successfully developed to screen polyphenol compounds (Lim et al., 2017). ZIKV RdRp can synthetize RNA de novo, which can then be quantified in-real time using a fluorescent dye. Saez-Alvarez and colleagues used SYTO9 fluorescent dye that allows the detection of double-stranded RNA generated after the copy of single-stranded homopolymeric RNA (Sáez-Álvarez, Arias, Del Águila, & Agudo, 2019). This approach allows for assay to scale up to high-throughput screening and for rapid testing of anti-ZIKV compound libraries.

Plant-derived antivirals against ZIKV

Natural antivirals are molecules isolated from natural sources capable of disrupting the viral life cycle, thus blocking infection (Wang & Shi, 2015). According to their mechanism of action on the viral cycle, antivirals can be classified into two groups: those that act on the early phase (e.g., direct action on the particle and entry level) and those that act during the late phase (replication, maturation, and exocytosis). Nowadays, there is no licensed antiviral against ZIKV or any other flavivirus. However, during a short period, many studies have been conducted worldwide to assess the antiviral activity against ZIKV of different molecules (natural or synthetic) or plant extracts by using different approaches and molecular tools presented previously (Mottin et al., 2018; Saiz, 2019). This section is designed to provide a nonexhaustive overview of medicinal plants as a promising source of molecular diversity that are useful for the discovery of effective antivirals against ZIKV.

ZIKV infectivity inhibitors

Antivirals acting on the viral particle are defined as being virostatic or virucidal. A virostatic substance allows the induction of a disturbance in the structure of the viral particle, which results in an inhibition of the recognition of binding sites at cellular receptors leading to a loss of infectivity. A virucidal substance induce, in addition to the morphological disturbance, an irreversible damage to the viral genome by lysis of the particle. Natural virostatic antivirals including EGCG and delphinidin have been identified to inhibit ZIKV infectivity (Carneiro, Batista, Braga, Nogueira, & Rahal, 2016; Sharma, Murali, Singh, & Giri, 2017; Vázquez-Calvo, Jiménez De Oya, Martín-Acebes, Garcia-Moruno, & Saiz, 2017). EGCG is a flavonoid of the flavonol class (Fig. 1). Found in large quantities in green tea (*Camellia sinensis*), this flavonol is suspected of binding to the envelope proteins of the ZIKV and DENV to prevent the attachment of virions to the target cell. This mechanism of action has also been described for delphinidin, an antocyanidol found in field dolphinelle (*Delphinium consolida*) (Vázquez-Calvo et al., 2017). In an equivalent manner, berberine, an alkaloid (Fig. 1) found mainly in two genera of plants, *Coptis* sp. and *Berberis* sp., acts on the preadsorption phase of the viral replication cycle of ZIKV (Batista et al., 2019). It induces modifications in the viral envelope and consequently causes a structural disorder that prevents the adsorption of the virus into the host cell. Similarly, curcumin, another polyphenol (Fig. 1) mainly found in turmeric (*Curcuma longa*), has anti-ZIKV activity by direct action on the viral particle, thus preventing its attachment to receptors in the host cell (Mounce, Cesaro, Carrau, Vallet, & Vignuzzi, 2017). The antiviral properties of curcumin were observed on different cell lines such as HeLa, BHK-21, and Vero cells infected with the African strain (HD78) of ZIKV. This study showed that curcumin has only antiviral activity against enveloped viruses. This observation suggests that curcumin causes a change in the dynamics of the viral envelope that no longer allows the interaction between ZIKV particle and cellular receptors (Mounce et al., 2017).

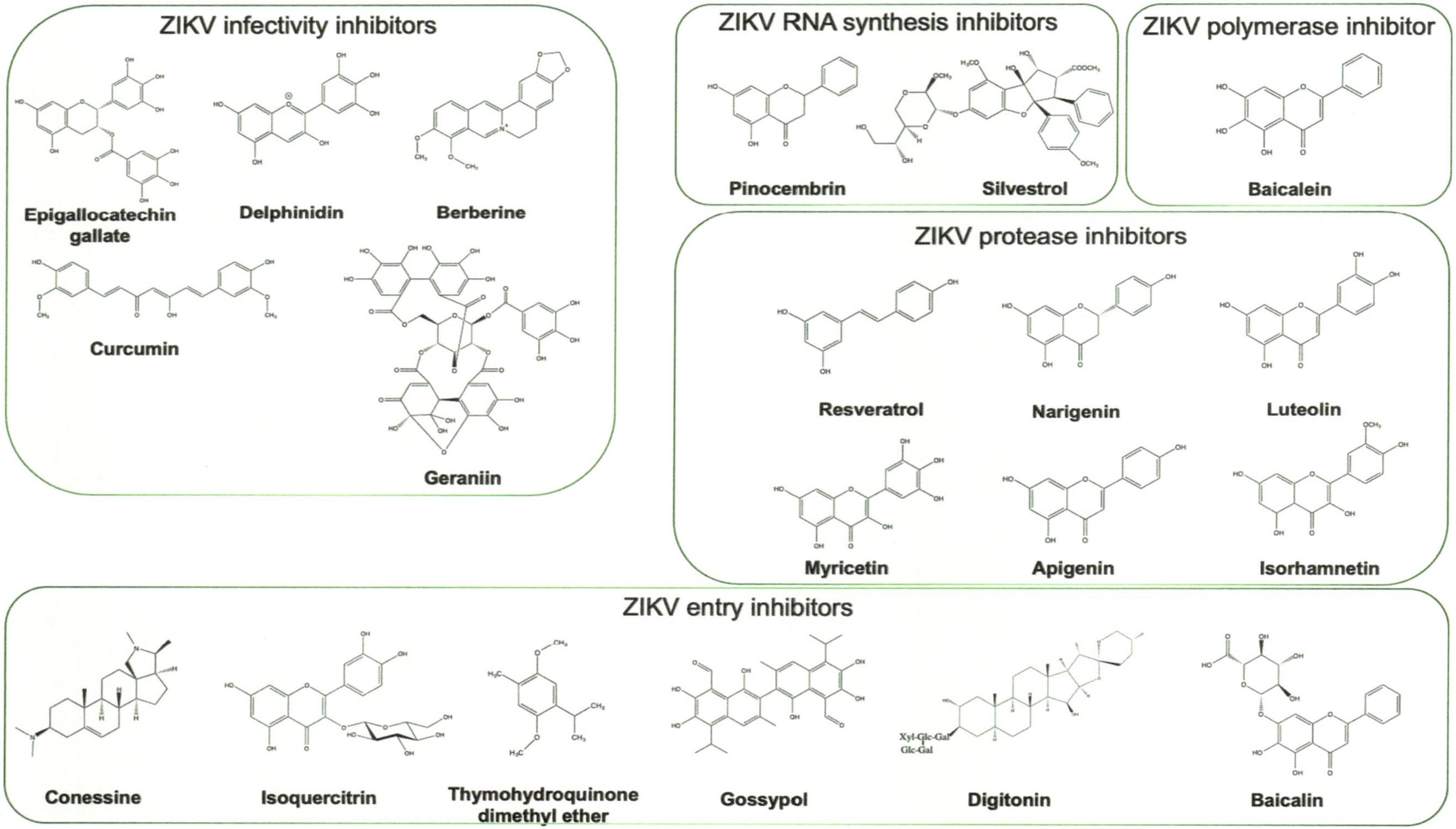

FIG. 1 Chemical structures of selected plant-derived ZIKV inhibitors. Chemical structures of natural compounds derived from plants and classified according to their antiviral mode of action inhibiting different stage of ZIKV life cycle: infectivity, entry, RNA synthesis, polymerase, and protease.

Several recent studies have highlighted the potential of medicinal plants from the Mascarene Islands as a source of antiviral agents against ZIKV. Indeed, *Aphloia theiformis*, *Psiloxylon mauritianum*, *Doratoxylon apetalum*, and *Phyllanthus phillyreifolius* are endemic or indigenous plants native from the Mascarenes: Reunion, Rodrigues, and Mauritius islands. The antiviral activity of *A. theiformis*, *P. mauritianum*, *D. apetalum*, and *P. phillyreifolius* leaf extracts was described for the first time by using a chimeric molecular clone of the African strain of ZIKV MR766 expressing a GFP reporter gene ($ZIKV_{GFP}$) (Gadea et al., 2016). Same results were obtained when testing these plant extracts against the Asian strain of ZIKV PF13 as well as the four DENV serotypes. In all cases, *A. theiformis*, *P. mauritianum*, *D. apetalum*, and *P. phillyreifolius* inhibit virus binding onto the host cell surface as well as the four DENV serotypes (Clain et al., 2019; Clain et al., 2018; Haddad et al., 2020; Haddad, Koishi, et al., 2019). *A. theiformis* has virucidal activity by deforming and inactivating the viral envelope particle. These properties have been shown by negative staining electron microscopy (Clain et al., 2018). The antiviral activity of *P. phillyreifolius* was related to its richness in geraniin (Fig. 1). It is important to note that *P. phillyreifolius* as well as its major compound geraniin did not present any toxicity in vivo using a zebrafish model at the effective antiviral doses (Haddad et al., 2020). These results highlight the safe and effective aspects of plant-derived phytochemicals.

ZIKV entry inhibitors

Antivirals can also act on the entry level of the viral cycle, that is, when the virion binds to the cell membrane; at the entry of the virus by endocytosis; or at the fusion of the viral membrane with the endosome. Many natural molecules have been studied for their ability to inhibit the attachment or entry of flaviviruses into host cells (Saiz, 2019). ZIKV binding inhibition was observed with gossypol, a natural polyphenol (Fig. 1) from plants belonging to the genus *Gossypium*. The antiviral properties of gossypol have been demonstrated on Vero and LLC-MK2 cells. In a recent study, gossypol has been shown to have antiviral activity on the preentry stage of the ZIKV. It has been described that gossypol binds to ZIKV E protein present on the viral envelope and more precisely to its domain III (EDIII), thus preventing the binding of ZIKV to the host cell membrane receptors (Gao et al., 2019). Other naturally derived compounds have also been shown to inhibit ZIKV attachment to the host cell surface such as digitonin, a steroidal saponin (Fig. 1) obtained from *Digitalis purpurea*, and conessine, a steroidal alkaloid found in several plant species belonging to the *Apocynaceae* family. Digitonin and conessine inhibit the entry step of ZIKV infection. These two molecules do not have the same mode of action on the ZIKV replication cycle. Digitonin interacts with host cell receptors and prevents adsorption of the virus to the host cell, while conessine intervenes at a late step of ZIKV entry by inhibiting the endocytosis phase or by targeting virus-cell fusion process (Gao et al., 2019). Plant-derived compounds have also been identified for their ability to interfere with ZIKV internalization step. For example, Q3G, a flavonoid (Fig. 1) present in a wide variety of plants such as St. John's wort (*Hypericum perforatum*) and mango fruits, has been described to inhibit ZIKV internalization step in A549, Vero, Huh7, and SH-SY5Y (Gaudry et al., 2018). To note, other flavonoids structurally close to Q3G like quercetin, kaempferol, and hyperoside have no antiviral activity against ZIKV. This implies an essential role for the glycosylated part of Q3G. This inhibition of internalization step has also been reported with thymohydroquinone dimethyl ether (THQ) (Fig. 1) isolated from the essential oil of *Ayapana triplinervis*. The antiviral properties of THQ have been investigated on A549 and Vero cells (Haddad, Picard, et al., 2019). THQ was tested on a zebrafish model in vivo to confirm its innocuity at the effective antiviral doses. The fish did not undergo any obvious stress or alteration in locomotor and eating behaviors (Haddad, Picard, et al., 2019). A recent study has demonstrated that phytochemicals from the leaves of *Ocimum basilicum* block the entry step of the virus by acting directly at the attachment and/or the endocytosis phases of the viral life cycle. The precise mode of action of these bioactive compounds is not specified in this study (Singh, Chakraborty, He, & Mergia, 2019).

If antivirals do not act on the viral particle or on the early phases, then they can act on one of the late phases of the viral cycle. Viral RNA synthesis, viral proteases, or viral polymerase are an attractive target because of their essential roles in processing the viral polypeptide and generating structural and nonstructural proteins during viral replication cycle.

ZIKV RNA synthesis inhibitors

Pinocembrin is a flavone (Fig. 1) that is a subfamily of flavonoids belonging to the family of polyphenols. This molecule is mainly present in honey, tea, and red wine. The antiviral activity of pinocembrin was evaluated on *Aedes albopictus* mosquito cells, BHK-21, JEG-3, HeLa, Huh7, HEK293T, and DR cell lines. This flavone is suggested to inhibit the viral RNA synthesis process and the production of ZIKV structural and nonstructural proteins during the replication cycle of ZIKV (Lee, Loe, Lee, & Chu, 2019). Silvestrol, a natural compound derived from *Rocaglates*, has been described to inhibit the ZIKV infection during the translation phase of the viral RNA within the A549 cells. This compound blocks the eukaryotic initiation factor 4A (eIF4A), which is required for the binding of the viral RNA to the small ribosomal subunit 40s. This prevents the synthesis of proteins involved in the viral replication and the assembly of ZIKV particles (Elgner et al., 2018).

ZIKV protease inhibitors

Resveratrol is a natural polyphenol belonging to the class of stilbenes. It is present in red wine, nuts, and berries. The antiviral properties of resveratrol are reported at different levels of ZIKV replication cycle on Vero and Huh7 cell lines (Mohd, Zainal, Tan, & AbuBakar, 2019). A research study described the ability of resveratrol to inhibit the attachment of ZIKV particles to the host cell receptors by interacting with the extracellular proteins of the viral envelope. Another study showed the inhibitory activity of resveratrol on the postentry steps of ZIKV life cycle. Resveratrol is able to bind to the NS3 protein of the virus and cause structural changes to this helicase, resulting in a lower affinity between NS3 and the viral RNA and a strong decrease in viral RNA synthesis, viral replication, and protein synthesis (Devnarain & Soliman, 2019). Additional natural flavonoids like naringenin, luteolin, myricetin, apigenin, and isorhamnetin inhibit the N2B-NS3 trypsin-like serine protease, which contributes to viral polyprotein processing (Cataneo et al., 2019; Lim et al., 2017).

ZIKV polymerase inhibitors

Baicalein and baicalin are compounds belonging to the flavone subfamily of flavonoids (Fig. 1). Baicalin is the glucuronide form of baicalein. Recently the antiviral activity of baicalin and baicalein was demonstrated on Vero cells (Oo, Teoh, Sam, Bakar, & Zandi, 2019). These two molecules have antiviral activity at different stages of ZIKV replication cycle, entry, and replication step. To note, baicalin has stronger inhibition during the early stages of the ZIKV replication cycle than baicalein. It has been shown that baicalin inhibits the entry of the virus into the host cell. However, the antiviral activity of baicalein is much more important than that of baicalin at the late stages of ZIKV replication cycle. This study showed, by molecular modeling, that baicalein interacts with the nonstructural protein NS5 polymerase of ZIKV and therefore prevents the replication of the viral RNA (Oo et al., 2019).

Concluding remarks

Despite the fact that scientific research has devoted considerable efforts to the research of antiviral drugs to prevent and treat ZIKV infections, there is still no approved treatment for this flavivirus. Plants have developed a tremendous molecular diversity over hundreds of millions of years of evolution. This molecular diversity is illustrated by the innumerable secondary metabolites that have been isolated from plants and that makes them a promising source of naturally derived antiviral compounds (Table 1). In this review, we focused on few phytochemicals used to exemplify the great diversity of antiviral molecules from plants that could be used against ZIKV (Table 1). These antiviral secondary metabolites can either target viral proteins or the lipid envelope or even ZIKV genome with resulting consequences on viral replication (Fig. 2 and Table 1). Most polyphenols or polyphenol-rich plant extracts act on the early stages of the viral cycle either by neutralizing ZIKV infectivity or by inhibiting its binding and internalization (Fig. 2 and Table 1). Flavonoids have been identified for their ability to interfere with the late stage of ZIKV replication cycle inhibiting protease or polymerase activity (Table 1).

There is still a long way to go before therapeutic measures against ZIKV can be implemented. Despite dozens of identified natural antiviral candidates acting at different stages of the viral infectious cycle, only a few of them have been tested in animal models. However, as a green light, the safety of some natural compounds has been highlighted in a small number of studies using a zebrafish model. Such results are encouraging and reassure if needed on the therapeutic potential of these natural compounds. Although the number of ZIKV infection cases has significantly decreased worldwide since 2018, the search for antivirals against this flavivirus must keep on, since it shares many biological characteristics with other flaviviruses of medical concern such as dengue virus. Thus the implementation of antivirals against ZIKV could also benefit to therapeutic approaches against other currently circulating flaviviruses.

Policy and procedures: Method to assess plant-derived antivirals in vivo

All animal experiments in zebrafish (Haddad et al., 2020; Haddad, Picard, et al., 2019) were conducted in accordance with the French and European Community Guidelines for the Use of Animals in Research (86/609/EEC and 2010/63/EU) and approved by the local ethics committee for animal experimentation of CYROI (APAFIS #2018072310507016_v5) and (APAFIS #2019052910002738_v4). For intraperitoneal injections, fish were anesthetized with 0.02% tricaine and injected with appropriate concentrations of plant extract, phytochemical, or vehicle. After injection, fish were immediately placed

TABLE 1 Selected plant-derived phytochemicals active against ZIKV.

Source	Phytochemical	Chemical classification	Mode of action	References
Camellia sinensis	EGCG	Polyphenol/ flavanol	Infectivity	Sharma et al. (2017)
Delphinium consolida	Delphinidin	Polyphenol/ anthocyanidin	Infectivity	Vázquez-Calvo, Jiménez De Oya, Martín-Acebes, Garcia-Moruno, & Saiz, 2017
Berberis sp.	Berberine	Alkaloid	Infectivity	Batista et al. (2019)
Curcuma longa	Curcumin	Polyphenol	Infectivity	Mounce et al. (2017)
Aphloia theiformis	ni	–	Infectivity	Clain et al. (2018)
Psiloxylon mauritianum	ni	Polyphenol	Infectivity	Clain et al. (2019)
Doratoxylon apetalum	ni	–	Infectivity	Haddad, Koishi, et al. (2019)
Phyllanthus phillyreifolius	Geraniin	Polyphenol/ ellagitannin	Infectivity	Haddad et al. (2020)
Gossypium sp.	Gossypol	Polyphenol	Entry	Gao et al. (2019)
Digitalis purpurea	Digitonin	Saponin	Entry	Gao et al. (2019)
Holarrhena floribunda	Conessine	Alkaloid	Entry	Gao et al. (2019)
Hypericum perforatum	Isoquercitrin	Polyphenol/ flavonoid	Entry	Gaudry et al. (2018)
Ayapana triplinervis	THQ	Terpene	Entry	Haddad, Picard, et al. (2019)
Ocimum basilicum	ni	Polyphenol	Entry	Singh et al. (2019)
Scutellaria baicalensis	Baicalin	Polyphenol/ flavone	Entry	Oo et al. (2019)
Turnera diffusa	Pinocembrin	Polyphenol/ flavanone	RNA synthesis	Lee et al. (2019)
Aglaia silvestris	Silvestrol	Flavaglines	RNA synthesis	Elgner et al. (2018)
Vitis vinifera	Resveratrol	Polyphenol/ stilbenes	Protease	Mohd et al. (2019)
Scutellaria baicalensis	Baicalein	Polyphenol/ flavone	Polymerase	Oo et al. (2019)

Antivirals derived from medicinal plants classified according to their mode of action (infectivity, entry, RNA synthesis, protease, and polymerase). The majority of the phytochemicals belong to the family of polyphenols and more specifically the group of flavonoids (flavanol, anthocyanidin, flavone, and flavanone). *ni*, nonidentified active compound.

back in fish water and carefully observed for detecting any striking signs of stress, locomotor activity, and feeding behavior issues. At the end of the procedure, the animals were sacrificed with an overdose of tricaine.

ZIKV experiments performed in mice (Chen, Liang, et al., 2017) were carried out in accordance with the recommendations in the Guide for the Care and Use of Laboratory Animals of the National Institutes of Health. All animal procedures were performed in compliance with the Institutional Animal Care and Use Committee (protocol # 1606029) at the

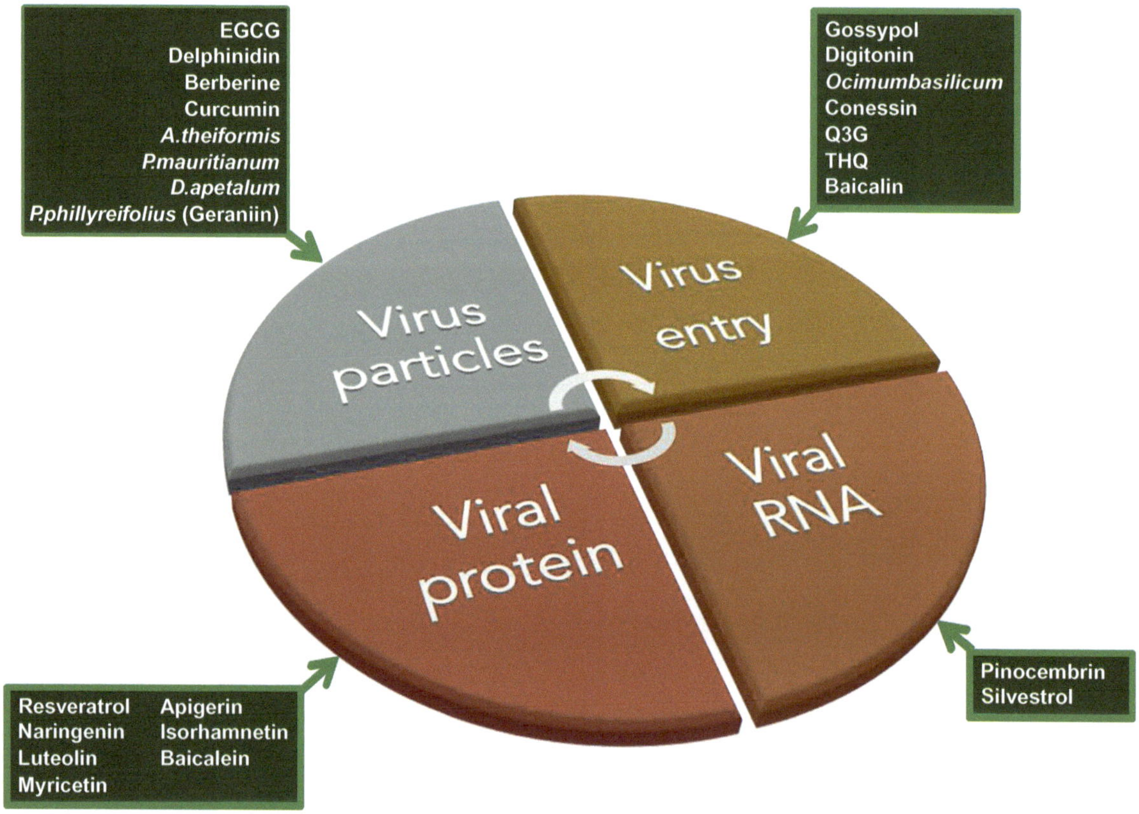

FIG. 2 Schematic representation of different successive ZIKV replication step. In the *green boxes* are the names of the phytochemicals or plant extracts that can inhibit infection at different stages of ZIKV infectious cycle.

University of Texas Medical Branch (UTMB) in Galveston, TX. In vivo experiments carried out in mice (Chen, Yang, et al., 2017) were approved by Fudan University and ShanghaiTech University. Animal studies were carried out in strict accordance with the Institutional Animal Care and Use Committee approved protocols. ZIKV studies realized in mice (Duggal et al., 2017) were conducted under approved Institutional Animal Care and Use Committee (IACUC) protocols at the Centers for Disease Control and prevention.

The cohort study conducted in Paz-Bailey et al. (2018) was reviewed and approved by the institutional review boards at the Centers for Disease Control and Prevention (CDC) and Ponce Health Sciences University.

Mini-dictionary of terms

Fluorescence resonance energy transfer. FRET stands for fluorescence resonance energy transfer and refers to two molecules within a close proximity that are able to transfer energy. This technology is commonly used in microscopy to study interaction between partners or in biochemistry to follow enzymatic reaction in real time (quantitative polymerase chain reactions or catalytic cleavages).

Replicons. DNA or RNA fragments able to replicate autonomously also known as replication unit. In virology, replicons are nucleic acid fragments covering the region of the nonstructural proteins so viral replicons are not capable to generate infectious viral particles.

Polyphenols. Diverse range of molecules (more than 8000), abundant in many vegetables and fruits, biosynthesized from the shikimate-derived phenylpropanoid and polyketide pathways and are chemically composed by more than one phenolic ring and are devoid of any nitrogen-based functional group.

Flavonoid. Among the 8000 polyphenols, a subclass of more than 4000 flavonoids have been identified. Flavonoids have the C6–C3–C6 general structural backbone in which the two C6 units (Ring A and Ring B) are of phenolic nature. The hydroxylation pattern and variations in the chromane ring (Ring C) give rise to numerous flavonoid subclasses such as flavone, flavanol, flavan-3-ol, flavanone, and anthocyanidin.

Alkaloid. Natural organic compounds that mostly contain basic nitrogen atoms usually heterocyclic. Alkaloids are derivatives of amino acids. They are found in the form of complex mixtures, often based on several or even dozens of different alkaloid molecules, with their precursors as secondary metabolites, mainly in plants.
Virucidal. Antivirals acting on the viral particle are defined as being virucidal. A virucidal effect allows the induction of a disturbance in the structure of the viral particle that results in an inhibition of the recognition of binding sites at cellular receptors leading to a loss of infectivity. In addition to the morphological disturbance, an irreversible damage to the viral genetic material could be induced by a virucidal substance.

Keys facts of phytochemicals evaluated in vitro and in vivo

- Asiatic and African strains of Zika virus are sensitive to the antiviral action of medicinal plants extracts and their active phytocompounds.
- Polyphenols target the early stages of Zika virus replication throughout ZIKV-envelope interaction or deforming and inactivating the viral envelope particle.
- Flavonoids mainly target the late stages of Zika virus replication.
- Small animals such as rodents and zebrafish are used as models for testing toxicity of anti-Zika compounds in vivo.
- Mouse model is used for assessing the efficacy of natural anti-Zika compounds.

Summary points

- Nowadays, we are still lacking of effective antivirals against ZIKV.
- Some broad-spectrum drugs are not suitable for use against ZIKV because they can be harmful to pregnant women.
- Secondary metabolites isolated from medicinal plants such as polyphenols, flavonoids, alkaloids, coumarins, and curcuminoids have been shown to inhibit infection, without toxic effect, at different stages of ZIKV infectious cycle.
- Most polyphenols or polyphenol-rich plant extracts act on the early stages of the viral cycle either by neutralizing ZIKV infectivity or by inhibiting its binding and internalization.
- Flavonoids have been identified for their ability to interfere with the late stage of ZIKV replication cycle inhibiting protease or polymerase activity.

References

Andersen, R. J., Ntie-Kang, F., & Tietjen, I. (2018). Natural product-derived compounds in HIV suppression, remission, and eradication strategies. *Antiviral Research, 158*, 63–77.

Arias-Arias, J. L., MacPherson, D. J., Hill, M. E., Hardy, J. A., & Mora-Rodríguez, R. (2020). A fluorescence-activatable reporter of flavivirus NS2B-NS3 protease activity enables live imaging of infection in single cells and viral plaques. *Journal of Biological Chemistry, 295*(8), 2212–2226.

Batista, M. N., Braga, A. C. S., Campos, G. R. F., Souza, M. M., Matos, R. P. A., Lopes, T. Z., … Rahal, P. (2019). Natural products isolated from oriental medicinal herbs inactivate Zika virus. *Viruses, 11*(1), 49.

Byler, K. G., Ogungbe, I. V., & Setzer, W. N. (2016). In-silico screening for anti-Zika virus phytochemicals. *Journal of Molecular Graphics and Modelling, 69*, 78–91.

Cao-Lormeau, V. M., Blake, A., Mons, S., Lastère, S., Roche, C., Vanhomwegen, J., … Ghawché, F. (2016). Guillain-Barré syndrome outbreak associated with Zika virus infection in French Polynesia: A case-control study. *The Lancet, 387*(10027), 1531–1539.

Carneiro, B. M., Batista, M. N., Braga, A. C. S., Nogueira, M. L., & Rahal, P. (2016). The green tea molecule EGCG inhibits Zika virus entry. *Virology, 496*, 215–218.

Cataneo, A. H. D., Kuczera, D., Koishi, A. C., Zanluca, C., Silveira, G. F., Arruda, T. B., … Bordignon, J. (2019). The citrus flavonoid naringenin impairs the in vitro infection of human cells by Zika virus. *Scientific Reports, 9*(1), 16348.

Chambers, T. J., Hahn, C. S., Galler, R., & Rice, C. M. (1990). Flavivirus genome organization, expression, and replication. *Annual Review of Microbiology, 44*(1), 649–688.

Chan, J. F., Choi, G. K., Yip, C. C., Cheng, V. C., & Yuen, K. Y. (2016). Zika fever and congenital Zika syndrome: An unexpected emerging arboviral disease. *Journal of Infection, 72*(5), 507–524.

Chen, J., Liang, Y., Yi, P., Xu, L., Hawkins, H. K., Rossi, S. L., … Sun, J. (2017). Outcomes of congenital Zika disease depend on timing of infection and maternal-fetal interferon action. *Cell Reports, 21*(6), 1588–1599.

Chen, J., Yang, Y. F., Chen, J., Zhou, X., Dong, Z., Chen, T., … Zhu, T. (2017). Zika virus infects renal proximal tubular epithelial cells with prolonged persistency and cytopathic effects. *Emerging Microbes & Infections, 6*(8). e77.

Clain, E., Haddad, J. G., Koishi, A. C., Sinigaglia, L., Rachidi, W., Desprès, P., … El Kalamouni, C. (2019). The polyphenol-rich extract from Psiloxylon mauritianum, an endemic medicinal plant from Reunion Island, inhibits the early stages of dengue and Zika virus infection. *International Journal of Molecular Sciences, 20*(8), 1860.

Clain, E., Sinigaglia, L., Koishi, A. C., Gorgette, O., Gadea, G., Viranaicken, W., … El Kalamouni, C. (2018). Extract from Aphloia theiformis, an edible indigenous plant from Reunion Island, impairs Zika virus attachment to the host cell surface. *Scientific Reports*, *8*(1), 10856.

Devnarain, N., & Soliman, M. (2019). Molecular mechanism of resveratrol inhibition of Zika virus NS3 helicase: Behind the scenes. *Future Virology*, *14* (2), 73–84.

Duggal, N. K., Ritter, J. M., Pestorius, S. E., Zaki, S. R., Davis, B. S., Chang, G. J., … Brault, A. C. (2017). Frequent Zika virus sexual transmission and prolonged viral RNA shedding in an immunodeficient mouse model. *Cell Reports*, *18*(7), 1751–1760.

Elgner, F., Sabino, C., Basic, M., Ploen, D., Grünweller, A., & Hildt, E. (2018). Inhibition of Zika virus replication by Silvestrol. *Viruses*, *10*(4), 149.

Gadea, G., Bos, S., Krejbich-Trotot, P., Clain, E., Viranaicken, W., El-Kalamouni, C., … Despres, P. (2016). A robust method for the rapid generation of recombinant Zika virus expressing the GFP reporter gene. *Virology*, *497*, 157–162.

Gao, Y., Tai, W., Wang, N., Li, X., Jiang, S., Debnath, A. K., … Chen, S. (2019). Identification of novel natural products as effective and broad-spectrum anti-Zika virus inhibitors. *Viruses*, *11*(11), 1019.

Gaudry, A., Bos, S., Viranaicken, W., Roche, M., Krejbich-Trotot, P., Gadea, G., … El-Kalamouni, C. (2018). The flavonoid isoquercitrin precludes initiation of Zika virus infection in human cells. *International Journal of Molecular Sciences*, *19*(4), 1093.

Haddad, J. G., Grauzdytė, D., Koishi, A. C., Viranaicken, W., Venskutonis, P. R., Nunes Duarte Dos Santos, C., … El Kalamouni, C. (2020). The geraniin-rich extract from Reunion Island endemic medicinal plant Phyllanthus phillyreifolius inhibits Zika and dengue virus infection at non-toxic effect doses in zebrafish. *Molecules (Basel, Switzerland)*, *25*(10), 2316.

Haddad, J. G., Koishi, A. C., Gaudry, A., Nunes Duarte Dos Santos, C., Viranaicken, W., Desprès, P., & El Kalamouni, C. (2019). Doratoxylon apetalum, an indigenous medicinal plant from Mascarene Islands, is a potent inhibitor of Zika and dengue virus infection in human cells. *International Journal of Molecular Sciences*, *20*(10), 2382.

Haddad, J. G., Picard, M., Bénard, S., Desvignes, C., Desprès, P., Diotel, N., & El Kalamouni, C. (2019). Ayapana triplinervis essential oil and its main component thymohydroquinone dimethyl ether inhibit Zika virus at doses devoid of toxicity in zebrafish. *Molecules (Basel, Switzerland)*, *24*(19), 3447.

Kang, C., Keller, T. H., & Luo, D. (2017). Zika virus protease: An antiviral drug target. *Trends in Microbiology*, *25*(10), 797–808.

Laureti, M., Narayanan, D., Rodriguez-Andres, J., Fazakerley, J. K., & Kedzierski, L. (2018). Flavivirus receptors: Diversity, identity, and cell entry. *Frontiers in Immunology*, *9*, 2180.

Lee, K. H. (2010). Discovery and development of natural product-derived chemotherapeutic agents based on a medicinal chemistry approach. *Journal of Natural Products*, *73*(3), 500–516.

Lee, J. L., Loe, M. W. C., Lee, R. C. H., & Chu, J. J. H. (2019). Antiviral activity of pinocembrin against Zika virus replication. *Antiviral Research*, *167*, 13–24.

Lee, J. K., & Shin, O. S. (2019). Advances in Zika virus-host cell interaction: Current knowledge and future perspectives. *International Journal of Molecular Sciences*, *20*(5), 1101.

Li, J. Q., Deng, C. L., Gu, D., Li, X., Shi, L., He, J., … Ye, H. Q. (2018). Development of a replicon cell line-based high throughput antiviral assay for screening inhibitors of Zika virus. *Antiviral Research*, *150*, 148–154.

Lim, H. J., Nguyen, T. T., Kim, N. M., Park, J. S., Jang, T. S., & Kim, D. (2017). Inhibitory effect of flavonoids against NS2B-NS3 protease of ZIKA virus and their structure activity relationship. *Biotechnology Letters*, *39*(3), 415–421.

Lu, C. Y., Lin, C. S., Lai, H. C., Yu, Y. W., Liao, C. Y., Su, W. C., … Lin, C. W. (2019). The rescue and characterization of recombinant, microcephaly-associated Zika viruses as single-round infectious particles. *Viruses*, *11*(11), 1005.

Mani, J. S., Johnson, J. B., Steel, J. C., Broszczak, D. A., Neilsen, P. M., Walsh, K. B., & Naiker, M. (2020). Natural product-derived phytochemicals as potential agents against coronaviruses: A review. *Virus Research*, *284*, 197989.

McFadden, M. J., Mitchell-Dick, A., Vazquez, C., Roder, A. E., Labagnara, K. F., McMahon, J. J., … Horner, S. M. (2018). A fluorescent cell-based system for imaging Zika virus infection in real-time. *Viruses*, *10*(2), 95.

Mittal, R., Nguyen, D., Debs, L. H., Patel, A. P., Liu, G., Jhaveri, V. M., … Liu, X. Z. (2017). Zika virus: An emerging global health threat. *Frontiers in Cellular and Infection Microbiology*, *7*, 486.

Mohd, A., Zainal, N., Tan, K. K., & AbuBakar, S. (2019). Resveratrol affects Zika virus replication in vitro. *Scientific Reports*, *9*(1), 14336.

Mottin, M., Borba, J., Braga, R. C., Torres, P. H. M., Martini, M. C., Proenca-Modena, J. L., … Horta Andrade, C. (2018). The A-Z of Zika drug discovery. *Drug Discovery Today*, *23*(11), 1833–1847.

Mounce, B. C., Cesaro, T., Carrau, L., Vallet, T., & Vignuzzi, M. (2017). Curcumin inhibits Zika and chikungunya virus infection by inhibiting cell binding. *Antiviral Research*, *142*, 148–157.

Münster, M., Płaszczyca, A., Cortese, M., Neufeldt, C. J., Goellner, S., Long, G., & Bartenschlager, R. (2018). A reverse genetics system for Zika virus based on a simple molecular cloning strategy. *Viruses*, *10*(7), 368.

Mutso, M., Saul, S., Rausalu, K., Susova, O., Žusinaite, E., Mahalingam, S., & Merits, A. (2017). Reverse genetic system, genetically stable reporter viruses and packaged subgenomic replicon based on a Brazilian Zika virus isolate. *Journal of General Virology*, *98*(11), 2712–2724.

Oo, A., Teoh, B. T., Sam, S. S., Bakar, S. A., & Zandi, K. (2019). Baicalein and baicalin as Zika virus inhibitors. *Archives of Virology*, *164*(2), 585–593.

Parra, B., Lizarazo, J., Jiménez-Arango, J. A., Zea-Vera, A. F., González-Manrique, G., Vargas, J., … Pardo, C. A. (2016). Guillain-Barré syndrome associated with Zika virus infection in Colombia. *New England Journal of Medicine*, *375*(16), 1513–1523.

Paz-Bailey, G., Rosenberg, E. S., Doyle, K., Munoz-Jordan, J., Santiago, G. A., Klein, L., … Sharp, T. M. (2018). Persistence of Zika virus in body fluids – Final report. *New England Journal of Medicine*, *379*(13), 1234–1243.

Perera-Lecoin, M., Meertens, L., Carnec, X., & Amara, A. (2013). Flavivirus entry receptors: An update. *Viruses*, *6*(1), 69–88.

Sáez-Álvarez, Y., Arias, A., Del Águila, C., & Agudo, R. (2019). Development of a fluorescence-based method for the rapid determination of Zika virus polymerase activity and the screening of antiviral drugs. *Scientific Reports*, *9*(1), 5397.

Saiz, J. C. (2019). Therapeutic advances against ZIKV: A quick response, a long way to go. *Pharmaceuticals*, *12*(3), 127.

Sharma, N., Murali, A., Singh, S. K., & Giri, R. (2017). Epigallocatechin gallate, an active green tea compound inhibits the Zika virus entry into host cells via binding the envelope protein. *International Journal of Biological Macromolecules*, *104*, 1046–1054.

Singh, P., Chakraborty, P., He, D. H., & Mergia, A. (2019). Extract prepared from the leaves of Ocimum basilicum inhibits the entry of Zika virus. *Acta Virologica*, *63*(3), 316–321.

Vázquez-Calvo, Á., Jiménez De Oya, N., Martín-Acebes, M. A., Garcia-Moruno, E., & Saiz, J.-C. (2017). Antiviral properties of the natural polyphenols delphinidin and epigallocatechin gallate against the Flaviviruses West Nile virus, Zika virus, and dengue virus. *Frontiers in Microbiology*, *8*, 1314.

Wang, Q. Y., & Shi, P. Y. (2015). Flavivirus entry inhibitors. *ACS Infectious Diseases*, *1*(9), 428–434.

Xie, X., Zou, J., Shan, C., Yang, Y., Kum, D. B., Dallmeier, K., … Shi, P. Y. (2016). Zika virus replicons for drug discovery. *eBioMedicine*, *12*, 156–160.

Yun, S. I., & Lee, Y. M. (2017). Zika virus: An emerging flavivirus. *Journal of Microbiology*, *55*(3), 204–219.

Chapter 39

Protein kinase C as a target in the control of viruses and implication for Zika virus

A.B. Blázquez and J.C. Saiz
Department of Biotechnology, National Agricultural and Food Research and Technology Institute (INIA), Madrid, Spain

Abbreviations

aPCK atypical protein kinase C
cPKC classical protein kinase C
DAG diacylglycerol
DENV dengue virus
HCV hepatitis C virus
JEV Japanese encephalitis virus
kDa kilodalton
nPKC novel protein kinase C
PKCs protein kinases C
PS phosphatidylserine
WNV West Nile virus
ZIKV Zika virus

Introduction

Protein kinases C (PKCs) constitute a family of multifunctional enzymes that phosphorylates serine and threonine residues in many cellular proteins. The first enzyme of this family was identified in bovine cerebellum as a phospholipid and calcium-dependent protein kinase (Takai et al., 1979) that phosphorylated histone and protamine (Webb, Hirst, & Giembycz, 2000). A few years later, Huang and coworkers demonstrated that this brain-isolated PKC resolved into three different species on hydroxylapatite columns, suggesting that PKC comprised multiple isoforms (Huang, Nakabayashi, & Huang, 1986). Since then, the role of PKC has been related to a wide variety of cellular processes such as apoptosis, differentiation, proliferation, cellular transformation, motility, adhesion, and angiogenesis (Brodie & Blumberg, 2003; Isakov, 2018; Mackay & Twelves, 2007; Sontag, Sontag, & Garcia, 1997), being also involved in different steps during viral infections (Blazquez, Vazquez-Calvo, Martin-Acebes, & Saiz, 2018). Even though most PKC isoenzymes are ubiquitous and many cells are able to co-express different PKC isoforms, some of them are expressed in a tissue-specific manner.

PKC structure and classification

The structure of PKC consists of a single polypeptide, with an N-terminal regulatory region 20–40 kDa in size and a C-terminal catalytic region of approximately 45 kDa (Newton, 1995), which presents four conserved domains: C1–C4 (Coussens et al., 1986) (Fig. 1). These domains show different functional features: C1 forms the DAG/phorbol ester binding site (Bell & Burns, 1991), and contains a Cys-rich motif; C2 contains the recognition site for acidic lipids. C1 and C2 comprise the regulatory domain, which is linked to the catalytic one by a hinge region. C3 and C4 form the ATP- and substrate-binding lobes of the catalytic domain (Newton, 1995). In addition, other regions of lower homology, called variable domains, are inserted among the conserved ones (Webb et al., 2000). Besides, all PKCs present a pseudo-substrate sequence preceding the C1 domain, which maintains the protein in an inactive state and thus regulates its activation (Nishizuka, 1988).

The classification of PKCs relies on structural and cofactor activation characteristics. There are currently 11 different PKC isozymes divided into three groups (Table 1): classical or conventional PKCs (cPKCs: PKCα, PKCβ, and PKCγ),

Zika Virus Impact, Diagnosis, Control, and Models. https://doi.org/10.1016/B978-0-12-820267-8.00039-X

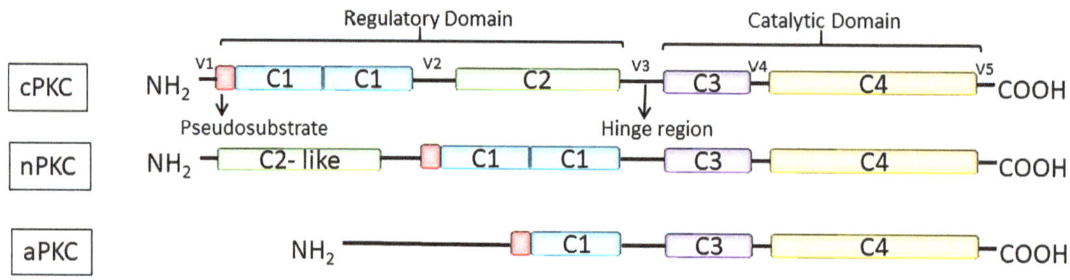

FIG. 1 Schematic representation of protein kinase C isoenzyme structure and classification. The PKC family is divided into three groups: classical (cPKC), novel (nPKC), and atypical (aPKC). PKC has four conserved domains (C1–C4) and five variable domains (V1–V5). This figure has not been previously published anywhere.

TABLE 1 Classification of PKCs.

	Cofactor requirement			Isoform
	Calcium	DAG	PS	
Classical/conventional PKC	Yes	Yes	Yes	PKCα PKCβ PKCϒ
Novel PKC	No	Yes	Yes	PKCσ PKCδ PKCε PKCη PKCθ
Atypical PKC	No	No	Yes	PKCζ PKCλ PKCι

Classification of PKCs has been determined according to their cofactor requirement. Table has not been previously published anywhere.

which are calcium-dependent PKCs and are activated by both PS and DAG; novel PKCs (nPKCs: PKCσ, PKCδ, PKCε, PKCη, and PKCθ), which are calcium independent and are regulated by DAG and PS; and atypical PKCs (aPKCs: PKCζ, PKCλ, and PKCι), which are calcium independent and do not require DAG for activation, although PS can regulate their activity (Mackay & Twelves, 2007). Domain structures are slightly different and define the subfamily of PKC isozymes (Fig. 1). cPKC and nPKC isoforms contain a tandem C1 domain that shows different affinity for DAG according to the PKC subfamily; cPKCs have a calcium sensitive C2 domain, while nPKCs have a C2-like domain that lacks calcium sensor ability (Newton, 2018b). Atypical PKCs are significantly different from the other two classes, and it has been reported that, contrary to them, they do not respond to phorbol esters either in vivo or in vitro (Newton, 1995). On the other hand, the aPKCs mechanism of phosphorylation differs from that of conventional (cPKC) and novel (nPKC) isozymes (Tobias et al., 2016).

PKC activation and functions

As previously mentioned, PKC is a family of protein kinase enzymes that mediates the activity of other proteins through phosphorylating serine and threonine amino acid residues on these ones (Wang, Liu, & Jia, 2018). Regulation of PKCs can be performed by two different mechanisms: phosphorylation and second messengers. PKC can be subjected to sequential phosphorylation steps in order to achieve its complete catalytic ability, and, in addition, it has also been described that in vitro autophosphorylation is a prerequisite for downregulation (Ohno, Konno, Akita, Yano, & Suzuki, 1990). Furthermore, PKCs can be activated by other physiological stimuli such as growth factors, hormones, cytokines, or antigens (Merida, Arranz-Nicolas, Rodriguez-Rodriguez, & Avila-Flores, 2019). Several posttranslational modifications also regulate the function of PKC isozymes in an agonist-dependent manner (Reyland, 2009; Steinberg, 2004). Along with this role in catalyzing phosphorylation reactions, PKCs have additional enzymatic functions such as ATPase and phosphatase activity (Newton, 1995).

Moreover, and given the myriad of substrates susceptible to the action of PKCs, a wide variety of functions have been ascribed to PKCs, so that the finding that PKC is the major receptor for phorbol esters, a potent tumor promoter, leads to its close association with cancer (An et al., 2019; Newton, 2018a). PKCs are also involved in autophagy (Wang et al., 2018), cell proliferation and differentiation, apoptosis (Singh et al., 2017), transcription regulation, memory (Webb et al., 2000), and many other pathophysiological conditions like Alzheimer, ischemic heart disease, Parkinson's disease, diabetes, bipolar disease, and psoriasis in humans (Mochly-Rosen, Das, & Grimes, 2012).

Role of PKC on viral infections

A considerable number of studies have focused on the role of PKCs during viral infection. PKC isoforms are involved in different replication steps, and PKC signaling is regulated during cell entry and infection in many unrelated viruses. For instance, herpes simplex virus type 1 needs PKC both for cell entry and nuclear egress of the viral capsid (Leach & Roller, 2010); PKCδ is required for porcine reproductive and respiratory syndrome virus replication (Zhao, Guo, Bi, Zhu, & Feng, 2014); influenza A virus requires PKC for endosomal escape, nuclear entry, and assembly (Mondal et al., 2017; Root, Wills, McNair, & Whittaker, 2000; Sieczkarski, Brown, & Whittaker, 2003); minute virus of mice replication is regulated by aPKC (Nuesch, Christensen, & Rommelaere, 2001); respiratory syncytial virus (Monick, Staber, Thomas, & Hunninghake, 2001), human herpesvirus (Naranatt, Akula, Zien, Krishnan, & Chandran, 2003), rabies virus (Gupta, Blondel, Choudhary, & Banerjee, 2000), and human parainfluenza virus (De, Gupta, Gupta, & Banerjee, 1995) replication is also influenced by different PKC isoforms. Additionally, cellular entry of some human adenovirus serotypes has been described to be sensitive to PKC inhibition (Yousuf et al., 2016). In this regard, PKC has been reported to be critical for the entry of enveloped viruses by receptor-mediated endocytosis (Constantinescu, Cernescu, & Popescu, 1991).

However, PKC can result in essential cell host factors restricting some virus replication. In this sense, hepatitis E virus replication was enhanced when conventional PKCα was inhibited either by pharmacological inhibition or by Sh-RNA mediated knockdown of the protein, and viral replication was inhibited when the protein was pharmacologically activated (Wang et al., 2017). Accordingly, activation of PKCs has been reported to inhibit replication in viruses like chikungunya (Remy & Litaudon, 2019) or spring viremia of carp virus (Dang, Qiu, Song, & Liu, 2019). Works on human immunodeficiency virus presented some controversial results, with studies reporting the importance of phosphorylated PKC for viral fusion and assembly (Contreras, Mzoughi, Gaston, Peterlin, & Bahraoui, 2012; Kudoh et al., 2014; Yu et al., 1995); and reports showing in vitro inhibition of virus replication by PKC natural activators (Huang et al., 2014; Matsuya et al., 2005; Nothias-Scaglia et al., 2015).

PKC implications for Zika virus and other related flaviviruses

Flaviviruses are relevant animal and human pathogens that include Zika virus (ZIKV), West Nile virus (WNV), and dengue virus (DENV), among others, and for which treatments are not currently available.

At present, as far as we know, no reports concerning the role of PKC in ZIKV infections are available in the literature. Only one study associates the loss of Ankle2, a protein linked to microcephaly in humans and known to interact with the NS4A protein of ZIKV, with the reduction of aPKC phosphorylation in *Drosophila* neuroblasts (Link et al., 2019). However, studies performed with other related flaviviruses may help to understand the possible role of PKCs in ZIKV infection. As happened with viruses of other families, diverse results have been reported with flaviviruses, but the specific role of PKCs has not yet been assessed. For instance, it has been described that inhibition of all PKC isoforms affected WNV entry into mosquito cells (Chu, Leong, & Ng, 2006) and that pharmacological inhibition of atypical isoenzymes decreased WNV in vitro infection, while inhibition of classical and novel PKCs did not show any effect (Blazquez et al., 2018). Also, PKC phosphorylation of the C viral protein has been described as the key regulator of apoptosis induction during WNV infection (Bhuvanakantham, Cheong, & Ng, 2010), and involvement of PKCs on JEV infection has also been described (Chen et al., 2012). In this line, treatment with the PKC inhibitor bisindolylmaleimide I showed a decrease in HCV replication (Murakami et al., 2009). However, conversely, DENV replication was enhanced when the same pharmacological inhibitor was used, even though viral entry to the host cells was not increased (Noppakunmongkolchai et al., 2016). At present, DENV is the only described member of this family in which PKC seems to act as a restricting mechanism that modulates replication.

Modulation of PKC for the control of viruses

As previously mentioned, PKCs can be exploited by viruses for efficient replication by facilitating viral entry and infection. However, the same kinases can be involved in the antiviral response in host cells. This different behavior requires a careful knowledge of the specific relationships that occur between PKC isoforms and multiplication of viruses, as modulation of the PKC pathway can be a suitable mechanism for the design of antiviral drugs. In fact, pharmacological modulation of PKCs has been achieved with a wide range of compounds including activators, small-molecule inhibitors, and peptides, as well as by the genetic manipulation of PKC (Wu-Zhang & Newton, 2013).

In the case of ZIKV, since there are still no specific investigations on the effect of PKC on its infection, this could be inferred from other flaviviruses. However, even though most of the studied members of the family show a direct effect of PKC at multiple stages during infection, in the case of DENV, PKC activation plays a role in the antiviral response in host cells, thus making it necessary to precisely assess the role that modulation of PKC has for the control of ZIKV.

Conclusions

PKC isoenzymes have been related to a wide variety of cellular processes, and the involvement of multiple PKC isoforms at different stages during viral infections has also been studied in a considerable range of viruses. In fact, the association of PKCs with a variety of both physio and pathological processes, either upregulating or downregulating them, makes this signaling pathway attractive as drug targets for broad-spectrum antiviral therapy against flavivirus, including ZIKV. Therefore, a deeper analysis deciphering the role of PKCs on ZIKV infection deserves further investigations.

Policy and procedures

PKCs structure and isoform roles

The differences in the PKC domains structure are crucial for their mechanism of action and the myriad of functions ascribed to PKCs. The presence of three diverse isoforms of PKC in rat brain was first demonstrated by the hydroxylapatite column chromatography technique, and later on two-dimensional peptide mapping analysis revealed the different sites of autophosphorylation shown by these kinases. Understanding the three-dimensional structure of PKC is a main target for an exhaustive knowledge of the activation mechanisms of PKC. Crystal structures of PKCs have been determined by X-ray diffraction and other similar techniques. To see details of already published crystal structures of some PKC domains, please consult Protein Data Bank: http://www.rcsb.org/pdb/results/results.do?tabtoshow=Current&qrid=B820B8C1.

Mini-dictionary of terms

Kinases. Enzymes that catalyze the phosphorylation of diverse molecules.
Isoenzymes. Enzymes with a different structure but the same function.
Cofactor. Nonproteic chemical component that helps in the activation of a chemical reaction.
Viral replication. Production of new viral genomes during an infection.
Flaviviruses. A family of single-stranded positive RNA viruses transmitted by arthropods, many of which cause severe human and animal illnesses.

Key facts of protein kinases C

- Protein kinase C family comprises 11 different isoforms, divided into three groups.
- Protein kinases C can be activated by different mechanisms: phosphorylation or second messengers.
- Protein kinases C are implicated in a wide variety of cellular processes.
- Protein kinases C play a role in viral replication.
- Protein kinases C are being studied as a target for antiviral therapy.

Summary points

- The role of protein kinase C has been related to a wide range of cellular processes, as well as in different steps during viral replication.
- Protein kinases C classification is highly related to their structure and cofactor activation requirements.
- A wide variety of substrates are susceptible to the action of this family of proteins, and modulation of protein kinase C activity is important for different physiological and pathological processes.
- The role of protein kinase C on viral infections has been studied for some viral families, but little is known about it during flaviviral infections.
- The association of this signaling pathway with a variety of processes makes it attractive as drug targets for broad-spectrum antiviral therapy against ZIKV and other viruses.

References

An, B., Pan, T., Hu, J., Pang, Y., Huang, L., Chan, A. S. C., ... Yan, J. (2019). The discovery of a potent and selective third-generation EGFR kinase inhibitor as a therapy for EGFR L858R/T790M double mutant non-small cell lung cancer. *European Journal of Medicinal Chemistry, 183*. https://doi.org/10.1016/j.ejmech.2019.111709, 111709.

Bell, R. M., & Burns, D. J. (1991). Lipid activation of protein kinase C. *The Journal of Biological Chemistry, 266*(8), 4661–4664.

Bhuvanakantham, R., Cheong, Y. K., & Ng, M. L. (2010). West Nile virus capsid protein interaction with importin and HDM2 protein is regulated by protein kinase C-mediated phosphorylation. *Microbes and Infection, 12*(8–9), 615–625. https://doi.org/10.1016/j.micinf.2010.04.005.

Blazquez, A. B., Vazquez-Calvo, A., Martin-Acebes, M. A., & Saiz, J. C. (2018). Pharmacological inhibition of protein kinase C reduces West Nile virus replication. *Viruses, 10*(2). https://doi.org/10.3390/v10020091.

Brodie, C., & Blumberg, P. M. (2003). Regulation of cell apoptosis by protein kinase C delta. *Apoptosis, 8*(1), 19–27. https://doi.org/10.1023/a:1021640817208.

Chen, C. J., Ou, Y. C., Chang, C. Y., Pan, H. C., Liao, S. L., Chen, S. Y., ... Lai, C. Y. (2012). Glutamate released by Japanese encephalitis virus-infected microglia involves TNF-alpha signaling and contributes to neuronal death. *Glia, 60*(3), 487–501. https://doi.org/10.1002/glia.22282.

Chu, J. J., Leong, P. W., & Ng, M. L. (2006). Analysis of the endocytic pathway mediating the infectious entry of mosquito-borne flavivirus West Nile into Aedes albopictus mosquito (C6/36) cells. *Virology, 349*(2), 463–475. https://doi.org/10.1016/j.virol.2006.01.022.

Constantinescu, S. N., Cernescu, C. D., & Popescu, L. M. (1991). Effects of protein kinase C inhibitors on viral entry and infectivity. *FEBS Letters, 292* (1–2), 31–33. https://doi.org/10.1016/0014-5793(91)80826-o.

Contreras, X., Mzoughi, O., Gaston, F., Peterlin, M. B., & Bahraoui, E. (2012). Protein kinase C-delta regulates HIV-1 replication at an early post-entry step in macrophages. *Retrovirology, 9*, 37. https://doi.org/10.1186/1742-4690-9-37.

Coussens, L., Parker, P. J., Rhee, L., Yang-Feng, T. L., Chen, E., Waterfield, M. D., ... Ullrich, A. (1986). Multiple, distinct forms of bovine and human protein kinase C suggest diversity in cellular signaling pathways. *Science, 233*(4766), 859–866. https://doi.org/10.1126/science.3755548.

Dang, Y. F., Qiu, T. X., Song, D. W., & Liu, L. (2019). PMA-triggered PKCepsilon activity enhances Nrf2-mediated antiviral response on fish rhabdovirus infection. *Fish & Shellfish Immunology, 87*, 871–878. https://doi.org/10.1016/j.fsi.2019.02.033.

De, B. P., Gupta, S., Gupta, S., & Banerjee, A. K. (1995). Cellular protein kinase C isoform zeta regulates human parainfluenza virus type 3 replication. *Proceedings of the National Academy of Sciences of the United States of America, 92*(11), 5204–5208. https://doi.org/10.1073/pnas.92.11.5204.

Gupta, A. K., Blondel, D., Choudhary, S., & Banerjee, A. K. (2000). The phosphoprotein of rabies virus is phosphorylated by a unique cellular protein kinase and specific isomers of protein kinase C. *Journal of Virology, 74*(1), 91–98. https://doi.org/10.1128/jvi.74.1.91-98.2000.

Huang, K. P., Nakabayashi, H., & Huang, F. L. (1986). Isozymic forms of rat brain Ca^{2+}-activated and phospholipid-dependent protein kinase. *Proceedings of the National Academy of Sciences of the United States of America, 83*(22), 8535–8539. https://doi.org/10.1073/pnas.83.22.8535.

Huang, S. Z., Zhang, X., Ma, Q. Y., Peng, H., Zheng, Y. T., Hu, J. M., ... Zhao, Y. X. (2014). Anti-HIV-1 tigliane diterpenoids from Excoecaria acertiflia Didr. *Fitoterapia, 95*, 34–41. https://doi.org/10.1016/j.fitote.2014.02.018.

Isakov, N. (2018). Protein kinase C (PKC) isoforms in cancer, tumor promotion and tumor suppression. *Seminars in Cancer Biology, 48*, 36–52. https://doi.org/10.1016/j.semcancer.2017.04.012.

Kudoh, A., Takahama, S., Sawasaki, T., Ode, H., Yokoyama, M., Okayama, A., ... Ryo, A. (2014). The phosphorylation of HIV-1 Gag by atypical protein kinase C facilitates viral infectivity by promoting Vpr incorporation into virions. *Retrovirology, 11*, 9. https://doi.org/10.1186/1742-4690-11-9.

Leach, N. R., & Roller, R. J. (2010). Significance of host cell kinases in herpes simplex virus type 1 egress and lamin-associated protein disassembly from the nuclear lamina. *Virology, 406*(1), 127–137. https://doi.org/10.1016/j.virol.2010.07.002.

Link, N., Chung, H., Jolly, A., Withers, M., Tepe, B., Arenkiel, B. R., ... Bellen, H. J. (2019). Mutations in ANKLE2, a ZIKA virus target, disrupt an asymmetric cell division pathway in drosophila neuroblasts to cause microcephaly. *Developmental Cell, 51*(6), 713–729. e716 https://doi.org/10.1016/j.devcel.2019.10.009.

Mackay, H. J., & Twelves, C. J. (2007). Targeting the protein kinase C family: Are we there yet? *Nature Reviews. Cancer, 7*(7), 554–562. https://doi.org/10.1038/nrc2168.

Matsuya, Y., Yu, Z., Yamamoto, N., Mori, M., Saito, H., Takeuchi, M., … Nemoto, H. (2005). Synthesis of new phorbol derivatives having ethereal side chain and evaluation of their anti-HIV activity. *Bioorganic & Medicinal Chemistry*, *13*(14), 4383–4388. https://doi.org/10.1016/j.bmc.2005.04.056.

Merida, I., Arranz-Nicolas, J., Rodriguez-Rodriguez, C., & Avila-Flores, A. (2019). Diacylglycerol kinase control of protein kinase C. *The Biochemical Journal*, *476*(8), 1205–1219. https://doi.org/10.1042/BCJ20180620.

Mochly-Rosen, D., Das, K., & Grimes, K. V. (2012). Protein kinase C, an elusive therapeutic target? *Nature Reviews. Drug Discovery*, *11*(12), 937–957. https://doi.org/10.1038/nrd3871.

Mondal, A., Dawson, A. R., Potts, G. K., Freiberger, E. C., Baker, S. F., Moser, L. A., … Mehle, A. (2017). Influenza virus recruits host protein kinase C to control assembly and activity of its replication machinery. *eLife*, *6*. https://doi.org/10.7554/eLife.26910.

Monick, M., Staber, J., Thomas, K., & Hunninghake, G. (2001). Respiratory syncytial virus infection results in activation of multiple protein kinase C isoforms leading to activation of mitogen-activated protein kinase. *Journal of Immunology*, *166*(4), 2681–2687. https://doi.org/10.4049/jimmunol.166.4.2681.

Murakami, Y., Noguchi, K., Yamagoe, S., Suzuki, T., Wakita, T., & Fukazawa, H. (2009). Identification of bisindolylmaleimides and indolocarbazoles as inhibitors of HCV replication by tube-capture-RT-PCR. *Antiviral Research*, *83*(2), 112–117. https://doi.org/10.1016/j.antiviral.2009.03.008.

Naranatt, P. P., Akula, S. M., Zien, C. A., Krishnan, H. H., & Chandran, B. (2003). Kaposi's sarcoma-associated herpesvirus induces the phosphatidylinositol 3-kinase-PKC-zeta-MEK-ERK signaling pathway in target cells early during infection: Implications for infectivity. *Journal of Virology*, *77*(2), 1524–1539. https://doi.org/10.1128/jvi.77.2.1524-1539.2003.

Newton, A. C. (1995). Protein kinase C: Structure, function, and regulation. *The Journal of Biological Chemistry*, *270*(48), 28495–28498. https://doi.org/10.1074/jbc.270.48.28495.

Newton, A. C. (2018a). Protein kinase C as a tumor suppressor. *Seminars in Cancer Biology*, *48*, 18–26. https://doi.org/10.1016/j.semcancer.2017.04.017.

Newton, A. C. (2018b). Protein kinase C: Perfectly balanced. *Critical Reviews in Biochemistry and Molecular Biology*, *53*(2), 208–230. https://doi.org/10.1080/10409238.2018.1442408.

Nishizuka, Y. (1988). The molecular heterogeneity of protein kinase C and its implications for cellular regulation. *Nature*, *334*(6184), 661–665. https://doi.org/10.1038/334661a0.

Noppakunmongkolchai, W., Poyomtip, T., Jittawuttipoka, T., Luplertlop, N., Sakuntabhai, A., Chimnaronk, S., … Tohtong, R. (2016). Inhibition of protein kinase C promotes dengue virus replication. *Virology Journal*, *13*, 35. https://doi.org/10.1186/s12985-016-0494-6.

Nothias-Scaglia, L. F., Pannecouque, C., Renucci, F., Delang, L., Neyts, J., Roussi, F., … Paolini, J. (2015). Antiviral activity of diterpene esters on chikungunya virus and HIV replication. *Journal of Natural Products*, *78*(6), 1277–1283. https://doi.org/10.1021/acs.jnatprod.5b00073.

Nuesch, J. P., Christensen, J., & Rommelaere, J. (2001). Initiation of minute virus of mice DNA replication is regulated at the level of origin unwinding by atypical protein kinase C phosphorylation of NS1. *Journal of Virology*, *75*(13), 5730–5739. https://doi.org/10.1128/JVI.75.13.5730-5739.2001.

Ohno, S., Konno, Y., Akita, Y., Yano, A., & Suzuki, K. (1990). A point mutation at the putative ATP-binding site of protein kinase C alpha abolishes the kinase activity and renders it down-regulation-insensitive. A molecular link between autophosphorylation and down-regulation. *The Journal of Biological Chemistry*, *265*(11), 6296–6300.

Remy, S., & Litaudon, M. (2019). Macrocyclic diterpenoids from Euphorbiaceae as a source of potent and selective inhibitors of chikungunya virus replication. *Molecules*, *24*(12). https://doi.org/10.3390/molecules24122336.

Reyland, M. E. (2009). Protein kinase C isoforms: Multi-functional regulators of cell life and death. *Frontiers in Bioscience (Landmark Edition)*, (14), 2386–2399. https://doi.org/10.2741/3385.

Root, C. N., Wills, E. G., McNair, L. L., & Whittaker, G. R. (2000). Entry of influenza viruses into cells is inhibited by a highly specific protein kinase C inhibitor. *The Journal of General Virology*, *81*(Pt 11), 2697–2705. https://doi.org/10.1099/0022-1317-81-11-2697.

Sieczkarski, S. B., Brown, H. A., & Whittaker, G. R. (2003). Role of protein kinase C betaII in influenza virus entry via late endosomes. *Journal of Virology*, *77*(1), 460–469. https://doi.org/10.1128/jvi.77.1.460-469.2003.

Singh, R. K., Kumar, S., Gautam, P. K., Tomar, M. S., Verma, P. K., Singh, S. P., … Acharya, A. (2017). Protein kinase C-alpha and the regulation of diverse cell responses. *Biomolecular Concepts*, *8*(3–4), 143–153. https://doi.org/10.1515/bmc-2017-0005.

Sontag, E., Sontag, J. M., & Garcia, A. (1997). Protein phosphatase 2A is a critical regulator of protein kinase C zeta signaling targeted by SV40 small t to promote cell growth and NF-kappaB activation. *The EMBO Journal*, *16*(18), 5662–5671. https://doi.org/10.1093/emboj/16.18.5662.

Steinberg, S. F. (2004). Distinctive activation mechanisms and functions for protein kinase Cdelta. *The Biochemical Journal*, *384*(Pt 3), 449–459. https://doi.org/10.1042/BJ20040704.

Takai, Y., Kishimoto, A., Iwasa, Y., Kawahara, Y., Mori, T., & Nishizuka, Y. (1979). Calcium-dependent activation of a multifunctional protein kinase by membrane phospholipids. *The Journal of Biological Chemistry*, *254*(10), 3692–3695.

Tobias, I. S., Kaulich, M., Kim, P. K., Simon, N., Jacinto, E., Dowdy, S. F., … Newton, A. C. (2016). Protein kinase Czeta exhibits constitutive phosphorylation and phosphatidylinositol-3,4,5-triphosphate-independent regulation. *The Biochemical Journal*, *473*(4), 509–523. https://doi.org/10.1042/BJ20151013.

Wang, T., Liu, C., & Jia, L. (2018). The roles of PKCs in regulating autophagy. *Journal of Cancer Research and Clinical Oncology*, *144*(12), 2303–2311. https://doi.org/10.1007/s00432-018-2731-4.

Wang, W., Wang, Y., Debing, Y., Zhou, X., Yin, Y., Xu, L., … Pan, Q. (2017). Biological or pharmacological activation of protein kinase C alpha constrains hepatitis E virus replication. *Antiviral Research*, *140*, 1–12. https://doi.org/10.1016/j.antiviral.2017.01.005.

Webb, B. L., Hirst, S. J., & Giembycz, M. A. (2000). Protein kinase C isoenzymes: A review of their structure, regulation and role in regulating airways smooth muscle tone and mitogenesis. *British Journal of Pharmacology*, *130*(7), 1433–1452. https://doi.org/10.1038/sj.bjp.0703452.

Wu-Zhang, A. X., & Newton, A. C. (2013). Protein kinase C pharmacology: Refining the toolbox. *The Biochemical Journal*, *452*(2), 195–209. https://doi.org/10.1042/BJ20130220.

Yousuf, M. A., Lee, J. S., Zhou, X., Ramke, M., Lee, J. Y., Chodosh, J., & Rajaiya, J. (2016). Protein kinase C signaling in adenoviral infection. *Biochemistry*, *55*(42), 5938–5946. https://doi.org/10.1021/acs.biochem.6b00858.

Yu, G., Shen, F. S., Sturch, S., Aquino, A., Glazer, R. I., & Felsted, R. L. (1995). Regulation of HIV-1 gag protein subcellular targeting by protein kinase C. *The Journal of Biological Chemistry*, *270*(9), 4792–4796. https://doi.org/10.1074/jbc.270.9.4792.

Zhao, H., Guo, X. K., Bi, Y., Zhu, Y., & Feng, W. H. (2014). PKCdelta is required for porcine reproductive and respiratory syndrome virus replication. *Virology*, *468–470*, 96–103. https://doi.org/10.1016/j.virol.2014.07.040.

Chapter 40

Nanotechnology applied in the control and diagnosis of *Zika virus* and its vectors

Gabriel Augusto Pires de Souza[a], Lívia Sacchetto[a], Betânia Paiva Drumond[a], Jônatas Santos Abrahão[a], Túlio César Rodrigues Leite[b], Breno de Mello Silva[c], Anna Carolina Toledo da Cunha Pereira[d], Gustavo Portela Ferreira[d], Luiz Cosme Cotta Malaquias[e], and Luiz Felipe Leomil Coelho[e]

[a]*Department of Microbiology, Federal University of Minas Gerais, Belo Horizonte, Minas Gerais, Brazil,* [b]*Biological Sciences Research Center, Federal University of Ouro Preto, Ouro Preto, Minas Gerais, Brazil,* [c]*Department of Biological Sciences, Federal University of Ouro Preto, Ouro Preto, Minas Gerais, Brazil,* [d]*Federal University of the Delta of Parnaíba, Parnaíba, Piauí, Brazil,* [e]*Vaccine Laboratory, Federal University of Alfenas, Alfenas, Minas Gerais, Brazil*

Abbreviations

Ag-NPs	silver nanoparticles
Au-NPs	gold nanoparticles
EPS	microbial exopolysaccharide
NPs	nanoparticles
NS1	nonstructural protein 1
OMVs	outer membrane vesicles
VLPs	virus-like particles
ZIKV	*Zika virus*
ZnO-NPs	zinc oxide nanoparticles

Introduction

Nanomedicine is a new and emerging subject of research that aims to use nanotechnology with the pharmaceutical and biomedical sciences. The main purpose of nanomedicine is to use nanoparticles to find new ways to deliver drugs and antigens, perform diagnosis, and improve safety and toxicity (Ventola, 2012). Nanoparticles (NPs) are defined by particles made of organic or inorganic polymers/materials with length of 1–1000 nm in at least one dimension. There is no universal definition about NPs, and the definition of what is NPs can vary between country legislations (Jeevanandam, Barhoum, Chan, Dufresne, & Danquah, 2018). The use of NPs to develop new vaccines, diagnostic tests, and treatments to infectious diseases is a new alternative to solve important problems in some diseases caused by viruses. The use of NPs could improve the antiviral activity of some compounds and improve the diagnostic and the immune response triggered by vaccination (Singh, Kruger, Maguire, Govender, & Parboosing, 2017; Sposito, Kurdekar, Zhao, & Hewlett, 2018; Sulczewski, Liszbinski, Romão, & Rodrigues Junior, 2018).

Nanotechnological strategies for *Zika virus* diagnosis

The use of nanotechnology to develop new and robust diagnostic devices is an alternative to *Zika virus* diagnosis. Accurate diagnosis of the *Zika virus* (ZIKV) is very important because some viral infections such as chikungunya, dengue, and yellow fever have similar symptoms, such as joint pain, rash, and fever. Therefore the diagnosis of ZIKV-infected individuals is a hard task due to nonspecific clinical symptoms. More than 80% of cases have clinical symptoms similar to other arboviruses (Martinez, de la Garza, & Cuellar-Barboza, 2019). Several diagnostic techniques were employed for ZIKV detection based on molecular and immunological techniques (Singh et al., 2018). However, some of these techniques

Zika Virus Impact, Diagnosis, Control, and Models. https://doi.org/10.1016/B978-0-12-820267-8.00040-6

are expensive, highly technical, and time consuming. These aspects turn the ZIKV diagnosis a challenge in countries with poor healthcare infrastructure. Therefore the search for new diagnostic alternatives for ZIKV is very important. Nanotechnology is a promising science that can help researchers to develop more sensitive, accurate, fast, and affordable methods to diagnose ZIKV infections. Several nanobiotechnology approaches were tested to improve the ZIKV diagnosis on biological samples (Table 1). All nanotechnology proposed strategies aim to improve ZIKV detection using methods that are able to detect the viral particle, viral genome, NS1 protein, or antibodies against ZIKV (Fig. 1).

Surface-modified microbeads and platinum NPs both coupled with monoclonal antibodies against ZIKV envelope protein were used in a sandwich assay for rapid detection of ZIKV particles in a microchip. These NPs are known for their catalytic properties in H_2O_2 that enable self-propelling by converting chemical energy in mechanical motion due to bubbling formation in a microchip. Such bubbles were detected by an application in a nanomotor-based bead motion cellphone system. The system was able to detect ZIKV on artificial urine and saliva containing ZIKV-spiked viruses and also on real serum samples (Draz, Lakshminaraasimulu, et al., 2018). A microchip assay using platinum NPs uses electrical signals in a paper microchip assay after laser stimulation. Virus particles were captured on the surface of microbeads with specific antibodies against the envelope protein of ZIKV, and a sandwich-like complex is formed by adding platinum NPs. Then, these complexes are isolated, lysated, and added on the paper microchip to perform the diagnosis (Draz, Venkataramani, et al., 2018).

Regarding ZIKV viral genome, a biosensor for detection of ZIKV genome using oxidized glassy carbon electrode modified with silsesquioxane-functionalized gold nanoparticles coupled to single-strand DNA complementary to ZIKV genome was described (Steinmetz et al., 2019). In this case, ZIKV genome bound to the device and the signal was detected by a Multipotenciostat. Another biosensor based on DNA origami is described as a platform to detect ZIKV genome. This optical DNA nanoantenna technology is composed of a DNA origami coupled with silver NPs and a molecular beacon-like

TABLE 1 New diagnostic tests for *Zika virus* based on nanotechnology.

Study	Nanotechnology	Target	Limit of detection
Draz, Lakshminaraasimulu, et al. (2018)	A microchip-based on surface modified microbeads and platinum nanoparticles both coupled with monoclonal antibodies against ZIKV	Envelope protein	1 particle/μL
Draz, Venkataramani, et al. (2018)	A paper microchip assay using platinum nanoparticles to detect ZIKV using electrical signals after laser stimulation	Envelope protein	10^1–10^2 particles/μL
Ochmann et al. (2017)	An optical DNA nanoantenna technology composed by a DNA origami coupled to silver nanoparticles and a molecular beacon-like structure	Viral genome	1 nM
Steinmetz et al. (2019)	A DNA biosensor based on an oxidized glassy carbon electrode modified with silsesquioxane-functionalized gold nanoparticles	Viral genome	0.82 pmol/L
Takemura, Adegoke, Suzuki, and Park (2019)	An immunofluorescence biosensor based on localized surface plasmon resonance signal from gold nanoparticles was used to amplify the fluorescence intensity signal of quantum dots	NS1	8.2–32 copies/mL
Chu et al. (2019)	A sandwich immunoassay based on laser desorption/ionization mass spectroscopy principle using gold nanoparticles	NS1	1 nM
Camacho, Sobral-Filho, Aoki, Constantino, and Brolo (2018)	Immunoassay based on surface-enhanced Raman scattering nanoprobes made by gold nanoshells covered with anti-NS1 antibodies	NS1	12.5 ng/mL
Zhang et al. (2017)	A multiplexed assay based on surface plasmon resonance signal from gold nanoparticles	IgG and IgA antibodies against NS1	Not reported
Ribeiro et al. (2019)	Immunofluorescence assay for detection of anti-ZIKV antibodies based on quantum-dots	IgG antibodies against envelope protein	1–2 μg/wel

NS1, nonstructural protein 1.

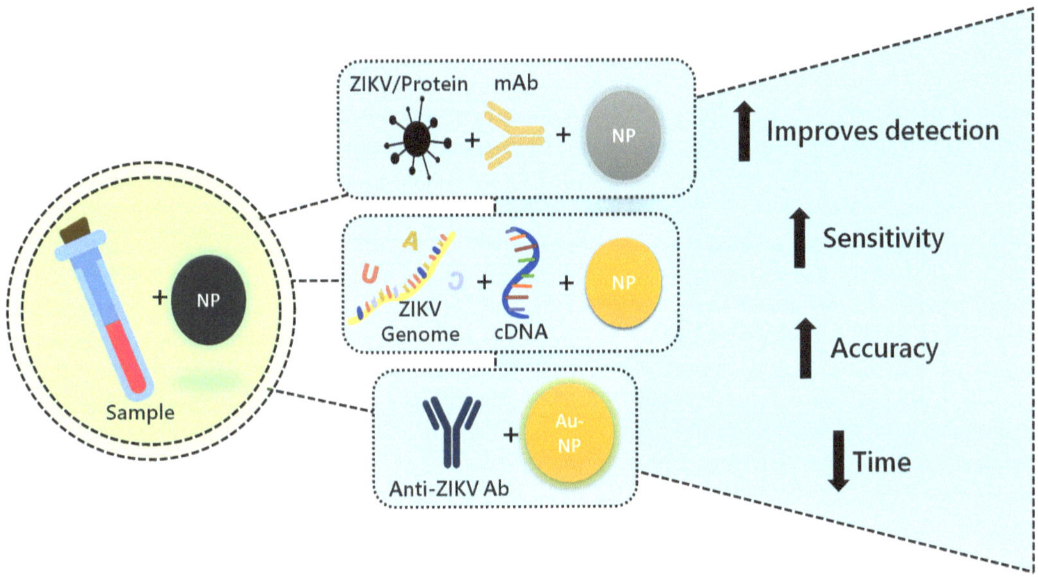

FIG. 1 Nanoparticles and earlier detection and diagnosis of ZIKV. Diagnostic tests based on nanoparticles could be used to improve the detection of ZIKV antigens (envelope or NS1 proteins), ZIKV genome or anti-ZIKV antibodies.

structure. In this case, ZIKV RNA can be detected in biological samples with no need for molecular multiplication. The high fluorescence signal is the result of the fluorescence signals present on a molecular beacon-like structure enhanced by plasmonic effects (Ochmann et al., 2017).

For nonstructural protein 1 (NS1), the main approach is to use NPs to increase the fluorescence signal derived from positive reactions. An immunofluorescence biosensor based on gold NPs and surface plasmon resonance-amplified immunofluorescence showed a rapid and sensitive detection of NS1 of ZIKV (Takemura et al., 2019). In this case the fluorescence signal of quantum dots was amplified using the gold NPs and surface plasmon resonance. NS1 antigen from ZIKV was also detected using the Raman effect from gold nanoshells. These NPs consist of a core of gold covered by a thin layer of silica, Nile blue as Raman reporter, and then a final shell of silica. Anti-NS1 antibodies placed onto coverslips act as a support to capture NS1 antigens. After the addition of nanoprobes covered with anti-NS1 and under excitation with a 633-nm laser, surface-enhanced resonance Raman scattering signal was obtained (Camacho et al., 2018). Another immunoassay to detect ZIKV NS1 was described using laser desorption/ionization mass spectroscopy principle. In this case the cellulose acetate membrane strips were coated with antibodies against NS1. Then, the strips were placed into a culture plate, and a sandwich-like assay was performed by adding samples containing 1.0×10^5 PFU/mL of ZIKV followed by the addition of gold NPs coupled with monoclonal antibodies against ZIKV (Chu et al., 2019).

Antibodies against ZIKV are important biomarkers for ZIKV infection. A multiplexed assay on a plasmonic gold platform using NS1 antigen from ZIKV and *dengue virus* for measuring IgG and IgA antibodies in real samples from ZIKV- and DENV-infected individuals was developed (Zhang et al., 2017). This test was able to differentiate between these two infections, especially on regions with cocirculation of these two arboviruses. In a similar way, indirect immunoassay based on fluorescent quantum dots showed a high sensibility to detect anti-ZIKV antibodies (Ribeiro et al., 2019). Besides the great performance of all new diagnostic tests described earlier, most of them fail to use human samples derived from ZIKV-infected patients. Future studies should be done to validate the diagnostic test using samples derived from ZIKV-infected individuals.

Nanotechnological strategies for control of ZIKV vectors

Mosquito control is critical in the prevention of outbreaks of Zika and other mosquito-borne diseases. In endemic countries, preventive measures aiming at control of ZIKV vectors, like *Aedes aegypti* (primary vector) and *Aedes albopictus* (secondary vector) are unsuccessfully used given the geographic expansion and high *A. aegypti* infestation (Boyer, Calvez, Chouin-Carneiro, Diallo, & Failloux, 2018). These procedures include the use of air conditioning, window/door screens, insect repellent, and insecticides and the removal of household debris, water containers, or other mosquito breeding sites (Benelli, 2016). Additionally the application of insecticides has created many health and environmental issues, such as mosquito resistance, environmental pollution, nontarget organism toxicity, and impact on aquatic species (Benelli, 2016, 2018).

Nanotechnology represents a revolutionary strategy to address the challenges in the control of mosquitoes. Several studies describing nanoparticle toxicity for various economically important arthropod species have been published, many based on the so-called green synthesis (Benelli, 2018). The green synthesis of NPs is advantageous over chemical and physical methods since it is cheap and single step and does not require high pressure, energy, temperature, or the use of highly toxic chemicals (Abinaya, Vaseeharan, Divya, Vijayakumar, et al., 2018). Green sources such as microorganisms (fungi and bacteria), enzymes, and plants can be used for stabilization and as reducing agents for the synthesis of metallic or metallic oxides NPs (Ishwarya et al., 2018; Selvan et al., 2018) (Fig. 2).

Silver nanoparticles (Ag-NPs) have been used in the majority of the studies followed by gold nanoparticles (Au-NPs). Silver nanostructures produced with seed extract of *Pedalium murex*, an ancient Indian medicinal plant, induced severe damages in hindgut and muscles of *A. aegypti* fourth instar larvae (Ishwarya et al., 2017). *A. aegypti* fourth instar larvae exposed for 24h to *Hedychium coronarium* Ag-NPs led to partial lyses of the midgut epithelial cells (Kalimuthu et al., 2017). In a similar study, Ag-NPs, biosynthesized using leaf extract of *Hugonia mystax*, exhibited high larvicidal action against *A. aegypti* third instar larvae (Govindarajan, Kadaikunnan, Alharbi, & Benelli, 2017). Ag-NPs fabricated with *Artemisia herba-alba* showed high larvicidal and adulticidal toxicity against Indian and Saudi Arabian strains of *A. aegypti* (Aziz et al., 2018). A simple fabrication process of polydisperse and stable Ag-NPs using the leaf extract of *Ichnocarpus frutescens* was also reported. The acute toxicity of these Ag-NPs was higher against *A. albopictus* compared with the leaf extract alone (Govindarajan, Hoti, & Benelli, 2016). Ag-NPs synthesized using the fruit pulp extract of *Cassia fistula* led to a significant decrease in total protein levels of *A. albopictus* fourth instar larvae (Fouad et al., 2018). A salicylic acid and 3,5-dinitrosalicylic acid-synthesized Ag-NPs showed a larvicidal effect on *A. albopictus* fourth instar larvae. The results showed that larvae treated with this formulation presented a relative decrease in total proteins, acetylcholinesterase, and phosphatase enzymes (Ga'al et al., 2018). Ag-NPs synthesized using *Habenaria plantaginea* leaf extract showed larvicidal activity against *A. aegypti* and *A. albopictus* third instar larvae (Aarthi et al., 2018).

In the literature, little has been described as the mode of action of Au-NPs against insects when compared with Ag-NPs. Treating *A. aegypti* larvae (third and fourth instar) with Au-NPs fabricated with the *Artemisia vulgaris* leaf extract led to the midgut, epithelial cell, and cortex damages, with an accumulation of Au-NPs in the midgut region (Sundararajan & Ranjitha Kumari, 2017). Au-NPs fabricated with *Cymbopogon citratus* leaf extract have been reported as toxic to *A. aegypti* (Murugan et al., 2015). Zein-coated Au-NPs had high toxic activity on *A. aegypti* larvae, with microscopy and histology

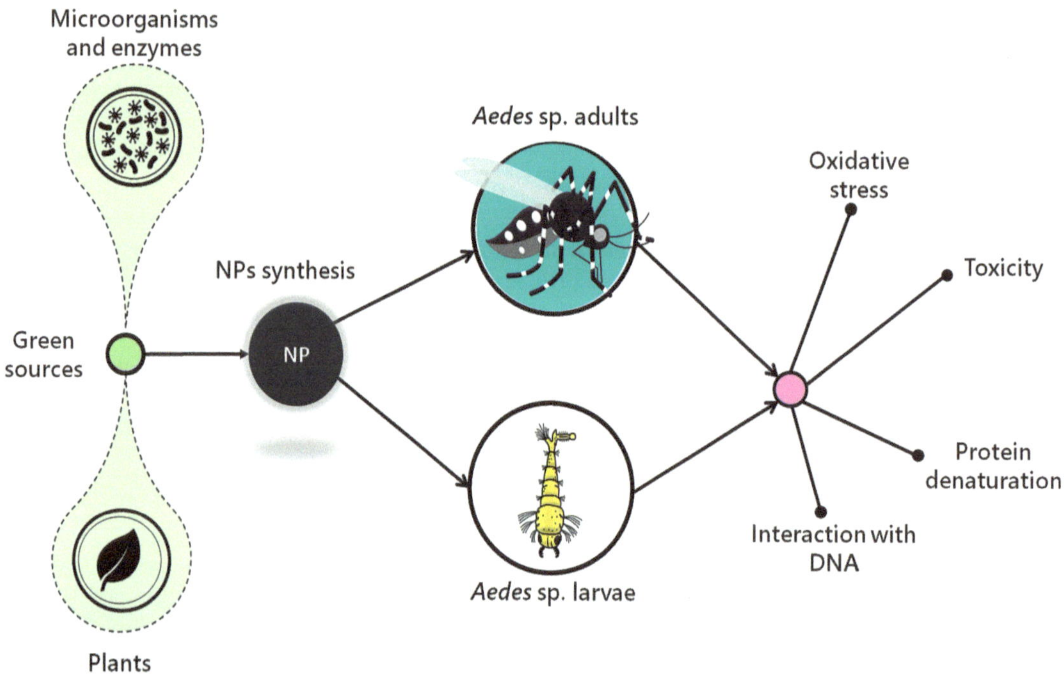

FIG. 2 Nanoparticles and control of ZIKV vectors. Metallic nanoparticles produced with green synthesis technology could induce toxic effects on larvae and adult *Aedes* sp. mosquitoes.

analyses showing the disintegration of the abdominal region (midgut and caeca) and the loss of lateral and caudal hairs (Suganya et al., 2017). Au-NPs can also be synthesized using microbial filtrates or biopolymers. Au-NP produced with microbial filtrate of the pathogenic fungus *Chrysosporium tropicum* induced 100% of mortality after 24h of exposure of *A. aegypti* first instar larvae (Soni & Prakash, 2012).

Copper-based nanostructured materials have efficient biocidal properties, commonly used in catalysis, pesticide formulations, and various health-related applications. Copper oxide NPs produced with *Tridax procumbens* leaf extract showed 100% of mortality against *A. aegypti* fourth instar larvae when administered at concentrations of 10μg/mL for 24h (Selvan et al., 2018). Zinc oxide nanoparticles (ZnO-NPs) produced with *Ulva lactuca* deposited in *A. aegypti* fourth instar larvae induce disintegration of the epithelial layer and outer cuticle, while the gastric ceca, muscles, and ganglia of the nervous cord seemed to be damaged and disorganized (Ishwarya et al., 2018).

Microbial exopolysaccharide (EPS) is a high molecular weight macromolecule that exhibits NP binding properties, promoting suspension stability by inhibiting aggregation. ZnO-NPs produced from coprecipitation with EPS induced 99% mortality of *A. aegypti* fourth instar larvae with the occurrence of the disintegration of the epithelial layer and outer cuticle, darken body, and destruction of the stigmal plate on the siphon apex (Abinaya, Vaseeharan, Divya, Sharmili, et al., 2018). *A. aegypti* third instar larvae treated with selenium nanowires associated with EPS presented 100% mortality with the observation of darkening of body, the disintegration of the epithelial layer and outer cuticle, and enlarged body structures at siphon apex. Histology analyses revealed that midgut epithelial cells and muscles were the most affected tissues (Abinaya et al., 2019). Another feature that makes selenium nanowire material also promising for insecticide applications is that its high antioxidant potential extends the life of commercial insecticides (Rekha et al., 2019). Besides the potential of the NPs in the control of larvae of mosquitoes, some of them also presented an ovicidal potential and effects on oviposition of several mosquitoes. In experiments conducted with *Anopheles stephensi, A. aegypti,* and *Culex quinquefasciatus,* the hatchability of eggs was reduced by 100% after a single exposure to 30ppm Ag-NPs synthesized with *Sargassum muticum* (Benelli, 2016).

Despite the promising results for the use of NPs as eco-friendly insecticides, accurate information on the mode of action of NPs on insects remains limited. Ag-NPs are the principal model for elucidation of toxic mechanism, and studies suggest two main pathways responsible for toxicity. The first describes that NPs could induce toxicity by triggering oxidative stress in arthropod tissues. The second suggests that NPs could penetrate through the exoskeleton and upon reaching intracellular space would bind to protein or DNA, leading to rapid denaturation of organelles and enzymes, leading to loss of function and cell death.

Several studies revealed that the green biological synthesis of NPs and their use as insecticides are a promising field in research. However, little is known about the potential environmental risks in terrestrial and aquatic ecosystems. Exposure to silver NP can cause morphological changes and DNA damage in both invertebrate and vertebrate organisms and represent a detrimental impact on enzymatic activities of various nontarget species. Thus efforts are still needed to validate nanopesticides, monitoring their stability and selectivity and their impact on the environment.

Nanotechnological strategies for antiviral and vaccine development

Complications including neurological fetal syndromes during pregnancy and Guillain-Barré syndrome associated with ZIKV infections turn the development of antivirals and vaccines an urgent measure to control the disease. There are no antivirals approved for therapeutic use in ZIKV infections. Many studies have aimed to find molecules with antiviral potential against ZIKV, including approaches based on repurposing antiviral drugs and discovery of molecules from natural products and antibodies (Wang, Yan, & Gao, 2017). Among the studies described in the literature, only a few studies use nanotechnology to developed antivirals and vaccines against ZIKV. One of these studies showed that carbon dots derived from benzoxazine monomers are able to reduce the capacity of ZIKV to infect Vero cells. The mechanism of action has not been determined but probably could be related to binding of these carbon dots to the viral surface, which could impair the early stages of ZIKV infection on these cells (Huang et al., 2019). Another study described the use of a synthetic nanoformulation of ivermectin for oral use. The ivermectin is a known antiparasitic drug with antiviral activity against ZIKV. NPs with encapsulated ivermectin can cross the intestinal epithelial barrier to enter the bloodstream. The ivermectin nanoformulation was able to inhibit NS1 expression on ZIKV-infected HEK293T cells and do not cross the placental barrier, turning this nanoformulation safe to treat pregnant ZIKV-infected population (Surnar et al., 2019) (Fig. 3).

Regarding the use of nanotechnology to develop vaccines against ZIKV, there are some nanovaccine candidates under development, all using different strategies to provide protection. NPs carrying RNA and prokaryotic outer membrane vesicles and virus-like particles (VLPs) were tested on several animal models. A nanoparticle RNA vaccine that generates a 54-kDa polypeptide from prM and E proteins was tested in a mouse model. In transfected cells, this polypeptide results in

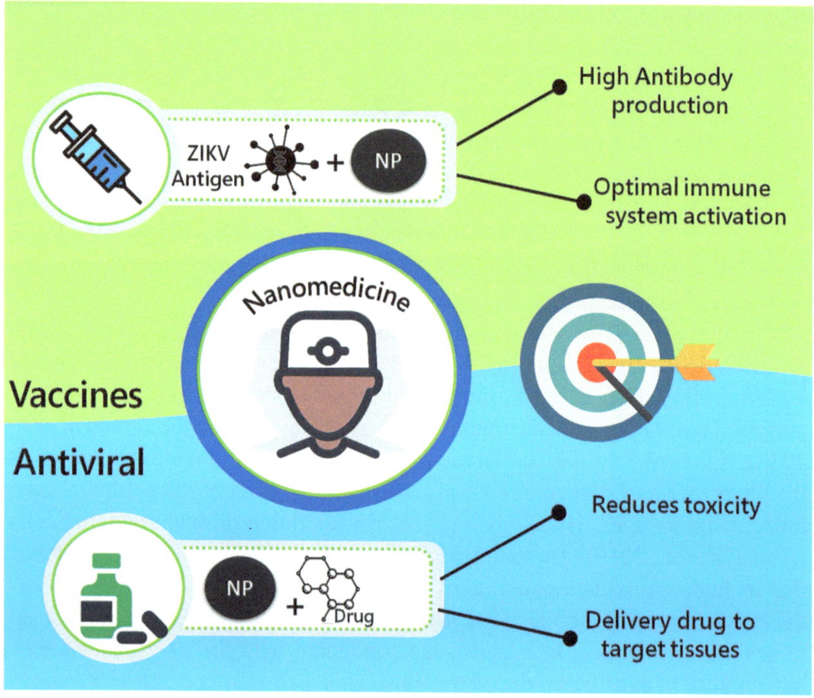

FIG. 3 Nanomedicine approaches to develop nanoparticles as antivirals and vaccines against ZIKV. Nanoparticles could be used to delivery antivirals to infected cells more efficiently or to delivery ZIKV antigens to the immune system.

the production and release of subviral particles that are antigenically similar to infectious viral particles. Immunized mice with this nanoparticle showed an antibody and CD8+ T-cell responses (Chahal et al., 2017). Another study used a nanostructured lipid carrier to deliver a replicative viral RNA encoding ZIKV antigens using an attenuated strain of Venezuelan equine encephalitis virus genetic backbone. This formulation was able to completely protect mice against a lethal ZIKV challenge after a single dose (Erasmus et al., 2018). In a similar way a mRNA lipid NP encoding prM-E genes that are able to produced VLPs induced an effective immune response to ZIKV. To prevent cross-reactive antibody responses and antibody-dependent enhancement in DENV infections, these VLPs do not have an immunodominant cross-reactive epitope in domain II of the envelope protein. Immunized mice with these nanolipids induced protective antibody responses against ZIKV (Richner et al., 2017). In another study, lipid nanoparticle-encapsulated nucleoside-modified mRNA encoding prM-*E*-ZIKV glycoproteins were tested in mouse and nonhuman primates. This nanovaccine protected mice and nonhuman primates after ZIKV infection even after 5 months post vaccination. The single low dose was able to induce long-lasting immunity based on neutralizing antibodies (Pardi et al., 2017).

VLPs are noninfectious protein cages derived from structural proteins of viruses. They are formed from self-assembling protein subunits and are considered a promising tool on nanotechnology. Use of VLPs as NPs to deliver antigens to the immune system is one of the most promising approaches on vaccinology. VLPs made of structural proteins of ZIKV (membrane, envelope, and/or capsid proteins) are produced on several cell systems, such as mammalian cells and plants. After production, they are purified and used to immunized mice. In all studies the VLPs are able to induce high levels of neutralizing antibodies on immunized animals (Boigard et al., 2017; Espinosa et al., 2018; Garg, Sedano, Plata, Punke, & Joshi, 2017; Yang, Lai, Sun, & Chen, 2017). In a similar way the use of outer membrane vesicles (OMVs) could induce a good immune response to antigens associated with this microbial structure. OMVs are produced by gram-negative bacteria and are able to activate the innate and adaptive immunity due to their adjuvant activity (van der Pol, Stork, & van der Ley, 2015). In this case, OMV from N*eisseria meningitidis* was fusioned with ZIKV particles to produce OMV-ZIKV fusion NPs. Immunized mice with this nanoparticle showed cellular immune response and neutralizing antibodies (Martins et al., 2018).

Overall, there are still few studies using nanotechnology tools to develop antivirals and vaccines to treat or prevent ZIKV infections, respectively. Much remains to be explored in this field, once there are several types of materials to construct NPs that could deliver antiviral drugs to infected cells or antigens to the immune system more efficiently.

Conclusion

ZIKV is an emerging flavivirus that has caused epidemic outbreaks in the Pacific islands and several countries of the Americas. In parallel, severe manifestations, including microcephaly and Guillain-Barré syndrome, have increased in incidence during the ZIKV outbreak. This scenario contributed to the establishment of a state of emergency by the World Health Organization. Therefore developing new methods of diagnostic, control of vector, antivirals, and vaccines is very important to control the disease. Several nanotechnological approaches were used to develop new diagnostic tests for ZIKV, new ways to control mosquito proliferation, new ways to deliver antiviral drugs, and new ways to deliver ZIKV antigens. Therefore NPs had unique biological and physical properties that can make them ideal candidates to overcome challenges associated with the control and diagnosis of ZIKV and its vectors.

Policy and procedures

- Nanoparticle technology—basic concepts:

Information about the theoretical aspects of nanoparticles and practical applications of nanotechnology can be obtained from this link: https://www.sciencedirect.com/book/9780444563361/nanoparticle-technology-handbook.

Hosokawa, M., Nogi, K., Naito, M. & Yokoyama, T. (2012). Nanoparticle Technology Handbook. (Second Edition). Netherlands: Elsevier.

- Green synthesis:

Green synthesis is defined as a new area of nanotechnology. Nanoparticles produced by green synthesis could provide economic and environmental benefits. In general, there is no requirement for toxic chemicals, high pressure, or high temperature to produced nanoparticles by green synthesis methods. A range of methodologies about green synthesis of nanoparticles can be obtained from this link: https://www.sciencedirect.com/book/9780081025796/green-synthesis-characterization-and-applications-of-nanoparticles.

Shukla, A. K. & Iravani, S. (2019). Micro and Nano Technologies, Green Synthesis, Characterization and Applications of Nanoparticles. (First Edition). Netherlands: Elsevier.

Mini-dictionary of terms

Nanoparticles. Nanoparticles are defined by particles made of organic or inorganic with length of 1–1000nm in at least one dimension. In general, the nanoparticle could have other physical or biological activities than the original organic or inorganic material.

Quantum dots. A particular type of semiconducting nanoparticles with length of 2–10nm. Due to small size they can transport electrons. After exposure to UV light they can emit light of various colors. Therefore quantum dots could be used in several applications, such as labeling cells or proteins.

Plasmonic effects. A physical reaction based on the interaction between free electrons in metal nanoparticles and an incident light that generates electron charge density waves called plasmons. This change can enhance the light emission of fluorescent emitters.

Green synthesis of nanoparticles. A set of methodologies applied to the production of nanoparticles with the main objective to minimize waste and implementing sustainable processes.

Raman scattering signal. An optical process where an excitation light will interact with a sample. This interaction produces scattered light that has less energy due to the vibrational modes of the chemical bonds of the specimen.

Neutralizing antibodies. Antibodies that are able to bind to the surface of ZIKV and inhibit cellular infection.

Key facts of nanomedicine and nanoparticles

- Nanomedicine is the use of nanotechnology for the development of new technologies applied to healthcare. This new field of research could improve the therapy, diagnostics, and prevention of infectious and noninfectious diseases.
- Nanoparticles are diverse and can be classified according to their physical (size and shape) and material (organic and inorganic) properties.
- The International Organization for Standardization defined nanoparticles as microscopic particles with at least one dimension less than 100nm.

- In general, nanoparticles should have different properties than the same bulk materials.
- All nanoparticles should be tested on several models to determine the toxicity to humans and environment.

Summary points

- Nanoparticles could be important tools to develop new ways to control and to perform the diagnosis of *Zika virus* and its vectors.
- Nanoparticles could be used to improve the detection of viral particle, viral genome, NS1 protein, or antibodies against ZIKV.
- Several nanoparticles could show larvicidal and adulticidal toxicity on *Aedes* sp. mosquitoes.
- Nanoparticles can also be used to increase the antiviral activity of drugs.
- Nanoparticles carrying RNA, prokaryotic outer membrane vesicles and virus-like particles (VLPs) were tested on several animal models and showed a good immune response.

References

Aarthi, C., Govindarajan, M., Rajaraman, P., Alharbi, N. S., Kadaikunnan, S., Khaled, J. M., … Benelli, G. (2018). Eco-friendly and cost-effective Ag nanocrystals fabricated using the leaf extract of *Habenaria plantaginea*: Toxicity on six mosquito vectors and four non-target species. *Environmental Science and Pollution Research*, *25*(11), 10317–10327. https://doi.org/10.1007/s11356-017-9203-2.

Abinaya, M., Vaseeharan, B., Divya, M., Sharmili, A., Govindarajan, M., Alharbi, N. S., … Benelli, G. (2018). Bacterial exopolysaccharide (EPS)-coated ZnO nanoparticles showed high antibiofilm activity and larvicidal toxicity against malaria and Zika virus vectors. *Journal of Trace Elements in Medicine and Biology*, *45*(August 2017), 93–103. https://doi.org/10.1016/j.jtemb.2017.10.002.

Abinaya, M., Vaseeharan, B., Divya, M., Vijayakumar, S., Govindarajan, M., Alharbi, N. S., … Benelli, G. (2018). Structural characterization of Bacillus licheniformis Dahb1 exopolysaccharide—Antimicrobial potential and larvicidal activity on malaria and Zika virus mosquito vectors. *Environmental Science and Pollution Research*, *25*(19), 18604–18619. https://doi.org/10.1007/s11356-018-2002-6.

Abinaya, M., Vaseeharan, B., Rekha, R., Shanthini, S., Govindarajan, M., Alharbi, N. S., … Al-Anbr, M. N. (2019). Microbial exopolymer-capped selenium nanowires—Towards new antibacterial, antibiofilm and arbovirus vector larvicides? *Journal of Photochemistry and Photobiology B: Biology*, *192*(2018), 55–67. https://doi.org/10.1016/j.jphotobiol.2019.01.009.

Aziz, A. T., Alshehri, M. A., Panneerselvam, C., Murugan, K., Trivedi, S., Mahyoub, J. A., … Benelli, G. (2018). The desert wormwood (*Artemisia herba-alba*)—From Arabian folk medicine to a source of green and effective nanoinsecticides against mosquito vectors. *Journal of Photochemistry and Photobiology B: Biology*, *180*(2017), 225–234. https://doi.org/10.1016/j.jphotobiol.2018.02.012.

Benelli, G. (2016). Green synthesized nanoparticles in the fight against mosquito-borne diseases and cancer—A brief review. *Enzyme and Microbial Technology*, *95*, 58–68. https://doi.org/10.1016/j.enzmictec.2016.08.022.

Benelli, G. (2018). Mode of action of nanoparticles against insects. *Environmental Science and Pollution Research*, *25*(13), 12329–12341. https://doi.org/10.1007/s11356-018-1850-4.

Boigard, H., Alimova, A., Martin, G. R., Katz, A., Gottlieb, P., & Galarza, J. M. (2017). Zika virus-like particle (VLP) based vaccine. *PLoS Neglected Tropical Diseases*, *11*(5), 1–20. https://doi.org/10.1371/journal.pntd.0005608.

Boyer, S., Calvez, E., Chouin-Carneiro, T., Diallo, D., & Failloux, A. B. (2018). An overview of mosquito vectors of Zika virus. *Microbes and Infection*, *20*(11–12), 646–660. https://doi.org/10.1016/j.micinf.2018.01.006.

Camacho, S. A., Sobral-Filho, R. G., Aoki, P. H. B., Constantino, C. J. L., & Brolo, A. G. (2018). Zika immunoassay based on surface-enhanced Raman scattering nanoprobes. *ACS Sensors*, *3*(3), 587–594. https://doi.org/10.1021/acssensors.7b00639.

Chahal, J. S., Fang, T., Woodham, A. W., Khan, O. F., Ling, J., Anderson, D. G., & Ploegh, H. L. (2017). An RNA nanoparticle vaccine against Zika virus elicits antibody and CD8+ T cell responses in a mouse model. *Scientific Reports*, *7*(1), 1–9. https://doi.org/10.1038/s41598-017-00193-w.

Chu, H. W., Lai, C. S., Ko, J. Y., Harroun, S. G., Chuang, C. I., Wang, R. Y. L., … Huang, C. C. (2019). Nanoparticle-based LDI-MS immunoassay for the multiple diagnosis of viral infections [research-article]. *ACS Sensors*, *4*(6), 1543–1551. https://doi.org/10.1021/acssensors.9b00054.

Draz, M. S., Lakshminaraasimulu, N. K., Krishnakumar, S., Battalapalli, D., Vasan, A., Kanakasabapathy, M. K., … Shafiee, H. (2018). Motion-based immunological detection of Zika virus using Pt-nanomotors and a cellphone. *ACS Nano*, *12*(6), 5709–5718. https://doi.org/10.1021/acsnano.8b01515.

Draz, M. S., Venkataramani, M., Lakshminarayanan, H., Saygili, E., Moazeni, M., Vasan, A., … Shafiee, H. (2018). Nanoparticle-enhanced electrical detection of Zika virus on paper microchips. *Nanoscale*, *10*(25), 11841–11849. https://doi.org/10.1039/c8nr01646a.

Erasmus, J. H., Khandhar, A. P., Guderian, J., Granger, B., Archer, J., Archer, M., … Van Hoeven, N. (2018). A nanostructured lipid carrier for delivery of a replicating viral RNA provides single, low-dose protection against Zika. *Molecular Therapy*, *26*(10), 2507–2522. https://doi.org/10.1016/j.ymthe.2018.07.010.

Espinosa, D., Mendy, J., Manayani, D., Vang, L., Wang, C., Richard, T., … Alexander, J. (2018). Passive transfer of immune sera induced by a Zika virus-like particle vaccine protects AG129 mice against lethal Zika virus challenge. *eBioMedicine*, *27*, 61–70. https://doi.org/10.1016/j.ebiom.2017.12.010.

Fouad, H., Hongjie, L., Hosni, D., Wei, J., Abbas, G., Ga'al, H., & Jianchu, M. (2018). Controlling *Aedes albopictus* and *Culex pipiens pallens* using silver nanoparticles synthesized from aqueous extract of *Cassia fistula* fruit pulp and its mode of action. *Artificial Cells, Nanomedicine and Biotechnology*, *46*(3), 558–567. https://doi.org/10.1080/21691401.2017.1329739.

Ga'al, H., Fouad, H., Tian, J., Hu, Y., Abbas, G., & Mo, J. (2018). Synthesis, characterization and efficacy of silver nanoparticles against Aedes albopictus larvae and pupae. *Pesticide Biochemistry and Physiology*, *144*, 49–56. https://doi.org/10.1016/j.pestbp.2017.11.004.

Garg, H., Sedano, M., Plata, G., Punke, E. B., & Joshi, A. (2017). Development of virus-like-particle vaccine and reporter assay for Zika virus. *Journal of Virology*, *91*(20), 1–16. https://doi.org/10.1128/jvi.00834-17.

Govindarajan, M., Hoti, S. L., & Benelli, G. (2016). Facile fabrication of eco-friendly nano-mosquitocides: Biophysical characterization and effectiveness on neglected tropical mosquito vectors. *Enzyme and Microbial Technology*, *95*, 155–163. https://doi.org/10.1016/j.enzmictec.2016.05.005.

Govindarajan, M., Kadaikunnan, S., Alharbi, N. S., & Benelli, G. (2017). Single-step biological fabrication of colloidal silver nanoparticles using *Hugonia mystax*: Larvicidal potential against Zika virus, dengue, and malaria vector mosquitoes. *Artificial Cells, Nanomedicine and Biotechnology*, *45*(7), 1317–1325. https://doi.org/10.1080/21691401.2016.1228664.

Huang, S., Gu, J., Ye, J., Fang, B., Wan, S., Wang, C., … Cao, S. (2019). Benzoxazine monomer derived carbon dots as a broad-spectrum agent to block viral infectivity. *Journal of Colloid and Interface Science*, *542*, 198–206. https://doi.org/10.1016/j.jcis.2019.02.010.

Ishwarya, R., Vaseeharan, B., Anuradha, R., Rekha, R., Govindarajan, M., Alharbi, N. S., … Benelli, G. (2017). Eco-friendly fabrication of Ag nanostructures using the seed extract of Pedalium murex, an ancient Indian medicinal plant: Histopathological effects on the Zika virus vector Aedes aegypti and inhibition of biofilm-forming pathogenic bacteria. *Journal of Photochemistry and Photobiology B: Biology*, *174*, 133–143. https://doi.org/10.1016/j.jphotobiol.2017.07.026.

Ishwarya, R., Vaseeharan, B., Kalyani, S., Banumathi, B., Govindarajan, M., Alharbi, N. S., … Benelli, G. (2018). Facile green synthesis of zinc oxide nanoparticles using Ulva lactuca seaweed extract and evaluation of their photocatalytic, antibiofilm and insecticidal activity. *Journal of Photochemistry and Photobiology B: Biology*, *178*(October 2017), 249–258. https://doi.org/10.1016/j.jphotobiol.2017.11.006.

Jeevanandam, J., Barhoum, A., Chan, Y. S., Dufresne, A., & Danquah, M. K. (2018). Review on nanoparticles and nanostructured materials: History, sources, toxicity and regulations. *Beilstein Journal of Nanotechnology*, *9*(1), 1050–1074. https://doi.org/10.3762/bjnano.9.98.

Kalimuthu, K., Panneerselvam, C., Chou, C., Tseng, L. C., Murugan, K., Tsai, K. H., … Benelli, G. (2017). Control of dengue and Zika virus vector Aedes aegypti using the predatory copepod Megacyclops formosanus: Synergy with *Hedychium coronarium*-synthesized silver nanoparticles and related histological changes in targeted mosquitoes. *Process Safety and Environmental Protection*, *109*, 82–96. https://doi.org/10.1016/j.psep.2017.03.027.

Martinez, J. D., de la Garza, J. A. C., & Cuellar-Barboza, A. (2019). Going viral 2019: Zika, chikungunya, and dengue. *Dermatologic Clinics*, *37*(1), 95–105. https://doi.org/10.1016/j.det.2018.07.008.

Martins, P., Machado, D., Theizen, T. H., Guarnieri, J. P. O., Bernardes, B. G., Gomide, G. P., … Lancellotti, M. (2018). Outer membrane vesicles from Neisseria meningitidis (proteossome) used for nanostructured Zika virus vaccine production. *Scientific Reports*, *8*(1), 1–10. https://doi.org/10.1038/s41598-018-26508-z.

Murugan, K., Benelli, G., Panneerselvam, C., Subramaniam, J., Jeyalalitha, T., Dinesh, D., … Madhiyazhagan, P. (2015). *Cymbopogon citratus*-synthesized gold nanoparticles boost the predation efficiency of copepod Mesocyclops aspericornis against malaria and dengue mosquitoes. *Experimental Parasitology*, *153*, 129–138. https://doi.org/10.1016/j.exppara.2015.03.017.

Ochmann, S. E., Vietz, C., Trofymchuk, K., Acuna, G. P., Lalkens, B., & Tinnefeld, P. (2017). Optical nanoantenna for single molecule-based detection of Zika virus nucleic acids without molecular multiplication. *Analytical Chemistry*, *89*(23), 13000–13007. https://doi.org/10.1021/acs.analchem.7b04082.

Pardi, N., Hogan, M. J., Pelc, R. S., Muramatsu, H., Andersen, H., DeMaso, C. R., … Weissman, D. (2017). Zika virus protection by a single low-dose nucleoside-modified mRNA vaccination. *Nature*, *543*(7644), 248–251. https://doi.org/10.1038/nature21428.

Rekha, R., Vaseeharan, B., Vijayakumar, S., Abinaya, M., Govindarajan, M., Alharbi, N. S., … Al-anbr, M. N. (2019). Crustin-capped selenium nanowires against microbial pathogens and Japanese encephalitis mosquito vectors—Insights on their toxicity and internalization. *Journal of Trace Elements in Medicine and Biology*, *51*, 191–203. https://doi.org/10.1016/j.jtemb.2018.10.017.

Ribeiro, J. F. F., Pereira, M. I. A., Assis, L. G., Cabral Filho, P. E., Santos, B. S., Pereira, G. A. L., … Fontes, A. (2019). Quantum dots-based fluoroimmunoassay for anti-Zika virus IgG antibodies detection. *Journal of Photochemistry and Photobiology B: Biology*, *194*(March), 135–139. https://doi.org/10.1016/j.jphotobiol.2019.03.019.

Richner, J. M., Himansu, S., Dowd, K. A., Butler, S. L., Salazar, V., Fox, J. M., … Diamond, M. S. (2017). Modified mRNA vaccines protect against Zika virus infection. *Cell*, *168*(6), 1114–1125.e10. https://doi.org/10.1016/j.cell.2017.02.017.

Selvan, S. M., Anand, K. V., Govindaraju, K., Tamilselvan, S., Kumar, V. G., Subramanian, K. S., … Raja, K. (2018). Green synthesis of copper oxide nanoparticles and mosquito larvicidal activity against dengue, zika and chikungunya causing vector *Aedes aegypti*. *IET Nanobiotechnology*, *12*(8), 1042–1046. https://doi.org/10.1049/iet-nbt.2018.5083.

Singh, R. K., Dhama, K., Karthik, K., Tiwari, R., Khandia, R., Munjal, A., … Bueno-Marí, R. (2018). Advances in diagnosis, surveillance, and monitoring of Zika virus: An update. *Frontiers in Microbiology*, *8*(January), 1–17. https://doi.org/10.3389/fmicb.2017.02677.

Singh, L., Kruger, H. G., Maguire, G. E. M., Govender, T., & Parboosing, R. (2017). The role of nanotechnology in the treatment of viral infections. *Therapeutic Advances in Infectious Disease*, *4*(4), 105–131. https://doi.org/10.1177/2049936117713593.

Soni, N., & Prakash, S. (2012). Efficacy of fungus mediated silver and gold nanoparticles against Aedes aegypti larvae. *Parasitology Research*, *110*(1), 175–184. https://doi.org/10.1007/s00436-011-2467-4.

Sposito, A. J., Kurdekar, A., Zhao, J., & Hewlett, I. (2018). Application of nanotechnology in biosensors for enhancing pathogen detection. *Wiley Interdisciplinary Reviews: Nanomedicine and Nanobiotechnology*, *10*(5), 1–22. https://doi.org/10.1002/wnan.1512.

Steinmetz, M., Lima, D., Viana, A. G., Fujiwara, S. T., Pessôa, C. A., Etto, R. M., & Wohnrath, K. (2019). A sensitive label-free impedimetric DNA biosensor based on silsesquioxane-functionalized gold nanoparticles for Zika virus detection. *Biosensors and Bioelectronics*, *141*(May), 111351. https://doi.org/10.1016/j.bios.2019.111351.

Suganya, P., Vaseeharan, B., Vijayakumar, S., Balan, B., Govindarajan, M., Alharbi, N. S., … Benelli, G. (2017). Biopolymer zein-coated gold nanoparticles: Synthesis, antibacterial potential, toxicity and histopathological effects against the Zika virus vector Aedes aegypti. *Journal of Photochemistry and Photobiology B: Biology*, *173*(January), 404–411. https://doi.org/10.1016/j.jphotobiol.2017.06.004.

Sulczewski, F. B., Liszbinski, R. B., Romão, P. R. T., & Rodrigues Junior, L. C. (2018). Nanoparticle vaccines against viral infections. *Archives of Virology*, *163*(9), 2313–2325. https://doi.org/10.1007/s00705-018-3856-0.

Sundararajan, B., & Ranjitha Kumari, B. D. (2017). Novel synthesis of gold nanoparticles using *Artemisia vulgaris* L. leaf extract and their efficacy of larvicidal activity against dengue fever vector *Aedes aegypti* L. *Journal of Trace Elements in Medicine and Biology*, *43*, 187–196. https://doi.org/10.1016/j.jtemb.2017.03.008.

Surnar, B., Kamran, M. Z., Shah, A. S., Basu, U., Kolishetti, N., Deo, S., … Dhar, S. (2019). Orally administrable therapeutic synthetic nanoparticle for Zika virus [research-article]. *ACS Nano*, *13*(10), 11034–11048. https://doi.org/10.1021/acsnano.9b02807.

Takemura, K., Adegoke, O., Suzuki, T., & Park, E. Y. (2019). A localized surface plasmon resonance-amplified immunofluorescence biosensor for ultrasensitive and rapid detection of nonstructural protein 1 of Zika virus. *PLoS One*, *14*(1), 1–14. https://doi.org/10.1371/journal.pone.0211517.

van der Pol, L., Stork, M., & van der Ley, P. (2015). Outer membrane vesicles as platform vaccine technology. *Biotechnology Journal*, *10*(11), 1689–1706. https://doi.org/10.1002/biot.201400395.

Ventola, C. L. (2012). Nanomedicine—Part 1 emerging concepts. *P & T*, *37*(9), 512–518.

Wang, Q., Yan, J., & Gao, G. F. (2017). Monoclonal antibodies against Zika virus: Therapeutics and their implications for vaccine design. *Journal of Virology*, *91*(20), 1–6. https://doi.org/10.1128/jvi.01049-17.

Yang, M., Lai, H., Sun, H., & Chen, Q. (2017). Virus-like particles that display Zika virus envelope protein domain III induce potent neutralizing immune responses in mice. *Scientific Reports*, *7*(1), 1–12. https://doi.org/10.1038/s41598-017-08247-9.

Zhang, B., Pinsky, B. A., Ananta, J. S., Zhao, S., Arulkumar, S., Wan, H., … Dai, H. (2017). Diagnosis of Zika virus infection on a nanotechnology platform. *Nature Medicine*, *23*(5), 548–550. https://doi.org/10.1038/nm.4302.

Section G

Models and modeling

Chapter 41

Neonatal microcephaly and humanized mouse models for Zika viral pathogenesis and immunity

Kimberly Schmitt, Tawfik Aboellail, and Ramesh Akkina

Department of Microbiology, Immunology and Pathology, Colorado State University, Fort Collins, CO, United States

Abbreviations

FNT flow cytometry-based neutralization test
Hu-BLT humanized bone marrow, liver, and thymus mice
Hu-HSC hemopoietic stem cell mice
Hu-mice humanized mice
PrM-E viral premembrane (prM) and envelope (E) proteins

Introduction

The worldwide impact of Zika virus (ZIKV) was devastating during the last severe outbreak that affected millions of people. Higher risk/incidence of Guillain-Barré syndrome (GBS) in adults and severe birth defects including fetal microcephaly [now collectively termed as congenital Zika Syndrome (CZS)] were among the more serious complications of ZIKV infection (Calvet et al., 2016; do Rosario et al., 2016; Mlakar et al., 2016; Oliveira Melo et al., 2016; Paploski et al., 2016; Vesnaver et al., 2017). Unusual for a typical arbovirus, sexual transmission was noted during infection (Foy et al., 2011). Persistence of the virus during pregnancy and in the male genital tract (GT) was seen despite a neutralizing antibody response (Barzon et al., 2018; Calvet et al., 2018; Paz-Bailey et al., 2018). Factors involved in fetal infection and the associated pathologies are not well understood. Animal models have an important role in answering these critical questions. Rapid progress was made using murine and nonhuman primate (NHP) animal models; however, many aspects of viral pathogenesis remain to be studied. Since wild-type mice support ZIKV infection but do not show Zika viral pathogenesis, IFN-deficient mouse models were the first widely used animal model (Caine, Jagger, & Diamond, 2018; Nazerai, Pravsgaard Christensen, & Randrup Thomsen, 2019). However, a significant drawback with these models is the lack of IFN-mediated innate immune response, the first line of antiviral defense. Several nonhuman primate (NHP) models have been employed for Zika studies providing important information on various aspects of ZIKV pathogenesis (Caine et al., 2018; Morrison & Diamond, 2017; Pena et al., 2018). A major advantage of these models is the genetic similarity of NHPs to humans relative to murine models. Many aspects of ZIKV pathology were reproduced in this system including congenital neurological abnormalities following infection of pregnant animals (Adams Waldorf et al., 2016; Caine et al., 2018; Nguyen et al., 2017). Cell-mediated T-cell responses and neutralizing antibody generation was seen in NHPs, thus permitting vaccine testing (Aliota et al., 2016, 2017; Dowd et al., 2016; Morrison & Diamond, 2017). Other observations in NHPs include detection of virus in body fluids, namely, saliva, urine, CSF, and GT secretions, in addition to multiple tissue sites (Aliota et al., 2017; Morrison & Diamond, 2017; Pena et al., 2018). However, NHP models are expensive, and the use of large numbers is cost prohibitive. Furthermore, these conventional animal models do not permit direct evaluation of the viral effects on human target cells in vivo and fail to assess human-specific immune responses.

Humanized mice

An in vivo animal model providing the physiological setting to support human cells and tissues would be an ideal human surrogate system to evaluate many important questions directly related to virus-human cell interactions. In this regard,

Zika Virus Impact, Diagnosis, Control, and Models. https://doi.org/10.1016/B978-0-12-820267-8.00041-8

humanized mice (hu-mice) prepared by transplanting various human tissues and cells can be used to overcome the previous limitations seen with traditional animal models. Immunodeficient mice such as NSG and Rag2$^{-/-}\gamma c^{-/-}$ mice that permit xenografts are used for human cell engraftment (Akkina, 2013). Many hu-mouse models currently exist with two specific models being widely used (Table 1). Of these models, the hu-HSC mice are prepared by transplanting human hematopoietic stem cells (HSC), whereas hu-BLT mice (bone marrow, liver, and thymus mice) are generated by engraftment of HSC and transplantation of autologous human fetal liver and thymic tissues under the kidney capsule of immunodeficient mice (Fig. 1A) (Akkina, 2013; Lan, Tonomura, Shimizu, Wang, & Yang, 2006; Melkus et al., 2006; Shultz, Brehm, Garcia-Martinez, & Greiner, 2012; Wege, Melkus, Denton, Estes, & Garcia, 2008). Post transplantation, both primary and secondary lymphoid organs develop with human hematopoietic cells generated de novo (Akkina, 2013; Shultz et al., 2012; Wege et al., 2008). The human cells populate systemically to various tissues including thymus, bone marrow, spleen, liver, lymph nodes, gut, and reproductive organs (Akkina, 2013; Akkina et al., 2016). Human T cells, B cells, macrophages, and dendritic cells, which are essential components of the human immune system, are present, thus reconstituting a functional human immune system (Fig. 1A and B) (Akkina, 2013; Akkina et al., 2016; Ito, Takahashi, Katano, & Ito, 2012; Traggiai et al., 2004; Wege et al., 2008). Therefore, hu-mice can respond with both antibody- and cell-mediated human immune responses after antigenic exposure (Akkina, 2013; Akkina et al., 2016; Melkus et al., 2006; Seung & Tager, 2013;

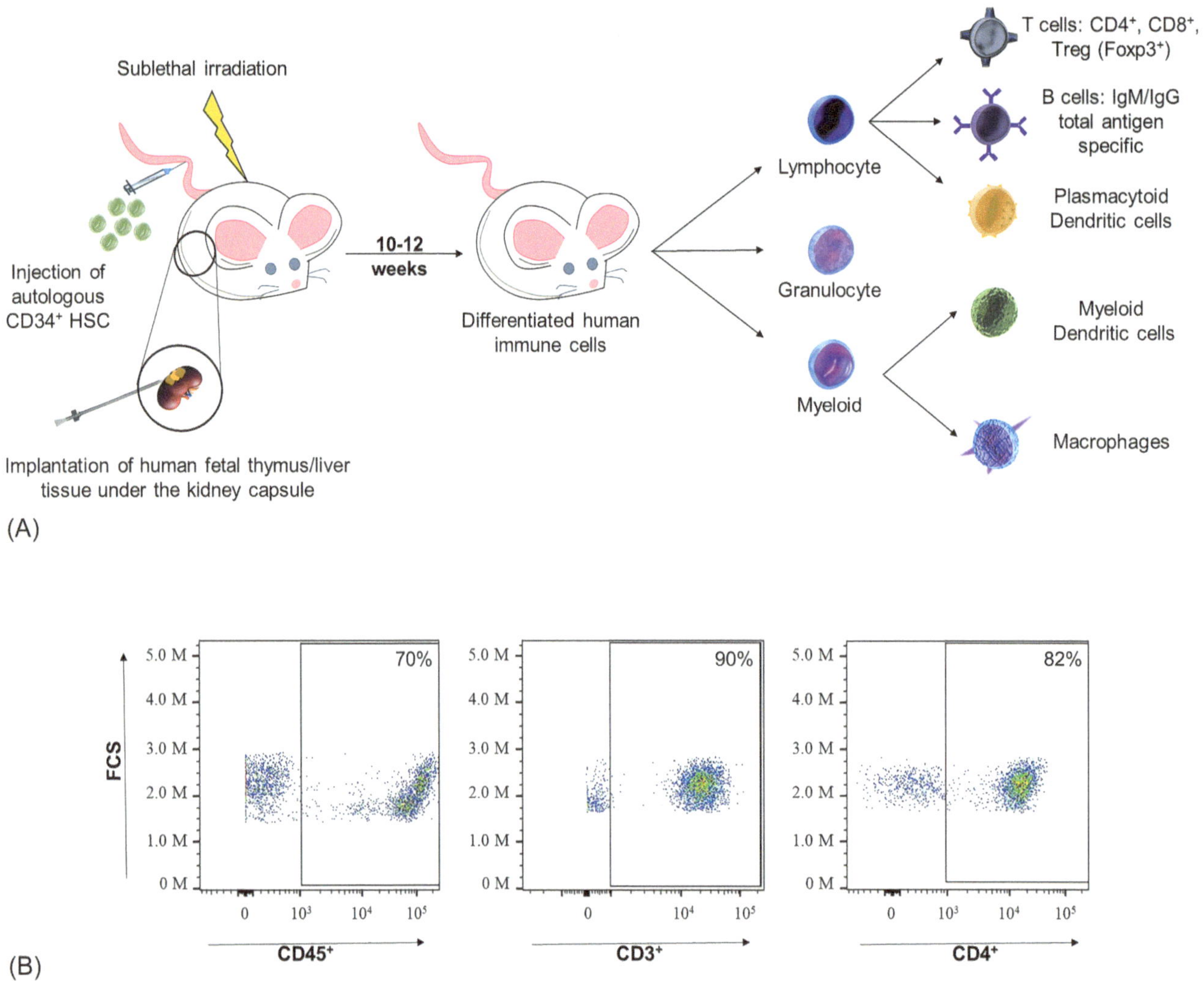

FIG. 1 Generation of hu-BLT mice with a human immune system. (A) Adult Balb/c-Rag2$^{-/-}\gamma c^{-/-}$ mice are preconditioned by irradiation. Mice were surgically implanted with human fetal liver and thymic fragments underneath the kidney capsule followed by i/v injection of autologous CD34^{+} hematopoietic stem cells. These mice generate a broad spectrum of human hematopoietic cells that encompass T cells, B cells, macrophages, NK cells, and dendritic cells. The presence of an autologous human thymus allows for appropriate T-cell education and restriction. (B) Human cell engraftment and cell subsets. At 10–20 weeks postreconstitution, the peripheral blood of mice was assessed using flow cytometry for cells positive for CD45 pan-leukocyte marker and followed by identification of different cell subset such as CD3 and CD4 cells (unpublished data).

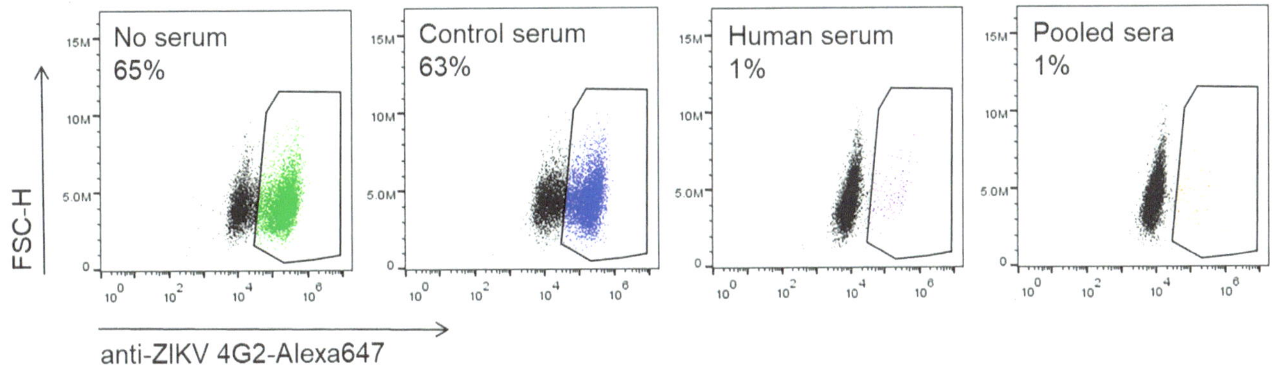

FIG. 2 Generation of anti-ZIKV neutralizing antibody response in BLT mice. Infected mouse plasma samples were evaluated by a flow cytometry-based virus neutralization assay in Vero cells. Note the drastic reduction of infected cells in pooled mouse sera and human positive sera compared with control uninfected mouse serum (about 1% vs 63%). Please see reference for more details on methodology (Schmitt et al., 2018) (unpublished data).

Shultz et al., 2012). A version of hu-HSC mice, the DRAG-hu mice with a human class II transgenic background, improved antibody production (Allam et al., 2015; Mendoza et al., 2018). Recent work with ZIKV involved the use of both the hu-BLT and DRAG-hu mouse models, the results of which are discussed next.

Zika viral human immune responses in hu-BLT mice

Studies in our laboratory as described by Schmitt et al. (2018) employed BLT mice to evaluate the effects of ZIKV on human hematopoietic cells and immune responses. Mice were challenged with ZIKV (Puerto Rico strain PRVABC59) by i/p and s/c routes to assess if hu-mice are susceptible to ZIKV. As controls, nonengrafted immunodeficient $Rag2^{-/-}\gamma c^{-/-}$ (non-hu) mice were similarly infected. It was found that all mice became readily infected with viremia detected by day 2. Viral loads increased gradually during the 14- to 15-day period. While virus-challenged non-hu mice, used as a control, also showed productive infection, BLT mice maintained significantly higher viral loads (one log) presumably due to the presence of human target cells. In mice that were studied long term, it was found that the virus persisted in some mice for as long as 133 days consistent with that described in NHPs and human cases, wherein viral persistence was detected for as long as 3–6 months (Aliota et al., 2017; Barzon et al., 2018; Calvet et al., 2018; Paz-Bailey et al., 2018; Pena et al., 2018; Schmitt et al., 2018). Hu-mice were previously shown to generate virus-specific human antibodies to flaviviruses such as dengue and other viruses like HIV and CMV (Akkina, 2013; Brainard et al., 2009; Crawford et al., 2017; Frias-Staheli et al., 2014; Jaiswal et al., 2012; Long & Stoddart, 2012). Therefore infected plasma from these mice was analyzed at different timepoints postinfection to detect human antibodies. To evaluate the neutralization capacity, a flow cytometry-based assay was used with ZIKV-infected Vero cells (Fig. 2). It was found that hu-mouse sera were fully capable of neutralizing the viral infectivity at the dilutions indicated and the degree of viral neutralization by BLT mouse antisera was comparable with that of a human antiserum sample from a patient that was previously infected with ZIKV (antibody titer of 1:240) and was employed as positive control (Schmitt et al., 2018). Whether or not the hu-mouse generated antibodies that react to ZIKV viral proteins was determined by a radioimmunoprecipitation assay. These results also showed that sera from ZIKV-infected BLT mice recognize virus-specific envelope and capsid polypeptides. This showed that BLT mice are also capable of generating ZIKV-specific antibody responses with neutralizing capacity (Schmitt et al., 2018). These findings open many possibilities for future ZIKV immunity studies. For example, many questions including the cross-reactivity of monospecific ZIKV sera to related viruses such as dengue viral antigens and the potential for antibody-dependent enhancement of infection can also be studied.

ZIKV viral infection and human immune responses in DRAG-hu mice

A common observation with many humanized immune system models is that while they are capable of generating human antibody responses, generation of IgG antibody responses is suboptimal due to inefficient class switching (Akkina, 2013; Akkina et al., 2016). In contrast, the humanized DRAG mouse model displayed improved antibody responses with the generation of human IgG antibodies shown with various antigens (Allam et al., 2015; Mendoza et al., 2018). Thus this model is better suited to evaluate vaccine candidates especially with those delivered by plasmid-based constructs that express various antigens. The DRAG mice lack murine class II molecules but are transgenically programmed to express human HLA-DR4 class II molecules (Allam et al., 2015; Mendoza et al., 2018). Engraftment of matched human HSC expressing human HLA DR4+ gives rise to higher human cell engraftment and improved T and B cell interactions, resulting in greater antibody responses. Studies of Yi et al. (2017) exploited this model for ZIKV in immunity studies and looked at what blood cells are targeted by the virus. Upon infection with ZIKV the DRAG mice did not show any adverse symptoms but showed viremia during an 8-day observation period with viral levels shown on day 3, which appeared to differ from the infection of BLT mice wherein the peak viral levels were higher and occurred later. To determine the capacity of DRAG mice to generate immune responses, they were inoculated with a DNA vaccine i/m expressing PrM-E proteins of ZIKV. Two versions containing either partial (AAs 216-794) or full (AAs 126-794) PrM-E sequences were used, and two doses were used 3 weeks apart. Sera collected 3 weeks postsecond dose were tested by ELISA and microneutralization assays. Results showed the presence of all IgG isotypes and IgM antibodies and were all found to be neutralizing. Later, the control and immunized mice were challenged with the virus to determine if any protection resulted. Whereas the control animals showed viremia, the vaccinated animals appeared to be protected due to almost undetectable viral loads at day 3 postchallenge. Overall, these data from DRAG-hu mice showed that the mice can be productively infected by ZIKV and that DNA vaccines encoding the PrM-E generate a neutralizing antibody response (Yi et al., 2017). While there is some degree of protection in vaccinated mice, it is unclear if the protection is complete since the animals were sacrificed within 3 days postchallenge and no later timepoints could be collected. Nevertheless, the use of DRAG-hu mice demonstrated the proof of concept that ZIKV vaccines can be tested using this system.

ZIKV infection of human hematopoietic cells and implications for viral persistence

While it is known that ZIKV has broad cell tropism and infects an extensive variety of cells including those in the nervous system, studies on viral infection of cells in human hematopoietic system have been sparse. While infection of monocytes/ macrophages was confirmed by analyzing human blood samples, experimental ZIKV infection of BLT mice provided an opportunity to examine cells from other blood lineages (Foo et al., 2017; Michlmayr, Andrade, Gonzalez, Balmaseda, & Harris, 2017). Schmitt et al. (2018) looked at human cell populations from spleen, bone marrow, and lymph nodes of chronically infected BLT mice at 16 weeks post-ZIKV infection. The results of flow cytometry analysis showed that human myeloid cells, B cells, and $CD34^+$ HSC were infected with ZIKV. In contrast, no T cell infection could be detected in any of those compartments (Schmitt et al., 2018). Studies of Yi et al. (2017) using humanized DRAG mice also evaluated hematopoietic cell infection by ZIKV. They found that in addition to myeloid cells, human B cells were also infected and further confirmed B cell infection by in vitro studies using $CD19^+$ B cells isolated from human PBMCs (Yi et al., 2017). In our recent studies (unpublished), we found that ZIKV infection of B cells leads to two outcomes. While a subpopulation of cells became activated and proliferated following infection, there was a decrease in another subset of cells indicative of cell death. Both outcomes have important consequences regarding ZIKV pathogenesis. For example, cell proliferation and polyclonal expansion can contribute to autoimmune phenomenon such GBS seen in a significant number of individuals recovering from Zika virus, whereas B cell loss can result in suboptimal immune responses and delayed recovery. Infection of human HSC, as shown in the BLT mouse model, also points to its potential significance to pathogenesis and viral persistence. Since HSC occupy a specialized niche in the bone marrow and give rise to all blood cell lineages, ZIKV infection of these cells may have additional consequences such as defects in their production. Future studies are needed to answer these important questions.

A neonatal $Rag2^{-/-}\gamma c^{-/-}$ mouse model for ZIKV microcephaly

Infection of the human fetus across different stages of development during gestation often resulted in CZS, with microcephaly as a prominent feature, clearly showing that ZIKV can cross the placental barrier and cause intrauterine infection (Calvet et al., 2016; Driggers et al., 2016). Severe disruption of the brain structures such as cortical calcifications, hypogyration, ventriculomegaly, and white matter brain abnormalities have been observed using head computed tomography and magnetic resonance imaging analysis (de Fatima Vasco Aragao et al., 2016; Hazin et al., 2016). Autopsy and neuropathological analysis revealed numerous calcifications in the fetal brain and severe microcephaly (Mlakar et al., 2016). Important questions regarding ZIKV pathogenesis and vertical transmission such as how the virus induces damage in the developing brain and the role of the immune system in mediating this damage continue to be intensively studied using both immunocompetent and immunocompromised murine models. Thus far, both wild-type mice and mice with a deficiency in interferon (IFN) signaling have been used most frequently to model ZIKV infection-mediated microcephaly during pregnancy (Xu, Li, Qin, & Xu, 2019). These murine pregnancy models confirmed that ZIKV compromises the integrity of the

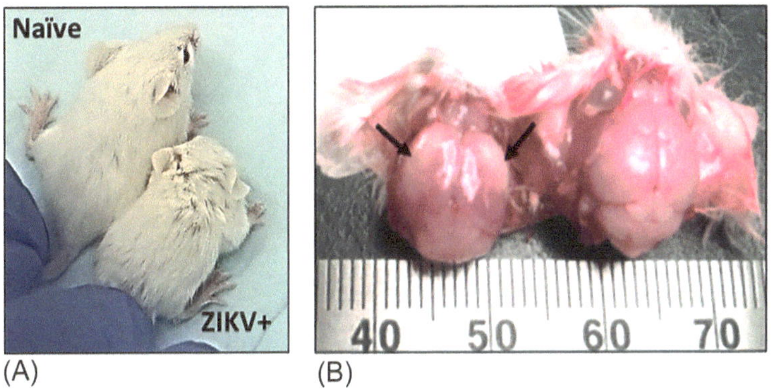

FIG. 3 ZIKV-induced growth retardation and microcephaly. Two-day-old Balb/c-Rag2$^{-/-}\gamma c^{-/-}$ mice were infected with ZIKV and assessed to determine viral effects. (A) ZIKV-infected mice display ruffled fur, hunched posture, and smaller stature (unpublished data). (B) ZIKV-infected humanized mice display gross microcephaly by 17 days post-ZIKV inoculation. Whether these neonatal Rag2$^{-/-}\gamma c^{-/-}$ mice were inoculated with cell-free ZIKV or ZIKV-infected CD34^{+} HSC intrahepatically, the results were similar (unpublished data). Brain from a ZIKV-infected mouse (left) compared with the control mouse (right). There is a significant diminution in the overall size of the cerebral hemispheres, which appear relatively paler due to thinning of the neocortex, neuronal necrosis, and mineralization (indicated by *black arrows*) (unpublished data).

placenta and that the timing of ZIKV infection during pregnancy correlates with the severity of congenital infection in the fetus (Miner et al., 2016; Szaba et al., 2018; Yockey et al., 2016). Mimicking the human CZS characteristics in experimental systems using mice has often been cumbersome, and no model thus far has proven to be optimal. Drawbacks include the need for direct injection of the virus into the developing fetal brain and a lack of survival of the pups soon after birth.

Since interferon-deficient mice lack a critical component of the innate immune system and only a subset of immunocompetent mice survive ZIKV postnatally, we developed a Rag2$^{-/-}\gamma c^{-/-}$ neonatal mouse model for CZS. Our model does not require pregnant dams or viral injection into fetal brains but rather parenteral ZIKV inoculation of 2- to 3-day-old neonatal Rag2$^{-/-}\gamma c^{-/-}$ mouse pups. Within a 14-day period, infected mice displayed stunted growth, and discernible clinical signs of ZIKV infection include hunched posture, ruffled fur, difficulty walking/hindleg paralysis, underdeveloped eyes, smaller stature, and weight loss (Fig. 3). Gross pathology of the brains from the ZIKV-infected neonatal mice showed overall significant diminution in the size of the cerebral hemispheres, which appear paler due to the thinning of the neocortex, neuronal necrosis, and mineralization, when compared with the age-matched control neonatal mice (Fig. 3). We also observed visible periarticular hemorrhages in the hindlegs and occasionally calvarial hemorrhage upon brain removal (data not shown). At the time of euthanasia, all neonatal ZIKV-infected mice displayed high viral titers in the plasma (10^6–10^7 copies/mL). Histologically, we were able to identify mineralization in the hippocampus, midbrain, thalamus, striatum, and cerebellum as well as diffuse necrosis of pyramidal neurons and subsequent collapse of the hippocampal formation (HCF). The cerebellum displays extensive underpopulation to complete absence of the external granular cell layer with focal loss of the internal granular cell layer, which leads to an overall decline in cerebellar size (unpublished data). Immunohistochemically, ZIKV antigen was seen in the cerebral cortex, hippocampus formation, dentate gyrus, thalamus, striatum, and cerebellum (unpublished data). Taken all together, these data not only documented occurrence of microcephaly, but more importantly, CZS in the Rag2$^{-/-}\gamma c^{-/-}$ neonatal mouse model with ZIKV infection is readily reproducible and easy to study. Although the Rag2$^{-/-}\gamma c^{-/-}$ neonatal mice do not have an adaptive immune system, one benefit to this model is that it is capable of being humanized. These Rag2$^{-/-}\gamma c^{-/-}$ neonatal mice have the ability to be xenografted with human blood forming CD34^{+} cells (hu-HSC) to produce a broad spectrum of hematopoietic stem cells that encompass T cells, B cells, monocyte/macrophages, and dendritic cells at various stages of development and viral susceptibility as described earlier (Akkina, 2013). The hu-HSC model would allow us to infect CD34^{+} HSC with ZIKV prior to intrahepatic injection to determine which human immune cells traffic ZIKV to the various organs such as the brain, spleen, and bone marrow, thus providing a unique platform to evaluate ZIKV pathogenesis and immunity to test novel vaccine and therapeutic candidates against CZS.

Policies and procedures

Generating humanized mice for the study of ZIKV

Among the new generation of humanized mouse models, hu-HSC are prepared by engrafting CD34^{+} hematopoietic stem cells (HSC) intrahepatically into 2- to 3-day-old immunodeficient neonatal Rag2$^{-/-}\gamma c^{-/-}$ mice, while hu-BLT (bone marrow, liver, and thymus) mice involve surgical engraftment of the human fetal liver and thymus under the kidney capsule followed by intravenous reconstitution with autologous $CD34^+$ HSC purified from the same fetal liver source into adult 5- to 8-week-old $Rag2^{-/-}\gamma c^{-/-}$ mice as previously described (Akkina, 2013; Akkina et al., 2016; Lan et al., 2006; Melkus et al., 2006; Shultz et al., 2012; Wege et al., 2008). Around 10–12 weeks postreconstitution, human cell engraftment is determined using flow cytometry and antibody-conjugated fluorophores for human immune cells (Akkina et al., 2016; Lan et al., 2006; Schmitt et al., 2018; Shultz et al., 2012; Wege et al., 2008). DRAG-hu mice, an advanced hu-BLT model, are prepared as hu-BLT mice with transgenes that express human HLA-DR4+ class II molecules. DRAG-hu mice are injected with HLA-matched human $CD34^+$ HSC (Allam et al., 2015; Mendoza et al., 2018; Yi et al., 2017).

Flow cytometry-based ZIKV neutralization assay (FNT)

ZIKV-infected humanized mouse plasma can be assessed for ZIKV neutralization in an FNT assay to assess virus-specific human antibody responses. Briefly, infected and uninfected hu-BLT mouse sera was serially diluted and mixed with ZIKV at a known concentration. The diluted virus-sera mixture was incubated and then used to infect Vero cells. Sera that caused a 50% reduction in Vero-infected cells detected by flow cytometry relative to the number of infected cells seen in the untreated virus control were scored as neutralizing (Kuruvilla, Troyer, Devi, & Akkina, 2007; Schmitt et al., 2018). End-point neutralization titers can be compared with ZIKV-infected human subject sera to demonstrate that hu-BLT mice can produce comparable human neutralizing antibody responses.

$Rag2^{-/-}\gamma c^{-/-}$ neonatal mouse model for microcephaly

The $Rag2^{-/-}\gamma c^{-/-}$ neonatal mouse model for CZS utilizes 2- to 3-day-old $Rag2^{-/-}\gamma c^{-/-}$ pups that are intrahepatically injected with cell-free ZIKV. Within 10 days, these mice display overt signs of CZS like ruffled fur, hunched posture, small stature, weight loss, and hindleg paralysis. By days 14–21, gross brain ZIKV pathology is observed (unpublished data). Since these $Rag2^{-/-}\gamma c^{-/-}$ mice lack an adaptive immune system, they are capable of being humanized. $CD34^+$ HSC can be infected with ZIKV, incubated with human ZIKV antibody to remove free ZIKV, prior to intrahepatic injection into 2- to 3-day-old $Rag2^{-/-}\gamma c^{-/-}$ pups. These ZIKV-infected human blood forming $CD34^+$ HSCs will produce de novo T cells, B cells, monocytes/macrophages, and dendritic cells at various stages of immune development, thus allowing the visualization of human immune responses to ZIKV and trafficking of ZIKV-infected cells throughout the model.

Mini-dictionary of terms

Bone marrow, liver, and thymus (Hu-BLT) mice. These mice are prepared by xenografting human fetal liver and thymus tissue under the kidney capsule of adult $Rag2^{-/-}\gamma c^{-/-}$ mice followed by reconstitution intravenously with autologous $CD34^+$ hematopoietic stem cells purified from the same fetal liver source. These mice generate a full complement of human immune cells such as T cells, B cells, monocytes/macrophages, dendritic cells, and NK cells in primary and secondary lymphoid compartments. This mouse model has been used to study human immunodeficiency virus, dengue virus, cytomegalovirus, and Zika virus.

Hematopoietic stem cell (Hu-HSC) mice. These mice are prepared by intrahepatically engrafting human blood forming $CD34^+$ hematopoietic stem cells into 2- to 3-day-old neonatal $Rag2^{-/-}\gamma c^{-/-}$ immunodeficient mice. A broad spectrum of human immune cells such as T cells, B cells, monocytes/macrophages, and dendritic cells are present at various stages of development and confer viral susceptibility.

DRAG-hu mice. More extensive than the hu-BLT mouse model in that these mice lack the murine class II molecules but are transgenically reconstituted to express human HLA-DR4 class II molecules. DRAG-hu mice give rise to higher rate of human immune cell reconstitution with longer lived B and T cells, all four classes of human immunoglobulins, and subclasses of IgG.

$Rag2^{-/-}\gamma c^{-/-}$ mice. Rag2 and interleukin 2 receptor gamma chain knockout mice. These mice exhibit T cell, B cell, and NK cell immunodeficiencies that make them effective xenograft hosts for human immune cell reconstitution.

Neutralizing antibody response. Endpoint flow cytometry-based neutralization assay (FNT) to determine the titer of ZIKV-specific human antibodies in hu-BLT mice. Hu-BLT sera that caused a 50% reduction in the number of infected cells relative to the number of infected cells seen in the untreated virus control are scored as neutralizing.

Puerto Rico strain (PRVABC59) Zika virus. Infectious viral clone isolated from a serum specimen by the CDC from a human infected in Puerto Rico in 2015.

Key facts

TABLE 1 Key facts of humanized mice.

• Various viruses pathogenic to humans have been studied in humanized mice, which include human immunodeficiency virus type 1 (HIV-1), Zika virus (ZIKV), Epstein-Barr virus (EBV), dengue virus (DENV), and hepatitis C virus (HCV)
• Humanized mice bridge the gap that exists between nonhuman primate (NHP) studies and the natural human hosts. Humanized mice allow for human-specific immune responses to identify immunogenic epitopes that trigger the human immune system
• The key feature of humanized mouse models is the transplantation of human cells/tissues into immunodeficient mice like SCID, NOG, NSG, or Rag1/2$^{-/-}\gamma c^{-/-}$
• Humanized mice are an ideal human surrogate system to evaluate virus-human cell interactions and are widely used for studies on vaccines and immunity
• Transplantation of hematopoietic stem cells (HSC) into a variety of immunodeficient mice yields hu-HSC mice. This results in human immune cell engraftment with a full complement of T cells, B cells, macrophage, NK cells, and dendritic cells allowing for the generation of human immune responses
• Hu-BLT (bone marrow, liver, and thymus) mice are created by HSC transplantation with autologous engraftment of human fetal liver and thymus tissue under the kidney capsule of immunodeficient mice. They develop both primary and secondary lymphoid organs with multilineage production of human T cells, B cells, macrophages, NK cells, and dendritic cells. The autologous thymus tissue results in T-cell education and restriction occurs
• DRAG-hu mice lack murine class II, but transgenically express human HLA-DR4. These mice are capable of HLA-restricted T-cell functions with improved cytokine secretion and IgG antibody production

This table lists the key facts regarding humanized mice, which includes different types of humanized mice, how they are produced, human immune cells generated, and their utility.

Future prospects: Compared with standard/conventional animal models, humanized mice have the advantage of studying virus-cell interactions and human immune responses more directly in a human surrogate system. The new neonatal microcephaly model we described here is not only unique but easy to establish and glean important data from.

Summary points

- Humanized bone marrow, liver, and thymus (hu-BLT) mice are permissive to ZIKV infection.
- Hu-BLT mice generate neutralizing human antibody responses.
- ZIKV infection of human B cells and hematopoietic stem cells (HSC), as determined in hu-BLT mice, has many implications for viral pathogenesis and persistence.
- NRG humanized mice encoding the HLA class II antigen binding domain molecules (DRAG-hu mice) and hu-BLT mice are useful models for ZIKV immunity and vaccine testing studies.
- The neonatal microcephaly model using newborn Rag2$^{-/-}\gamma c^{-/-}$ mice are useful for studying Zika neuropathology.

Acknowledgments

Work done in Akkina laboratory was supported by NIH and US grants RO1 AI123234 and ROI AI120021 to R. Akkina. The CCTSI work was partially funded by CTSA Grant 5UL1TR002535-02.

References

Adams Waldorf, K. M., Stencel-Baerenwald, J. E., Kapur, R. P., Studholme, C., Boldenow, E., Vornhagen, J., … Rajagopal, L. (2016). Fetal brain lesions after subcutaneous inoculation of Zika virus in a pregnant nonhuman primate. *Nature Medicine*, *22*(11), 1256–1259. https://doi.org/10.1038/nm.4193.

Akkina, R. (2013). New generation humanized mice for virus research: Comparative aspects and future prospects. *Virology*, *435*(1), 14–28. https://doi.org/10.1016/j.virol.2012.10.007.

Akkina, R., Allam, A., Balazs, A. B., Blankson, J. N., Burnett, J. C., Casares, S., … Zack, J. A. (2016). Improvements and limitations of humanized mouse models for HIV research: NIH/NIAID "meet the experts" 2015 workshop summary. *AIDS Research and Human Retroviruses*, *32*(2), 109–119. https://doi.org/10.1089/AID.2015.0258.

Aliota, M. T., Bassit, L., Bradrick, S. S., Cox, B., Garcia-Blanco, M. A., Gavegnano, C., … Weaver, S. C. (2017). Zika in the Americas, year 2: What have we learned? What gaps remain? A report from the global virus network. *Antiviral Research*, *144*, 223–246. https://doi.org/10.1016/j.antiviral.2017.06.001.

Aliota, M. T., Dudley, D. M., Newman, C. M., Mohr, E. L., Gellerup, D. D., Breitbach, M. E., … O'Connor, D. H. (2016). Heterologous protection against Asian Zika virus challenge in rhesus macaques. *PLoS Neglected Tropical Diseases*, *10*(12), e0005168. https://doi.org/10.1371/journal.pntd.0005168.

Allam, A., Majji, S., Peachman, K., Jagodzinski, L., Kim, J., Ratto-Kim, S., … Rao, M. (2015). TFH cells accumulate in mucosal tissues of humanized-DRAG mice and are highly permissive to HIV-1. *Scientific Reports*, *5*, 10443. https://doi.org/10.1038/srep10443.

Barzon, L., Percivalle, E., Pacenti, M., Rovida, F., Zavattoni, M., Del Bravo, P., … Baldanti, F. (2018). Virus and antibody dynamics in travelers with acute Zika virus infection. *Clinical Infectious Diseases*, *66*(8), 1173–1180. https://doi.org/10.1093/cid/cix967.

Brainard, D. M., Seung, E., Frahm, N., Cariappa, A., Bailey, C. C., Hart, W. K., … Tager, A. M. (2009). Induction of robust cellular and humoral virus-specific adaptive immune responses in human immunodeficiency virus-infected humanized BLT mice. *Journal of Virology*, *83*(14), 7305–7321. https://doi.org/10.1128/JVI.02207-08.

Caine, E. A., Jagger, B. W., & Diamond, M. S. (2018). Animal models of Zika virus infection during pregnancy. *Viruses*, *10*(11). https://doi.org/10.3390/v10110598.

Calvet, G., Aguiar, R. S., Melo, A. S. O., Sampaio, S. A., de Filippis, I., Fabri, A., … de Filippis, A. M. B. (2016). Detection and sequencing of Zika virus from amniotic fluid of fetuses with microcephaly in Brazil: A case study. *The Lancet Infectious Diseases*, *16*(6), 653–660. https://doi.org/10.1016/S1473-3099(16)00095-5.

Calvet, G. A., Kara, E. O., Giozza, S. P., Botto-Menezes, C. H. A., Gaillard, P., de Oliveira Franca, R. F., … Team, Z. S. (2018). Study on the persistence of Zika virus (ZIKV) in body fluids of patients with ZIKV infection in Brazil. *BMC Infectious Diseases*, *18*(1), 49. https://doi.org/10.1186/s12879-018-2965-4.

Crawford, L. B., Tempel, R., Streblow, D. N., Kreklywich, C., Smith, P., Picker, L. J., … Caposio, P. (2017). Human cytomegalovirus induces cellular and humoral virus-specific immune responses in humanized BLT mice. *Scientific Reports*, *7*(1), 937. https://doi.org/10.1038/s41598-017-01051-5.

de Fatima Vasco Aragao, M., van der Linden, V., Brainer-Lima, A. M., Coeli, R. R., Rocha, M. A., Sobral da Silva, P., … Valenca, M. M. (2016). Clinical features and neuroimaging (CT and MRI) findings in presumed Zika virus related congenital infection and microcephaly: Retrospective case series study. *BMJ*, *i1901*, 353. https://doi.org/10.1136/bmj.i1901.

do Rosario, M. S., de Jesus, P. A., Vasilakis, N., Farias, D. S., Novaes, M. A., Rodrigues, S. G., … de Siqueira, I. C. (2016). Guillain-Barre syndrome after Zika virus infection in Brazil. *The American Journal of Tropical Medicine and Hygiene*, *95*(5), 1157–1160. https://doi.org/10.4269/ajtmh.16-0306.

Dowd, K. A., Ko, S. Y., Morabito, K. M., Yang, E. S., Pelc, R. S., DeMaso, C. R., … Graham, B. S. (2016). Rapid development of a DNA vaccine for Zika virus. *Science*, *354*(6309), 237–240. https://doi.org/10.1126/science.aai9137.

Driggers, R. W., Ho, C. Y., Korhonen, E. M., Kuivanen, S., Jaaskelainen, A. J., Smura, T., … Vapalahti, O. (2016). Zika virus infection with prolonged maternal viremia and fetal brain abnormalities. *The New England Journal of Medicine*, *374*(22), 2142–2151. https://doi.org/10.1056/NEJMoa1601824.

Foo, S. S., Chen, W., Chan, Y., Bowman, J. W., Chang, L. C., Choi, Y., … Jung, J. U. (2017). Asian Zika virus strains target CD14(+) blood monocytes and induce M2-skewed immunosuppression during pregnancy. *Nature Microbiology*, *2*(11), 1558–1570. https://doi.org/10.1038/s41564-017-0016-3.

Foy, B. D., Kobylinski, K. C., Chilson Foy, J. L., Blitvich, B. J., Travassos da Rosa, A., Haddow, A. D., … Tesh, R. B. (2011). Probable non-vector-borne transmission of Zika virus, Colorado, USA. *Emerging Infectious Diseases*, *17*(5), 880–882. https://doi.org/10.3201/eid1705.101939.

Frias-Staheli, N., Dorner, M., Marukian, S., Billerbeck, E., Labitt, R. N., Rice, C. M., & Ploss, A. (2014). Utility of humanized BLT mice for analysis of dengue virus infection and antiviral drug testing. *Journal of Virology*, *88*(4), 2205–2218. https://doi.org/10.1128/JVI.03085-13.

Hazin, A. N., Poretti, A., Di Cavalcanti Souza Cruz, D., Tenorio, M., van der Linden, A., Pena, L. J., … Microcephaly Epidemic Research Group. (2016). Computed tomographic findings in microcephaly associated with Zika virus. *The New England Journal of Medicine*, *374*(22), 2193–2195. https://doi.org/10.1056/NEJMc1603617.

Ito, R., Takahashi, T., Katano, I., & Ito, M. (2012). Current advances in humanized mouse models. *Cellular & Molecular Immunology*, *9*(3), 208–214. https://doi.org/10.1038/cmi.2012.2.

Jaiswal, S., Pazoles, P., Woda, M., Shultz, L. D., Greiner, D. L., Brehm, M. A., & Mathew, A. (2012). Enhanced humoral and HLA-A2-restricted dengue virus-specific T-cell responses in humanized BLT NSG mice. *Immunology*, *136*(3), 334–343. https://doi.org/10.1111/j.1365-2567.2012.03585.x.

Kuruvilla, J. G., Troyer, R. M., Devi, S., & Akkina, R. (2007). Dengue virus infection and immune response in humanized RAG2(−/−)gamma(c)(−/−) (RAG-hu) mice. *Virology*, *369*(1), 143–152. https://doi.org/10.1016/j.virol.2007.06.005.

Lan, P., Tonomura, N., Shimizu, A., Wang, S., & Yang, Y. G. (2006). Reconstitution of a functional human immune system in immunodeficient mice through combined human fetal thymus/liver and CD34 + cell transplantation. *Blood*, *108*(2), 487–492. https://doi.org/10.1182/blood-2005-11-4388.

Long, B. R., & Stoddart, C. A. (2012). Alpha interferon and HIV infection cause activation of human T cells in NSG-BLT mice. *Journal of Virology*, *86*(6), 3327–3336. https://doi.org/10.1128/JVI.06676-11.

Melkus, M. W., Estes, J. D., Padgett-Thomas, A., Gatlin, J., Denton, P. W., Othieno, F. A., … Garcia, J. V. (2006). Humanized mice mount specific adaptive and innate immune responses to EBV and TSST-1. *Nature Medicine*, *12*(11), 1316–1322. https://doi.org/10.1038/nm1431.

Mendoza, M., Ballesteros, A., Qiu, Q., Pow Sang, L., Shashikumar, S., Casares, S., & Brumeanu, T. D. (2018). Generation and testing anti-influenza human monoclonal antibodies in a new humanized mouse model (DRAGA: HLA-A2. HLA-DR4. Rag1 KO. IL-2Rgammac KO. NOD). *Human Vaccines & Immunotherapeutics*, *14*(2), 345–360. https://doi.org/10.1080/21645515.2017.1403703.

Michlmayr, D., Andrade, P., Gonzalez, K., Balmaseda, A., & Harris, E. (2017). CD14(+)CD16(+) monocytes are the main target of Zika virus infection in peripheral blood mononuclear cells in a paediatric study in Nicaragua. *Nature Microbiology*, *2*(11), 1462–1470. https://doi.org/10.1038/s41564-017-0035-0.

Miner, J. J., Cao, B., Govero, J., Smith, A. M., Fernandez, E., Cabrera, O. H., … Diamond, M. S. (2016). Zika virus infection during pregnancy in mice causes placental damage and fetal demise. *Cell*, *165*(5), 1081–1091. https://doi.org/10.1016/j.cell.2016.05.008.

Mlakar, J., Korva, M., Tul, N., Popovic, M., Poljsak-Prijatelj, M., Mraz, J., … Avsic Zupanc, T. (2016). Zika virus associated with microcephaly. *The New England Journal of Medicine*, *374*(10), 951–958. https://doi.org/10.1056/NEJMoa1600651.

Morrison, T. E., & Diamond, M. S. (2017). Animal models of Zika virus infection, pathogenesis, and immunity. *Journal of Virology*, *91*(8). https://doi.org/10.1128/JVI.00009-17.

Nazerai, L., Pravsgaard Christensen, J., & Randrup Thomsen, A. (2019). A 'Furry-Tale' of Zika virus infection: What have we learned from animal models? *Viruses*, *11*(1). https://doi.org/10.3390/v11010029.

Nguyen, S. M., Antony, K. M., Dudley, D. M., Kohn, S., Simmons, H. A., Wolfe, B., … Golos, T. G. (2017). Highly efficient maternal-fetal Zika virus transmission in pregnant rhesus macaques. *PLoS Pathogens*, *13*(5), e1006378. https://doi.org/10.1371/journal.ppat.1006378.

Oliveira Melo, A. S., Malinger, G., Ximenes, R., Szejnfeld, P. O., Alves Sampaio, S., & Bispo de Filippis, A. M. (2016). Zika virus intrauterine infection causes fetal brain abnormality and microcephaly: Tip of the iceberg? *Ultrasound in Obstetrics & Gynecology*, *47*(1), 6–7. https://doi.org/10.1002/uog.15831.

Paploski, I. A., Prates, A. P., Cardoso, C. W., Kikuti, M., Silva, M. M., Waller, L. A., … Ribeiro, G. S. (2016). Time lags between exanthematous illness attributed to Zika virus, Guillain-Barre syndrome, and microcephaly, Salvador, Brazil. *Emerging Infectious Diseases*, *22*(8), 1438–1444. https://doi.org/10.3201/eid2208.160496.

Paz-Bailey, G., Rosenberg, E. S., Doyle, K., Munoz-Jordan, J., Santiago, G. A., Klein, L., … Sharp, T. M. (2018). Persistence of Zika virus in body fluids—Final report. *The New England Journal of Medicine*, *379*(13), 1234–1243. https://doi.org/10.1056/NEJMoa1613108.

Pena, L. J., Miranda Guarines, K., Duarte Silva, A. J., Sales Leal, L. R., Mendes Felix, D., Silva, A., … de Freitas, A. C. (2018). In vitro and in vivo models for studying Zika virus biology. *The Journal of General Virology*, *99*(12), 1529–1550. https://doi.org/10.1099/jgv.0.001153.

Schmitt, K., Charlins, P., Veselinovic, M., Kinner-Bibeau, L., Hu, S., Curlin, J., … Akkina, R. (2018). Zika viral infection and neutralizing human antibody response in a BLT humanized mouse model. *Virology*, *515*, 235–242. https://doi.org/10.1016/j.virol.2017.12.026.

Seung, E., & Tager, A. M. (2013). Humoral immunity in humanized mice: A work in progress. *The Journal of Infectious Diseases*, *208*(Suppl. 2), S155–S159. https://doi.org/10.1093/infdis/jit448.

Shultz, L. D., Brehm, M. A., Garcia-Martinez, J. V., & Greiner, D. L. (2012). Humanized mice for immune system investigation: Progress, promise and challenges. *Nature Reviews. Immunology*, *12*(11), 786–798. https://doi.org/10.1038/nri3311.

Szaba, F. M., Tighe, M., Kummer, L. W., Lanzer, K. G., Ward, J. M., Lanthier, P., … Lin, J. S. (2018). Zika virus infection in immunocompetent pregnant mice causes fetal damage and placental pathology in the absence of fetal infection. *PLoS Pathogens*, *14*(4), e1006994. https://doi.org/10.1371/journal.ppat.1006994.

Traggiai, E., Chicha, L., Mazzucchelli, L., Bronz, L., Piffaretti, J. C., Lanzavecchia, A., & Manz, M. G. (2004). Development of a human adaptive immune system in cord blood cell-transplanted mice. *Science*, *304*(5667), 104–107. https://doi.org/10.1126/science.1093933.

Vesnaver, T. V., Tul, N., Mehrabi, S., Parissone, F., Strafela, P., Mlakar, J., … Popovic, M. (2017). Zika virus associated microcephaly/micrencephaly-fetal brain imaging in comparison with neuropathology. *BJOG*, *124*(3), 521–525. https://doi.org/10.1111/1471-0528.14423.

Wege, A. K., Melkus, M. W., Denton, P. W., Estes, J. D., & Garcia, J. V. (2008). Functional and phenotypic characterization of the humanized BLT mouse model. *Current Topics in Microbiology and Immunology*, *324*, 149–165. https://doi.org/10.1007/978-3-540-75647-7_10.

Xu, D., Li, C., Qin, C. F., & Xu, Z. (2019). Update on the animal models and underlying mechanisms for ZIKV-induced microcephaly. *Annual Review of Virology*, *6*(1), 459–479. https://doi.org/10.1146/annurev-virology-092818-015740.

Yi, G., Xu, X., Abraham, S., Petersen, S., Guo, H., Ortega, N., … Manjunath, N. (2017). A DNA vaccine protects human immune cells against Zika virus infection in humanized mice. *eBioMedicine*, *25*, 87–94. https://doi.org/10.1016/j.ebiom.2017.10.006.

Yockey, L. J., Varela, L., Rakib, T., Khoury-Hanold, W., Fink, S. L., Stutz, B., … Iwasaki, A. (2016). Vaginal exposure to Zika virus during pregnancy leads to fetal brain infection. *Cell*, *166*(5), 1247–1256. e1244 https://doi.org/10.1016/j.cell.2016.08.004.

Chapter 42

Use of liver cells to discover novel peptides for anti-Zika strategies

Ahmad Suhail Khazali[a] and Rohana Yusof[b]

[a]*Faculty of Applied Sciences, MARA University of Technology (UiTM) Perlis Branch, Arau, Malaysia,* [b]*Department of Molecular Medicine, Faculty of Medicine, University of Malaya, Kuala Lumpur, Malaysia*

Abbreviations

ALB	albumin
ALT	alanine aminotransferase
AST	aspartate aminotransferase
DILI	drug-induced liver injury
Env	envelope
hPSC-HLC	human pluripotent stem cell-derived hepatocyte-like cells
MTase	methyltransferase
NS2B3pro	NS2B-NS3 protease
NS3-hel	NS3 helicase
RdRp	RNA-dependent RNA polymerase
SAM	*S*-adenosylmethionine

Introduction

Zika virus (ZIKV) is an arbovirus belonging to the *Flaviviridae* family that includes other notorious flaviviruses such as dengue virus (DENV), Japanese encephalitis virus, West Nile virus (WNV), and yellow fever virus (YFV). Prior to a series of ZIKV outbreaks, ZIKV infection in humans was rare and was typically associated with mild clinical symptoms such as fever, conjunctivitis, rash, and arthralgias. Investigations following the 2013 and 2015 ZIKV outbreaks established the causal relationship between ZIKV and neurological complications such as Guillain-Barré syndrome in adults and microcephaly in newborns (Cao-Lormeau et al., 2016; Rasmussen, Jamieson, Honein, & Petersen, 2016).

Given the severity of its clinical complications, extensive efforts were geared toward understanding the pathophysiology of ZIKV for vaccine and/or antiviral drug development. ZIKV infection has been detected not only in the brain and central nervous system but also in other organs and cell types (Chan et al., 2016; Sousa et al., 2017). This chapter will discuss the relevance of using liver cells in ZIKV studies in three subsections: hepatotoxicity of drugs, liver cells as the standard cell model in ZIKV studies, and liver complications in Zika patients. This chapter will also discuss the strategies employed in anti-ZIKV drug development using liver cells.

Hepatotoxicity is a major hurdle in drug development

The liver is crucial for many physiological processes including metabolism and detoxification, thus exposing it to hepatotoxic xenobiotics. Drug toxicity, typically observed in a subset of patients, is one of the main causes of clinical trial failure (Khetani et al., 2015). Among the toxicity observed, drug-induced liver injury (DILI) is the main cause of drug attrition at multiple stages of drug development and is also the leading cause of acute liver failure (60%) in the United States (Kaplowitz, 2005; Lee, 2013). DILI is also one of the main causes of black box warnings and drug withdrawal from the market (Lasser et al., 2002). Hepatotoxicity of antiviral drugs has been documented in human immunodeficiency virus (HIV) and hepatitis C virus (HCV) patients receiving their respective treatments (Dyson et al., 2016; Nunez, 2010). The risk of DILI is further increased in HIV patients with HCV or tuberculosis coinfection receiving the concomitant treatments (Nunez, 2010; Yimer et al., 2014).

Zika Virus Impact, Diagnosis, Control, and Models. https://doi.org/10.1016/B978-0-12-820267-8.00042-X

Current drug development process mandates drug testing in vitro and in two preclinical animal models: one rodent and one nonrodent species. The animal models, though sufficient in most cases, fail to completely model human response to drugs and predict drug toxicity in humans (Griffith, Wells, & Stolz, 2014). This is primarily due to the differences in human-animal physiology and cytochrome p450 profiles, a superfamily of enzymes important in xenobiotic metabolism (Bogaards et al., 2000). Animal models also fail to completely replicate human virus infections that sometimes demonstrate species-specific tropism and potentiate different host responses (Takaki, Oshiumi, Shingai, Matsumoto, & Seya, 2017). Similarly the standard 2D in vitro culture of human hepatocytes is not reflective of in vivo responses with numerous drawbacks (Edmondson, Broglie, Adcock, & Yang, 2014).

To better predict drug toxicity and response, investigators have opted to construct 3D human liver culture systems using various strategies [reviewed here by Griffith et al. (2014) and Khetani et al. (2015)] and complement the systems with the followings to recreate human liver microenvironment:

- Introduce medium perfusion or flow to replenish nutrients and remove waste from the cells (Beckwitt et al., 2018).
- Coculture human hepatocytes with liver nonparenchymal cells for cell signaling (Khazali, Clark, & Wells, 2017).
- Use soft scaffold to mimic liver stiffness (Clark et al., 2017).
- Improve architectural design of the system to imitate liver sinusoid and achieve partial hypoxia (Kang, Eo, Bulutoglu, Yarmush, & Usta, 2019; Vernetti et al., 2016).
- Recreate a universal culture medium containing physiologic and defined levels of metabolites to support the growth and viability of liver resident cells (Vorrink, Zhou, Ingelman-Sundberg, & Lauschke, 2018).

3D human liver culture systems have been successfully utilized to model sustained HBV infection and phenocopy host responses against HBV infection (Ortega-Prieto et al., 2018). Importantly, utilization of these technologies was able to predict the toxicity of DILI-causing drugs including fialuridine, a nucleoside analog that was developed to treat HBV but caused severe liver injury and mortality (Vorrink et al., 2018).

In short, assessing liver toxicity is one of the critical parameters in drug development pipeline. The use of 3D culture systems has not been reported in ZIKV research but may serve as alternative tools in anti-ZIKV drug safety testing due to decreasing ZIKV circulation post-2015 outbreak.

Liver cells as the model cell system for ZIKV studies

Liver cells (Huh7 cells) have been used as the standard cell system in flavivirus studies due to impaired innate immunity that permits ZIKV infection and replication. Specifically, ZIKV has been shown to alter the architecture of endoplasmic reticulum (ER) and mitochondria, modify cytoskeleton network, and affect physiological processes.

Cortese et al. (2017) reported that ZIKV infection in liver cells resulted in the formation of zippered rough ER and convoluted membrane (CM) of smooth ER with vesicle packets (VPs), which are the viral RNA replication factories, clustering at the periphery of the CM and in the perinuclear region. Interestingly the formation of CM and zippered rough ER was not observed in ZIKV-infected human neural progenitor cells (hNPCs). Additionally the VPs formed in the hNPCs are smaller than those in liver cells. These differences could be attributed to the availability of cell-specific factors and could be the reason for efficient ZIKV infection in human pluripotent stem cell-derived hepatocyte-like cells (hPSC-HLC) when compared with hPSC-derived neuronal cells (Tricot, Helsen, Kaptein, Neyts, & Verfaillie, 2018). However, regardless of virus strains and cell types, the pores of the VPs, through which neosynthesized viral RNA exits the VPs, were similar in size, implying that pore formation is due to conserved viral or host factors (Cortese et al., 2017). VP formation was also accompanied by an extensive cytoskeleton remodeling in ZIKV-infected Huh7 cells (Cortese et al., 2017).

In a separate study the same group observed a substantial mitochondrial elongation in ZIKV-infected Huh7 cells and demonstrated that the perturbation of mitochondria dynamics affected ZIKV replication (Chatel-Chaix et al., 2016). Specifically, knockdown of DRP1 (a fission factor) or mitofusin2 (a fusion factor) increased and decreased ZIKV replication, respectively, potentially by modulating innate immune responses.

ZIKV could also suppress innate immunity by repressing interferon-stimulated genes (ISGs) such as viperin. Overexpression of viperin restricted ZIKV-MR766 infection in Huh7 cells (Van der Hoek et al., 2017). ZIKV also disrupts cellular proteostasis in Huh7 cells by exploiting HSP70 chaperone to assemble viral replication complexes (Taguwa et al., 2019).

The impaired innate immunity in Huh7 cells might mask crucial cellular processes during ZIKV infection. hPSC-HLC have been shown to acquire sufficient innate immunity upon differentiation and could be useful in understanding host-pathogen interactions (Fischer, Hay, & O'Farrelly, 2018). Tricot et al. (2018) reported that hPSC-HLC supported ZIKV replication and infectious virion production but, unlike Huh7 cells, hPSC-HLC could mount immune responses. Interestingly, treatment with polymerase inhibitors produced contrasting results where effective anti-ZIKV effects were observed in Huh7 cells but not in immunocompetent hPSC-HLC. However, this result awaits validation, and better differentiation method may be required to improve drug uptake and metabolic functions of hPSC-HLC.

As a standard cell system for flaviviruses, investigators have utilized liver cells to screen and identify host factors that favor viral replication. Marceau et al. (2016) employed a pooled CRISPR genetic screening to generate gene knockout Huh7.5 cells and identified the catalytic subunits of oligosaccharyltransferase complex, namely, staurosporine and temperature-sensitive (STT) 3A and 3B, to be crucial for DENV genome replication. The enriched genes from the CRISPR screening were investigated in ZIKV infection and established ZIKV dependency on STT3A only. Similarly, CRISPR screening identified 12 factors that are required for WNV infection including signal peptidase complex subunit 1 (SPCS1). ZIKV dependency on SPCS1 was established as ZIKV failed to produce infectious virion in SPCS1$^{-/-}$ Huh7.5 cells (Zhang et al., 2016). Another study identified ZIKV dependency on HSP70 for viral entry and capsid assembly in Huh7 cells (Taguwa et al., 2019).

In conclusion, hepatocytes serve as an excellent model cell system for flavivirus infection and have generated extensive knowledge on ZIKV pathophysiology. These studies have identified some cell-specific differences that could be important in understanding ZIKV biology in different host cells. More importantly, these studies have shed light on potential anti-ZIKV targets.

Liver complications in ZIKV-infected patients

Liver complications in ZIKV-infected patients are rare. The first purported report on Zika infection during a jaundice outbreak in 1954 was a misidentification (Wikan & Smith, 2017), and liver involvement in ZIKV infection was not reported until the 2015 outbreak. Table 1 summarizes some of the clinical findings involving ZIKV infection in the liver including a case of severe liver injury and reduced liver function in a previously healthy individual. Although rare, these observations

TABLE 1 Liver involvement in ZIKV-infected patients.

Subject	ZIKV validation	Liver complications	Other complications
7 Neonates (Sousa et al., 2017)	ZIKV RNA detected in the cerebrospinal fluid and in organ samples (brain, liver, lung, heart, spleen, and kidney). One neonate was also positive for DENV. All are HIV negative	5 Neonates had moderate to severe hydropic liver degeneration, megamitochondria, steatosis, hepatocyte necrosis, and periportal lymphocytic inflammation	Microcephaly, severe depletion of neural progenitor cells, macroscopic calcification, brain lymphocytic inflammation, pulmonary hypoplasia, focal glomerular sclerosis
61-year-old man. History of ischemic heart and chronic kidney disease (Rajahram et al., 2019)	ZIKV NS5 and Env genes detected in kidney, spleen, liver, and heart samples. Negative for DENV, malaria, and *Leptospira*. ZIKV positive in kidney and heart tissues via ISH	Steatosis	Cardiomegaly, calcified coronary artery, bilateral lung edema, contracted right kidney, splenic congestion
29-year-old man. No prior medical history (Wu et al., 2017)	Virus isolated from serum was positive for ZIKV RNA, IgG, and IgM. Negative for DENV, CHIKV, HCMV, HSV-1, HIV, HCV, EBV, and *Treponema pallidum*	Prolonged elevation of AST and ALT levels and low ALB	Thrombocytopenia, heart injury
17 patients from a travel center after ZIKV PHEIC. 24–62 years old (Sokal et al., 2016)	ZIKV positive by RT-PCR or IgM in plasma or urine samples	1 Case of severe hepatic cytolysis	Most patients suffer mild symptoms. 2 cases of low platelet count
3 Atypical lethal cases associated with ZIKV infection. 59-, 61-, and 64-year-old males (Zonneveld et al., 2016)	Plasma samples were ZIKV positive by RT-PCR. Negative for DENV, CHIKV, HIV, leptospirosis, and bacteria by serology test or RT-PCR	2 Patients had elevated AST and ALT	All patients had multiple organ failure including respiratory, circulatory, and renal failure

warrant for vigilance in diagnosing ZIKV infection and assessing anti-ZIKV drug toxicity since both the virus and the drugs might exacerbate liver complications as observed in patients with HIV and HCV coinfection.

Anti-ZIKV drug development

Generally, antiviral drugs are divided into two modes of action: (i) direct-acting agents (DAAs) targeting active sites or conserved regions of viral proteins and (ii) host-targeting agents (HTAs) targeting host factors required for ZIKV infection and/or replication. Fig. 1 summarizes some of the targets and strategies to prevent or inhibit ZIKV. This section will discuss several key targets and summarize some of the studies utilizing liver cells in anti-ZIKV drug discovery.

The targets and the drugs

ZIKV envelope (ZIKV-env) protein mediates critical aspects of virus infection, making it an ideal target to prevent ZIKV infection (Fig. 2). The fusion loop epitope (FLE) on domain II is crucial for viral-endosomal membrane fusion under acidic conditions (Fig. 2; Dai et al., 2016). Another important target is a conserved region within the stem of the envelope protein (Yu et al., 2017).

ZIKV nonstructural (NS) proteins are important in viral RNA replication and thus may serve as ideal targets to block ZIKV genome replication. ZIKV NS5 contains two domains where the domain near the N-terminus encodes for a methyltransferase (MTase). MTase methylates guanosine N-7 and ribose 2′-O of viral RNA cap to enhance genome stability and translation efficiency. The MTase also interacts with the template RNA and contributes to RNA elongation (Zhao et al., 2017). SAM-binding site, GTP-cap-binding pocket, and RNA-binding site are the important and the most conserved regions of the MTase based on alignment analyses comparing 73 flaviviruses sequences (Fig. 3A; Duan et al., 2017). Other important and unique features of ZIKV MTase include (i) a conserved hydrophobic pocket next to SAM-binding site and (ii) interacts with guanosine cap differently than cellular MTase (Duan et al., 2017). These features may be utilized to develop specific anti-ZIKV without affecting host MTase.

The C-terminus of ZIKV NS5 encodes for an RNA-dependent RNA polymerase (RdRp) that synthesizes the viral genome. Structural and computational analyses identified several druggable sites of NS5-RdRp such as four flexible loops in the finger subdomain that modulate RdRp conformation during replication, the catalytic active residues (D535, D665, and D666), a priming loop that positions the nucleotides during replication and promotes the formation of the initiation complex, and a zinc-binding pocket near the NTP entry channel that stabilizes the incoming NTP (Fig. 3B; Duan et al., 2017).

ZIKV NS3 encodes for two enzymes: viral protease (NS3pro) on the N-terminus and viral helicase (NS3-hel) on the C-terminus. ZIKV protease requires about 40 residues from the NS2B domain to activate the NS3pro. The catalytic active residues (S135-H51-D75) reside within the NS3pro domain (Fig. 4; Lei et al., 2016). The apo conformation of ZIKV NS2B3pro assumes a preopen conformation in which the NS2B cofactor is disordered, rendering the enzyme incapable of binding to the substrate (Lee et al., 2017). Binding of substrates or inhibitors induces conformational changes to a closed conformation (Fig. 4B).

ZIKV NS3-hel contains motifs that are associated with ATP binding and hydrolysis, interdomain communication, and RNA binding (Jain, Coloma, Garcia-Sastre, & Aggarwal, 2016). FTMap analysis identified two druggable hotspots: site 1 is close to the ATP site (Fig. 5A), while site 2 is within the RNA-binding groove (Fig. 5B; Jain et al., 2016). Another study identified specific residues of ATP-binding sites (G199 and K200) and RNA-binding site (D410 and K431) that are crucial for NS3-hel activity (Xu et al., 2019). Other key targets are the conserved residues in domain III, especially N569 and E573, that interact with NS5 to maintain helicase activity during replication (Xu et al., 2019).

The conserved residues, active sites, and other key residues of these structural and NS proteins are the prime targets in DAA development. In-depth analyses of these viral proteins are reviewed here (Valente & Moraes, 2019), and examples of compounds or drugs that can inhibit these NS proteins are summarized here (Baz & Boivin, 2019; Wang et al., 2019), but reports on NS3-hel inhibitors are limited due to low hit rate of NS3-hel unwinding and ATPase assays (Abrams, Solis, & Nath, 2017).

HTA development is based on ZIKV pathophysiology (Fig. 1). Targeting the receptors for viral entry substantially reduced ZIKV infection (Hamel et al., 2015) but may require a combination of multiple agents as the ablation of AXL, one of the entry receptors, alone did not protect against ZIKV infection (Wells et al., 2016). Inhibiting clathrin with chlorpromazine and blocking endosomal acidification with chloroquine potently blocked ZIKV infection (Persaud, Martinez-Lopez, Buffone, Porcelli, & Diaz-Griffero, 2018). Other pharmacological interventions targeting viral attachment, endocytosis, and fusion are reviewed here (Abrams et al., 2017). Numerous host factors such as HSP70,

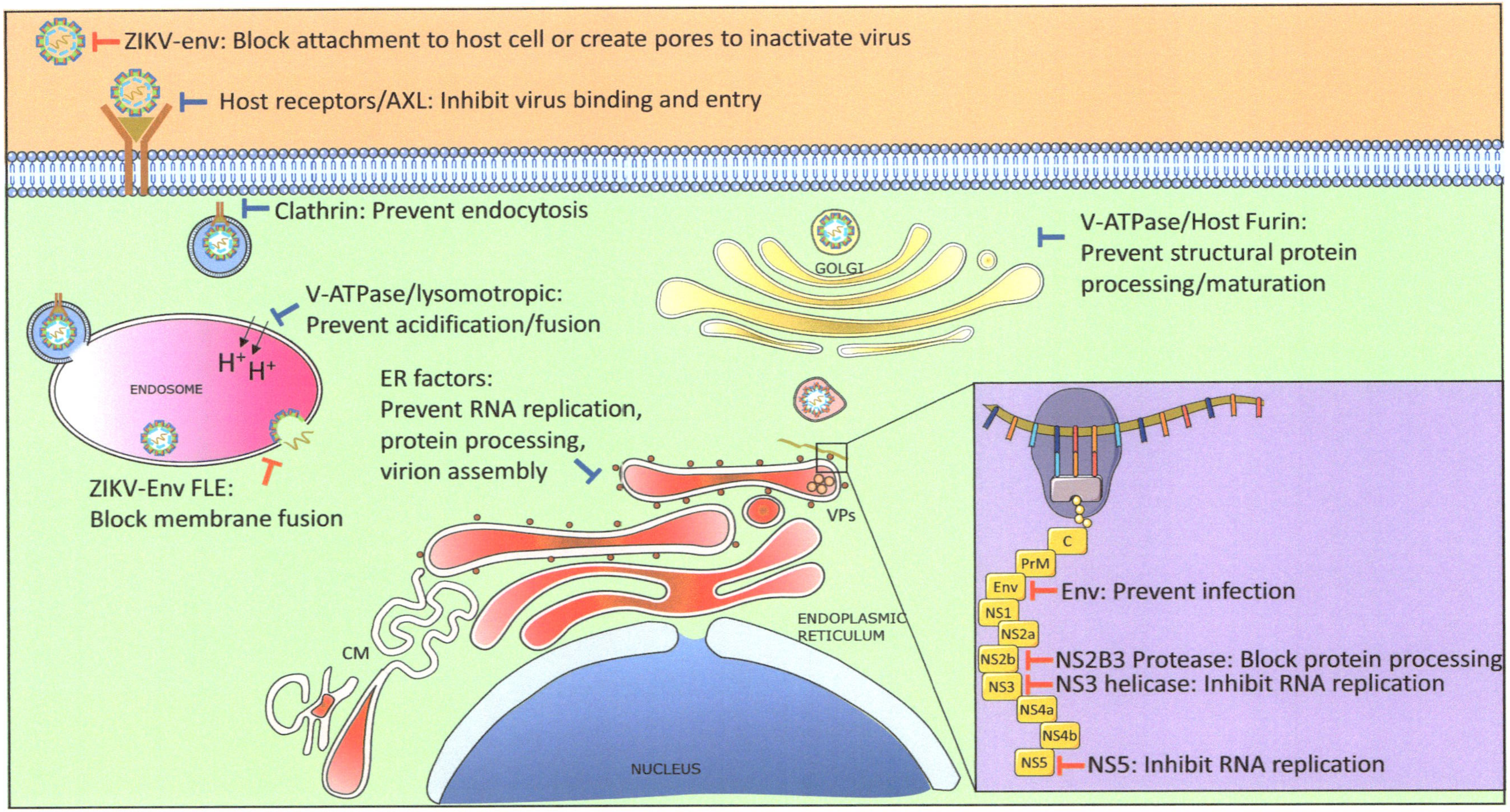

FIG. 1 Summary of anti-ZIKV target. *Black arrows* indicate the progression of viral life cycle. *Blunt-ended arrows* indicate inhibition of viral *(red)* and host *(blue)* targets. The artwork was prepared using materials from SMART website (Laureau, 2005).

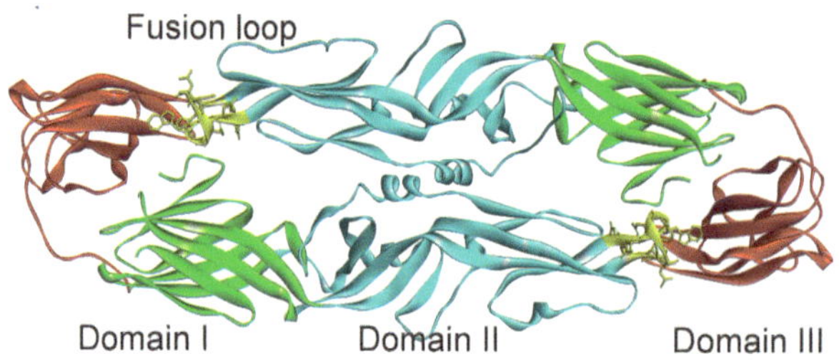

FIG. 2 Crystal structure of ZIKV-env monomer. ZIKV-env (PDB 5JHM) consists of three domains: B-barrel-shaped domain I *(green)*, finger-like domain II *(cyan)*, and immunoglobulin-like domain III *(red)* (Dai et al., 2016). Domain II contains the fusion loop epitope (FLE) *(yellow)*. The structure was visualized using BIOVIA Discovery Studio Visualizer software.

OST complex, reticulon, and ER membrane complex are crucial in viral RNA replication, viral protein processing, and virion assemblies (Aktepe, Liebscher, Prier, Simmons, & Mackenzie, 2017; Marceau et al., 2016; Savidis et al., 2016; Taguwa et al., 2019). Pharmacological inhibitors targeting these factors may be utilized to abrogate ZIKV RNA replication. A key virion maturation step is the cleavage of premembrane by host furin under low pH conditions. Inhibition of this step by directly inhibiting host furin (Imran et al., 2019) or by preventing the acidification of the Golgi microenvironment (Sabino et al., 2019) reduced the production of infectious ZIKV progenies. Several purine/pyrimidine inhibitors, inosine monophosphate dehydrogenase inhibitors, and other classes of drugs targeting other host factors have also been shown to exhibit potent anti-ZIKV activities (Baz & Boivin, 2019; Saiz & Martin-Acebes, 2017). However, the clinical efficacy and safety, especially in pregnant women of these agents, are yet to be determined.

There are several advantages and disadvantages of DAAs and HTAs. The DAAs are virus specific, thus reducing potential toxicity and side effects. However, DAA targets are prone to mutations due to the lack of proofreading ability of viral polymerase that can lead to reduced drug efficacy or resistance (Taguwa et al., 2019). Moreover, rapid ZIKV genome mutation rate (Logan, 2016) may limit the number of druggable targets and reduce the spectrum of drug action. The HTAs are free from the effects of error-prone viral polymerase but require comprehensive studies to prevent off-target effects. Viruses are also less likely to develop resistance against HTAs, but host factor redundancy and viral mutations may reduce ZIKV dependency on the targeted host factors.

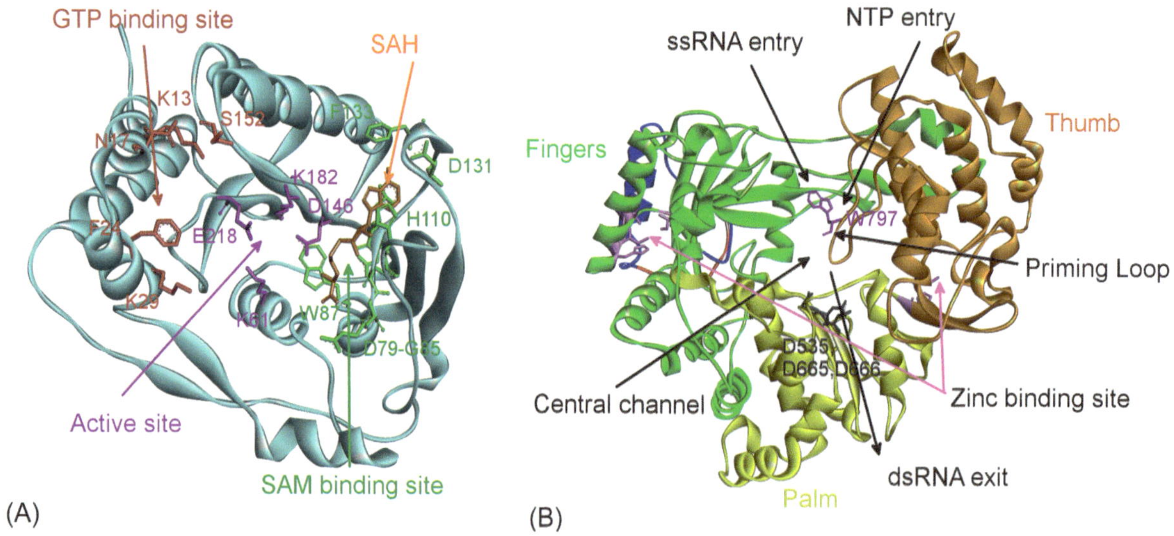

FIG. 3 Crystal structure of ZIKV NS5. (A) ZIKV MTase crystal structure *(cyan)* containing *S*-adenosylhomocysteine molecule *(orange stick)*. Important regions: SAM-binding site *(green)*, GTP-binding site *(red)*, and MTase active site *(pink)*. Key residues for each site are shown as stick. (B) ZIKV RdRp assumes a canonical right-hand conformation with three subdomains. $Zinc^{2+}$ binds to two sites *(pink arrows)*. The catalytic active residues are in the palm subdomain *(gray stick)*. A priming loop (*orange* with a *purple* key residue) extends from the thumb subdomain into the central channel. The RdRp is linked to MTase by a linker *(blue)* and an extension *(red)*. MTase and RdRp structures were prepared using PDB 5U0B (Zhao et al., 2017).

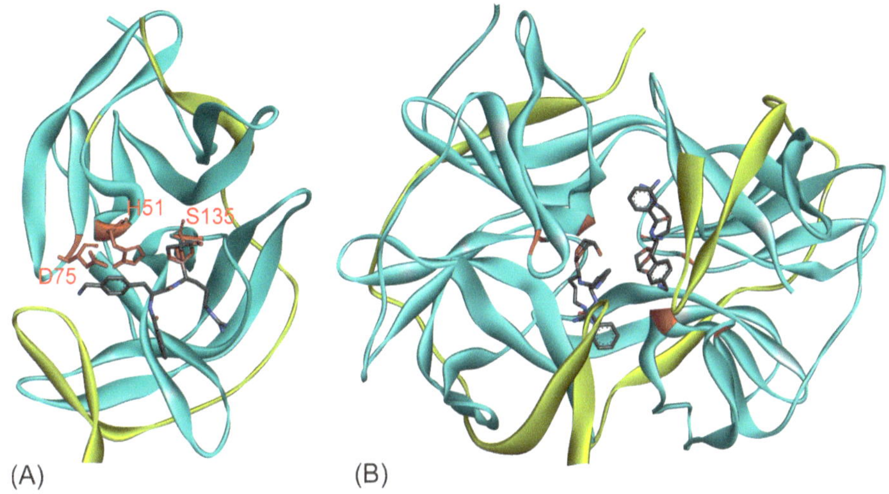

FIG. 4 Crystal structure of ZIKV NS2B3pro. ZIKV protease (PDB 5LC0) consists of NS2B cofactor *(yellow)* and NS3 protease domain *(cyan)* (Lei et al., 2016). (A) The catalytic triad residues (H51, D75 S135 in *red*) are within NS3 protease domain. Shown here is a boronic acid cn-716 inhibitor *(stick)* interacting with the triad. (B) Closed conformation of a NS2B3pro dimer with inhibitors binding to the catalytic triad *(stick)*.

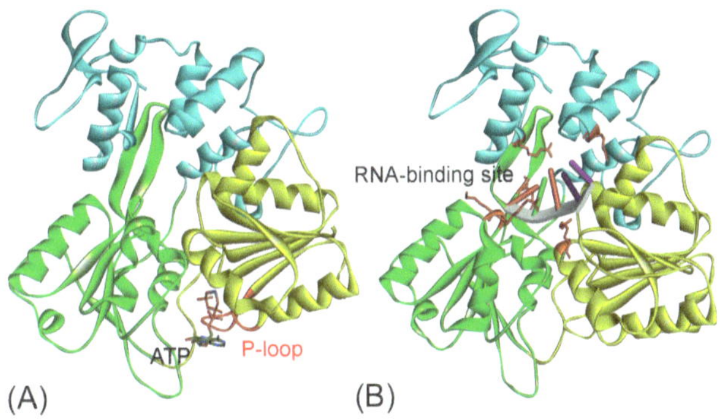

FIG. 5 Crystal structure of ZIKV NS3-hel. NS3-hel contains 3 domains: domain I *(yellow)*, domain II *(green)*, and domain III *(cyan)*. (A) ATP (stick) binds to the ATP-binding site containing the P-loop *(red)*. (B) ssRNA (structure) binds to the RNA-binding site, which is lined with positively charged residues *(orange)*. The structures were prepared using PDB 5GJB and 5GJC (Tian et al., 2016).

Utilizing liver cells in anti-ZIKV development strategies

Several anti-ZIKV development studies utilizing liver cells are illustrated here. Yu et al. (2017) performed an alignment analysis to identify a conserved stem region of ZIKV-env and designed a 33-amino acid peptide designated as Z2. Z2 inactivated ZIKV by disrupting the viral membrane, causing pore formation and the release of the viral genome. Importantly, Z2 peptide was not cytotoxic in vitro (Vero, BHK21, and Huh7 cells) and in vivo, could cross the placental barrier, blocked vertical transmission, and prevented ZIKV-induced death. This study represents a comprehensive preclinical drug development steps utilizing multiple in silico, in vitro, and in vivo techniques.

Yuan et al. (2017) performed a structure-based screening of 8227 entries and identified several approved drugs as potential anti-ZIKV agents targeting the catalytic triad of NS2B3pro. Novobiocin, which can be used during pregnancy, inhibited NS2B3pro in protease biochemical assay and significantly blocked ZIKV replication in Vero and Huh7 cells with selectivity indexes (SI) of 32.5 and 28.9, respectively. Novobiocin also improved the clinical outcome and the survival of infected mice and could be repurposed to treat ZIKV infection.

Abdulrahman, Khazali, Teoh, Rothan, and Yusof (2019) utilized a similar strategy in which HyperChem software was used to optimize a template peptide, producing short cationic peptides that could inhibit ZIKV NS2B3pro at low micromolar concentrations. HADDOCK docking simulation predicted peptide interaction with the catalytic triad. Potency was improved when the short peptides were fused to MAP30 peptide with minimal cytotoxic effects in human liver cells.

Barrows et al. (2016) screened 774 FDA-approved drugs in Huh7 cells and identified several drugs with potent anti-ZIKV activities such as ivermectin, mycophenolic acid, and daptomycin. However, some of these drugs are unsafe for pregnant women or inefficient in crossing the blood-brain barrier (Abrams et al., 2017).

Besides screening, investigators also tested specific compounds based on previous studies. Some of the tested compounds exhibited anti-ZIKV activity with minimal cytotoxic effect in Huh7 cells such as nucleoside analog sofosbuvir (Sacramento et al., 2017).

Concluding remarks

Antiviral drug development is a challenging process. Currently, antiviral drugs are approved to treat only nine types of virus infection excluding *Flavivirus* infection (De Clercq & Li, 2016). Liver toxicity is one of the main problems in drug development. Additional considerations in anti-ZIKV drug development include safety in pregnant women and the ability to cross the placenta and the blood-brain barrier. Several promising anti-ZIKV drugs have been developed but awaiting clinical validation. The erratic pattern of Zika outbreaks and the expansion of Zika vector geographic regions necessitate continuous research on vaccine and anti-ZIKV drug development. These studies should be complemented with vector control and other preventative studies.

Policy and procedures

A typical strategy in drug development is compound screening. Investigators typically employ computational, biochemical, or in vitro assays to screen thousands of compounds against a specific target. The computational approaches include ligand-based, structure-based, and fragment-based drug design and in silico drug repurposing. The methods and software to perform these approaches are summarized here (Mottin et al., 2018; Ramharack & Soliman, 2016). Structure-based methods such as protein-ligand docking, structure-based pharmacophore, and molecular dynamics simulations are widely used in anti-ZIKV drug design. Biochemical screening assays are performed by expressing and purifying key ZIKV enzymes, namely, protease, helicase, and RdRp. The assays are typically fluorescence or luminescence based that allow high-throughput analyses. In vitro assays are also microscopy, fluorescence, or luminescence based and typically used to assess the reduction of virus yield (VY) or cytopathic effect (CPE). VY reduction is detected via qRT-PCR or immunolabeling, whereas CPE reduction is quantified by labeling the viable cells with specific dyes or by measuring the level of apoptosis markers. Alves, Vielle, Thiel, and Pfaender (2018) summarized other tools and assays used in anti-Zika drug development. The subsequent steps of anti-ZIKV drug development are similar to other drug development pipeline, which include in vivo toxicity and efficacy testing, pharmacokinetic study, and clinical trials.

Mini-dictionary of terms

Hepatotoxicity. Liver injury or damage due to exposure to chemicals or nonpharmacological agents.
Steatosis. A fatty appearance of the liver due to lipid accumulation.
Hydropic liver degeneration. A foamy appearance of the hepatocytes due to ER swelling related to increased intracellular water.
S-adenosylmethionine (SAM). Methyl donor in ZIKV MTase activity.
Selectivity index (SI). The ratio of cytotoxicity to antiviral activity (CC_{50}/IC_{50}) where a higher SI indicates a more effective and safer drug.

Key facts: Drug-induced liver injury (DILI)

- 60% of acute liver failure patients in the United States.
- Divided into two: intrinsic DILI and idiosyncratic DILI.
- Intrinsic DILI (80% of DILI): predictable and dose-related injury. Hyperacute liver injury. Mainly caused by acetaminophen overdose.
- Idiosyncratic DILI (20% of DILI): dose-independent and unpredictable (in a subset of drug users). Subacute liver injury. Caused by prescribed drugs or complementary/alternative medicines.

Summary points

- The use of liver cells in anti-ZIKV drug development is primarily to assess hepatotoxicity.
- Liver cells are the standard cell system in ZIKV studies.
- Liver complications in ZIKV-infected patients may complicate anti-ZIKV treatment.
- Investigators employ in silico, in vitro, and in vivo techniques to target specific viral proteins or host factors.
- DAAs are virus specific, but mutation may cause resistance. Toxicity might be an issue for the HTAs.

References

Abdulrahman, A. Y., Khazali, A. S., Teoh, T. C., Rothan, H. A., & Yusof, R. (2019). Novel peptides inhibit Zika NS2B-NS3 serine protease and virus replication in human hepatic cell line. *International Journal of Peptide Research and Therapeutics*. https://doi.org/10.1007/s10989-019-09808-4.

Abrams, R. P. M., Solis, J., & Nath, A. (2017). Therapeutic approaches for Zika virus infection of the nervous system. *Neurotherapeutics*, *14*(4), 1027–1048. https://doi.org/10.1007/s13311-017-0575-2.

Aktepe, T. E., Liebscher, S., Prier, J. E., Simmons, C. P., & Mackenzie, J. M. (2017). The host protein reticulon 3.1A is utilized by flaviviruses to facilitate membrane remodelling. *Cell Reports*, *21*(6), 1639–1654. https://doi.org/10.1016/j.celrep.2017.10.055.

Alves, M. P., Vielle, N. J., Thiel, V., & Pfaender, S. (2018). Research models and tools for the identification of antivirals and therapeutics against Zika virus infection. *Viruses*, *10*(11). https://doi.org/10.3390/v10110593.

Barrows, N. J., Campos, R. K., Powell, S. T., Prasanth, K. R., Schott-Lerner, G., Soto-Acosta, R., … Garcia-Blanco, M. A. (2016). A screen of FDA-approved drugs for inhibitors of Zika virus infection. *Cell Host & Microbe*, *20*(2), 259–270. https://doi.org/10.1016/j.chom.2016.07.004.

Baz, M., & Boivin, G. (2019). Antiviral agents in development for Zika virus infections. *Pharmaceuticals (Basel)*, *12*(3). https://doi.org/10.3390/ph12030101.

Beckwitt, C. H., Clark, A. M., Wheeler, S., Taylor, D. L., Stolz, D. B., Griffith, L., & Wells, A. (2018). Liver 'organ on a chip. *Experimental Cell Research*, *363*(1), 15–25. https://doi.org/10.1016/j.yexcr.2017.12.023.

Bogaards, J. J., Bertrand, M., Jackson, P., Oudshoorn, M. J., Weaver, R. J., van Bladeren, P. J., & Walther, B. (2000). Determining the best animal model for human cytochrome P450 activities: A comparison of mouse, rat, rabbit, dog, micropig, monkey and man. *Xenobiotica*, *30*(12), 1131–1152. https://doi.org/10.1080/00498250010021684.

Cao-Lormeau, V. M., Blake, A., Mons, S., Lastere, S., Roche, C., Vanhomwegen, J., … Ghawche, F. (2016). Guillain-Barre Syndrome outbreak associated with Zika virus infection in French Polynesia: A case-control study. *Lancet*, *387*(10027), 1531–1539. https://doi.org/10.1016/S0140-6736(16)00562-6.

Chan, J. F., Yip, C. C., Tsang, J. O., Tee, K. M., Cai, J. P., Chik, K. K., … Yuen, K. Y. (2016). Differential cell line susceptibility to the emerging Zika virus: Implications for disease pathogenesis, non-vector-borne human transmission and animal reservoirs. *Emerging Microbes & Infections*, *5*, e93. https://doi.org/10.1038/emi.2016.99.

Chatel-Chaix, L., Cortese, M., Romero-Brey, I., Bender, S., Neufeldt, C. J., Fischl, W., … Bartenschlager, R. (2016). Dengue virus perturbs mitochondrial morphodynamics to dampen innate immune responses. *Cell Host & Microbe*, *20*(3), 342–356. https://doi.org/10.1016/j.chom.2016.07.008.

Clark, A. M., Wheeler, S. E., Young, C. L., Stockdale, L., Shepard Neiman, J., Zhao, W., … Wells, A. (2017). A liver microphysiological system of tumor cell dormancy and inflammatory responsiveness is affected by scaffold properties. *Lab on a Chip*, *17*(1), 156–168. https://doi.org/10.1039/c6lc01171c.

Cortese, M., Goellner, S., Acosta, E. G., Neufeldt, C. J., Oleksiuk, O., Lampe, M., … Bartenschlager, R. (2017). Ultrastructural characterization of Zika virus replication factories. *Cell Reports*, *18*(9), 2113–2123. https://doi.org/10.1016/j.celrep.2017.02.014.

Dai, L., Song, J., Lu, X., Deng, Y. Q., Musyoki, A. M., Cheng, H., … Gao, G. F. (2016). Structures of the Zika virus envelope protein and its complex with a flavivirus broadly protective antibody. *Cell Host & Microbe*, *19*(5), 696–704. https://doi.org/10.1016/j.chom.2016.04.013.

De Clercq, E., & Li, G. D. (2016). Approved antiviral drugs over the past 50 years. *Clinical Microbiology Reviews*, *29*(3), 695–747. https://doi.org/10.1128/Cmr.00102-15.

Duan, W., Song, H., Wang, H., Chai, Y., Su, C., Qi, J., … Gao, G. F. (2017). The crystal structure of Zika virus NS5 reveals conserved drug targets. *The EMBO Journal*, *36*(7), 919–933. https://doi.org/10.15252/embj.201696241.

Dyson, J. K., Hutchinson, J., Harrison, L., Rotimi, O., Tiniakos, D., Foster, G. R., … McPherson, S. (2016). Liver toxicity associated with sofosbuvir, an NS5A inhibitor and ribavirin use. *Journal of Hepatology*, *64*(1), 234–238. https://doi.org/10.1016/j.jhep.2015.07.041.

Edmondson, R., Broglie, J. J., Adcock, A. F., & Yang, L. (2014). Three-dimensional cell culture systems and their applications in drug discovery and cell-based biosensors. *Assay and Drug Development Technologies*, *12*(4), 207–218. https://doi.org/10.1089/adt.2014.573.

Fischer, L., Hay, D. C., & O'Farrelly, C. (2018). Innate immunity in stem cell-derived hepatocytes. *Philosophical Transactions of the Royal Society of London. Series B, Biological Sciences*, *373*(1750). https://doi.org/10.1098/rstb.2017.0220.

Griffith, L. G., Wells, A., & Stolz, D. B. (2014). Engineering liver. *Hepatology*, 1426–1434. https://doi.org/10.1002/hep.27150.

Hamel, R., Dejarnac, O., Wichit, S., Ekchariyawat, P., Neyret, A., Luplertlop, N., … Misse, D. (2015). Biology of Zika virus infection in human skin cells. *Journal of Virology*, *89*(17), 8880–8896. https://doi.org/10.1128/JVI.00354-15.

Imran, M., Saleemi, M. K., Chen, Z., Wang, X., Zhou, D., Li, Y., … Ye, J. (2019). Decanoyl-Arg-Val-Lys-Arg-chloromethylketone: An antiviral compound that acts against flaviviruses through the inhibition of furin-mediated prM cleavage. *Viruses*, *11*(11). https://doi.org/10.3390/v11111011.

Jain, R., Coloma, J., Garcia-Sastre, A., & Aggarwal, A. K. (2016). Structure of the NS3 helicase from Zika virus. *Nature Structural & Molecular Biology*, *23*(8), 752–754. https://doi.org/10.1038/nsmb.3258.

Kang, Y. B. A., Eo, J., Bulutoglu, B., Yarmush, M. L., & Usta, O. B. (2019). Progressive hypoxia-on-a-chip: An in vitro oxygen gradient model for capturing the effects of hypoxia on primary hepatocytes in health and disease. *Biotechnology and Bioengineering*. https://doi.org/10.1002/bit.27225.

Kaplowitz, N. (2005). Idiosyncratic drug hepatotoxicity. *Nature Reviews. Drug Discovery*, *4*(6), 489–499. https://doi.org/10.1038/nrd1750.

Khazali, A. S., Clark, A. M., & Wells, A. (2017). A pathway to personalizing therapy for metastases using liver-on-a-chip platforms. *Stem Cell Reviews*, *13* (3), 364–380. https://doi.org/10.1007/s12015-017-9735-3.

Khetani, S. R., Berger, D. R., Ballinger, K. R., Davidson, M. D., Lin, C., & Ware, B. R. (2015). Microengineered liver tissues for drug testing. *Journal of Laboratory Automation*, *20*(3), 216–250. https://doi.org/10.1177/2211068214566939.

Lasser, K. E., Allen, P. D., Woolhandler, S. J., Himmelstein, D. U., Wolfe, S. M., & Bor, D. H. (2002). Timing of new black box warnings and withdrawals for prescription medications. *JAMA*, *287*(17), 2215–2220. https://doi.org/10.1001/jama.287.17.2215.

Laureau, O. (2005). Servier medical art. Les Laboratoires Servier. Retrieved from https://smart.servier.com.

Lee, W. M. (2013). Drug-induced acute liver failure. *Clinics in Liver Disease*, *17*(4), 575–586. viii https://doi.org/10.1016/j.cld.2013.07.001.

Lee, H., Ren, J., Nocadello, S., Rice, A. J., Ojeda, I., Light, S., … Johnson, M. E. (2017). Identification of novel small molecule inhibitors against NS2B/NS3 serine protease from Zika virus. *Antiviral Research*, *139*, 49–58. https://doi.org/10.1016/j.antiviral.2016.12.016.

Lei, J., Hansen, G., Nitsche, C., Klein, C. D., Zhang, L., & Hilgenfeld, R. (2016). Crystal structure of Zika virus NS2B-NS3 protease in complex with a boronate inhibitor. *Science*, *353*(6298), 503–505. https://doi.org/10.1126/science.aag2419.

Logan, I. S. (2016). ZIKA—How fast does this virus mutate? *Dongwuxue Yanjiu*, *37*(2), 110–115. https://doi.org/10.13918/j.issn.2095-8137.2016.2.110.

Marceau, C. D., Puschnik, A. S., Majzoub, K., Ooi, Y. S., Brewer, S. M., Fuchs, G., … Carette, J. E. (2016). Genetic dissection of Flaviviridae host factors through genome-scale CRISPR screens. *Nature*, *535*(7610), 159–163. https://doi.org/10.1038/nature18631.

Mottin, M., Borba, J., B, V., Melo, C. C., Neves, B. J., Muratov, E., … Andrade, C.H. (2018). Computational drug discovery for the Zika virus. *Brazilian Journal of Pharmaceutical Sciences*, *54*. https://doi.org/10.1590/s2175-97902018000001002.

Nunez, M. (2010). Clinical syndromes and consequences of antiretroviral-related hepatotoxicity. *Hepatology*, *52*(3), 1143–1155. https://doi.org/10.1002/hep.23716.

Ortega-Prieto, A. M., Skelton, J. K., Wai, S. N., Large, E., Lussignol, M., Vizcay-Barrena, G., … Dorner, M. (2018). 3D microfluidic liver cultures as a physiological preclinical tool for hepatitis B virus infection. *Nature Communications*, *9*(1), 682. https://doi.org/10.1038/s41467-018-02969-8.

Persaud, M., Martinez-Lopez, A., Buffone, C., Porcelli, S. A., & Diaz-Griffero, F. (2018). Infection by Zika viruses requires the transmembrane protein AXL, endocytosis and low pH. *Virology*, *518*, 301–312. https://doi.org/10.1016/j.virol.2018.03.009.

Rajahram, G. S., Hale, G., Bhatnagar, J., Hiu, J., Thayan, R., William, T., … Yeo, T. W. (2019). Postmortem evidence of disseminated Zika virus infection in an adult patient. *International Journal of Infectious Diseases*, *83*, 163–166. https://doi.org/10.1016/j.ijid.2019.01.047.

Ramharack, P., & Soliman, M. E. S. (2016). Zika virus drug targets: A missing link in drug design and discovery—A route map to fill the gap. *RSC Advances*, *6*(73), 68719–68731. https://doi.org/10.1039/c6ra12142j.

Rasmussen, S. A., Jamieson, D. J., Honein, M. A., & Petersen, L. R. (2016). Zika virus and birth defects—Reviewing the evidence for causality. *New England Journal of Medicine*, *374*(20), 1981–1987. https://doi.org/10.1056/NEJMsr1604338.

Sabino, C., Basic, M., Bender, D., Elgner, F., Himmelsbach, K., & Hildt, E. (2019). Bafilomycin A1 and U18666A efficiently impair ZIKV infection. *Viruses*, *11*(6). https://doi.org/10.3390/v11060524.

Sacramento, C. Q., de Melo, G. R., de Freitas, C. S., Rocha, N., Hoelz, L. V., Miranda, M., … Souza, T. M. (2017). The clinically approved antiviral drug sofosbuvir inhibits Zika virus replication. *Scientific Reports*, *7*, 40920. https://doi.org/10.1038/srep40920.

Saiz, J. C., & Martin-Acebes, M. A. (2017). The race to find antivirals for Zika virus. *Antimicrobial Agents and Chemotherapy*, *61*(6). https://doi.org/10.1128/AAC.00411-17.

Savidis, G., McDougall, W. M., Meraner, P., Perreira, J. M., Portmann, J. M., Trincucci, G., … Brass, A. L. (2016). Identification of Zika virus and dengue virus dependency factors using functional genomics. *Cell Reports*, *16*(1), 232–246. https://doi.org/10.1016/j.celrep.2016.06.028.

Sokal, A., D'Ortenzio, E., Houhou-Fidouh, N., Brichler, S., Dorchies, J., Cabras, O., … Matheron, S. (2016). Zika virus infection: Report of the first imported cases in a Paris travel centre. *Journal of Travel Medicine*, *24*(1). https://doi.org/10.1093/jtm/taw066.

Sousa, A. Q., Cavalcante, D. I. M., Franco, L. M., Araujo, F. M. C., Sousa, E. T., Valenca-Junior, J. T., … Pompeu, M. M. L. (2017). Postmortem findings for 7 neonates with congenital Zika virus infection. *Emerging Infectious Diseases*, *23*(7), 1164–1167. https://doi.org/10.3201/eid2307.162019.

Taguwa, S., Yeh, M. T., Rainbolt, T. K., Nayak, A., Shao, H., Gestwicki, J. E., … Frydman, J. (2019). Zika virus dependence on host Hsp70 provides a protective strategy against infection and disease. *Cell Reports*, *26*(4), 906–920. e903 https://doi.org/10.1016/j.celrep.2018.12.095.

Takaki, H., Oshiumi, H., Shingai, M., Matsumoto, M., & Seya, T. (2017). Development of mouse models for analysis of human virus infections. *Microbiology and Immunology*, *61*(3–4), 107–113. https://doi.org/10.1111/1348-0421.12477.

Tian, H., Ji, X., Yang, X., Zhang, Z., Lu, Z., Yang, K., … Rao, Z. (2016). Structural basis of Zika virus helicase in recognizing its substrates. *Protein & Cell*, *7*(8), 562–570. https://doi.org/10.1007/s13238-016-0293-2.

Tricot, T., Helsen, N., Kaptein, S. J. F., Neyts, J., & Verfaillie, C. M. (2018). Human stem cell-derived hepatocyte-like cells support Zika virus replication and provide a relevant model to assess the efficacy of potential antivirals. *PLoS One*, *13*(12), e0209097. https://doi.org/10.1371/journal.pone.0209097.

Valente, A. P., & Moraes, A. H. (2019). Zika virus proteins at an atomic scale: How does structural biology help us to understand and develop vaccines and drugs against Zika virus infection? *Journal of Venomous Animals and Toxins Including Tropical Diseases*, *25*, e20190013. https://doi.org/10.1590/1678-9199-JVATITD-2019-0013.

Van der Hoek, K. H., Eyre, N. S., Shue, B., Khantisitthiporn, O., Glab-Ampi, K., Carr, J. M., … Beard, M. R. (2017). Viperin is an important host restriction factor in control of Zika virus infection. *Scientific Reports*, *7*(1), 4475. https://doi.org/10.1038/s41598-017-04138-1.

Vernetti, L. A., Senutovitch, N., Boltz, R., DeBiasio, R., Shun, T. Y., Gough, A., & Taylor, D. L. (2016). A human liver microphysiology platform for investigating physiology, drug safety, and disease models. *Experimental Biology and Medicine (Maywood, NJ)*, *241*(1), 101–114. https://doi.org/10.1177/1535370215592121.

Vorrink, S. U., Zhou, Y., Ingelman-Sundberg, M., & Lauschke, V. M. (2018). Prediction of drug-induced hepatotoxicity using long-term stable primary hepatic 3D spheroid cultures in chemically defined conditions. *Toxicological Sciences*, *163*(2), 655–665. https://doi.org/10.1093/toxsci/kfy058.

Wang, L., Liang, R., Gao, Y., Li, Y., Deng, X., Xiang, R., … Yu, F. (2019). Development of small-molecule inhibitors against Zika virus infection. *Frontiers in Microbiology*, *10*, 2725. https://doi.org/10.3389/fmicb.2019.02725.

Wells, M. F., Salick, M. R., Wiskow, O., Ho, D. J., Worringer, K. A., Ihry, R. J., … Eggan, K. (2016). Genetic ablation of AXL does not protect human neural progenitor cells and cerebral organoids from Zika virus infection. *Cell Stem Cell*, *19*(6), 703–708. https://doi.org/10.1016/j.stem.2016.11.011.

Wikan, N., & Smith, D. R. (2017). First published report of Zika virus infection in people: Simpson, not MacNamara. *Lancet Infectious Diseases*, *17*(1), 15–17. https://doi.org/10.1016/S1473-3099(16)30525-4.

Wu, Y., Cui, X., Wu, N., Song, R., Yang, W., Zhang, W., … An, J. (2017). A unique case of human Zika virus infection in association with severe liver injury and coagulation disorders. *Scientific Reports*, *7*(1), 11393. https://doi.org/10.1038/s41598-017-11568-4.

Xu, S., Ci, Y., Wang, L., Yang, Y., Zhang, L., Xu, C., … Shi, L. (2019). Zika virus NS3 is a canonical RNA helicase stimulated by NS5 RNA polymerase. *Nucleic Acids Research*, *47*(16), 8693–8707. https://doi.org/10.1093/nar/gkz650.

Yimer, G., Gry, M., Amogne, W., Makonnen, E., Habtewold, A., Petros, Z., … Aklillu, E. (2014). Evaluation of patterns of liver toxicity in patients on antiretroviral and anti-tuberculosis drugs: A prospective four arm observational study in ethiopian patients. *PLoS One*, *9*(4), e94271. https://doi.org/10.1371/journal.pone.0094271.

Yu, Y., Deng, Y. Q., Zou, P., Wang, Q., Dai, Y., Yu, F., … Lu, L. (2017). A peptide-based viral inactivator inhibits Zika virus infection in pregnant mice and fetuses. *Nature Communications*, *8*, 15672. https://doi.org/10.1038/ncomms15672.

Yuan, S. F., Chan, J. F. W., Den-Haan, H., Chik, K. K. H., Zhang, A. J., Chan, C. C. S., … Yuen, K. Y. (2017). Structure-based discovery of clinically approved drugs as Zika virus NS2B-NS3 protease inhibitors that potently inhibit Zika virus infection in vitro and in vivo. *Antiviral Research*, *145*, 33–43. https://doi.org/10.1016/j.antiviral.2017.07.007.

Zhang, R., Miner, J. J., Gorman, M. J., Rausch, K., Ramage, H., White, J. P., … Diamond, M. S. (2016). A CRISPR screen defines a signal peptide processing pathway required by flaviviruses. *Nature*, *535*(7610), 164–168. https://doi.org/10.1038/nature18625.

Zhao, B., Yi, G., Du, F., Chuang, Y. C., Vaughan, R. C., Sankaran, B., … Li, P. (2017). Structure and function of the Zika virus full-length NS5 protein. *Nature Communications*, *8*, 14762. https://doi.org/10.1038/ncomms14762.

Zonneveld, R., Roosblad, J., Staveren, J. W., Wilschut, J. C., Vreden, S. G., & Codrington, J. (2016). Three atypical lethal cases associated with acute Zika virus infection in Suriname. *IDCases*, *5*, 49–53. https://doi.org/10.1016/j.idcr.2016.07.009.

Chapter 43

In vivo mouse models to investigate the microcephaly associated with Zika virus

Raissa R. Christoff and Patricia P. Garcez
Institute of Biomedical Sciences, Federal University of Rio de Janeiro, Rio de Janeiro, Brazil

Abbreviations

CNS	central nervous system
CZS	congenital Zika syndrome
DPI	days post infection
E	embryonic day
ICV	intracerebroventricular
IFN	interferon
IP	intraperitoneal
IUGR	intrauterine growth restriction
IV	intravenous
KO	knockout
NHP	nonhuman primates
NPCs	neural progenitor cells
P	postnatal day
PFU	plaque-forming unit
SC	subcutaneous
WT	wild-type
ZIKV	Zika virus

Introduction

Zika virus (ZIKV) is a member of the Flaviviridae family belonging to the genus *Flavivirus* along with other pathogens of major public health concern such as dengue and yellow fever virus (Musso & Gubler, 2016). ZIKV is an arbovirus and *Aedes* mosquito is its main vector. The illness caused by ZIKV has mild to moderate symptoms that include fever, joint pain, conjunctivitis, and rash, which can be easily attributed to the other arboviruses leading to misdiagnosis. ZIKV was first isolated from a rhesus monkey in the Zika forest of Uganda in 1947 (Dick, 1952a, 1952b, 1952c). A few years later after, the first human case was reported (Simpson, 1964). In the following years, only sporadic cases emerged in humans restricted to the African continent (Wikan & Smith, 2016). However, in 2007, the first outbreak of the virus was reported in the Yap Islands of Micronesia, causing epidemiological interest (Lanciotti et al., 2008). In 2013 another virus outbreak occurred in French Polynesia, where the first severe neurological manifestations associated with ZIKV were observed in adults, such as Guillain-Barré syndrome (Cao-Lormeau et al., 2016). During the outbreak in Brazil, in 2015, ZIKV infection was followed by an increase in the number of reported cases of microcephaly, raising the hypothesis that the virus could be transmitted vertically from mother to fetus. ZIKV was found in the amniotic fluid, placenta, and brain tissues of compromised fetuses, leading the scientists to establish a correlation between the virus and microcephaly (Mlakar et al., 2016). Other brain abnormalities are also present in congenital Zika syndrome (CZS), including ventriculomegaly, calcifications, and corpus callosum dysgenesis (Chimelli et al., 2017; De Oliveira Melo et al., 2016). Infants with CZS also present seizures, motor problems, and visual impairments (Pessoa et al., 2018).

Zika Virus Impact, Diagnosis, Control, and Models. https://doi.org/10.1016/B978-0-12-820267-8.00043-1

The first studies to show the interaction between ZIKV and neural progenitor cells (NPCs) were performed in vitro. Tang, Hammack, et al. (2016) showed that after 3 days in vitro, ZIKV is able to infect NPCs preferentially. Also, ZIKV infection causes cell cycle deregulation and apoptosis. The same results were observed using 3D cellular models, like neurospheres and brain organoids, in which infection reduces proliferation, causes cell death, and produces a thinner cell layer resembling microcephaly (Garcez et al., 2016). These first publications were crucial in determining the causality between ZIKV and the correlated microcephaly, showing that the virus is able to infect, replicate in NPC, and disturb neuronal production. However, the brain abnormalities present in CSZ, and the outcome symptoms suggest that ZIKV pathogenesis is far more complicated than microcephaly. Indeed, pregnant women in all trimesters are susceptible to ZIKV vertical transmission, despite first trimester offering higher risk (Pool et al., 2019). Importantly, during ZIKV outbreak, some children have been born with an average cephalic perimeter, no microcephaly, but developed other more subtle CZS brain abnormalities such as language impairment and learning disabilities, detected later in their development (Nielsen-Saines et al., 2019).

Furthermore the geographic distribution of CZS cases is asymmetrical, as well as the variability of vertical transmission prevalence among human populations (Barbeito-Andrés, Schuler-Faccini, & Garcez, 2018). In this scenario, animal models are advantageous to understand the underlying mechanisms of ZIKV pathogenesis and the potential cofactors that modulate ZIKV vertical transmission (Barbeito-Andrés et al., 2020; Pedrosa et al., 2019). The development of the central nervous system (CNS) is well documented in many species, which enable researchers to study the ZIKV infection in different gestational stages (Semple, Blomgren, Gimlin, Ferriero, & Noble-Haeusslein, 2013). Also, they can control the route of infection and determine the viral replication in maternal and fetal tissues. Different animal models were developed to study the mechanisms by which ZIKV causes brain impairments. In this chapter, we summarize the most relevant ones used to model CZS.

Animal models of ZIKV infection

Different types of animal models were developed to study ZIKV pathogenesis. These include murine models like mice, hamsters, and guinea pigs, as well as chicken, piglets, and nonhuman primates (NHP). When ZIKV was discovered, wild-type (WT) mouse was the first model to investigate the pathogenesis of the African lineage. Using intraperitoneal (IP) route, Dick described that mice with 2 weeks old or older could not be infected. In contrast, using the same IP viral inoculation WT younger than 2 weeks displayed susceptibility and viremia (Dick, 1952a, 1952b, 1952c). The first neurological findings were described using intracerebral inoculation, resulting in hippocampal and cortical necrosis and viral replication in mouse neural tissue (Bell, Field, & Narang, 1971; Dick, 1952a, 1952b, 1952c). However, microcephaly was not observed at that time. Only 50 years later, when ZIKV gained worldwide attention, these results were reviewed, and new experiments were performed with more recently isolated Asian strains. Mice with deficiency in interferon signaling pathway have been proved to be good models for viral infections via peripheral route (Lazear et al., 2016). Viral RNA can be found in different types of tissues, and mice present signs of disease. Knockout (KO) mice lacking type I interferon (IFN) receptor (Ifnar1$^{-/-}$) or WT mice with administration of a blocking antitype I IFN receptor antibody were most used (Dowall et al., 2016; Rossi et al., 2016). Mice lacking both type I and II receptors are even more sensitive to ZIKV infection, showing that type II IFN signaling is also relevant for antiviral response (Aliota et al., 2016).

Vertical transmission of ZIKV

Since immunocompromised mice were more susceptible to ZIKV infection, vertical transmission studies were carried out using these infection models (Table 1). Miner et al. (2016) inject with ZIKV (strain H/PF/2013) Ifnar1$^{-/-}$ pregnant mice at embryonic day 6.5 (E6.5) via subcutaneous. Female mice were mated with WT male mice, so the offspring were Ifnar1 heterozygotes and still have IFN responses. ZIKV was able to cross the placenta and cause fetal demise. The virus has been found in both maternal and placenta. Also the embryonic brain exhibited signs of apoptosis. Blocking the IFN receptor using an antibody in WT animals also allows ZIKV vertical transmission but did not result in abortions. Instead, newborns had signs of intrauterine growth restriction (IUGR). This study demonstrated that ZIKV could successfully cross the placental barrier and vertically be transmitted from mother to fetus (Miner et al., 2016). Concomitantly, CZS was modeled in vivo using SJL pregnant mice, inoculated by intravenous route. Pups presented brain malformations, with reduction of cortical layer thickness (Cugola et al., 2016). Despite SJL mice being considered immunocompetent, it is known that they display elevated levels of T cells and are susceptible for autoimmune disease (Hutchings, Varey, & Cooke, 1986).

Other studies followed up investigating the importance of IFN signaling to vertical transmission. Type III interferon plays a role in antiviral response to ZIKV in the placenta, and when deficient, it can increase infection in placenta and

TABLE 1 Mouse models of vertical transmission.

Mice strain	ZIKV strain	Dosage (PFU/mL)	Administration route	Age of administration	Analysis	Major findings	Refs.
Ifnar1$^{-/-}$	H/PF/2013	10^3	SC	E6.5 or E7.5	E13.5, E15.5, or E16.5	Infection of placental cells and fetal demise	Miner et al. (2016)
SJL	ZIKVbr	4×10^{10}	IV	E10 to E13	P0	Intrauterine growth restriction; thinner cortex	Cugola et al. (2016)
C57BL/6	SZ01	3×10^5	IP	E13.5	E17.5 and P1	Infection of radial glial cells and thinner cortex	Wu et al. (2016)
C57BL/6 and FVB/NJ	ZIKVbr	10^5	Intravascular	E5.5 to E9.5	Multiple stages	Susceptibility window for ZIKV congenital infection; cortical malformations most commonly observed in early infection such as hydrocephalus	Xavier-Neto et al. (2017)
Ifnar1-blocking	ZIKVbr	10^3	SC	E9.5	E14	IFN role in autophagy and infection	Cao, Parnell, Diamond, and Mysorekar (2017)
WT e Ifnar1$^{-/-}$ mice	FSS13025	2.5×10^4 to 5×10^5	Vaginal	E4.5 and E8.5	E18.5	Vaginal mucosa infection; intrauterine growth restriction	Yockey et al. (2016)
Ifnar1$^{-/-}$	ZIKVbr	10^3	Vaginal	E6, E9, and E12	Multiple stages	Early ZIKV infection resulted in placental insufficiency and fetal demise; infections at midstage produces reduced cranial dimensions	Jagger et al. (2017)
Ifnar1-blocking	DAKAR	10^6	Vaginal	E0	Multiple stages	ZIKV vaginal infection is controlled by hormonal stage-dependent role for IFN-λ	Caine et al. (2019)
Ifng$^{-/-}$	MR766	10^5	IV	E12.5	E15 and P2	Neural progenitor and neuronal cell reduction and vascular development disruption	Garcez et al. (2018)
C57BL/6	PRVABC59	10^4 to10^5	IV	Multiple stages	Multiple stages	Immunocompetent pregnant mice showed limited viremia. Infected embryos developed profound placental pathology and high frequency of fetal demise in early stages of ZIKV infection	Szaba et al. (2018)
C57BL6 with low protein diet	ZIKVbr	10^5	IP	E12	E15 and P0	Immunocompetent pregnant mice subjected to malnutrition vertically transmitted ZIKV to offspring causing microcephaly	Barbeito-Andrés et al. (2018)
Ifnar$^{-/-}$	FSS13025	1×10^5	Vaginal	E0	Multiple stages	Mice infected during estrus-like phase were resistant to vaginal infection. In contrast, when infected during diestrum-like phase, shows viremia and persistent vaginal replication	Tang, Young, et al. (2016)

Representative studies of ZIKV infection during pregnancy, which establishes viral transmission from mother to fetus. *PFU*, plaque-forming unit; *SC*, subcutaneous; *IV*, intravenous; *IP*, intraperitoneal; *E*, embryonic day.

embryos (Jagger et al., 2017). Nevertheless, when it is administered, type III IFN can protect against infection (Caine et al., 2019). Pregnant mice lacking type II interferon are also susceptible to ZIKV infection. Infected embryos exhibit a reduction of NPCs, neurons, and brain vasculature is impaired (Garcez et al., 2018).

Although it is debatable in which extension ZIKV can be vertically transmitted by WT mice, in some models, the viral infection is not enough to cause illness to the pregnant dam but might be enough to disturb the developing embryo (Caine, Jagger, & Diamond, 2018). Some aspects should be taken into consideration like virus strain, viral inoculation load, administration route, and gestational age. Wu et al. (2016) successfully infect WT mice at E13.5 via intraperitoneal route, causing a transient viremia after 1 day postinfection. They were able to detect ZIKV RNA in 5 of 9 placentas 3 days postinfection, which lead to infection of radial glia cells of the dorsal ventricular zone of the developing brain. A thorough analysis done using WT mice injecting ZIKV IV in early gestational ages described a window of susceptibility to ZIKV infection and fetal malformations depending on the stage of development (Xavier-Neto et al., 2017). Using the Brazilian ZIKV strain, WT pregnant mice vertically transmitted the virus and produced distinct malformations when infected between E5.5 and E.9.5. In a similar study, using intravenous injection on pregnant WT mice, the outcome of ZIKV infection was shown to depend on the viral dose and the gestational age (Szaba et al., 2018). Also, placental pathology could contribute to fetal malformations such as IUGR. Putting these results together, it is clear that placental infection is crucial to the maternal-fetal viral vertical transmission and subsequently embryonic infection. Possibly, at the early stages of mice pregnancy, the placenta barrier is still developing and could be potentially more permissive to ZIKV infection (Vermillion et al., 2017).

ZIKV is not a classical arbovirus since it can also be sexually transmitted (D'Ortenzio et al., 2016). Case reports of viral persistence in semen and infection through sexual contact raised the possibility that ZIKV could have other routes of infection, which were confirmed later by animal models (Tang, Young, et al., 2016). ZIKV replicates in the vaginal tract of Ifnar$^{-/-}$ female mice, and it depends on the hormonal phase. ZIKV also infects the vaginal mucosa of WT mice but does not cause extremely viremia and illness (Yockey et al., 2016). In pregnant Ifnar1$^{-/-}$ mice, vaginal infection in the early gestational period can lead to a severe IUGR and fetal demise. In pregnant WT mice, IUGR was also observed, and ZIKV was found in fetal brain tissue (Tang, Young, et al., 2016).

Direct administration of ZIKV in utero

The IFN pathway seems to be essential to ZIKV infection, and a deficiency in IFN signaling helps to produce animal models of vertical transmission. However, as with any viral infection, the immune system is required for antiviral response and contributes to pathogenesis itself. During brain development, it is known that the immune system plays an important role (Dantzer, 2018). Also, KO mice are great to understand cellular and molecular mechanisms, but they might be challenging strains to set behavioral parameters. Since infection of WT pregnant mice does not produce a consistent infection in offspring, several groups developed animal models of CZS by direct injecting ZIKV in utero (Table 2 and Fig. 1).

To study the direct effect of ZIKV into developing, brain different groups inject the virus intracerebroventricular/lateral ventricle (ICV) or amniotic cavity in several gestational days. This methodology is very consistent in producing the CZS results. Infected embryos show impairments in NPC proliferation, apoptosis, reduced brain size, and cortical thickness. Li et al. (2016) was first to show that ZIKV could infect embryonic mouse brain and cause microcephaly. The same group in a later publication showed that ZIKV infection in later gestation (E15.5) target glial progenitor cells, leading to a disruption in oligodendrocyte proliferation and differentiation (Li et al., 2018). ICV infection also leads to massive neuronal death and extensive microglial activation and reactive gliosis (Li et al., 2016; Shao et al., 2016). Further ZIKV infection disturbs vasculature development, causing a leaky blood-brain barrier (Shao et al., 2016).

A limitation of this approach is that it causes a severe infection and survival rate is very low. Nevertheless, Cui et al. (2017) were able to grow infected mice into puberty. These mice exhibit severe motor deficits and paralysis, cerebellar defects, and calcifications. Another research group was able to grow in utero infected mice until adulthood. ZIKV infection continues to impair the postnatal brain and alter adult neurogenesis. Also, it leads to spinal cord dysfunction (Zhang et al., 2019).

In mice the first week after birth corresponds to the third trimester of the human embryonic brain development (Thompson, Levitt, & Stanwood, 2009). Therefore some of the research around CZS is performed in postnatal infection (Table 3). ZIKV injection in the first week postnatal leads to a widespread infection in mice brain followed by apoptosis (Huang et al., 2016). ZIKV exposure right after birth leads to infection of astrocytes and neurons. The virus was also found in the retina and spinal cord (van den Pol et al., 2017). When evaluated in a short and long term, postnatal ZIKV infection can lead to anatomical alteration in different brain structures, seizures, and behavioral deficits in adulthood (De Oliveira Souza et al., 2018)

TABLE 2 Mouse models of in utero infection.

Mice strain	ZIKV strain	Dosage (PFU/mL)	Administration route	Age of administration	Analysis	Major findings	Refs.
C57BL/6	MEX1-44	1.7×10^6	ICV	E14.5	E17.5	Cell cycle arrest and apoptosis of NPC; disruption of neurovascular development	Shao et al. (2016)
CD1	MR766	10^6	Intrauterine injection	E10 or E14	Multiple stages	Intrauterine inoculation at embryonic day (E) 10 causes microglia activation and reduces cortical thickness	Vermillion et al. (2017)
C57BL/6	SZ01	500	Amniotic fluid	E15	Adult	Adult mice show visual and motor deficits with defects in retina and cerebellum	Cui et al. (2017)
ICR	SZ01	6.5×10^5	ICV	E13.5	E16.5	Infection of NPC leading to cell cycle arrest and apoptosis	Li et al. (2016)
ICR	SZ01	6.5×10^5	ICV	E15.5	Afterbirth	ZIKV infection disturbs the proliferation and differentiation of the oligodendrocyte and progenitor cells	Li et al. (2018)
C57BL/6	SMGC-1	1500	Amniotic fluid	E13.5	Afterbirth	Multiple brain abnormalities such as microcephaly and hippocampal dysplasia; damage of visual system and cerebellar malformation	Shi et al. (2018)
C57BL/6	PRVABC59	500	ICV	E13.5	Multiple stages	ZIKV infection of neural stem cells leads to a thinner cortex and a smaller brain	Zhang et al. (2019)

Representative studies performed with in utero infection to study ZIKV pathogenesis. *PFU*, plaque-forming unit; *ICV*, intracerebroventricular; *E*, embryonic day.

Other animal models to study ZIKV infection

Mice are standard models used to study brain development and infectious disease. Most of the ZIKV infection research is done in mice. However, some studies used different animal models to understand ZIKV pathogenesis (Table 4). Chicken embryos are frequently used to study embryogenesis. Goodfellow et al. (2016) demonstrated that chicken embryos are susceptible to ZIKV infection. Guinea pigs and hamster may also be used as models of ZIKV infection in adults. Infected guinea pigs are permissive to viral infection and show a sizeable antiviral response; however, it did not disturb with fetal development (Bierle et al., 2017). In hamster, STAT-2 KO pregnant animals were susceptible to ZIKV infection and able to transmit ZIKV to embryos (Siddharthan et al., 2017). Pigs have high similarity to human immune system, so they are good models to study different types of infection. When pregnant pigs are challenged with ZIKV, pups showed smaller brains and increased susceptibility to seizures (Darbellay, Lai, et al., 2017).

NHP are most used for preclinical research and vaccine development due to their proximity to humans. Also, they are natural hosts of ZIKV, which enables the viral infection and immune response research. In addition, NHP pregnancy

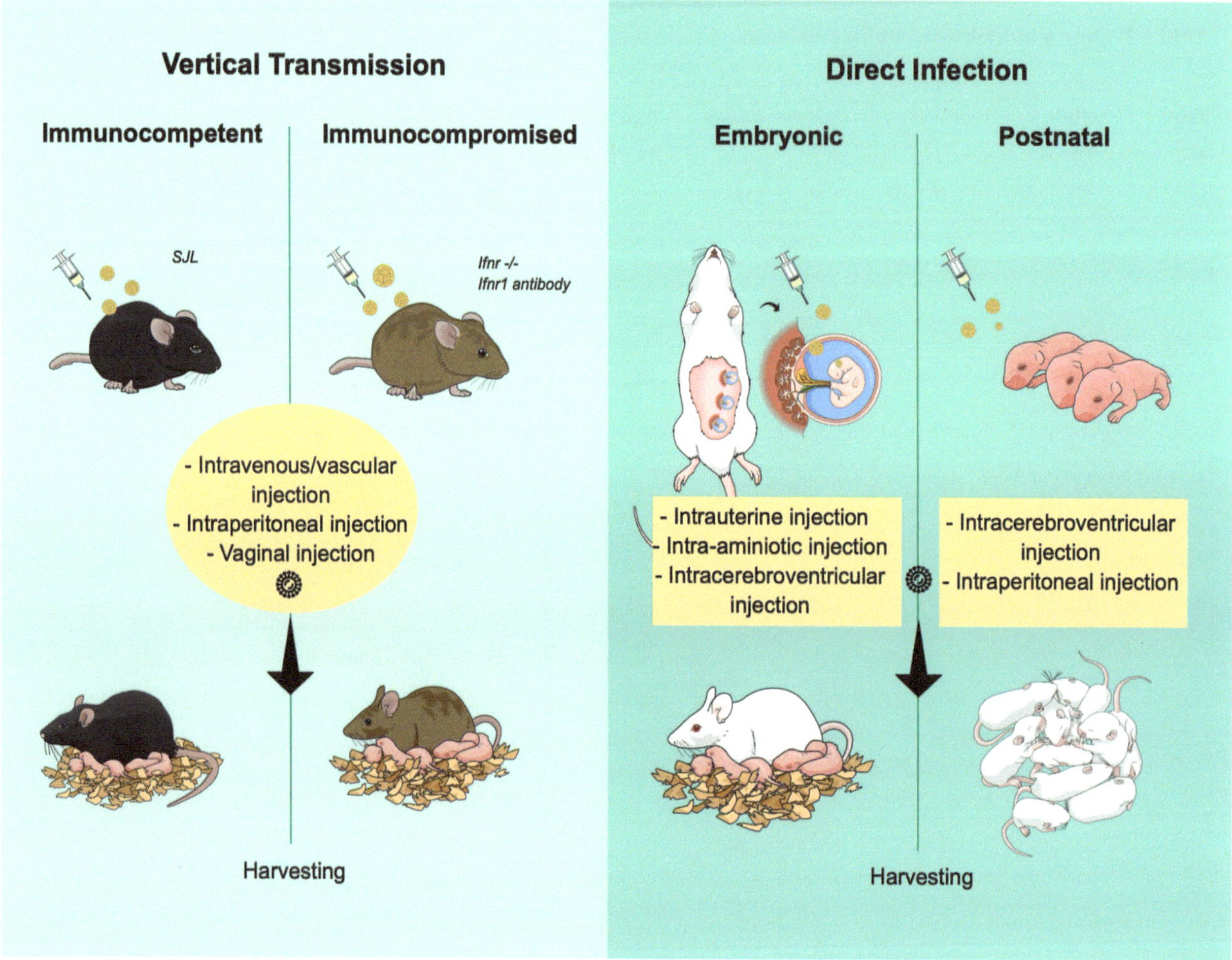

FIG. 1 Mouse models of ZIKV infection to produce congenital Zika syndrome in vivo. Mouse models of ZIKV infection were developed in immunocompetent and immunocompromised mice. For vertical transmission, mice lacking interferon signaling were the most successful in reproducing the CZS phenotype. Wild-type mice were also used for vertical transmission. For vertical transmission models, pregnant mice were injected in different gestational ages. Injections performed in early stages of development may cause fetal demise. To study the direct effect of the virus in brain development and ensure the infection, ZIKV was injected directly to the fetal brain in different gestational stages or neonatal mice. WT neonatal mice have a restricted susceptibility to ZIKV infection when the injection is via intraperitoneal (IP) or subcutaneous (SC).

resembles human gestational development and placental barrier. Pregnant rhesus macaques infected by SC route were capable to transmit vertically ZIKV to fetus and produce microcephaly. Moreover a robust immune response was observed (Nguyen et al., 2017). In another work, pregnant pigtail macaques were inoculated with ZIKV, which led to several brain abnormalities. Fetal brains showed signs of white matter hypoplasia, gliosis, and axonal injury (Adams Waldorf et al., 2018).

Conclusions

Different animal models were developed to understand ZIKV pathogenesis (Tables 1–4). As a result, it is possible to begin to explain which mechanisms ZIKV infection leads to microcephaly. Despite the different ZIKV or mouse strain used, animal age, and inoculation route, ZIKV infection in the developing brain produces a set of characteristics that is common to all approaches (Fig. 2). ZIKV infects NPCs disturbing the cell cycle and causing their apoptosis. The viral infection leads to an activation of the immune system with reactive gliosis and neuronal death. Infected animals show signs of microcephaly associated with behavioral impairments such as motor difficulties and seizures. All the features are comparable with those found in CZS children. CZS it is a complex disorder in which not only ZIKV, but also other factors could be contributing to enhancing viral susceptibility. Therefore animal models are crucial for understanding ZIKV pathogenesis.

TABLE 3 Mouse models of postnatal infection.

Mice strain	ZIKV strain	Dosage (PFU/mL)	Administration route	Age of administration	Analysis	Major findings	Refs.
ICR	FSS13025 and VEN/2016	100	ICV	P0	Multiple stages	The VEN/2016 strain causes more severe brain damage followed by several neuronal loss and oligodendrocyte developmental impairment	Zhang et al. (2017)
Swiss	ZIKVbr	10^6	SC	P3	Multiple stage; until adulthood	ZIKV persists into adulthood causing several behavior deficits	De Oliveira Souza et al. (2018)
C57BL/6	PRVABC59	$2 \times 10^{3-4}$	SC	P1 and P10	Until 2 weeks	Infiltration of lymphocytes into the brain and ZIKV infection in cerebellum; paralysis; macrophage infiltration in the brain	Manangeeswaran, Ireland, and Verthelyi (2016)
C57BL/6 and Swiss	FSS13025	6×10^4 and 4×10^5	IP	P0	1 to 10 dpi	Postnatal infection lead to infection first in astrocytes and later in neurons	van den Pol, Mao, Yang, Ornaghi, and Davis (2017)
C57BL/6	MR766	1.9×10^7	ICV	P7 and P21	4 dpi	Increased apoptosis trough the brain targeting different cell types	Huang, Abraham, Shim, Choe, and Page (2016)
Swiss	ZIKVbr		ICV and SC	P0	Multiple stages	Apoptosis and increased reactive astrocytes	Fernandes et al. (2017)
Balb/c	FSS13025 and SZ01	10	ICV	P0	26 dpi	Higher mortality rate when neonatal mice were injected with Asian lineage	Yuan et al. (2017)

Representative studies of ZIKV infection in postnatal mice. *PFU*, plaque-forming unit; *ICV*, intracerebroventricular; *SC*, subcutaneous; *P*, postnatal day; *DPI*, days post infection.

Policy and procedures: Producing mouse models of ZIKV infection

- The most common mouse models to study ZIKV vertical transmission are type 1 IFN KO mice or WT mice treated with a blocking antibody to IFN receptors.
- It is possible to obtain CZS in WT mice if ZIKV infection is associated with other cofactors that enhance viral vertical transmission, for instance, combined with a maternal protein restriction diet or maternal administration of water with saxitoxin.

TABLE 4 Alternative models of ZIKV infection.

Animal model	ZIKV strain	Dosage (PFU/mL)	Administration route	Age of administration	Analysis	Major findings	Refs.
Chicken	MEX1-44	2×10 to 2×10^4	Amniotic fluid	E2.5 and E5	E15 and E20	Chicken embryos are permissive to Zikv and show microcephaly-like features with enlarged ventricles and thinner cortex	Goodfellow et al. (2016)
STAT2 KO hamster	Malasyan (P6-740)	50 TCID50	SC	E8.5	E14.5 and postnatal	STAT2 KO pregnant hamsters are susceptible to infection, and ZIKV was detected into brain and placentas	Siddharthan et al. (2017)
Guinea pig	H/PF/2013	1×10^7 or 1×10^8	SC	E18 and E21	Multiple stages	Pup development is undisturbed; nevertheless, antibody response against ZIKV was detected in both the pups and dams	Bierle et al. (2017)
Porcine	PRVABC59	5 log10 TCID50	Intrauterine injection	E50	Afterbirth	ZIKV was found in fetal tissues, and some pups presented undersized brains and showed seizures	Darbellay, Cox, et al. (2017)
Porcine	PRVABC59	$10^{5.6}$ TCID50	Intrauterine injection	E50	Multiple stages	Mild to severe neuropathology, characterized by a depletion of cortical neurons	Wichgers Schreur, Van Keulen, Anjema, Kant, and Kortekaas (2018)

Different animal models were used to study ZIKV pathogenesis. *PFU*, plaque-forming unit; *SC*, subcutaneous; *E*, embryonic day.

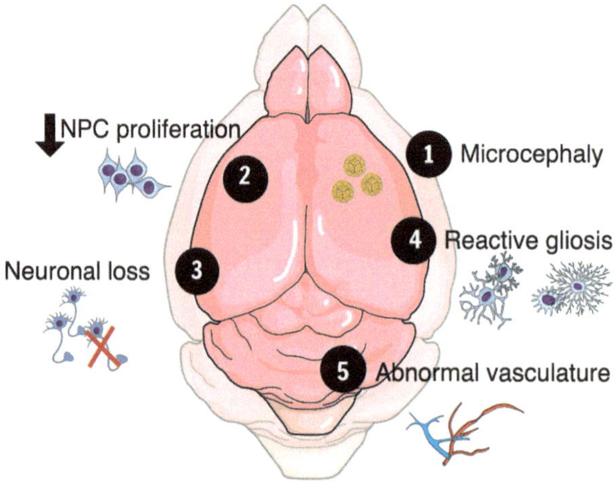

FIG. 2 Neurological aspects of ZIKV infection in newborn mice. ZIKV infection in developing mouse brain produces microcephaly followed by many cellular abnormalities such as the reduction of NPC proliferation and increase in apoptosis, neuronal loss, microglia reaction and astrocyte activation, and disruption of the vasculature.

- Injection of ZIKV early in the development (before E11) has a higher risk of abortion, and the main pathologies are associated with the placenta.
- Mouse injections of 10^3–10^6 PFU/mL of ZIKV are considered pathophysiological, mimicking the viral load in humans.
- Direct injection of ZIKV into the developing brain produces robust results, but inoculations with a low amount of viral titer are needed to analyze the effect of the virus after 7 dpi.

Mini-dictionary of terms

Arboviruses. Classification of viruses that are transmitted by arthropods, such as mosquitoes.
Ventriculomegaly. It is a condition that the brain lateral ventricles are larger than average.
Corpus callosum dysgenesis. It is a brain malformation that affects the size of the major commissure connecting both hemispheres (the corpus callosum) during development.
Interferon signaling pathway. It is an innate immune system pathway that plays a crucial role in antiviral response.
Reactive gliosis. It is a reaction to brain injury performed mostly by microglia and astrocytes, which can be observed by a change in cell morphology and molecular pattern.

Key facts of brain development

- The neocortex expanded region of the mammalian brain.
- The steps by which brain develops are very similar in different species that can be divided in proliferation of NPCs, migration of immature neurons, neuronal differentiation, axonal extension and pathfinding, gliogenesis, synapse formation, regressive events, and myelinization.
- In mice, some of these events happen postnatally, corresponding to the third trimester of human brain development. In humans, some developmental events, such as myelinization, will only be completed in the brain of adolescents and young adults.
- Despite the differences in the developmental timeline and cell composition, the key events of molecular and cellular changes are preserved between human and mice.
- The advantage using the mouse as an animal model to research the infectious diseases that affect brain formation is to understand not only the direct impact of pathogens in neural cells but also the effects of the immune system responding to the infection.

Summary points

- The Zika virus can be vertically transmitted from mother to fetus during pregnancy during the three trimesters and cause brain abnormalities known as congenital Zika syndrome.
- The hallmark of congenital Zika syndrome is microcephaly, but other brain abnormalities are also present such as ventriculomegaly, calcifications, and corpus callosum dysgenesis.
- Mouse models were developed to explain how Zika virus could cause microcephaly and other brain malformations, as well as to test different cofactors that can increase viral vertical transmission.
- Immunocompromised mice are more susceptible to ZIKV infection, especially those with IFN1 pathway deficiency that highlights the importance of IFN1 signaling controlling ZIKV infection.
- Mouse models of vertical transmission were developed both after antibody blockade of type 1 IFN signaling in WT mice and using IFN1 pathway deficiency mice.
- Direct injection into embryonic and postnatal brain leads to a microcephaly phenotype modeling the disease.
- ZIKV has a tropism to NPCs disturbing their proliferation and differentiation but also infects other types of neural cells and causes apoptosis and reactive gliosis.
- ZIKV embryonic infection leads to small brain size and thinner cortex at birth. In a long term, mice present multiple behavior abnormalities, such as motor, learning, and memory disabilities.

References

Adams Waldorf, K. M., Nelson, B. R., Stencel-Baerenwald, J. E., Studholme, C., Kapur, R. P., Armistead, B., … Rajagopal, L. (2018). Congenital Zika virus infection as a silent pathology with loss of neurogenic output in the fetal brain. *Nature Medicine*. https://doi.org/10.1038/nm.4485.

Aliota, M. T., Caine, E. A., Walker, E. C., Larkin, K. E., Camacho, E., & Osorio, J. E. (2016). Characterization of lethal Zika virus infection in AG129 mice. *PLoS Neglected Tropical Diseases*, *10*(4), 1–11. https://doi.org/10.1371/journal.pntd.0004682.

Barbeito-Andrés, J., Pezzuto, P., Higa, L. M., Dias, A. A., Vasconcelos, J. M., Santos, T. M. P., … Garcez, P. P. (2020). Congenital Zika syndrome is associated with maternal protein malnutrition. *Science Advances*, *6*(2), eaaw6284. https://doi.org/10.1126/sciadv.aaw6284.

Barbeito-Andrés, J., Schuler-Faccini, L., & Garcez, P. P. (2018). Why is congenital Zika syndrome asymmetrically distributed among human populations? *PLoS Biology*, *16*(8), 1–11. https://doi.org/10.1371/journal.pbio.2006592.

Bell, T. M., Field, E. J., & Narang, H. K. (1971). Zika virus infection of the central nervous system of mice. *Archiv für die Gesamte Virusforschung*, *35*(2–3), 183–193. https://doi.org/10.1007/BF01249709.

Bierle, C. J., Fernández-Alarcón, C., Hernandez-Alvarado, N., Zabeli, J. C., Janus, B. C., Putri, D. S., & Schleiss, M. R. (2017). Assessing Zika virus replication and the development of Zika-specific antibodies after a mid-gestation viral challenge in guinea pigs. *PLoS One*, *12*(11), 1–14. https://doi.org/10.1371/journal.pone.0187720.

Caine, E. A., Jagger, B. W., & Diamond, M. S. (2018). Animal models of Zika virus infection during pregnancy. *Viruses*, *10*(11), 1–21. https://doi.org/10.3390/v10110598.

Caine, E. A., Scheaffer, S. M., Arora, N., Zaitsev, K., Artyomov, M. N., Coyne, C. B., … Diamond, M. S. (2019). Interferon lambda protects the female reproductive tract against Zika virus infection. *Nature Communications*, *10*(1). https://doi.org/10.1038/s41467-018-07993-2.

Cao, B., Parnell, L. A., Diamond, M. S., & Mysorekar, I. U. (2017). Inhibition of autophagy limits vertical transmission of Zika virus in pregnant mice. *The Journal of Experimental Medicine*, *214*(8), 2303–2313. https://doi.org/10.1084/jem.20170957.

Cao-Lormeau, V. M., Blake, A., Mons, S., Lastère, S., Roche, C., Vanhomwegen, J., … Ghawché, F. (2016). Guillain-Barré Syndrome outbreak associated with Zika virus infection in French Polynesia: A case-control study. *The Lancet*. https://doi.org/10.1016/S0140-6736(16)00562-6.

Chimelli, L., Melo, A. S. O., Avvad-Portari, E., Wiley, C. A., Camacho, A. H. S., Lopes, V. S., … Tanuri, A. (2017). The spectrum of neuropathological changes associated with congenital Zika virus infection. *Acta Neuropathologica*, *133*(6), 983–999. https://doi.org/10.1007/s00401-017-1699-5.

Cugola, F. R., Fernandes, I. R., Russo, F. B., Freitas, B. C., Dias, J. L. M., Guimarães, K. P., … Beltrao-Braga, P. C. B. B. (2016). The Brazilian Zika virus strain causes birth defects in experimental models. *Nature*, *534*(7606), 267–271. https://doi.org/10.1038/nature18296.

Cui, L., Zou, P., Chen, E., Yao, H., Zheng, H., Wang, Q., … Zhang, J. (2017). Visual and motor deficits in grown-up mice with congenital Zika virus infection. *eBioMedicine*, *20*, 193–201. https://doi.org/10.1016/j.ebiom.2017.04.029.

D'Ortenzio, E., Matheron, S., de Lamballerie, X., Hubert, B., Piorkowski, G., Maquart, M., … Leparc-Goffart, I. (2016). Evidence of sexual transmission of Zika virus. *New England Journal of Medicine*, *374*(22), 2195–2198. https://doi.org/10.1056/NEJMc1604449.

Dantzer, R. (2018). Neuroimmune interactions: From the brain to the immune system and vice versa. *Physiological Reviews*. https://doi.org/10.1152/physrev.00039.2016.

Darbellay, J., Cox, B., Lai, K., Delgado-Ortega, M., Wheler, C., Wilson, D., … Karniychuk, U. (2017). Zika virus causes persistent infection in porcine conceptuses and may impair health in offspring. *eBioMedicine*, *25*, 73–86. https://doi.org/10.1016/j.ebiom.2017.09.021.

Darbellay, J., Lai, K., Babiuk, S., Berhane, Y., Ambagala, A., Wheler, C., … Karniychuk, U. (2017). Neonatal pigs are susceptible to experimental Zika virus infection. *Emerging Microbes and Infections*, *6*(2). https://doi.org/10.1038/emi.2016.133. e6-4.

De Oliveira Melo, A. S., Aguiar, R. S., Amorim, M. M. R., Arruda, M. B., De Oliveira Melo, F., Ribeiro, S. T. C., … Tanuri, A. (2016). Congenital Zika virus infection: Beyond neonatal microcephaly. *JAMA Neurology*, *73*(12), 1407–1416. https://doi.org/10.1001/jamaneurol.2016.3720.

De Oliveira Souza, I. N., Frost, P. S., França, J. V., Nascimento-Viana, J. B., Neris, R. L. S., Freitas, L., … Clarke, J. R. (2018). Acute and chronic neurological consequences of early-life Zika virus infection in mice. *Science Translational Medicine*, *10*(444). https://doi.org/10.1126/scitranslmed.aar2749.

Dick, G. W. (1952a). Zika virus. II. Pathogenicity and physical properties. *Transactions of the Royal Society of Tropical Medicine and Hygiene*, *46*(5), 521–534. https://doi/10.1016/0035-9203(52)90043-6.

Dick, G. W. A. (1952b). Pathogenicity and physical properties. *Transactions of the Royal Society of Tropical Medicine and Hygiene*, *46*(5), 521–534.

Dick, G. (1952c). Zika isolation and serological specificity. *Transactions of the Royal Society of Tropical Medicine and Hygiene*, *46*(5), 509–520.

Dowall, S. D., Graham, V. A., Rayner, E., Atkinson, B., Hall, G., Watson, R. J., … Hewson, R. (2016). A susceptible mouse model for Zika virus infection. *PLoS Neglected Tropical Diseases*, *10*(5), 1–13. https://doi.org/10.1371/journal.pntd.0004658.

Fernandes, N. C. C. A., Nogueira, J. S., Réssio, R. A., Cirqueira, C. S., Kimura, L. M., Fernandes, K. R., … Guerra, J. M. (2017). Experimental Zika virus infection induces spinal cord injury and encephalitis in newborn Swiss mice. *Experimental and Toxicologic Pathology*, *69*(2), 63–71. https://doi.org/10.1016/j.etp.2016.11.004.

Garcez, P. P., Loiola, E. C., Madeiro da Costa, R., Higa, L. M., Trindade, P., Delvecchio, R., … Rehen, S. K. (2016). Zika virus impairs growth in human neurospheres and brain organoids. *Science*, *352*(6287), 816–818. https://doi.org/10.1126/science.aaf6116.

Garcez, P. P., Stolp, H. B., Sravanam, S., Christoff, R. R., Ferreira, J. C. C. G., Dias, A. A., … Molnár, Z. (2018). Zika virus impairs the development of blood vessels in a mouse model of congenital infection. *Scientific Reports*, *8*(1), 1–10. https://doi.org/10.1038/s41598-018-31149-3.

Goodfellow, F. T., Tesla, B., Simchick, G., Zhao, Q., Hodge, T., Brindley, M. A., & Stice, S. L. (2016). Zika virus induced mortality and microcephaly in chicken embryos. *Stem Cells and Development*, *25*(22), 1691–1697. https://doi.org/10.1089/scd.2016.0231.

Huang, W. C., Abraham, R., Shim, B. S., Choe, H., & Page, D. T. (2016). Zika virus infection during the period of maximal brain growth causes microcephaly and corticospinal neuron apoptosis in wild type mice. *Scientific Reports*, *6*(October), 1–8. https://doi.org/10.1038/srep34793.

Hutchings, P. R., Varey, A. M., & Cooke, A. (1986). Immunological defects in SJL mice. *Immunology*, *59*(3), 445–450.

Jagger, B. W., Miner, J. J., Cao, B., Arora, N., Smith, A. M., Kovacs, A., … Diamond, M. S. (2017). Gestational stage and IFN-λ signaling regulate ZIKV infection in utero. *Cell Host and Microbe*, *22*(3). https://doi.org/10.1016/j.chom.2017.08.012. 366–376.e3.

Lanciotti, R. S., Kosoy, O. L., Laven, J. J., Velez, J. O., Lambert, A. J., Johnson, A. J., … Duffy, M. R. (2008). Genetic and serologic properties of Zika virus associated with an epidemic, Yap State, Micronesia, 2007. *Emerging Infectious Diseases*. https://doi.org/10.3201/eid1408.080287.

Lazear, H. M., Govero, J., Smith, A. M., Platt, D. J., Fernandez, E., Miner, J. J., & Diamond, M. S. (2016). A mouse model of Zika virus pathogenesis. *Cell Host & Microbe*, *19*(5), 720–730. https://doi.org/10.1016/j.chom.2016.03.010.

Li, C., Wang, Q., Jiang, Y., Ye, Q., Xu, D., Gao, F., … Xu, Z. (2018). Disruption of glial cell development by Zika virus contributes to severe microcephalic newborn mice. *Cell Discovery*, *4*(1), 1–12. https://doi.org/10.1038/s41421-018-0042-1.

Li, C., Xu, D., Ye, Q., Hong, S., Jiang, Y., Liu, X., … Xu, Z. (2016). Zika virus disrupts neural progenitor development and leads to microcephaly in mice. *Cell Stem Cell*, *19*(1), 120–126. https://doi.org/10.1016/j.stem.2016.04.017.

Manangeeswaran, M., Ireland, D. D. C., & Verthelyi, D. (2016). Zika (PRVABC59) infection is associated with t cell infiltration and neurodegeneration in CNS of immunocompetent neonatal C57Bl/6 mice. *PLoS Pathogens*, *12*(11), 1–20. https://doi.org/10.1371/journal.ppat.1006004.

Miner, J. J., Cao, B., Govero, J., Smith, A. M., Fernandez, E., Cabrera, O. H., … Diamond, M. S. (2016). Zika virus infection during pregnancy in mice causes placental damage and fetal demise. *Cell*, *165*(5), 1081–1091. https://doi.org/10.1016/j.cell.2016.05.008.

Mlakar, J., Korva, M., Tul, N., Popović, M., Poljšak-Prijatelj, M., Mraz, J., … Zupanc, T. A. (2016). Zika virus associated with microcephaly. *New England Journal of Medicine*, *374*(10), 951–958. https://doi.org/10.1056/NEJMoa1600651.

Musso, D., & Gubler, D. J. (2016). Zika virus. *Clinical Microbiology Reviews*, *29*(3), 487–524. https://doi.org/10.1128/CMR.00072-15.

Nguyen, S. M., Antony, K. M., Dudley, D. M., Kohn, S., Simmons, H. A., Wolfe, B., … Golos, T. G. (2017). Highly efficient maternal-fetal Zika virus transmission in pregnant rhesus macaques. *PLoS Pathogens*. https://doi.org/10.1371/journal.ppat.1006378.

Nielsen-Saines, K., Brasil, P., Kerin, T., Vasconcelos, Z., Gabaglia, C. R., Damasceno, L., … Moreira, M. E. (2019). Delayed childhood neurodevelopment and neurosensory alterations in the second year of life in a prospective cohort of ZIKV-exposed children. *Nature Medicine*, *25*(8), 1213–1217. https://doi.org/10.1038/s41591-019-0496-1.

Pedrosa, C.d. S. G., Souza, L. R. Q., de Lima, C. V. F., Ledur, P. F., Karmirian, K., Gomes, T. A., … Rehen, S. K. (2019). The cyanobacterial saxitoxin exacerbates neural cell death and brain malformations induced by Zika virus. *BioRxiv*. https://doi.org/10.1101/755066. 755066.

Pessoa, A., Van Der Linden, V., Yeargin-Allsopp, M., Carvalho, M. D. C. G., Ribeiro, E. M., Van Naarden Braun, K., … Moore, C. A. (2018). Motor abnormalities and epilepsy in infants and children with evidence of congenital Zika virus infection. *Pediatrics*. https://doi.org/10.1542/peds.2017-2038F.

Pool, K. L., Adachi, K., Karnezis, S., Salamon, N., Romero, T., Nielsen-Saines, K., … Pone, M. (2019). Association between neonatal neuroimaging and clinical outcomes in Zika-exposed infants from Rio de Janeiro, Brazil. *JAMA Network Open*. https://doi.org/10.1001/jamanetworkopen.2019.8124.

Rossi, S. L., Tesh, R. B., Azar, S. R., Muruato, A. E., Hanley, K. A., Auguste, A. J., … Weaver, S. C. (2016). Characterization of a novel murine model to study Zika virus. *American Journal of Tropical Medicine and Hygiene*, *94*(6), 1362–1369. https://doi.org/10.4269/ajtmh.16-0111.

Semple, B. D., Blomgren, K., Gimlin, K., Ferriero, D. M., & Noble-Haeusslein, L. J. (2013). Brain development in rodents and humans: Identifying benchmarks of maturation and vulnerability to injury across species. *Progress in Neurobiology*. https://doi.org/10.1016/j.pneurobio.2013.04.001.

Shao, Q., Herrlinger, S., Yang, S.-L., Lai, F., Moore, J. M., Brindley, M. A., & Chen, J.-F. (2016). Zika virus infection disrupts neurovascular development and results in postnatal microcephaly with brain damage. *Development*, *143*(22), 4127–4136. https://doi.org/10.1242/dev.143768.

Shi, Y., Li, S., Wu, Q., Sun, L., Zhang, J., Pan, N., … Wang, X. (2018). Vertical transmission of the Zika virus causes neurological disorders in mouse offspring. *Scientific Reports*, *8*(1), 1–14. https://doi.org/10.1038/s41598-018-21894-w.

Siddharthan, V., Van Wettere, A. J., Li, R., Miao, J., Wang, Z., Morrey, J. D., & Julander, J. G. (2017). Zika virus infection of adult and fetal STAT2 knock-out hamsters. *Virology*, *507*(April), 89–95. https://doi.org/10.1016/j.virol.2017.04.013.

Simpson, D. I. H. (1964). Zika virus infection in man. *Transactions of the Royal Society of Tropical Medicine and Hygiene*, *58*(4), 335–338. https://doi.org/10.1016/0035-9203(64)90200-7.

Szaba, F. M., Tighe, M., Kummer, L. W., Lanzer, K. G., Ward, J. M., Lanthier, P., … Lin, J. S. (2018). Zika virus infection in immunocompetent pregnant mice causes fetal damage and placental pathology in the absence of fetal infection. *PLoS Pathogens*, *14*(4), e1006994. https://doi.org/10.1371/journal.ppat.1006994.

Tang, H., Hammack, C., Ogden, S. C., Wen, Z., Qian, X., Li, Y., … Ming, G. L. (2016). Zika virus infects human cortical neural progenitors and attenuates their growth. *Cell Stem Cell*, *18*(5), 587–590. https://doi.org/10.1016/j.stem.2016.02.016.

Tang, W. W., Young, M. P., Mamidi, A., Regla-Nava, J. A., Kim, K., & Shresta, S. (2016). A mouse model of Zika virus sexual transmission and vaginal viral replication. *Cell Reports*, *17*(12), 3091–3098. https://doi.org/10.1016/j.celrep.2016.11.070.

Thompson, B. L., Levitt, P., & Stanwood, G. D. (2009). Prenatal exposure to drugs: Effects on brain development and implications for policy and education. *Nature Reviews Neuroscience*, *10*(4), 303–312. https://doi.org/10.1038/nrn2598.

van den Pol, A. N., Mao, G., Yang, Y., Ornaghi, S., & Davis, J. N. (2017). Zika virus targeting in the developing brain. *The Journal of Neuroscience*, *37*(8), 2161–2175. https://doi.org/10.1523/JNEUROSCI.3124-16.2017.

Vermillion, M. S., Lei, J., Shabi, Y., Baxter, V. K., Crilly, N. P., McLane, M., … Burd, I. (2017). Intrauterine Zika virus infection of pregnant immunocompetent mice models transplacental transmission and adverse perinatal outcomes. *Nature Communications*, *8*, 1–14. https://doi.org/10.1038/ncomms14575.

Wichgers Schreur, P. J., Van Keulen, L., Anjema, D., Kant, J., & Kortekaas, J. (2018). Microencephaly in fetal piglets following in utero inoculation of Zika virus article. *Emerging Microbes and Infections*, *7*(1). https://doi.org/10.1038/s41426-018-0044-y.

Wikan, N., & Smith, D. R. (2016). Zika virus: History of a newly emerging arbovirus. *The Lancet Infectious Diseases*. https://doi.org/10.1016/S1473-3099(16)30010-X.

Wu, K. Y., Zuo, G. L., Li, X. F., Ye, Q., Deng, Y. Q., Huang, X. Y., … Luo, Z. G. (2016). Vertical transmission of Zika virus targeting the radial glial cells affects cortex development of offspring mice. *Cell Research*, *26*(6), 645–654. https://doi.org/10.1038/cr.2016.58.

Xavier-Neto, J., Carvalho, M., Pascoalino, B.d. S., Cardoso, A. C., Costa, Â. M. S., Pereira, A. H. M., … Franchini, K. (2017). Hydrocephalus and arthrogryposis in an immunocompetent mouse model of ZIKA teratogeny: A developmental study. *PLoS Neglected Tropical Diseases*, *11*(2), 1–29. https://doi.org/10.1371/journal.pntd.0005363.

Yockey, L. J., Varela, L., Rakib, T., Khoury-Hanold, W., Fink, S. L., Stutz, B., … Iwasaki, A. (2016). Vaginal exposure to Zika virus during pregnancy leads to fetal brain infection. *Cell*, *166*(5). https://doi.org/10.1016/j.cell.2016.08.004. 1247–1256.e4.

Yuan, L., Huang, X. Y., Liu, Z. Y., Zhang, F., Zhu, X. L., Yu, J. Y., … Qin, C. F. (2017). A single mutation in the prM protein of Zika virus contributes to fetal microcephaly. *Science*. https://doi.org/10.1126/science.aam7120.

Zhang, W., Tan, Y. W., Yam, W. K., Tu, H., Qiu, L., Tan, E. K., … Zeng, L. (2019). In utero infection of Zika virus leads to abnormal central nervous system development in mice. *Scientific Reports*, *9*(1), 7298. https://doi.org/10.1038/s41598-019-43303-6.

Zhang, F., Wang, H. J., Wang, Q., Liu, Z. Y., Yuan, L., Huang, X. Y., … Xu, Z. (2017). American strain of Zika virus causes more severe microcephaly than an old Asian strain in neonatal mice. *eBioMedicine*, *25*, 95–105. https://doi.org/10.1016/j.ebiom.2017.10.019.

Chapter 44

Zika virus infection with primates: Fetal outcomes

Sunam Gurung[a], Rachel Jordan[b], James Papin[c], and Dean Myers[d]

[a]*Dean Myers Lab, Department of Obstetrics and Gynecology, The University of Oklahoma Health Sciences Center, Oklahoma City, OK, United States,* [b]*The University of Oklahoma Health Sciences Center, College of Medicine, Oklahoma City, OK, United States,* [c]*Division of Comparative Medicine, The University of Oklahoma Health Sciences Center, Oklahoma City, OK, United States,* [d]*Harold Hamm Diabetes Center, Department of Obstetrics and Gynecology, The University of Oklahoma Health Sciences Center, Oklahoma City, OK, United States*

Abbreviations

CNS	central nervous system
CZS	congenital Zika syndrome
GFAP	glial fibrillary acidic protein
Iba-1	ionized calcium binding adaptor molecule 1
IL-6	interleukin 6
IUGR	intrauterine growth restriction
IZ	intermediate zone
NeuN	neuronal nuclei
NHP	nonhuman primate
NPC	neuronal progenitor cells
O1	oligodendrocyte
Preol	premature oligodendrocyte
RG	radial glia
SC	subcutaneous
SGZ	subgranular zone
SP	subplate
ZIKV	Zika virus

Introduction

Zika virus (ZIKV) belongs to the Flaviviridae family, genus *Flavivirus*, which includes other mosquito-borne viruses including dengue (DENV), West Nile (WNV), yellow fever (YFV), and Japanese encephalitis virus (JEV) (Dick, 1952; Dick, Kitchen, & Haddow, 1952). While ZIKV symptoms are typically mild in most individuals, with occasional rash, mild fever, and arthralgia, ZIKV infection during pregnancy is firmly linked to an increased incidence of newborns with microcephaly and a variety of other congenital anomalies, collectively referred to as congenital Zika syndrome (CZS) (Alvarado & Schwartz, 2017; Brasil et al., 2016; Chan, Choi, Yip, Cheng, & Yuen, 2016; Mlakar et al., 2016). In addition to the spectrum of congenital malformations, ZIKV infection in pregnancy is associated with intrauterine growth restriction (IUGR), increased incidence of miscarriage, and fetal demise (Brasil et al., 2016; Chan et al., 2016). There is also growing evidence of occurrence of microcephaly and developmental impairments postnatal in infants exposed to ZIKV during pregnancy regardless of gestational age and without obvious abnormalities at birth (Aragao et al., 2017; van der Linden et al., 2016). One longitudinal study following a cohort of infants for 8 months after birth found deleterious effects of CZS in infants such as increased incidences of irritability, altered muscle tone and movement, epileptic seizures, and dysphagia with abnormal brain imaging seen in all these infants (Moura da Silva et al., 2016). Another longitudinal study followed CZS-affected children from Brazil for 19–24 months that reported severe motor impairment, seizure disorders, hearing and vision abnormalities, and sleep difficulties, resulting in significant delays in achieving developmental milestones (Satterfield-Nash et al., 2017).

Zika Virus Impact, Diagnosis, Control, and Models. https://doi.org/10.1016/B978-0-12-820267-8.00044-3

Nonhuman primate animal models for ZIKV study

The development of animal models that faithfully recapitulate the complex pathogenesis of ZIKV infection is essential for understanding how this virus transverses the antiviral placental barrier in ~15% of seemingly healthy pregnancies resulting in CZS and critically, in which cell types in the fetal CNS are targeted at the most vulnerable windows of gestation at times that are comparable with human pregnancy. Therefore it is important to develop animal models for studying long-term neurodevelopmental outcomes of in utero exposure to the ZIKV infection to develop therapeutic interventions to improve the clinical outcomes of such births. Nonhuman primates (NHPs) are evolutionarily closer to humans and therefore share similar placentation, immune responses, and brain development. NHPs are also the best-documented animal reservoirs for Zika and related flaviviruses. ZIKV infection has been characterized in male and nonpregnant female rhesus macaques [*Macaca mulatta* (Aid et al., 2017; Coffey et al., 2017; Dudley et al., 2016; George et al., 2017; Hirsch et al., 2017; Pantoja et al., 2017; Silveira et al., 2017)] cynomolgus macaques [*Macacca fascicularis* (Koide et al., 2016; Osuna et al., 2016)], and baboons [*Papio anubis* (Gurung et al., 2018, 2019, 2020; Peregrine et al., 2019)] following the standard subcutaneous (SC) route of inoculation. Successful infection of rhesus macaques and baboons has also been described following intravaginal/intrarectal (Carroll et al., 2017; Gurung et al., 2020; Haddow et al., 2017) and in macaques oropharyngeal mucosal (Newman et al., 2017) or mosquito bite (Dudley et al., 2017) routes of inoculation, and seroprevalence of ZIKV has been reported in wild African green monkeys (*Chlorocebus aethiops*) and baboons, which are endemic to the African continent (Buechler et al., 2017).

In utero ZIKV exposure and neuropathology in primate fetus

Zika virus infection of pregnant baboons (Gurung et al., 2019, 2020), rhesus macaques (Coffey et al., 2018; Dudley et al., 2016; Hirsch et al., 2018; Magnani et al., 2018; Martinot et al., 2018; Mohr et al., 2018; Nguyen et al., 2017), pigtail macaques [*Macacca nemestrina* (Adams Waldorf et al., 2016, 2018)], and marmosets [*Callithrix jacchus* (Seferovic et al., 2018)] have been achieved to model pregnancy outcomes and fetoplacental pathologies in NHPs. Similar to humans, intrauterine fetal death and/or miscarriage has been reported as a common (26%) outcome following ZIKV infection in macaques and baboons (Dudley et al., 2018; Gurung et al., 2019). While microcephaly per se has not been reported in macaques infected with ZIKV during gestation, a range of fetal or infant neuropathologies has been documented. An initial study of four rhesus macaques, SC inoculation (mid-1st or early 3rd trimesters) with the French Polynesian ZIKV isolate (Nguyen et al., 2017), yielded no overt fetal brain pathologies by late gestation, although ocular and lung pathology was observed. However, significant fetal neuropathology was reported in a single pigtail macaque following multisite SC infection (Adams Waldorf et al., 2016) with the Cambodian strain of ZIKV, albeit the dose of ZIKV was artificially high (5×10^7 pfu). In this study the most consistent CNS pathology was the loss of fetal noncortical brain volume with white matter and periventricular gliosis and ependymal loss. However, the authors noted normal cortical folding and found no evidence of cortical malformations. A subsequent study from this group using the same inoculating dose of either the Cambodian (2 macaques) or Brazilian ZIKV isolates (3 macaques) confirmed the initial findings and additionally reported a decrease in late neuroprogenitor cells (NPCs) in the subventricular zone (SVZ) of the temporal cortex and disorganization of the subgranular zone (SGZ) of the hippocampal dentate gyrus (Adams Waldorf et al., 2018). They also reported ependymal fusion and periventricular gliosis in the fetal brain. Interestingly the Cambodian ZIKV isolate is not associated with adverse pregnancy outcome or CZS in humans. Another study showed (Hirsch et al., 2018) subtle effects on fetal brain structure such as thinner postcentral gyrus and missing secondary sulcus identified on MRI imaging in 2 of 5 fetuses from pregnant rhesus macaques following SC inoculation (1st or 2nd trimesters) with the Puerto Rican isolate of ZIKV. This study also found aberrant placental function secondary to ZIKV infection resulting in uteroplacental pathology. They suggested that the placental injury and dysfunction due to ZIKV infection leads to decreased oxygen permeability and oxygen transport across the villi that may contribute (via hypoxia) to the neurodevelopmental deficits and fetal loss reported in ZIKV-induced CZS. The most severe fetal neuropathology in NHPs was reported in a study of six rhesus macaques following SC inoculation with the Brazilian strain of ZIKV (1×10^3 pfu) early in pregnancy, which resulted in one in utero fetal death/abortion while the remaining five infants (at birth) exhibited smaller brain size and CNS lesions including microcalcifications, hemorrhage, neuronal necrosis, vasculitis, gliosis, and abnormal increase in NPCs in the frontal cortex and basal ganglia and disorganization of NPC migration (Martinot et al., 2018). In addition, these authors found significant placental pathology, potentially contributing to the more extensive infant CNS pathology observed in this study. In this study, they showed 66% of the ZIKV-infected macaque dams exhibited severe placental pathology such as placental infarctions, neutrophilic chorioamnionitis, and histiocytic villitis. Their data suggested that the ZIKV-induced inflammation and damage to the placental vasculature and the resulting placental insufficiency may have contributed to the extensive neuropathology with vasculitis and vascular compromise of the CNS as hallmarks of the CZS observed in the fetuses.

Olive baboon as a nonhuman primate model for ZIKV pathogenesis

There are some clear differences between ZIKV pathogenesis in macaques and humans that need to be addressed. One of the significant difference is the prolonged maternal viremia seen in pregnant rhesus dams that typically lasts 30–55 days (Dudley et al., 2016; Nguyen et al., 2017), which is not commonly seen in the human population. Vertical transmission is also highly efficient in the macaques (100%). It is unclear whether the high rate of vertical transmission in macaques is inherent to these species of primate or due to the prolonged maternal viremia seen in pregnant macaques, which could be exposing the fetus to the virus repeatedly over the long gestation period despite development of the neutralizing antibodies. One of the key questions regarding ZIKV infection during pregnancy is whether adverse pregnancy outcomes requires vertical transmission or whether placental dysfunction and damage without vertical transmission is sufficient for fetal neuropathologies and ZIKV-induced CZS. Our lab developed and established olive baboons (*Papio anubis*) as an alternative NHP model to study ZIKV infection and pathogenicity during pregnancy that can be compared/contrasted with ZIKV infection in human and other NHP pregnancies. The olive baboon is similar to humans in terms of placentation (monodiscoid hemomonochorial placenta), brain development, and immune repertoire, which makes the baboon an excellent translational NHP model to study ZIKV infection and for vaccine and therapeutics development (Shearer et al., 1997; Shearer, Dark, Chodosh, & Kennedy, 1999; Wolf et al., 2006). The baboon is permissive to flavivirus infection and replication, including ZIKV, and produces a virus-specific immune response (Gurung et al., 2018; Wolf et al., 2006).

So far, studies in macaques have focused on late gestation fetal or infant ZIKV-induced neuropathologies. For our early studies in baboons, we chose to study the early events of ZIKV infection to shed light on the timing of maternal ZIKV infection, vertical transmission, and CNS pathology since late gestation fetal or infant CNS outcome is likely cumulative effect of ZIKV targeting of cells, chronic inflammation, and the long-term effects of placental dysfunction (e.g., chronic fetal hypoxia, hypoxia/ischemia, and nutrient restriction). Therefore we studied the acute effects of ZIKV infection during pregnancy in pregnant baboons ($n = 4$) infected with the French Polynesian isolate of ZIKV (H/PF/2013) at midgestation and terminate pregnancies at 7, 14, and 21 days post infection (dpi) to study the immediate effects of ZIKV infection during pregnancy.

Cortical neuropathology in the fetal baboon

Our acute study showed efficient maternal infection, with vertical transmission in two of four fetuses and one case of fetal demise. ZIKV RNA was detected in the brains of two fetal baboons (14 dpi; 21 dpi); however, the brain was unavailable for analysis due to necrosis, from the one fetus that died in utero. One of the major findings of our study was the noted neuropathology in fetal frontal cortex at 21 dpi. To our knowledge, this is the earliest description of cortical neuropathology post-ZIKV infection in a primate in a viable fetus that was not aborted or underwent in utero death. A pronounced difference in immunostaining for GFAP, a marker for radial glia (RG), was noted in the ZIKV-infected frontal cortex compared with the control fetus (and the two fetuses from ZIKV dams with no detectable vertical transfer of virus to the fetus). In the ZIKV-infected frontal cortex, a substantial loss of radial glial fibers was observed (~90% reduction in density), in particular in the intermediate zone (IZ)/subplate (SP) region of the frontal cortex (Fig. 1). Radial glial fibers serve as scaffolding for migrating neurons and neuronal precursors to the cortical plate to form the characteristic six-layered cortical structure (Malatesta, Appolloni, & Calzolari, 2008). To our knowledge, this is the first description of loss of RG fibers in primates in response to ZIKV. Besides the loss of RG fibers, we also observed an ~50% decrease in the number of NeuN neurons in the cortical plate of this fetus compared with the control fetus or the two fetuses without vertical transfer of ZIKV. In addition, NeuN neurons in the cortical plate of these fetuses were organized into the typical pattern of rows following RG fibers, while in the ZIKV-infected frontal cortex, the NeuN neurons appeared disorganized.

Radial glial fibers, neuronal migration, and differentiation in the fetal brain

In addition to serving as neural stem cells and providing scaffolding for neuronal migration in the cortex, RG also differentiate to astrocytes and preoligodendrocytes (Howard, Chen, & Zecevic, 2006; Mo & Zecevic, 2009; Noctor, Flint, Weissman, Dammerman, & Kriegstein, 2001; Rakic, 2003). Normal transformation of RG to astrocytes in the human fetal frontal cortex takes place gradually over several weeks, mainly in the second half of gestation (deAzevedo et al., 2003). Concurrent with the loss of the RG fibers, a dramatic increase (~fivefold) in RG/astrocytes was observed in the ZIKV-infected frontal cortex of the

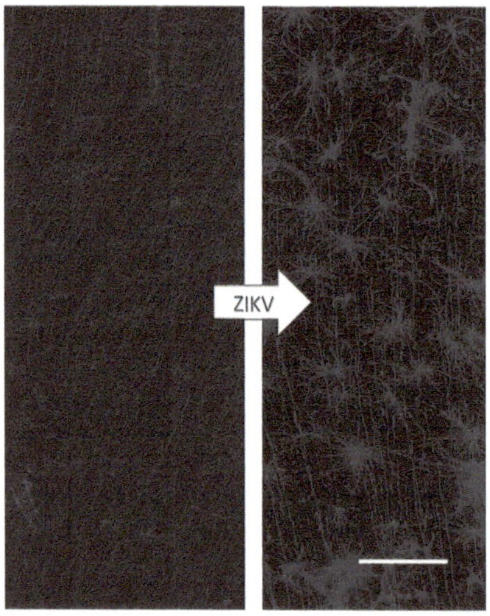

FIG. 1 Loss of radial glial fibers in ZIKV-infected fetal frontal cortex. GFAP staining of control and ZIKV-infected fetal frontal cortex. Left image showing normal midgestation fetal baboon frontal cortex with dense GFAP+ radial glial fibers in the subplate and intermediate zone (SP/IZ). Right image showing 21-day ZIKV-infected fetal baboon frontal cortex with marked loss of GFAP+ radial glial fibers and a significant increase in density of astrocytes compared with the control fetus.

fetus in the IZ/SP (Fig. 1). In the control fetus and the two fetuses lacking vertical transfer of virus, GFAP staining followed the expected pattern of RG/astrocyte distribution. The expanded population of astrocytes in the ZIKV-infected cortex suggested a rapid induction of differentiation of RG to astrocytes (away from neuronal differentiation) coupled with potential migration of astrocytes into this region, rather than death of the RG per se. In the prior study in the pigtail macaque, vertical transfer of the Cambodian strain of ZIKV also resulted in an increase in GFAP-stained astrocytes in the white matter of the cortex as observed at 6 weeks after infection (at near-term gestation) (Adams Waldorf et al., 2016). The findings of our study support that an early event following ZIKV penetration of the fetal CNS may have accelerated RG differentiation to astrocytes, in particular in the IZ/SP where the RG are primarily located during this stage of development coupled with a loss of RG fibers.

Gyri/sulci formation in the fetal cortex

It should also be emphasized that the RG fibers provide an additional function in the formation of gyri and sulci (Stahl et al., 2013). Loss of RG fibers in the ZIKV brain would seemingly predict a less folded brain as gestation progresses. Loss of gyri/sulci is a hallmark in ZIKV-infected cases of human microcephaly (Hazin et al., 2016; Mlakar et al., 2016). This was evident in our longitudinal study with timed pregnant baboons infected early or midgestation and taken to near term. In this study, we examined the fetal neuropathologies and manifestations of CZS as a result of ZIKV infection during critical time of pregnancy for fetal brain development. In these late gestation fetuses, we observed significant gyral/sulcal malformations including absence of major gyri/sulci, enlarged gyri, and abnormal sulcal length and depth (Fig. 2A (1,2,3)). We also observed gross and histological cerebellar hypoplasia including significant loss and distortion of Purkinje cells, loss of cellular volume in the molecular layer, increased apoptotic cells, and increased inflammation in different regions of the cerebellum compared with the control cerebellum (Fig. 2A (4) and 2B (4, 5)). Cerebellar hypoplasia and gyri/sulci malformations are manifestations of CZS seen in ZIKV-infected newborns (Musso, Ko, & Baud, 2019). Histological staining of the ZIKV-exposed fetal frontal cortices showed significant disruption in the formation of the six-layered neocortex mainly in one of the early gestation and one midgestation fetus where significant disorganization of the layers, lack of layer-specific cell populations, loss of pyramidal neurons, and focal hemorrhage in the gray and white matter were seen, which also corresponded to the increase in apoptosis and inflammatory microglia in these areas compared with the control brains (Fig. 2B (1, 2, 3)). To our knowledge, this is the most detailed emulation in an NHP model of ZIKV-induced brain abnormalities seen in human infants manifesting as cortical atrophy, cerebellar hypoplasia, and neuronal migration defect seen as gyri/sulci malformation possibly due to the loss or disorganization of RG fibers as reported by our acute studies in baboons (Gurung et al., 2019).

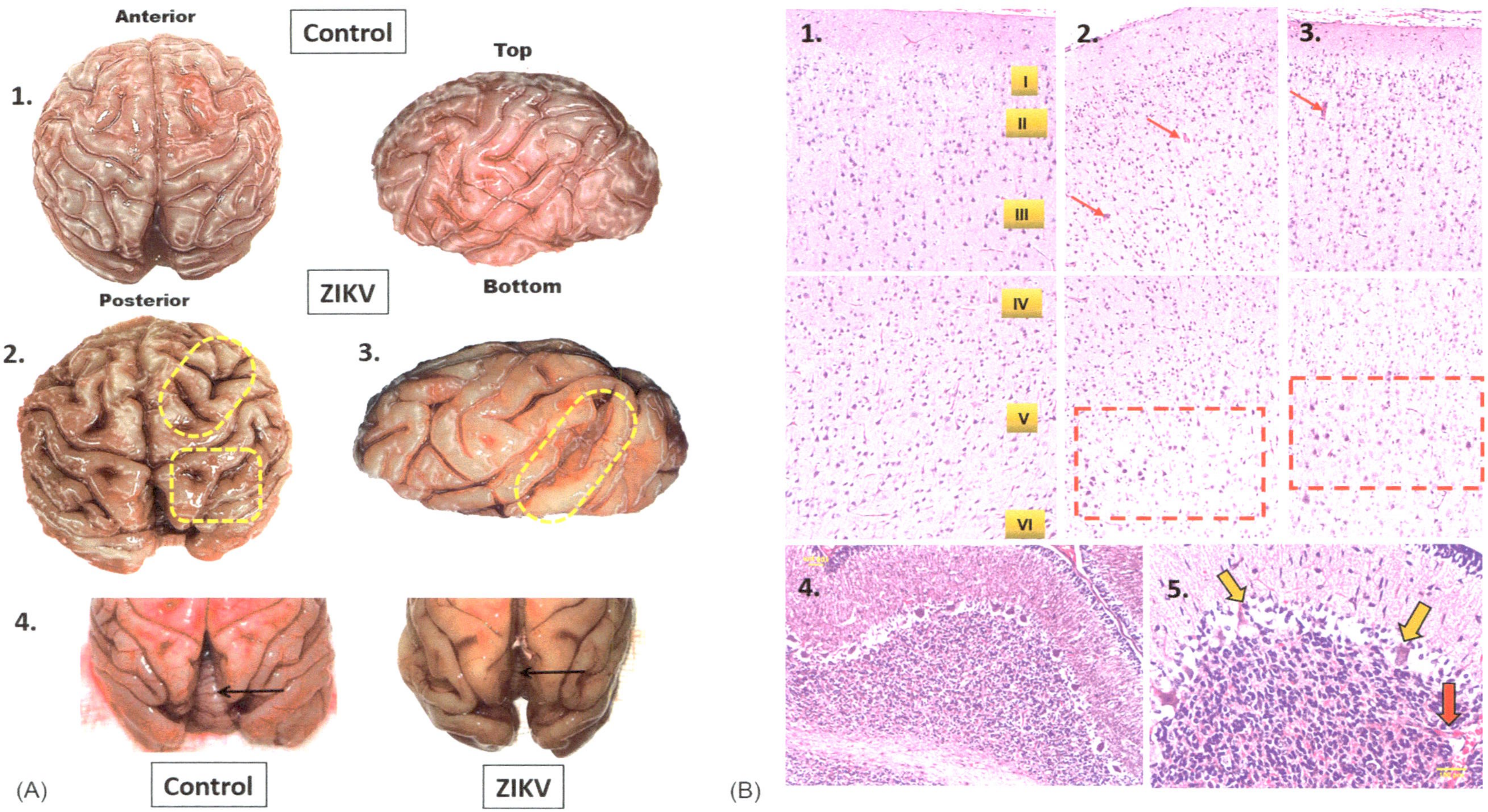

FIG. 2 Gross and histological fetal brain malformations. (A) Top and side view of control (1), early gest (2), and mid-gest (3) fetal brains showing major gyri and sulci formations. Areas marked by *yellow dotted ovals* highlight significant gyral/sulcal malformations in various areas of the cortex. Top view of fetal brains (A 4) highlighting major cerebellar atrophy in mid-gest fetus compared with the control brain. (B) Representative images from H&E staining of control (1), early (2), and mid-gest (3) brain showing six layers of cortex. *Red arrows* represent focal hemorrhage, and *red dotted rectangle* represents substantial lack of pyramidal cells in Layer V of the cortical plate. (B) Representative images from H&E staining of control (4), and early gest (5) cerebellar sections. Loss and/or distorted Purkinje cells are marked by *yellow arrows*, and *red arrow* shows focal hemorrhage (unpublished data).

Oligodendrocyte differentiation, maturation, and myelination

Astrogliosis is a normal response to viral infection and brain injury, and our results are in agreement with this. In humans, RG differentiate into preoligodendrocytes (preOL) (Back et al., 2001; Back, Luo, Borenstein, Volpe, & Kinney, 2002). Thus a loss of RG could conceivably reduce subsequent formation of OL and reduce myelination consistent with the observations in the pigtail macaque and human fetuses obtained from ZIKV-infected pregnancies (Adams Waldorf et al., 2016, 2018). In our acute studies with the baboons, in the 21-day ZIKV-infected fetal cortex, we found reduced numbers of $O1^+$ (immature oligodendrocytes) as well as without their characteristic multibranched processes with the appearance of undergoing degeneration compared with the control fetus where an abundant population of $O1^+$ immature oligodendrocytes was observed in a gradient from the IZ/SP through the CP and developing white matter similar to midgestation human fetus. In primates, cortical white matter forms within and eventually replaces the IZ/SP, and as such, the effects of ZIKV infection observed in our study may disrupt normal white matter development (Mo & Zecevic, 2009).

Neuroinflammation in the fetal brain

In addition to increased astrocytes, in the acute studies, the 21-day ZIKV-infected cortex exhibited other indices of neuroinflammation with increased Iba-1 (microglia) and IL-6 (proinflammatory cytokine) immunostaining. Similar increase in Iba-1 staining was seen in the white matter of the cerebellum and cortical layer of the frontal cortex in ZIKV-exposed fetuses during early and midgestation period in our longitudinal studies. While astrogliosis was described in the fetal CNS of pigtail macaque fetuses when analyzed at near-term gestation, there are no other reports of active neuroinflammation in studies of rhesus macaques in response to ZIKV infection. However, brain lesions have been described consisting of necrosis and gliosis in infant rhesus macaques whose mothers were infected during early gestation (Martinot et al., 2018). We observed noted increase in both Iba1 and IL-6 in the frontal cortex of the day 14 postinfection fetus (compared with the control fetus and day 7 fetus) despite not detecting ZIKV in the cortex of this fetus. This implicates that the increased neuroinflammation may be in response to maternal or placental inflammation or that ZIKV had transferred to the fetus in a tissue not sampled or in an adjacent region of the fetal brain not sampled. Increased neuroinflammation may also be a hallmark of impending ZIKV infiltration into the fetal CNS (Fig. 3). The potential implications of increased fetal neuroinflammation in the absence of vertical transfer have implications for human infants from ZIKV-infected mothers without notable gross CNS pathologies that may lead to subtle neurobehavioral or cognitive deficits postbirth.

Placental inflammation, vascular deficiency, and neuropathology

Placental insufficiency is recognized as a major contributor to IUGR, birth defects, preterm birth, and stillbirth in human maternal-fetal medicine (American College of Obstetricians & Gynecologists, 2013). In our acute baboon study, we observed one dam with intrauterine fetal death that exhibited significant placental pathology, with extensive fibrin deposition in the intervillous space with nearly uniform degenerated villi with frequent necrosis and acute inflammation. While we can't determine if loss of placental function contributed to fetal demise in this dam, placental inflammation is associated with pregnancy loss in women, and this dam exhibited the highest systemic cytokine response. The placentas of two other baboons (7 and 21 day) exhibited minor evidence of inflammation, and the placenta of 14-day dam was histologically similar to the control placenta, despite having one cotyledon positive for ZIKV RNA. Immunofluorescence for ZIKV (pan flavivirus) verified the presence of ZIKV in the placenta of all three dams with ZIKV RNA mainly in the syncytiotrophoblast layer of all three placentas to varying intensities with the day 14 postinfection placenta exhibiting the most uniform IF in the syncytiotrophoblast layer, while there was restriction of the IF signal by 21 dpi indicating potential resolution of placental infection by this time postinfection.

Conclusion

In conclusion, the consequences of ZIKV exposure to a fetus in utero are enduring and devastating. NHP models are critical in understanding the short- and long-term consequences of ZIKV infection during pregnancy and therapeutic interventions. Our study in baboons and other macaque studies have highlighted the fact that any exposure to the virus in utero can lead to neurodevelopmental abnormalities as seen in human infants facing significant developmental, cognitive, and behavioral challenges postbirth. Even though ZIKV may not seem as an immediate risk anymore, in many parts of the globe, the virus still poses a significant threat to the health and well-being of pregnant women and their pregnancy outcomes. The lifetime cost of caring for a baby born with ZIKV-induced CZS is expected to exceed 3 million USD. Therefore diseases like ZIKV inflict a huge economic and personal burden on societies if left unchecked. Understanding the disease pathogenesis is

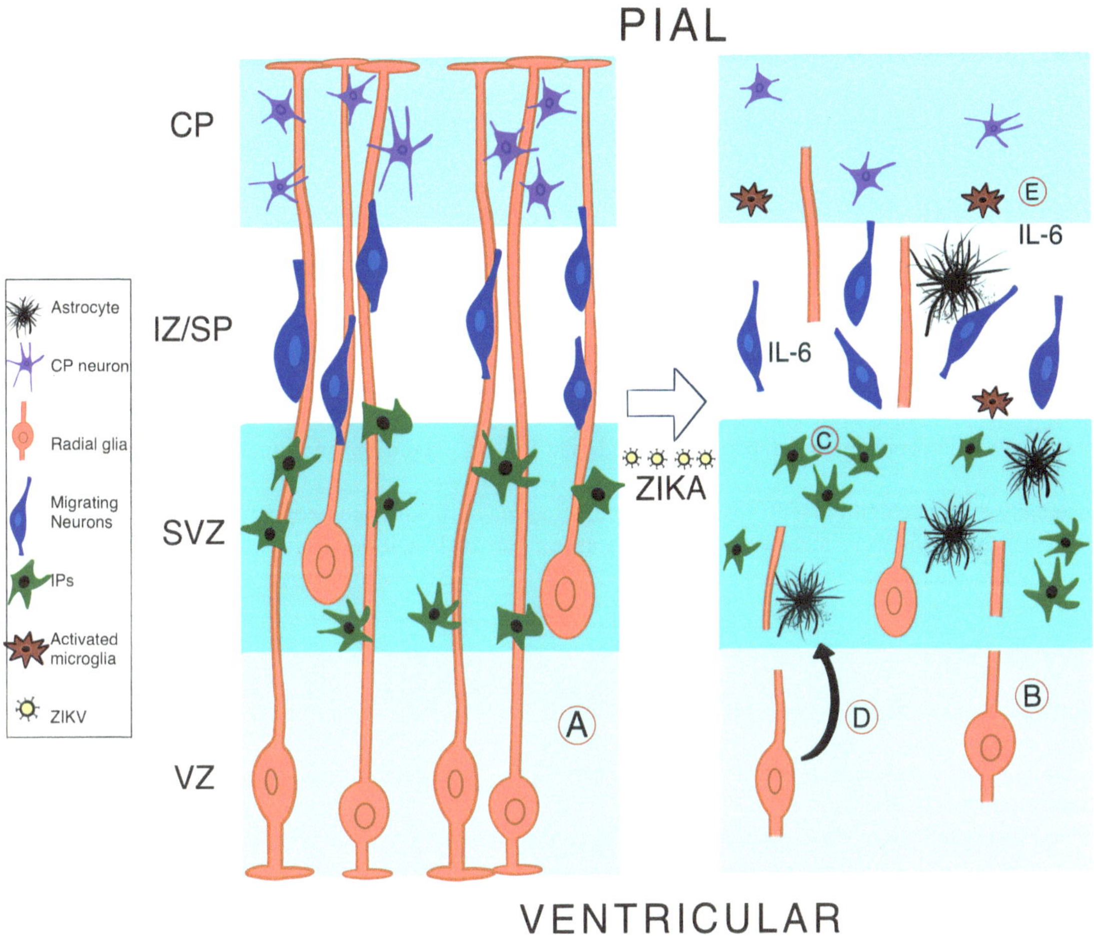

FIG. 3 Summary figure showing potential mechanisms for ZIKV pathogenesis in fetal primate brain. (A) Radial glial (RG) fibers serve as scaffolding for migrating neurons and intermediate precursors as they make their way to the cortical plate (CP) to form the six-layered cortex during normal development of the fetal primate cortex. Based on studies in nonhuman primates, it is known that ZIKV infection of the fetal cortex results in loss of RG fibers (B) resulting in disorganized precursor and migrating neurons (C) due to the loss of RG fiber scaffolding. (D) An increase in astrogliosis is also observed, which may be due to premature differentiation of RG to astrocytes and/or due to noted neuroinflammatory response to ZIKV seen as increase in activated microglial population and proinflammatory cytokine such as IL-6 (E). *VZ*, ventricular zone; *SVZ*, subventricular zone; *IZ*, intermediate zone; *SP*, subplate; *CP*, cortical plate.

critical in preparing for future outbreaks or outbreaks of new and reemerging viruses that transfer vertically like ZIKV and cause ZIKV like CZS.

Policy and procedures

Fetal CNS histology and immunostaining: Following removal, fetal brains were divided midsagittal with one-half fixed in 4% paraformaldehyde for 48h. For cryoembedding, brains were cryoprotected in 30% sucrose until sunk and frozen in embedding molds in OCT on dry ice. For paraffin embedding, after fixation, brains were moved to 70% EtOH and paraffin processed. The identical region of the frontal cortex was selected from each fetus and subjected to serial coronal sectioning. H&E staining was performed on brain sections for morphology, and immunostaining utilizing various antibodies was used for immunolabeling.

Mini-dictionary of terms

Congenital Zika syndrome. Various neurologic, ocular, and musculoskeletal manifestations of fetal malformations associated with maternal ZIKV infection including intrauterine growth restriction, preterm birth, miscarriage, and fetal demise.

Microcephaly. Rare neurological condition in infants born with head circumference smaller than average due to abnormal head development usually in utero but could also be postnatal as seen in ZIKV cases.

Neuropathology. Pathogenesis of nervous system usually characterized by atrophy of various brain structures, lesions, inflammation, and loss of cellular structure and integrity using histochemical and immunohistochemical stains.
Placental insufficiency. Pregnancy complication due to placental vascular deficiency resulting in insufficient nutrient and oxygen transfer to the developing fetus resulting in fetal growth restriction.

Key facts of microcephaly

- Children born with small head circumference than average at birth are termed microcephalic usually confirmed through CT or MRI to look at abnormal brain development.
- Microcephaly can occur either at birth or postnatally and ranges from mild to severe.
- Microcephaly is known to occur due to various factors such as genetic and chromosomal disorders, infections passed to the fetus during pregnancy (e.g., ZIKA virus), cerebral anoxia, severe malnutrition, and exposure to drugs, alcohol, and toxins in utero.
- Children with microcephaly have permanent and debilitating cognitive and intellectual disability and growth impairment.
- Microcephaly is a rare condition with 1 baby in several thousand born with microcephaly. However, significant increase in microcephalic births were recorded in the northeastern part of Brazil between 2015 and 2016 (5.9 per 1000 live births), which studies later directly associated with Zika virus infection during pregnancy.

Summary points

- ZIKV infection during pregnancy causes severe fetal neuropathology resulting in significant developmental delay and trauma.
- Placental damage, inflammation, and insufficiency may alone cause ZIKV-induced CZS in fetuses in absence of vertical transmission.
- Disruption/disorganization of the radial glial fibers and subsequent migration of neuronal cells to the outer layers of the cortex may result in low cortical volume and collapse of the skull seen in microcephalic fetuses.
- Damage to the RG fibers may also cause defect in gyri/sulci formation.
- Changes in neuronal progenitor cell proliferation, differentiation, and apoptosis due to the targeting of the NPCs by ZIKV directly or ZIKV infection-related inflammation may also result in disruption of formation of the six-layered neocortex.
- Early and midgestation infection may result in the most severe ZIKV-induced CZS; however, any exposure to the virus at any gestational age can lead to neurodevelopmental abnormalities, and some CZS have been seen in babies postnatal.

References

Adams Waldorf, K. M., Nelson, B. R., Stencel-Baerenwald, J. E., Studholme, C., Kapur, R. P., Armistead, B., … Rajagopal, L. (2018). Congenital Zika virus infection as a silent pathology with loss of neurogenic output in the fetal brain. *Nature Medicine*, *24*(3), 368–374. https://doi.org/10.1038/nm.4485.

Adams Waldorf, K. M., Stencel-Baerenwald, J. E., Kapur, R. P., Studholme, C., Boldenow, E., Vornhagen, J., … Rajagopal, L. (2016). Fetal brain lesions after subcutaneous inoculation of Zika virus in a pregnant nonhuman primate. *Nature Medicine*, *22*(11), 1256–1259. https://doi.org/10.1038/nm.4193.

Aid, M., Abbink, P., Larocca, R. A., Boyd, M., Nityanandam, R., Nanayakkara, O., … Barouch, D. H. (2017). Zika virus persistence in the central nervous system and lymph nodes of rhesus monkeys. *Cell*, *169*(4), 610–620. e614 https://doi.org/10.1016/j.cell.2017.04.008.

Alvarado, M. G., & Schwartz, D. A. (2017). Zika virus infection in pregnancy, microcephaly, and maternal and fetal health: What we think, what we know, and what we think we know. *Archives of Pathology & Laboratory Medicine*, *141*(1), 26–32. https://doi.org/10.5858/arpa.2016-0382-RA.

American College of Obstetricians & Gynecologists. (2013). ACOG practice bulletin no. 134: Fetal growth restriction. *Obstetrics and Gynecology*, *121*(5), 1122–1133. https://doi.org/10.1097/01.AOG.0000429658.85846.f9.

Aragao, M. F. V. V., Holanda, A. C., Brainer-Lima, A. M., Petribu, N. C. L., Castillo, M., van der Linden, V., … Costello, A. (2017). Nonmicrocephalic infants with congenital Zika syndrome suspected only after neuroimaging evaluation compared with those with microcephaly at birth and postnatally: How large is the Zika virus "iceberg"? *AJNR. American Journal of Neuroradiology*, *38*(7), 1427–1434. https://doi.org/10.3174/ajnr.A5216.

Back, S. A., Luo, N. L., Borenstein, N. S., Levine, J. M., Volpe, J. J., & Kinney, H. C. (2001). Late oligodendrocyte progenitors coincide with the developmental window of vulnerability for human perinatal white matter injury. *The Journal of Neuroscience*, *21*(4), 1302–1312.

Back, S. A., Luo, N. L., Borenstein, N. S., Volpe, J. J., & Kinney, H. C. (2002). Arrested oligodendrocyte lineage progression during human cerebral white matter development: Dissociation between the timing of progenitor differentiation and myelinogenesis. *Journal of Neuropathology and Experimental Neurology*, *61*(2), 197–211.

Brasil, P., Pereira, J. P., Jr., Moreira, M. E., Ribeiro Nogueira, R. M., Damasceno, L., Wakimoto, M., … Nielsen-Saines, K. (2016). Zika virus infection in pregnant women in Rio de Janeiro. *The New England Journal of Medicine*, *375*(24), 2321–2334. https://doi.org/10.1056/NEJMoa1602412.

Buechler, C. R., Bailey, A. L., Weiler, A. M., Barry, G. L., Breitbach, M. E., Stewart, L. M., … O'Connor, D. H. (2017). Seroprevalence of Zika virus in wild African green monkeys and baboons. *mSphere*, *2*(2). https://doi.org/10.1128/mSphere.00392-16.

Carroll, T., Lo, M., Lanteri, M., Dutra, J., Zarbock, K., Silveira, P., … Miller, C. J. (2017). Zika virus preferentially replicates in the female reproductive tract after vaginal inoculation of rhesus macaques. *PLoS Pathogens*, *13*(7), e1006537. https://doi.org/10.1371/journal.ppat.1006537.

Chan, J. F., Choi, G. K., Yip, C. C., Cheng, V. C., & Yuen, K. Y. (2016). Zika fever and congenital Zika syndrome: An unexpected emerging arboviral disease. *The Journal of Infection*, *72*(5), 507–524. https://doi.org/10.1016/j.jinf.2016.02.011.

Coffey, L. L., Keesler, R. I., Pesavento, P. A., Woolard, K., Singapuri, A., Watanabe, J., … Van Rompay, K. K. A. (2018). Intraamniotic Zika virus inoculation of pregnant rhesus macaques produces fetal neurologic disease. *Nature Communications*, *9*(1), 2414. https://doi.org/10.1038/s41467-018-04777-6.

Coffey, L. L., Pesavento, P. A., Keesler, R. I., Singapuri, A., Watanabe, J., Watanabe, R., … Van Rompay, K. K. (2017). Zika virus tissue and blood compartmentalization in acute infection of rhesus macaques. *PLoS One*, *12*(1), e0171148. https://doi.org/10.1371/journal.pone.0171148.

deAzevedo, L. C., Fallet, C., Moura-Neto, V., Daumas-Duport, C., Hedin-Pereira, C., & Lent, R. (2003). Cortical radial glial cells in human fetuses: Depth-correlated transformation into astrocytes. *Journal of Neurobiology*, *55*(3), 288–298. https://doi.org/10.1002/neu.10205.

Dick, G. W. (1952). Zika virus. II. Pathogenicity and physical properties. *Transactions of the Royal Society of Tropical Medicine and Hygiene*, *46*(5), 521–534.

Dick, G. W., Kitchen, S. F., & Haddow, A. J. (1952). Zika virus. I. Isolations and serological specificity. *Transactions of the Royal Society of Tropical Medicine and Hygiene*, *46*(5), 509–520.

Dudley, D. M., Aliota, M. T., Mohr, E. L., Weiler, A. M., Lehrer-Brey, G., Weisgrau, K. L., … O'Connor, D. H. (2016). A rhesus macaque model of Asian-lineage Zika virus infection. *Nature Communications*, *7*, 12204. https://doi.org/10.1038/ncomms12204.

Dudley, D. M., Newman, C. M., Lalli, J., Stewart, L. M., Koenig, M. R., Weiler, A. M., … Aliota, M. T. (2017). Infection via mosquito bite alters Zika virus tissue tropism and replication kinetics in rhesus macaques. *Nature Communications*, *8*(1), 2096. https://doi.org/10.1038/s41467-017-02222-8.

Dudley, D. M., Van Rompay, K. K., Coffey, L. L., Ardeshir, A., Keesler, R. I., Bliss-Moreau, E., … O'Connor, D. H. (2018). Miscarriage and stillbirth following maternal Zika virus infection in nonhuman primates. *Nature Medicine*, *24*(8), 1104–1107. https://doi.org/10.1038/s41591-018-0088-5.

George, J., Valiant, W. G., Mattapallil, M. J., Walker, M., Huang, Y. S., Vanlandingham, D. L., … Mattapallil, J. J. (2017). Prior exposure to Zika virus significantly enhances peak dengue-2 viremia in rhesus macaques. *Scientific Reports*, *7*(1), 10498. https://doi.org/10.1038/s41598-017-10901-1.

Gurung, S., Nadeau, H., Maxted, M., Peregrine, J., Reuter, D., Norris, A., … Myers, D. A. (2020). Maternal Zika virus (Zikv) infection following vaginal inoculation with Zikv-infected semen in the timed-pregnant olive baboon. *Journal of Virology*. https://doi.org/10.1128/JVI.00058-20.

Gurung, S., Preno, A. N., Dubaut, J. P., Nadeau, H., Hyatt, K., Reuter, N., … Papin, J. F. (2018). Translational model of Zika virus disease in baboons. *Journal of Virology*. https://doi.org/10.1128/JVI.00186-18.

Gurung, S., Reuter, N., Preno, A., Dubaut, J., Nadeau, H., Hyatt, K., … Myers, D. A. (2019). Zika virus infection at mid-gestation results in fetal cerebral cortical injury and fetal death in the olive baboon. *PLoS Pathogens*, *15*(1), e1007507. https://doi.org/10.1371/journal.ppat.1007507.

Haddow, A. D., Nalca, A., Rossi, F. D., Miller, L. J., Wiley, M. R., Perez-Sautu, U., … Nasar, F. (2017). High infection rates for adult macaques after intravaginal or intrarectal inoculation with Zika virus. *Emerging Infectious Diseases*, *23*(8), 1274–1281. https://doi.org/10.3201/eid2308.170036.

Hazin, A. N., Poretti, A., Di Cavalcanti Souza Cruz, D., Tenorio, M., van der Linden, A., Pena, L. J., … Microcephaly Epidemic Research Group. (2016). Computed tomographic findings in microcephaly associated with Zika virus. *The New England Journal of Medicine*, *374*(22), 2193–2195. https://doi.org/10.1056/NEJMc1603617.

Hirsch, A. J., Roberts, V. H. J., Grigsby, P. L., Haese, N., Schabel, M. C., Wang, X., … Streblow, D. N. (2018). Zika virus infection in pregnant rhesus macaques causes placental dysfunction and immunopathology. *Nature Communications*, *9*(1), 263. https://doi.org/10.1038/s41467-017-02499-9.

Hirsch, A. J., Smith, J. L., Haese, N. N., Broeckel, R. M., Parkins, C. J., Kreklywich, C., … Streblow, D. N. (2017). Zika virus infection of rhesus macaques leads to viral persistence in multiple tissues. *PLoS Pathogens*, *13*(3), e1006219. https://doi.org/10.1371/journal.ppat.1006219.

Howard, B., Chen, Y., & Zecevic, N. (2006). Cortical progenitor cells in the developing human telencephalon. *Glia*, *53*(1), 57–66. https://doi.org/10.1002/glia.20259.

Koide, F., Goebel, S., Snyder, B., Walters, K. B., Gast, A., Hagelin, K., … Rayner, J. (2016). Development of a Zika virus infection model in cynomolgus macaques. *Frontiers in Microbiology*, *7*, 2028. https://doi.org/10.3389/fmicb.2016.02028.

Magnani, D. M., Rogers, T. F., Maness, N. J., Grubaugh, N. D., Beutler, N., Bailey, V. K., … Watkins, D. I. (2018). Fetal demise and failed antibody therapy during Zika virus infection of pregnant macaques. *Nature Communications*, *9*(1), 1624. https://doi.org/10.1038/s41467-018-04056-4.

Malatesta, P., Appolloni, I., & Calzolari, F. (2008). Radial glia and neural stem cells. *Cell and Tissue Research*, *331*(1), 165–178. https://doi.org/10.1007/s00441-007-0481-8.

Martinot, A. J., Abbink, P., Afacan, O., Prohl, A. K., Bronson, R., Hecht, J. L., … Barouch, D. H. (2018). Fetal neuropathology in Zika virus-infected pregnant female rhesus monkeys. *Cell*, *173*(5), 1111–1122. e1110 https://doi.org/10.1016/j.cell.2018.03.019.

Mlakar, J., Korva, M., Tul, N., Popovic, M., Poljsak-Prijatelj, M., Mraz, J., … Avsic Zupanc, T. (2016). Zika virus associated with microcephaly. *The New England Journal of Medicine*, *374*(10), 951–958. https://doi.org/10.1056/NEJMoa1600651.

Mo, Z., & Zecevic, N. (2009). Human fetal radial glia cells generate oligodendrocytes in vitro. *Glia*, *57*(5), 490–498. https://doi.org/10.1002/glia.20775.

Mohr, E. L., Block, L. N., Newman, C. M., Stewart, L. M., Koenig, M., Semler, M., … Golos, T. G. (2018). Ocular and uteroplacental pathology in a macaque pregnancy with congenital Zika virus infection. *PLoS One*, *13*(1), e0190617. https://doi.org/10.1371/journal.pone.0190617.

Moura da Silva, A. A., Ganz, J. S., Sousa, P. D., Doriqui, M. J., Ribeiro, M. R., Branco, M. D., … Alves, M.T. (2016). Early growth and neurologic outcomes of infants with probable congenital Zika virus syndrome. *Emerging Infectious Diseases*, *22*(11), 1953–1956. https://doi.org/10.3201/eid2211.160956.

Musso, D., Ko, A. I., & Baud, D. (2019). Zika virus infection—After the pandemic. *New England Journal of Medicine*, *381*(15), 1444–1457. https://doi.org/10.1056/NEJMra1808246.

Newman, C. M., Dudley, D. M., Aliota, M. T., Weiler, A. M., Barry, G. L., Mohns, M. S., … Friedrich, T. C. (2017). Oropharyngeal mucosal transmission of Zika virus in rhesus macaques. *Nature Communications*, *8*(1), 169. https://doi.org/10.1038/s41467-017-00246-8.

Nguyen, S. M., Antony, K. M., Dudley, D. M., Kohn, S., Simmons, H. A., Wolfe, B., … Golos, T. G. (2017). Highly efficient maternal-fetal Zika virus transmission in pregnant rhesus macaques. *PLoS Pathogens*, *13*(5), e1006378. https://doi.org/10.1371/journal.ppat.1006378.

Noctor, S. C., Flint, A. C., Weissman, T. A., Dammerman, R. S., & Kriegstein, A. R. (2001). Neurons derived from radial glial cells establish radial units in neocortex. *Nature*, *409*(6821), 714–720. https://doi.org/10.1038/35055553.

Osuna, C. E., Lim, S. Y., Deleage, C., Griffin, B. D., Stein, D., Schroeder, L. T., … Whitney, J. B. (2016). Zika viral dynamics and shedding in rhesus and cynomolgus macaques. *Nature Medicine*, *22*(12), 1448–1455. https://doi.org/10.1038/nm.4206.

Pantoja, P., Perez-Guzman, E. X., Rodriguez, I. V., White, L. J., Gonzalez, O., Serrano, C., … Sariol, C. A. (2017). Zika virus pathogenesis in rhesus macaques is unaffected by pre-existing immunity to dengue virus. *Nature Communications*, *8*, 15674. https://doi.org/10.1038/ncomms15674.

Peregrine, J., Gurung, S., Lindgren, M. C., Husain, S., Zavy, M. T., Myers, D. A., & Papin, J. F. (2019). Zika virus infection, reproductive organ targeting, and semen transmission in the male olive baboon. *Journal of Virology*, *94*(1). https://doi.org/10.1128/JVI.01434-19.

Rakic, P. (2003). Developmental and evolutionary adaptations of cortical radial glia. *Cerebral Cortex*, *13*(6), 541–549.

Satterfield-Nash, A., Kotzky, K., Allen, J., Bertolli, J., Moore, C. A., Pereira, I. O., … Peacock, G. (2017). Health and development at age 19-24 months of 19 children who were born with microcephaly and laboratory evidence of congenital Zika virus infection during the 2015 Zika virus outbreak—Brazil, 2017. *MMWR. Morbidity and Mortality Weekly Report*, *66*(49), 1347–1351. https://doi.org/10.15585/mmwr.mm6649a2.

Seferovic, M., Martin, C. S., Tardif, S. D., Rutherford, J., Castro, E. C. C., Li, T., … Patterson, J. L. (2018). Experimental Zika virus infection in the pregnant common marmoset induces spontaneous fetal loss and neurodevelopmental abnormalities. *Scientific Reports*, *8*(1), 6851. https://doi.org/10.1038/s41598-018-25205-1.

Shearer, M. H., Dark, R. D., Chodosh, J., & Kennedy, R. C. (1999). Comparison and characterization of immunoglobulin G subclasses among primate species. *Clinical and Diagnostic Laboratory Immunology*, *6*(6), 953–958.

Shearer, M. H., Lucas, A. H., Anderson, P. W., Carey, K. D., Jenson, H. B., Chanh, T. C., … Kennedy, R. C. (1997). The baboon as a nonhuman primate model for assessing the effects of maternal immunization with Haemophilus influenzae type b polysaccharide vaccines. *Infection and Immunity*, *65*(8), 3267–3270.

Silveira, E. L. V., Rogers, K. A., Gumber, S., Amancha, P., Xiao, P., Woollard, S. M., … Villinger, F. (2017). Immune cell dynamics in rhesus macaques infected with a Brazilian strain of Zika virus. *Journal of Immunology*, *199*(3), 1003–1011. https://doi.org/10.4049/jimmunol.1700256.

Stahl, R., Walcher, T., De Juan Romero, C., Pilz, G. A., Cappello, S., Irmler, M., … Gotz, M. (2013). Trnp1 regulates expansion and folding of the mammalian cerebral cortex by control of radial glial fate. *Cell*, *153*(3), 535–549. https://doi.org/10.1016/j.cell.2013.03.027.

van der Linden, V., Pessoa, A., Dobyns, W., Barkovich, A. J., Júnior, H. V., Filho, E. L., … Moore, C. A. (2016). Description of 13 infants born during October 2015-January 2016 with congenital Zika virus infection without microcephaly at birth—Brazil. *MMWR. Morbidity and Mortality Weekly Report*, *65*(47), 1343–1348. https://doi.org/10.15585/mmwr.mm6547e2.

Wolf, R. F., Papin, J. F., Hines-Boykin, R., Chavez-Suarez, M., White, G. L., Sakalian, M., & Dittmer, D. P. (2006). Baboon model for West Nile virus infection and vaccine evaluation. *Virology*, *355*(1), 44–51. https://doi.org/10.1016/j.virol.2006.06.033.

Index

Note: Page numbers followed by *f* indicate figures, and *t* indicate tables.

H

I

J

K

L

M

N